Unit Conversions (Equivalents)

Length

1 in. = 2.54 cm (defined)
1 cm = 0.3937 in.
1 ft = 30.48 cm
1 m = 39.37 in. = 3.281 ft
1 mi = 5280 ft = 1.609 km
1 km = 0.6214 mi
1 nautical mile (U.S.) = 1.151 mi = 6076 ft = 1.852 km
1 fermi = 1 femtometer (fm) = 10^{-15} m
1 angstrom (Å) = 10^{-10} m = 0.1 nm
1 light-year (ly) = 9.461×10^{15} m
1 parsec = 3.26 ly = 3.09×10^{16} m

Volume

1 liter (L) = 1000 mL = 1000 cm^3 = 1.0×10^{-3} m^3 = 1.057 qt (U.S.) = 61.02 in.^{3}
1 gal (U.S.) = 4 qt (U.S.) = 231 in.^{3} = 3.785 L = 0.8327 gal (British)
1 quart (U.S.) = 2 pints (U.S.) = 946 mL
1 pint (British) = 1.20 pints (U.S.) = 568 mL
1 m^3 = 35.31 ft^3

Speed

1 mi/h = 1.4667 ft/s = 1.6093 km/h = 0.4470 m/s
1 km/h = 0.2778 m/s = 0.6214 mi/h
1 ft/s = 0.3048 m/s (exact) = 0.6818 mi/h = 1.0973 km/h
1 m/s = 3.281 ft/s = 3.600 km/h = 2.237 mi/h
1 knot = 1.151 mi/h = 0.5144 m/s

Angle

1 radian (rad) = 57.30° = 57°18′
1° = 0.01745 rad
1 rev/min (rpm) = 0.1047 rad/s

Time

1 day = 8.640×10^4 s
1 year = 3.15581×10^7 s

Mass

1 atomic mass unit (u) = 1.6605×10^{-27} kg
1 kg = 0.06852 slug
[1 kg has a weight of 2.20 lb where $g = 9.80$ m/s^2.]

Force

1 lb = 4.44822 N
1 N = 10^5 dyne = 0.2248 lb

Energy and Work

1 J = 10^7 ergs = 0.7376 ft·lb
1 ft·lb = 1.356 J = 1.29×10^{-3} Btu = 3.24×10^{-4} kcal
1 kcal = 4.19×10^3 J = 3.97 Btu
1 eV = 1.6022×10^{-19} J
1 kWh = 3.600×10^6 J = 860 kcal
1 Btu = 1.056×10^3 J

Power

1 W = 1 J/s = 0.7376 ft·lb/s = 3.41 Btu/h
1 hp = 550 ft·lb/s = 746 W

Pressure

1 atm = 1.01325 bar = 1.01325×10^5 N/m^2 = 14.7 lb/in.^{2} = 760 torr
1 lb/in.^{2} = 6.895×10^3 N/m^2
1 Pa = 1 N/m^2 = 1.450×10^{-4} lb/in.^{2}

SI Derived Units and Their Abbreviations

Quantity	Unit	Abbreviation	In Terms of Base Units[†]
Force	newton	N	kg·m/s^2
Energy and work	joule	J	kg·m^2/s^2
Power	watt	W	kg·m^2/s^3
Pressure	pascal	Pa	kg/(m·s^2)
Frequency	hertz	Hz	s^{-1}
Electric charge	coulomb	C	A·s
Electric potential	volt	V	kg·m^2/(A·s^3)
Electric resistance	ohm	Ω	kg·m^2/(A^2·s^3)
Capacitance	farad	F	A^2·s^4/(kg·m^2)
Magnetic field	tesla	T	kg/(A·s^2)
Magnetic flux	weber	Wb	kg·m^2/(A·s^2)
Inductance	henry	H	kg·m^2/(s^2·A^2)

[†] kg = kilogram (mass), m = meter (length), s = second (time), A = ampere (electric current).

Metric (SI) Multipliers

Prefix	Abbreviation	Value
yotta	Y	10^{24}
zeta	Z	10^{21}
exa	E	10^{18}
peta	P	10^{15}
tera	T	10^{12}
giga	G	10^9
mega	M	10^6
kilo	k	10^3
hecto	h	10^2
deka	da	10^1
deci	d	10^{-1}
centi	c	10^{-2}
milli	m	10^{-3}
micro	μ	10^{-6}
nano	n	10^{-9}
pico	p	10^{-12}
femto	f	10^{-15}
atto	a	10^{-18}
zepto	z	10^{-21}
yocto	y	10^{-24}

FOURTH EDITION

Volume II

PHYSICS

for

SCIENTISTS & ENGINEERS

with Modern Physics

DOUGLAS C. GIANCOLI

PEARSON

Prentice
Hall

Upper Saddle River, New Jersey 07458

Library of Congress Cataloging-in-Publication Data

Giancoli, Douglas C.
    Physics for scientists and engineers with modern physics / Douglas C.
Giancoli.—4th ed.
        p. cm.
    Includes bibliographical references and index.
    ISBN 0-13-227359-4
    1. Physics—Textbooks.   I. Title.
    QC21.3.G539 2008
    530—dc22

                                                    2006039431

President, ESM: Paul Corey
Sponsoring Editor: Christian Botting
Production Editor: Clare Romeo
Senior Managing Editor: Kathleen Schiaparelli
Art Director and Interior & Cover Designer: John Christiana
Manager, Art Production: Sean Hogan
Senior Development Editor: Karen Karlin
Copy Editor: Jocelyn Phillips
Proofreaders: Marne Evans, Jennie Kaufman, Karen Bosch, Colleen Brosnan,
    Gina Cheselka, and Traci Douglas
Senior Operations Specialist: Alan Fischer
Art Production Editor: Connie Long
Illustrators: Audrey Simonetti and Mark Landis
Photo Researchers: Mary Teresa Giancoli and Truitt & Marshall
Senior Administrative Coordinator: Trisha Tarricone
Composition: Emilcomp/Prepare Inc.; Pearson Education/Clara Bartunek,
    Karen Stephens, Lissette Quinones, Julita Nazario
Photo credits appear on page A-54 which constitutes
    a continuation of the copyright page.

© 2008, 2000, 1989, 1984 by Douglas C. Giancoli

Published by Pearson Education, Inc.
Pearson Prentice Hall
Pearson Education, Inc.
Upper Saddle River, NJ 07458

Pearson Prentice Hall™ is a trademark of Pearson Education, Inc.

Printed in the United States of America
10

ISBN-13: 978-0-13-227359-6
ISBN-10:   0-13-227359-4

Pearson Education LTD., *London*
Pearson Education Australia PTY, Limited, *Sydney*
Pearson Education Singapore, Pte. Ltd.
Pearson Education North Asia Ltd., *Hong Kong*
Pearson Education Canada, Ltd., *Toronto*
Pearson Educación de Mexico, S.A. de C.V.
Pearson Education—Japan, *Tokyo*
Pearson Education Malaysia, Pte. Ltd.

Contents

Volume 1

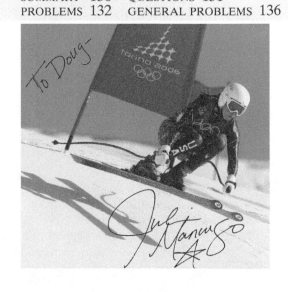

Volume 2

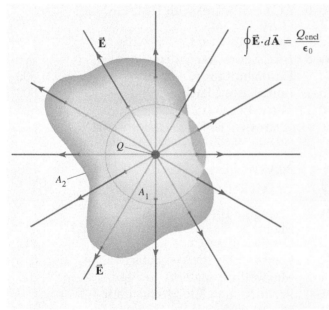

$$\oint \vec{E} \cdot d\vec{A} = \frac{Q_{encl}}{\epsilon_0}$$

subtends a _____ As shown in Fig. 33–33a, the object is placed at the focal poi____ __agnifying gla__, converging lens produces a virtual image, which _____ _f the eye is to focus on it. If the eye is relaxed, comparison of part (a) __ __se the object is exactly at the focal point. __iewed at the near point with __ when you "focus" on the object __ __bject subtends at the eye is much la__ part (b), in which _____ __gular **magnification** or **magnifying power**, __ eye, re__ _____ __e ratio of the angle subtended by an object when using __ __e subtended using the unaided eye, with the object at the nea__ __ the eye ($N = 25$ cm for a normal eye):

$$M = \frac{\theta'}{\theta}, \qquad (33-5)$$

where θ and θ' are shown in Fig. 33–33__ write M in terms of the focal length by noting that $\theta = h/N$ (Fig. 33–__ $\theta' = h/d_o$ (Fig. 33–33a), where h is the height of the object and we ass__ __gles are small so θ and θ' equal __heir sines and tangents. If the eye is __ __ least eye strain), the image will __ at infinity and the object will b__ __ at the focal point; see Fig. 33–34. ___ $d_o = f$ and $\theta' = h/f$. Th__

Volume 3

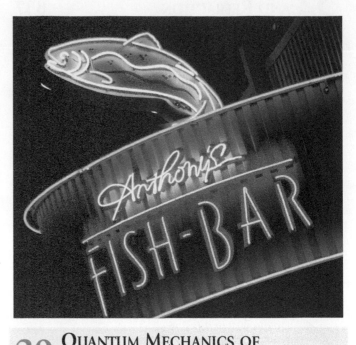

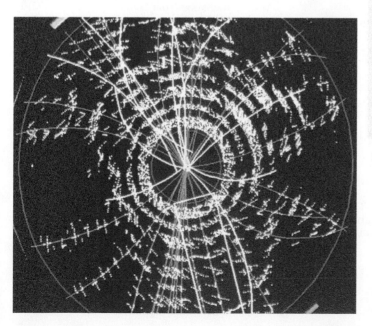

APPLICATIONS (SELECTED)

Preface

I was motivated from the beginning to write a textbook different from others that present physics as a sequence of facts, like a Sears catalog: "here are the facts and you better learn them." Instead of that approach in which topics are begun formally and dogmatically, I have sought to begin each topic with concrete observations and experiences students can relate to: start with specifics and only then go to the great generalizations and the more formal aspects of a topic, showing *why* we believe what we believe. This approach reflects how science is actually practiced.

Why a Fourth Edition?

Two recent trends in physics texbooks are disturbing: (1) their revision cycles have become short—they are being revised every 3 or 4 years; (2) the books are getting larger, some over 1500 pages. I don't see how either trend can be of benefit to students. My response: (1) It has been 8 years since the previous edition of this book. (2) This book makes use of physics education research, although it avoids the detail a Professor may need to say in class but in a book shuts down the reader. And this book still remains among the shortest.

This new edition introduces some important new pedagogic tools. It contains new physics (such as in cosmology) and many new appealing applications (list on previous page). Pages and page breaks have been carefully formatted to make the physics easier to follow: no turning a page in the middle of a derivation or Example. Great efforts were made to make the book attractive so students will want to *read* it.

Some of the new features are listed below.

What's New

Chapter-Opening Questions: Each Chapter begins with a multiple-choice question, whose responses include common misconceptions. Students are asked to answer before starting the Chapter, to get them involved in the material and to get any preconceived notions out on the table. The issues reappear later in the Chapter, usually as Exercises, after the material has been covered. The Chapter-Opening Questions also show students the power and usefulness of Physics.

APPROACH paragraph in worked-out numerical Examples: A short introductory paragraph before the Solution, outlining an approach and the steps we can take to get started. Brief NOTES after the Solution may remark on the Solution, may give an alternate approach, or mention an application.

Step-by-Step Examples: After many Problem Solving Strategies (more than 20 in the book), the next Example is done step-by-step following precisely the steps just seen.

Exercises within the text, after an Example or derivation, give students a chance to see if they have understood enough to answer a simple question or do a simple calculation. Many are multiple choice.

Greater clarity: No topic, no paragraph in this book was overlooked in the search to improve the clarity and conciseness of the presentation. Phrases and sentences that may slow down the principal argument have been eliminated: keep to the essentials at first, give the elaborations later.

$\vec{F}, \vec{v}, \vec{B}$ ***Vector notation, arrows***: The symbols for vector quantities in the text and Figures now have a tiny arrow over them, so they are similar to what we write by hand.

Cosmological Revolution: With generous help from top experts in the field, readers have the latest results.

Page layout: more than in the previous edition, serious attention has been paid to how each page is formatted. Examples and all important derivations and arguments are on facing pages. Students then don't have to turn back and forth. Throughout, readers see, on two facing pages, an important slice of physics.

New Applications: LCDs, digital cameras and electronic sensors (CCD, CMOS), electric hazards, GFCIs, photocopiers, inkjet and laser printers, metal detectors, underwater vision, curve balls, airplane wings, DNA, how we actually *see* images. (Turn back a page to see a longer list.)

Examples modified: more math steps are spelled out, and many new Examples added. About 10% of all Examples are Estimation Examples.

This Book is Shorter than other complete full-service books at this level. Shorter explanations are easier to understand and more likely to be read.

Content and Organizational Changes

- **Rotational Motion**: Chapters 10 and 11 have been reorganized. All of angular momentum is now in Chapter 11.

- **First law of thermodynamics**, in Chapter 19, has been rewritten and extended. The full form is given: $\Delta K + \Delta U + \Delta E_{int} = Q - W$, where internal energy is E_{int}, and U is potential energy; the form $Q - W$ is kept so that $dW = P\,dV$.

- Kinematics and Dynamics of Circular Motion are now treated together in Chapter 5.

- Work and Energy, Chapters 7 and 8, have been carefully revised.

- Work done by friction is discussed now with energy conservation (energy terms due to friction).

- Chapters on Inductance and AC Circuits have been combined into one: Chapter 30.

- Graphical Analysis and Numerical Integration is a new optional Section 2–9. Problems requiring a computer or graphing calculator are found at the end of most Chapters.

- Length of an object is a script ℓ rather than normal l, which looks like 1 or I (moment of inertia, current), as in $F = I\ell B$. Capital L is for angular momentum, latent heat, inductance, dimensions of length $[L]$.

- Newton's law of gravitation remains in Chapter 6. Why? Because the $1/r^2$ law is too important to relegate to a late chapter that might not be covered at all late in the semester; furthermore, it is one of the basic forces in nature. In Chapter 8 we can treat real gravitational potential energy and have a fine instance of using $U = -\int \vec{\mathbf{F}} \cdot d\vec{\boldsymbol{\ell}}$.

- New Appendices include the differential form of Maxwell's equations and more on dimensional analysis.

- Problem Solving Strategies are found on pages 30, 58, 64, 96, 102, 125, 166, 198, 229, 261, 314, 504, 551, 571, 600, 685, 716, 740, 763, 849, 871, and 913.

Organization

Some instructors may find that this book contains more material than can be covered in their courses. The text offers great flexibility. Sections marked with a star * are considered optional. These contain slightly more advanced physics material, or material not usually covered in typical courses and/or interesting applications; they contain no material needed in later Chapters (except perhaps in later optional Sections). For a brief course, all optional material could be dropped as well as major parts of Chapters 1, 13, 16, 26, 30, and 35, and selected parts of Chapters 9, 12, 19, 20, 33, and the modern physics Chapters. Topics not covered in class can be a valuable resource for later study by students. Indeed, this text can serve as a useful reference for years because of its wide range of coverage.

Versions of this Book

Complete version: 44 Chapters including 9 Chapters of modern physics.

Classic version: 37 Chapters including one each on relativity and quantum theory.

3 Volume version: Available separately or packaged together (Vols. 1 & 2 or all 3 Volumes):

Volume 1: Chapters 1–20 on mechanics, including fluids, oscillations, waves, plus heat and thermodynamics.

Volume 2: Chapters 21–35 on electricity and magnetism, plus light and optics.

Volume 3: Chapters 36–44 on modern physics: relativity, quantum theory, atomic physics, condensed matter, nuclear physics, elementary particles, cosmology and astrophysics.

Thanks

Many physics professors provided input or direct feedback on every aspect of this textbook. They are listed below, and I owe each a debt of gratitude.

Mario Affatigato, Coe College
Lorraine Allen, United States Coast Guard Academy
Zaven Altounian, McGill University
Bruce Barnett, Johns Hopkins University
Michael Barnett, Lawrence Berkeley Lab
Anand Batra, Howard University
Cornelius Bennhold, George Washington University
Bruce Birkett, University of California Berkeley
Dr. Robert Boivin, Auburn University
Subir Bose, University of Central Florida
David Branning, Trinity College
Meade Brooks, Collin County Community College
Bruce Bunker, University of Notre Dame
Grant Bunker, Illinois Institute of Technology
Wayne Carr, Stevens Institute of Technology
Charles Chiu, University of Texas Austin
Robert Coakley, University of Southern Maine
David Curott, University of North Alabama
Biman Das, SUNY Potsdam
Bob Davis, Taylor University
Kaushik De, University of Texas Arlington
Michael Dennin, University of California Irvine
Kathy Dimiduk, University of New Mexico
John DiNardo, Drexel University
Scott Dudley, United States Air Force Academy
John Essick, Reed College
Cassandra Fesen, Dartmouth College
Alex Filippenko, University of California Berkeley
Richard Firestone, Lawrence Berkeley Lab
Mike Fortner, Northern Illinois University
Tom Furtak, Colorado School of Mines
Edward Gibson, California State University Sacramento
John Hardy, Texas A&M
J. Erik Hendrickson, University of Wisconsin Eau Claire
Laurent Hodges, Iowa State University
David Hogg, New York University
Mark Hollabaugh, Normandale Community College
Andy Hollerman, University of Louisiana at Lafayette
William Holzapfel, University of California Berkeley
Bob Jacobsen, University of California Berkeley
Teruki Kamon, Texas A&M
Daryao Khatri, University of the District of Columbia
Jay Kunze, Idaho State University

Jim LaBelle, Dartmouth College
M.A.K. Lodhi, Texas Tech
Bruce Mason, University of Oklahoma
Dan Mazilu, Virginia Tech
Linda McDonald, North Park College
Bill McNairy, Duke University
Raj Mohanty, Boston University
Giuseppe Molesini, Istituto Nazionale di Ottica Florence
Lisa K. Morris, Washington State University
Blaine Norum, University of Virginia
Alexandria Oakes, Eastern Michigan University
Michael Ottinger, Missouri Western State University
Lyman Page, Princeton and WMAP
Bruce Partridge, Haverford College
R. Daryl Pedigo, University of Washington
Robert Pelcovitz, Brown University
Vahe Peroomian, UCLA
James Rabchuk, Western Illinois University
Michele Rallis, Ohio State University
Paul Richards, University of California Berkeley
Peter Riley, University of Texas Austin
Larry Rowan, University of North Carolina Chapel Hill
Cindy Schwarz, Vassar College
Peter Sheldon, Randolph-Macon Woman's College
Natalia A. Sidorovskaia, University of Louisiana at Lafayette
James Siegrist, UC Berkeley, Director Physics Division LBNL
George Smoot, University of California Berkeley
Mark Sprague, East Carolina University
Michael Strauss, University of Oklahoma
Laszlo Takac, University of Maryland Baltimore Co.
Franklin D. Trumpy, Des Moines Area Community College
Ray Turner, Clemson University
Som Tyagi, Drexel University
John Vasut, Baylor University
Robert Webb, Texas A&M
Robert Weidman, Michigan Technological University
Edward A. Whittaker, Stevens Institute of Technology
John Wolbeck, Orange County Community College
Stanley George Wojcicki, Stanford University
Edward Wright, UCLA
Todd Young, Wayne State College
William Younger, College of the Albemarle
Hsiao-Ling Zhou, Georgia State University

I owe special thanks to Prof. Bob Davis for much valuable input, and especially for working out all the Problems and producing the Solutions Manual for all Problems, as well as for providing the answers to odd-numbered Problems at the end of this book. Many thanks also to J. Erik Hendrickson who collaborated with Bob Davis on the solutions, and to the team they managed (Profs. Anand Batra, Meade Brooks, David Currott, Blaine Norum, Michael Ottinger, Larry Rowan, Ray Turner, John Vasut, William Younger). I am grateful to Profs. John Essick, Bruce Barnett, Robert Coakley, Biman Das, Michael Dennin, Kathy Dimiduk, John DiNardo, Scott Dudley, David Hogg, Cindy Schwarz, Ray Turner, and Som Tyagi, who inspired many of the Examples, Questions, Problems, and significant clarifications.

Crucial for rooting out errors, as well as providing excellent suggestions, were Profs. Kathy Dimiduk, Ray Turner, and Lorraine Allen. A huge thank you to them and to Prof. Giuseppe Molesini for his suggestions and his exceptional photographs for optics.

For Chapters 43 and 44 on Particle Physics and Cosmology and Astrophysics, I was fortunate to receive generous input from some of the top experts in the field, to whom I owe a debt of gratitude: George Smoot, Paul Richards, Alex Filippenko, James Siegrist, and William Holzapfel (UC Berkeley), Lyman Page (Princeton and WMAP), Edward Wright (UCLA and WMAP), and Michael Strauss (University of Oklahoma).

I especially wish to thank Profs. Howard Shugart, Chair Frances Hellman, and many others at the University of California, Berkeley, Physics Department for helpful discussions, and for hospitality. Thanks also to Prof. Tito Arecchi and others at the Istituto Nazionale di Ottica, Florence, Italy.

Finally, I am grateful to the many people at Prentice Hall with whom I worked on this project, especially Paul Corey, Karen Karlin, Christian Botting, John Christiana, and Sean Hogan.

The final responsibility for all errors lies with me. I welcome comments, corrections, and suggestions as soon as possible to benefit students for the next reprint.

D.C.G.

email: Paul.Corey@Pearson.com

Post: Paul Corey
    One Lake Street
    Upper Saddle River, NJ 07458

About the Author

Douglas C. Giancoli obtained his BA in physics (summa cum laude) from the University of California, Berkeley, his MS in physics at the Massachusetts Institute of Technology, and his PhD in elementary particle physics at the University of California, Berkeley. He spent 2 years as a post-doctoral fellow at UC Berkeley's Virus lab developing skills in molecular biology and biophysics. His mentors include Nobel winners Emilio Segrè and Donald Glaser.

He has taught a wide range of undergraduate courses, traditional as well as innovative ones, and continues to update his textbooks meticulously, seeking ways to better provide an understanding of physics for students.

Doug's favorite spare-time activity is the outdoors, especially climbing peaks (here he is heading for the top). He says climbing peaks is like learning physics: it takes effort and the rewards are great.

Online Supplements (partial list)

MasteringPhysics™ (www.masteringphysics.com)
is a sophisticated online tutoring and homework system developed specially for courses using calculus-based physics. Originally developed by David Pritchard and collaborators at MIT, MasteringPhysics provides **students** with individualized online tutoring by responding to their wrong answers and providing hints for solving multi-step problems when they get stuck. It gives them immediate and up-to-date assessment of their progress, and shows where they need to practice more. MasteringPhysics provides **instructors** with a fast and effective way to assign tried-and-tested online homework assignments that comprise a range of problem types. The powerful post-assignment diagnostics allow instructors to assess the progress of their class as a whole as well as individual students, and quickly identify areas of difficulty.

WebAssign (www.webassign.com)
CAPA and LON-CAPA (www.lon-capa.org)

Student Supplements (partial list)

Student Study Guide & Selected Solutions Manual (Volume I: 0-13-227324-1, Volumes II & III: 0-13-227325-X) by Frank Wolfs

Student Pocket Companion (0-13-227326-8) by Biman Das

Tutorials in Introductory Physics (0-13-097069-7)
by Lillian C. McDermott, Peter S. Schaffer, and the Physics Education Group at the University of Washington

Physlet® Physics (0-13-101969-4)
by Wolfgang Christian and Mario Belloni

Ranking Task Exercises in Physics, Student Edition (0-13-144851-X)
by Thomas L. O'Kuma, David P. Maloney, and Curtis J. Hieggelke

E&M TIPERs: Electricity & Magnetism Tasks Inspired by Physics Education Research (0-13-185499-2) by Curtis J. Hieggelke, David P. Maloney, Stephen E. Kanim, and Thomas L. O'Kuma

Mathematics for Physics with Calculus (0-13-191336-0)
by Biman Das

To Students

HOW TO STUDY

1. Read the Chapter. Learn new vocabulary and notation. Try to respond to questions and exercises as they occur.
2. Attend all class meetings. Listen. Take notes, especially about aspects you do not remember seeing in the book. Ask questions (everyone else wants to, but maybe you will have the courage). You will get more out of class if you read the Chapter first.
3. Read the Chapter again, paying attention to details. Follow derivations and worked-out Examples. Absorb their logic. Answer Exercises and as many of the end of Chapter Questions as you can.
4. Solve 10 to 20 end of Chapter Problems (or more), especially those assigned. In doing Problems you find out what you learned and what you didn't. Discuss them with other students. Problem solving is one of the great learning tools. Don't just look for a formula—it won't cut it.

NOTES ON THE FORMAT AND PROBLEM SOLVING

1. Sections marked with a star (*) are considered **optional**. They can be omitted without interrupting the main flow of topics. No later material depends on them except possibly later starred Sections. They may be fun to read, though.
2. The customary **conventions** are used: symbols for quantities (such as m for mass) are italicized, whereas units (such as m for meter) are not italicized. Symbols for vectors are shown in boldface with a small arrow above: $\vec{F}$.
3. Few equations are valid in all situations. Where practical, the **limitations** of important equations are stated in square brackets next to the equation. The equations that represent the great laws of physics are displayed with a tan background, as are a few other indispensable equations.
4. At the end of each Chapter is a set of **Problems** which are ranked as Level I, II, or III, according to estimated difficulty. Level I Problems are easiest, Level II are standard Problems, and Level III are "challenge problems." These ranked Problems are arranged by Section, but Problems for a given Section may depend on earlier material too. There follows a group of General Problems, which are not arranged by Section nor ranked as to difficulty. Problems that relate to optional Sections are starred (*). Most Chapters have 1 or 2 Computer/Numerical Problems at the end, requiring a computer or graphing calculator. Answers to odd-numbered Problems are given at the end of the book.
5. Being able to solve **Problems** is a crucial part of learning physics, and provides a powerful means for understanding the concepts and principles. This book contains many aids to problem solving: (a) worked-out **Examples** and their solutions in the text, which should be studied as an integral part of the text; (b) some of the worked-out Examples are **Estimation Examples**, which show how rough or approximate results can be obtained even if the given data are sparse (see Section 1–6); (c) special **Problem Solving Strategies** placed throughout the text to suggest a step-by-step approach to problem solving for a particular topic—but remember that the basics remain the same; most of these "Strategies" are followed by an Example that is solved by explicitly following the suggested steps; (d) special problem-solving Sections; (e) "Problem Solving" marginal notes which refer to hints within the text for solving Problems; (f) **Exercises** within the text that you should work out immediately, and then check your response against the answer given at the bottom of the last page of that Chapter; (g) the Problems themselves at the end of each Chapter (point 4 above).
6. **Conceptual Examples** pose a question which hopefully starts you to think and come up with a response. Give yourself a little time to come up with your own response before reading the Response given.
7. **Math** review, plus some additional topics, are found in Appendices. Useful data, conversion factors, and math formulas are found inside the front and back covers.

USE OF COLOR

Vectors

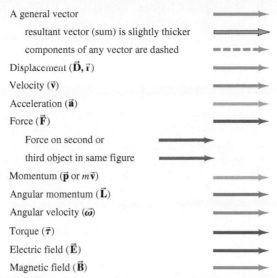

A general vector

    resultant vector (sum) is slightly thicker

    components of any vector are dashed

Displacement ($\vec{\mathbf{D}}, \vec{\mathbf{r}}$)

Velocity ($\vec{\mathbf{v}}$)

Acceleration ($\vec{\mathbf{a}}$)

Force ($\vec{\mathbf{F}}$)

    Force on second or

    third object in same figure

Momentum ($\vec{\mathbf{p}}$ or $m\vec{\mathbf{v}}$)

Angular momentum ($\vec{\mathbf{L}}$)

Angular velocity ($\vec{\boldsymbol{\omega}}$)

Torque ($\vec{\boldsymbol{\tau}}$)

Electric field ($\vec{\mathbf{E}}$)

Magnetic field ($\vec{\mathbf{B}}$)

Electricity and magnetism

Electric field lines

Equipotential lines

Magnetic field lines

Electric charge (+)   +   or   ● +

Electric charge (−)   −   or   ● −

Electric circuit symbols

Wire, with switch S

Resistor

Capacitor

Inductor

Battery

Ground

Optics

Light rays

Object

Real image (dashed)

Virtual image (dashed and paler)

Other

Energy level (atom, etc.)

Measurement lines |←1.0 m→|

Path of a moving object

Direction of motion or current

This comb has acquired a static electric charge, either from passing through hair, or being rubbed by a cloth or paper towel. The electrical charge on the comb induces a polarization (separation of charge) in scraps of paper, and thus attracts them.

Our introduction to electricity in this Chapter covers conductors and insulators, and Coulomb's law which relates the force between two point charges as a function of their distance apart. We also introduce the powerful concept of electric field.

CHAPTER 21

Electric Charge and Electric Field

CHAPTER-OPENING QUESTION—Guess now!

Two identical tiny spheres have the same electric charge. If the electric charge on each of them is doubled, and their separation is also doubled, the force each exerts on the other will be

(a) half.
(b) double.
(c) four times larger.
(d) one-quarter as large.
(e) unchanged.

The word "electricity" may evoke an image of complex modern technology: lights, motors, electronics, and computers. But the electric force plays an even deeper role in our lives. According to atomic theory, electric forces between atoms and molecules hold them together to form liquids and solids, and electric forces are also involved in the metabolic processes that occur within our bodies. Many of the forces we have dealt with so far, such as elastic forces, the normal force, and friction and other contact forces (pushes and pulls), are now considered to result from electric forces acting at the atomic level. Gravity, on the other hand, is a separate force.[†]

[†] As we discussed in Section 6–7, physicists in the twentieth century came to recognize four different fundamental forces in nature: (1) gravitational force, (2) electromagnetic force (we will see later that electric and magnetic forces are intimately related), (3) strong nuclear force, and (4) weak nuclear force. The last two forces operate at the level of the nucleus of an atom. Recent theory has combined the electromagnetic and weak nuclear forces so they are now considered to have a common origin known as the electroweak force. We will discuss these forces in later Chapters.

CONTENTS

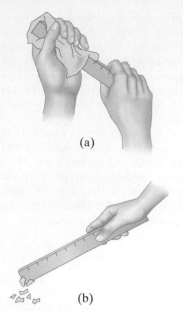

(a)

(b)

FIGURE 21-1 (a) Rub a plastic ruler and (b) bring it close to some tiny pieces of paper.

FIGURE 21-2 Like charges repel one another; unlike charges attract. (Note color coding: positive and negative charged objects are often colored pink and blue-green, respectively, when we want to emphasize them. We use these colors especially for point charges, but not often for real objects.)

(a) Two charged plastic rulers repel

(b) Two charged glass rods repel

(c) Charged glass rod attracts charged plastic ruler

LAW OF CONSERVATION OF ELECTRIC CHARGE

The earliest studies on electricity date back to the ancients, but only in the past two centuries has electricity been studied in detail. We will discuss the development of ideas about electricity, including practical devices, as well as its relation to magnetism, in the next eleven Chapters.

21-1 Static Electricity; Electric Charge and Its Conservation

The word *electricity* comes from the Greek word *elektron*, which means "amber." Amber is petrified tree resin, and the ancients knew that if you rub a piece of amber with a cloth, the amber attracts small pieces of leaves or dust. A piece of hard rubber, a glass rod, or a plastic ruler rubbed with a cloth will also display this "amber effect," or **static electricity** as we call it today. You can readily pick up small pieces of paper with a plastic comb or ruler that you have just vigorously rubbed with even a paper towel. See the photo on the previous page and Fig. 21-1. You have probably experienced static electricity when combing your hair or when taking a synthetic blouse or shirt from a clothes dryer. And you may have felt a shock when you touched a metal doorknob after sliding across a car seat or walking across a nylon carpet. In each case, an object becomes "charged" as a result of rubbing, and is said to possess a net **electric charge**.

Is all electric charge the same, or is there more than one type? In fact, there are *two* types of electric charge, as the following simple experiments show. A plastic ruler suspended by a thread is vigorously rubbed with a cloth to charge it. When a second plastic ruler, which has been charged in the same way, is brought close to the first, it is found that one ruler *repels* the other. This is shown in Fig. 21-2a. Similarly, if a rubbed glass rod is brought close to a second charged glass rod, again a repulsive force is seen to act, Fig. 21-2b. However, if the charged glass rod is brought close to the charged plastic ruler, it is found that they *attract* each other, Fig. 21-2c. The charge on the glass must therefore be different from that on the plastic. Indeed, it is found experimentally that all charged objects fall into one of two categories. Either they are attracted to the plastic and repelled by the glass; or they are repelled by the plastic and attracted to the glass. Thus there seem to be two, and only two, types of electric charge. Each type of charge repels the same type but attracts the opposite type. That is: **unlike charges attract; like charges repel**.

The two types of electric charge were referred to as *positive* and *negative* by the American statesman, philosopher, and scientist Benjamin Franklin (1706–1790). The choice of which name went with which type of charge was arbitrary. Franklin's choice set the charge on the rubbed glass rod to be positive charge, so the charge on a rubbed plastic ruler (or amber) is called negative charge. We still follow this convention today.

Franklin argued that whenever a certain amount of charge is produced on one object, an equal amount of the opposite type of charge is produced on another object. The positive and negative are to be treated *algebraically*, so during any process, the net change in the amount of charge produced is zero. For example, when a plastic ruler is rubbed with a paper towel, the plastic acquires a negative charge and the towel acquires an equal amount of positive charge. The charges are separated, but the sum of the two is zero.

This is an example of a law that is now well established: the **law of conservation of electric charge**, which states that

> **the net amount of electric charge produced in any process is zero;**

or, said another way,

> **no net electric charge can be created or destroyed.**

If one object (or a region of space) acquires a positive charge, then an equal amount of negative charge will be found in neighboring areas or objects. No violations have ever been found, and this conservation law is as firmly established as those for energy and momentum.

21–2 Electric Charge in the Atom

century has it become clear that an understanding of
inside the atom itself. In later Chapters we will discuss atomic
that led to our present view of the atom in more detail. But
standing of electricity if we discuss it briefly now.

del of an atom shows it as having a tiny but heavy, positively
ounded by one or more negatively charged electrons (Fig. 21–3).
s protons, which are positively charged, and neutrons, which
charge. All protons and all electrons have exactly the same
ic charge; but their signs are opposite. Hence neutral atoms,
e, contain equal numbers of protons and electrons. Sometimes
ne or more of its electrons, or may gain extra electrons, in which
et positive or negative charge and is called an **ion**.

als the nuclei tend to remain close to fixed positions, whereas
ns may move quite freely. When an object is *neutral*, it contains
ositive and negative charge. The charging of a solid object by
ained by the transfer of electrons from one object to the other.
r becomes negatively charged by rubbing with a paper towel,
the transfer of electrons from the towel to the plastic leaves the towel with a
positive charge equal in magnitude to the negative charge acquired by the plastic.
In liquids and gases, nuclei or ions can move as well as electrons.

Normally when objects are charged by rubbing, they hold their charge only for
a limited time and eventually return to the neutral state. Where does the charge
go? Usually the charge "leaks off" onto water molecules in the air. This is because
water molecules are **polar**—that is, even though they are neutral, their charge is not
distributed uniformly, Fig. 21–4. Thus the extra electrons on, say, a charged plastic ruler
can "leak off" into the air because they are attracted to the positive end of water molecules. A positively charged object, on the other hand, can be neutralized by transfer of
loosely held electrons from water molecules in the air. On dry days, static electricity is
much more noticeable since the air contains fewer water molecules to allow leakage.
On humid or rainy days, it is difficult to make any object hold a net charge for long.

21–3 Insulators and Conductors

Suppose we have two metal spheres, one highly charged and the other electrically
neutral (Fig. 21–5a). If we now place a metal object, such as a nail, so that it
touches both spheres (Fig. 21–5b), the previously uncharged sphere quickly
becomes charged. If, instead, we had connected the two spheres by a wooden rod
or a piece of rubber (Fig. 21–5c), the uncharged ball would not become noticeably
charged. Materials like the iron nail are said to be **conductors** of electricity,
whereas wood and rubber are **nonconductors** or **insulators**.

Metals are generally good conductors, whereas most other materials are insulators (although even insulators conduct electricity very slightly). Nearly all natural
materials fall into one or the other of these two very distinct categories. However,
a few materials (notably silicon and germanium) fall into an intermediate category
known as **semiconductors**.

From the atomic point of view, the electrons in an insulating material are
bound very tightly to the nuclei. In a good conductor, on the other hand, some of
the electrons are bound very loosely and can move about freely within the
material (although they cannot *leave* the object easily) and are often referred to as
free electrons or *conduction electrons*. When a positively charged object is brought
close to or touches a conductor, the free electrons in the conductor are attracted
by this positively charged object and move quickly toward it. On the other hand,
the free electrons move swiftly away from a negatively charged object that is
brought close to the conductor. In a semiconductor, there are many fewer free
electrons, and in an insulator, almost none.

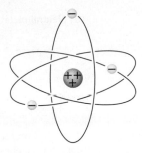

FIGURE 21–3 Simple model of the atom.

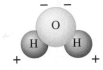

FIGURE 21–4 Diagram of a water molecule. Because it has opposite charges on different ends, it is called a "polar" molecule.

FIGURE 21–5 (a) A charged metal sphere and a neutral metal sphere. (b) The two spheres connected by a conductor (a metal nail), which conducts charge from one sphere to the other. (c) The original two spheres connected by an insulator (wood); almost no charge is conducted.

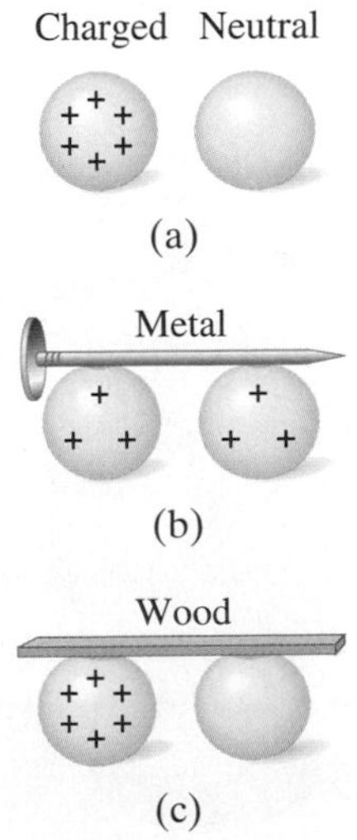

21–4 Induced Charge; the Electroscope

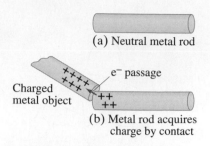
(a) Neutral metal rod

Charged metal object
e^- passage
(b) Metal rod acquires charge by contact

FIGURE 21–6 A neutral metal rod in (a) will acquire a positive charge if placed in contact (b) with a positively charged metal object. (Electrons move as shown by the orange arrow.) This is called charging by conduction.

Suppose a positively charged metal object is brought close to an uncharged metal object. If the two touch, the free electrons in the neutral one are attracted to the positively charged object and some will pass over to it, Fig. 21–6. Since the second object, originally neutral, is now missing some of its negative electrons, it will have a net positive charge. This process is called "charging by conduction," or "by contact," and the two objects end up with the same sign of charge.

Now suppose a positively charged object is brought close to a neutral metal rod, but does not touch it. Although the free electrons of the metal rod do not leave the rod, they still move within the metal toward the external positive charge, leaving a positive charge at the opposite end of the rod (Fig. 21–7). A charge is said to have been *induced* at the two ends of the metal rod. No net charge has been created in the rod: charges have merely been *separated*. The net charge on the metal rod is still zero. However, if the metal is separated into two pieces, we would have two charged objects: one charged positively and one charged negatively.

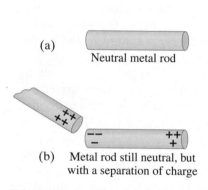
(a) Neutral metal rod

(b) Metal rod still neutral, but with a separation of charge

FIGURE 21–7 Charging by induction.

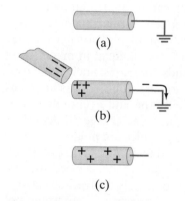
(a)

(b)

(c)

FIGURE 21–8 Inducing a charge on an object connected to ground.

Another way to induce a net charge on a metal object is to first connect it with a conducting wire to the ground (or a conducting pipe leading into the ground) as shown in Fig. 21–8a (the symbol $\equiv$ means connected to "ground"). The object is then said to be "grounded" or "earthed." The Earth, because it is so large and can conduct, easily accepts or gives up electrons; hence it acts like a reservoir for charge. If a charged object—say negative this time—is brought up close to the metal object, free electrons in the metal are repelled and many of them move down the wire into the Earth, Fig. 21–8b. This leaves the metal positively charged. If the wire is now cut, the metal object will have a positive induced charge on it (Fig. 21–8c). If the wire were cut after the negative object was moved away, the electrons would all have moved back into the metal object and it would be neutral.

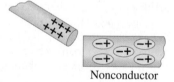
Nonconductor

FIGURE 21–9 A charged object brought near an insulator causes a charge separation within the insulator's molecules.

FIGURE 21–10 Electroscope.

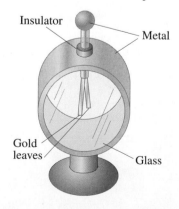
Insulator
Metal
Gold leaves
Glass

Charge separation can also be done in nonconductors. If you bring a positively charged object close to a neutral nonconductor as shown in Fig. 21–9, almost no electrons can move about freely within the nonconductor. But they can move slightly within their own atoms and molecules. Each oval in Fig. 21–9 represents a molecule (not to scale); the negatively charged electrons, attracted to the external positive charge, tend to move in its direction within their molecules. Because the negative charges in the nonconductor are nearer to the external positive charge, the nonconductor as a whole is attracted to the external positive charge (see the Chapter-Opening Photo, page 559).

An **electroscope** is a device that can be used for detecting charge. As shown in Fig. 21–10, inside of a case are two movable metal leaves, often made of gold, connected to a metal knob on the outside. (Sometimes only one leaf is movable.)

If a positively charged object is brought close to the knob, a separation of charge is induced: electrons are attracted up into the knob, leaving the leaves positively charged, Fig. 21–11a. The two leaves repel each other as shown, because they are both positively charged. If, instead, the knob is charged by conduction, the whole apparatus acquires a net charge as shown in Fig. 21–11b. In either case, the greater the amount of charge, the greater the separation of the leaves.

Note that you cannot tell the sign of the charge in this way since negative charge will cause the leaves to separate just as much as an equal amount of positive charge; in either case, the two leaves repel each other. An electroscope can, however, be used to determine the sign of the charge if it is first charged by conduction, say, negatively, as in Fig. 21–12a. Now if a negative object is brought close, as in Fig. 21–12b, more electrons are induced to move down into the leaves and they separate further. If a positive charge is brought close instead, the electrons are induced to flow upward, leaving the leaves less negative and their separation is reduced, Fig. 21–12c.

The electroscope was used in the early studies of electricity. The same principle, aided by some electronics, is used in much more sensitive modern **electrometers**.

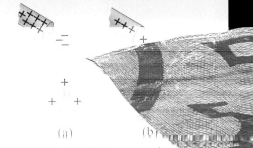

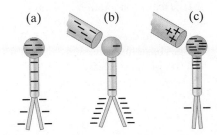

FIGURE 21–11 Electroscope charged (a) by induction, (b) by conduction.

FIGURE 21–12 A previously charged electroscope can be used to determine the sign of a charged object.

21–5 Coulomb's Law

We have seen that an electric charge exerts a force of attraction or repulsion on other electric charges. What factors affect the magnitude of this force? To find an answer, the French physicist Charles Coulomb (1736–1806) investigated electric forces in the 1780s using a torsion balance (Fig. 21–13) much like that used by Cavendish for his studies of the gravitational force (Chapter 6).

Precise instruments for the measurement of electric charge were not available in Coulomb's time. Nonetheless, Coulomb was able to prepare small spheres with different magnitudes of charge in which the *ratio* of the charges was known.[†] Although he had some difficulty with induced charges, Coulomb was able to argue that the force one tiny charged object exerts on a second tiny charged object is directly proportional to the charge on each of them. That is, if the charge on either one of the objects is doubled, the force is doubled; and if the charge on both of the objects is doubled, the force increases to four times the original value. This was the case when the distance between the two charges remained the same. If the distance between them was allowed to increase, he found that the force decreased with the *square of the distance* between them. That is, if the distance was doubled, the force fell to one-fourth of its original value. Thus, Coulomb concluded, the force one small charged object exerts on a second one is proportional to the product of the magnitude of the charge on one, Q_1, times the magnitude of the charge on the other, Q_2, and inversely proportional to the square of the distance r between them (Fig. 21–14). As an equation, we can write **Coulomb's law** as

$$F = k\frac{Q_1 Q_2}{r^2}, \qquad \text{[magnitudes]} \quad \textbf{(21–1)}$$

where k is a proportionality constant.[‡]

FIGURE 21–13 (below) Coulomb used a torsion balance to investigate how the electric force varies as a function of the magnitude of the charges and of the distance between them. When an external charged sphere is placed close to the charged one on the suspended bar, the bar rotates slightly. The suspending fiber resists the twisting motion, and the angle of twist is proportional to the electric force.

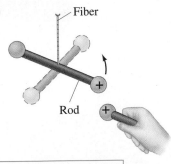

COULOMB'S LAW

FIGURE 21–14 Coulomb's law, Eq. 21–1, gives the force between two point charges, Q_1 and Q_2, a distance r apart.

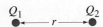

[†]Coulomb reasoned that if a charged conducting sphere is placed in contact with an identical uncharged sphere, the charge on the first would be shared equally by the two of them because of symmetry. He thus had a way to produce charges equal to $\frac{1}{2}, \frac{1}{4}$, and so on, of the original charge.

[‡]The validity of Coulomb's law today rests on precision measurements that are much more sophisticated than Coulomb's original experiment. The exponent 2 in Coulomb's law has been shown to be accurate to 1 part in 10^{16} [that is, $2 \pm (1 \times 10^{-16})$].

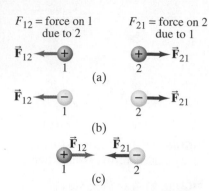

F_{12} = force on 1
due to 2

F_{21} = force on 2
due to 1

$\vec{\mathbf{F}}_{12}$ ⟵ ⊕   ⊕ ⟶ $\vec{\mathbf{F}}_{21}$
    1   2

(a)

$\vec{\mathbf{F}}_{12}$ ⟵ ⊖   ⊖ ⟶ $\vec{\mathbf{F}}_{21}$
    1   2

(b)

⊕ $\vec{\mathbf{F}}_{12}$ ⟶ $\vec{\mathbf{F}}_{21}$ ⟵ ⊖
1   2

(c)

FIGURE 21–15 The direction of the static electric force one point charge exerts on another is always along the line joining the two charges, and depends on whether the charges have the same sign as in (a) and (b), or opposite signs (c).

As we just saw, Coulomb's law,

$$F = k\frac{Q_1 Q_2}{r^2},$$

[magnitudes] **(21–1)**

gives the *magnitude* of the electric force that either charge exerts on the other. The *direction* of the electric force *is always along the line joining the two charges*. If the two charges have the same sign, the force on either charge is directed away from the other (they repel each other). If the two charges have opposite signs, the force on one is directed toward the other (they attract). See Fig. 21–15. Notice that the force one charge exerts on the second is equal but opposite to that exerted by the second on the first, in accord with Newton's third law.

The SI unit of charge is the **coulomb** (C).[†] The precise definition of the coulomb today is in terms of electric current and magnetic field, and will be discussed later (Section 28–3). In SI units, the constant k in Coulomb's law has the value

$$k = 8.99 \times 10^9 \, \text{N} \cdot \text{m}^2/\text{C}^2$$

or, when we only need two significant figures,

$$k \approx 9.0 \times 10^9 \, \text{N} \cdot \text{m}^2/\text{C}^2.$$

Thus, 1 C is that amount of charge which, if placed on each of two point objects that are 1.0 m apart, will result in each object exerting a force of $(9.0 \times 10^9 \, \text{N} \cdot \text{m}^2/\text{C}^2)(1.0 \, \text{C})(1.0 \, \text{C})/(1.0 \, \text{m})^2 = 9.0 \times 10^9 \, \text{N}$ on the other. This would be an enormous force, equal to the weight of almost a million tons. We rarely encounter charges as large as a coulomb.

Charges produced by rubbing ordinary objects (such as a comb or plastic ruler) are typically around a microcoulomb $(1 \, \mu\text{C} = 10^{-6} \, \text{C})$ or less. Objects that carry a positive charge have a deficit of electrons, whereas negatively charged objects have an excess of electrons. The charge on one electron has been determined to have a magnitude of about $1.602 \times 10^{-19} \, \text{C}$, and is negative. This is the smallest charge found in nature,[‡] and because it is fundamental, it is given the symbol e and is often referred to as the *elementary charge*:

$$e = 1.602 \times 10^{-19} \, \text{C}.$$

Note that e is defined as a positive number, so the charge on the electron is $-e$. (The charge on a proton, on the other hand, is $+e$.) Since an object cannot gain or lose a fraction of an electron, the net charge on any object must be an integral multiple of this charge. Electric charge is thus said to be **quantized** (existing only in discrete amounts: $1e$, $2e$, $3e$, etc.). Because e is so small, however, we normally do not notice this discreteness in macroscopic charges (1 μC requires about 10^{13} electrons), which thus seem continuous.

Coulomb's law looks a lot like the *law of universal gravitation*, $F = Gm_1 m_2/r^2$, which expresses the gravitational force a mass m_1 exerts on a mass m_2 (Eq. 6–1). Both are **inverse square laws** $(F \propto 1/r^2)$. Both also have a proportionality to a property of each object—mass for gravity, electric charge for electricity. And both act over a distance (that is, there is no need for contact). A major difference between the two laws is that gravity is always an attractive force, whereas the electric force can be either attractive or repulsive. Electric charge comes in two types, positive and negative; gravitational mass is only positive.

[†]In the once common cgs system of units, k is set equal to 1, and the unit of electric charge is called the *electrostatic unit* (esu) or the statcoulomb. One esu is defined as that charge, on each of two point objects 1 cm apart, that gives rise to a force of 1 dyne.

[‡]According to the standard model of elementary particle physics, subnuclear particles called quarks have a smaller charge than that on the electron, equal to $\frac{1}{3}e$ or $\frac{2}{3}e$. Quarks have not been detected directly as isolated objects, and theory indicates that free quarks may not be detectable.

The constant k in Eq. 21–1 is often written in terms of another constant, ϵ_0, called the **permittivity of free space**. It is related to k by $k = 1/4\pi\epsilon_0$. Coulomb's law can then be written

$$F = \frac{1}{4\pi\epsilon_0}\frac{Q_1 Q_2}{r^2}. \qquad (21\text{–}2)$$

where

$$\epsilon_0 = \frac{1}{4\pi k} = 8.85 \times 10^{-12}\ C^2/N\cdot m^2.$$

Equation 21–2 looks more complicated than Eq. 21–1, but other fundamental equations we haven't seen yet are simpler in terms of ϵ_0 rather than k. It doesn't matter which form we use since Eqs. 21–1 and 21–2 are equivalent. (The latest precise values of e and ϵ_0 are given inside the front cover.)

[Our convention for units, such as $C^2/N\cdot m^2$ for ϵ_0, means m^2 is in the denominator. That is, $C^2/N\cdot m^2$ does *not* mean $(C^2/N)\cdot m^2 = C^2\cdot m^2/N$.]

Equations 21–1 and 21–2 apply to objects whose size is much smaller than the distance between them. Ideally, it is precise for **point charges** (spatial size negligible compared to other distances). For finite-sized objects, it is not always clear what value to use for r, particularly since the charge may not be distributed uniformly on the objects. If the two objects are spheres and the charge is known to be distributed uniformly on each, then r is the distance between their centers.

Coulomb's law describes the force between two charges when they are at rest. Additional forces come into play when charges are in motion, and will be discussed in later Chapters. In this Chapter we discuss only charges at rest, the study of which is called **electrostatics**, and Coulomb's law gives the **electrostatic force**.

When calculating with Coulomb's law, we usually ignore the signs of the charges and determine the direction of a force separately based on whether the force is attractive or repulsive.

PROBLEM SOLVING

Use magnitudes in Coulomb's law; find force direction from signs of charges

EXERCISE A Return to the Chapter-Opening Question, page 559, and answer it again now. Try to explain why you may have answered differently the first time.

CONCEPTUAL EXAMPLE 21–1 **Which charge exerts the greater force?** Two positive point charges, $Q_1 = 50\ \mu C$ and $Q_2 = 1\ \mu C$, are separated by a distance ℓ, Fig. 21–16. Which is larger in magnitude, the force that Q_1 exerts on Q_2, or the force that Q_2 exerts on Q_1?

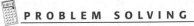

FIGURE 21–16   Example 21–1.

RESPONSE From Coulomb's law, the force on Q_1 exerted by Q_2 is

$$F_{12} = k\frac{Q_1 Q_2}{\ell^2}.$$

The force on Q_2 exerted by Q_1 is

$$F_{21} = k\frac{Q_2 Q_1}{\ell^2}$$

which is the same magnitude. The equation is symmetric with respect to the two charges, so $F_{21} = F_{12}$.

NOTE Newton's third law also tells us that these two forces must have equal magnitude.

EXERCISE B What is the magnitude of F_{12} (and F_{21}) in Example 21–1 if $\ell = 30\ cm$?

Keep in mind that Eq. 21–2 (or 21–1) gives the force on a charge due to only *one* other charge. If several (or many) charges are present, the *net force on any one of them will be the vector sum of the forces due to each of the others*. This **principle of superposition** is based on experiment, and tells us that electric force vectors add like any other vector. For continuous distributions of charge, the sum becomes an integral.

EXAMPLE 21–2 **Three charges in a line.** Three charged particles are arranged in a line, as shown in Fig. 21–17. Calculate the net electrostatic force on particle 3 (the $-4.0\,\mu C$ on the right) due to the other two charges.

APPROACH The net force on particle 3 is the vector sum of the force $\vec{F}_{31}$ exerted on 3 by particle 1 and the force $\vec{F}_{32}$ exerted on 3 by particle 2: $\vec{F} = \vec{F}_{31} + \vec{F}_{32}$.

SOLUTION The magnitudes of these two forces are obtained using Coulomb's law, Eq. 21–1:

$$F_{31} = k\frac{Q_3 Q_1}{r_{31}^2}$$

$$= \frac{(9.0 \times 10^9\,N\cdot m^2/C^2)(4.0 \times 10^{-6}\,C)(8.0 \times 10^{-6}\,C)}{(0.50\,m)^2} = 1.2\,N,$$

where $r_{31} = 0.50\,m$ is the distance from Q_3 to Q_1. Similarly,

$$F_{32} = k\frac{Q_3 Q_2}{r_{32}^2}$$

$$= \frac{(9.0 \times 10^9\,N\cdot m^2/C^2)(4.0 \times 10^{-6}\,C)(3.0 \times 10^{-6}\,C)}{(0.20\,m)^2} = 2.7\,N.$$

Since we were calculating the magnitudes of the forces, we omitted the signs of the charges. But we must be aware of them to get the direction of each force. Let the line joining the particles be the x axis, and we take it positive to the right. Then, because $\vec{F}_{31}$ is repulsive and $\vec{F}_{32}$ is attractive, the directions of the forces are as shown in Fig. 21–17b: F_{31} points in the positive x direction and F_{32} points in the negative x direction. The net force on particle 3 is then

$$F = -F_{32} + F_{31} = -2.7\,N + 1.2\,N = -1.5\,N.$$

The magnitude of the net force is 1.5 N, and it points to the left.

NOTE Charge Q_1 acts on charge Q_3 just as if Q_2 were not there (this is the principle of superposition). That is, the charge in the middle, Q_2, in no way blocks the effect of charge Q_1 acting on Q_3. Naturally, Q_2 exerts its own force on Q_3.

| **EXERCISE C** Determine the magnitude and direction of the net force on Q_1 in Fig. 21–17a.

EXAMPLE 21–3 **Electric force using vector components.** Calculate the net electrostatic force on charge Q_3 shown in Fig. 21–18a due to the charges Q_1 and Q_2.

APPROACH We use Coulomb's law to find the magnitudes of the individual forces. The direction of each force will be along the line connecting Q_3 to Q_1 or Q_2. The forces $\vec{F}_{31}$ and $\vec{F}_{32}$ have the directions shown in Fig. 21–18a, since Q_1 exerts an attractive force on Q_3, and Q_2 exerts a repulsive force. The forces $\vec{F}_{31}$ and $\vec{F}_{32}$ are *not* along the same line, so to find the resultant force on Q_3 we resolve $\vec{F}_{31}$ and $\vec{F}_{32}$ into x and y components and perform the vector addition.

SOLUTION The magnitudes of $\vec{F}_{31}$ and $\vec{F}_{32}$ are (ignoring signs of the charges since we know the directions)

$$F_{31} = k\frac{Q_3 Q_1}{r_{31}^2} = \frac{(9.0 \times 10^9\,N\cdot m^2/C^2)(6.5 \times 10^{-5}\,C)(8.6 \times 10^{-5}\,C)}{(0.60\,m)^2} = 140\,N,$$

$$F_{32} = k\frac{Q_3 Q_2}{r_{32}^2} = \frac{(9.0 \times 10^9\,N\cdot m^2/C^2)(6.5 \times 10^{-5}\,C)(5.0 \times 10^{-5}\,C)}{(0.30\,m)^2} = 330\,N.$$

We resolve $\vec{F}_{31}$ into its components along the x and y axes, as shown in Fig. 21–18a:

$$F_{31x} = F_{31}\cos 30° = (140\,N)\cos 30° = 120\,N,$$

$$F_{31y} = -F_{31}\sin 30° = -(140\,N)\sin 30° = -70\,N.$$

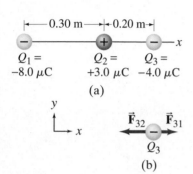

|←—0.30 m—→|←0.20 m→|

$Q_1 = -8.0\,\mu C$ $Q_2 = +3.0\,\mu C$ $Q_3 = -4.0\,\mu C$

(a)

$\vec{F}_{32}$ $\vec{F}_{31}$

Q_3

(b)

FIGURE 21–17 Example 21–2.

⚠ **CAUTION**

Each charge exerts its own force. No charge blocks the effect of the others

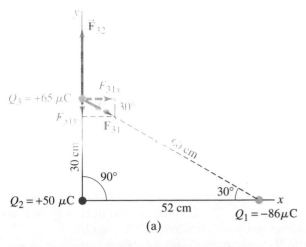

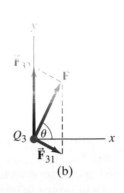

FIGURE 21–18 Determining the forces for Example 21–3. (a) The directions of the individual forces are as shown because $\vec{F}_{32}$ is repulsive (the force on Q_3 is in the direction away from Q_2 because Q_3 and Q_2 are both positive) whereas $\vec{F}_{31}$ is attractive (Q_3 and Q_1 have opposite signs), so $\vec{F}_{31}$ points toward Q_1. (b) Adding $\vec{F}_{32}$ to $\vec{F}_{31}$ to obtain the net force $\vec{F}$.

The force $\vec{F}_{32}$ has only a y component. So the net force $\vec{F}$ on Q_3 has components

$$F_x = F_{31x} = 120\,\text{N},$$
$$F_y = F_{32} + F_{31y} = 330\,\text{N} - 70\,\text{N} = 260\,\text{N}.$$

The magnitude of the net force is

$$F = \sqrt{F_x^2 + F_y^2} = \sqrt{(120\,\text{N})^2 + (260\,\text{N})^2} = 290\,\text{N};$$

and it acts at an angle θ (see Fig. 21–18b) given by

$$\tan\theta = \frac{F_y}{F_x} = \frac{260\,\text{N}}{120\,\text{N}} = 2.2,$$

so $\theta = \tan^{-1}(2.2) = 65°$.

NOTE Because $\vec{F}_{31}$ and $\vec{F}_{32}$ are not along the same line, the magnitude of $\vec{F}_3$ is not equal to the sum (or difference as in Example 21–2) of the separate magnitudes.

CONCEPTUAL EXAMPLE 21–4 Make the force on Q_3 zero. In Fig. 21–18, where could you place a fourth charge, $Q_4 = -50\,\mu\text{C}$, so that the net force on Q_3 would be zero?

RESPONSE By the principle of superposition, we need a force in exactly the opposite direction to the resultant $\vec{F}$ due to Q_2 and Q_1 that we calculated in Example 21–3, Fig. 21–18b. Our force must have magnitude 290 N, and must point down and to the left of Q_3 in Fig. 21–18b. So Q_4 must be along this line. See Fig. 21–19.

EXERCISE D (a) Consider two point charges of the same magnitude but opposite sign ($+Q$ and $-Q$), which are fixed a distance d apart. Can you find a location where a third positive charge Q could be placed so that the net electric force on this third charge is zero? (b) What if the first two charges were both $+Q$?

*Vector Form of Coulomb's Law

Coulomb's law can be written in vector form (as we did for Newton's law of universal gravitation in Chapter 6, Section 6–2) as

$$\vec{F}_{12} = k\frac{Q_1 Q_2}{r_{21}^2}\,\hat{r}_{21},$$

where $\vec{F}_{12}$ is the vector force on charge Q_1 due to Q_2 and $\hat{r}_{21}$ is the unit vector pointing from Q_2 toward Q_1. That is, $\hat{r}_{21}$ points from the "source" charge (Q_2) toward the charge on which we want to know the force (Q_1). See Fig. 21–20. The charges Q_1 and Q_2 can be either positive or negative, and this will affect the direction of the electric force. If Q_1 and Q_2 have the same sign, the product $Q_1 Q_2 > 0$ and the force on Q_1 points away from Q_2—that is, it is repulsive. If Q_1 and Q_2 have opposite signs, $Q_1 Q_2 < 0$ and $\vec{F}_{12}$ points toward Q_2—that is, it is attractive.

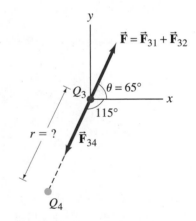

FIGURE 21–19 Example 21–4: Q_4 exerts force ($\vec{F}_{34}$) that makes the net force on Q_3 zero.

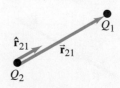

FIGURE 21–20 Determining the force on Q_1 due to Q_2, showing the direction of the unit vector $\hat{r}_{21}$.

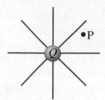

FIGURE 21–21 An electric field surrounds every charge. P is an arbitrary point.

FIGURE 21–22 Force exerted by charge $+Q$ on a small test charge, q, placed at points A, B, and C.

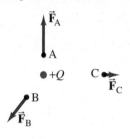

FIGURE 21–23 (a) Electric field at a given point in space. (b) Force on a positive charge at that point. (c) Force on a negative charge at that point.

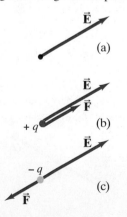

21–6 The Electric Field

Many common forces might be referred to as "contact forces," such as your hands pushing or pulling a cart, or a tennis racket hitting a tennis ball.

In contrast, both the gravitational force and the electrical force act over a distance: there is a force between two objects even when the objects are not touching. The idea of a force *acting at a distance* was a difficult one for early thinkers. Newton himself felt uneasy with this idea when he published his law of universal gravitation. A helpful way to look at the situation uses the idea of the **field**, developed by the British scientist Michael Faraday (1791–1867). In the electrical case, according to Faraday, an *electric field* extends outward from every charge and permeates all of space (Fig. 21–21). If a second charge (call it Q_2) is placed near the first charge, it feels a force exerted by the electric field that is there (say, at point P in Fig. 21–21). The electric field at point P is considered to interact directly with charge Q_2 to produce the force on Q_2.

We can in principle investigate the electric field surrounding a charge or group of charges by measuring the force on a small positive **test charge** at rest. By a test charge we mean a charge so small that the force it exerts does not significantly affect the charges that create the field. If a tiny positive test charge q is placed at various locations in the vicinity of a single positive charge Q as shown in Fig. 21–22 (points A, B, C), the force exerted on q is as shown. The force at B is less than at A because B's distance from Q is greater (Coulomb's law); and the force at C is smaller still. In each case, the force on q is directed radially away from Q. The electric field is defined in terms of the force on such a positive test charge. In particular, the **electric field**, $\vec{E}$, at any point in space is defined as the force $\vec{F}$ exerted on a tiny positive test charge placed at that point divided by the magnitude of the test charge q:

$$\vec{E} = \frac{\vec{F}}{q}. \qquad (21\text{–}3)$$

More precisely, $\vec{E}$ is defined as the limit of $\vec{F}/q$ as q is taken smaller and smaller, approaching zero. That is, q is so tiny that it exerts essentially no force on the other charges which created the field. From this definition (Eq. 21–3), we see that the electric field at any point in space is a vector whose direction is the direction of the force on a tiny positive test charge at that point, and whose magnitude is the *force per unit charge*. Thus $\vec{E}$ has SI units of newtons per coulomb (N/C).

The reason for defining $\vec{E}$ as $\vec{F}/q$ (with $q \to 0$) is so that $\vec{E}$ does not depend on the magnitude of the test charge q. This means that $\vec{E}$ describes only the effect of the charges creating the electric field at that point.

The electric field at any point in space can be measured, based on the definition, Eq. 21–3. For simple situations involving one or several point charges, we can calculate $\vec{E}$. For example, the electric field at a distance r from a single point charge Q would have magnitude

$$E = \frac{F}{q} = \frac{kqQ/r^2}{q}$$

$$E = k\frac{Q}{r^2}; \qquad \text{[single point charge]} \quad (21\text{–}4a)$$

or, in terms of ϵ_0 as in Eq. 21–2 $(k = 1/4\pi\epsilon_0)$:

$$E = \frac{1}{4\pi\epsilon_0}\frac{Q}{r^2}. \qquad \text{[single point charge]} \quad (21\text{–}4b)$$

Notice that E is independent of the test charge q—that is, E depends only on the charge Q which produces the field, and not on the value of the test charge q. Equations 21–4 are referred to as the electric field form of Coulomb's law.

If we are given the electric field $\vec{E}$ at a given point in space, then we can calculate the force $\vec{F}$ on any charge q placed at that point by writing (see Eq. 21–3):

$$\vec{F} = q\vec{E}. \qquad (21\text{–}5)$$

This is valid even if q is not small as long as q does not cause the charges creating $\vec{E}$ to move. If q is positive, $\vec{F}$ and $\vec{E}$ point in the same direction. If q is negative, $\vec{F}$ and $\vec{E}$ point in opposite directions. See Fig. 21–23.

EXAMPLE 21–5 **Photocopy machine.** A photocopy machine works by arranging positive charges (in the pattern to be copied) on the surface of a drum, then gently sprinkling negatively charged dry toner (ink) particles onto the drum. The toner particles temporarily stick to the pattern on the drum (Fig. 21–24) and are later transferred to paper and "melted" to produce the copy. Suppose each toner particle has a mass of 9.0×10^{-16} kg and carries an average of 20 extra electrons to provide an electric charge. Assuming that the electric force on a toner particle must exceed twice its weight in order to ensure sufficient attraction, compute the required electric field strength near the surface of the drum.

APPROACH The electric force on a toner particle of charge $q = 20e$ is $F = qE$, where E is the needed electric field. This force needs to be at least as great as twice the weight (mg) of the particle.

SOLUTION The minimum value of electric field satisfies the relation

$$qE = 2mg$$

where $q = 20e$. Hence

$$E = \frac{2mg}{q} = \frac{2(9.0 \times 10^{-16}\,\text{kg})(9.8\,\text{m/s}^2)}{20(1.6 \times 10^{-19}\,\text{C})} = 5.5 \times 10^3\,\text{N/C}.$$

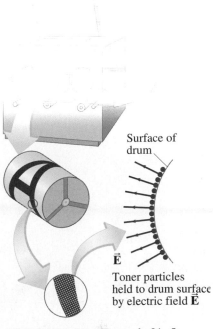

Surface of drum

Ē

Toner particles held to drum surface by electric field **Ē**

FIGURE 21–24 Example 21–5.

EXAMPLE 21–6 **Electric field of a single point charge.** Calculate the magnitude and direction of the electric field at a point P which is 30 cm to the right of a point charge $Q = -3.0 \times 10^{-6}$ C.

APPROACH The magnitude of the electric field due to a single point charge is given by Eq. 21–4. The direction is found using the sign of the charge Q.

SOLUTION The magnitude of the electric field is:

$$E = k\frac{Q}{r^2} = \frac{(9.0 \times 10^9\,\text{N}\cdot\text{m}^2/\text{C}^2)(3.0 \times 10^{-6}\,\text{C})}{(0.30\,\text{m})^2} = 3.0 \times 10^5\,\text{N/C}.$$

The direction of the electric field is *toward* the charge Q, to the left as shown in Fig. 21–25a, since we defined the direction as that of the force on a positive test charge which here would be attractive. If Q had been positive, the electric field would have pointed away, as in Fig. 21–25b.

NOTE There is no electric charge at point P. But there is an electric field there. The only real charge is Q.

FIGURE 21–25 Example 21–6. Electric field at point P (a) due to a negative charge Q, and (b) due to a positive charge Q, each 30 cm from P.

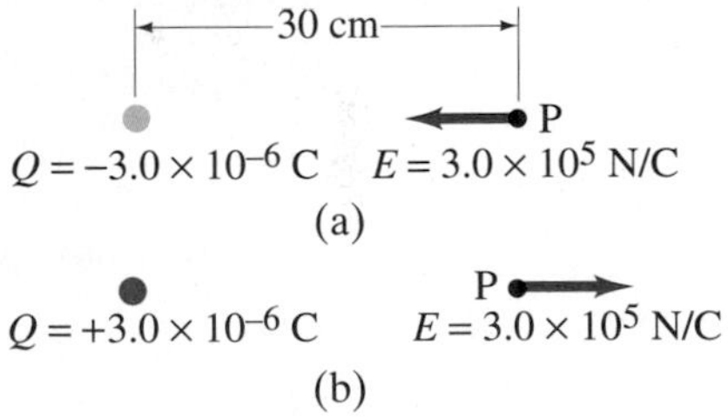

|←——— 30 cm ———→|

$Q = -3.0 \times 10^{-6}$ C $E = 3.0 \times 10^5$ N/C
(a)

$Q = +3.0 \times 10^{-6}$ C   P $E = 3.0 \times 10^5$ N/C
(b)

This Example illustrates a general result: The electric field $\vec{E}$ due to a positive charge points away from the charge, whereas $\vec{E}$ due to a negative charge points toward that charge.

EXERCISE E Four charges of equal magnitude, but possibly different sign, are placed on the corners of a square. What arrangement of charges will produce an electric field with the greatest magnitude at the center of the square? (*a*) All four positive charges; (*b*) all four negative charges; (*c*) three positive and one negative; (*d*) two positive and two negative; (*e*) three negative and one positive.

If the electric field at a given point in space is due to more than one charge, the individual fields (call them $\vec{E}_1, \vec{E}_2$, etc.) due to each charge are added vectorially to get the total field at that point:

$$\vec{E} = \vec{E}_1 + \vec{E}_2 + \cdots.$$

The validity of this **superposition principle** for electric fields is fully confirmed by experiment.

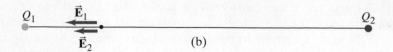

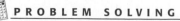

EXAMPLE 21–7 E **at a point between two charges.** Two point charges are separated by a distance of 10.0 cm. One has a charge of $-25\,\mu$C and the other $+50\,\mu$C. (a) Determine the direction and magnitude of the electric field at a point P between the two charges that is 2.0 cm from the negative charge (Fig. 21–26a). (b) If an electron (mass $= 9.11 \times 10^{-31}$ kg) is placed at rest at P and then released, what will be its initial acceleration (direction and magnitude)?

APPROACH The electric field at P will be the vector sum of the fields created separately by Q_1 and Q_2. The field due to the negative charge Q_1 points toward Q_1, and the field due to the positive charge Q_2 points away from Q_2. Thus both fields point to the left as shown in Fig. 21–26b and we can add the magnitudes of the two fields together algebraically, ignoring the signs of the charges. In (b) we use Newton's second law ($F = ma$) to determine the acceleration, where $F = qE$ (Eq. 21–5).

SOLUTION (a) Each field is due to a point charge as given by Eq. 21–4, $E = kQ/r^2$. The total field is

$$E = k\frac{Q_1}{r_1^2} + k\frac{Q_2}{r_2^2} = k\left(\frac{Q_1}{r_1^2} + \frac{Q_2}{r_2^2}\right)$$

$$= (9.0 \times 10^9\,\text{N·m}^2/\text{C}^2)\left(\frac{25 \times 10^{-6}\,\text{C}}{(2.0 \times 10^{-2}\,\text{m})^2} + \frac{50 \times 10^{-6}\,\text{C}}{(8.0 \times 10^{-2}\,\text{m})^2}\right)$$

$$= 6.3 \times 10^8\,\text{N/C}.$$

(b) The electric field points to the left, so the electron will feel a force to the *right* since it is negatively charged. Therefore the acceleration $a = F/m$ (Newton's second law) will be to the right. The force on a charge q in an electric field E is $F = qE$ (Eq. 21–5). Hence the magnitude of the acceleration is

$$a = \frac{F}{m} = \frac{qE}{m} = \frac{(1.60 \times 10^{-19}\,\text{C})(6.3 \times 10^8\,\text{N/C})}{9.11 \times 10^{-31}\,\text{kg}} = 1.1 \times 10^{20}\,\text{m/s}^2.$$

NOTE By carefully considering the directions of *each* field ($\vec{E}_1$ and $\vec{E}_2$) before doing any calculations, we made sure our calculation could be done simply and correctly.

EXAMPLE 21–8 $\vec{E}$ **above two point charges.** Calculate the total electric field (a) at point A and (b) at point B in Fig. 21–27 due to both charges, Q_1 and Q_2.

APPROACH The calculation is much like that of Example 21–3, except now we are dealing with electric fields instead of force. The electric field at point A is the vector sum of the fields $\vec{E}_{A1}$ due to Q_1, and $\vec{E}_{A2}$ due to Q_2. We find the magnitude of the field produced by each point charge, then we add their components to find the total field at point A. We do the same for point B.

SOLUTION (a) The magnitude of the electric field produced at point A by each of the charges Q_1 and Q_2 is given by $E = kQ/r^2$, so

$$E_{A1} = \frac{(9.0 \times 10^9\,\text{N·m}^2/\text{C}^2)(50 \times 10^{-6}\,\text{C})}{(0.60\,\text{m})^2} = 1.25 \times 10^6\,\text{N/C},$$

$$E_{A2} = \frac{(9.0 \times 10^9\,\text{N·m}^2/\text{C}^2)(50 \times 10^{-6}\,\text{C})}{(0.30\,\text{m})^2} = 5.0 \times 10^6\,\text{N/C}.$$

The direction of E_{A1} points from A toward Q_1 (negative charge), whereas E_{A2} points

PROBLEM SOLVING

Ignore signs of charges and determine direction physically, showing directions on diagram

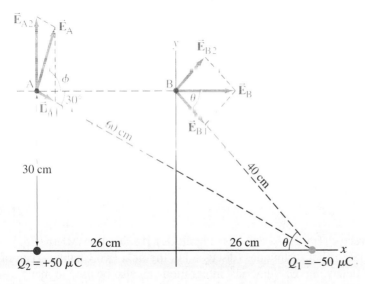

FIGURE 21-27 Calculation of the electric field at points A and B for Example 21-9.

from A away from Q_2, as shown; so the total electric field at A, $\vec{E}_A$, has components

$$E_{Ax} = E_{A1} \cos 30° = 1.1 \times 10^6 \, \text{N/C},$$
$$E_{Ay} = E_{A2} - E_{A1} \sin 30° = 4.4 \times 10^6 \, \text{N/C}.$$

Thus the magnitude of $\vec{E}_A$ is

$$E_A = \sqrt{(1.1)^2 + (4.4)^2} \times 10^6 \, \text{N/C} = 4.5 \times 10^6 \, \text{N/C},$$

and its direction is ϕ given by $\tan \phi = E_{Ay}/E_{Ax} = 4.4/1.1 = 4.0$, so $\phi = 76°$.

(b) Because B is equidistant from the two equal charges (40 cm by the Pythagorean theorem), the magnitudes of E_{B1} and E_{B2} are the same; that is,

$$E_{B1} = E_{B2} = \frac{kQ}{r^2} = \frac{(9.0 \times 10^9 \, \text{N·m}^2/\text{C}^2)(50 \times 10^{-6} \, \text{C})}{(0.40 \, \text{m})^2}$$
$$= 2.8 \times 10^6 \, \text{N/C}.$$

Also, because of the symmetry, the y components are equal and opposite, and so cancel out. Hence the total field E_B is horizontal and equals $E_{B1} \cos \theta + E_{B2} \cos \theta = 2E_{B1} \cos \theta$. From the diagram, $\cos \theta = 26 \, \text{cm}/40 \, \text{cm} = 0.65$. Then

$$E_B = 2E_{B1} \cos \theta = 2(2.8 \times 10^6 \, \text{N/C})(0.65)$$
$$= 3.6 \times 10^6 \, \text{N/C},$$

and the direction of $\vec{E}_B$ is along the $+x$ direction.

NOTE We could have done part (b) in the same way we did part (a). But symmetry allowed us to solve the problem with less effort.

PROBLEM SOLVING

Use symmetry to save work, when possible

Electrostatics: Electric Forces and Electric Fields

Solving electrostatics problems follows, to a large extent, the general problem-solving procedure discussed in Section 4–8. Whether you use electric field or electrostatic forces, the procedure is similar:

1. **Draw** a careful **diagram**—namely, a free-body diagram for each object, showing all the forces acting on that object, or showing the electric field at a point due to all significant charges present. Determine the **direction** of each force or electric field physically: like charges repel each other, unlike charges attract; fields point away from a + charge, and toward

a − charge. Show and label each vector force or field on your diagram.

2. **Apply Coulomb's law** to calculate the magnitude of the force that each contributing charge exerts on a charged object, or the magnitude of the electric field each charge produces at a given point. Deal only with magnitudes of charges (leaving out minus signs), and obtain the magnitude of each force or electric field.

3. **Add vectorially** all the forces on an object, or the contributing fields at a point, to get the resultant. Use **symmetry** (say, in the geometry) whenever possible.

4. **Check** your answer. Is it **reasonable**? If a function of distance, does it give reasonable results in limiting cases?

21–7 Electric Field Calculations for Continuous Charge Distributions

In many cases we can treat charge as being distributed continuously.[†] We can divide up a charge distribution into infinitesimal charges dQ, each of which will act as a tiny point charge. The contribution to the electric field at a distance r from each dQ is

$$dE = \frac{1}{4\pi\epsilon_0}\frac{dQ}{r^2}. \qquad (21\text{–}6a)$$

Then the electric field, $\vec{\mathbf{E}}$, at any point is obtained by summing over all the infinitesimal contributions, which is the integral

$$\vec{\mathbf{E}} = \int d\vec{\mathbf{E}}. \qquad (21\text{–}6b)$$

Note that $d\vec{\mathbf{E}}$ is a vector (Eq. 21–6a gives its magnitude). [In situations where Eq. 21–6b is difficult to evaluate, other techniques (discussed in the next two Chapters) can often be used instead to determine $\vec{\mathbf{E}}$. Numerical integration can also be used in many cases.]

EXAMPLE 21–9 **A ring of charge.** A thin, ring-shaped object of radius a holds a total charge $+Q$ distributed uniformly around it. Determine the electric field at a point P on its axis, a distance x from the center. See Fig. 21–28. Let λ be the charge per unit length (C/m).

APPROACH AND SOLUTION We explicitly follow the steps of the Problem Solving Strategy on page 571.

1. **Draw** a careful **diagram**. The **direction** of the electric field due to one infinitesimal length $d\ell$ of the charged ring is shown in Fig. 21–28.
2. **Apply Coulomb's law**. The electric field, $d\vec{\mathbf{E}}$, due to this particular segment of the ring of length $d\ell$ has magnitude

$$dE = \frac{1}{4\pi\epsilon_0}\frac{dQ}{r^2}.$$

The whole ring has length (circumference) of $2\pi a$, so the charge on a length $d\ell$ is

$$dQ = Q\left(\frac{d\ell}{2\pi a}\right) = \lambda\, d\ell$$

where $\lambda = Q/2\pi a$ is the charge per unit length. Now we write dE as

$$dE = \frac{1}{4\pi\epsilon_0}\frac{\lambda\, d\ell}{r^2}.$$

3. **Add vectorially** and use **symmetry**: The vector $d\vec{\mathbf{E}}$ has components dE_x along the x axis and $dE_\perp$ perpendicular to the x axis (Fig. 21–28). We are going to sum (integrate) around the entire ring. We note that an equal-length segment diametrically opposite the $d\ell$ shown will produce a $d\vec{\mathbf{E}}$ whose component perpendicular to the x axis will just cancel the $dE_\perp$ shown. This is true for all segments of the ring, so by symmetry $\vec{\mathbf{E}}$ will have zero y component, and so we need only sum the x components, dE_x. The total field is then

$$E = E_x = \int dE_x = \int dE\cos\theta = \frac{1}{4\pi\epsilon_0}\lambda\int\frac{d\ell}{r^2}\cos\theta.$$

Since $\cos\theta = x/r$, where $r = (x^2 + a^2)^{\frac{1}{2}}$, we have

$$E = \frac{\lambda}{(4\pi\epsilon_0)}\frac{x}{(x^2 + a^2)^{\frac{3}{2}}}\int_0^{2\pi a} d\ell = \frac{1}{4\pi\epsilon_0}\frac{\lambda x(2\pi a)}{(x^2 + a^2)^{\frac{3}{2}}} = \frac{1}{4\pi\epsilon_0}\frac{Qx}{(x^2 + a^2)^{\frac{3}{2}}}.$$

4. To **check reasonableness**, note that at great distances, $x \gg a$, this result reduces to $E = Q/(4\pi\epsilon_0 x^2)$. We would expect this result because at great distances the ring would appear to be a point charge ($1/r^2$ dependence). Also note that our result gives $E = 0$ at $x = 0$, as we might expect because all components will cancel at the center of the circle.

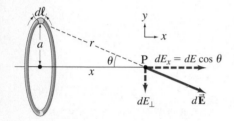

FIGURE 21–28 Example 21–9.

PROBLEM SOLVING

Use symmetry when possible

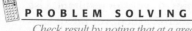

PROBLEM SOLVING

Check result by noting that at a great distance the ring looks like a point charge

---

[†]Because we believe there is a minimum charge (e), the treatment here is only for convenience; it is nonetheless useful and accurate since e is usually very much smaller than macroscopic charges.

Note in this Example three important problem-solving techniques that can be used elsewhere: (1) using symmetry to reduce the complexity of the problem; (2) expressing the charge dQ in terms of a charge density (here linear, $\lambda = Q/2\pi a$); and (3) checking the answer at the limit of large r, which serves as an indication (but not proof) of the correctness of the answer—if the result does not check at large r, your result has to be wrong.

CONCEPTUAL EXAMPLE 21–10 | **Charge at the center of a ring.** Imagine a small positive charge placed at the center of a nonconducting ring carrying a uniformly distributed negative charge. Is the positive charge in equilibrium if it is displaced slightly from the center along the axis of the ring, and if so is it stable? What if the small charge is negative? Neglect gravity, as it is much smaller than the electrostatic forces.

RESPONSE The positive charge is in equilibrium because there is no net force on it, by *symmetry*. If the positive charge moves away from the center of the ring along the axis in either direction, the net force will be back towards the center of the ring and so the charge is in *stable* equilibrium. A negative charge at the center of the ring would feel no net force, but is in *unstable* equilibrium because if it moved along the ring's axis, the net force would be away from the ring and the charge would be pushed farther away.

EXAMPLE 21–11 **Long line of charge.** Determine the magnitude of the electric field at any point P a distance x from the midpoint 0 of a very long line (a wire, say) of uniformly distributed positive charge, Fig. 21–29. Assume x is much smaller than the length of the wire, and let λ be the charge per unit length (C/m).

APPROACH We set up a coordinate system so the wire is on the y axis with origin 0 as shown. A segment of wire dy has charge $dQ = \lambda\, dy$. The field $d\vec{E}$ at point P due to this length dy of wire (at y) has magnitude

$$dE = \frac{1}{4\pi\epsilon_0}\frac{dQ}{r^2} = \frac{1}{4\pi\epsilon_0}\frac{\lambda\, dy}{(x^2 + y^2)},$$

where $r = (x^2 + y^2)^{\frac{1}{2}}$ as shown in Fig. 21–29. The vector $d\vec{E}$ has components dE_x and dE_y as shown where $dE_x = dE\cos\theta$ and $dE_y = dE\sin\theta$.

SOLUTION Because 0 is at the midpoint of the wire, the y component of $\vec{E}$ will be zero since there will be equal contributions to $E_y = \int dE_y$ from above and below point 0:

$$E_y = \int dE\sin\theta = 0.$$

Thus we have

$$E = E_x = \int dE\cos\theta = \frac{\lambda}{4\pi\epsilon_0}\int \frac{\cos\theta\, dy}{x^2 + y^2}.$$

The integration here is over y, along the wire, with x treated as constant. We must now write θ as a function of y, or y as a function of θ. We do the latter: since $y = x\tan\theta$, then $dy = x\, d\theta/\cos^2\theta$. Furthermore, because $\cos\theta = x/\sqrt{x^2 + y^2}$, then $1/(x^2 + y^2) = \cos^2\theta/x^2$ and our integrand above is $(\cos\theta)(x\, d\theta/\cos^2\theta)(\cos^2\theta/x^2) = \cos\theta\, d\theta/x$. Hence

$$E = \frac{\lambda}{4\pi\epsilon_0}\frac{1}{x}\int_{-\pi/2}^{\pi/2}\cos\theta\, d\theta = \frac{\lambda}{4\pi\epsilon_0 x}(\sin\theta)\Big|_{-\pi/2}^{\pi/2} = \frac{1}{2\pi\epsilon_0}\frac{\lambda}{x},$$

where we have assumed the wire is extremely long in both directions ($y \to \pm\infty$) which corresponds to the limits $\theta = \pm\pi/2$. Thus the field near a long straight wire of uniform charge decreases inversely as the first power of the distance from the wire.

NOTE This result, obtained for an infinite wire, is a good approximation for a wire of finite length as long as x is small compared to the distance of P from the ends of the wire.

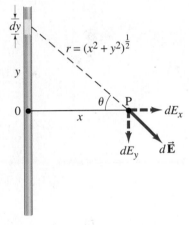

FIGURE 21–29 Example 21–11.

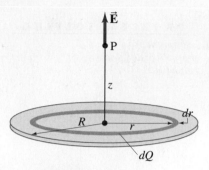

FIGURE 21–30 Example 21–12; a uniformly charged flat disk of radius R.

EXAMPLE 21–12 **Uniformly charged disk.** Charge is distributed uniformly over a thin circular disk of radius R. The charge per unit area (C/m^2) is σ. Calculate the electric field at a point P on the axis of the disk, a distance z above its center, Fig. 21–30.

APPROACH We can think of the disk as a set of concentric rings. We can then apply the result of Example 21–9 to each of these rings, and then sum over all the rings.

SOLUTION For the ring of radius r shown in Fig. 21–30, the electric field has magnitude (see result of Example 21–9)

$$dE = \frac{1}{4\pi\epsilon_0} \frac{z\,dQ}{(z^2 + r^2)^{\frac{3}{2}}}$$

where we have written dE (instead of E) for this thin ring of total charge dQ. The ring has area $(dr)(2\pi r)$ and charge per unit area $\sigma = dQ/(2\pi r\,dr)$. We solve this for dQ $(= \sigma\,2\pi r\,dr)$ and insert it in the equation above for dE:

$$dE = \frac{1}{4\pi\epsilon_0} \frac{z\sigma 2\pi r\,dr}{(z^2 + r^2)^{\frac{3}{2}}} = \frac{z\sigma r\,dr}{2\epsilon_0(z^2 + r^2)^{\frac{3}{2}}}.$$

Now we sum over all the rings, starting at $r = 0$ out to the largest with $r = R$:

$$E = \frac{z\sigma}{2\epsilon_0} \int_0^R \frac{r\,dr}{(z^2 + r^2)^{\frac{3}{2}}} = \frac{z\sigma}{2\epsilon_0} \left[-\frac{1}{(z^2 + r^2)^{\frac{1}{2}}} \right]_0^R$$

$$= \frac{\sigma}{2\epsilon_0} \left[1 - \frac{z}{(z^2 + R^2)^{\frac{1}{2}}} \right].$$

This gives the magnitude of $\vec{E}$ at any point z along the axis of the disk. The direction of each $d\vec{E}$ due to each ring is along the z axis (as in Example 21–9), and therefore the direction of $\vec{E}$ is along z. If Q (and σ) are positive, $\vec{E}$ points away from the disk; if Q (and σ) are negative, $\vec{E}$ points toward the disk.

If the radius of the disk in Example 21–12 is much greater than the distance of our point P from the disk (i.e., $z \ll R$) then we can obtain a very useful result: the second term in the solution above becomes very small, so

$$E = \frac{\sigma}{2\epsilon_0}. \qquad\qquad \text{[infinite plane]} \quad \textbf{(21–7)}$$

This result is valid for any point above (or below) an infinite plane of any shape holding a uniform charge density σ. It is also valid for points close to a finite plane, as long as the point is close to the plane compared to the distance to the edges of the plane. Thus the field near a large uniformly charged plane is uniform, and directed outward if the plane is positively charged.

It is interesting to compare here the distance dependence of the electric field due to a point charge $\left(E \sim 1/r^2\right)$, due to a very long uniform line of charge $\left(E \sim 1/r\right)$, and due to a very large uniform plane of charge (E does not depend on r).

EXAMPLE 21–13 **Two parallel plates.** Determine the electric field between two large parallel plates or sheets, which are very thin and are separated by a distance d which is small compared to their height and width. One plate carries a uniform surface charge density σ and the other carries a uniform surface charge density $-\sigma$, as shown in Fig. 21–31 (the plates extend upward and downward beyond the part shown).

APPROACH From Eq. 21–7, each plate sets up an electric field of magnitude $\sigma/2\epsilon_0$. The field due to the positive plate points away from that plate whereas the field due to the negative plate points toward that plate.

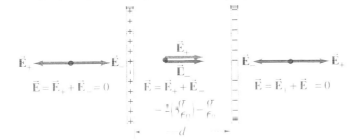

SOLUTION In the region between the plates, the fields add together as shown:

$$E = E_+ + E_- = \frac{\sigma}{2\epsilon_0} + \frac{\sigma}{2\epsilon_0} = \frac{\sigma}{\epsilon_0}.$$

The field is uniform, since the plates are very large compared to their separation, so this result is valid for any point, whether near one or the other of the plates, or midway between them as long as the point is far from the ends. Outside the plates, the fields cancel,

$$E = E_+ + E_- = \frac{\sigma}{2\epsilon_0} - \frac{\sigma}{2\epsilon_0} = 0,$$

as shown in Fig. 21-31. These results are valid ideally for infinitely large plates; they are a good approximation for finite plates if the separation is much less than the dimensions of the plate and for points not too close to the edge.

NOTE: These useful and extraordinary results illustrate the principle of superposition and its great power.

21-8 Field Lines

Since the electric field is a vector, it is sometimes referred to as a **vector field**. We could indicate the electric field with arrows at various points in a given situation, such as at A, B, and C in Fig. 21-32. The directions of $\vec{E}_A$, $\vec{E}_B$, and $\vec{E}_C$ are the same as that of the forces shown earlier in Fig. 21-22, but the magnitudes (arrow lengths) are different since we divide $\vec{F}$ in Fig. 21-22 by q to get $\vec{E}$. However, the relative lengths of $\vec{E}_A$, $\vec{E}_B$, and $\vec{E}_C$ are the same as for the forces since we divide by the same q each time. To indicate the electric field in such a way at *many* points, however, would result in many arrows, which might appear complicated or confusing. To avoid this, we use another technique, that of field lines.

To visualize the electric field, we draw a series of lines to indicate the direction of the electric field at various points in space. These **electric field lines** (sometimes called **lines of force**) are drawn so that they indicate the direction of the force due to the given field on a positive test charge. The lines of force due to a single isolated positive charge are shown in Fig. 21-33a, and for a single isolated negative charge in Fig. 21-33b. In part (a) the lines point radially outward from the charge, and in part (b) they point radially inward toward the charge because that is the direction the force would be on a positive test charge in each case (as in Fig. 21-25). Only a few representative lines are shown. We could just as well draw lines in between those shown since the electric field exists there as well. We can draw the lines so that the *number of lines starting on a positive charge, or ending on a negative charge, is proportional to the magnitude of the charge*. Notice that nearer the charge, where the electric field is greater $(F \propto 1/r^2)$, the lines are closer together. This is a general property of electric field lines: *the closer together the lines are, the stronger the electric field in that region.* In fact, field lines can be drawn so that the number of lines crossing unit area perpendicular to $\vec{E}$ is proportional to the magnitude of the electric field.

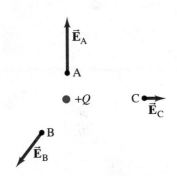

FIGURE 21-32 Electric field vector shown at three points, due to a single point charge Q. (Compare to Fig. 21-22.)

FIGURE 21-33 Electric field lines (a) near a single positive point charge, (b) near a single negative point charge.

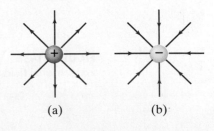

(a)   (b)

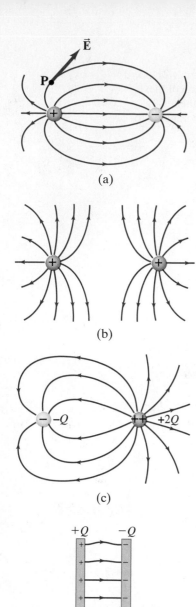

(a)

(b)

$-Q$ $+2Q$

(c)

$+Q$ $-Q$

(d)

FIGURE 21–34 Electric field lines for four arrangements of charges.

Figure 21–34a shows the electric field lines due to two equal charges of opposite sign, a combination known as an **electric dipole**. The electric field lines are curved in this case and are directed from the positive charge to the negative charge. The direction of the electric field at any point is tangent to the field line at that point as shown by the vector arrow $\vec{E}$ at point P. To satisfy yourself that this is the correct pattern for the electric field lines, you can make a few calculations such as those done in Example 21–8 for just this case (see Fig. 21–27). Figure 21–34b shows the electric field lines for two equal positive charges, and Fig. 21–34c for unequal charges, $-Q$ and $+2Q$. Note that twice as many lines leave $+2Q$, as enter $-Q$ (number of lines is proportional to magnitude of Q). Finally, in Fig. 21–34d, we see the field lines between two parallel plates carrying equal but opposite charges. Notice that the electric field lines between the two plates start out perpendicular to the surface of the metal plates (we will see why this is true in the next Section) and go directly from one plate to the other, as we expect because a positive test charge placed between the plates would feel a strong repulsion from the positive plate and a strong attraction to the negative plate. The field lines between two close plates are parallel and equally spaced in the central region, but fringe outward near the edges. Thus, in the central region, the electric field has the same magnitude at all points, and we can write (see Example 21–13)

$$E = \text{constant} = \frac{\sigma}{\epsilon_0}. \quad \begin{bmatrix} \text{between two closely spaced,} \\ \text{oppositely charged, parallel plates} \end{bmatrix} \quad \textbf{(21–8)}$$

The fringing of the field near the edges can often be ignored, particularly if the separation of the plates is small compared to their height and width.

We summarize the properties of field lines as follows:

1. Electric field lines indicate the direction of the electric field; the field points in the direction tangent to the field line at any point.
2. The lines are drawn so that the magnitude of the electric field, E, is proportional to the number of lines crossing unit area perpendicular to the lines. The closer together the lines, the stronger the field.
3. Electric field lines start on positive charges and end on negative charges; and the number starting or ending is proportional to the magnitude of the charge.

Also note that field lines never cross. Why not? Because the electric field can not have two directions at the same point, nor exert more than one force on a test charge.

Gravitational Field

The field concept can also be applied to the gravitational force as mentioned in Chapter 6. Thus we can say that a **gravitational field** exists for every object that has mass. One object attracts another by means of the gravitational field. The Earth, for example, can be said to possess a gravitational field (Fig. 21–35) which is responsible for the gravitational force on objects. The *gravitational field* is defined as the *force per unit mass*. The magnitude of the Earth's gravitational field at any point above the Earth's surface is thus (GM_E/r^2), where M_E is the mass of the Earth, r is the distance of the point from the Earth's center, and G is the gravitational constant (Chapter 6). At the Earth's surface, r is the radius of the Earth and the gravitational field is equal to g, the acceleration due to gravity. Beyond the Earth, the gravitational field can be calculated at any point as a sum of terms due to Earth, Sun, Moon, and other bodies that contribute significantly.

FIGURE 21–35 The Earth's gravitational field.

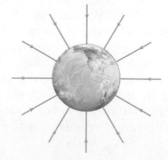

Electric Fields and Conductors

We now discuss some properties of conductors. First, *the electric field inside a conductor is zero in the static situation*—that is, when the charges are at rest. If there were an electric field within a conductor, there would be a force on the free electrons. The electrons would move until they reached positions where the electric field, and therefore the electric force on them, did become zero.

This reasoning has some interesting consequences. For one, *any net charge on a conductor distributes itself on the surface.* (If there were charges inside, there would be an electric field.) For a negatively charged conductor, you can imagine that the negative charges repel one another and race to the surface to get as far from one another as possible. Another consequence is the following. Suppose that a positive charge Q is surrounded by an isolated uncharged metal conductor whose shape is a spherical shell, Fig. 21–36. Because there can be no field within the metal, the lines leaving the central positive charge must end on negative charges on the inner surface of the metal. Thus an equal amount of negative charge, $-Q$, is induced on the inner surface of the spherical shell. Then, since the shell is neutral, a positive charge of the same magnitude, $+Q$, must exist on the outer surface of the shell. Thus, although no field exists in the metal itself, an electric field exists outside of it, as shown in Fig. 21–36, as if the metal were not even there.

A related property of static electric fields and conductors is that *the electric field is always perpendicular to the surface outside of a conductor.* If there were a component of $\vec{E}$ parallel to the surface (Fig. 21–37), it would exert a force on free electrons at the surface, causing the electrons to move along the surface until they reached positions where no net force was exerted on them parallel to the surface—that is, until the electric field was perpendicular to the surface.

These properties apply only to conductors. Inside a nonconductor, which does not have free electrons, a static electric field can exist as we will see in the next Chapter. Also, the electric field outside a nonconductor does not necessarily make an angle of 90° to the surface.

FIGURE 21–36 A charge inside a neutral spherical metal shell induces charge on its surfaces. The electric field exists even beyond the shell, but not within the conductor itself.

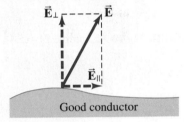

FIGURE 21–37 If the electric field $\vec{E}$ at the surface of a conductor had a component parallel to the surface, $\vec{E}_\parallel$, the latter would accelerate electrons into motion. In the static case, $\vec{E}_\parallel$ must be zero, and the electric field must be perpendicular to the conductor's surface: $\vec{E} = \vec{E}_\perp$.

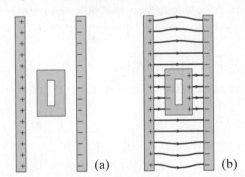

FIGURE 21–38 Example 21–14.

(a)　　　(b)

FIGURE 21–39 A strong electric field exists in the vicinity of this "Faraday cage," so strong that stray electrons in the atmosphere are accelerated to the kinetic energy needed to knock electrons out of air atoms, causing an avalanche of charge which flows to (or from) the metal cage. Yet the person inside the cage is not affected.

CONCEPTUAL EXAMPLE 21–14 | **Shielding, and safety in a storm.** A neutral hollow metal box is placed between two parallel charged plates as shown in Fig. 21–38a. What is the field like inside the box?

RESPONSE If our metal box had been solid, and not hollow, free electrons in the box would have redistributed themselves along the surface until all their individual fields would have canceled each other inside the box. The net field inside the box would have been zero. For a hollow box, the external field is not changed since the electrons in the metal can move just as freely as before to the surface. Hence the field inside the hollow metal box is also zero, and the field lines are shown in Fig. 21–38b. A conducting box used in this way is an effective device for shielding delicate instruments and electronic circuits from unwanted external electric fields. We also can see that a relatively safe place to be during a lightning storm is inside a parked car, surrounded by metal. See also Fig. 21–39, where a person inside a porous "cage" is protected from a strong electric discharge.

(人) **P H Y S I C S   A P P L I E D**
Electrical shielding

21–10 Motion of a Charged Particle in an Electric Field

If an object having an electric charge q is at a point in space where the electric field is $\vec{E}$, the force on the object is given by

$$\vec{F} = q\vec{E}$$

(see Eq. 21–5). In the past few Sections we have seen how to determine $\vec{E}$ for some particular situations. Now let us suppose we know $\vec{E}$ and we want to find the force on a charged object and the object's subsequent motion. (We assume no other forces act.)

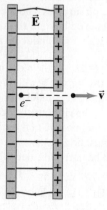

FIGURE 21–40 Example 21–15.

EXAMPLE 21–15 **Electron accelerated by electric field.** An electron (mass $m = 9.1 \times 10^{-31}$ kg) is accelerated in the uniform field $\vec{E}$ ($E = 2.0 \times 10^4$ N/C) between two parallel charged plates. The separation of the plates is 1.5 cm. The electron is accelerated from rest near the negative plate and passes through a tiny hole in the positive plate, Fig. 21–40. (a) With what speed does it leave the hole? (b) Show that the gravitational force can be ignored. Assume the hole is so small that it does not affect the uniform field between the plates.

APPROACH We can obtain the electron's velocity using the kinematic equations of Chapter 2, after first finding its acceleration from Newton's second law, $F = ma$. The magnitude of the force on the electron is $F = qE$ and is directed to the right.

SOLUTION (a) The magnitude of the electron's acceleration is

$$a = \frac{F}{m} = \frac{qE}{m}.$$

Between the plates $\vec{E}$ is uniform so the electron undergoes uniformly accelerated motion with acceleration

$$a = \frac{(1.6 \times 10^{-19}\,\text{C})(2.0 \times 10^4\,\text{N/C})}{(9.1 \times 10^{-31}\,\text{kg})} = 3.5 \times 10^{15}\,\text{m/s}^2.$$

It travels a distance $x = 1.5 \times 10^{-2}$ m before reaching the hole, and since its initial speed was zero, we can use the kinematic equation, $v^2 = v_0^2 + 2ax$ (Eq. 2–12c), with $v_0 = 0$:

$$v = \sqrt{2ax} = \sqrt{2(3.5 \times 10^{15}\,\text{m/s}^2)(1.5 \times 10^{-2}\,\text{m})} = 1.0 \times 10^7\,\text{m/s}.$$

There is no electric field outside the plates, so after passing through the hole, the electron moves with this speed, which is now constant.

(b) The magnitude of the electric force on the electron is

$$qE = (1.6 \times 10^{-19}\,\text{C})(2.0 \times 10^4\,\text{N/C}) = 3.2 \times 10^{-15}\,\text{N}.$$

The gravitational force is

$$mg = (9.1 \times 10^{-31}\,\text{kg})(9.8\,\text{m/s}^2) = 8.9 \times 10^{-30}\,\text{N},$$

which is 10^{14} times smaller! Note that the electric field due to the electron does not enter the problem (since a particle cannot exert a force on itself).

EXAMPLE 21–16 **Electron moving perpendicular to $\vec{E}$.** Suppose an electron traveling with speed v_0 enters a uniform electric field $\vec{E}$, which is at right angles to $\vec{v}_0$ as shown in Fig. 21–41. Describe its motion by giving the equation of its path while in the electric field. Ignore gravity.

APPROACH Again we use Newton's second law, with $F = qE$, and the kinematic equations from Chapter 2.

SOLUTION When the electron enters the electric field (at $x = y = 0$) it has velocity $\vec{v}_0 = v_0\hat{i}$ in the x direction. The electric field $\vec{E}$, pointing vertically upward, imparts a uniform vertical acceleration to the electron of

$$a_y = \frac{F}{m} = \frac{qE}{m} = -\frac{eE}{m},$$

where we set $q = -e$ for the electron.

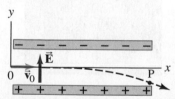

FIGURE 21–41 Example 21–16.

The electron's vertical position is given by Eq. 2–12b.

$$y = \frac{1}{2} a_y t^2 = -\frac{eE}{2m} t^2$$

since the motion is at constant acceleration. The horizontal position is given by

$$x = v_0 t$$

since $a_x = 0$. We eliminate t between these two equations and obtain

$$y = -\frac{eE}{2mv_0^2} x^2,$$

which is the equation of a parabola (just as in projectile motion, Section 3–7).

21–11 Electric Dipoles

The combination of two equal charges of opposite sign, $+Q$ and $-Q$, separated by a distance ℓ, is referred to as an **electric dipole**. The quantity $Q\ell$ is called the **dipole moment** and is represented[†] by the symbol p. The dipole moment can be considered to be a vector $\vec{p}$, of magnitude $Q\ell$, that points from the negative to the positive charge as shown in Fig. 21–42. Many molecules, such as the diatomic molecule CO, have a dipole moment (C has a small positive charge and O a small negative charge of equal magnitude), and are referred to as **polar molecules**. Even though the molecule as a whole is neutral, there is a separation of charge that results from an uneven sharing of electrons by the two atoms.[‡] (Symmetric diatomic molecules, like O_2, have no dipole moment.) The water molecule, with its uneven sharing of electrons (O is negative, the two H are positive), also has a dipole moment—see Fig. 21–43.

Dipole in an External Field

First let us consider a dipole, of dipole moment $p = Q\ell$, that is placed in a uniform electric field $\vec{E}$, as shown in Fig. 21–44. If the field is uniform, the force $Q\vec{E}$ on the positive charge and the force $-Q\vec{E}$ on the negative charge result in no net force on the dipole. There will, however, be a *torque* on the dipole (Fig. 21–44) which has magnitude (calculated about the center, 0, of the dipole)

$$\tau = QE\frac{\ell}{2}\sin\theta + QE\frac{\ell}{2}\sin\theta = pE\sin\theta. \qquad \textbf{(21–9a)}$$

This can be written in vector notation as

$$\vec{\tau} = \vec{p} \times \vec{E}. \qquad \textbf{(21–9b)}$$

The effect of the torque is to try to turn the dipole so $\vec{p}$ is parallel to $\vec{E}$. The work done on the dipole by the electric field to change the angle θ from θ_1 to θ_2 is (see Eq. 10–22)

$$W = \int_{\theta_1}^{\theta_2} \tau \, d\theta.$$

We need to write the torque as $\tau = -pE\sin\theta$ because its direction is opposite to the direction of increasing θ (right-hand rule). Then

$$W = \int_{\theta_1}^{\theta_2} \tau \, d\theta = -pE \int_{\theta_1}^{\theta_2} \sin\theta \, d\theta = pE\cos\theta \Big|_{\theta_1}^{\theta_2} = pE(\cos\theta_2 - \cos\theta_1).$$

Positive work done by the field decreases the potential energy, U, of the dipole in this field. (Recall the relation between work and potential energy, Eq. 8–4, $\Delta U = -W$.) If we choose $U = 0$ when $\vec{p}$ is perpendicular to $\vec{E}$ (that is, choosing $\theta_1 = 90°$ so $\cos\theta_1 = 0$), and setting $\theta_2 = \theta$, then

$$U = -W = -pE\cos\theta = -\vec{p} \cdot \vec{E}. \qquad \textbf{(21–10)}$$

If the electric field is *not* uniform, the force on the $+Q$ of the dipole may not have the same magnitude as on the $-Q$, so there may be a net force as well as a torque.

[†]Be careful not to confuse this p for dipole moment with p for momentum.

[‡]The value of the separated charges may be a fraction of e (say $\pm 0.2e$ or $\pm 0.4e$) but note that such charges do not violate what we said about e being the smallest charge. These charges less than e cannot be isolated and merely represent how much time electrons spend around one atom or the other.

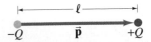

FIGURE 21–42 A dipole consists of equal but opposite charges, $+Q$ and $-Q$, separated by a distance ℓ. The dipole moment is $\vec{p} = Q\vec{\ell}$ and points from the negative to the positive charge.

FIGURE 21–43 In the water molecule (H_2O), the electrons spend more time around the oxygen atom than around the two hydrogen atoms. The net dipole moment $\vec{p}$ can be considered as the vector sum of two dipole moments $\vec{p}_1$ and $\vec{p}_2$ that point from the O toward each H as shown: $\vec{p} = \vec{p}_1 + \vec{p}_2$.

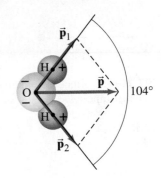

FIGURE 21–44 (below) An electric dipole in a uniform electric field.

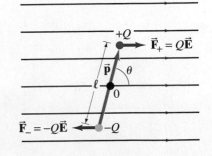

EXAMPLE 21–17 **Dipole in a field.** The dipole moment of a water molecule is 6.1×10^{-30} C·m. A water molecule is placed in a uniform electric field with magnitude 2.0×10^5 N/C. (*a*) What is the magnitude of the maximum torque that the field can exert on the molecule? (*b*) What is the potential energy when the torque is at its maximum? (*c*) In what position will the potential energy take on its greatest value? Why is this different than the position where the torque is maximum?

APPROACH The torque is given by Eq. 21–9 and the potential energy by Eq. 21–10.

SOLUTION (*a*) From Eq. 21–9 we see that τ is maximized when θ is 90°. Then $\tau = pE = (6.1 \times 10^{-30}$ C·m$)(2.0 \times 10^5$ N/C$) = 1.2 \times 10^{-24}$ N·m.
(*b*) The potential energy for $\theta = 90°$ is zero (Eq. 21–10). Note that the potential energy is negative for smaller values of θ, so U is not a minimum for $\theta = 90°$.
(*c*) The potential energy U will be a maximum when $\cos\theta = -1$ in Eq. 21–10, so $\theta = 180°$, meaning $\vec{\mathbf{E}}$ and $\vec{\mathbf{p}}$ are antiparallel. The potential energy is maximized when the dipole is oriented so that it has to rotate through the largest angle, 180°, to reach the equilibrium position at $\theta = 0°$. The torque on the other hand is maximized when the electric forces are perpendicular to $\vec{\mathbf{p}}$.

Electric Field Produced by a Dipole

We have just seen how an external electric field affects an electric dipole. Now let us suppose that there is no external field, and we want to determine the electric field produced *by* the dipole. For brevity, we restrict ourselves to points that are on the perpendicular bisector of the dipole, such as point P in Fig. 21–45 which is a distance r above the midpoint of the dipole. Note that r in Fig. 21–45 is not the distance from either charge to point P; the latter distance is $(r^2 + \ell^2/4)^{\frac{1}{2}}$ and this is what must be used in Eq. 21–4. The total field at P is

$$\vec{\mathbf{E}} = \vec{\mathbf{E}}_+ + \vec{\mathbf{E}}_-,$$

where $\vec{\mathbf{E}}_+$ and $\vec{\mathbf{E}}_-$ are the fields due to the $+$ and $-$ charges respectively. The magnitudes E_+ and E_- are equal:

$$E_+ = E_- = \frac{1}{4\pi\epsilon_0}\frac{Q}{r^2 + \ell^2/4}.$$

Their y components cancel at point P (*symmetry* again), so the magnitude of the total field $\vec{\mathbf{E}}$ is

$$E = 2E_+\cos\phi = \frac{1}{2\pi\epsilon_0}\left(\frac{Q}{r^2 + \ell^2/4}\right)\frac{\ell}{2(r^2 + \ell^2/4)^{\frac{1}{2}}}$$

or, setting $Q\ell = p$,

$$E = \frac{1}{4\pi\epsilon_0}\frac{p}{(r^2 + \ell^2/4)^{\frac{3}{2}}}. \qquad \left[\begin{array}{c}\text{on perpendicular bisector}\\ \text{of dipole}\end{array}\right] \quad \textbf{(21–11)}$$

Far from the dipole, $r \gg \ell$, this reduces to

$$E = \frac{1}{4\pi\epsilon_0}\frac{p}{r^3}. \qquad \left[\begin{array}{c}\text{on perpendicular bisector}\\ \text{of dipole; } r \gg \ell\end{array}\right] \quad \textbf{(21–12)}$$

So the field decreases more rapidly for a dipole than for a single point charge ($1/r^3$ versus $1/r^2$), which we expect since at large distances the two opposite charges appear so close together as to neutralize each other. This $1/r^3$ dependence also applies for points not on the perpendicular bisector (see Problem 67).

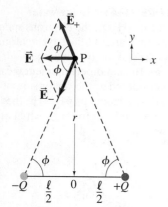

FIGURE 21–45 Electric field due to an electric dipole.

Electric Forces in Molecular Biology; DNA

The interior of every biological cell is mainly water. We can imagine a cell as a vast sea of molecules continually in motion (kinetic theory, Chapter 18), colliding with one another with various amounts of kinetic energy. These molecules interact with one another because of *electrostatic attraction between molecules*.

Indeed, cellular processes are now considered to be the result of *random ("thermal") molecular motion plus the ordering effect of the electrostatic force*. As an example, we look at DNA structure and replication. The picture we present has not been seen "in action." Rather, it is a model of what happens based on physical theories and experiment.

The genetic information that is passed on from generation to generation in all living cells is contained in the chromosomes, which are made up of genes. Each gene contains the information needed to produce a particular type of protein molecule, and that information is built into the principal molecule of a chromosome, DNA (deoxyribonucleic acid), Fig. 21–46. DNA molecules are made up of many small molecules known as nucleotide bases which are each polar due to unequal sharing of electrons. There are four types of nucleotide bases in DNA: adenine (A), cytosine (C), guanine (G), and thymine (T).

The DNA of a chromosome generally consists of two long DNA strands wrapped about one another in the shape of a "double helix." The genetic information is contained in the specific order of the four bases (A, C, G, T) along the strand. As shown in Fig. 21–47, the two strands are attracted by electrostatic forces—that is, by the attraction of positive charges to negative charges that exist on parts of the molecules. We see in Fig. 21–47a that an A (adenine) on one strand is always opposite a T on the other strand; similarly, a G is always opposite a C. This important ordering effect occurs because the shapes of A, T, C, and G are such that a T fits closely only into an A, and a G into a C; and only in the case of this close proximity of the charged portions is the electrostatic force great enough to hold them together even for a short time (Fig. 21–47b), forming what are referred to as "weak bonds."

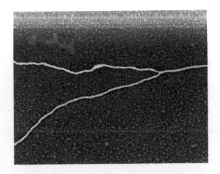

FIGURE 21–46 DNA replicating in a human HeLa cancer cell. This is a false-color image made by a transmission electron microscope (TEM; discussed in Chapter 37).

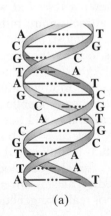

(a)

FIGURE 21–47 (a) Section of a DNA double helix. (b) "Close-up" view of the helix, showing how A and T attract each other and how G and C attract each other through electrostatic forces. The + and − signs represent net charges, usually a fraction of *e*, due to uneven sharing of electrons. The red dots indicate the electrostatic attraction (often called a "weak bond" or "hydrogen bond"—Section 40–3). Note that there are two weak bonds between A and T, and three between C and G.

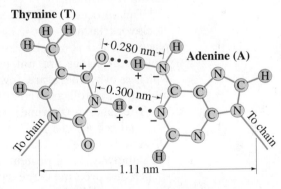

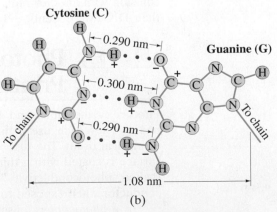

(b)

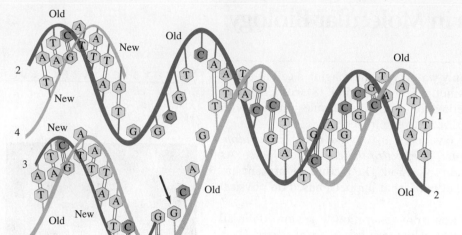

FIGURE 21–48 Replication of DNA.

When the DNA replicates (duplicates) itself just before cell division, the arrangement of A opposite T and G opposite C is crucial for ensuring that the genetic information is passed on accurately to the next generation, Fig. 21–48. The two strands of DNA separate (with the help of enzymes, which also operate via the electrostatic force), leaving the charged parts of the bases exposed. Once replication starts, let us see how the correct order of bases occurs by looking at the G molecule indicated by the red arrow in Fig. 21–48. Many unattached nucleotide bases of all four kinds are bouncing around in the cellular fluid, and the only type that will experience attraction to our G, if it bounces close to it, will be a C. The charges on the other three bases can not get close enough to those on the G to provide a significant attractive force—remember that the force decreases rapidly with distance ($\propto 1/r^2$). Because the G does not attract an A, T, or G appreciably, an A, T, or G will be knocked away by collisions with other molecules before enzymes can attach it to the growing chain (number 3). But the electrostatic force will often hold a C opposite our G long enough so that an enzyme can attach the C to the growing end of the new chain. Thus we see that electrostatic forces are responsible for selecting the bases in the proper order during replication.

This process of DNA replication is often presented as if it occurred in clockwork fashion—as if each molecule knew its role and went to its assigned place. But this is not the case. The forces of attraction are rather weak, and if the molecular shapes are not just right, there is almost no electrostatic attraction, which is why there are few mistakes. Thus, out of the random motion of the molecules, the electrostatic force acts to bring order out of chaos.

The random (thermal) velocities of molecules in a cell affect *cloning*. When a bacterial cell divides, the two new bacteria have nearly identical DNA. Even if the DNA were perfectly identical, the two new bacteria would not end up behaving in the same way. Long protein, DNA, and RNA molecules get bumped into different shapes, and even the expression of genes can thus be different. Loosely held parts of large molecules such as a methyl group (CH_3) can also be knocked off by a strong collision with another molecule. Hence, cloned organisms are not identical, even if their DNA were identical. Indeed, there can not really be genetic determinism.

*21–13 Photocopy Machines and Computer Printers Use Electrostatics

Photocopy machines and laser printers use electrostatic attraction to print an image. They each use a different technique to project an image onto a special cylindrical drum. The drum is typically made of aluminum, a good conductor; its surface is coated with a thin layer of selenium, which has the interesting property (called "photoconductivity") of being an electrical nonconductor in the dark, but a conductor when exposed to light.

In a *photocopier*, lenses and mirrors focus an image of the original sheet of paper onto the drum, much like a camera lens focuses an image on film. Step 1 is

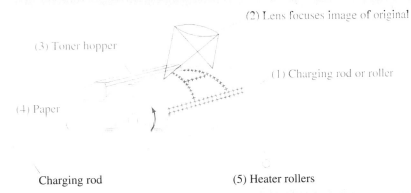

(2) Lens focuses image of original

(3) Toner hopper

(1) Charging rod or roller

(4) Paper

Charging rod

(5) Heater rollers

FIGURE 21-49 Inside a photocopy machine: (1) the selenium drum is given a + charge; (2) the lens focuses image on drum—only dark spots stay charged; (3) toner particles (negatively charged) are attracted to positive areas on drum; (4) the image is transferred to paper; (5) heat binds the image to the paper.

the placing of a uniform positive charge on the drum's selenium layer by a charged rod or roller, done in the dark. In step 2, the image to be copied or printed is projected onto the drum. For simplicity, let us assume the image is a dark letter A on a white background (as on the page of a book) as shown in Fig. 21–49. The letter A on the drum is dark, but all around it is light. At all these light places, the selenium becomes conducting and electrons flow in from the aluminum beneath, neutralizing those positive areas. In the dark areas of the letter A, the selenium is nonconducting and so retains a positive charge, Fig. 21–49. In step 3, a fine dark powder known as *toner* is given a negative charge, and brushed on the drum as it rotates. The negatively charged toner particles are attracted to the positive areas on the drum (the A in our case) and stick only there. In step 4, the rotating drum presses against a piece of paper which has been positively charged more strongly than the selenium, so the toner particles are transferred to the paper, forming the final image. Finally, step 5, the paper is heated to fix the toner particles firmly on the paper.

In a color copier (or printer), this process is repeated for each color—black, cyan (blue), magenta (red), and yellow. Combining these four colors in different proportions produces any desired color.

A *laser printer*, on the other hand, uses a computer output to program the intensity of a laser beam onto the selenium-coated drum. The thin beam of light from the laser is scanned (by a movable mirror) from side to side across the drum in a series of horizontal lines, each line just below the previous line. As the beam sweeps across the drum, the intensity of the beam is varied by the computer output, being strong for a point that is meant to be white or bright, and weak or zero for points that are meant to come out dark. After each sweep, the drum rotates very slightly for additional sweeps, Fig. 21–50, until a complete image is formed on it. The light parts of the selenium become conducting and lose their electric charge, and the toner sticks only to the dark, electrically charged areas. The drum then transfers the image to paper, as in a photocopier.

An *inkjet printer* does not use a drum. Instead nozzles spray tiny droplets of ink directly at the paper. The nozzles are swept across the paper, each sweep just above the previous one as the paper moves down. On each sweep, the ink makes dots on the paper, except for those points where no ink is desired, as directed by the computer. The image consists of a huge number of very tiny dots. The quality or resolution of a printer is usually specified in dots per inch (dpi) in each (linear) direction.

PHYSICS APPLIED
Photocopy machines

PHYSICS APPLIED
Laser printer

PHYSICS APPLIED
Inkjet printer

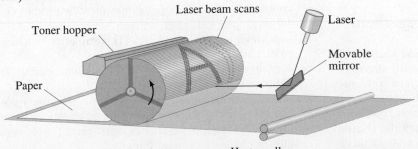

Laser beam scans

Toner hopper

Laser

Movable mirror

Paper

Heater rollers

FIGURE 21–50 Inside a laser printer: A movable mirror sweeps the laser beam in horizontal lines across the drum.

Summary

There are two kinds of **electric charge**, positive and negative. These designations are to be taken algebraically—that is, any charge is plus or minus so many coulombs (C), in SI units.

Electric charge is **conserved**: if a certain amount of one type of charge is produced in a process, an equal amount of the opposite type is also produced; thus the *net* charge produced is zero.

According to atomic theory, electricity originates in the atom, each consisting of a positively charged nucleus surrounded by negatively charged electrons. Each electron has a charge $-e = -1.6 \times 10^{-19}$ C.

Electric **conductors** are those materials in which many electrons are relatively free to move, whereas electric **insulators** are those in which very few electrons are free to move.

An object is negatively charged when it has an excess of electrons, and positively charged when it has less than its normal amount of electrons. The charge on any object is thus a whole number times $+e$ or $-e$. That is, charge is **quantized**.

An object can become charged by rubbing (in which electrons are transferred from one material to another), by conduction (which is transfer of charge from one charged object to another by touching), or by induction (the separation of charge within an object because of the close approach of another charged object but without touching).

Electric charges exert a force on each other. If two charges are of opposite types, one positive and one negative, they each exert an attractive force on the other. If the two charges are the same type, each repels the other.

The magnitude of the force one point charge exerts on another is proportional to the product of their charges, and inversely proportional to the square of the distance between them:

$$F = k\frac{Q_1 Q_2}{r^2} = \frac{1}{4\pi\epsilon_0}\frac{Q_1 Q_2}{r^2}; \qquad \textbf{(21–1, 21–2)}$$

this is **Coulomb's law**. In SI units, k is often written as $1/4\pi\epsilon_0$.

We think of an **electric field** as existing in space around any charge or group of charges. The force on another charged object is then said to be due to the electric field present at its location.

The *electric field*, $\vec{\mathbf{E}}$, at any point in space due to one or more charges, is defined as the force per unit charge that would act on a positive test charge q placed at that point:

$$\vec{\mathbf{E}} = \frac{\vec{\mathbf{F}}}{q}. \qquad \textbf{(21–3)}$$

The magnitude of the electric field a distance r from a point charge Q is

$$E = k\frac{Q}{r^2}. \qquad \textbf{(21–4a)}$$

The total electric field at a point in space is equal to the vector sum of the individual fields due to each contributing charge (**principle of superposition**).

Electric fields are represented by **electric field lines** that start on positive charges and end on negative charges. Their direction indicates the direction the force would be on a tiny positive test charge placed at each point. The lines can be drawn so that the number per unit area is proportional to the magnitude of E.

The static electric field inside a conductor is zero, and the electric field lines just outside a charged conductor are perpendicular to its surface.

An **electric dipole** is a combination of two equal but opposite charges, $+Q$ and $-Q$, separated by a distance ℓ. The **dipole moment** is $p = Q\ell$. A dipole placed in a uniform electric field feels no net force but does feel a net torque (unless $\vec{\mathbf{p}}$ is parallel to $\vec{\mathbf{E}}$). The electric field produced by a dipole decreases as the third power of the distance r from the dipole $\left(E \propto 1/r^3\right)$ for r large compared to ℓ.

[*In the replication of DNA, the electrostatic force plays a crucial role in selecting the proper molecules so the genetic information is passed on accurately from generation to generation.]

Questions

1. If you charge a pocket comb by rubbing it with a silk scarf, how can you determine if the comb is positively or negatively charged?

2. Why does a shirt or blouse taken from a clothes dryer sometimes cling to your body?

3. Explain why fog or rain droplets tend to form around ions or electrons in the air.

4. A positively charged rod is brought close to a neutral piece of paper, which it attracts. Draw a diagram showing the separation of charge in the paper, and explain why attraction occurs.

5. Why does a plastic ruler that has been rubbed with a cloth have the ability to pick up small pieces of paper? Why is this difficult to do on a humid day?

6. Contrast the *net charge* on a conductor to the "free charges" in the conductor.

7. Figures 21–7 and 21–8 show how a charged rod placed near an uncharged metal object can attract (or repel) electrons. There are a great many electrons in the metal, yet only some of them move as shown. Why not all of them?

8. When an electroscope is charged, the two leaves repel each other and remain at an angle. What balances the electric force of repulsion so that the leaves don't separate further?

9. The form of Coulomb's law is very similar to that for Newton's law of universal gravitation. What are the differences between these two laws? Compare also gravitational mass and electric charge.

10. We are not normally aware of the gravitational or electric force between two ordinary objects. What is the reason in each case? Give an example where we are aware of each one and why.

11. Is the electric force a conservative force? Why or why not? (See Chapter 8.)

12. What experimental observations mentioned in the text rule out the possibility that the numerator in Coulomb's law contains the sum $(Q_1 + Q_2)$ rather than the product $Q_1 Q_2$?

13. When a charged ruler attracts small pieces of paper, sometimes a piece jumps quickly away after touching the ruler. Explain.

14. Explain why the test charges we use when measuring electric fields must be small.

15. When determining an electric field, must we use a *positive* test charge, or would a negative one do as well? Explain.

16. Draw the electric field lines surrounding two negative electric charges a distance ℓ apart.

17. Assume that the two opposite charges in Fig. 21–34a are 12.0 cm apart. Consider the magnitude of the electric field 2.5 cm from the positive charge. On which side of this charge—top, bottom, left, or right—is the electric field the strongest? The weakest?

18. Consider the electric field at the three points indicated by the letters A, B, and C in Fig. 21–51. First draw an arrow at each point indicating the direction of the net force that a positive test charge would experience if placed at that point, then list the letters in order of *decreasing* field strength (strongest first).

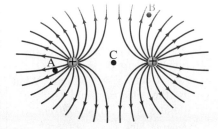

FIGURE 21–51
Question 18.

19. Why can electric field lines never cross?

20. Show, using the three rules for field lines given in Section 21–8, that the electric field lines starting or ending on a single point charge must be symmetrically spaced around the charge.

21. Given two point charges, Q and $2Q$, a distance ℓ apart, is there a point along the straight line that passes through them where $E = 0$ when their signs are (a) opposite, (b) the same? If yes, state roughly where this point will be.

22. Suppose the ring of Fig. 21–28 has a uniformly distributed negative charge Q. What is the magnitude and direction of $\vec{E}$ at point P?

23. Consider a small positive test charge located on an electric field line at some point, such as point P in Fig. 21–34a. Is the direction of the velocity and/or acceleration of the test charge along this line? Discuss.

24. We wish to determine the electric field at a point near a positively charged metal sphere (a good conductor). We do so by bringing a small test charge, q_0, to this point and measure the force F_0 on it. Will F_0/q_0 be greater than, less than, or equal to the electric field $\vec{E}$ as it was at that point before the test charge was present?

25. In what ways does the electron motion in Example 21–16 resemble projectile motion (Section 3–7)? In which ways not?

26. Describe the motion of the dipole shown in Fig. 21–44 if it is released from rest at the position shown.

27. Explain why there can be a net force on an electric dipole placed in a nonuniform electric field.

Problems

21–5 Coulomb's Law

$[1\,\mathrm{mC} = 10^{-3}\,\mathrm{C},\ 1\,\mu\mathrm{C} = 10^{-6}\,\mathrm{C},\ 1\,\mathrm{nC} = 10^{-9}\,\mathrm{C}.]$

1. (I) What is the magnitude of the electric force of attraction between an iron nucleus ($q = +26e$) and its innermost electron if the distance between them is $1.5 \times 10^{-12}\,\mathrm{m}$?

2. (I) How many electrons make up a charge of $-38.0\,\mu\mathrm{C}$?

3. (I) What is the magnitude of the force a $+25\,\mu\mathrm{C}$ charge exerts on a $+2.5\,\mathrm{mC}$ charge 28 cm away?

4. (I) What is the repulsive electrical force between two protons $4.0 \times 10^{-15}\,\mathrm{m}$ apart from each other in an atomic nucleus?

5. (II) When an object such as a plastic comb is charged by rubbing it with a cloth, the net charge is typically a few microcoulombs. If that charge is $3.0\,\mu\mathrm{C}$, by what percentage does the mass of a 35-g comb change during charging?

6. (II) Two charged dust particles exert a force of $3.2 \times 10^{-2}\,\mathrm{N}$ on each other. What will be the force if they are moved so they are only one-eighth as far apart?

7. (II) Two charged spheres are 8.45 cm apart. They are moved, and the force on each of them is found to have been tripled. How far apart are they now?

8. (II) A person scuffing her feet on a wool rug on a dry day accumulates a net charge of $-46\,\mu\mathrm{C}$. How many excess electrons does she get, and by how much does her mass increase?

9. (II) What is the total charge of all the electrons in a 15-kg bar of gold? What is the net charge of the bar? (Gold has 79 electrons per atom and an atomic mass of 197 u.)

10. (II) Compare the electric force holding the electron in orbit $(r = 0.53 \times 10^{-10}\,\mathrm{m})$ around the proton nucleus of the hydrogen atom, with the gravitational force between the same electron and proton. What is the ratio of these two forces?

11. (II) Two positive point charges are a fixed distance apart. The sum of their charges is Q_T. What charge must each have in order to (a) maximize the electric force between them, and (b) minimize it?

12. (II) Particles of charge $+75$, $+48$, and $-85\,\mu\mathrm{C}$ are placed in a line (Fig. 21–52). The center one is 0.35 m from each of the others. Calculate the net force on each charge due to the other two.

FIGURE 21–52
Problem 12.

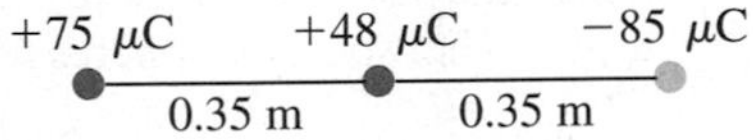

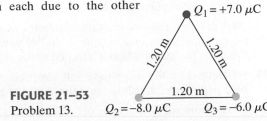

$+75\,\mu\mathrm{C}$ $+48\,\mu\mathrm{C}$ $-85\,\mu\mathrm{C}$
0.35 m   0.35 m

13. (II) Three charged particles are placed at the corners of an equilateral triangle of side 1.20 m (Fig. 21–53). The charges are $+7.0\,\mu\mathrm{C}$, $-8.0\,\mu\mathrm{C}$, and $-6.0\,\mu\mathrm{C}$. Calculate the magnitude and direction of the net force on each due to the other two.

$Q_1 = +7.0\,\mu\mathrm{C}$
1.20 m   1.20 m

FIGURE 21–53
Problem 13. $Q_2 = -8.0\,\mu\mathrm{C}$ 1.20 m $Q_3 = -6.0\,\mu\mathrm{C}$

14. (II) Two small nonconducting spheres have a total charge of $90.0\,\mu\mathrm{C}$. (a) When placed 1.16 m apart, the force each exerts on the other is 12.0 N and is repulsive. What is the charge on each? (b) What if the force were attractive?

15. (II) A charge of 4.15 mC is placed at each corner of a square 0.100 m on a side. Determine the magnitude and direction of the force on each charge.

16. (II) Two negative and two positive point charges (magnitude $Q = 4.15$ mC) are placed on opposite corners of a square as shown in Fig. 21–54. Determine the magnitude and direction of the force on each charge.

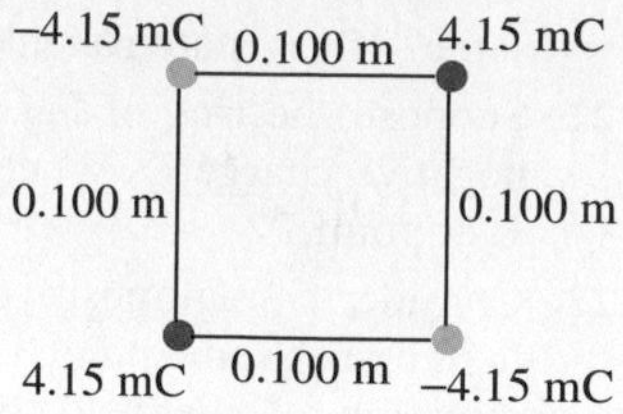

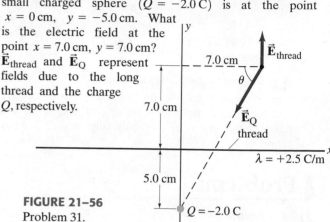

FIGURE 21–54
Problem 16.

17. (II) A charge Q is transferred from an initially uncharged plastic ball to an identical ball 12 cm away. The force of attraction is then 17 mN. How many electrons were transferred from one ball to the other?

18. (III) Two charges, $-Q_0$ and $-4Q_0$, are a distance ℓ apart. These two charges are free to move but do not because there is a third charge nearby. What must be the magnitude of the third charge and its placement in order for the first two to be in equilibrium?

19. (III) Two positive charges $+Q$ are affixed rigidly to the x axis, one at $x = +d$ and the other at $x = -d$. A third charge $+q$ of mass m, which is constrained to move only along the x axis, is displaced from the origin by a small distance $s \ll d$ and then released from rest. (*a*) Show that (to a good approximation) $+q$ will execute simple harmonic motion and determine an expression for its oscillation period T. (*b*) If these three charges are each singly ionized sodium atoms $(q = Q = +e)$ at the equilibrium spacing $d = 3 \times 10^{-10}$ m typical of the atomic spacing in a solid, find T in picoseconds.

20. (III) Two small charged spheres hang from cords of equal length ℓ as shown in Fig. 21–55 and make small angles θ_1 and θ_2 with the vertical. (*a*) If $Q_1 = Q$, $Q_2 = 2Q$, and $m_1 = m_2 = m$, determine the ratio θ_1/θ_2. (*b*) If $Q_1 = Q$, $Q_2 = 2Q$, $m_1 = m$, and $m_2 = 2m$, determine the ratio θ_1/θ_2. (*c*) Estimate the distance between the spheres for each case.

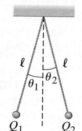

FIGURE 21–55
Problem 20.

21–6 to 21–8 Electric Field, Field Lines

21. (I) What are the magnitude and direction of the electric force on an electron in a uniform electric field of strength 1920 N/C that points due east?

22. (I) A proton is released in a uniform electric field, and it experiences an electric force of 2.18×10^{-14} N toward the south. What are the magnitude and direction of the electric field?

23. (I) Determine the magnitude and direction of the electric field 16.4 cm directly above an isolated 33.0×10^{-6} C charge.

24. (I) A downward electric force of 8.4 N is exerted on a $-8.8\,\mu$C charge. What are the magnitude and direction of the electric field at the position of this charge?

25. (I) The electric force on a $+4.20$-μC charge is $\vec{\mathbf{F}} = (7.22 \times 10^{-4}\,\text{N})\hat{\mathbf{j}}$. What is the electric field at the position of the charge?

26. (I) What is the electric field at a point when the force on a 1.25-μC charge placed at that point is $\vec{\mathbf{F}} = (3.0\hat{\mathbf{i}} - 3.9\hat{\mathbf{j}}) \times 10^{-3}$ N?

27. (II) Determine the magnitude of the acceleration experienced by an electron in an electric field of 576 N/C. How does the direction of the acceleration depend on the direction of the field at that point?

28. (II) Determine the magnitude and direction of the electric field at a point midway between a $-8.0\,\mu$C and a $+5.8\,\mu$C charge 8.0 cm apart. Assume no other charges are nearby.

29. (II) Draw, approximately, the electric field lines about two point charges, $+Q$ and $-3Q$, which are a distance ℓ apart.

30. (II) What is the electric field strength at a point in space where a proton experiences an acceleration of 1.8 million "g's"?

31. (II) A long uniformly charged thread (linear charge density $\lambda = 2.5$ C/m) lies along the x axis in Fig. 21–56. A small charged sphere $(Q = -2.0$ C) is at the point $x = 0$ cm, $y = -5.0$ cm. What is the electric field at the point $x = 7.0$ cm, $y = 7.0$ cm? $\vec{\mathbf{E}}_{\text{thread}}$ and $\vec{\mathbf{E}}_Q$ represent fields due to the long thread and the charge Q, respectively.

FIGURE 21–56
Problem 31.

32. (II) The electric field midway between two equal but opposite point charges is 586 N/C, and the distance between the charges is 16.0 cm. What is the magnitude of the charge on each?

33. (II) Calculate the electric field at one corner of a square 1.22 m on a side if the other three corners are occupied by 2.25×10^{-6} C charges.

34. (II) Calculate the electric field at the center of a square 52.5 cm on a side if one corner is occupied by a $-38.6\,\mu$C charge and the other three are occupied by $-27.0\,\mu$C charges.

35. (II) Determine the direction and magnitude of the electric field at the point P in Fig. 21–57. The charges are separated by a distance $2a$, and point P is a distance x from the midpoint between the two charges. Express your answer in terms of Q, x, a, and k.

FIGURE 21–57
Problem 35.

36. (II) Two point charges, $Q_1 = -25\,\mu$C and $Q_2 = +45\,\mu$C, are separated by a distance of 12 cm. The electric field at the point P (see Fig. 21–58) is zero. How far from Q_1 is P?

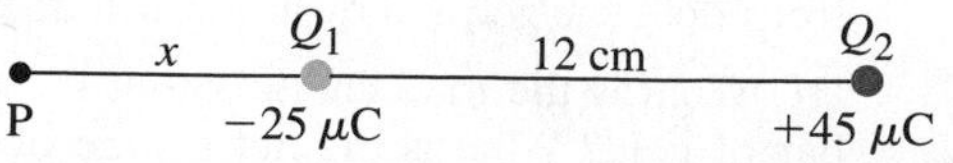

FIGURE 21–58
Problem 36.

37. (II) A very thin line of charge lies along the x axis from $x = -\infty$ to $x = +\infty$. Another similar line of charge lies along the y axis from $y = -\infty$ to $y = +\infty$. Both lines have a uniform charge per length λ. Determine the resulting electric field magnitude and direction (relative to the x axis) at a point (x, y) in the first quadrant of the xy plane.

38. (II) (a) Determine the electric field $\vec{E}$ at the origin 0 in Fig. 21–59 due to the two charges at A and B. (b) Repeat, but let the charge at B be reversed in sign.

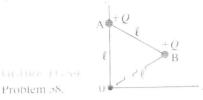

FIGURE 21–59
Problem 38.

39. (II) Draw, approximately, the electric field lines emanating from a uniformly charged straight wire whose length ℓ is not great. The spacing between lines near the wire should be much less than ℓ. [*Hint*: Also consider points very far from the wire.]

40. (II) Two parallel circular rings of radius R have their centers on the x axis separated by a distance ℓ as shown in Fig. 21–60. If each ring carries a uniformly distributed charge Q, find the electric field, $\vec{E}(x)$, at points along the x axis.

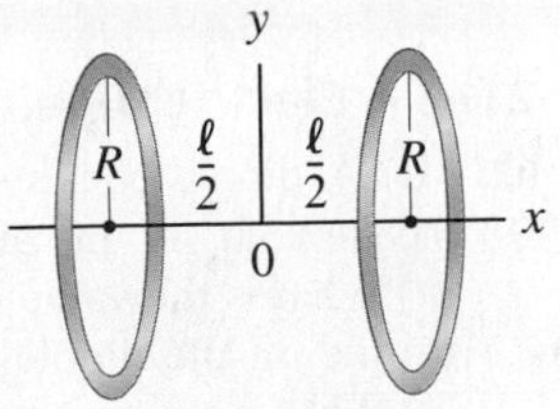

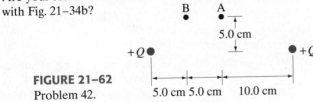

FIGURE 21–60
Problem 40.

41. (II) You are given two unknown point charges, Q_1 and Q_2. At a point on the line joining them, one-third of the way from Q_1 to Q_2, the electric field is zero (Fig. 21–61). What is the ratio Q_1/Q_2?

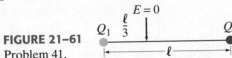

FIGURE 21–61
Problem 41.

42. (II) Use Coulomb's law to determine the magnitude and direction of the electric field at points A and B in Fig. 21–62 due to the two positive charges $(Q = 5.7\,\mu\text{C})$ shown. Are your results consistent with Fig. 21–34b?

FIGURE 21–62
Problem 42.

43. (II) (a) Two equal charges Q are positioned at points $(x = \ell, y = 0)$ and $(x = -\ell, y = 0)$. Determine the electric field as a function of y for points along the y axis. (b) Show that the field is a maximum at $y = \pm\ell/\sqrt{2}$.

44. (II) At what position, $x = x_\text{M}$, is the magnitude of the electric field along the axis of the ring of Example 21–9 a maximum?

45. (II) Estimate the electric field at a point 2.40 cm perpendicular to the midpoint of a uniformly charged 2.00-m-long thin wire carrying a total charge of $4.75\,\mu\text{C}$.

46. (II) The uniformly charged straight wire in Fig. 21–29 has the length ℓ, where point 0 is at the midpoint. Show that the field at point P, a perpendicular distance x from 0, is given by

$$E = \frac{\lambda}{2\pi\epsilon_0}\frac{\ell}{x(\ell^2 + 4x^2)^{1/2}},$$

where λ is the charge per unit length.

47. (II) Use your result from Problem 46 to find the electric field (magnitude and direction) a distance z above the center of a square loop of wire, each of whose sides has length ℓ and uniform charge per length λ (Fig. 21–63).

FIGURE 21–63
Problem 47.

48. (II) Determine the direction and magnitude of the electric field at the point P shown in Fig. 21–64. The two charges are separated by a distance of $2a$. Point P is on the perpendicular bisector of the line joining the charges, a distance x from the midpoint between them. Express your answers in terms of $Q, x, a,$ and k.

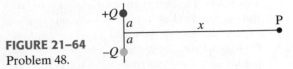

FIGURE 21–64
Problem 48.

49. (III) A thin rod bent into the shape of an arc of a circle of radius R carries a uniform charge per unit length λ. The arc subtends a total angle $2\theta_0$, symmetric about the x axis, as shown in Fig. 21–65. Determine the electric field $\vec{E}$ at the origin 0.

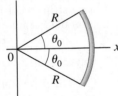

FIGURE 21–65
Problem 49.

50. (III) A thin glass rod is a semicircle of radius R, Fig. 21–66. A charge is nonuniformly distributed along the rod with a linear charge density given by $\lambda = \lambda_0 \sin\theta$, where λ_0 is a positive constant. Point P is at the center of the semicircle. (a) Find the electric field $\vec{E}$ (magnitude and direction) at point P. [*Hint*: Remember $\sin(-\theta) = -\sin\theta$, so the two halves of the rod are oppositely charged.] (b) Determine the acceleration (magnitude and direction) of an electron placed at point P, assuming $R = 1.0\,\text{cm}$ and $\lambda_0 = 1.0\,\mu\text{C/m}$.

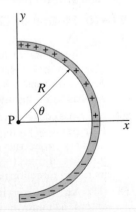

FIGURE 21–66
Problem 50.

51. (III) Suppose a uniformly charged wire starts at point 0 and rises vertically along the positive y axis to a length ℓ. (a) Determine the components of the electric field E_x and E_y at point $(x, 0)$. That is, calculate $\vec{E}$ near one end of a long wire, in the plane perpendicular to the wire. (b) If the wire extends from $y = 0$ to $y = \infty$, so that $\ell = \infty$, show that $\vec{E}$ makes a 45° angle to the horizontal for any x. [*Hint*: See Example 21–11 and Fig. 21–29.]

52. (III) Suppose in Example 21–11 that $x = 0.250\,\text{m}$, $Q = 3.15\,\mu\text{C}$, and that the uniformly charged wire is only 6.50 m long and extends along the y axis from $y = -4.00\,\text{m}$ to $y = +2.50\,\text{m}$. (a) Calculate E_x and E_y at point P. (b) Determine what the error would be if you simply used the result of Example 21–11, $E = \lambda/2\pi\epsilon_0 x$. Express this error as $(E_x - E)/E$ and E_y/E.

53. (III) A thin rod of length ℓ carries a total charge Q distributed uniformly along its length. See Fig. 21–67. Determine the electric field along the axis of the rod starting at one end—that is, find $E(x)$ for $x \geq 0$ in Fig. 21–67.

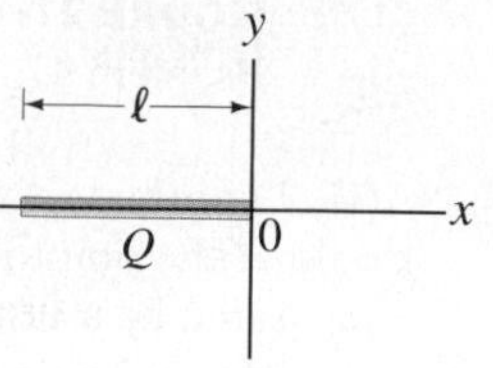

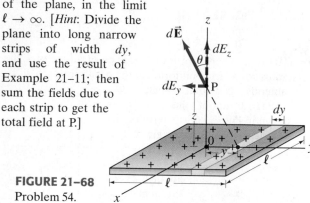

FIGURE 21–67
Problem 53.

54. (III) *Uniform plane of charge.* Charge is distributed uniformly over a large square plane of side ℓ, as shown in Fig. 21–68. The charge per unit area (C/m^2) is σ. Determine the electric field at a point P a distance z above the center of the plane, in the limit $\ell \to \infty$. [*Hint*: Divide the plane into long narrow strips of width dy, and use the result of Example 21–11; then sum the fields due to each strip to get the total field at P.]

FIGURE 21–68
Problem 54.

55. (III) Suppose the charge Q on the ring of Fig. 21–28 was all distributed uniformly on only the upper half of the ring, and no charge was on the lower half. Determine the electric field $\vec{E}$ at P. (Take y vertically upward.)

21–10 Motion of Charges in an Electric Field

56. (II) An electron with speed $v_0 = 27.5 \times 10^6\,\text{m/s}$ is traveling parallel to a uniform electric field of magnitude $E = 11.4 \times 10^3\,\text{N/C}$. (a) How far will the electron travel before it stops? (b) How much time will elapse before it returns to its starting point?

57. (II) An electron has an initial velocity $\vec{v}_0 = (8.0 \times 10^4\,\text{m/s})\hat{\mathbf{j}}$. It enters a region where $\vec{E} = (2.0\hat{\mathbf{i}} + 8.0\hat{\mathbf{j}}) \times 10^4\,\text{N/C}$. (a) Determine the vector acceleration of the electron as a function of time. (b) At what angle θ is it moving (relative to its initial direction) at $t = 1.0\,\text{ns}$?

58. (II) An electron moving to the right at $7.5 \times 10^5\,\text{m/s}$ enters a uniform electric field parallel to its direction of motion. If the electron is to be brought to rest in the space of 4.0 cm, (a) what direction is required for the electric field, and (b) what is the strength of the field?

59. (II) At what angle will the electrons in Example 21–16 leave the uniform electric field at the end of the parallel plates (point P in Fig. 21–41)? Assume the plates are 4.9 cm long, $E = 5.0 \times 10^3\,\text{N/C}$, and $v_0 = 1.00 \times 10^7\,\text{m/s}$. Ignore fringing of the field.

60. (II) An electron is traveling through a uniform electric field. The field is constant and given by $\vec{E} = (2.00 \times 10^{-11}\,\text{N/C})\hat{\mathbf{i}} - (1.20 \times 10^{-11}\,\text{N/C})\hat{\mathbf{j}}$. At $t = 0$, the electron is at the origin and traveling in the x direction with a speed of 1.90 m/s. What is its position 2.00 s later?

61. (II) A positive charge q is placed at the center of a circular ring of radius R. The ring carries a uniformly distributed negative charge of total magnitude $-Q$. (a) If the charge q is displaced from the center a small distance x as shown in Fig. 21–69, show that it will undergo simple harmonic motion when released. (b) If its mass is m, what is its period?

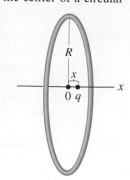

FIGURE 21–69
Problem 61.

21–11 Electric Dipoles

62. (II) A dipole consists of charges $+e$ and $-e$ separated by 0.68 nm. It is in an electric field $E = 2.2 \times 10^4\,\text{N/C}$. (a) What is the value of the dipole moment? (b) What is the torque on the dipole when it is perpendicular to the field? (c) What is the torque on the dipole when it is at an angle of 45° to the field? (d) What is the work required to rotate the dipole from being oriented parallel to the field to being antiparallel to the field?

63. (II) The HCl molecule has a dipole moment of about $3.4 \times 10^{-30}\,\text{C·m}$. The two atoms are separated by about $1.0 \times 10^{-10}\,\text{m}$. (a) What is the net charge on each atom? (b) Is this equal to an integral multiple of e? If not, explain. (c) What maximum torque would this dipole experience in a $2.5 \times 10^4\,\text{N/C}$ electric field? (d) How much energy would be needed to rotate one molecule 45° from its equilibrium position of lowest potential energy?

64. (II) Suppose both charges in Fig. 21–45 (for a dipole) were positive. (a) Show that the field on the perpendicular bisector, for $r \gg \ell$, is given by $(1/4\pi\epsilon_0)(2Q/r^2)$. (b) Explain why the field decreases as $1/r^2$ here whereas for a dipole it decreases as $1/r^3$.

65. (II) An electric dipole, of dipole moment p and moment of inertia I, is placed in a uniform electric field $\vec{E}$. (a) If displaced by an angle θ as shown in Fig. 21–44 and released, under what conditions will it oscillate in simple harmonic motion? (b) What will be its frequency?

66. (III) Suppose a dipole $\vec{p}$ is placed in a nonuniform electric field $\vec{E} = E\hat{\mathbf{i}}$ that points along the x axis. If $\vec{E}$ depends only on x, show that the net force on the dipole is

$$\vec{F} = \left(\vec{p} \cdot \frac{d\vec{E}}{dx}\right)\hat{\mathbf{i}},$$

where $d\vec{E}/dx$ is the gradient of the field in the x direction.

67. (III) (a) Show that at points along the axis of a dipole (along the same line that contains $+Q$ and $-Q$), the electric field has magnitude

$$E = \frac{1}{4\pi\epsilon_0} \frac{2p}{r^3}$$

for $r \gg \ell$ (Fig. 21–45), where r is the distance from a point to the center of the dipole. (b) In what direction does $\vec{E}$ point?

General Problems

68. How close must two electrons be if the electric force between them is equal to the weight of either at the Earth's surface?

69. Given that the human body is mostly made of water, estimate the total amount of positive charge in a 65-kg person.

70. A 3.0-g copper penny has a positive charge of $38\,\mu C$. What fraction of its electrons has it lost?

71. Measurements indicate that there is an electric field surrounding the Earth. Its magnitude is about $150\,N/C$ at the Earth's surface and points inward toward the Earth's center. What is the magnitude of the electric charge on the Earth? Is it positive or negative? [*Hint*: The electric field outside a uniformly charged sphere is the same as if all the charge were concentrated at its center.]

72. (*a*) The electric field near the Earth's surface has magnitude of about $150\,N/C$. What is the acceleration experienced by an electron near the surface of the Earth? (*b*) What about a proton? (*c*) Calculate the ratio of each acceleration to $g = 9.8\,m/s^2$.

73. A water droplet of radius 0.018 mm remains stationary in the air. If the downward-directed electric field of the Earth is $150\,N/C$, how many excess electron charges must the water droplet have?

74. Estimate the net force between the CO group and the HN group shown in Fig. 21–70. The C and O have charges $\pm 0.40e$, and the H and N have charges $\pm 0.20e$, where $e = 1.6 \times 10^{-19}\,C$. [*Hint*: Do not include the "internal" forces between C and O, or between H and N.]

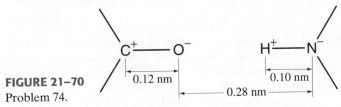

FIGURE 21–70
Problem 74.

75. Suppose that electrical attraction, rather than gravity, were responsible for holding the Moon in orbit around the Earth. If equal and opposite charges Q were placed on the Earth and the Moon, what should be the value of Q to maintain the present orbit? Use data given on the inside front cover of this book. Treat the Earth and Moon as point particles.

76. In a simple model of the hydrogen atom, the electron revolves in a circular orbit around the proton with a speed of $2.2 \times 10^6\,m/s$. Determine the radius of the electron's orbit. [*Hint*: See Chapter 5 on circular motion.]

77. A positive point charge $Q_1 = 2.5 \times 10^{-5}\,C$ is fixed at the origin of coordinates, and a negative point charge $Q_2 = -5.0 \times 10^{-6}\,C$ is fixed to the x axis at $x = +2.0\,m$. Find the location of the place(s) along the x axis where the electric field due to these two charges is zero.

78. When clothes are removed from a dryer, a 40-g sock is stuck to a sweater, even with the sock clinging to the sweater's underside. Estimate the minimum attractive force between the sock and the sweater. Then estimate the minimum charge on the sock and the sweater. Assume the charging came entirely from the sock rubbing against the sweater so that they have equal and opposite charges, and approximate the sweater as a flat sheet of uniform charge.

79. A small lead sphere is encased in insulating plastic and suspended vertically from an ideal spring (spring constant $k = 126\,N/m$) as in Fig. 21–71. The total mass of the coated sphere is 0.650 kg, and its center lies 15.0 cm above a tabletop when in equilibrium. The sphere is pulled down 5.00 cm below equilibrium, an electric charge $Q = -5.00 \times 10^{-6}\,C$ is deposited on it, and then it is released. Using what you know about harmonic oscillation, write an expression for the electric field strength as a function of time that would be measured at the point on the tabletop (P) directly below the sphere.

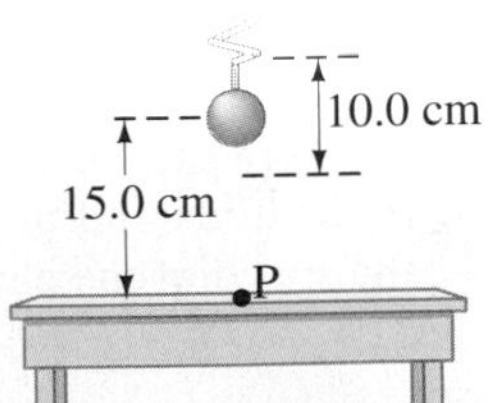

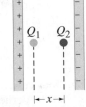

FIGURE 21–71
Problem 79.

80. A large electroscope is made with "leaves" that are 78-cm-long wires with tiny 24-g spheres at the ends. When charged, nearly all the charge resides on the spheres. If the wires each make a 26° angle with the vertical (Fig. 21–72), what total charge Q must have been applied to the electroscope? Ignore the mass of the wires.

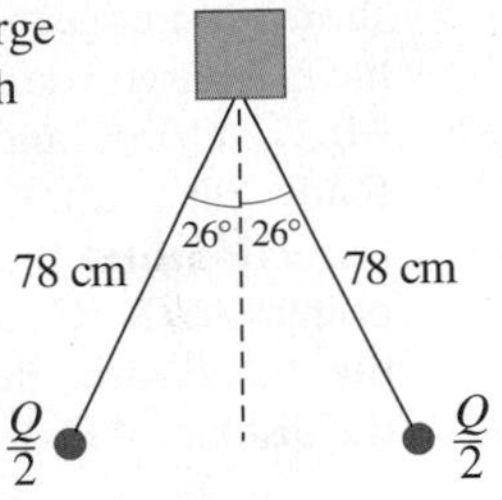

FIGURE 21–72
Problem 80.

81. Dry air will break down and generate a spark if the electric field exceeds about $3 \times 10^6\,N/C$. How much charge could be packed onto a green pea (diameter 0.75 cm) before the pea spontaneously discharges? [*Hint*: Eqs. 21–4 work outside a sphere if r is measured from its center.]

82. Two point charges, $Q_1 = -6.7\,\mu C$ and $Q_2 = 1.8\,\mu C$, are located between two oppositely charged parallel plates, as shown in Fig. 21–73. The two charges are separated by a distance of $x = 0.34\,m$. Assume that the electric field produced by the charged plates is uniform and equal to $E = 73{,}000\,N/C$. Calculate the net electrostatic force on Q_1 and give its direction.

FIGURE 21–73
Problem 82.

83. Packing material made of pieces of foamed polystyrene can easily become charged and stick to each other. Given that the density of this material is about $35\,kg/m^3$, estimate how much charge might be on a 2.0-cm-diameter foamed polystyrene sphere, assuming the electric force between two spheres stuck together is equal to the weight of one sphere.

84. One type of *electric quadrupole* consists of two dipoles placed end to end with their negative charges (say) overlapping; that is, in the center is $-2Q$ flanked (on a line) by a $+Q$ to either side (Fig. 21–74). Determine the electric field $\vec{E}$ at points along the perpendicular bisector and show that E decreases as $1/r^4$. Measure r from the $-2Q$ charge and assume $r \gg \ell$.

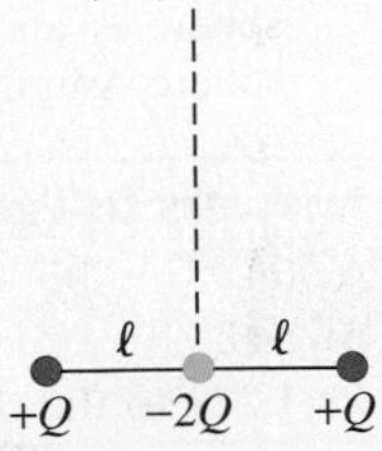

FIGURE 21–74
Problem 84.

85. Suppose electrons enter a uniform electric field midway between two plates at an angle θ_0 to the horizontal, as shown in Fig. 21–75. The path is symmetrical, so they leave at the same angle θ_0 and just barely miss the top plate. What is θ_0? Ignore fringing of the field.

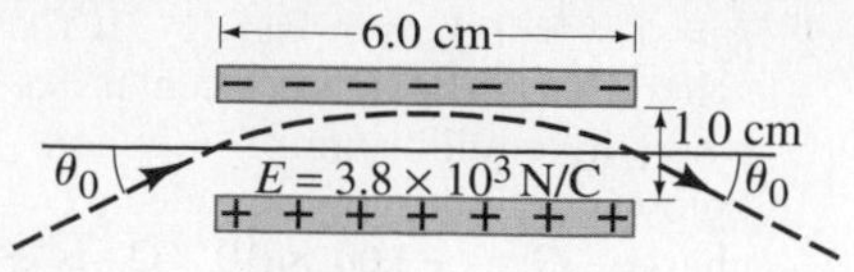

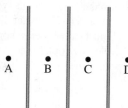

FIGURE 21–75
Problem 85.

$|\leftarrow$ —6.0 cm—$\rightarrow|$

$\uparrow$1.0 cm

$E = 3.8 \times 10^3$ N/C

θ_0

86. An electron moves in a circle of radius r around a very long uniformly charged wire in a vacuum chamber, as shown in Fig. 21–76. The charge density on the wire is $\lambda = 0.14\,\mu$C/m. (a) What is the electric field at the electron (magnitude and direction in terms of r and λ)? (b) What is the speed of the electron?

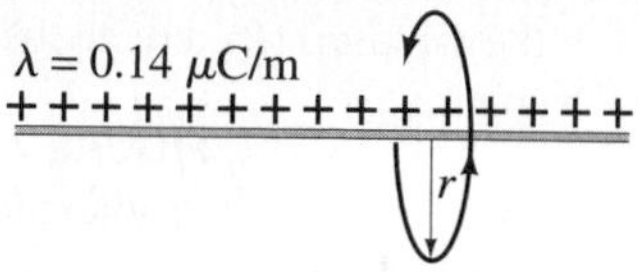

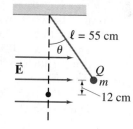

$\lambda = 0.14\,\mu$C/m

FIGURE 21–76
Problem 86.

r

87. Three very large square planes of charge are arranged as shown (on edge) in Fig. 21–77. From left to right, the planes have charge densities per unit area of $-0.50\,\mu$C/m^2, $+0.25\,\mu$C/m^2, and $-0.35\,\mu$C/m^2. Find the total electric field (direction and magnitude) at the points A, B, C, and D. Assume the plates are much larger than the distance AD.

A   B   C   D

FIGURE 21–77
Problem 87.

88. A point charge ($m = 1.0$ g) at the end of an insulating cord of length 55 cm is observed to be in equilibrium in a uniform horizontal electric field of 15,000 N/C, when the pendulum's position is as shown in Fig. 21–78, with the charge 12 cm above the lowest (vertical) position. If the field points to the right in Fig. 21–78, determine the magnitude and sign of the point charge.

$\ell = 55$ cm

θ

$\vec{E}$

Q

m

12 cm

FIGURE 21–78
Problem 88.

89. Four equal positive point charges, each of charge 8.0 μC, are at the corners of a square of side 9.2 cm. What charge should be placed at the center of the square so that all charges are at equilibrium? Is this a stable or unstable equilibrium (Section 12–3) in the plane?

90. Two small, identical conducting spheres A and B are a distance R apart; each carries the same charge Q. (a) What is the force sphere B exerts on sphere A? (b) An identical sphere with zero charge, sphere C, makes contact with sphere B and is then moved very far away. What is the net force now acting on sphere A? (c) Sphere C is brought back and now makes contact with sphere A and is then moved far away. What is the force on sphere A in this third case?

91. A point charge of mass 0.210 kg, and net charge $+0.340\,\mu$C, hangs at rest at the end of an insulating cord above a large sheet of charge. The horizontal sheet of fixed uniform charge creates a uniform vertical electric field in the vicinity of the point charge. The tension in the cord is measured to be 5.18 N. (a) Calculate the magnitude and direction of the electric field due to the sheet of charge (Fig. 21–79). (b) What is the surface charge density σ (C/m^2) on the sheet?

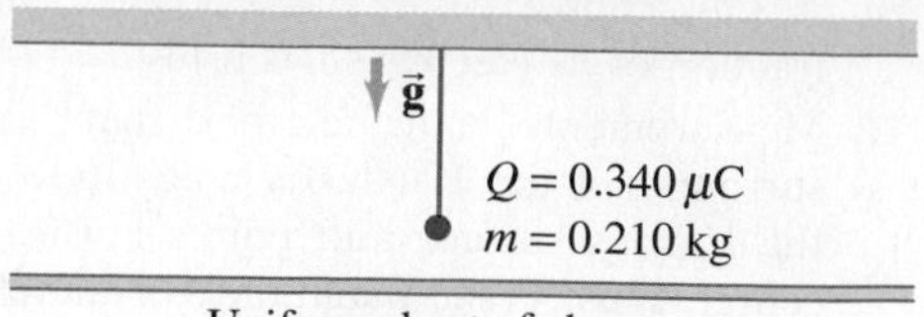

$\vec{g}$

$Q = 0.340\,\mu$C
$m = 0.210$ kg

FIGURE 21–79
Problem 91.   Uniform sheet of charge

92. A one-dimensional row of positive ions, each with charge $+Q$ and separated from its neighbors by a distance d, occupies the right-hand half of the x axis. That is, there is a $+Q$ charge at $x = 0$, $x = +d$, $x = +2d$, $x = +3d$, and so on out to ∞. (a) If an electron is placed at the position $x = -d$, determine F, the magnitude of force that this row of charges exerts on the electron. (b) If the electron is instead placed at $x = -3d$, what is the value of F? [Hint: The infinite sum $\sum_{n=1}^{n=\infty} \frac{1}{n^2} = \frac{\pi^2}{6}$, where n is a positive integer.]

Numerical/Computer

*93. (III) A thin ring-shaped object of radius a contains a total charge Q uniformly distributed over its length. The electric field at a point on its axis a distance x from its center is given in Example 21–9 as

$$E = \frac{1}{4\pi\epsilon_0} \frac{Qx}{(x^2 + a^2)^{\frac{3}{2}}}.$$

(a) Take the derivative to find where on the x axis ($x > 0$) E_x is a maximum. Assume $Q = 6.00\,\mu$C and $a = 10.0$ cm. (b) Calculate the electric field for $x = 0$ to $x = +12.0$ cm in steps of 0.1 cm, and make a graph of the electric field. Does the maximum of the graph coincide with the maximum of the electric field you obtained analytically? Also, calculate and graph the electric field (c) due to the ring, and (d) due to a point charge $Q = 6.00\,\mu$C at the center of the ring. Make a single graph, from $x = 0$ (or $x = 1.0$ cm) out to $x = 50.0$ cm in 1.0 cm steps, with two curves of the electric fields, and show that both fields converge at large distances from the center. (e) At what distance does the electric field of the ring differ from that of the point charge by 10%?

*94. (III) An 8.00 μC charge is on the x axis of a coordinate system at $x = +5.00$ cm. A $-2.00\,\mu$C charge is at $x = -5.00$ cm. (a) Plot the x component of the electric field for points on the x axis from $x = -30.0$ cm to $x = +30.0$ cm. The sign of E_x is positive when $\vec{E}$ points to the right and negative when it points to the left. (b) Make a plot of E_x and E_y for points on the y axis from $y = -30.0$ to $+30.0$ cm.

Answers to Exercises

A: (e).

B: 5 N.

C: 1.2 N, to the right.

D: (a) No; (b) yes, midway between them.

E: (d), if the two + charges are not at opposite corners (use symmetry).

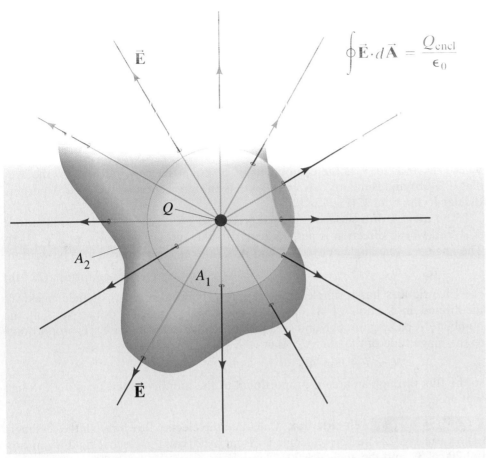

$$\oint \vec{E} \cdot d\vec{A} = \frac{Q_{encl}}{\epsilon_0}$$

$\vec{E}$

Q

A_2

A_1

$\vec{E}$

Gauss's law is an elegant relation between electric charge and electric field. It is more general than Coulomb's law. Gauss's law involves an integral of the electric field $\vec{E}$ at each point on a closed surface. The surface is only imaginary, and we choose the shape and placement of the surface so that we can evaluate the integral. In this drawing, two different 3-D surfaces are shown (one green, one blue), both enclosing a point charge Q. Gauss's law states that the product $\vec{E} \cdot d\vec{A}$, where $d\vec{A}$ is an infinitesimal area of the surface, integrated over the entire surface, equals the charge enclosed by the surface Q_{encl} divided by ϵ_0. Both surfaces here enclose the same charge Q. Hence $\oint \vec{E} \cdot d\vec{A}$ will give the same result for both surfaces.

Gauss's Law

CHAPTER 22

CHAPTER-OPENING QUESTION—Guess now!

A nonconducting sphere has a uniform charge density throughout. How does the magnitude of the electric field vary inside with distance from the center?

(a) The electric field is zero throughout.
(b) The electric field is constant but nonzero throughout.
(c) The electric field is linearly increasing from the center to the outer edge.
(d) The electric field is exponentially increasing from the center to the outer edge.
(e) The electric field increases quadratically from the center to the outer edge.

The great mathematician Karl Friedrich Gauss (1777–1855) developed an important relation, now known as Gauss's law, which we develop and discuss in this Chapter. It is a statement of the relation between electric charge and electric field and is a more general and elegant form of Coulomb's law.

We can, in principle, determine the electric field due to any given distribution of electric charge using Coulomb's law. The total electric field at any point will be the vector sum (or integral) of contributions from all charges present (see Eq. 21–6). Except for some simple cases, the sum or integral can be quite complicated to evaluate. For situations in which an analytic solution (such as we carried out in the Examples of Sections 21–6 and 21–7) is not possible, a computer can be used.

In some cases, however, the electric field due to a given charge distribution can be calculated more easily or more elegantly using Gauss's law, as we shall see in this Chapter. But the major importance of Gauss's law is that it gives us additional insight into the nature of electrostatic fields, and a more general relationship between charge and field.

Before discussing Gauss's law itself, we first discuss the concept of *flux*.

CONTENTS

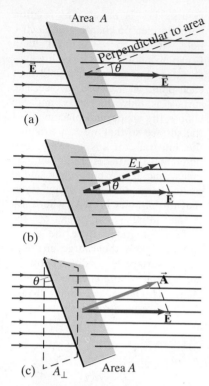

Area A

Perpendicular to area

θ

$\vec{E}$

$\vec{E}$

(a)

$E_\perp$

θ

$\vec{E}$

(b)

θ

$\vec{A}$

$\vec{E}$

(c) $A_\perp$ Area A

FIGURE 22–1 (a) A uniform electric field $\vec{E}$ passing through a flat area A. (b) $E_\perp = E\cos\theta$ is the component of $\vec{E}$ perpendicular to the plane of area A. (c) $A_\perp = A\cos\theta$ is the projection (dashed) of the area A perpendicular to the field $\vec{E}$.

FIGURE 22–2 Electric flux through a curved surface. One small area of the surface, $\Delta\vec{A}_i$, is indicated.

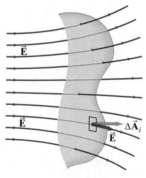

$\vec{E}$

$\vec{E}$

$\Delta\vec{A}_i$

$\vec{E}$

FIGURE 22–3 Electric flux through a closed surface.

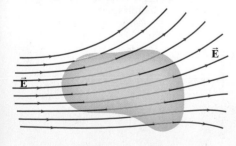

$\vec{E}$

$\vec{E}$

22–1 Electric Flux

Gauss's law involves the concept of **electric flux**, which refers to the electric field passing through a given area. For a uniform electric field $\vec{E}$ passing through an area A, as shown in Fig. 22–1a, the electric flux Φ_E is defined as

$$\Phi_E = EA\cos\theta,$$

where θ is the angle between the electric field direction and a line drawn perpendicular to the area. The flux can be written equivalently as

$$\Phi_E = E_\perp A = EA_\perp = EA\cos\theta, \qquad [\vec{E}\text{ uniform}] \quad (22\text{–}1a)$$

where $E_\perp = E\cos\theta$ is the component of $\vec{E}$ along the perpendicular to the area (Fig. 22–1b) and, similarly, $A_\perp = A\cos\theta$ is the projection of the area A perpendicular to the field $\vec{E}$ (Fig. 22–1c).

The area A of a surface can be represented by a vector $\vec{A}$ whose magnitude is A and whose direction is perpendicular to the surface, as shown in Fig. 22–1c. The angle θ is the angle between $\vec{E}$ and $\vec{A}$, so the electric flux can also be written

$$\Phi_E = \vec{E} \cdot \vec{A}. \qquad [\vec{E}\text{ uniform}] \quad (22\text{–}1b)$$

Electric flux has a simple intuitive interpretation in terms of field lines. We mentioned in Section 21–8 that field lines can always be drawn so that the number (N) passing through unit area perpendicular to the field ($A_\perp$) is proportional to the magnitude of the field (E): that is, $E \propto N/A_\perp$. Hence,

$$N \propto EA_\perp = \Phi_E,$$

so the flux through an area is proportional to the number of lines passing through that area.

> **EXAMPLE 22–1** **Electric flux.** Calculate the electric flux through the rectangle shown in Fig. 22–1a. The rectangle is 10 cm by 20 cm, the electric field is uniform at 200 N/C, and the angle θ is 30°.
>
> **APPROACH** We use the definition of flux, $\Phi_E = \vec{E} \cdot \vec{A} = EA\cos\theta$.
>
> **SOLUTION** The electric flux is
> $$\Phi_E = (200\,\text{N/C})(0.10\,\text{m} \times 0.20\,\text{m})\cos 30° = 3.5\,\text{N}\cdot\text{m}^2/\text{C}.$$

> **EXERCISE A** Which of the following would cause a change in the electric flux through a circle lying in the xz plane where the electric field is $(10\,\text{N})\hat{j}$? (a) Changing the magnitude of the electric field. (b) Changing the size of the circle. (c) Tipping the circle so that it is lying in the xy plane. (d) All of the above. (e) None of the above.

In the more general case, when the electric field $\vec{E}$ is not uniform and the surface is not flat, Fig. 22–2, we divide up the chosen surface into n small elements of surface whose areas are $\Delta A_1, \Delta A_2, \cdots, \Delta A_n$. We choose the division so that each ΔA_i is small enough that (1) it can be considered flat, and (2) the electric field varies so little over this small area that it can be considered uniform. Then the electric flux through the entire surface is approximately

$$\Phi_E \approx \sum_{i=1}^{n} \vec{E}_i \cdot \Delta\vec{A}_i,$$

where $\vec{E}_i$ is the field passing through $\Delta\vec{A}_i$. In the limit as we let $\Delta\vec{A}_i \to 0$, the sum becomes an integral over the entire surface and the relation becomes mathematically exact:

$$\Phi_E = \int \vec{E} \cdot d\vec{A}. \qquad (22\text{–}2)$$

Gauss's law involves the *total* flux through a *closed* surface—a surface of any shape that completely encloses a volume of space, such as that shown in Fig. 22–3. In this case, the net flux through the enclosing surface is given by

$$\Phi_E = \oint \vec{E} \cdot d\vec{A}, \qquad (22\text{–}3)$$

where the integral sign is written $\oint$ to indicate that the integral is over the value of $\vec{E}$ at every point on an enclosing surface.

Up to now we have not been concerned with an ambiguity in the direction of the vector $\vec{A}$ or $d\vec{A}$ that represents a surface. For example, in Fig. 22–1c, the vector $\vec{A}$ could point upward and to the right (as shown) or downward to the left and still be perpendicular to the surface. For a closed surface, we define (arbitrarily) the direction of $\vec{A}$, or of $d\vec{A}$, to point *outward* from the enclosed volume, Fig. 22–4. For an electric field line leaving the enclosed volume (on the right in Fig. 22–4), the angle θ between $\vec{E}$ and $d\vec{A}$ must be less than $\pi/2$ (= 90°); hence $\cos\theta > 0$. For a line entering the volume (on the left in Fig. 22–4) $\theta > \pi/2$; hence $\cos\theta < 0$. Hence, *flux entering the enclosed volume is negative* ($\int E \cos\theta \, dA < 0$), whereas *flux leaving the volume is positive*. Consequently, Eq. 22–3 gives the net flux *out* of the volume. If Φ_E is negative, there is a net flux *into* the volume.

In Figs. 22–3 and 22–4, each field line that enters the volume also leaves the volume. Hence $\Phi_E = \oint \vec{E} \cdot d\vec{A} = 0$. There is no net flux into or out of this enclosed surface. The flux, $\oint \vec{E} \cdot d\vec{A}$, will be nonzero only if one or more lines start or end within the surface. Since electric field lines start and stop only on electric charges, the flux will be nonzero only if the surface encloses a net charge. For example, the surface labeled A_1 in Fig. 22–5 encloses a positive charge and there is a net outward flux through this surface ($\Phi_E > 0$). The surface A_2 encloses an equal magnitude negative charge and there is a net inward flux ($\Phi_E < 0$). For the configuration shown in Fig. 22–6, the flux through the surface shown is negative (count the lines). The value of Φ_E depends on the charge enclosed by the surface, and this is what Gauss's law is all about.

22–2 Gauss's Law

The precise relation between the electric flux through a closed surface and the net charge Q_{encl} enclosed within that surface is given by **Gauss's law**:

$$\oint \vec{E} \cdot d\vec{A} = \frac{Q_{\text{encl}}}{\epsilon_0}, \tag{22–4}$$

where ϵ_0 is the same constant (permittivity of free space) that appears in Coulomb's law. The integral on the left is over the value of $\vec{E}$ on any closed surface, and we choose that surface for our convenience in any given situation. The charge Q_{encl} is the net charge *enclosed* by that surface. It doesn't matter where or how the charge is distributed within the surface. Any charge outside this surface must not be included. A charge outside the chosen surface may affect the position of the electric field lines, but will not affect the net number of lines entering or leaving the surface. For example, Q_{encl} for the gaussian surface A_1 in Fig. 22–5 would be the positive charge enclosed by A_1; the negative charge does contribute to the electric field at A_1 but it is *not* enclosed by surface A_1 and so is not included in Q_{encl}.

Now let us see how Gauss's law is related to Coulomb's law. First, we show that Coulomb's law follows from Gauss's law. In Fig. 22–7 we have a single isolated charge Q. For our "gaussian surface," we choose an imaginary sphere of radius r centered on the charge. Because Gauss's law is supposed to be valid for any surface, we have chosen one that will make our calculation easy. Because of the *symmetry* of this (imaginary) sphere about the charge at its center, we know that $\vec{E}$ must have the same magnitude at any point on the surface, and that $\vec{E}$ points radially outward (inward for a negative charge) parallel to $d\vec{A}$, an element of the surface area. Hence, we write the integral in Gauss's law as

$$\oint \vec{E} \cdot d\vec{A} = \oint E \, dA = E \oint dA = E(4\pi r^2)$$

since the surface area of a sphere of radius r is $4\pi r^2$, and the magnitude of $\vec{E}$ is the same at all points on this spherical surface. Then Gauss's law becomes, with $Q_{\text{encl}} = Q$,

$$\frac{Q}{\epsilon_0} = \oint \vec{E} \cdot d\vec{A} = E(4\pi r^2)$$

because $\vec{E}$ and $d\vec{A}$ are both perpendicular to the surface at each point, and $\cos\theta = 1$. Solving for E we obtain

$$E = \frac{Q}{4\pi\epsilon_0 r^2},$$

which is the electric field form of Coulomb's law, Eq. 21–4b.

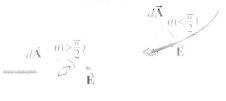

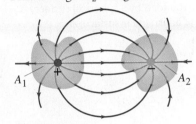

FIGURE 22–4 The direction of an element of area $d\vec{A}$ is taken to point outward from an enclosed surface.

FIGURE 22–5 An electric dipole. Flux through surface A_1 is positive. Flux through A_2 is negative.

FIGURE 22–6 Net flux through surface A is negative.

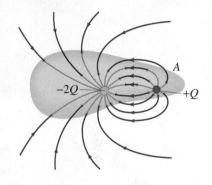

FIGURE 22–7 A single point charge Q at the center of an imaginary sphere of radius r (our "gaussian surface"—that is, the closed surface we choose to use for applying Gauss's law in this case).

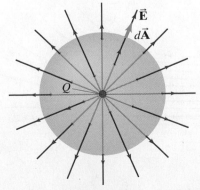

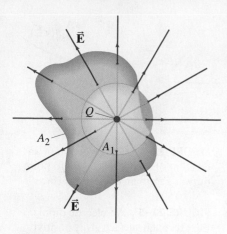

FIGURE 22–8 A single point charge surrounded by a spherical surface, A_1, and an irregular surface, A_2.

Now let us do the reverse, and derive Gauss's law from Coulomb's law for static electric charges[†]. First we consider a single point charge Q surrounded by an imaginary spherical surface as in Fig. 22–7 (and shown again, green, in Fig. 22–8). Coulomb's law tells us that the electric field at the spherical surface is $E = (1/4\pi\epsilon_0)(Q/r^2)$. Reversing the argument we just used, we have

$$\oint \vec{E} \cdot d\vec{A} = \oint \frac{1}{4\pi\epsilon_0} \frac{Q}{r^2} dA = \frac{Q}{4\pi\epsilon_0 r^2}(4\pi r^2) = \frac{Q}{\epsilon_0}.$$

This is Gauss's law, with $Q_{encl} = Q$, and we derived it for the special case of a spherical surface enclosing a point charge at its center. But what about some other surface, such as the irregular surface labeled A_2 in Fig. 22–8? The same number of field lines (due to our charge Q) pass through surface A_2, as pass through the spherical surface, A_1. Therefore, because the flux through a surface is proportional to the number of lines through it as we saw in Section 22–1, the flux through A_2 is the same as through A_1:

$$\oint_{A_2} \vec{E} \cdot d\vec{A} = \oint_{A_1} \vec{E} \cdot d\vec{A} = \frac{Q}{\epsilon_0}.$$

Hence, we can expect that

$$\oint \vec{E} \cdot d\vec{A} = \frac{Q}{\epsilon_0}$$

would be valid for *any* surface surrounding a single point charge Q.

Finally, let us look at the case of more than one charge. For each charge, Q_i, enclosed by the chosen surface,

$$\oint \vec{E}_i \cdot d\vec{A} = \frac{Q_i}{\epsilon_0},$$

where $\vec{E}_i$ refers to the electric field produced by Q_i alone. By the superposition principle for electric fields (Section 21–6), the total field $\vec{E}$ is equal to the sum of the fields due to each separate charge, $\vec{E} = \Sigma\vec{E}_i$. Hence

$$\oint \vec{E} \cdot d\vec{A} = \oint (\Sigma\vec{E}_i) \cdot d\vec{A} = \Sigma \frac{Q_i}{\epsilon_0} = \frac{Q_{encl}}{\epsilon_0},$$

where $Q_{encl} = \Sigma Q_i$ is the total net charge enclosed within the surface. Thus we see, based on this simple argument, that Gauss's law follows from Coulomb's law for any distribution of static electric charge enclosed within a closed surface of any shape.

The derivation of Gauss's law from Coulomb's law is valid for electric fields produced by static electric charges. We will see later that electric fields can also be produced by changing magnetic fields. Coulomb's law cannot be used to describe such electric fields. But Gauss's law *is* found to hold also for electric fields produced in any of these ways. Hence *Gauss's law is a more general law than Coulomb's law*. It holds for any electric field whatsoever.

Even for the case of static electric fields that we are considering in this Chapter, it is important to recognize that $\vec{E}$ on the left side of Gauss's law is not necessarily due only to the charge Q_{encl} that appears on the right. For example, in Fig. 22–9 there is an electric field $\vec{E}$ at all points on the imaginary gaussian surface, but it is not due to the charge enclosed by the surface (which is $Q_{encl} = 0$ in this case). The electric field $\vec{E}$ which appears on the left side of Gauss's law is the *total* electric field at each point, on the gaussian surface chosen, not just that due to the charge Q_{encl}, which appears on the right side. Gauss's law has been found to be valid for the total field at any surface. It tells us that any *difference* between the input and output flux of the electric field over any surface is due to charge within that surface.

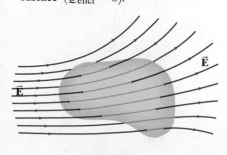

FIGURE 22–9 Electric flux through a closed surface. (Same as Fig. 22–3.) No electric charge is enclosed by this surface ($Q_{encl} = 0$).

[†]Note that Gauss's law would look more complicated in terms of the constant $k = 1/4\pi\epsilon_0$ that we originally used in Coulomb's law (Eqs. 21–1 or 21–4a):

Coulomb's law	*Gauss's law*
$E = k\dfrac{Q}{r^2}$	$\oint \vec{E} \cdot d\vec{A} = 4\pi kQ$
$E = \dfrac{1}{4\pi\epsilon_0}\dfrac{Q}{r^2}$	$\oint \vec{E} \cdot d\vec{A} = \dfrac{Q}{\epsilon_0}.$

Gauss's law has a simpler form using ϵ_0; Coulomb's law is simpler using k. The normal convention is to use ϵ_0 rather than k because Gauss's law is considered more general and therefore it is preferable to have it in simpler form.

Flux from Gauss's law. Consider the two gaussian surfaces. A_1 and A_2, shown in Fig. 22–10. The only charge present is the charge Q at the center of surface A_1. What is the net flux through each surface, A_1 and A_2?

RESPONSE The surface A_1 encloses the charge $+Q$. By Gauss's law, the net flux through A_1 is then Q/ϵ_0. For surface A_2, the charge $+Q$ is outside the surface. Surface A_2 encloses zero net charge, so the net electric flux through A_2 is zero, by Gauss's law. Note that all field lines that enter the volume enclosed by surface A_2 also leave it.

EXERCISE B A point charge Q is at the center of a spherical gaussian surface A. When a second charge Q is placed just outside A, the total flux through this spherical surface A is (a) unchanged, (b) doubled, (c) halved, (d) none of these.

EXERCISE C Three 2.95 μC charges are in a small box. What is the net flux leaving the box? (a) $3.3 \times 10^{12}\,\text{N}\cdot\text{m}^2/\text{C}$, (b) $3.3 \times 10^{5}\,\text{N}\cdot\text{m}^2/\text{C}$, (c) $1.0 \times 10^{12}\,\text{N}\cdot\text{m}^2/\text{C}$, (d) $1.0 \times 10^{6}\,\text{N}\cdot\text{m}^2/\text{C}$, (e) $6.7 \times 10^{6}\,\text{N}\cdot\text{m}^2/\text{C}$.

We note that the integral in Gauss's law is often rather difficult to carry out in practice. We rarely need to do it except for some fairly simple situations that we now discuss.

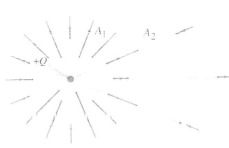

FIGURE 22–10 Example 22–2. Two gaussian surfaces.

22–3 Applications of Gauss's Law

Gauss's law is a very compact and elegant way to write the relation between electric charge and electric field. It also offers a simple way to determine the electric field when the charge distribution is simple and/or possesses a high degree of *symmetry*. In order to apply Gauss's law, however, we must choose the "gaussian" surface very carefully (for the integral on the left side of Gauss's law) so we can determine $\vec{\mathbf{E}}$. We normally try to think of a surface that has just the symmetry needed so that E will be constant on all or on parts of its surface. Sometimes we choose a surface so the flux through part of the surface is zero.

EXAMPLE 22–3 **Spherical conductor.** A thin spherical shell of radius r_0 possesses a total net charge Q that is uniformly distributed on it (Fig. 22–11). Determine the electric field at points (a) outside the shell, and (b) inside the shell. (c) What if the conductor were a solid sphere?

APPROACH Because the charge is distributed symmetrically, the electric field must also be *symmetric*. Thus the field outside the sphere must be directed radially outward (inward if $Q < 0$) and must depend only on r, not on angle (spherical coordinates).

SOLUTION (a) The electric field will have the same magnitude at all points on an imaginary gaussian surface, if we choose that surface as a sphere of radius r ($r > r_0$) concentric with the shell, and shown in Fig. 22–11 as the dashed circle A_1. Because $\vec{\mathbf{E}}$ is perpendicular to this surface, the cosine of the angle between $\vec{\mathbf{E}}$ and $d\vec{\mathbf{A}}$ is always 1. Gauss's law then gives (with $Q_{\text{encl}} = Q$ in Eq. 22–4)

$$\oint \vec{\mathbf{E}} \cdot d\vec{\mathbf{A}} = E(4\pi r^2) = \frac{Q}{\epsilon_0},$$

where $4\pi r^2$ is the surface area of our sphere (gaussian surface) of radius r. Thus

$$E = \frac{1}{4\pi\epsilon_0}\frac{Q}{r^2}. \qquad [r > r_0]$$

Thus the field outside a uniformly charged spherical shell is the same as if all the charge were concentrated at the center as a point charge.

(b) Inside the shell, the electric field must also be symmetric. So E must again have the same value at all points on a spherical gaussian surface (A_2 in Fig. 22–11) concentric with the shell. Thus E can be factored out of the integral and, with $Q_{\text{encl}} = 0$ because the charge enclosed within the sphere A_2 is zero, we have

$$\oint \vec{\mathbf{E}} \cdot d\vec{\mathbf{A}} = E(4\pi r^2) = 0.$$

Hence

$$E = 0 \qquad [r < r_0]$$

inside a uniform spherical shell of charge.

(c) These same results also apply to a uniformly charged solid spherical conductor, since all the charge would lie in a thin layer at the surface.

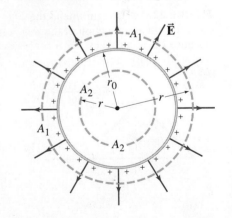

FIGURE 22–11 Cross-sectional drawing of a thin spherical shell of radius r_0, carrying a net charge Q uniformly distributed. A_1 and A_2 represent two gaussian surfaces we use to determine $\vec{\mathbf{E}}$. Example 22–3.

EXERCISE D A charge Q is placed on a hollow metal ball. We saw in Chapter 21 that the charge is all on the surface of the ball because metal is a conductor. How does the charge distribute itself on the ball? (*a*) Half on the inside surface and half on the outside surface. (*b*) Part on each surface in inverse proportion to the two radii. (*c*) Part on each surface but with a more complicated dependence on the radii than in answer (*b*). (*d*) All on the inside surface. (*e*) All on the outside surface.

EXAMPLE 22–4 Solid sphere of charge.

An electric charge Q is distributed uniformly throughout a nonconducting sphere of radius r_0, Fig. 22–12. Determine the electric field (*a*) outside the sphere $(r > r_0)$ and (*b*) inside the sphere $(r < r_0)$.

APPROACH Since the charge is distributed symmetrically in the sphere, the electric field at all points must again be *symmetric*. $\vec{E}$ depends only on r and is directed radially outward (or inward if $Q < 0$).

SOLUTION (*a*) For our gaussian surface we choose a sphere of radius r $(r > r_0)$, labeled A_1 in Fig. 22–12. Since E depends only on r, Gauss's law gives, with $Q_{encl} = Q$,

$$\oint \vec{E} \cdot d\vec{A} = E(4\pi r^2) = \frac{Q}{\epsilon_0}$$

or

$$E = \frac{1}{4\pi\epsilon_0}\frac{Q}{r^2}.$$

Again, the field outside a spherically symmetric distribution of charge is the same as that for a point charge of the same magnitude located at the center of the sphere.

(*b*) Inside the sphere, we choose for our gaussian surface a concentric sphere of radius r $(r < r_0)$, labeled A_2 in Fig. 22–12. From symmetry, the magnitude of $\vec{E}$ is the same at all points on A_2, and $\vec{E}$ is perpendicular to the surface, so

$$\oint \vec{E} \cdot d\vec{A} = E(4\pi r^2).$$

We must equate this to Q_{encl}/ϵ_0 where Q_{encl} is the charge enclosed by A_2. Q_{encl} is not the total charge Q but only a portion of it. We define the **charge density**, ρ_E, as the charge per unit volume $(\rho_E = dQ/dV)$, and here we are given that $\rho_E = $ constant. So the charge enclosed by the gaussian surface A_2, a sphere of radius r, is

$$Q_{encl} = \left(\frac{\frac{4}{3}\pi r^3 \rho_E}{\frac{4}{3}\pi r_0^3 \rho_E}\right)Q = \frac{r^3}{r_0^3}Q.$$

Hence, from Gauss's law,

$$E(4\pi r^2) = \frac{Q_{encl}}{\epsilon_0} = \frac{r^3}{r_0^3}\frac{Q}{\epsilon_0}$$

or

$$E = \frac{1}{4\pi\epsilon_0}\frac{Q}{r_0^3}r. \qquad [r < r_0]$$

Thus the field increases linearly with r, until $r = r_0$. It then decreases as $1/r^2$, as plotted in Fig. 22–13.

EXERCISE E Return to the Chapter-Opening Question, page 591, and answer it again now. Try to explain why you may have answered differently the first time.

The results in Example 22–4 would have been difficult to obtain from Coulomb's law by integrating over the sphere. Using Gauss's law and the *symmetry* of the situation, this result is obtained rather easily, and shows the great power of Gauss's law. However, its use in this way is limited mainly to cases where the charge distribution has a high degree of symmetry. In such cases, we *choose* a simple surface on which $E = $ constant, so the integration is simple. Gauss's law holds, of course, for any surface.

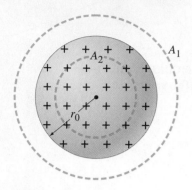

FIGURE 22–12 A solid sphere of uniform charge density. Example 22–4.

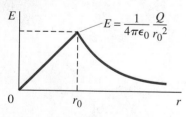

FIGURE 22–13 Magnitude of the electric field as a function of the distance r from the center of a uniformly charged solid sphere.

EXAMPLE 22–5 **Nonuniformly charged solid sphere.** Suppose the charge density of the solid sphere in Fig. 22–12, Example 22–4, is given by $\rho_E = \alpha r^2$, where α is a constant. (a) Find α in terms of the total charge Q on the sphere and its radius r_0. (b) Find the electric field as a function of r inside the sphere.

APPROACH We divide the sphere up into concentric thin shells of thickness dr as shown in Fig. 22–14, and integrate (a) setting $Q = \int \rho_1 \, dV$ and (b) using Gauss's law.

SOLUTION (a) A thin shell of radius r and thickness dr (Fig. 22–14) has volume $dV = 4\pi r^2 \, dr$. The total charge is given by

$$Q = \int \rho_E \, dV = \int_0^{r_0} (\alpha r^2)(4\pi r^2 \, dr) = 4\pi\alpha \int_0^{r_0} r^4 \, dr = \frac{4\pi\alpha}{5} r_0^5.$$

Thus $\alpha = 5Q/4\pi r_0^5$.

(b) To find E inside the sphere at distance r from its center, we apply Gauss's law to an imaginary sphere of radius r which will enclose a charge

$$Q_{encl} = \int_0^r \rho_E \, dV = \int_0^r (\alpha r^2) \, 4\pi r^2 \, dr = \int_0^r \left(\frac{5Q}{4\pi r_0^5} r^2\right) 4\pi r^2 \, dr = Q \frac{r^5}{r_0^5}.$$

By *symmetry*, E will be the same at all points on the surface of a sphere of radius r, so Gauss's law gives

$$\oint \vec{E} \cdot d\vec{A} = \frac{Q_{encl}}{\epsilon_0}$$

$$(E)(4\pi r^2) = Q \frac{r^5}{\epsilon_0 r_0^5},$$

so

$$E = \frac{Qr^3}{4\pi\epsilon_0 r_0^5}.$$

FIGURE 22–14 Example 22–5.

EXAMPLE 22–6 **Long uniform line of charge.** A very long straight wire possesses a uniform positive charge per unit length, λ. Calculate the electric field at points near (but outside) the wire, far from the ends.

APPROACH Because of the *symmetry*, we expect the field to be directed radially outward and to depend only on the perpendicular distance, R, from the wire. Because of the cylindrical symmetry, the field will be the same at all points on a gaussian surface that is a cylinder with the wire along its axis, Fig. 22–15. $\vec{E}$ is perpendicular to this surface at all points. For Gauss's law, we need a closed surface, so we include the flat ends of the cylinder. Since $\vec{E}$ is parallel to the ends, there is no flux through the ends (the cosine of the angle between $\vec{E}$ and $d\vec{A}$ on the ends is $\cos 90° = 0$).

SOLUTION For our chosen gaussian surface Gauss's law gives

$$\oint \vec{E} \cdot d\vec{A} = E(2\pi R\ell) = \frac{Q_{encl}}{\epsilon_0} = \frac{\lambda\ell}{\epsilon_0},$$

where ℓ is the length of our chosen gaussian surface ($\ell \ll$ length of wire), and $2\pi R$ is its circumference. Hence

$$E = \frac{1}{2\pi\epsilon_0} \frac{\lambda}{R}.$$

NOTE This is the same result we found in Example 21–11 using Coulomb's law (we used x there instead of R), but here it took much less effort. Again we see the great power of Gauss's law.[†]

NOTE Recall from Chapter 10, Fig. 10–2, that we use R for the distance of a particle from an axis (cylindrical symmetry), but lower case r for the distance from a point (usually the origin 0).

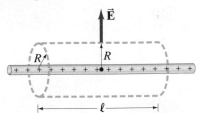

FIGURE 22–15 Calculation of $\vec{E}$ due to a very long line of charge. Example 22–6.

[†]But note that the method of Example 21–11 allows calculation of E also for a short line of charge by using the appropriate limits for the integral, whereas Gauss's law is not readily adapted due to lack of symmetry.

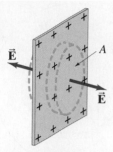

FIGURE 22–16 Calculation of the electric field outside a large uniformly charged nonconducting plane surface. Example 22–7.

EXAMPLE 22–7 **Infinite plane of charge.** Charge is distributed uniformly, with a surface charge density σ (σ = charge per unit area = dQ/dA), over a very large but very thin nonconducting flat plane surface. Determine the electric field at points near the plane.

APPROACH We choose as our gaussian surface a small closed cylinder whose axis is perpendicular to the plane and which extends through the plane as shown in Fig. 22–16. Because of the symmetry, we expect $\vec{E}$ to be directed perpendicular to the plane on both sides as shown, and to be uniform over the end caps of the cylinder, each of whose area is A.

SOLUTION Since no flux passes through the curved sides of our chosen cylindrical surface, all the flux is through the two end caps. So Gauss's law gives

$$\oint \vec{E} \cdot d\vec{A} = 2EA = \frac{Q_{encl}}{\epsilon_0} = \frac{\sigma A}{\epsilon_0},$$

where $Q_{encl} = \sigma A$ is the charge enclosed by our gaussian cylinder. The electric field is then

$$E = \frac{\sigma}{2\epsilon_0}.$$

NOTE This is the same result we obtained much more laboriously in Chapter 21, Eq. 21–7. The field is uniform for points far from the ends of the plane, and close to its surface.

EXAMPLE 22–8 **Electric field near any conducting surface.** Show that the electric field just outside the surface of any good conductor of arbitrary shape is given by

$$E = \frac{\sigma}{\epsilon_0},$$

where σ is the surface charge density on the conductor's surface at that point.

APPROACH We choose as our gaussian surface a small cylindrical box, as we did in the previous Example. We choose the cylinder to be very small in height, so that one of its circular ends is just above the conductor (Fig. 22–17). The other end is just below the conductor's surface, and the sides are perpendicular to it.

FIGURE 22–17 Electric field near surface of a conductor. Example 22–8.

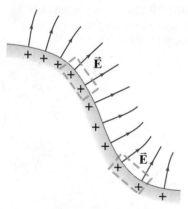

SOLUTION The electric field is zero inside a conductor and is perpendicular to the surface just outside it (Section 21–9), so electric flux passes only through the outside end of our cylindrical box; no flux passes through the short sides or inside end. We choose the area A (of the flat cylinder end) small enough so that E is essentially uniform over it. Then Gauss's law gives

$$\oint \vec{E} \cdot d\vec{A} = EA = \frac{Q_{encl}}{\epsilon_0} = \frac{\sigma A}{\epsilon_0},$$

so that

$$E = \frac{\sigma}{\epsilon_0}. \qquad \text{[at surface of conductor]} \quad \textbf{(22–5)}$$

NOTE This useful result applies for a conductor of any shape.

⚠ **CAUTION**
When is $E = \sigma/\epsilon_0$ and when is $E = \sigma/2\epsilon_0$

Why is it that the field outside a large plane nonconductor is $E = \sigma/2\epsilon_0$ (Example 22–7) whereas outside a conductor it is $E = \sigma/\epsilon_0$ (Example 22–8)? The reason for the factor of 2 comes not from conductor verses nonconductor. It comes instead from how we define charge per unit area σ. For a thin flat nonconductor, Fig. 22–16, the charge may be distributed throughout the volume (not only on the surface, as for a conductor). The charge per unit area σ represents all the charge throughout the thickness of the thin nonconductor. Also our gaussian surface has its ends outside the nonconductor on each side, so as to include all this charge.

For a conductor, on the other hand, the charge accumulates on the outer surfaces only. For a thin flat conductor, as shown in Fig. 22–18, the charge accumulates on both surfaces, and using the same small gaussian surface we did in Fig. 22–17, with one end inside and the other end outside the conductor, we came up with the result, $E = \sigma/\epsilon_0$. If we defined σ for a conductor, as we did for a nonconductor, σ would represent the charge per area for the entire conductor. Then Fig. 22–18 would show $\sigma/2$ as the surface charge on each surface, and Gauss's law would give $\int \mathbf{E} \cdot d\mathbf{A} = EA = (\sigma/2)A/\epsilon_0 = \sigma A/2\epsilon_0$, so $E = \sigma/2\epsilon_0$, just as for a nonconductor. We need to be careful about how we define charge per unit area σ.

We saw in Section 21–9 that in the static situation, the electric field inside any conductor must be zero even if it has a net charge. (Otherwise, the free charges in the conductor would move—until the net force on each, and hence $\vec{E}$, were zero.) We also mentioned there that any net electric charge on a conductor must all reside on its outer surface. This is readily shown using Gauss's law. Consider any charged conductor of any shape, such as that shown in Fig. 22–19, which carries a net charge Q. We choose the gaussian surface, shown dashed in the diagram, so that it all lies just below the surface of the conductor and encloses essentially the whole volume of the conductor. Our gaussian surface can be arbitrarily close to the surface, but still *inside* the conductor. The electric field is zero at all points on this gaussian surface since it is inside the conductor. Hence, from Gauss's law, Eq. 22–4, the net charge within the surface must be zero. Thus, there can be no net charge within the conductor. Any net charge must lie on the surface of the conductor.

If there is an empty cavity inside a conductor, can charge accumulate on that (inner) surface too? As shown in Fig. 22–20, if we imagine a gaussian surface (shown dashed) just inside the conductor above the cavity, we know that $\vec{E}$ must be zero everywhere on this surface since it is inside the conductor. Hence, by Gauss's law, *there can be no net charge at the surface of the cavity*.

But what if the cavity is not empty and there is a charge inside it?

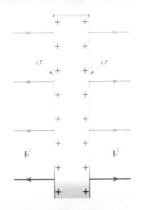

FIGURE 22–18 Thin flat charged conductor with surface charge density σ at each surface. For the conductor as a whole, the charge density is $\sigma' = 2\sigma$.

FIGURE 22–19 An insulated charged conductor of arbitrary shape, showing a gaussian surface (dashed) just below the surface of the conductor.

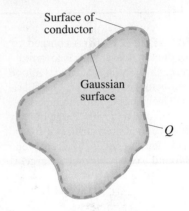

Surface of conductor

Gaussian surface

Q

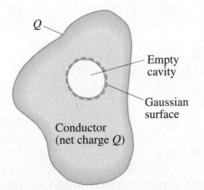

Q

Empty cavity

Gaussian surface

Conductor (net charge Q)

FIGURE 22–20 An empty cavity inside a charged conductor carries zero net charge.

FIGURE 22–21 Example 22–9.

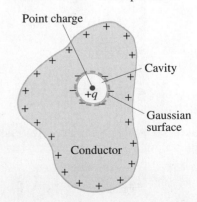

Point charge

Cavity

$+q$

Gaussian surface

Conductor

CONCEPTUAL EXAMPLE 22–9 | **Conductor with charge inside a cavity.** Suppose a conductor carries a net charge $+Q$ and contains a cavity, inside of which resides a point charge $+q$. What can you say about the charges on the inner and outer surfaces of the conductor?

RESPONSE As shown in Fig. 22–21, a gaussian surface just inside the conductor surrounding the cavity must contain zero net charge ($E = 0$ in a conductor). Thus a net charge of $-q$ must exist on the cavity surface. The conductor itself carries a net charge $+Q$, so its outer surface must carry a charge equal to $Q + q$. These results apply to a cavity of any shape.

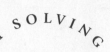

EXERCISE F Which of the following statements about Gauss's law is correct? (*a*) If we know the charge enclosed by a surface, we always know the electric field everywhere at the surface. (*b*) When finding the electric field with Gauss's law, we always use a sphere for the gaussian surface. (*c*) If we know the total flux through a surface, we also know the total charge inside the surface. (*d*) We can only use Gauss's law if the electric field is constant in space.

Gauss's Law for Symmetric Charge Distributions

1. First identify the **symmetry** of the charge distribution: spherical, cylindrical, planar. This identification should suggest a gaussian surface for which $\vec{E}$ will be constant and/or zero on all or on parts of the surface: a sphere for spherical symmetry, a cylinder for cylindrical symmetry and a small cylinder or "pillbox" for planar symmetry.

2. Draw the appropriate gaussian surface making sure it passes through the point where you want to know the electric field.

3. Use the symmetry of the charge distribution to determine the direction of $\vec{E}$ at points on the gaussian surface.

4. Evaluate the flux, $\oint \vec{E} \cdot d\vec{A}$. With an appropriate gaussian surface, the dot product $\vec{E} \cdot d\vec{A}$ should be zero or equal to $\pm E\, dA$, with the magnitude of E being constant over all or parts of the surface.

5. Calculate the charge *enclosed* by the gaussian surface. Remember it's the enclosed charge that matters. Ignore all the charge outside the gaussian surface.

6. Equate the flux to the enclosed charge and solve for E.

FIGURE 22–22 (a) A charged conductor (metal ball) is lowered into an insulated metal can (a good conductor) carrying zero net charge. (b) The charged ball is touched to the can and all of its charge quickly flows to the outer surface of the can. (c) When the ball is then removed, it is found to carry zero net charge.

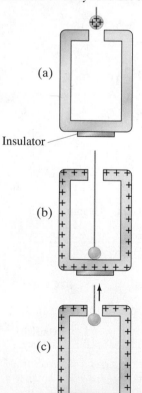

*22–4 Experimental Basis of Gauss's and Coulomb's Laws

Gauss's law predicts that any net charge on a conductor must lie only on its surface. But is this true in real life? Let us see how it can be verified experimentally. And in confirming this prediction of Gauss's law, Coulomb's law is also confirmed since the latter follows from Gauss's law, as we saw in Section 22–2. Indeed, the earliest observation that charge resides only on the outside of a conductor was recorded by Benjamin Franklin some 30 years before Coulomb stated his law.

A simple experiment is illustrated in Fig. 22–22. A metal can with a small opening at the top rests on an insulator. The can, a conductor, is initially uncharged (Fig. 22–22a). A charged metal ball (also a conductor) is lowered by an insulating thread into the can, and is allowed to touch the can (Fig. 22–22b). The ball and can now form a single conductor. Gauss's law, as discussed above, predicts that all the charge will flow to the outer surface of the can. (The flow of charge in such situations does not occur instantaneously, but the time involved is usually negligible). These predictions are confirmed in experiments by (1) connecting an electroscope to the can, which will show that the can is charged, and (2) connecting an electroscope to the ball after it has been withdrawn from the can (Fig. 22–22c), which will show that the ball carries zero charge.

The precision with which Coulomb's and Gauss's laws hold can be stated quantitatively by writing Coulomb's law as

$$ F = k\frac{Q_1 Q_2}{r^{2+\delta}}. $$

For a perfect inverse-square law, $\delta = 0$. The most recent and precise experiments (1971) give $\delta = (2.7 \pm 3.1) \times 10^{-16}$. Thus Coulomb's and Gauss's laws are found to be valid to an extremely high precision!

Summary

The **electric flux** passing through a flat area A for a uniform electric field $\vec{E}$ is

$$\Phi_E = \vec{E} \cdot \vec{A}. \qquad (22\text{–}1b)$$

If the field is not uniform, the flux is determined from the integral

$$\Phi_E = \int \vec{E} \cdot d\vec{A}. \qquad (22\text{–}2)$$

The direction of the vector $\vec{A}$ or $d\vec{A}$ is chosen to be perpendicular to the surface whose area is A or dA, and points outward from an enclosed surface. The flux through a surface is proportional to the number of field lines passing through it.

Gauss's law states that the net flux passing through any closed surface is equal to the net charge Q_{encl} enclosed by the surface divided by ϵ_0:

$$\oint \vec{E} \cdot d\vec{A} = \frac{Q_{encl}}{\epsilon_0}. \qquad (22\text{–}4)$$

Gauss's law can in principle be used to determine the electric field due to a given charge distribution, but its usefulness is mainly limited to a small number of cases, usually where the charge distribution displays much symmetry. The real importance of Gauss's law is that it is a more general and elegant statement (than Coulomb's law) for the relation between electric charge and electric field. It is one of the basic equations of electromagnetism.

Questions

1. If the electric flux through a closed surface is zero, is the electric field necessarily zero at all points on the surface? Explain. What about the converse: If $\vec{E} = 0$ at all points on the surface is the flux through the surface zero?

2. Is the electric field $\vec{E}$ in Gauss's law, $\oint \vec{E} \cdot d\vec{A} = Q_{encl}/\epsilon_0$, created only by the charge Q_{encl}?

3. A point charge is surrounded by a spherical gaussian surface of radius r. If the sphere is replaced by a cube of side r, will Φ_E be larger, smaller, or the same? Explain.

4. What can you say about the flux through a closed surface that encloses an electric dipole?

5. The electric field $\vec{E}$ is zero at all points on a closed surface; is there necessarily no net charge within the surface? If a surface encloses zero net charge, is the electric field necessarily zero at all points on the surface?

6. Define gravitational flux in analogy to electric flux. Are there "sources" and "sinks" for the gravitational field as there are for the electric field? Discuss.

7. Would Gauss's law be helpful in determining the electric field due to an electric dipole?

8. A spherical basketball (a nonconductor) is given a charge Q distributed uniformly over its surface. What can you say about the electric field inside the ball? A person now steps on the ball, collapsing it, and forcing most of the air out without altering the charge. What can you say about the field inside now?

9. In Example 22–6, it may seem that the electric field calculated is due only to the charge on the wire that is enclosed by the cylinder chosen as our gaussian surface. In fact, the entire charge along the whole length of the wire contributes to the field. Explain how the charge outside the cylindrical gaussian surface of Fig. 22–15 contributes to E at the gaussian surface. [*Hint*: Compare to what the field would be due to a short wire.]

10. Suppose the line of charge in Example 22–6 extended only a short way beyond the ends of the cylinder shown in Fig. 22–15. How would the result of Example 22–6 be altered?

11. A point charge Q is surrounded by a spherical surface of radius r_0, whose center is at C. Later, the charge is moved to the right a distance $\frac{1}{2}r_0$, but the sphere remains where it was, Fig. 22–23. How is the electric flux Φ_E through the sphere changed? Is the electric field at the surface of the sphere changed? For each "yes" answer, describe the change.

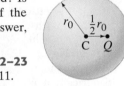

FIGURE 22–23
Question 11.

12. A solid conductor carries a net positive charge Q. There is a hollow cavity within the conductor, at whose center is a negative point charge $-q$ (Fig. 22–24). What is the charge on (a) the outer surface of the conductor and (b) the inner surface of the conductor's cavity?

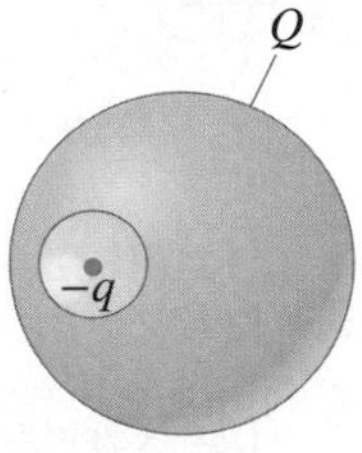

FIGURE 22–24
Question 12.

13. A point charge q is placed at the center of the cavity of a thin metal shell which is neutral. Will a charge Q placed outside the shell feel an electric force? Explain.

14. A small charged ball is inserted into a balloon. The balloon is then blown up slowly. Describe how the flux through the balloon's surface changes as the balloon is blown up. Consider both the total flux and the flux per unit surface area of the balloon.

Problems

22–1 Electric Flux

1. (I) A uniform electric field of magnitude 5.8×10^2 N/C passes through a circle of radius 13 cm. What is the electric flux through the circle when its face is (a) perpendicular to the field lines, (b) at 45° to the field lines, and (c) parallel to the field lines?

2. (I) The Earth possesses an electric field of (average) magnitude 150 N/C near its surface. The field points radially inward. Calculate the net electric flux outward through a spherical surface surrounding, and just beyond, the Earth's surface.

3. (II) A cube of side ℓ is placed in a uniform field E_0 with edges parallel to the field lines. (*a*) What is the net flux through the cube? (*b*) What is the flux through each of its six faces?

4. (II) A uniform field $\vec{\mathbf{E}}$ is parallel to the axis of a hollow hemisphere of radius r, Fig. 22–25. (*a*) What is the electric flux through the hemispherical surface? (*b*) What is the result if $\vec{\mathbf{E}}$ is instead perpendicular to the axis?

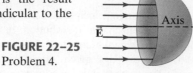

FIGURE 22–25
Problem 4.

22–2 Gauss's Law

5. (I) The total electric flux from a cubical box 28.0 cm on a side is $1.84 \times 10^3\,\mathrm{N \cdot m^2/C}$. What charge is enclosed by the box?

6. (I) Figure 22–26 shows five closed surfaces that surround various charges in a plane, as indicated. Determine the electric flux through each surface, S_1, S_2, S_3, S_4, and S_5. The surfaces are flat "pillbox" surfaces that extend only slightly above and below the plane in which the charges lie.

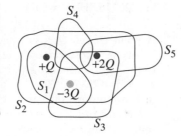

FIGURE 22–26
Problem 6.

7. (II) In Fig. 22–27, two objects, O_1 and O_2, have charges $+1.0\,\mu\mathrm{C}$ and $-2.0\,\mu\mathrm{C}$ respectively, and a third object, O_3, is electrically neutral. (*a*) What is the electric flux through the surface A_1 that encloses all the three objects? (*b*) What is the electric flux through the surface A_2 that encloses the third object only?

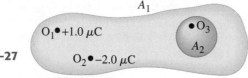

FIGURE 22–27
Problem 7.

8. (II) A ring of charge with uniform charge density is completely enclosed in a hollow donut shape. An exact copy of the ring is completely enclosed in a hollow sphere. What is the ratio of the flux out of the donut shape to that out of the sphere?

9. (II) In a certain region of space, the electric field is constant in direction (say horizontal, in the x direction), but its magnitude decreases from $E = 560\,\mathrm{N/C}$ at $x = 0$ to $E = 410\,\mathrm{N/C}$ at $x = 25\,\mathrm{m}$. Determine the charge within a cubical box of side $\ell = 25\,\mathrm{m}$, where the box is oriented so that four of its sides are parallel to the field lines (Fig. 22–28).

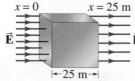

FIGURE 22–28
Problem 9.

10. (II) A point charge Q is placed at the center of a cube of side ℓ. What is the flux through one face of the cube?

11. (II) A 15.0-cm-long uniformly charged plastic rod is sealed inside a plastic bag. The total electric flux leaving the bag is $7.3 \times 10^5\,\mathrm{N \cdot m^2/C}$. What is the linear charge density on the rod?

22–3 Applications of Gauss's Law

12. (I) Draw the electric field lines around a negatively charged metal egg.

13. (I) The field just outside a 3.50-cm-radius metal ball is $6.25 \times 10^2\,\mathrm{N/C}$ and points toward the ball. What charge resides on the ball?

14. (I) Starting from the result of Example 22–3, show that the electric field just outside a uniformly charged spherical conductor is $E = \sigma/\epsilon_0$, consistent with Example 22–8.

15. (I) A long thin wire, hundreds of meters long, carries a uniformly distributed charge of $-7.2\,\mu\mathrm{C}$ per meter of length. Estimate the magnitude and direction of the electric field at points (*a*) 5.0 m and (*b*) 1.5 m perpendicular from the center of the wire.

16. (I) A metal globe has 1.50 mC of charge put on it at the north pole. Then -3.00 mC of charge is applied to the south pole. Draw the field lines for this system after it has come to equilibrium.

17. (II) A nonconducting sphere is made of two layers. The innermost section has a radius of 6.0 cm and a uniform charge density of $-5.0\,\mathrm{C/m^3}$. The outer layer has a uniform charge density of $+8.0\,\mathrm{C/m^3}$ and extends from an inner radius of 6.0 cm to an outer radius of 12.0 cm. Determine the electric field for (*a*) $0 < r < 6.0$ cm, (*b*) 6.0 cm $< r < 12.0$ cm, and (*c*) 12.0 cm $< r < 50.0$ cm. (*d*) Plot the magnitude of the electric field for $0 < r < 50.0$ cm. Is the field continuous at the edges of the layers?

18. (II) A solid metal sphere of radius 3.00 m carries a total charge of $-5.50\,\mu\mathrm{C}$. What is the magnitude of the electric field at a distance from the sphere's center of (*a*) 0.250 m, (*b*) 2.90 m, (*c*) 3.10 m, and (*d*) 8.00 m? How would the answers differ if the sphere were (*e*) a thin shell, or (*f*) a solid nonconductor uniformly charged throughout?

19. (II) A 15.0-cm-diameter nonconducting sphere carries a total charge of $2.25\,\mu\mathrm{C}$ distributed uniformly throughout its volume. Graph the electric field E as a function of the distance r from the center of the sphere from $r = 0$ to $r = 30.0$ cm.

20. (II) A flat square sheet of thin aluminum foil, 25 cm on a side, carries a uniformly distributed 275 nC charge. What, approximately, is the electric field (*a*) 1.0 cm above the center of the sheet and (*b*) 15 m above the center of the sheet?

21. (II) A spherical cavity of radius 4.50 cm is at the center of a metal sphere of radius 18.0 cm. A point charge $Q = 5.50\,\mu\mathrm{C}$ rests at the very center of the cavity, whereas the metal conductor carries no net charge. Determine the electric field at a point (*a*) 3.00 cm from the center of the cavity, (*b*) 6.00 cm from the center of the cavity, (*c*) 30.0 cm from the center.

22. (II) A point charge Q rests at the center of an uncharged thin spherical conducting shell. What is the electric field E as a function of r (*a*) for r less than the radius of the shell, (*b*) inside the shell, and (*c*) beyond the shell? (*d*) Does the shell affect the field due to Q alone? Does the charge Q affect the shell?

23. (II) A solid metal cube has a spherical cavity at its center as shown in Fig. 22–29. At the center of the cavity there is a point charge $Q = +8.00\,\mu C$. The metal cube carries a net charge $q = -6.10\,\mu C$ (not including Q). Determine (a) the total charge on the surface of the spherical cavity and (b) the total charge on the outer surface of the cube.

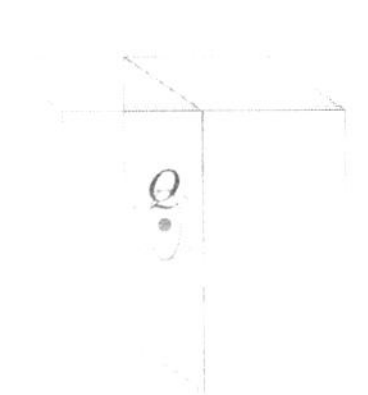

FIGURE 22–29
Problem 23

24. (II) Two large, flat metal plates are separated by a distance that is very small compared to their height and width. The conductors are given equal but opposite uniform surface charge densities $\pm\sigma$. Ignore edge effects and use Gauss's law to show (a) that for points far from the edges, the electric field between the plates is $E = \sigma/\epsilon_0$ and (b) that outside the plates on either side the field is zero. (c) How would your results be altered if the two plates were nonconductors? (See Fig. 22–30).

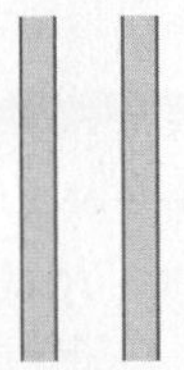

FIGURE 22–30
Problems 24, 25, and 26.
$+\sigma$ $-\sigma$

25. (II) Suppose the two conducting plates in Problem 24 have the *same* sign and magnitude of charge. What then will be the electric field (a) between them and (b) outside them on either side? (c) What if the plates are nonconducting?

26. (II) The electric field between two square metal plates is $160\,N/C$. The plates are $1.0\,m$ on a side and are separated by $3.0\,cm$, as in Fig. 22–30. What is the charge on each plate? Neglect edge effects.

27. (II) Two thin concentric spherical shells of radii r_1 and r_2 $(r_1 < r_2)$ contain uniform surface charge densities σ_1 and σ_2, respectively (see Fig. 22–31). Determine the electric field for (a) $0 < r < r_1$, (b) $r_1 < r < r_2$, and (c) $r > r_2$. (d) Under what conditions will $E = 0$ for $r > r_2$? (e) Under what conditions will $E = 0$ for $r_1 < r < r_2$? Neglect the thickness of the shells.

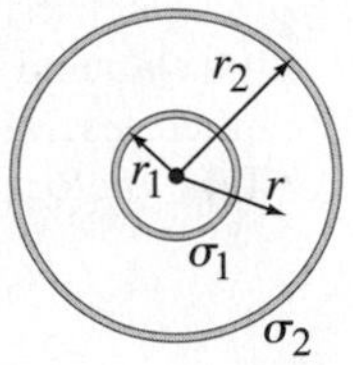

FIGURE 22–31 Two spherical shells (Problem 27).

28. (II) A spherical rubber balloon carries a total charge Q uniformly distributed on its surface. At $t = 0$ the nonconducting balloon has radius r_0 and the balloon is then slowly blown up so that r increases linearly to $2r_0$ in a time T. Determine the electric field as a function of time (a) just outside the balloon surface and (b) at $r = 3.2r_0$.

29. (II) Suppose the nonconducting sphere of Example 22–4 has a spherical cavity of radius r_1 centered at the sphere's center (Fig. 22–32). Assuming the charge Q is distributed uniformly in the "shell" (between $r = r_1$ and $r = r_0$), determine the electric field as a function of r for (a) $0 < r < r_1$, (b) $r_1 < r < r_0$, and (c) $r > r_0$.

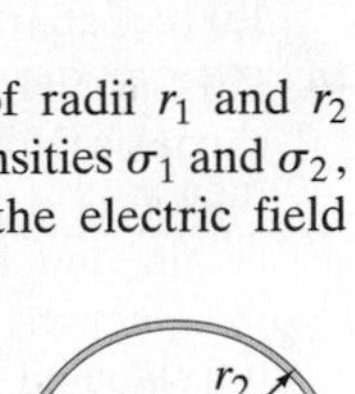

FIGURE 22–32
Problems 29, 30, 31, and 44.

30. (II) Suppose in Fig. 22–32, Problem 29, there is also a charge q at the center of the cavity. Determine the electric field for (a) $0 < r < r_1$, (b) $r_1 < r < r_0$, and (c) $r > r_0$.

31. (II) Suppose the thick spherical shell of Problem 29 is a conductor. It carries a total net charge Q and at its center there is a point charge q. What total charge is found on (a) the inner surface of the shell and (b) the outer surface of the shell? Determine the electric field for (c) $0 < r < r_1$, (d) $r_1 < r < r_0$, and (e) $r > r_0$.

32. (II) Suppose that at the center of the cavity inside the shell (charge Q) of Fig. 22–11 (and Example 22–3) there is a point charge $q\,(\neq \pm Q)$. Determine the electric field for (a) $0 < r < r_0$ and for (b) $r > r_0$. What are your answers if (c) $q = Q$ and (d) $q = -Q$?

33. (II) A long cylindrical shell of radius R_0 and length ℓ $(R_0 \ll \ell)$ possesses a uniform surface charge density (charge per unit area) σ (Fig. 22–33). Determine the electric field at points (a) outside the cylinder $(R > R_0)$ and (b) inside the cylinder $(0 < R < R_0)$; assume the points are far from the ends and not too far from the shell $(R \ll \ell)$. (c) Compare to the result for a long line of charge, Example 22–6. Neglect the thickness of shell.

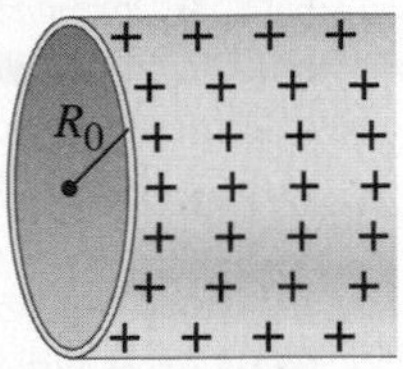

FIGURE 22–33
Problem 33.

34. (II) A very long solid nonconducting cylinder of radius R_0 and length ℓ $(R_0 \ll \ell)$ possesses a uniform volume charge density $\rho_E\,(C/m^3)$, Fig. 22–34. Determine the electric field at points (a) outside the cylinder $(R > R_0)$ and (b) inside the cylinder $(R < R_0)$. Do only for points far from the ends and for which $R \ll \ell$.

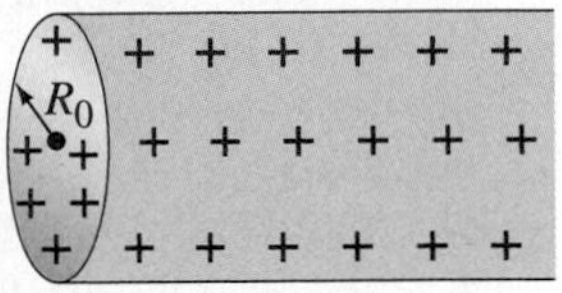

FIGURE 22–34
Problem 34.

35. (II) A thin cylindrical shell of radius R_1 is surrounded by a second concentric cylindrical shell of radius R_2 (Fig. 22–35). The inner shell has a total charge $+Q$ and the outer shell $-Q$. Assuming the length ℓ of the shells is much greater than R_1 or R_2, determine the electric field as a function of R (the perpendicular distance from the common axis of the cylinders) for (a) $0 < R < R_1$, (b) $R_1 < R < R_2$, and (c) $R > R_2$. (d) What is the kinetic energy of an electron if it moves between (and concentric with) the shells in a circular orbit of radius $(R_1 + R_2)/2$? Neglect thickness of shells.

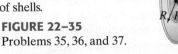

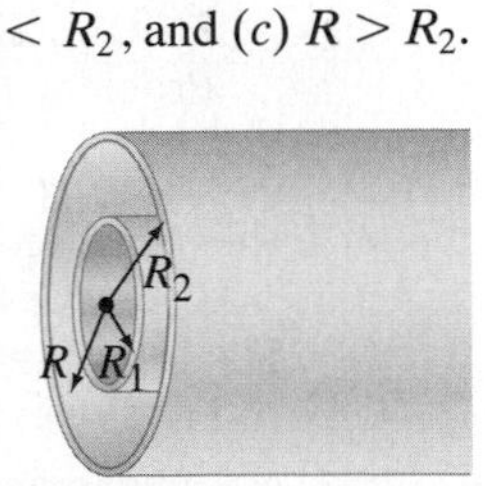

FIGURE 22–35
Problems 35, 36, and 37.

36. (II) A thin cylindrical shell of radius $R_1 = 6.5\,cm$ is surrounded by a second cylindrical shell of radius $R_2 = 9.0\,cm$, as in Fig. 22–35. Both cylinders are $5.0\,m$ long and the inner one carries a total charge $Q_1 = -0.88\,\mu C$ and the outer one $Q_2 = +1.56\,\mu C$. For points far from the ends of the cylinders, determine the electric field at a radial distance R from the central axis of (a) $3.0\,cm$, (b) $7.0\,cm$, and (c) $12.0\,cm$.

37. (II) (a) If an electron $(m = 9.1 \times 10^{-31}\,kg)$ escaped from the surface of the inner cylinder in Problem 36 (Fig. 22–35) with negligible speed, what would be its speed when it reached the outer cylinder? (b) If a proton $(m = 1.67 \times 10^{-27}\,kg)$ revolves in a circular orbit of radius $R = 7.0\,cm$ about the axis (i.e., between the cylinders), what must be its speed?

38. (II) A very long solid nonconducting cylinder of radius R_1 is uniformly charged with a charge density ρ_E. It is surrounded by a concentric cylindrical tube of inner radius R_2 and outer radius R_3 as shown in Fig. 22–36, and it too carries a uniform charge density ρ_E. Determine the electric field as a function of the distance R from the center of the cylinders for (a) $0 < R < R_1$, (b) $R_1 < R < R_2$, (c) $R_2 < R < R_3$, and (d) $R > R_3$. (e) If $\rho_E = 15\,\mu\text{C/m}^3$ and $R_1 = \frac{1}{2}R_2 = \frac{1}{3}R_3 = 5.0\,\text{cm}$, plot E as a function of R from $R = 0$ to $R = 20.0\,\text{cm}$. Assume the cylinders are very long compared to R_3.

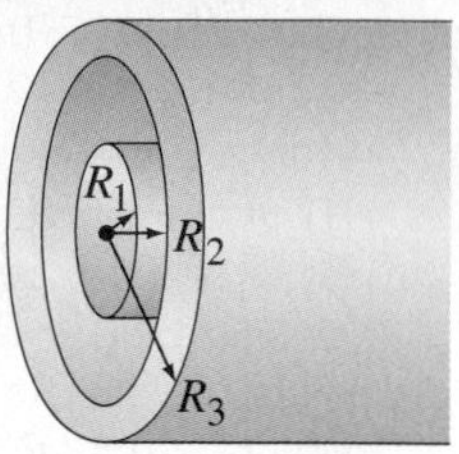

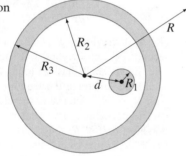

FIGURE 22–36
Problem 38.

39. (II) A nonconducting sphere of radius r_0 is uniformly charged with volume charge density ρ_E. It is surrounded by a concentric metal (conducting) spherical shell of inner radius r_1 and outer radius r_2, which carries a net charge $+Q$. Determine the resulting electric field in the regions (a) $0 < r < r_0$, (b) $r_0 < r < r_1$, (c) $r_1 < r < r_2$, and (d) $r > r_2$ where the radial distance r is measured from the center of the nonconducting sphere.

40. (II) A very long solid nonconducting cylinder of radius R_1 is uniformly charged with charge density ρ_E. It is surrounded by a cylindrical metal (conducting) tube of inner radius R_2 and outer radius R_3, which has no net charge (cross-sectional view shown in Fig. 22–37). If the axes of the two cylinders are parallel, but displaced from each other by a distance d, determine the resulting electric field in the region $R > R_3$, where the radial distance R is measured from the metal cylinder's axis. Assume $d < (R_2 - R_1)$.

FIGURE 22–37
Problem 40.

41. (II) A flat ring (inner radius R_0, outer radius $4R_0$) is uniformly charged. In terms of the total charge Q, determine the electric field on the axis at points (a) $0.25R_0$ and (b) $75R_0$ from the center of the ring. [*Hint:* The ring can be replaced with two oppositely charged superposed disks.]

42. (II) An uncharged solid conducting sphere of radius r_0 contains two spherical cavities of radii r_1 and r_2, respectively. Point charge Q_1 is then placed within the cavity of radius r_1 and point charge Q_2 is placed within the cavity of radius r_2 (Fig. 22–38). Determine the resulting electric field (magnitude and direction) at locations outside the solid sphere $(r > r_0)$, where r is the distance from its center.

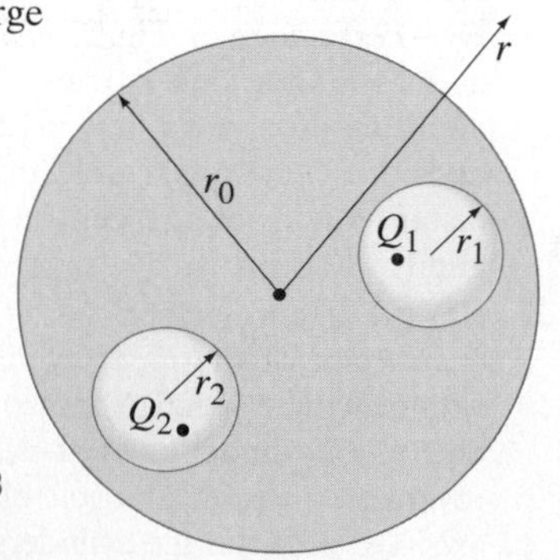

FIGURE 22–38
Problem 42.

43. (III) A very large (i.e., assume infinite) flat slab of nonconducting material has thickness d and a uniform volume charge density $+\rho_E$. (a) Show that a uniform electric field exists outside of this slab. Determine its magnitude E and its direction (relative to the slab's surface). (b) As shown in Fig. 22–39, the slab is now aligned so that one of its surfaces lies on the line $y = x$. At time $t = 0$, a pointlike particle (mass m, charge $+q$) is located at position $\vec{r} = +y_0\hat{j}$ and has velocity $\vec{v} = v_0\hat{i}$. Show that the particle will collide with the slab if $v_0 \geq \sqrt{\sqrt{2}qy_0\rho_E d/\epsilon_0 m}$. Ignore gravity.

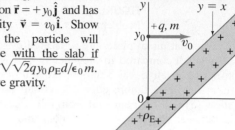

FIGURE 22–39
Problem 43.

44. (III) Suppose the density of charge between r_1 and r_0 of the hollow sphere of Problem 29 (Fig. 22–32) varies as $\rho_E = \rho_0 r_1/r$. Determine the electric field as a function of r for (a) $0 < r < r_1$, (b) $r_1 < r < r_0$, and (c) $r > r_0$. (d) Plot E versus r from $r = 0$ to $r = 2r_0$.

45. (III) Suppose two thin flat plates measure $1.0\,\text{m} \times 1.0\,\text{m}$ and are separated by $5.0\,\text{mm}$. They are oppositely charged with $\pm 15\,\mu\text{C}$. (a) Estimate the total force exerted by one plate on the other (ignore edge effects). (b) How much work would be required to move the plates from $5.0\,\text{mm}$ apart to $1.00\,\text{cm}$ apart?

46. (III) A flat slab of nonconducting material (Fig. 22–40) carries a uniform charge per unit volume, ρ_E. The slab has thickness d which is small compared to the height and breadth of the slab. Determine the electric field as a function of x (a) inside the slab and (b) outside the slab (at distances much less than the slab's height or breadth). Take the origin at the center of the slab.

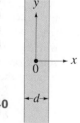

FIGURE 22–40
Problem 46.

47. (III) A flat slab of nonconducting material has thickness $2d$, which is small compared to its height and breadth. Define the x axis to be along the direction of the slab's thickness with the origin at the center of the slab (Fig. 22–41). If the slab carries a volume charge density $\rho_E(x) = -\rho_0$ in the region $-d \leq x < 0$, and $\rho_E(x) = +\rho_0$ in the region $0 < x \leq +d$, determine the electric field $\vec{E}$ as a function of x in the regions (a) outside the slab, (b) $0 < x \leq +d$, and (c) $-d \leq x < 0$. Let ρ_0 be a positive constant.

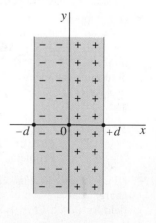

FIGURE 22–41
Problem 47.

48. (III) An extremely long, solid nonconducting cylinder has a radius R_0. The charge density within the cylinder is a function of the distance R from the axis, given by $\rho_1(R) = \rho_0(R/R_0)^2$. What is the electric field everywhere inside and outside the cylinder (far away from the ends) in terms of ρ_0 and R_0?

49. (III) Charge is distributed within a solid sphere of radius r_0 in such a way that the charge density is a function of the radial position within the sphere of the form: $\rho_1(r) = \rho_0(r/r_0)$. If the total charge within the sphere is Q (and positive), what is the electric field everywhere within the sphere in terms of Q, r_0, and the radial position r?

General Problems

50. A point charge Q is on the axis of a short cylinder at its center. The diameter of the cylinder is equal to its length ℓ (Fig. 22–42). What is the total flux through the curved sides of the cylinder? [*Hint*: First calculate the flux through the ends.]

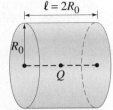

FIGURE 22–42
Problem 50.

51. Write Gauss's law for the gravitational field $\vec{g}$ (see Section 6–6).

52. The Earth is surrounded by an electric field, pointing inward at every point, of magnitude $E \approx 150\,\text{N/C}$ near the surface. (*a*) What is the net charge on the Earth? (*b*) How many excess electrons per square meter on the Earth's surface does this correspond to?

53. A cube of side ℓ has one corner at the origin of coordinates, and extends along the positive x, y, and z axes. Suppose the electric field in this region is given by $\vec{E} = (ay + b)\hat{j}$. Determine the charge inside the cube.

54. A solid nonconducting sphere of radius r_0 has a total charge Q which is distributed according to $\rho_E = br$, where ρ_E is the charge per unit volume, or charge density (C/m^3), and b is a constant. Determine (*a*) b in terms of Q, (*b*) the electric field at points inside the sphere, and (*c*) the electric field at points outside the sphere.

55. A point charge of $9.20\,\text{nC}$ is located at the origin and a second charge of $-5.00\,\text{nC}$ is located on the x axis at $x = 2.75\,\text{cm}$. Calculate the electric flux through a sphere centered at the origin with radius $1.00\,\text{m}$. Repeat the calculation for a sphere of radius $2.00\,\text{m}$.

56. A point charge produces an electric flux of $+235\,\text{N}\cdot\text{m}^2/\text{C}$ through a gaussian sphere of radius $15.0\,\text{cm}$ centered on the charge. (*a*) What is the flux through a gaussian sphere with a radius $27.5\,\text{cm}$? (*b*) What is the magnitude and sign of the charge?

57. A point charge Q is placed a distance $r_0/2$ above the surface of an imaginary spherical surface of radius r_0 (Fig. 22–43). (*a*) What is the electric flux through the sphere? (*b*) What range of values does E have at the surface of the sphere? (*c*) Is $\vec{E}$ perpendicular to the sphere at all points? (*d*) Is Gauss's law useful for obtaining E at the surface of the sphere?

$Q\bullet\!\begin{matrix}\}\frac{1}{2}r_0\end{matrix}$

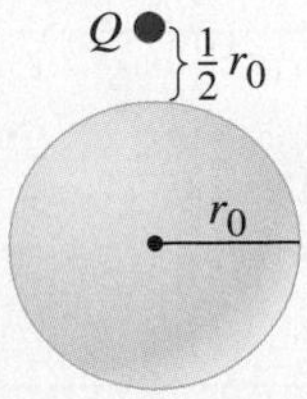

FIGURE 22–43
Problem 57.

58. Three large but thin charged sheets are parallel to each other as shown in Fig. 22–44. Sheet I has a total surface charge density of $6.5\,\text{nC/m}^2$, sheet II a charge of $-2.0\,\text{nC/m}^2$, and sheet III a charge of $5.0\,\text{nC/m}^2$. Estimate the force per unit area on each sheet, in N/m^2.

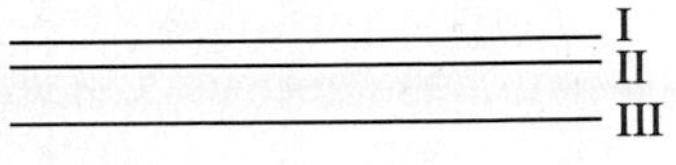

FIGURE 22–44
Problem 58.

I
II
III

59. Neutral hydrogen can be modeled as a positive point charge $+1.6 \times 10^{-19}\,\text{C}$ surrounded by a distribution of negative charge with volume density given by $\rho_E(r) = -Ae^{-2r/a_0}$ where $a_0 = 0.53 \times 10^{-10}\,\text{m}$ is called the *Bohr radius*, A is a constant such that the total amount of negative charge is $-1.6 \times 10^{-19}\,\text{C}$, and $e = 2.718\cdots$ is the base of the natural log. (*a*) What is the net charge inside a sphere of radius a_0? (*b*) What is the strength of the electric field at a distance a_0 from the nucleus? [*Hint*: Do not confuse the exponential number e with the elementary charge e which uses the same symbol but has a completely different meaning and value $(e = 1.6 \times 10^{-19}\,\text{C})$.]

60. A very large thin plane has uniform surface charge density σ. Touching it on the right (Fig. 22–45) is a long wide slab of thickness d with uniform volume charge density ρ_E. Determine the electric field (*a*) to the left of the plane, (*b*) to the right of the slab, and (*c*) everywhere inside the slab.

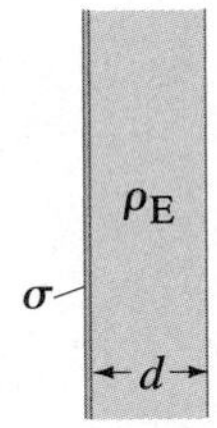

ρ_E

σ

FIGURE 22–45
Problem 60.

$\leftarrow d \rightarrow$

61. A sphere of radius r_0 carries a volume charge density ρ_E (Fig. 22–46). A spherical cavity of radius $r_0/2$ is then scooped out and left empty, as shown. (*a*) What is the magnitude and direction of the electric field at point A? (*b*) What is the direction and magnitude of the electric field at point B? Points A and C are at the centers of the respective spheres.

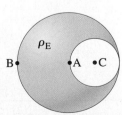

FIGURE 22–46
Problem 61.

62. Dry air will break down and generate a spark if the electric field exceeds about $3 \times 10^6\,\text{N/C}$. How much charge could be packed onto the surface of a green pea (diameter $0.75\,\text{cm}$) before the pea spontaneously discharges?

63. Three very large sheets are separated by equal distances of 15.0 cm (Fig. 22–47). The first and third sheets are very thin and nonconducting and have charge per unit area σ of $+5.00\,\mu C/m^2$ and $-5.00\,\mu C/m^2$ respectively. The middle sheet is conducting but has no net charge. (a) What is the electric field inside the middle sheet? What is the electric field (b) between the left and middle sheets, and (c) between the middle and right sheets? (d) What is the charge density on the surface of the middle sheet facing the left sheet, and (e) on the surface facing the right sheet?

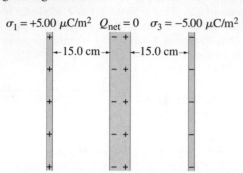

FIGURE 22–47   Problem 63.

64. In a cubical volume, 0.70 m on a side, the electric field is

$$\vec{E} = E_0\left(1 + \frac{z}{a}\right)\hat{i} + E_0\left(\frac{z}{a}\right)\hat{j}$$

where $E_0 = 0.125\,N/C$ and $a = 0.70\,m$. The cube has its sides parallel to the coordinate axes, Fig. 22–48. Determine the net charge within the cube.

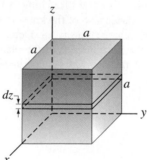

FIGURE 22–48
Problem 64.

65. A conducting spherical shell (Fig. 22–49) has inner radius $= 10.0\,cm$, outer radius $= 15.0\,cm$, and has a $+3.0\,\mu C$ point charge at the center. A charge of $-3.0\,\mu C$ is put on the conductor. (a) Where on the conductor does the $-3.0\,\mu C$ end up? (b) What is the electric field both inside and outside the shell?

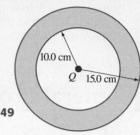

FIGURE 22–49
Problem 65.

66. A hemisphere of radius R is placed in a charge-free region of space where a uniform electric field exists of magnitude E directed perpendicular to the hemisphere's circular base (Fig. 22–50). (a) Using the definition of Φ_E through an "open" surface, calculate (via explicit integration) the electric flux through the hemisphere. [Hint: In Fig. 22–50 you can see that, on the surface of a sphere, the infinitesimal area located between the angles θ and $\theta + d\theta$ is $dA = (2\pi R \sin\theta)(R\,d\theta) = 2\pi R^2 \sin\theta\,d\theta$.] (b) Choose an appropriate gaussian surface and use Gauss's law to much more easily obtain the same result for the electric flux through the hemisphere.

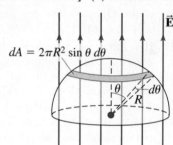

FIGURE 22–50
Problem 66.

*Numerical/Computer

*67. (III) An electric field is given by

$$\mathbf{E} = E_{x0}e^{-\left(\frac{x+y}{a}\right)^2}\hat{i} + E_{y0}e^{-\left(\frac{x+y}{a}\right)^2}\hat{j},$$

where $E_{x0} = 50\,N/C$, $E_{y0} = 25\,N/C$, and $a = 1.0\,m$. Given a cube with sides parallel to the coordinate axes, with one corner at the origin (as in Fig. 22–48), and with sides of length 1.0 m, estimate the flux out of the cube using a spreadsheet or other numerical method. How much total charge is enclosed by the cube?

Answers to Exercises

A: (d).

B: (a).

C: (d).

D: (e).

E: (c).

F: (c).

We are used to voltage in our lives—a 12-volt car battery, 110 V or 220 V at home, 1.5 volt flashlight batteries, and so on. Here we see a Van de Graaff generator, whose voltage may reach 50,000 V or more. Voltage is the same as electric potential difference between two points. Electric potential is defined as the potential energy per unit charge.

The children here, whose hair stands on end because each hair has received the same sign of charge, are not harmed by the voltage because the Van de Graaff cannot provide much current before the voltage drops. (It is current through the body that is harmful, as we will see later.)

C H A P T E R

23

Electric Potential

CHAPTER-OPENING QUESTION—Guess now!

Consider a pair of parallel plates with equal and opposite charge densities, σ. Which of the following actions will increase the voltage between the plates (assuming fixed charge density)?

(a) Moving the plates closer together.
(b) Moving the plates apart.
(c) Doubling the area of the plates.
(d) Halving the area of the plates.

We saw in Chapters 7 and 8 that the concept of energy was extremely useful in dealing with the subject of mechanics. The energy point of view is especially useful for electricity. It not only extends the law of conservation of energy, but it gives us another way to view electrical phenomena. Energy is also a powerful tool for solving Problems more easily in many cases than by using forces and electric fields.

23–1 Electric Potential Energy and Potential Difference

Electric Potential Energy

To apply conservation of energy, we need to define electric potential energy as we did for other types of potential energy. As we saw in Chapter 8, potential energy can be defined only for a conservative force. The work done by a conservative force in moving an object between any two positions is independent of the path taken. The electrostatic force between any two charges (Eq. 21–1, $F = kQ_1Q_2/r^2$) is conservative since the dependence on position is just like the gravitational force, $1/r^2$, which we saw in Section 8–7 is conservative. Hence we can define potential energy U for the electrostatic force.

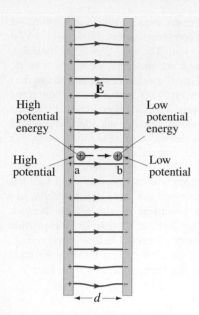

FIGURE 23–1 Work is done by the electric field in moving the positive charge q from position a to position b.

We saw in Chapter 8 that the change in potential energy between two points, a and b, equals the negative of the work done by the conservative force as an object moves from a to b: $\Delta U = -W$.

Thus we define the change in electric potential energy, $U_b - U_a$, when a point charge q moves from some point a to another point b, as the negative of the work done by the electric force as the charge moves from a to b. For example, consider the electric field between two equally but oppositely charged parallel plates; we assume their separation is small compared to their width and height, so the field $\vec{E}$ will be uniform over most of the region, Fig. 23–1. Now consider a tiny positive point charge q placed at point a very near the positive plate as shown. This charge q is so small it has no effect on $\vec{E}$. If this charge q at point a is released, the electric force will do work on the charge and accelerate it toward the negative plate. The work W done by the electric field E to move the charge a distance d is

$$W = Fd = qEd$$

where we used Eq. 21–5, $F = qE$. The change in electric potential energy equals the negative of the work done by the electric force:

$$U_b - U_a = -W = -qEd \qquad \text{[uniform } \vec{E}\text{]} \quad \textbf{(23–1)}$$

for this case of uniform electric field $\vec{E}$. In the case illustrated, the potential energy decreases (ΔU is negative); and as the charged particle accelerates from point a to point b in Fig. 23–1, the particle's kinetic energy K increases—by an equal amount. In accord with the conservation of energy, electric potential energy is transformed into kinetic energy, and the total energy is conserved. Note that the positive charge q has its greatest potential energy at point a, near the positive plate.[†] The reverse is true for a negative charge: its potential energy is greatest near the negative plate.

Electric Potential and Potential Difference

In Chapter 21, we found it useful to define the electric field as the force per unit charge. Similarly, it is useful to define the **electric potential** (or simply the **potential** when "electric" is understood) as the *electric potential energy per unit charge*. Electric potential is given the symbol V. If a positive test charge q in an electric field has electric potential energy U_a at some point a (relative to some zero potential energy), the electric potential V_a at this point is

$$V_a = \frac{U_a}{q}. \qquad \textbf{(23–2a)}$$

As we discussed in Chapter 8, only differences in potential energy are physically meaningful. Hence only the **difference in potential**, or the **potential difference**, between two points a and b (such as between a and b in Fig. 23–1) is measurable. When the electric force does positive work on a charge, the kinetic energy increases and the potential energy decreases. The difference in potential energy, $U_b - U_a$, is equal to the negative of the work, W_{ba}, done by the electric field as the charge moves from a to b; so the potential difference V_{ba} is

$$V_{ba} = \Delta V = V_b - V_a = \frac{U_b - U_a}{q} = -\frac{W_{ba}}{q}. \qquad \textbf{(23–2b)}$$

Note that electric potential, like electric field, does not depend on our test charge q. V depends on the other charges that create the field, not on q; q acquires potential energy by being in the potential V due to the other charges.

We can see from our definition that the positive plate in Fig. 23–1 is at a higher potential than the negative plate. Thus a positively charged object moves naturally from a high potential to a low potential. A negative charge does the reverse.

The unit of electric potential, and of potential difference, is joules/coulomb and is given a special name, the **volt**, in honor of Alessandro Volta (1745–1827) who is best known for inventing the electric battery. The volt is abbreviated V, so $1\ \text{V} = 1\ \text{J/C}$. Potential difference, since it is measured in volts, is often referred to as **voltage**.

[†]At this point the charge has its greatest ability to do work (on some other object or system).

If we wish to speak of the potential V_a at some point a, we must be aware that V_a depends on where the potential is chosen to be zero. The zero for electric potential in a given situation can be chosen arbitrarily, just as for potential energy, because only differences in potential energy can be measured. Often the ground, or a conductor connected directly to the ground (the Earth), is taken as zero potential, and other potentials are given with respect to ground. (Thus, a point where the voltage is 50 V is one where the difference of potential between it and ground is 50 V.) In other cases, as we shall see, we may choose the potential to be zero at an infinite distance ($r = \infty$).

FIGURE 23–2 Central part of Fig. 23–1, showing a negative point charge near the negative plate, where its potential energy (PE) is high. Example 23–1.

> **CONCEPTUAL EXAMPLE 23–1** **A negative charge.** Suppose a negative charge, such as an electron, is placed near the negative plate in Fig. 23–1, at point b, shown here in Fig. 23–2. If the electron is free to move, will its electric potential energy increase or decrease? How will the electric potential change?
>
> **RESPONSE** An electron released at point b will move toward the positive plate. As the electron moves toward the positive plate, its potential energy *decreases* as its kinetic energy gets larger, so $U_a < U_b$ and $\Delta U = U_a - U_b < 0$. But note that the electron moves from point b at low potential to point a at higher potential: $V_{ab} = V_a - V_b > 0$. (Potentials V_a and V_b are due to the charges on the plates, not due to the electron.) The sign of ΔU and ΔV are opposite because of the negative charge.

⚠ **CAUTION**

A negative charge has high potential energy when potential V is low

Because the electric potential difference is defined as the potential energy difference per unit charge, then the change in potential energy of a charge q when moved between two points a and b is

$$\Delta U = U_b - U_a = q(V_b - V_a) = qV_{ba}. \qquad (23\text{--}3)$$

That is, if an object with charge q moves through a potential difference V_{ba}, its potential energy changes by an amount qV_{ba}. For example, if the potential difference between the two plates in Fig. 23–1 is 6 V, then a $+1$ C charge moved (say by an external force) from point b to point a will gain $(1\,\text{C})(6\,\text{V}) = 6\,\text{J}$ of electric potential energy. (And it will lose 6 J of electric potential energy if it moves from a to b.) Similarly, a $+2$ C charge will gain 12 J, and so on. Thus, electric potential difference is a measure of how much energy an electric charge can acquire in a given situation. And, since energy is the ability to do work, the electric potential difference is also a measure of how much work a given charge can do. The exact amount depends both on the potential difference and on the charge.

To better understand electric potential, let's make a comparison to the gravitational case when a rock falls from the top of a cliff. The greater the height, h, of a cliff, the more potential energy ($= mgh$) the rock has at the top of the cliff, relative to the bottom, and the more kinetic energy it will have when it reaches the bottom. The actual amount of kinetic energy it will acquire, and the amount of work it can do, depends both on the height of the cliff and the mass m of the rock. A large rock and a small rock can be at the same height h (Fig. 23–3a) and thus have the same "gravitational potential," but the larger rock has the greater potential energy (it has more mass). The electrical case is similar (Fig. 23–3b): the potential energy change, or the work that can be done, depends both on the potential difference (corresponding to the height of the cliff) and on the charge (corresponding to mass), Eq. 23–3. But note a significant difference: electric charge comes in two types, $+$ and $-$, whereas gravitational mass is always $+$.

Sources of electrical energy such as batteries and electric generators are meant to maintain a potential difference. The actual amount of energy transformed by such a device depends on how much charge flows, as well as the potential difference (Eq. 23–3). For example, consider an automobile headlight connected to a 12.0-V battery. The amount of energy transformed (into light and thermal energy) is proportional to how much charge flows, which depends on how long the light is on. If over a given period of time 5.0 C of charge flows through the light, the total energy transformed is $(5.0\,\text{C})(12.0\,\text{V}) = 60\,\text{J}$. If the headlight is left on twice as long, 10.0 C of charge will flow and the energy transformed is $(10.0\,\text{C})(12.0\,\text{V}) = 120\,\text{J}$. Table 23–1 presents some typical voltages.

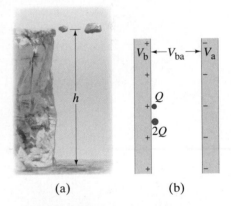

FIGURE 23–3 (a) Two rocks are at the same height. The larger rock has more potential energy. (b) Two charges have the same electric potential. The $2Q$ charge has more potential energy.

TABLE 23–1 Some Typical Potential Differences (Voltages)

Source	Voltage (approx.)
Thundercloud to ground	10^8 V
High-voltage power line	10^5–10^6 V
Power supply for TV tube	10^4 V
Automobile ignition	10^4 V
Household outlet	10^2 V
Automobile battery	12 V
Flashlight battery	1.5 V
Resting potential across nerve membrane	10^{-1} V
Potential changes on skin (EKG and EEG)	10^{-4} V

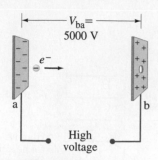

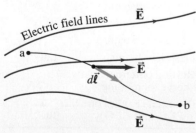

FIGURE 23-4 Electron accelerated in CRT. Example 23-2.

EXAMPLE 23-2 **Electron in CRT.** Suppose an electron in a cathode ray tube (Section 23–9) is accelerated from rest through a potential difference $V_b - V_a = V_{ba} = +5000\,V$ (Fig. 23–4). (*a*) What is the change in electric potential energy of the electron? (*b*) What is the speed of the electron $(m = 9.1 \times 10^{-31}\,kg)$ as a result of this acceleration?

APPROACH The electron, accelerated toward the positive plate, will decrease in potential energy by an amount $\Delta U = qV_{ba}$ (Eq. 23–3). The loss in potential energy will equal its gain in kinetic energy (energy conservation).

SOLUTION (*a*) The charge on an electron is $q = -e = -1.6 \times 10^{-19}\,C$. Therefore its change in potential energy is

$$\Delta U = qV_{ba} = (-1.6 \times 10^{-19}\,C)(+5000\,V) = -8.0 \times 10^{-16}\,J.$$

The minus sign indicates that the potential energy decreases. The potential difference V_{ba} has a positive sign since the final potential V_b is higher than the initial potential V_a. Negative electrons are attracted toward a positive electrode and repelled away from a negative electrode.

(*b*) The potential energy lost by the electron becomes kinetic energy K. From conservation of energy (Eq. 8–9a), $\Delta K + \Delta U = 0$, so

$$\Delta K = -\Delta U$$
$$\tfrac{1}{2}mv^2 - 0 = -q(V_b - V_a) = -qV_{ba},$$

where the initial kinetic energy is zero since we are given that the electron started from rest. We solve for v:

$$v = \sqrt{-\frac{2qV_{ba}}{m}} = \sqrt{-\frac{2(-1.6 \times 10^{-19}\,C)(5000\,V)}{9.1 \times 10^{-31}\,kg}} = 4.2 \times 10^7\,m/s.$$

NOTE The electric potential energy does not depend on the mass, only on the charge and voltage. The speed *does* depend on m.

23-2 Relation between Electric Potential and Electric Field

The effects of any charge distribution can be described either in terms of electric field or in terms of electric potential. Electric potential is often easier to use because it is a scalar, as compared to electric field which is a vector. There is a crucial connection between the electric potential produced by a given arrangement of charges and the electric field due to those charges, which we now examine.

We start by recalling the relation between a conservative force $\vec{F}$ and the potential energy U associated with that force. As discussed in Section 8–2, the difference in potential energy between any two points in space, a and b, is given by Eq. 8–4:

$$U_b - U_a = -\int_a^b \vec{F} \cdot d\vec{\ell},$$

FIGURE 23-5 To find V_{ba} in a nonuniform electric field $\vec{E}$, we integrate $\vec{E} \cdot d\vec{\ell}$ from point a to point b.

where $d\vec{\ell}$ is an infinitesimal increment of displacement, and the integral is taken along any path in space from point a to point b. For the electrical case, we are more interested in the potential difference, given by Eq. 23–2b, $V_{ba} = V_b - V_a = (U_b - U_a)/q$, rather than in the potential energy itself. Also, the electric field $\vec{E}$ at any point in space is defined as the force per unit charge (Eq. 21–3): $\vec{E} = \vec{F}/q$. Putting these two relations in the above equation gives us

$$V_{ba} = V_b - V_a = -\int_a^b \vec{E} \cdot d\vec{\ell}. \qquad \textbf{(23-4a)}$$

This is the general relation between electric field and potential difference. See Fig. 23–5. If we are given the electric field due to some arrangement of electric charge, we can use Eq. 23–4a to determine V_{ba}.

A simple special case is a uniform field. In Fig. 23–1, for example, a path parallel to the electric field lines from point a at the positive plate to point b at the negative plate gives (since $\vec{E}$ and $d\vec{\ell}$ are in the same direction at each point),

$$V_{ba} = V_b - V_a = -\int_a^b \vec{E} \cdot d\vec{\ell} = -E\int_a^b d\ell = -Ed$$

or

$$V_{ba} = -Ed \qquad \text{[only if } E \text{ is uniform]} \quad (23\text{–}4b)$$

where d is the distance, parallel to the field lines, between points a and b. Be careful not to use Eq. 23–4b unless you are sure the electric field is uniform.

From either of Eqs. 23–4 we can see that the units for electric field intensity can be written as volts per meter (V/m) as well as newtons per coulomb (N/C). These are equivalent in general, since $1\,\text{N/C} = 1\,\text{N·m/C·m} = 1\,\text{J/C·m} = 1\,\text{V/m}$.

EXERCISE A Return to the Chapter-Opening Question, page 607, and answer it again now. Try to explain why you may have answered differently the first time.

EXAMPLE 23–3 **Electric field obtained from voltage.** Two parallel plates are charged to produce a potential difference of 50 V. If the separation between the plates is 0.050 m, calculate the magnitude of the electric field in the space between the plates (Fig. 23–6).

APPROACH We apply Eq. 23–4b to obtain the magnitude of E, assumed uniform.

SOLUTION The electric field magnitude is $E = V_{ba}/d = (50\,\text{V}/0.050\,\text{m}) = 1000\,\text{V/m}$.

EXAMPLE 23–4 **Charged conducting sphere.** Determine the potential at a distance r from the center of a charged conducting sphere of radius r_0 for (a) $r > r_0$, (b) $r = r_0$, (c) $r < r_0$. The total charge on the sphere is Q.

APPROACH The charge Q is distributed over the surface of the sphere since it is a conductor. We saw in Example 22–3 that the electric field outside a conducting sphere is

$$E = \frac{1}{4\pi\epsilon_0}\frac{Q}{r^2} \qquad [r > r_0]$$

and points radially outward (inward if $Q < 0$). Since we know $\vec{E}$, we can start by using Eq. 23–4a.

SOLUTION (a) We use Eq. 23–4a and integrate along a radial line with $d\vec{\ell}$ parallel to $\vec{E}$ (Fig. 23–7) between two points which are distances r_a and r_b from the sphere's center:

$$V_b - V_a = -\int_{r_a}^{r_b} \vec{E} \cdot d\vec{\ell} = -\frac{Q}{4\pi\epsilon_0}\int_{r_a}^{r_b}\frac{dr}{r^2} = \frac{Q}{4\pi\epsilon_0}\left(\frac{1}{r_b} - \frac{1}{r_a}\right)$$

and we set $d\ell = dr$. If we let $V = 0$ for $r = \infty$ (let's choose $V_b = 0$ at $r_b = \infty$), then at any other point r (for $r > r_0$) we have

$$V = \frac{1}{4\pi\epsilon_0}\frac{Q}{r}. \qquad [r > r_0]$$

We will see in the next Section that this same equation applies for the potential a distance r from a single point charge. Thus the electric potential outside a spherical conductor with a uniformly distributed charge is the same as if all the charge were at its center.

(b) As r approaches r_0, we see that

$$V = \frac{1}{4\pi\epsilon_0}\frac{Q}{r_0} \qquad [r = r_0]$$

at the surface of the conductor.

(c) For points within the conductor, $E = 0$. Thus the integral, $\int\vec{E} \cdot d\vec{\ell}$, between $r = r_0$ and any point within the conductor gives zero change in V. Hence V is constant within the conductor:

$$V = \frac{1}{4\pi\epsilon_0}\frac{Q}{r_0}. \qquad [r \le r_0]$$

The whole conductor, not just its surface, is at this same potential. Plots of both E and V as a function of r are shown in Fig. 23–8 for a positively charged conducting sphere.

FIGURE 23–6 Example 23–3.

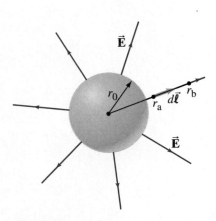

FIGURE 23–7 Example 23–4. Integrating $\vec{E} \cdot d\vec{\ell}$ for the field outside a spherical conductor.

FIGURE 23–8 (a) E versus r, and (b) V versus r, for a positively charged solid conducting sphere of radius r_0 (the charge distributes itself evenly on the surface); r is the distance from the center of the sphere.

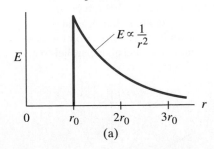

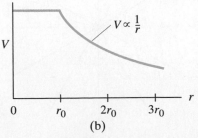

EXAMPLE 23–5 Breakdown voltage. In many kinds of equipment, very high voltages are used. A problem with high voltage is that the air can become ionized due to the high electric fields: free electrons in the air (produced by cosmic rays, for example) can be accelerated by such high fields to speeds sufficient to ionize O_2 and N_2 molecules by collision, knocking out one or more of their electrons. The air then becomes conducting and the high voltage cannot be maintained as charge flows. The breakdown of air occurs for electric fields of about 3×10^6 V/m. (*a*) Show that the breakdown voltage for a spherical conductor in air is proportional to the radius of the sphere, and (*b*) estimate the breakdown voltage in air for a sphere of diameter 1.0 cm.

APPROACH The electric potential at the surface of a spherical conductor of radius r_0 (Example 23–4), and the electric field just outside its surface, are

$$V = \frac{1}{4\pi\epsilon_0} \frac{Q}{r_0} \quad \text{and} \quad E = \frac{1}{4\pi\epsilon_0} \frac{Q}{r_0^2}.$$

SOLUTION (*a*) We combine these two equations and obtain

$$V = r_0 E. \qquad \text{[at surface of spherical conductor]}$$

(*b*) For $r_0 = 5 \times 10^{-3}$ m, the breakdown voltage in air is

$$V = (5 \times 10^{-3}\,\text{m})(3 \times 10^6\,\text{V/m}) \approx 15,000\,\text{V}.$$

When high voltages are present, a glow may be seen around sharp points, known as a **corona discharge**, due to the high electric fields at these points which ionize air molecules. The light we see is due to electrons jumping down to empty lower states. **Lightning rods**, with their sharp tips, are intended to ionize the surrounding air when a storm cloud is near, and to provide a conduction path to discharge a dangerous high-voltage cloud slowly, over a period of time. Thus lightning rods, connected to the ground, are intended to draw electric charge off threatening clouds before a large buildup of charge results in a swift destructive lightning bolt.

EXERCISE B On a dry day, a person can become electrically charged by rubbing against rugs and other ordinary objects. Suppose you notice a small shock as you reach for a metal doorknob, noting that the shock occurs along with a tiny spark when your hand is about 3.0 mm from the doorknob. As a rough estimate, use Eq. 23–4b to estimate the potential difference between your hand and the doorknob. (*a*) 9 V; (*b*) 90 V; (*c*) 900 V; (*d*) 9000 V; (*e*) none of these.

23–3 Electric Potential Due to Point Charges

The electric potential at a distance r from a single point charge Q can be derived directly from Eq. 23–4a, $V_b - V_a = -\int\vec{E} \cdot d\vec{\ell}$. The electric field due to a single point charge has magnitude (Eq. 21–4)

$$E = \frac{1}{4\pi\epsilon_0} \frac{Q}{r^2} \quad \text{or} \quad E = k\frac{Q}{r^2}$$

(where $k = 1/4\pi\epsilon_0 = 8.99 \times 10^9\,\text{N·m}^2/\text{C}^2$), and is directed radially outward from a positive charge (inward if $Q < 0$). We take the integral in Eq. 23–4a along a (straight) field line (Fig. 23–9) from point a, a distance r_a from Q, to point b, a distance r_b from Q. Then $d\vec{\ell}$ will be parallel to $\vec{E}$ and $d\ell = dr$. Thus

$$V_b - V_a = -\int_{r_a}^{r_b}\vec{E} \cdot d\vec{\ell} = -\frac{Q}{4\pi\epsilon_0}\int_{r_a}^{r_b} \frac{1}{r^2}\, dr = \frac{1}{4\pi\epsilon_0}\left(\frac{Q}{r_b} - \frac{Q}{r_a}\right).$$

As mentioned earlier, only differences in potential have physical meaning. We are free, therefore, to choose the value of the potential at some one point to

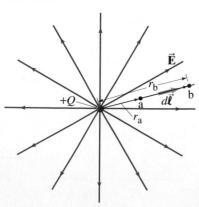

FIGURE 23–9 We integrate Eq. 23–4a along the straight line (shown in black) from point a to point b. The line ab is parallel to a field line.

be whatever we please. It is common to choose the potential to be zero at infinity (let $V_b = 0$ at $r_b = \infty$). Then the electric potential V at a distance r from a single point charge is

$$V = \frac{1}{4\pi\epsilon_0}\frac{Q}{r}. \qquad \left[\begin{array}{c}\text{single point charge:} \\ V = 0 \text{ at } r = \infty\end{array}\right] \qquad \textbf{(23–5)}$$

We can think of V here as representing the absolute potential, where $V = 0$ at $r = \infty$, or we can think of V as the potential difference between r and infinity. Notice that the potential V decreases with the first power of the distance, whereas the electric field (Eq. 21–4) decreases as the *square* of the distance. The potential near a positive charge is large, and it decreases toward zero at very large distances (Fig. 23–10). For a negative charge, the potential is negative and increases toward zero at large distances (Fig. 23–11). Equation 23–5 is sometimes called the **Coulomb potential** (it has its origin in Coulomb's law).

In Example 23–4 we found that the potential due to a uniformly charged sphere is given by the same relation, Eq. 23–5, for points outside the sphere. Thus we see that the potential outside a uniformly charged sphere is the same as if all the charge were concentrated at its center.

EXERCISE C What is the potential at a distance of 3.0 cm from a point charge $Q = -2.0 \times 10^{-9}$ C? (a) 600 V; (b) 60 V; (c) 6 V; (d) -600 V; (e) -60 V; (f) -6 V.

EXAMPLE 23–6 **Work required to bring two positive charges close together.** What minimum work must be done by an external force to bring a charge $q = 3.00\,\mu$C from a great distance away (take $r = \infty$) to a point 0.500 m from a charge $Q = 20.0\,\mu$C?

APPROACH To find the work we cannot simply multiply the force times distance because the force is not constant. Instead we can set the change in potential energy equal to the (positive of the) work required of an *external* force (Chapter 8), and Eq. 23–3: $W = \Delta U = q(V_b - V_a)$. We get the potentials V_b and V_a using Eq. 23–5.

SOLUTION The work required is equal to the change in potential energy:

$$W = q(V_b - V_a)$$

$$= q\left(\frac{kQ}{r_b} - \frac{kQ}{r_a}\right),$$

where $r_b = 0.500$ m and $r_a = \infty$. The right-hand term within the parentheses is zero $(1/\infty = 0)$ so

$$W = (3.00 \times 10^{-6}\,\text{C})\frac{(8.99 \times 10^9\,\text{N}\cdot\text{m}^2/\text{C}^2)(2.00 \times 10^{-5}\,\text{C})}{(0.500\,\text{m})} = 1.08\,\text{J}.$$

NOTE We could not use Eq. 23–4b here because it applies *only* to uniform fields. But we did use Eq. 23–3 because it is always valid.

To determine the electric field at points near a collection of two or more point charges requires adding up the electric fields due to each charge. Since the electric field is a vector, this can be time consuming or complicated. To find the electric potential at a point due to a collection of point charges is far easier, since the electric potential is a scalar, and hence you only need to add numbers (with appropriate signs) without concern for direction. This is a major advantage in using electric potential for solving Problems.

FIGURE 23–10 Potential V as a function of distance r from a single point charge Q when the charge is positive.

FIGURE 23–11 Potential V as a function of distance r from a single point charge Q when the charge is negative.

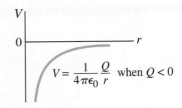

⚠ **CAUTION**

We cannot use $W = Fd$ *when F is not constant*

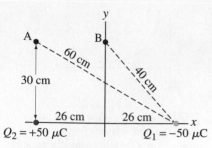

FIGURE 23–12 Example 23–7. (See also Example 21–8, Fig. 21–27.)

> ⚠ **CAUTION**
>
> *Potential is a scalar and has no components*

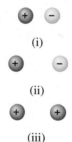

FIGURE 23–13 Exercise D.

EXAMPLE 23–7 **Potential above two charges.** Calculate the electric potential (*a*) at point A in Fig. 23–12 due to the two charges shown, and (*b*) at point B. [This is the same situation as Example 21–8, Fig. 21–27, where we calculated the electric field at these points.]

APPROACH The total potential at point A (or at point B) is the sum of the potentials at that point due to each of the two charges Q_1 and Q_2. The potential due to each single charge is given by Eq. 23–5. We do not have to worry about directions because electric potential is a scalar quantity. But we do have to keep track of the signs of charges.

SOLUTION (*a*) We add the potentials at point A due to each charge Q_1 and Q_2, and we use Eq. 23–5 for each:

$$V_A = V_{A2} + V_{A1}$$
$$= k\frac{Q_2}{r_{2A}} + k\frac{Q_1}{r_{1A}}$$

where $r_{1A} = 60\text{ cm}$ and $r_{2A} = 30\text{ cm}$. Then

$$V_A = \frac{(9.0 \times 10^9\,\text{N·m}^2/\text{C}^2)(5.0 \times 10^{-5}\,\text{C})}{0.30\,\text{m}}$$
$$+ \frac{(9.0 \times 10^9\,\text{N·m}^2/\text{C}^2)(-5.0 \times 10^{-5}\,\text{C})}{0.60\,\text{m}}$$
$$= 1.50 \times 10^6\,\text{V} - 0.75 \times 10^6\,\text{V}$$
$$= 7.5 \times 10^5\,\text{V}.$$

(*b*) At point B, $r_{1B} = r_{2B} = 0.40\text{ m}$, so

$$V_B = V_{B2} + V_{B1}$$
$$= \frac{(9.0 \times 10^9\,\text{N·m}^2/\text{C}^2)(5.0 \times 10^{-5}\,\text{C})}{0.40\,\text{m}}$$
$$+ \frac{(9.0 \times 10^9\,\text{N·m}^2/\text{C}^2)(-5.0 \times 10^{-5}\,\text{C})}{0.40\,\text{m}}$$
$$= 0\,\text{V}.$$

NOTE The two terms in the sum in (*b*) cancel for any point equidistant from Q_1 and Q_2 ($r_{1B} = r_{2B}$). Thus the potential will be zero everywhere on the plane equidistant between the two opposite charges. This plane where V is constant is called an equipotential surface.

Simple summations like these can easily be performed for any number of point charges.

> **EXERCISE D** Consider the three pairs of charges, Q_1 and Q_2, in Fig. 23–13. (*a*) Which set has a positive potential energy? (*b*) Which set has the most negative potential energy? (*c*) Which set requires the most work to separate the charges to infinity? Assume the charges all have the same magnitude.

23–4 Potential Due to Any Charge Distribution

If we know the electric field in a region of space due to any distribution of electric charge, we can determine the difference in potential between two points in the region using Eq. 23–4a, $V_{ba} = -\int_a^b \vec{E} \cdot d\vec{\ell}$. In many cases we don't know $\vec{E}$ as a function of position, and it may be difficult to calculate. We can calculate the potential V due to a given charge distribution in another way, using the potential due to a single point charge, Eq. 23–5:

$$V = \frac{1}{4\pi\epsilon_0}\frac{Q}{r},$$

where we choose $V = 0$ at $r = \infty$. Then we can sum over all the charges.

If we have n individual point charges, the potential at some point a (relative to $V = 0$ at $r = \infty$) is

$$V_a = \sum_{i=1}^{n} V_i = \frac{1}{4\pi\epsilon_0} \sum_{i=1}^{n} \frac{Q_i}{r_{ia}}.$$ (23–6a)

where r_{ia} is the distance from the ith charge (Q_i) to the point a. (We already used this approach in Example 23–7.) If the charge distribution can be considered continuous, then

$$V = \frac{1}{4\pi\epsilon_0} \int \frac{dq}{r}.$$ (23–6b)

where r is the distance from a tiny element of charge, dq, to the point where V is being determined.

EXAMPLE 23–8 **Potential due to a ring of charge.** A thin circular ring of radius R has a uniformly distributed charge Q. Determine the electric potential at a point P on the axis of the ring a distance x from its center, Fig. 23–14.

APPROACH We integrate over the ring using Eq. 23–6b.

SOLUTION Each point on the ring is equidistant from point P, and this distance is $(x^2 + R^2)^{\frac{1}{2}}$. So the potential at P is:

$$V = \frac{1}{4\pi\epsilon_0} \int \frac{dq}{r} = \frac{1}{4\pi\epsilon_0} \frac{1}{(x^2 + R^2)^{\frac{1}{2}}} \int dq = \frac{1}{4\pi\epsilon_0} \frac{Q}{(x^2 + R^2)^{\frac{1}{2}}}.$$

NOTE For points very far away from the ring, $x \gg R$, this result reduces to $(1/4\pi\epsilon_0)(Q/x)$, the potential of a point charge, as we should expect.

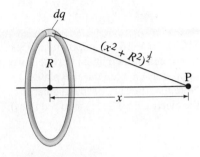

FIGURE 23–14 Example 23–8. Calculating the potential at point P, a distance x from the center of a uniform ring of charge.

EXAMPLE 23–9 **Potential due to a charged disk.** A thin flat disk, of radius R_0, has a uniformly distributed charge Q, Fig. 23–15. Determine the potential at a point P on the axis of the disk, a distance x from its center.

APPROACH We divide the disk into thin rings of radius R and thickness dR and use the result of Example 23–8 to sum over the disk.

SOLUTION The charge Q is distributed uniformly, so the charge contained in each ring is proportional to its area. The disk has area πR_0^2 and each thin ring has area $dA = (2\pi R)(dR)$. Hence

$$\frac{dq}{Q} = \frac{2\pi R\, dR}{\pi R_0^2}$$

so

$$dq = Q \frac{(2\pi R)(dR)}{\pi R_0^2} = \frac{2QR\, dR}{R_0^2}.$$

Then the potential at P, using Eq. 23–6b in which r is replaced by $(x^2 + R^2)^{\frac{1}{2}}$, is

$$V = \frac{1}{4\pi\epsilon_0} \int \frac{dq}{(x^2 + R^2)^{\frac{1}{2}}} = \frac{2Q}{4\pi\epsilon_0 R_0^2} \int_0^{R_0} \frac{R\, dR}{(x^2 + R^2)^{\frac{1}{2}}} = \frac{Q}{2\pi\epsilon_0 R_0^2} (x^2 + R^2)^{\frac{1}{2}} \Big|_{R=0}^{R=R_0}$$

$$= \frac{Q}{2\pi\epsilon_0 R_0^2} \left[(x^2 + R_0^2)^{\frac{1}{2}} - x \right].$$

NOTE For $x \gg R_0$, this formula reduces to

$$V \approx \frac{Q}{2\pi\epsilon_0 R_0^2} \left[x\left(1 + \frac{1}{2}\frac{R_0^2}{x^2}\right) - x \right] = \frac{Q}{4\pi\epsilon_0 x}.$$

This is the formula for a point charge, as we expect.

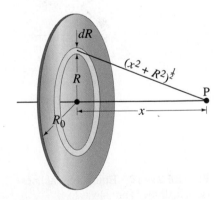

FIGURE 23–15 Example 23–9. Calculating the electric potential at point P on the axis of a uniformly charged thin disk.

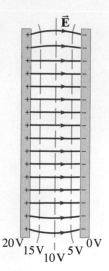

FIGURE 23–16 Equipotential lines (the green dashed lines) between two oppositely charged parallel plates. Note that they are perpendicular to the electric field lines (solid red lines).

FIGURE 23–17 Example 23–10. Electric field lines and equipotential surfaces for a point charge.

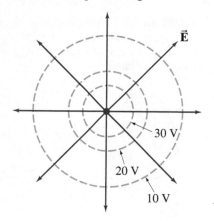

FIGURE 23–18 Equipotential lines (green, dashed) are always perpendicular to the electric field lines (solid red) shown here for two equal but oppositely charged particles.

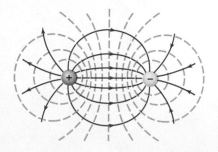

23–5 Equipotential Surfaces

The electric potential can be represented graphically by drawing **equipotential lines** or, in three dimensions, **equipotential surfaces**. An equipotential surface is one on which all points are at the same potential. That is, the potential difference between any two points on the surface is zero, and no work is required to move a charge from one point to the other. An *equipotential surface must be perpendicular to the electric field* at any point. If this were not so—that is, if there were a component of $\vec{E}$ parallel to the surface—it would require work to move the charge along the surface against this component of $\vec{E}$; and this would contradict the idea that it is an equipotential surface. This can also be seen from Eq. 23–4a, $\Delta V = -\int \vec{E} \cdot d\vec{\ell}$. On a surface where V is constant, $\Delta V = 0$, so we must have either $\vec{E} = 0$, $d\vec{\ell} = 0$, or $\cos \theta = 0$ where θ is the angle between $\vec{E}$ and $d\vec{\ell}$. Thus in a region where $\vec{E}$ is not zero, the path $d\vec{\ell}$ along an equipotential must have $\cos \theta = 0$, meaning $\theta = 90°$ and $\vec{E}$ is perpendicular to the equipotential.

The fact that the electric field lines and equipotential surfaces are mutually perpendicular helps us locate the equipotentials when the electric field lines are known. In a normal two-dimensional drawing, we show equipotential *lines*, which are the intersections of equipotential surfaces with the plane of the drawing. In Fig. 23–16, a few of the equipotential lines are drawn (dashed green lines) for the electric field (red lines) between two parallel plates at a potential difference of 20 V. The negative plate is arbitrarily chosen to be zero volts and the potential of each equipotential line is indicated. Note that $\vec{E}$ points toward lower values of V.

EXAMPLE 23–10 **Point charge equipotential surfaces.** For a single point charge with $Q = 4.0 \times 10^{-9}$ C, sketch the equipotential surfaces (or lines in a plane containing the charge) corresponding to $V_1 = 10$ V, $V_2 = 20$ V, and $V_3 = 30$ V.

APPROACH The electric potential V depends on the distance r from the charge (Eq. 23–5).

SOLUTION The electric field for a positive point charge is directed radially outward. Since the equipotential surfaces must be perpendicular to the lines of electric field, they will be spherical in shape, centered on the point charge, Fig. 23–17. From Eq. 23–5 we have $r = (1/4\pi\epsilon_0)(Q/V)$, so that for $V_1 = 10$ V, $r_1 = (9.0 \times 10^9 \text{ N} \cdot \text{m}^2/\text{C}^2)(4.0 \times 10^{-9} \text{ C})/(10 \text{ V}) = 3.6$ m, for $V_2 = 20$ V, $r_2 = 1.8$ m, and for $V_3 = 30$ V, $r_3 = 1.2$ m, as shown.

NOTE The equipotential surface with the largest potential is closest to the positive charge. How would this change if Q were negative?

The equipotential lines for the case of two equal but oppositely charged particles are shown in Fig. 23–18 as green dashed lines. Equipotential lines and surfaces, unlike field lines, are always continuous and never end, and so continue beyond the borders of Figs. 23–16 and 23–18.

We saw in Section 21–9 that there can be no electric field within a conductor in the static case, for otherwise the free electrons would feel a force and would move. Indeed, the entire volume of *a conductor must be entirely at the same potential in the static case*, and the surface of a conductor is then an equipotential surface. (If it weren't, the free electrons at the surface would move, since whenever there is a potential difference between two points, free charges will move.) This is fully consistent with our result, discussed earlier, that the electric field at the surface of a conductor must be perpendicular to the surface.

A useful analogy for equipotential lines is a topographic map: the contour lines are essentially gravitational equipotential lines (Fig. 23–19).

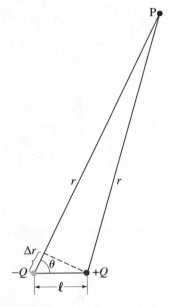

FIGURE 23–19 A topographic map (here, a portion of the Sierra Nevada in California) shows continuous contour lines, each of which is at a fixed height above sea level. Here they are at 80 ft (25 m) intervals. If you walk along one contour line, you neither climb nor descend. If you cross lines, and especially if you climb perpendicular to the lines, you will be changing your gravitational potential (rapidly, if the lines are close together).

23–6 Electric Dipole Potential

Two equal point charges Q, of opposite sign, separated by a distance ℓ, are called an **electric dipole**, as we saw in Section 21–11. Also, the two charges we saw in Figs. 23–12 and 23–18 constitute an electric dipole, and the latter shows the electric field lines and equipotential surfaces for a dipole. Because electric dipoles occur often in physics, as well as in other fields, it is useful to examine them more closely.

The electric potential at an arbitrary point P due to a dipole, Fig. 23–20, is the sum of the potentials due to each of the two charges (we take $V = 0$ at $r = \infty$):

$$V = \frac{1}{4\pi\epsilon_0}\frac{Q}{r} + \frac{1}{4\pi\epsilon_0}\frac{(-Q)}{(r + \Delta r)} = \frac{1}{4\pi\epsilon_0}Q\left(\frac{1}{r} - \frac{1}{r + \Delta r}\right) = \frac{Q}{4\pi\epsilon_0}\frac{\Delta r}{r(r + \Delta r)},$$

where r is the distance from P to the positive charge and $r + \Delta r$ is the distance to the negative charge. This equation becomes simpler if we consider points P whose distance from the dipole is much larger than the separation of the two charges— that is, for $r \gg \ell$. From Fig. 23–20 we see that $\Delta r \approx \ell \cos\theta$; since $r \gg \Delta r = \ell \cos\theta$, we can neglect Δr in the denominator as compared to r. Therefore, we obtain

$$V = \frac{1}{4\pi\epsilon_0}\frac{Q\ell\cos\theta}{r^2} = \frac{1}{4\pi\epsilon_0}\frac{p\cos\theta}{r^2} \qquad \text{[dipole; } r \gg \ell\text{]} \quad \textbf{(23–7)}$$

where $p = Q\ell$ is called the **dipole moment**. We see that the potential decreases as the *square* of the distance from the dipole, whereas for a single point charge the potential decreases with the first power of the distance (Eq. 23–5). It is not surprising that the potential should fall off faster for a dipole; for when you are far from a dipole, the two equal but opposite charges appear so close together as to tend to neutralize each other.

Table 23–2 gives the dipole moments for several molecules. The + and − signs indicate on which atoms these charges lie. The last two entries are a part of many organic molecules and play an important role in molecular biology. A dipole moment has units of coulomb-meters (C·m), although for molecules a smaller unit called a *debye* is sometimes used: 1 debye = 3.33×10^{-30} C·m.

FIGURE 23–20 Electric dipole. Calculation of potential V at point P.

TABLE 23–2 Dipole Moments of Selected Molecules

Molecule	Dipole Moment (C · m)
$H_2^{(+)}O^{(-)}$	6.1×10^{-30}
$H^{(+)}Cl^{(-)}$	3.4×10^{-30}
$N^{(-)}H_3^{(+)}$	5.0×10^{-30}
$>N^{(-)}-H^{(+)}$	$\approx 3.0^{\dagger} \times 10^{-30}$
$>C^{(+)}=O^{(-)}$	$\approx 8.0^{\dagger} \times 10^{-30}$

† These groups often appear on larger molecules; hence the value for the dipole moment will vary somewhat, depending on the rest of the molecule.

23–7 $\vec{E}$ Determined from V

We can use Eq. 23–4a, $V_b - V_a = -\int_a^b \vec{E} \cdot d\vec{\ell}$, to determine the difference in potential between two points if the electric field is known in the region between those two points. By inverting Eq. 23–4a, we can write the electric field in terms of the potential. Then the electric field can be determined from a knowledge of V. Let us see how to do this.

We write Eq. 23-4a in differential form as

$$dV = -\vec{E} \cdot d\vec{\ell} = -E_\ell \, d\ell,$$

where dV is the infinitesimal difference in potential between two points a distance $d\ell$ apart, and E_ℓ is the component of the electric field in the direction of the infinitesimal displacement $d\vec{\ell}$. We can then write

$$E_\ell = -\frac{dV}{d\ell}. \qquad \text{(23-8)}$$

Thus *the component of the electric field in any direction is equal to the negative of the rate of change of the electric potential with distance in that direction.* The quantity $dV/d\ell$ is called the gradient of V in a particular direction. If the direction is not specified, the term *gradient* refers to that direction in which V changes most rapidly; this would be the direction of $\vec{E}$ at that point, so we can write

$$E = -\frac{dV}{d\ell}. \qquad \text{[if } d\vec{\ell} \parallel \vec{E}]$$

If $\vec{E}$ is written as a function of x, y, and z, and we let ℓ refer to the x, y, and z axes, then Eq. 23-8 becomes

$$E_x = -\frac{\partial V}{\partial x}, \qquad E_y = -\frac{\partial V}{\partial y}, \qquad E_z = -\frac{\partial V}{\partial z}. \qquad \text{(23-9)}$$

Here, $\partial V/\partial x$ is the "partial derivative" of V with respect to x, with y and z held constant.[†] For example, if $V(x, y, z) = (2\text{ V/m}^2)x^2 + (8\text{ V/m}^3)y^2 z + (2\text{ V/m}^2)z^2$, then

$$E_x = -\partial V/\partial x = -(4\text{ V/m}^2)x,$$
$$E_y = -\partial V/\partial y = -(16\text{ V/m}^3)yz,$$

and

$$E_z = -\partial V/\partial z = -(8\text{ V/m}^3)y^2 - (4\text{ V/m}^2)z.$$

EXAMPLE 23-11 $\vec{E}$ **for ring and disk.** Use electric potential to determine the electric field at point P on the axis of (*a*) a circular ring of charge (Fig. 23-14) and (*b*) a uniformly charged disk (Fig. 23-15).

APPROACH We obtained V as a function of x in Examples 23-8 and 23-9, so we find E by taking derivatives (Eqs. 23-9).

SOLUTION (*a*) From Example 23-8,

$$V = \frac{1}{4\pi\epsilon_0} \frac{Q}{(x^2 + R^2)^{\frac{1}{2}}}.$$

Then

$$E_x = -\frac{\partial V}{\partial x} = \frac{1}{4\pi\epsilon_0} \frac{Qx}{(x^2 + R^2)^{\frac{3}{2}}}.$$

This is the same result we obtained in Example 21-9.

(*b*) From Example 23-9,

$$V = \frac{Q}{2\pi\epsilon_0 R_0^2}[(x^2 + R_0^2)^{\frac{1}{2}} - x],$$

so

$$E_x = -\frac{\partial V}{\partial x} = \frac{Q}{2\pi\epsilon_0 R_0^2}\left[1 - \frac{x}{(x^2 + R_0^2)^{\frac{1}{2}}}\right].$$

For points very close to the disk, $x \ll R_0$, this can be approximated by

$$E_x \approx \frac{Q}{2\pi\epsilon_0 R_0^2} = \frac{\sigma}{2\epsilon_0}$$

where $\sigma = Q/\pi R_0^2$ is the surface charge density. We also obtained these results in Chapter 21, Example 21-12 and Eq. 21-7.

[†]Equation 23-9 can be written as a vector equation,

$$\vec{E} = -\text{grad } V = -\vec{\nabla} V = -\left(\hat{i}\frac{\partial}{\partial x} + \hat{j}\frac{\partial}{\partial y} + \hat{k}\frac{\partial}{\partial z}\right)V$$

where the symbol $\vec{\nabla}$ is called the *del* or *gradient operator*: $\vec{\nabla} = \hat{i}\frac{\partial}{\partial x} + \hat{j}\frac{\partial}{\partial y} + \hat{k}\frac{\partial}{\partial z}$.

If we compare this last Example with Examples 21–9 and 21–12, we see that here. as for many charge distributions, it is easier to calculate V first, and then $\vec{E}$ from Eq. 23–9, rather than to calculate $\vec{E}$ due to each charge from Coulomb's law. This is because V due to many charges is a scalar sum, whereas $\vec{E}$ is a vector sum.

Electrostatic Potential Energy; the Electron Volt

Suppose a point charge q is moved between two points in space, a and b, where the electric potential due to other charges is V_a and V_b, respectively. The change in electrostatic potential energy of q in the field of these other charges is, according to Eq. 23–2b,

$$\Delta U = U_b - U_a = q(V_b - V_a).$$

Now suppose we have a system of several point charges. What is the electrostatic potential energy of the system? It is most convenient to choose the electric potential energy to be zero when the charges are very far (ideally infinitely far) apart. A single point charge, Q_1, in isolation, has no potential energy, because if there are no other charges around, no electric force can be exerted on it. If a second point charge Q_2 is brought close to Q_1, the potential due to Q_1 at the position of this second charge is

$$V = \frac{1}{4\pi\epsilon_0} \frac{Q_1}{r_{12}},$$

where r_{12} is the distance between the two. The potential energy of the two charges, relative to $V = 0$ at $r = \infty$, is

$$U = Q_2 V = \frac{1}{4\pi\epsilon_0} \frac{Q_1 Q_2}{r_{12}}. \qquad \textbf{(23–10)}$$

This represents the work that needs to be done by an external force to bring Q_2 from infinity $(V = 0)$ to a distance r_{12} from Q_1. It is also the negative of the work needed to separate them to infinity.

If the system consists of three charges, the total potential energy will be the work needed to bring all three together. Equation 23–10 represents the work needed to bring Q_2 close to Q_1; to bring a third charge Q_3 so that it is a distance r_{13} from Q_1 and r_{23} from Q_2 requires work equal to

$$\frac{1}{4\pi\epsilon_0} \frac{Q_1 Q_3}{r_{13}} + \frac{1}{4\pi\epsilon_0} \frac{Q_2 Q_3}{r_{23}}.$$

So the potential energy of a system of three point charges is

$$U = \frac{1}{4\pi\epsilon_0} \left(\frac{Q_1 Q_2}{r_{12}} + \frac{Q_1 Q_3}{r_{13}} + \frac{Q_2 Q_3}{r_{23}} \right). \qquad [V = 0 \text{ at } r = \infty]$$

For a system of four charges, the potential energy would contain six such terms, and so on. (Caution must be used when making such sums to avoid double counting of the different pairs.)

The Electron Volt Unit

The joule is a very large unit for dealing with energies of electrons, atoms, or molecules (see Example 23–2), and for this purpose, the unit **electron volt** (eV) is used. One electron volt is defined as the energy acquired by a particle carrying a charge whose magnitude equals that on the electron $(q = e)$ as a result of moving through a potential difference of 1 V. Since $e = 1.60 \times 10^{-19}$ C, and since the change in potential energy equals qV, 1 eV is equal to $(1.60 \times 10^{-19} \text{ C})(1.00 \text{ V}) = 1.60 \times 10^{-19}$ J:

$$1 \text{ eV} = 1.60 \times 10^{-19} \text{ J}.$$

An electron that accelerates through a potential difference of 1000 V will lose 1000 eV of potential energy and will thus gain 1000 eV or 1 keV (kiloelectron volt) of kinetic energy. On the other hand, if a particle with a charge equal to twice the magnitude of the charge on the electron $(= 2e = 3.20 \times 10^{-19} \text{ C})$ moves through a potential difference of 1000 V, its energy will change by 2000 eV $= 2$ keV.

Although the electron volt is handy for *stating* the energies of molecules and elementary particles, it is not a proper SI unit. For calculations it should be converted to joules using the conversion factor given above. In Example 23–2, for example, the electron acquired a kinetic energy of 8.0×10^{-16} J. We normally would quote this energy as 5000 eV $(= 8.0 \times 10^{-16}$ J$/1.6 \times 10^{-19}$ J/eV$)$. But when determining the speed of a particle in SI units, we must use the kinetic energy in J.

EXERCISE E What is the kinetic energy of a He^{2+} ion released from rest and accelerated through a potential difference of 1.0 kV? (a) 1000 eV, (b) 500 eV, (c) 2000 eV, (d) 4000 eV, (e) 250 eV.

EXAMPLE 23–12 **Disassembling a hydrogen atom.** Calculate the work needed to "disassemble" a hydrogen atom. Assume that the proton and electron are initially separated by a distance equal to the "average" radius of the hydrogen atom in its ground state, 0.529×10^{-10} m, and that they end up an infinite distance apart from each other.

APPROACH The work necessary will be equal to the total energy, kinetic plus potential, of the electron and proton as an atom, compared to their total energy when infinitely far apart.

SOLUTION From Eq. 23–10 we have initially

$$U = \frac{1}{4\pi\epsilon_0} \frac{Q_1 Q_2}{r} = \frac{1}{4\pi\epsilon_0} \frac{(e)(-e)}{r} = \frac{-(8.99 \times 10^9 \, \text{N} \cdot \text{m}^2/\text{C}^2)(1.60 \times 10^{-19} \, \text{C})^2}{(0.529 \times 10^{-10} \, \text{m})}$$
$$= -27.2(1.60 \times 10^{-19}) \, \text{J} = -27.2 \, \text{eV}.$$

This represents the potential energy. The total energy must include also the kinetic energy of the electron moving in an orbit of radius $r = 0.529 \times 10^{-10}$ m. From $F = ma$ for centripetal acceleration, we have

$$\frac{1}{4\pi\epsilon_0}\left(\frac{e^2}{r^2}\right) = \frac{mv^2}{r}.$$

Then

$$K = \tfrac{1}{2}mv^2 = \tfrac{1}{2}\left(\frac{1}{4\pi\epsilon_0}\right)\frac{e^2}{r}$$

which equals $-\tfrac{1}{2}U$ (as calculated above), so $K = +13.6$ eV. The total energy initially is $E = K + U = 13.6 \, \text{eV} - 27.2 \, \text{eV} = -13.6 \, \text{eV}$. To separate a stable hydrogen atom into a proton and an electron at rest very far apart ($U = 0$ at $r = \infty$, $K = 0$ because $v = 0$) requires $+13.6$ eV. This is, in fact, the measured ionization energy for hydrogen.

NOTE To treat atoms properly, we need to use quantum theory (Chapters 37 to 39). But our "classical" calculation does give the correct answer here.

EXERCISE F The kinetic energy of a 1000-kg automobile traveling 20 m/s (70 km/h) would be about (a) 100 GeV, (b) 1000 TeV, (c) 10^6 TeV, (d) 10^{12} TeV, (e) 10^{18} TeV.

FIGURE 23–21 If the cathode inside the evacuated glass tube is heated to glowing, negatively charged "cathode rays" (electrons) are "boiled off" and flow across to the anode (+) to which they are attracted.

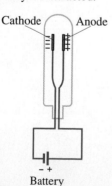

Cathode   Anode

Battery

*23–9 Cathode Ray Tube: TV and Computer Monitors, Oscilloscope

An important device that makes use of voltage, and that allows us to "visualize" how a voltage changes in time, is the *cathode ray tube* (CRT). A CRT used in this way is an *oscilloscope*. The CRT has also been used for many years as the picture tube of television sets and computer monitors, but LCD (Chapter 35) and other screens are now more common.

The operation of a CRT depends on the phenomenon of **thermionic emission** discovered by Thomas Edison (1847–1931). Consider two small plates (electrodes) inside an evacuated "bulb" or "tube" as shown in Fig. 23–21, to which is applied a potential difference. The negative electrode is called the **cathode**, the positive one the **anode**. If the negative cathode is heated (usually by an electric current, as in a lightbulb) so that it becomes hot and glowing, it is found that negative charge leaves the cathode and flows to the positive anode. These negative charges are now called electrons, but originally they were called **cathode rays** since they seemed to come from the cathode (see Section 27–7 on the discovery of the electron).

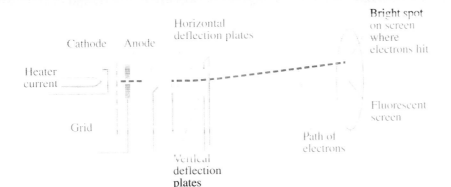

Heater current

Cathode   Anode

Horizontal deflection plates

Grid

Vertical deflection plates

Path of electrons

Bright spot on screen where electrons hit

Fluorescent screen

The **cathode ray tube** (CRT) derives its name from the fact that inside an evacuated glass tube, a beam of cathode rays (electrons) is directed to various parts of a screen to produce a "picture." A simple CRT is diagrammed in Fig. 23–22. Electrons emitted by the heated cathode are accelerated by a high voltage (5000–50,000 V) applied between the anode and cathode. The electrons pass out of this "electron gun" through a small hole in the anode. The inside of the tube face is coated with a fluorescent material that glows when struck by electrons. A tiny bright spot is thus visible where the electron beam strikes the screen. Two horizontal and two vertical plates can deflect the beam of electrons when a voltage is applied to them. The electrons are deflected toward whichever plate is positive. By varying the voltage on the deflection plates, the bright spot can be placed at any point on the screen. Many CRTs use magnetic deflection coils (Chapter 27) instead of electric plates.

In the picture tube or monitor for a computer or television set, the electron beam is made to sweep over the screen in the manner shown in Fig. 23–23 by changing voltages applied to the deflection plates. For standard television in the United States, 525 lines constitutes a complete sweep in $\frac{1}{30}$ s, over the entire screen. High-definition TV provides more than double this number of lines (1080), giving greater picture sharpness. We see a picture because the image is retained by the fluorescent screen and by our eyes for about $\frac{1}{20}$ s. The picture we see consists of the varied brightness of the spots on the screen, controlled by the grid (a "porous" electrode, such as a wire grid, that allows passage of electrons). The grid limits the flow of electrons by means of the voltage (the "video signal") applied to it: the more negative this voltage, the more electrons are repelled and the fewer pass through. This video signal sent out by the TV station, and received by the TV set, is accompanied by signals that synchronize the grid voltage to the horizontal and vertical sweeps. (More in Chapter 31.)

An **oscilloscope** is a device for amplifying, measuring, and visually observing an electrical signal as a function of time on a CRT or LCD screen (a "signal" is usually a time-varying voltage). The electron beam is swept horizontally at a uniform rate in time by the horizontal deflection plates. The signal to be displayed is applied (after amplification) to the vertical deflection plates. The visible "trace" on the screen, which could be an electrocardiogram (Fig. 23–24), or a signal from an experiment on nerve conduction, is a plot of the signal voltage (vertically) versus time (horizontally).

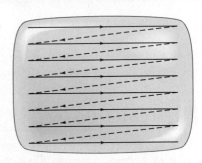

FIGURE 23–23 Electron beam sweeps across a television screen in a succession of horizontal lines. Each horizontal sweep is made by varying the voltage on the horizontal deflection plates. Then the electron beam is moved down a short distance by a change in voltage on the vertical deflection plates, and the process is repeated.

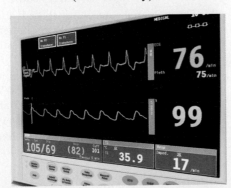

FIGURE 23–24 An electrocardiogram (ECG) trace.

Summary

Electric potential is defined as electric potential energy per unit charge. That is, the **electric potential difference** between any two points in space is defined as the difference in potential energy of a test charge q placed at those two points, divided by the charge q:

$$V_{ba} = \frac{U_b - U_a}{q}. \qquad (23\text{-}2b)$$

Potential difference is measured in volts ($1\,V = 1\,J/C$) and is sometimes referred to as **voltage**.

The change in potential energy of a charge q when it moves through a potential difference V_{ba} is

$$\Delta U = q V_{ba}. \qquad (23\text{-}3)$$

The potential difference V_{ba} between two points, a and b, is given by the relation

$$V_{ba} = V_b - V_a = -\int_a^b \vec{E} \cdot d\vec{\ell}. \qquad (23\text{-}4a)$$

Thus V_{ba} can be found in any region where $\vec{E}$ is known. If the electric field is uniform, the integral is easy: $V_{ba} = -Ed$,

where d is the distance (parallel to the field lines) between the two points.

An **equipotential line** or **surface** is all at the same potential, and is perpendicular to the electric field at all points.

The electric potential due to a single point charge Q, relative to zero potential at infinity, is given by

$$V = \frac{1}{4\pi\epsilon_0}\frac{Q}{r}. \qquad (23\text{-}5)$$

The potential due to any charge distribution can be obtained by summing (or integrating) over the potentials for all the charges.

The potential due to an **electric dipole** drops off as $1/r^2$. The **dipole moment** is $p = Q\ell$, where ℓ is the distance between the two equal but opposite charges of magnitude Q.

When V is known, the components of $\vec{E}$ can be found from the inverse of Eq. 23–4a, namely

$$E_x = -\frac{\partial V}{\partial x}, \qquad E_y = -\frac{\partial V}{\partial y}, \qquad E_z = -\frac{\partial V}{\partial z}. \qquad (23\text{-}9)$$

[*Television and computer monitors traditionally use a **cathode ray tube** (CRT) that accelerates electrons by high voltage, and sweeps them across the screen in a regular way using deflection plates.]

Questions

1. If two points are at the same potential, does this mean that no work is done in moving a test charge from one point to the other? Does this imply that no force must be exerted? Explain.

2. If a negative charge is initially at rest in an electric field, will it move toward a region of higher potential or lower potential? What about a positive charge? How does the potential energy of the charge change in each instance?

3. State clearly the difference (a) between electric potential and electric field, (b) between electric potential and electric potential energy.

4. An electron is accelerated from rest by a potential difference of, say, 0.10 V. How much greater would its final speed be if it is accelerated with four times as much voltage? Explain.

5. Can a particle ever move from a region of low electric potential to one of high potential and yet have its electric potential energy decrease? Explain.

6. If $V = 0$ at a point in space, must $\vec{E} = 0$? If $\vec{E} = 0$ at some point, must $V = 0$ at that point? Explain. Give examples for each.

7. When dealing with practical devices, we often take the ground (the Earth) to be 0 V. (a) If instead we said the ground was $-10\,V$, how would this affect V and E at other points? (b) Does the fact that the Earth carries a net charge affect the choice of V at its surface?

8. Can two equipotential lines cross? Explain.

9. Draw in a few equipotential lines in Fig. 21–34b and c.

10. What can you say about the electric field in a region of space that has the same potential throughout?

11. A satellite orbits the Earth along a gravitational equipotential line. What shape must the orbit be?

12. Suppose the charged ring of Example 23–8 was not uniformly charged, so that the density of charge was twice as great near the top as near the bottom. Assuming the total charge Q is unchanged, would this affect the potential at point P on the axis (Fig. 23–14)? Would it affect the value of $\vec{E}$ at that point? Is there a discrepancy here? Explain.

13. Consider a metal conductor in the shape of a football. If it carries a total charge Q, where would you expect the charge density σ to be greatest, at the ends or along the flatter sides? Explain. [*Hint*: Near the surface of a conductor, $E = \sigma/\epsilon_0$.]

14. If you know V at a point in space, can you calculate $\vec{E}$ at that point? If you know $\vec{E}$ at a point can you calculate V at that point? If not, what else must be known in each case?

15. A conducting sphere carries a charge Q and a second identical conducting sphere is neutral. The two are initially isolated, but then they are placed in contact. (a) What can you say about the potential of each when they are in contact? (b) Will charge flow from one to the other? If so, how much? (c) If the spheres do not have the same radius, how are your answers to parts (a) and (b) altered?

16. At a particular location, the electric field points due north. In what direction(s) will the rate of change of potential be (a) greatest, (b) least, and (c) zero?

17. Equipotential lines are spaced 1.00 V apart. Does the distance between the lines in different regions of space tell you anything about the relative strengths of $\vec{E}$ in those regions? If so, what?

18. If the electric field $\vec{E}$ is uniform in a region, what can you infer about the electric potential V? If V is uniform in a region of space, what can you infer about $\vec{E}$?

19. Is the electric potential energy of two unlike charges positive or negative? What about two like charges? What is the significance of the sign of the potential energy in each case?

Problems

1. (I) What potential difference is needed to stop an electron that has an initial velocity $v = 5.0 \times 10^5 \, \text{m/s}$?

2. (I) How much work does the electric field do in moving a proton from a point with a potential of $+185$ V to a point where it is -55 V?

3. (I) An electron acquires 5.25×10^{-16} J of kinetic energy when it is accelerated by an electric field from plate A to plate B. What is the potential difference between the plates, and which plate is at the higher potential?

4. (II) The work done by an external force to move a $-9.10 \, \mu\text{C}$ charge from point a to point b is 7.00×10^{-4} J. If the charge was started from rest and had 2.10×10^{-4} J of kinetic energy when it reached point b, what must be the potential difference between a and b?

23–2 Potential Related to Electric Field

5. (I) Thunderclouds typically develop voltage differences of about 1×10^8 V. Given that an electric field of 3×10^6 V/m is required to produce an electrical spark within a volume of air, estimate the length of a thundercloud lightning bolt. [Can you see why, when lightning strikes from a cloud to the ground, the bolt actually has to propagate as a sequence of steps?]

6. (I) The electric field between two parallel plates connected to a 45-V battery is 1300 V/m. How far apart are the plates?

7. (I) What is the maximum amount of charge that a spherical conductor of radius 6.5 cm can hold in air?

8. (I) What is the magnitude of the electric field between two parallel plates 4.0 mm apart if the potential difference between them is 110 V?

9. (I) What minimum radius must a large conducting sphere of an electrostatic generating machine have if it is to be at 35,000 V without discharge into the air? How much charge will it carry?

10. (II) A manufacturer claims that a carpet will not generate more than 5.0 kV of static electricity. What magnitude of charge would have to be transferred between a carpet and a shoe for there to be a 5.0-kV potential difference between the shoe and the carpet? Approximate the shoe and the carpet as large sheets of charge separated by a distance $d = 1.0$ mm.

11. (II) A uniform electric field $\vec{E} = -4.20 \, \text{N/C} \, \hat{i}$ points in the negative x direction as shown in Fig. 23–25. The x and y coordinates of points A, B, and C are given on the diagram (in meters). Determine the differences in potential (a) V_{BA}, (b) V_{CB}, and (c) V_{CA}.

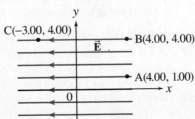

FIGURE 23–25
Problem 11.

12. (II) The electric potential of a very large isolated flat metal plate is V_0. It carries a uniform distribution of charge of surface density σ (C/m^2), or $\sigma/2$ on each surface. Determine V at a distance x from the plate. Consider the point x to be far from the edges and assume x is much smaller than the plate dimensions.

13. (II) The Earth produces an inwardly directed electric field of magnitude 150 V/m near its surface. (a) What is the potential of the Earth's surface relative to $V = 0$ at $r = \infty$? (b) If the potential of the Earth is chosen to be zero, what is the potential at infinity? (Ignore the fact that positive charge in the ionosphere approximately cancels the Earth's net charge; how would this affect your answer?)

14. (II) A 32-cm-diameter conducting sphere is charged to 680 V relative to $V = 0$ at $r = \infty$. (a) What is the surface charge density σ? (b) At what distance will the potential due to the sphere be only 25 V?

15. (II) An insulated spherical conductor of radius r_1 carries a charge Q. A second conducting sphere of radius r_2 and initially uncharged is then connected to the first by a long conducting wire. (a) After the connection, what can you say about the electric potential of each sphere? (b) How much charge is transferred to the second sphere? Assume the connected spheres are far apart compared to their radii. (Why make this assumption?)

16. (II) Determine the difference in potential between two points that are distances R_a and R_b from a very long ($\gg R_a$ or R_b) straight wire carrying a uniform charge per unit length λ.

17. (II) Suppose the end of your finger is charged. (a) Estimate the breakdown voltage in air for your finger. (b) About what surface charge density would have to be on your finger at this voltage?

18. (II) Estimate the electric field in the membrane wall of a living cell. Assume the wall is 10 nm thick and has a potential of 0.10 V across it.

19. (II) A nonconducting sphere of radius r_0 carries a total charge Q distributed uniformly throughout its volume. Determine the electric potential as a function of the distance r from the center of the sphere for (a) $r > r_0$ and (b) $r < r_0$. Take $V = 0$ at $r = \infty$. (c) Plot V versus r and E versus r.

20. (III) Repeat Problem 19 assuming the charge density ρ_E increases as the square of the distance from the center of the sphere, and $\rho_E = 0$ at the center.

21. (III) The volume charge density ρ_E within a sphere of radius r_0 is distributed in accordance with the following spherically symmetric relation

$$\rho_E(r) = \rho_0 \left[1 - \frac{r^2}{r_0^2} \right]$$

where r is measured from the center of the sphere and ρ_0 is a constant. For a point P inside the sphere ($r < r_0$), determine the electric potential V. Let $V = 0$ at infinity.

22. (III) A hollow spherical conductor, carrying a net charge $+Q$, has inner radius r_1 and outer radius $r_2 = 2r_1$ (Fig. 23–26). At the center of the sphere is a point charge $+Q/2$. (a) Write the electric field strength E in all three regions as a function of r. Then determine the potential as a function of r, the distance from the center, for (b) $r > r_2$, (c) $r_1 < r < r_2$, and (d) $0 < r < r_1$. (e) Plot both V and E as a function of r from $r = 0$ to $r = 2r_2$.

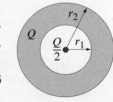

FIGURE 23–26
Problem 22.

23. (III) A very long conducting cylinder (length ℓ) of radius R_0 ($R_0 \ll \ell$) carries a uniform surface charge density σ (C/m²). The cylinder is at an electric potential V_0. What is the potential, at points far from the end, at a distance R from the center of the cylinder? Determine for (a) $R > R_0$ and (b) $R < R_0$. (c) Is $V = 0$ at $R = \infty$ (assume $\ell = \infty$)? Explain.

23–3 Potential Due to Point Charges

24. (I) A point charge Q creates an electric potential of $+185$ V at a distance of 15 cm. What is Q (let $V = 0$ at $r = \infty$) ?

25. (I) (a) What is the electric potential 0.50×10^{-10} m from a proton (charge $+e$)? Let $V = 0$ at $r = \infty$. (b) What is the potential energy of an electron at this point?

26. (II) Two point charges, $3.4\,\mu$C and $-2.0\,\mu$C, are placed 5.0 cm apart on the x axis. At what points along the x axis is (a) the electric field zero and (b) the potential zero? Let $V = 0$ at $r = \infty$.

27. (II) A $+25\,\mu$C point charge is placed 6.0 cm from an identical $+25\,\mu$C point charge. How much work would be required by an external force to move a $+0.18\,\mu$C test charge from a point midway between them to a point 1.0 cm closer to either of the charges?

28. (II) Point a is 26 cm north of a $-3.8\,\mu$C point charge, and point b is 36 cm west of the charge (Fig. 23–27). Determine (a) $V_b - V_a$, and (b) $\vec{\mathbf{E}}_b - \vec{\mathbf{E}}_a$ (magnitude and direction).

FIGURE 23–27
Problem 28.

26 cm
36 cm
$Q = -3.8\,\mu$C

29. (II) How much voltage must be used to accelerate a proton (radius 1.2×10^{-15} m) so that it has sufficient energy to just "touch" a silicon nucleus? A silicon nucleus has a charge of $+14e$ and its radius is about 3.6×10^{-15} m. Assume the potential is that for point charges.

30. (II) Two identical $+5.5\,\mu$C point charges are initially spaced 6.5 cm from each other. If they are released at the same instant from rest, how fast will they be moving when they are very far away from each other? Assume they have identical masses of 1.0 mg.

31. (II) An electron starts from rest 42.5 cm from a fixed point charge with $Q = -0.125$ nC. How fast will the electron be moving when it is very far away?

32. (II) Two equal but opposite charges are separated by a distance d, as shown in Fig. 23–28. Determine a formula for $V_{BA} = V_B - V_A$ for points B and A on the line between the charges situated as shown.

FIGURE 23–28
Problem 32.

$+q$ A   B $-q$

23–4 Potential Due to Charge Distribution

33. (II) A thin circular ring of radius R (as in Fig. 23–14) has charge $+Q/2$ uniformly distributed on the top half, and $-Q/2$ on the bottom half. (a) What is the value of the electric potential at a point a distance x along the axis through the center of the circle? (b) What can you say about the electric field $\vec{\mathbf{E}}$ at a distance x along the axis? Let $V = 0$ at $r = \infty$.

34. (II) Three point charges are arranged at the corners of a square of side ℓ as shown in Fig. 23–29. What is the potential at the fourth corner (point A), taking $V = 0$ at a great distance?

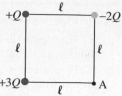

FIGURE 23–29
Problem 34.

35. (II) A flat ring of inner radius R_1 and outer radius R_2, Fig. 23–30, carries a uniform surface charge density σ. Determine the electric potential at points along the axis (the x axis). [*Hint*: Try substituting variables.]

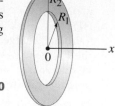

FIGURE 23–30
Problem 35.

36. (II) A total charge Q is uniformly distributed on a thread of length ℓ. The thread forms a semicircle. What is the potential at the center? (Assume $V = 0$ at large distances.)

37. (II) A 12.0-cm-radius thin ring carries a uniformly distributed $15.0\,\mu$C charge. A small 7.5-g sphere with a charge of $3.0\,\mu$C is placed exactly at the center of the ring and given a very small push so it moves along the ring axis ($+x$ axis). How fast will the sphere be moving when it is 2.0 m from the center of the ring (ignore gravity)?

38. (II) A thin rod of length 2ℓ is centered on the x axis as shown in Fig. 23–31. The rod carries a uniformly distributed charge Q. Determine the potential V as a function of y for points along the y axis. Let $V = 0$ at infinity.

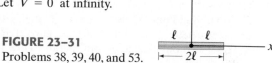

FIGURE 23–31
Problems 38, 39, 40, and 53.

39. (II) Determine the potential $V(x)$ for points along the x axis outside the rod of Fig. 23–31 (Problem 38).

40. (III) The charge on the rod of Fig. 23–31 has a nonuniform linear charge distribution, $\lambda = ax$. Determine the potential V for (a) points along the y axis and (b) points along the x axis outside the rod.

41. (III) Suppose the flat circular disk of Fig. 23–15 (Example 23–9) has a nonuniform surface charge density $\sigma = aR^2$, where R is measured from the center of the disk. Find the potential $V(x)$ at points along the x axis, relative to $V = 0$ at $x = \infty$.

23–5 Equipotentials

42. (I) Draw a conductor in the shape of a football. This conductor carries a net negative charge, $-Q$. Draw in a dozen or so electric field lines and equipotential lines.

43. (II) Equipotential surfaces are to be drawn 100 V apart near a very large uniformly charged metal plate carrying a surface charge density $\sigma = 0.75\,\mu$C/m². How far apart (in space) are the equipotential surfaces?

44. (II) A metal sphere of radius $r_0 = 0.44$ m carries a charge $Q = 0.50\,\mu$C. Equipotential surfaces are to be drawn for 100-V intervals outside the sphere. Determine the radius r of (a) the first, (b) the tenth, and (c) the 100th equipotential from the surface.

45. (II) Calculate the electric potential due to a tiny dipole whose dipole moment is 4.8×10^{-30} C·m at a point 4.1×10^{-9} m away if this point is (a) along the axis of the dipole nearer the positive charge; (b) 45° above the axis but nearer the positive charge; (c) 45° above the axis but nearer the negative charge. Let $V = 0$ at $r = \infty$.

46. (III) The dipole moment, considered as a vector, points from the negative to the positive charge. The water molecule, Fig. 23–32, has a dipole moment $\vec{\mathbf{p}}$ which can be considered as the vector sum of the two dipole moments $\vec{\mathbf{p}}_1$ and $\vec{\mathbf{p}}_2$ as shown. The distance between each H and the O is about 0.96×10^{-10} m; the lines joining the center of the O atom with each H atom make an angle of 104° as shown, and the net dipole moment has been measured to be $p = 6.1 \times 10^{-30}$ C·m. (a) Determine the effective charge q on each H atom. (b) Determine the electric potential, far from the molecule, due to each dipole, $\vec{\mathbf{p}}_1$ and $\vec{\mathbf{p}}_2$, and show that

$$V = \frac{1}{4\pi\epsilon_0} \frac{p \cos\theta}{r^2},$$

where p is the magnitude of the net dipole moment, $\vec{\mathbf{p}} = \vec{\mathbf{p}}_1 + \vec{\mathbf{p}}_2$, and V is the total potential due to both $\vec{\mathbf{p}}_1$ and $\vec{\mathbf{p}}_2$. Take $V = 0$ at $r = \infty$.

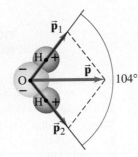

FIGURE 23–32
Problem 46.

23–7 $\vec{\mathbf{E}}$ Determined from V

47. (I) Show that the electric field of a single point charge (Eq. 21–4) follows from Eq. 23–5, $V = (1/4\pi\epsilon_0)(Q/r)$.

48. (I) What is the potential gradient just outside the surface of a uranium nucleus $(Q = +92e)$ whose diameter is about 15×10^{-15} m?

49. (II) The electric potential between two parallel plates is given by $V(x) = (8.0 \text{ V/m})x + 5.0 \text{ V}$, with $x = 0$ taken at one of the plates and x positive in the direction toward the other plate. What is the charge density on the plates?

50. (II) The electric potential in a region of space varies as $V = by/(a^2 + y^2)$. Determine $\vec{\mathbf{E}}$.

51. (II) In a certain region of space, the electric potential is given by $V = y^2 + 2.5xy - 3.5xyz$. Determine the electric field vector, $\vec{\mathbf{E}}$, in this region.

52. (II) A dust particle with mass of 0.050 g and a charge of 2.0×10^{-6} C is in a region of space where the potential is given by $V(x) = (2.0 \text{ V/m}^2)x^2 - (3.0 \text{ V/m}^3)x^3$. If the particle starts at $x = 2.0$ m, what is the initial acceleration of the charge?

53. (III) Use the results of Problems 38 and 39 to determine the electric field due to the uniformly charged rod of Fig. 23–31 for points (a) along the y axis and (b) along the x axis.

23–8 Electrostatic Potential Energy; Electron Volt

54. (I) How much work must be done to bring three electrons from a great distance apart to within 1.0×10^{-10} m from one another (at the corners of an equilateral triangle)?

55. (I) What potential difference is needed to give a helium nucleus $(Q = 3.2 \times 10^{-19}$ C) 125 keV of kinetic energy?

56. (I) What is the speed of (a) a 1.5-keV (kinetic energy) electron and (b) a 1.5-keV proton?

57. (II) Many chemical reactions release energy. Suppose that at the beginning of a reaction, an electron and proton are separated by 0.110 nm, and their final separation is 0.100 nm. How much electric potential energy was lost in this reaction (in units of eV)?

58. (II) An alpha particle (which is a helium nucleus, $Q = +2e$, $m = 6.64 \times 10^{-27}$ kg) is emitted in a radioactive decay with kinetic energy 5.53 MeV. What is its speed?

59. (II) Write the total electrostatic potential energy, U, for (a) four point charges and (b) five point charges. Draw a diagram defining all quantities.

60. (II) Four equal point charges, Q, are fixed at the corners of a square of side b. (a) What is their total electrostatic potential energy? (b) How much potential energy will a fifth charge, Q, have at the center of the square (relative to $V = 0$ at $r = \infty$)? (c) If constrained to remain in that plane, is the fifth charge in stable or unstable equilibrium? (d) If a negative $(-Q)$ charge is at the center, is it in stable equilibrium?

61. (II) An electron starting from rest acquires 1.33 keV of kinetic energy in moving from point A to point B. (a) How much kinetic energy would a proton acquire, starting from rest at B and moving to point A? (b) Determine the ratio of their speeds at the end of their respective trajectories.

62. (II) Determine the total electrostatic potential energy of a conducting sphere of radius r_0 that carries a total charge Q distributed uniformly on its surface.

63. (II) The **liquid-drop model** of the nucleus suggests that high-energy oscillations of certain nuclei can split ("fission") a large nucleus into two unequal fragments plus a few neutrons. Using this model, consider the case of a uranium nucleus fissioning into two spherical fragments, one with a charge $q_1 = +38e$ and radius $r_1 = 5.5 \times 10^{-15}$ m, the other with $q_2 = +54e$ and $r_2 = 6.2 \times 10^{-15}$ m. Calculate the electric potential energy (MeV) of these fragments, assuming that the charge is uniformly distributed throughout the volume of each spherical nucleus and that their surfaces are initially in contact at rest. The electrons surrounding the nuclei can be neglected. This electric potential energy will then be entirely converted to kinetic energy as the fragments repel each other. How does your predicted kinetic energy of the fragments agree with the observed value associated with uranium fission (approximately 200 MeV total)? [$1 \text{ MeV} = 10^6 \text{ eV}$.]

64. (III) Determine the total electrostatic potential energy of a nonconducting sphere of radius r_0 carrying a total charge Q distributed uniformly throughout its volume.

*23–9 CRT

*65. (I) Use the ideal gas as a model to estimate the rms speed of a free electron in a metal at 273 K, and at 2700 K (a typical temperature of the cathode in a CRT).

*66. (III) Electrons are accelerated by 6.0 kV in a CRT. The screen is 28 cm wide and is 34 cm from the 2.6-cm-long deflection plates. Over what range must the horizontally deflecting electric field vary to sweep the beam fully across the screen?

*67. (III) In a given CRT, electrons are accelerated horizontally by 7.2 kV. They then pass through a uniform electric field E for a distance of 2.8 cm which deflects them upward so they reach the top of the screen 22 cm away, 11 cm above the center. Estimate the value of E.

General Problems

68. If the electrons in a single raindrop, 3.5 mm in diameter, could be removed from the Earth (without removing the atomic nuclei), by how much would the potential of the Earth increase?

69. By rubbing a nonconducting material, a charge of 10^{-8} C can readily be produced. If this is done to a sphere of radius 15 cm, estimate the potential produced at the surface. Let $V = 0$ at $r = \infty$.

70. Sketch the electric field and equipotential lines for two charges of the same sign and magnitude separated by a distance d.

71. A $+33\,\mu$C point charge is placed 36 cm from an identical $+33\,\mu$C charge. A $-1.5\,\mu$C charge is moved from point a to point b, Fig. 23–33. What is the change in potential energy?

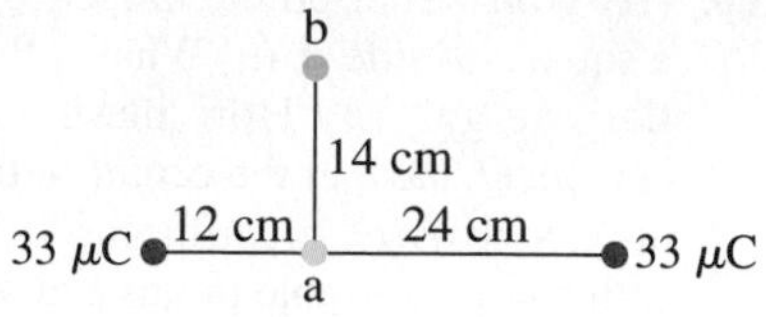

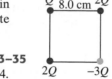

FIGURE 23–33
Problem 71.

72. At each corner of a cube of side ℓ there is a point charge Q, Fig. 23–34. (a) What is the potential at the center of the cube ($V = 0$ at $r = \infty$)? (b) What is the potential at each corner due to the other seven charges? (c) What is the total potential energy of this system?

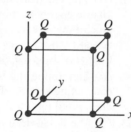

FIGURE 23–34
Problem 72.

73. In a television picture tube (CRT), electrons are accelerated by thousands of volts through a vacuum. If a television set is laid on its back, would electrons be able to move upward against the force of gravity? What potential difference, acting over a distance of 3.5 cm, would be needed to balance the downward force of gravity so that an electron would remain stationary? Assume that the electric field is uniform.

74. Four point charges are located at the corners of a square that is 8.0 cm on a side. The charges, going in rotation around the square, are Q, $2Q$, $-3Q$, and $2Q$, where $Q = 3.1\,\mu$C (Fig. 23–35). What is the total electric potential energy stored in the system, relative to $U = 0$ at infinite separation?

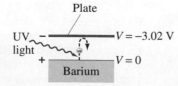

FIGURE 23–35
Problem 74.

75. In a **photocell**, ultraviolet (UV) light provides enough energy to some electrons in barium metal to eject them from the surface at high speed. See Fig. 23–36. To measure the maximum energy of the electrons, another plate above the barium surface is kept at a negative enough potential that the emitted electrons are slowed down and stopped, and return to the barium surface. If the plate voltage is -3.02 V (compared to the barium) when the fastest electrons are stopped, what was the speed of these electrons when they were emitted?

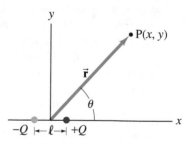

FIGURE 23–36
Problem 75.

76. An electron is accelerated horizontally from rest in a television picture tube by a potential difference of 5500 V. It then passes between two horizontal plates 6.5 cm long and 1.3 cm apart that have a potential difference of 250 V (Fig. 23–37). At what angle θ will the electron be traveling after it passes between the plates?

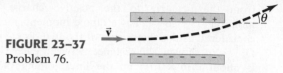

FIGURE 23–37
Problem 76.

77. Three charges are at the corners of an equilateral triangle (side ℓ) as shown in Fig. 23–38. Determine the potential at the midpoint of each of the sides. Let $V = 0$ at $r = \infty$.

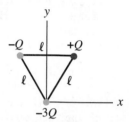

FIGURE 23–38
Problem 77.

78. Near the surface of the Earth there is an electric field of about 150 V/m which points downward. Two identical balls with mass $m = 0.340$ kg are dropped from a height of 2.00 m, but one of the balls is positively charged with $q_1 = 450\,\mu$C, and the second is negatively charged with $q_2 = -450\,\mu$C. Use conservation of energy to determine the difference in the speeds of the two balls when they hit the ground. (Neglect air resistance.)

79. A lightning flash transfers 4.0 C of charge and 4.8 MJ of energy to the Earth. (a) Between what potential difference did it travel? (b) How much water could this energy boil, starting from room temperature? [Hint: See Chapter 19.]

80. Determine the components of the electric field, E_x and E_y, as a function of x and y in the xy plane due to a dipole, Fig. 23–39, starting with Eq. 23–7. Assume $r = (x^2 + y^2)^{\frac{1}{2}} \gg \ell$.

FIGURE 23–39
Problem 80.

81. A nonconducting sphere of radius r_2 contains a concentric spherical cavity of radius r_1. The material between r_1 and r_2 carries a uniform charge density ρ_E (C/m³). Determine the electric potential V, relative to $V = 0$ at $r = \infty$, as a function of the distance r from the center for (a) $r > r_2$, (b) $r_1 < r < r_2$, and (c) $0 < r < r_1$. Is V continuous at r_1 and r_2?

82. A thin flat nonconducting disk, with radius R_0 and charge Q, has a hole with a radius $R_0/2$ in its center. Find the electric potential $V(x)$ at points along the symmetry (x) axis of the disk (a line perpendicular to the disk, passing through its center). Let $V = 0$ at $x = \infty$.

83. A **Geiger counter** is used to detect charged particles emitted by radioactive nuclei. It consists of a thin, positively charged central wire of radius R_a surrounded by a concentric conducting cylinder of radius R_b with an equal negative charge (Fig. 23–40). The charge per unit length on the inner wire is λ (units C/m). The interior space between wire and cylinder is filled with low-pressure inert gas. Charged particles ionize some of these gas atoms; the resulting free electrons are attracted toward the positive central wire. If the radial electric field is strong enough, the freed electrons gain enough energy to ionize other atoms, causing an "avalanche" of electrons to strike the central wire, generating an electric "signal." Find the expression for the electric field between the wire and the cylinder, and show that the potential difference between R_a and R_b is

$$V_a - V_b = \left(\frac{\lambda}{2\pi\epsilon_0}\right) \ln\left(\frac{R_b}{R_a}\right).$$

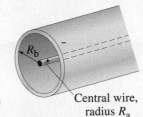

FIGURE 23–40
Problem 83.

Central wire, radius R_a

84. A **Van de Graaff generator** (Fig. 23–41) can develop a very large potential difference, even millions of volts. Electrons are pulled off the belt by the high voltage pointed electrode at A, leaving the belt positively charged. (Recall Example 23–5 where we saw that near sharp points the electric field is high and ionization can occur.) The belt carries the positive charge up inside the spherical shell where electrons from the large conducting sphere are attracted over to the pointed conductor at B, leaving the outer surface of the conducting sphere positively charged. As more charge is brought up, the sphere reaches extremely high voltage. Consider a Van de Graaff generator with a sphere of radius 0.20 m. (a) What is the electric potential on the surface of the sphere when electrical breakdown occurs? (Assume $V = 0$ at $r = \infty$.) (b) What is the charge on the sphere for the potential found in part (a)?

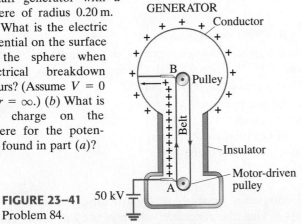

FIGURE 23–41
Problem 84.

85. The potential in a region of space is given by $V = B/(x^2 + R^2)^2$ where $B = 150\ \mathrm{V \cdot m^4}$ and $R = 0.20\ \mathrm{m}$. (a) Find V at $x = 0.20\ \mathrm{m}$. (b) Find $\vec{\mathbf{E}}$ as a function of x. (c) Find $\vec{\mathbf{E}}$ at $x = 0.20\ \mathrm{m}$.

86. A charge $-q_1$ of mass m rests on the y axis at a distance b above the x axis. Two positive charges of magnitude $+q_2$ are fixed on the x axis at $x = +a$ and $x = -a$, respectively (Fig. 23–42). If the $-q_1$ charge is given an initial velocity v_0 in the positive y direction, what is the minimum value of v_0 such that the charge escapes to a point infinitely far away from the two positive charges?

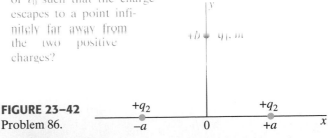

FIGURE 23–42
Problem 86.

Numerical/Computer

*87. (II) A dipole is composed of a $-1.0\ \mathrm{nC}$ charge at $x = -1.0\ \mathrm{cm}$ and a $+1.0\ \mathrm{nC}$ charge at $x = +1.0\ \mathrm{cm}$. (a) Make a plot of V along the x axis from $x = 2.0\ \mathrm{cm}$ to $x = 15\ \mathrm{cm}$. (b) On the same graph, plot the approximate V using Eq. 23–7 from $x = 2.0\ \mathrm{cm}$ to $x = 15\ \mathrm{cm}$. Let $V = 0$ at $x = \infty$.

*88. (II) A thin flat disk of radius R_0 carries a total charge Q that is distributed uniformly over its surface. The electric potential at a distance x on the x axis is given by

$$V(x) = \frac{Q}{2\pi\epsilon_0 R_0^2}\left[(x^2 + R_0^2)^{\frac{1}{2}} - x\right].$$

(See Example 23–9.) Show that the electric field at a distance x on the x axis is given by

$$E(x) = \frac{Q}{2\pi\epsilon_0 R_0^2}\left(1 - \frac{x}{(x^2 + R_0^2)^{\frac{1}{2}}}\right).$$

Make graphs of $V(x)$ and $E(x)$ as a function of x/R_0 for $x/R_0 = 0$ to 4. (Do the calculations in steps of 0.1.) Use $Q = 5.0\ \mu\mathrm{C}$ and $R_0 = 10\ \mathrm{cm}$ for the calculation and graphs.

*89. (III) You are trying to determine an unknown amount of charge using only a voltmeter and a ruler, knowing that it is either a single sheet of charge or a point charge that is creating it. You determine the direction of greatest change of potential, and then measure potentials along a line in that direction. The potential versus position (note that the zero of position is arbitrary, and the potential is measured relative to ground) is measured as follows:

x (cm)	0.0	1.0	2.0	3.0	4.0	5.0	6.0	7.0	8.0	9.0
V (volts)	3.9	3.0	2.5	2.0	1.7	1.5	1.4	1.4	1.2	1.1

(a) Graph V versus position. Do you think the field is caused by a sheet or a point charge? (b) Graph the data in such a way that you can determine the magnitude of the charge and determine that value. (c) Is it possible to determine where the charge is from this data? If so, give the position of the charge.

Answers to Exercises

A: (b).

B: (d).

C: (d).

D: (a) iii, (b) i, (c) i.

E: (c).

F: (d).

Capacitors come in a wide range of sizes and shapes, only a few of which are shown here. A capacitor is basically two conductors that do not touch, and which therefore can store charge of opposite sign on its two conductors. Capacitors are used in a wide variety of circuits, as we shall see in this and later Chapters.

24

Capacitance, Dielectrics, Electric Energy Storage

CHAPTER-OPENING QUESTION—Guess now!

A fixed potential difference V exists between a pair of close parallel plates carrying opposite charges $+Q$ and $-Q$. Which of the following would not increase the magnitude of charge that you could put on the plates?

(a) Increase the size of the plates.

(b) Move the plates farther apart.

(c) Fill the space between the plates with paper.

(d) Increase the fixed potential difference V.

(e) None of the above.

This Chapter will complete our study of electrostatics. It deals first of all with an important device, the capacitor, which is used in many electronic circuits. We will also discuss electric energy storage and the effects of an insulator, or dielectric, on electric fields and potential differences.

24–1 Capacitors

PHYSICS APPLIED

Uses of capacitors

A **capacitor** is a device that can store electric charge, and normally consists of two conducting objects (usually plates or sheets) placed near each other but not touching. Capacitors are widely used in electronic circuits. They store charge for later use, such as in a camera flash, and as energy backup in computers if the power fails. Capacitors also block surges of charge and energy to protect circuits.

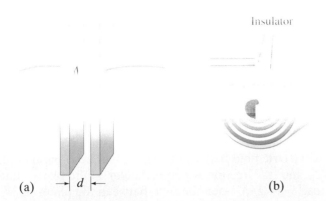

Insulator

(a) $\rightarrow | d | \leftarrow$ (b)

FIGURE 24-1 Capacitors; diagrams of (a) parallel plate, (b) cylindrical (rolled up parallel plate).

Very tiny capacitors serve as memory for the "ones" and "zeros" of the binary code in the random access memory (RAM) of computers. Capacitors serve many other applications, some of which we will discuss.

A simple capacitor consists of a pair of parallel plates of area A separated by a small distance d (Fig. 24-1a). Often the two plates are rolled into the form of a cylinder with plastic, paper, or other insulator separating the plates, Fig. 24-1b. In a diagram, the symbol

$$\dashv\vdash \quad \text{or} \quad \dashv\vdash \qquad \text{[capacitor symbol]}$$

represents a capacitor. A battery, which is a source of voltage, is indicated by the symbol:

$$\dashv\!\!\vdash \qquad \text{[battery symbol]}$$

with unequal arms.

If a voltage is applied across a capacitor by connecting the capacitor to a battery with conducting wires as in Fig. 24-2, the two plates quickly become charged: one plate acquires a negative charge, the other an equal amount of positive charge. Each battery terminal and the plate of the capacitor connected to it are at the same potential; hence the full battery voltage appears across the capacitor. For a given capacitor, it is found that the amount of charge Q acquired by each plate is proportional to the magnitude of the potential difference V between them:

$$Q = CV. \qquad \textbf{(24-1)}$$

The constant of proportionality, C, in the above relation is called the **capacitance** of the capacitor. The unit of capacitance is coulombs per volt and this unit is called a **farad** (F). Common capacitors have capacitance in the range of $1\,\text{pF}$ (picofarad $= 10^{-12}\,\text{F}$) to $10^3\,\mu\text{F}$ (microfarad $= 10^{-6}\,\text{F}$). The relation, Eq. 24-1, was first suggested by Volta in the late eighteenth century. The capacitance C does not in general depend on Q or V. Its value depends only on the size, shape, and relative position of the two conductors, and also on the material that separates them.

In Eq. 24-1, and from now on, we use simply V (in italics) to represent a potential difference, rather than V_{ba}, ΔV, or $V_b - V_a$, as previously. (Be sure not to confuse italic V and C which stand for voltage and capacitance, with non-italic V and C which stand for the units volts and coulombs).

| **EXERCISE A** Graphs for charge versus voltage are shown in Fig. 24-3 for three capacitors, A, B, and C. Which has the greatest capacitance?

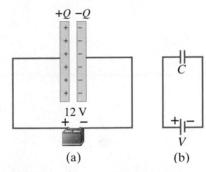
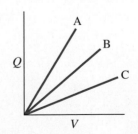

$+Q$ $-Q$

12 V

C

V

(a) (b)

FIGURE 24-2 (a) Parallel-plate capacitor connected to a battery. (b) Same circuit shown using symbols.

FIGURE 24-3 Exercise A.

Q

A

B

C

V

⚠ **CAUTION**

V = potential difference from here on

24–2 Determination of Capacitance

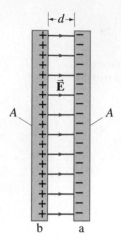

FIGURE 24–4 Parallel-plate capacitor, each of whose plates has area A. Fringing of the field is ignored.

The capacitance of a given capacitor can be determined experimentally directly from Eq. 24–1, by measuring the charge Q on either conductor for a given potential difference V.

For capacitors whose geometry is simple, we can determine C analytically, and in this Section we assume the conductors are separated by a vacuum or air. First, we determine C for a parallel-plate capacitor, Fig. 24–4. Each plate has area A and the two plates are separated by a distance d. We assume d is small compared to the dimensions of each plate so that the electric field $\vec{E}$ is uniform between them and we can ignore fringing (lines of $\vec{E}$ not straight) at the edges. We saw earlier (Example 21–13) that the electric field between two closely spaced parallel plates has magnitude $E = \sigma/\epsilon_0$ and its direction is perpendicular to the plates. Since σ is the charge per unit area, $\sigma = Q/A$, then the field between the plates is

$$E = \frac{Q}{\epsilon_0 A}.$$

The relation between electric field and electric potential, as given by Eq. 23–4a, is

$$V = V_{ba} = V_b - V_a = -\int_a^b \vec{E} \cdot d\vec{\ell}.$$

We can take the line integral along a path antiparallel to the field lines, from plate a to plate b; then $\theta = 180°$ and $\cos 180° = -1$, so

$$V = V_b - V_a = -\int_a^b E \, d\ell \cos 180° = +\int_a^b E \, d\ell = \frac{Q}{\epsilon_0 A} \int_a^b d\ell = \frac{Qd}{\epsilon_0 A}.$$

This relates Q to V, and from it we can get the capacitance C in terms of the geometry of the plates:

$$C = \frac{Q}{V} = \epsilon_0 \frac{A}{d}. \qquad \text{[parallel-plate capacitor]} \qquad \textbf{(24–2)}$$

Note from Eq. 24–2 that the value of C does not depend on Q or V, so Q is predicted to be proportional to V as is found experimentally.

EXAMPLE 24–1 **Capacitor calculations.** (*a*) Calculate the capacitance of a parallel-plate capacitor whose plates are 20 cm × 3.0 cm and are separated by a 1.0-mm air gap. (*b*) What is the charge on each plate if a 12-V battery is connected across the two plates? (*c*) What is the electric field between the plates? (*d*) Estimate the area of the plates needed to achieve a capacitance of 1 F, given the same air gap d.

APPROACH The capacitance is found by using Eq. 24–2, $C = \epsilon_0 A/d$. The charge on each plate is obtained from the definition of capacitance, Eq. 24–1, $Q = CV$. The electric field is uniform, so we can use Eq. 23–4b for the magnitude $E = V/d$. In (*d*) we use Eq. 24–2 again.

SOLUTION (*a*) The area $A = (20 \times 10^{-2}\,\text{m})(3.0 \times 10^{-2}\,\text{m}) = 6.0 \times 10^{-3}\,\text{m}^2$. The capacitance C is then

$$C = \epsilon_0 \frac{A}{d} = (8.85 \times 10^{-12}\,\text{C}^2/\text{N} \cdot \text{m}^2)\frac{6.0 \times 10^{-3}\,\text{m}^2}{1.0 \times 10^{-3}\,\text{m}} = 53\,\text{pF}.$$

(*b*) The charge on each plate is

$$Q = CV = (53 \times 10^{-12}\,\text{F})(12\,\text{V}) = 6.4 \times 10^{-10}\,\text{C}.$$

(*c*) From Eq. 23–4b for a uniform electric field, the magnitude of E is

$$E = \frac{V}{d} = \frac{12\,\text{V}}{1.0 \times 10^{-3}\,\text{m}} = 1.2 \times 10^4\,\text{V/m}.$$

(*d*) We solve for A in Eq. 24–2 and substitute $C = 1.0\,\text{F}$ and $d = 1.0\,\text{mm}$ to find that we need plates with an area

$$A = \frac{Cd}{\epsilon_0} \approx \frac{(1\,\text{F})(1.0 \times 10^{-3}\,\text{m})}{(9 \times 10^{-12}\,\text{C}^2/\text{N} \cdot \text{m}^2)} \approx 10^8\,\text{m}^2.$$

NOTE This is the area of a square 10^4 m or 10 km on a side. That is the size of a city like San Francisco or Boston! Large-capacitance capacitors will not be simple parallel plates.

EXERCISE B Two circular plates of radius 5.0 cm are separated by a 0.10-mm air gap. What is the magnitude of the charge on each plate when connected to a 12-V battery?

Not long ago, a capacitance greater than a few mF was unusual. Today capacitors are available that are 1 or 2 F, yet they are just a few cm on a side. Such capacitors are used as power backups, for example, in computer memory and electronics where the time and date can be maintained through tiny charge flow. [Capacitors are superior to rechargable batteries for this purpose because they can be recharged more than 10^5 times with no degradation.] Such high-capacitance capacitors can be made of "activated" carbon which has very high porosity so that the surface area is very large; one tenth of a gram of activated carbon can have a surface area of $100 \, m^2$. Furthermore, the equal and opposite charges exist in an electric "double layer" about $10^{-9} \, m$ thick. Thus, the capacitance of 0.1 g of activated carbon, whose internal area can be $10^2 \, m^2$, is equivalent to a parallel-plate capacitor with $C \approx \epsilon_0 A/d = (8.85 \times 10^{-12} \, C^2/N \cdot m^2)(10^2 \, m^2)/(10^{-9} \, m) \approx 1 \, F$.

One type of computer keyboard operates by capacitance. As shown in Fig. 24–5, each key is connected to the upper plate of a capacitor. The upper plate moves down when the key is pressed, reducing the spacing between the capacitor plates, and increasing the capacitance (Eq. 24–2: smaller d, larger C). The *change* in capacitance results in an electric signal that is detected by an electronic circuit.

The proportionality, $C \propto A/d$ in Eq. 24–2, is valid also for a parallel-plate capacitor that is rolled up into a spiral cylinder, as in Fig. 24–1b. However, the constant factor, ϵ_0, must be replaced if an insulator such as paper separates the plates, as is usual, and this is discussed in Section 24–5. For a true cylindrical capacitor—consisting of two long coaxial cylinders—the result is somewhat different as the next Example shows.

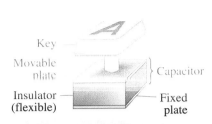

FIGURE 24–5 Key on a computer keyboard. Pressing the key reduces the capacitor spacing thus increasing the capacitance which can be detected electronically.

PHYSICS APPLIED
Computer key

EXAMPLE 24–2 **Cylindrical capacitor.** A cylindrical capacitor consists of a cylinder (or wire) of radius R_b surrounded by a coaxial cylindrical shell of inner radius R_a, Fig. 24–6a. Both cylinders have length ℓ which we assume is much greater than the separation of the cylinders, $R_a - R_b$, so we can neglect end effects. The capacitor is charged (by connecting it to a battery) so that one cylinder has a charge $+Q$ (say, the inner one) and the other one a charge $-Q$. Determine a formula for the capacitance.

APPROACH To obtain $C = Q/V$, we need to determine the potential difference V between the cylinders in terms of Q. We can use our earlier result (Example 21–11 or 22–6) that the electric field outside a long wire is directed radially outward and has magnitude $E = (1/2\pi\epsilon_0)(\lambda/R)$, where R is the distance from the axis and λ is the charge per unit length, Q/ℓ. Then $E = (1/2\pi\epsilon_0)(Q/\ell R)$ for points between the cylinders.

SOLUTION To obtain the potential difference V in terms of Q, we use this result for E in Eq. 23–4a, $V = V_b - V_a = -\int_a^b \vec{E} \cdot d\vec{\ell}$, and write the line integral from the outer cylinder to the inner one (so $V > 0$) along a radial line:[†]

$$V = V_b - V_a = -\int_a^b \vec{E} \cdot d\vec{\ell} = -\frac{Q}{2\pi\epsilon_0\ell} \int_{R_a}^{R_b} \frac{dR}{R}$$

$$= -\frac{Q}{2\pi\epsilon_0\ell} \ln\frac{R_b}{R_a} = \frac{Q}{2\pi\epsilon_0\ell} \ln\frac{R_a}{R_b}.$$

Q and V are proportional, and the capacitance C is

$$C = \frac{Q}{V} = \frac{2\pi\epsilon_0\ell}{\ln(R_a/R_b)}. \qquad \text{[cylindrical capacitor]}$$

NOTE If the space between cylinders, $R_a - R_b = \Delta R$ is small, we have $\ln(R_a/R_b) = \ln[(R_b + \Delta R)/R_b] = \ln[1 + \Delta R/R_b] \approx \Delta R/R_b$ (see Appendix A–3) so $C \approx 2\pi\epsilon_0\ell R_b/\Delta R = \epsilon_0 A/\Delta R$ because the area of cylinder b is $A = 2\pi R_b\ell$. This is just Eq. 24–2 ($d = \Delta R$), a nice check.

FIGURE 24–6 (a) Cylindrical capacitor consists of two coaxial cylindrical conductors. (b) The electric field lines are shown in cross-sectional view.

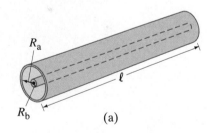

(a)

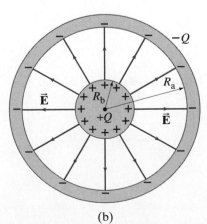

(b)

EXERCISE C What is the capacitance per unit length of a cylindrical capacitor with radii $R_a = 2.5 \, mm$ and $R_b = 0.40 \, mm$? (a) 30 pF/m; (b) −30 pF/m; (c) 56 pF/m; (d) −56 pF/m; (e) 100 pF/m; (f) −100 pF/m.

[†]Note that $\vec{E}$ points outward in Fig. 24–6b, but $d\vec{\ell}$ points inward for our chosen direction of integration; the angle between $\vec{E}$ and $d\vec{\ell}$ is 180° and cos 180° = −1. Also, $d\ell = -dr$ because dr increases outward. These two minus signs cancel.

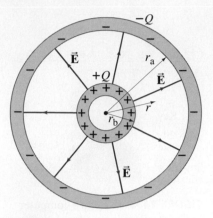

FIGURE 24–7 Cross section through the center of a spherical capacitor. The thin inner shell has radius r_b and the thin outer shell has radius r_a.

PROBLEM SOLVING

Checking with a limiting case

EXAMPLE 24–3 **Spherical capacitor.** A spherical capacitor consists of two thin concentric spherical conducting shells, of radius r_a and r_b as shown in Fig. 24–7. The inner shell carries a uniformly distributed charge Q on its surface, and the outer shell an equal but opposite charge $-Q$. Determine the capacitance of the two shells.

APPROACH In Example 22–3 we used Gauss's law to show that the electric field outside a uniformly charged conducting sphere is $E = Q/4\pi\epsilon_0 r^2$ as if all the charge were concentrated at the center. Now we use Eq. 23–4a, $V = -\int_a^b \vec{E} \cdot d\vec{\ell}$.

SOLUTION We integrate Eq. 23–4a along a radial line to obtain the potential difference between the two conducting shells:

$$V_{ba} = -\int_a^b \vec{E} \cdot d\vec{\ell} = -\frac{Q}{4\pi\epsilon_0} \int_{r_a}^{r_b} \frac{1}{r^2} dr$$

$$= \frac{Q}{4\pi\epsilon_0}\left(\frac{1}{r_b} - \frac{1}{r_a}\right) = \frac{Q}{4\pi\epsilon_0}\left(\frac{r_a - r_b}{r_a r_b}\right).$$

Finally,

$$C = \frac{Q}{V_{ba}} = 4\pi\epsilon_0\left(\frac{r_a r_b}{r_a - r_b}\right).$$

NOTE If the separation $\Delta r = r_a - r_b$ is very small, then $C = 4\pi\epsilon_0 r^2/\Delta r \approx \epsilon_0 A/\Delta r$ (since $A = 4\pi r^2$), which is the parallel-plate formula, Eq. 24–2.

A single isolated conductor can also be said to have a capacitance, C. In this case, C can still be defined as the ratio of the charge to absolute potential V on the conductor (relative to $V = 0$ at $r = \infty$), so that the relation

$$Q = CV$$

remains valid. For example, the potential of a single conducting sphere of radius r_b can be obtained from our results in Example 24–3 by letting r_a become infinitely large. As $r_a \to \infty$, then

$$V = \frac{Q}{4\pi\epsilon_0}\left(\frac{1}{r_b} - \frac{1}{r_a}\right) = \frac{1}{4\pi\epsilon_0}\frac{Q}{r_b};$$

so its capacitance is

$$C = \frac{Q}{V} = 4\pi\epsilon_0 r_b.$$

In practical cases, a single conductor may be near other conductors or the Earth (which can be thought of as the other "plate" of a capacitor), and these will affect the value of the capacitance.

EXAMPLE 24–4 **Capacitance of two long parallel wires.** Estimate the capacitance per unit length of two very long straight parallel wires, each of radius R, carrying uniform charges $+Q$ and $-Q$, and separated by a distance d which is large compared to R ($d \gg R$), Fig. 24–8.

APPROACH We calculate the potential difference between the wires by treating the electric field at any point between them as the superposition of the two fields created by each wire. (The electric field inside each wire conductor is zero.)

SOLUTION The electric field outside of a long straight conductor was found in Examples 21–11 and 22–6 to be radial and given by $E = \lambda/(2\pi\epsilon_0 x)$ where λ is the charge per unit length, $\lambda = Q/\ell$. The total electric field at distance x from the left-hand wire in Fig. 24–8 has magnitude

$$E = \frac{\lambda}{2\pi\epsilon_0 x} + \frac{\lambda}{2\pi\epsilon_0(d - x)},$$

and points to the left (from $+$ to $-$). Now we find the potential difference

FIGURE 24–8 Example 24–4.

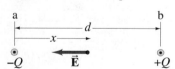

between the two wires using Eq. 23–4a and integrating along the straight line from the surface of the negative wire to the surface of the positive wire, noting that $\vec{E}$ and $d\vec{\ell}$ point in opposite directions ($\vec{E} \cdot d\vec{\ell} < 0$):

$$V = V_b - V_a = -\int_a^b \vec{E} \cdot d\vec{\ell} = \left(\frac{\lambda}{2\pi\epsilon_0}\right) \int_R^{d-R} \left[\frac{1}{x} + \frac{1}{(d-x)}\right] dx$$

$$= \left(\frac{\lambda}{2\pi\epsilon_0}\right) \Big[\ln(x) - \ln(d-x)\Big]\Big|_R^{d-R}$$

$$= \left(\frac{\lambda}{2\pi\epsilon_0}\right)\Big[\ln(d-R) - \ln R - \ln R + \ln(d-R)\Big]$$

$$= \left(\frac{\lambda}{\pi\epsilon_0}\right)\Big[\ln(d-R) - \ln(R)\Big] \approx \left(\frac{\lambda}{\pi\epsilon_0}\right)\Big[\ln(d) - \ln(R)\Big].$$

We are given that $d \gg R$, so

$$V \approx \left(\frac{Q}{\pi\epsilon_0 \ell}\right)\left[\ln\left(\frac{d}{R}\right)\right].$$

The capacitance from Eq. 24–1 is $C = Q/V \approx (\pi\epsilon_0 \ell)/\ln(d/R)$, so the capacitance per unit length is given approximately by

$$\frac{C}{\ell} \approx \frac{\pi\epsilon_0}{\ln(d/R)}.$$

24–3 Capacitors in Series and Parallel

Capacitors are found in many electric circuits. By electric circuit we mean a closed path of conductors, usually wires connecting capacitors and/or other devices, in which charge can flow and which includes a source of voltage such as a battery. The battery voltage is usually given the symbol V, which means that V represents a potential *difference*. Capacitors can be connected together in various ways. Two common ways are in *series*, or in *parallel*, and we now discuss both.

A circuit containing three capacitors connected in **parallel** is shown in Fig. 24–9. They are in "parallel" because when a battery of voltage V is connected to points a and b, this voltage $V = V_{ab}$ exists across each of the capacitors. That is, since the left-hand plates of all the capacitors are connected by conductors, they all reach the same potential V_a when connected to the battery; and the right-hand plates each reach potential V_b. Each capacitor plate acquires a charge given by $Q_1 = C_1 V$, $Q_2 = C_2 V$, and $Q_3 = C_3 V$. The total charge Q that must leave the battery is then

$$Q = Q_1 + Q_2 + Q_3 = C_1 V + C_2 V + C_3 V.$$

Let us try to find a single equivalent capacitor that will hold the same charge Q at the same voltage $V = V_{ab}$. It will have a capacitance C_{eq} given by

$$Q = C_{eq} V.$$

Combining the two previous equations, we have

$$C_{eq} V = C_1 V + C_2 V + C_3 V = (C_1 + C_2 + C_3)V$$

or

$$C_{eq} = C_1 + C_2 + C_3. \qquad \text{[parallel]} \quad \textbf{(24–3)}$$

The net effect of connecting capacitors in parallel is thus to *increase* the capacitance. This makes sense because we are essentially increasing the area of the plates where charge can accumulate (see, for example, Eq. 24–2).

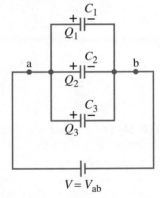

FIGURE 24–9 Capacitors in parallel: $C_{eq} = C_1 + C_2 + C_3$.

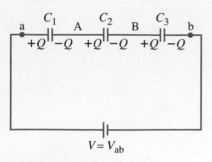

FIGURE 24–10 Capacitors in series:
$$\frac{1}{C_{eq}} = \frac{1}{C_1} + \frac{1}{C_2} + \frac{1}{C_3}.$$

Capacitors can also be connected in **series**: that is, end to end as shown in Fig. 24–10. A charge $+Q$ flows from the battery to one plate of C_1, and $-Q$ flows to one plate of C_3. The regions A and B between the capacitors were originally neutral; so the net charge there must still be zero. The $+Q$ on the left plate of C_1 attracts a charge of $-Q$ on the opposite plate. Because region A must have a zero net charge, there is thus $+Q$ on the left plate of C_2. The same considerations apply to the other capacitors, so we see the charge on each capacitor is the same value Q. A single capacitor that could replace these three in series without affecting the circuit (that is, Q and V the same) would have a capacitance C_{eq} where

$$Q = C_{eq}V.$$

Now the total voltage V across the three capacitors in series must equal the sum of the voltages across each capacitor:

$$V = V_1 + V_2 + V_3.$$

We also have for each capacitor $Q = C_1V_1$, $Q = C_2V_2$, and $Q = C_3V_3$, so we substitute for V, V_1, V_2, and V_3 into the last equation and get

$$\frac{Q}{C_{eq}} = \frac{Q}{C_1} + \frac{Q}{C_2} + \frac{Q}{C_3} = Q\left(\frac{1}{C_1} + \frac{1}{C_2} + \frac{1}{C_3}\right)$$

or

$$\frac{1}{C_{eq}} = \frac{1}{C_1} + \frac{1}{C_2} + \frac{1}{C_3}. \qquad \text{[series]} \quad \textbf{(24–4)}$$

Notice that the equivalent capacitance C_{eq} is smaller than the smallest contributing capacitance.

EXERCISE D Consider two identical capacitors $C_1 = C_2 = 10\,\mu F$. What are the minimum and maximum capacitances that can be obtained by connecting these in series or parallel combinations? (*a*) $0.2\,\mu F, 5\,\mu F$; (*b*) $0.2\,\mu F, 10\,\mu F$; (*c*) $0.2\,\mu F, 20\,\mu F$; (*d*) $5\,\mu F, 10\,\mu F$; (*e*) $5\,\mu F, 20\,\mu F$; (*f*) $10\,\mu F, 20\,\mu F$.

Other connections of capacitors can be analyzed similarly using charge conservation, and often simply in terms of series and parallel connections.

FIGURE 24–11 Examples 24–5 and 24–6.

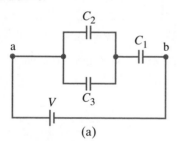

(a)

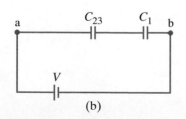

(b)

EXAMPLE 24–5 Equivalent capacitance. Determine the capacitance of a single capacitor that will have the same effect as the combination shown in Fig. 24–11a. Take $C_1 = C_2 = C_3 = C$.

APPROACH First we find the equivalent capacitance of C_2 and C_3 in parallel, and then consider that capacitance in series with C_1.

SOLUTION Capacitors C_2 and C_3 are connected in parallel, so they are equivalent to a single capacitor having capacitance

$$C_{23} = C_2 + C_3 = 2C.$$

This C_{23} is in series with C_1, Fig. 24–11b, so the equivalent capacitance of the entire circuit, C_{eq}, is given by

$$\frac{1}{C_{eq}} = \frac{1}{C_1} + \frac{1}{C_{23}} = \frac{1}{C} + \frac{1}{2C} = \frac{3}{2C}.$$

Hence the equivalent capacitance of the entire combination is $C_{eq} = \frac{2}{3}C$, and it is smaller than any of the contributing capacitors, $C_1 = C_2 = C_3 = C$.

EXAMPLE 24-6 **Charge and voltage on capacitors.** Determine the charge on each capacitor in Fig. 24–11a of Example 24–5 and the voltage across each, assuming $C = 3.0\,\mu F$ and the battery voltage is $V = 4.0\,V$.

APPROACH We have to work "backward" through Example 24–5. That is, we find the charge Q that leaves the battery, using the equivalent capacitance. Then we find the charge on each separate capacitor and the voltage across each. Each step uses Eq. 24–1, $Q = CV$.

SOLUTION The 4.0-V battery behaves as if it is connected to a capacitance $C_{eq} = \frac{3}{2}C = \frac{3}{2}(3.0\,\mu F) = 2.0\,\mu F$. Therefore the charge Q that leaves the battery, by Eq. 24–1, is

$$Q = CV = (2.0\,\mu F)(4.0\,V) = 8.0\,\mu C.$$

From Fig. 24–11a, this charge arrives at the negative plate of C_1, so $Q_1 = 8.0\,\mu C$. The charge Q that leaves the positive plate of the battery is split evenly between C_2 and C_3 (symmetry: $C_2 = C_3$) and is $Q_2 = Q_3 = \frac{1}{2}Q = 4.0\,\mu C$. Next, the voltages across C_2 and C_3 have to be the same. The voltage across each capacitor is obtained using $V = Q/C$. So

$$V_1 = Q_1/C_1 = (8.0\,\mu C)/(3.0\,\mu F) = 2.7\,V$$
$$V_2 = Q_2/C_2 = (4.0\,\mu C)/(3.0\,\mu F) = 1.3\,V$$
$$V_3 = Q_3/C_3 = (4.0\,\mu C)/(3.0\,\mu F) = 1.3\,V.$$

EXAMPLE 24-7 **Capacitors reconnected.** Two capacitors, $C_1 = 2.2\,\mu F$ and $C_2 = 1.2\,\mu F$, are connected in parallel to a 24-V source as shown in Fig. 24–12a. After they are charged they are disconnected from the source and from each other, and then reconnected directly to each other with plates of opposite sign connected together (see Fig. 24–12b). Find the charge on each capacitor and the potential across each after equilibrium is established.

APPROACH We find the charge $Q = CV$ on each capacitor initially. Charge is conserved, although rearranged after the switch. The two new voltages will have to be equal.

SOLUTION First we calculate how much charge has been placed on each capacitor after the power source has charged them fully, using Eq. 24–1:

$$Q_1 = C_1 V = (2.2\,\mu F)(24\,V) = 52.8\,\mu C,$$
$$Q_2 = C_2 V = (1.2\,\mu F)(24\,V) = 28.8\,\mu C.$$

Next the capacitors are connected in parallel, Fig. 24–12b, and the potential difference across each must quickly equalize. Thus, the charge cannot remain as shown in Fig. 24–12b, but the charge must rearrange itself so that the upper plates at least have the same sign of charge, with the lower plates having the opposite charge as shown in Fig. 24–12c. Equation 24–1 applies for each:

$$q_1 = C_1 V' \qquad \text{and} \qquad q_2 = C_2 V',$$

where V' is the voltage across each capacitor after the charges have rearranged themselves. We don't know q_1, q_2, or V', so we need a third equation. This is provided by charge conservation. The charges have rearranged themselves between Figs. 24–12b and c. The total charge on the upper plates in those two Figures must be the same, so we have

$$q_1 + q_2 = Q_1 - Q_2 = 24.0\,\mu C.$$

Combining the last three equations we find:

$$V' = (q_1 + q_2)/(C_1 + C_2) = 24.0\,\mu C/3.4\,\mu F = 7.06\,V \approx 7.1\,V$$
$$q_1 = C_1 V' = (2.2\,\mu F)(7.06\,V) = 15.5\,\mu C \approx 16\,\mu C$$
$$q_2 = C_2 V' = (1.2\,\mu F)(7.06\,V) = 8.5\,\mu C$$

where we have kept only two significant figures in our final answers.

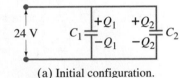

(a) Initial configuration.

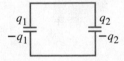

(b) At the instant of reconnection only.

(c) A short time later.

FIGURE 24-12 Example 24–7.

24–4 Electric Energy Storage

A charged capacitor stores electrical energy. The energy stored in a capacitor will be equal to the work done to charge it. The net effect of charging a capacitor is to remove charge from one plate and add it to the other plate. This is what a battery does when it is connected to a capacitor. A capacitor does not become charged instantly. It takes time (Section 26–5). Initially, when the capacitor is uncharged, it requires no work to move the first bit of charge over. When some charge is on each plate, it requires work to add more charge of the same sign because of the electric repulsion. The more charge already on a plate, the more work required to add additional charge. The work needed to add a small amount of charge dq, when a potential difference V is across the plates, is $dW = V \, dq$. Since $V = q/C$ at any moment (Eq. 24–1), where C is the capacitance, the work needed to store a total charge Q is

$$W = \int_0^Q V \, dq = \frac{1}{C} \int_0^Q q \, dq = \frac{1}{2} \frac{Q^2}{C}.$$

Thus we can say that the energy "stored" in a capacitor is

$$U = \frac{1}{2} \frac{Q^2}{C}$$

when the capacitor C carries charges $+Q$ and $-Q$ on its two conductors. Since $Q = CV$, where V is the potential difference across the capacitor, we can also write

$$U = \frac{1}{2} \frac{Q^2}{C} = \frac{1}{2} C V^2 = \frac{1}{2} QV. \qquad \textbf{(24–5)}$$

PHYSICS APPLIED
Camera flash

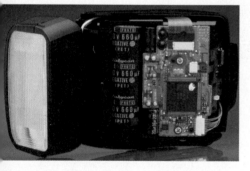

FIGURE 24–13 A camera flash unit.

EXAMPLE 24–8 **Energy stored in a capacitor.** A camera flash unit (Fig. 24–13) stores energy in a 150-μF capacitor at 200 V. (*a*) How much electric energy can be stored? (*b*) What is the power output if nearly all this energy is released in 1.0 ms?

APPROACH We use Eq. 24–5 in the form $U = \frac{1}{2} CV^2$ because we are given C and V.

SOLUTION The energy stored is

$$U = \frac{1}{2} C V^2 = \frac{1}{2} (150 \times 10^{-6} \, \text{F})(200 \, \text{V})^2 = 3.0 \, \text{J}.$$

If this energy is released in $\frac{1}{1000}$ of a second, the power output is $P = U/t = (3.0 \, \text{J})/(1.0 \times 10^{-3} \, \text{s}) = 3000 \, \text{W}$.

CONCEPTUAL EXAMPLE 24–9 **Capacitor plate separation increased.** A parallel-plate capacitor carries charge Q and is then disconnected from a battery. The two plates are initially separated by a distance d. Suppose the plates are pulled apart until the separation is $2d$. How has the energy stored in this capacitor changed?

RESPONSE If we increase the plate separation d, we decrease the capacitance according to Eq. 24–2, $C = \epsilon_0 A/d$, by a factor of 2. The charge Q hasn't changed. So according to Eq. 24–5, where we choose the form $U = \frac{1}{2} Q^2/C$ because we know Q is the same and C has been halved, the reduced C means the potential energy stored increases by a factor of 2.

NOTE We can see why the energy stored increases from a physical point of view: the two plates are charged equal and opposite, so they attract each other. If we pull them apart, we must do work, so we raise their potential energy.

EXAMPLE 24–10 **Moving parallel capacitor plates.** The plates of a parallel-plate capacitor have area A, separation x, and are connected to a battery with voltage V. While connected to the battery, the plates are pulled apart until they are separated by $3x$. (a) What are the initial and final energies stored in the capacitor? (b) How much work is required to pull the plates apart (assume constant speed)? (c) How much energy is exchanged with the battery?

APPROACH The stored energy is given by Eq. 24–5: $U = \frac{1}{2}CV^2$, where $C = \epsilon_0 A/x$. Unlike Example 24–9, here the capacitor remains connected to the battery. Hence charge and energy can flow to or from the battery, and we can not set the work $W = \Delta U$. Instead, the work can be calculated from Eq. 7–7, $W = \int \mathbf{F} \cdot d\boldsymbol{\ell}$.

SOLUTION (a) When the separation is x, the capacitance is $C_1 = \epsilon_0 A/x$ and the energy stored is

$$ U_1 = \frac{1}{2} C_1 V^2 = \frac{1}{2} \frac{\epsilon_0 A}{x} V^2. $$

When the separation is $3x$, $C_2 = \epsilon_0 A/3x$ and

$$ U_2 = \frac{1}{2} \frac{\epsilon_0 A}{3x} V^2. $$

Then

$$ \Delta U_{\text{cap}} = U_2 - U_1 = -\frac{\epsilon_0 A V^2}{3x}. $$

The potential energy decreases as the oppositely charged plates are pulled apart, which makes sense. The plates remain connected to the battery, so V does not change and C decreases; hence some charge leaves each plate $(Q = CV)$, causing U to decrease.

(b) The work done in pulling the plates apart is $W = \int_x^{3x} F \, d\ell = \int_x^{3x} QE \, d\ell$, where Q is the charge on one plate at a given moment when the plates are a distance ℓ apart, and E is the field due to the other plate at that instant. You might think we could use $E = V/\ell$ where ℓ is the separation of the plates (Eq. 23–4b). But we want the force on one plate (of charge Q) due to the electric field of the other plate only—which is half by *symmetry*: so we take $E = V/2\ell$. The charge at any separation ℓ is given by $Q = CV$, where $C = \epsilon_0 A/\ell$. Substituting, the work is

$$ W = \int_{\ell=x}^{\ell=3x} QE \, d\ell = \frac{\epsilon_0 A V^2}{2} \int_x^{3x} \frac{d\ell}{\ell^2} = -\frac{\epsilon_0 A V^2}{2\ell}\bigg|_{\ell=x}^{\ell=3x} = \frac{\epsilon_0 A V^2}{2}\left(\frac{-1}{3x} + \frac{1}{x}\right) = \frac{\epsilon_0 A V^2}{3x}. $$

As you might expect, the work required to pull these oppositely charged plates apart is positive.

(c) Even though the work done is positive, the potential energy decreased, which tells us that energy must have gone into the battery (as if charging it). Conservation of energy tells us that the work W done on the system must equal the change in potential energy of the capacitor plus that of the battery (kinetic energy can be assumed to be essentially zero):

$$ W = \Delta U_{\text{cap}} + \Delta U_{\text{batt}}. $$

Thus the battery experiences a change in energy of

$$ \Delta U_{\text{batt}} = W - \Delta U_{\text{cap}} = \frac{\epsilon_0 A V^2}{3x} + \frac{\epsilon_0 A V^2}{3x} = \frac{2\epsilon_0 A V^2}{3x}. $$

Thus charge flows back into the battery, raising its stored energy. In fact, the battery energy increase is double the work we do.

It is useful to think of the energy stored in a capacitor as being stored in the electric field between the plates. As an example let us calculate the energy stored in a parallel-plate capacitor in terms of the electric field.

We have seen (Eq. 23–4b) that the electric field $\vec{E}$ between two close parallel plates is (approximately) uniform and its magnitude is related to the potential difference by $V = Ed$ where d is the plate separation. Also, Eq. 24–2 tells us $C = \epsilon_0 A/d$ for a parallel-plate capacitor. Thus

$$U = \tfrac{1}{2}CV^2 = \tfrac{1}{2}\left(\frac{\epsilon_0 A}{d}\right)(E^2 d^2)$$
$$= \tfrac{1}{2}\epsilon_0 E^2 Ad.$$

The quantity Ad is the volume between the plates in which the electric field E exists. If we divide both sides by the volume, we obtain an expression for the energy per unit volume or **energy density**, u:

$$u = \text{energy density} = \tfrac{1}{2}\epsilon_0 E^2. \qquad \textbf{(24–6)}$$

The *electric energy stored per unit volume in any region of space is proportional to the square of the electric field* in that region. We derived Eq. 24–6 for the special case of a parallel-plate capacitor. But it can be shown to be true for any region of space where there is an electric field. Note that the units check: for $(\epsilon_0 E^2)$ we have $(C^2/N\cdot m^2)(N/C)^2 = N/m^2 = (N\cdot m)/m^3 = J/m^3$.

Health Effects

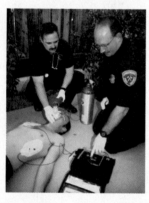

FIGURE 24–14 Heart defibrillator.

The energy stored in a large capacitance can do harm, giving you a burn or a shock. One reason you are warned not to touch a circuit, or the inside of electronic devices, is because capacitors may still be carrying charge even if the external power has been turned off.

On the other hand, the basis of a **heart defibrillator** is a capacitor charged to a high voltage. A heart attack can be characterized by fast irregular beating of the heart, known as *ventricular* (or *cardiac*) *fibrillation*. The heart then does not pump blood to the rest of the body properly, and if it lasts for long, death results. A sudden, brief jolt of charge through the heart from a defibrillator can cause complete heart stoppage, sometimes followed by a resumption of normal beating. The defibrillator capacitor is charged to a high voltage, typically a few thousand volts, and is allowed to discharge very rapidly through the heart via a pair of wide contacts known as "paddles" that spread out the current over the chest (Fig. 24–14).

24–5 Dielectrics

In most capacitors there is an insulating sheet of material, such as paper or plastic, called a **dielectric** between the plates. This serves several purposes. First of all, dielectrics break down (allowing electric charge to flow) less readily than air, so higher voltages can be applied without charge passing across the gap. Furthermore, a dielectric allows the plates to be placed closer together without touching, thus allowing an increased capacitance because d is smaller in Eq. 24–2. Finally, it is found experimentally that if the dielectric fills the space between the two conductors, it increases the capacitance by a factor K which is known as the **dielectric constant**. Thus

$$C = KC_0, \qquad \textbf{(24–7)}$$

where C_0 is the capacitance when the space between the two conductors of the capacitor is a vacuum, and C is the capacitance when the space is filled with a material whose dielectric constant is K.

The values of the dielectric constant for various materials are given in Table 24–1. Also shown in Table 24–1 is the **dielectric strength**, the maximum electric field before breakdown (charge flow) occurs.

For a parallel-plate capacitor (see Eq. 24–2),

$$C = K\epsilon_0 \frac{A}{d} \qquad \text{[parallel-plate capacitor]} \quad \textbf{(24–8)}$$

when the space between the plates is completely filled with a dielectric whose dielectric constant is K. (The situation when the dielectric only partially fills the space will be discussed shortly in Example 24–11.) The quantity $K\epsilon_0$ appears so

TABLE 24–1
Dielectric Constants (at 20°C)

Material	Dielectric constant K	Dielectric strength (V/m)
Vacuum	1.0000	
Air (1 atm)	1.0006	3×10^6
Paraffin	2.2	10×10^6
Polystyrene	2.6	24×10^6
Vinyl (plastic)	2–4	50×10^6
Paper	3.7	15×10^6
Quartz	4.3	8×10^6
Oil	4	12×10^6
Glass, Pyrex	5	14×10^6
Porcelain	6–8	5×10^6
Mica	7	150×10^6
Water (liquid)	80	
Strontium titanate	300	8×10^6

often in formulas that we define a new quantity

$$\epsilon = K\epsilon_0 \qquad (24\text{--}9)$$

called the **permittivity** of a material. Then the capacitance of a parallel-plate capacitor becomes

$$C = \epsilon \frac{A}{d}.$$

Note that ϵ_0 represents the permittivity of free space (a vacuum) as in Section 24-5.

The energy density stored in an electric field E (Section 24-4) in a dielectric is given by (see Eq. 24-6)

$$u = \tfrac{1}{2}K\epsilon_0 E^2 = \tfrac{1}{2}\epsilon E^2. \qquad [E \text{ in a dielectric}]$$

EXERCISE E Return to the Chapter-Opening Question, page 628, and answer it again now. Try to explain why you may have answered differently the first time.

Two simple experiments illustrate the effect of a dielectric. In the first, Fig. 24-15a, a battery of voltage V_0 is kept connected to a capacitor as a dielectric is inserted between the plates. If the charge on the plates without dielectric is Q_0, then when the dielectric is inserted, it is found experimentally (first by Faraday) that the charge Q on the plates is increased by a factor K,

$$Q = KQ_0. \qquad [\text{voltage constant}]$$

The capacitance has increased to $C = Q/V_0 = KQ_0/V_0 = KC_0$, which is Eq. 24-7. In a second experiment, Fig. 24-15b, a battery V_0 is connected to a capacitor C_0 which then holds a charge $Q_0 = C_0 V_0$. The battery is then disconnected, leaving the capacitor isolated with charge Q_0 and still at voltage V_0. Next a dielectric is inserted between the plates of the capacitor. The charge remains Q_0 (there is nowhere for the charge to go) but the voltage is found experimentally to drop by a factor K:

$$V = \frac{V_0}{K}. \qquad [\text{charge constant}]$$

Note that the capacitance changes to $C = Q_0/V = Q_0/(V_0/K) = KQ_0/V_0 = KC_0$, so this experiment too confirms Eq. 24-7.

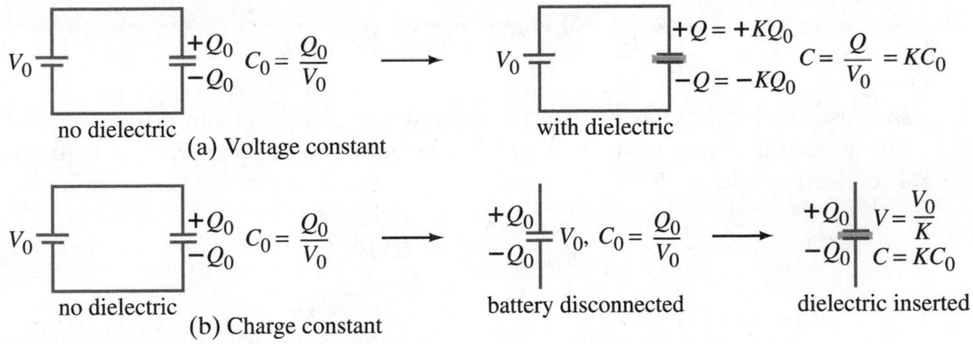

no dielectric
(a) Voltage constant
with dielectric

no dielectric
(b) Charge constant
battery disconnected
dielectric inserted

FIGURE 24-15 Two experiments with a capacitor. Dielectric inserted with (a) voltage held constant, (b) charge held constant.

The electric field when a dielectric is inserted is also altered. When no dielectric is present, the electric field between the plates of a parallel-plate capacitor is given by Eq. 23-4b:

$$E_0 = \frac{V_0}{d},$$

where V_0 is the potential difference between the plates and d is their separation. If the capacitor is isolated so that the charge remains fixed on the plates when a dielectric is inserted, filling the space between the plates, the potential difference drops to $V = V_0/K$. So the electric field in the dielectric is

$$E = E_{\mathrm{D}} = \frac{V}{d} = \frac{V_0}{Kd}$$

or

$$E_{\mathrm{D}} = \frac{E_0}{K}. \qquad [\text{in a dielectric}] \quad (24\text{--}10)$$

The electric field in a dielectric is reduced by a factor equal to the dielectric constant. The field in a dielectric (or insulator) is not reduced all the way to zero as in a conductor. Equation 24-10 is valid even if the dielectric's width is smaller than the gap between the capacitor plates.

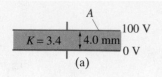

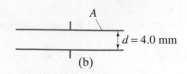

FIGURE 24–16 Example 24–11.

EXAMPLE 24–11 **Dielectric removal.** A parallel-plate capacitor, filled with a dielectric with $K = 3.4$, is connected to a 100-V battery (Fig. 24–16a). After the capacitor is fully charged, the battery is disconnected. The plates have area $A = 4.0 \, \text{m}^2$, and are separated by $d = 4.0 \, \text{mm}$. (a) Find the capacitance, the charge on the capacitor, the electric field strength, and the energy stored in the capacitor. (b) The dielectric is carefully removed, without changing the plate separation nor does any charge leave the capacitor (Fig. 24–16b). Find the new values of capacitance, electric field strength, voltage between the plates, and the energy stored in the capacitor.

APPROACH We use the formulas for parallel-plate capacitance and electric field with and without a dielectric.

SOLUTION (a) First we find the capacitance, with dielectric:

$$C = \frac{K\epsilon_0 A}{d} = \frac{3.4(8.85 \times 10^{-12} \, \text{C}^2/\text{N} \cdot \text{m}^2)(4.0 \, \text{m}^2)}{4.0 \times 10^{-3} \, \text{m}}$$
$$= 3.0 \times 10^{-8} \, \text{F}.$$

The charge Q on the plates is

$$Q = CV = (3.0 \times 10^{-8} \, \text{F})(100 \, \text{V}) = 3.0 \times 10^{-6} \, \text{C}.$$

The electric field between the plates is

$$E = \frac{V}{d} = \frac{100 \, \text{V}}{4.0 \times 10^{-3} \, \text{m}} = 25 \, \text{kV/m}.$$

Finally, the total energy stored in the capacitor is

$$U = \tfrac{1}{2}CV^2 = \tfrac{1}{2}(3.0 \times 10^{-8} \, \text{F})(100 \, \text{V})^2 = 1.5 \times 10^{-4} \, \text{J}.$$

(b) The capacitance without dielectric decreases by a factor $K = 3.4$:

$$C_0 = \frac{C}{K} = \frac{(3.0 \times 10^{-8} \, \text{F})}{3.4} = 8.8 \times 10^{-9} \, \text{F}.$$

Because the battery has been disconnected, the charge Q can not change; when the dielectric is removed, $V = Q/C$ increases by a factor $K = 3.4$ to 340 V. The electric field is

$$E = \frac{V}{d} = \frac{340 \, \text{V}}{4.0 \times 10^{-3} \, \text{m}} = 85 \, \text{kV/m}.$$

The energy stored is

$$U = \tfrac{1}{2}CV^2 = \tfrac{1}{2}(8.8 \times 10^{-9} \, \text{F})(340 \, \text{V})^2$$
$$= 5.1 \times 10^{-4} \, \text{J}.$$

NOTE Where did all this extra energy come from? The energy increased because work had to be done to remove the dielectric. The work required was $W = (5.1 \times 10^{-4} \, \text{J}) - (1.5 \times 10^{-4} \, \text{J}) = 3.6 \times 10^{-4} \, \text{J}$. (We will see in the next Section that work is required because of the force of attraction between induced charge on the dielectric and the charges on the plates, Fig. 24–17c.)

*24–6 Molecular Description of Dielectrics

Let us examine, from the molecular point of view, why the capacitance of a capacitor should be larger when a dielectric is between the plates. A capacitor whose plates are separated by an air gap has a charge $+Q$ on one plate and $-Q$ on the other (Fig. 24–17a). Assume it is isolated (not connected to a battery) so charge cannot flow to or from the plates. The potential difference between the plates, V_0, is given by Eq. 24–1:

$$Q = C_0 V_0,$$

where the subscripts refer to air between the plates. Now we insert a dielectric between the plates (Fig. 24–17b). Because of the electric field between the capacitor plates, the dielectric molecules will tend to become oriented as shown in Fig. 24–17b. If the dielectric molecules are *polar*, the positive end is attracted to the negative plate and vice versa. Even if the dielectric molecules are not polar, electrons within them will tend to move slightly toward the positive capacitor plate, so the effect is the same. The net effect of the aligned dipoles is a net negative charge on the outer edge of the dielectric facing the positive plate, and a net positive charge on the opposite side, as shown in Fig. 24–17c.

Some of the electric field lines, then, do not pass through the dielectric but instead end on charges induced on the surface of the dielectric as shown in Fig. 24–17c. Hence the electric field within the dielectric is less than in air. That is, the electric field between the capacitor plates, assumed filled by the dielectric, has been reduced by some factor K. The voltage across the capacitor is reduced by the same factor K because $V = Ed$ (Eq. 23–4b) and hence, by Eq. 24–1, $Q = CV$, the capacitance C must increase by that same factor K to keep Q constant.

As shown in Fig. 24–17d, the electric field within the dielectric E_D can be considered as the vector sum of the electric field $\vec{E}_0$ due to the "free" charges on the conducting plates, and the field $\vec{E}_{ind}$ due to the induced charge on the surfaces of the dielectric. Since these two fields are in opposite directions, the net field within the dielectric, $E_0 - E_{ind}$, is less than E_0. The precise relationship is given by Eq. 24–10, even if the dielectric does not fill the gap between the plates:

$$E_D = E_0 - E_{ind} = \frac{E_0}{K},$$

so

$$E_{ind} = E_0\left(1 - \frac{1}{K}\right).$$

The electric field between two parallel plates is related to the surface charge density, σ, by $E = \sigma/\epsilon_0$ (Example 21–13 or 22–8). Thus

$$E_0 = \sigma/\epsilon_0$$

where $\sigma = Q/A$ is the surface charge density on the conductor; Q is the net charge on the conductor and is often called the **free charge** (since charge is free to move in a conductor). Similarly, we define an equivalent induced surface charge density σ_{ind} on the dielectric; then

$$E_{ind} = \sigma_{ind}/\epsilon_0$$

where E_{ind} is the electric field due to the induced charge $Q_{ind} = \sigma_{ind} A$ on the surface of the dielectric, Fig. 24–17d. Q_{ind} is often called the **bound charge**, since it is on an insulator and is not free to move. Since $E_{ind} = E_0(1 - 1/K)$ as shown above, we now have

$$\sigma_{ind} = \sigma\left(1 - \frac{1}{K}\right) \qquad \textbf{(24–11a)}$$

and

$$Q_{ind} = Q\left(1 - \frac{1}{K}\right). \qquad \textbf{(24–11b)}$$

Since K is always greater than 1, we see that the charge induced on the dielectric is always less than the free charge on each of the capacitor plates.

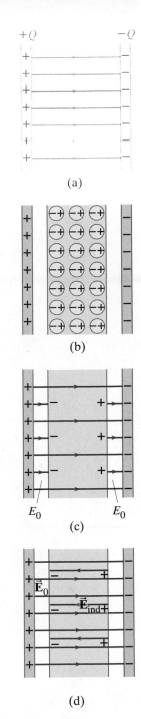

FIGURE 24–17 Molecular view of the effects of a dielectric.

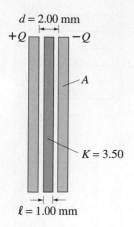

$d = 2.00$ mm

$+Q$ $-Q$

A

$K = 3.50$

$\ell = 1.00$ mm

FIGURE 24–18 Example 24–12.

EXAMPLE 24–12 **Dielectric partially fills capacitor.** A parallel-plate capacitor has plates of area $A = 250\,\text{cm}^2$ and separation $d = 2.00\,\text{mm}$. The capacitor is charged to a potential difference $V_0 = 150\,\text{V}$. Then the battery is disconnected (the charge Q on the plates then won't change), and a dielectric sheet ($K = 3.50$) of the same area A but thickness $\ell = 1.00\,\text{mm}$ is placed between the plates as shown in Fig. 24–18. Determine (*a*) the initial capacitance of the air-filled capacitor, (*b*) the charge on each plate before the dielectric is inserted, (*c*) the charge induced on each face of the dielectric after it is inserted, (*d*) the electric field in the space between each plate and the dielectric, (*e*) the electric field in the dielectric, (*f*) the potential difference between the plates after the dielectric is added, and (*g*) the capacitance after the dielectric is in place.

APPROACH We use the expressions for capacitance and charge developed in this Section plus (part *e*), Eq. 23–4a, $V = -\int \vec{\mathbf{E}} \cdot d\vec{\boldsymbol{\ell}}$.

SOLUTION (*a*) Before the dielectric is in place, the capacitance is

$$C_0 = \epsilon_0 \frac{A}{d} = \left(8.85 \times 10^{-12}\,\text{C}^2/\text{N} \cdot \text{m}^2\right)\left(\frac{2.50 \times 10^{-2}\,\text{m}^2}{2.00 \times 10^{-3}\,\text{m}}\right) = 111\,\text{pF}.$$

(*b*) The charge on each plate is

$$Q = C_0 V_0 = \left(1.11 \times 10^{-10}\,\text{F}\right)(150\,\text{V}) = 1.66 \times 10^{-8}\,\text{C}.$$

(*c*) Equations 24–10 and 24–11 are valid even when the dielectric does not fill the gap, so (Eq. 24–11b)

$$Q_{\text{ind}} = Q\left(1 - \frac{1}{K}\right) = \left(1.66 \times 10^{-8}\,\text{C}\right)\left(1 - \frac{1}{3.50}\right) = 1.19 \times 10^{-8}\,\text{C}.$$

(*d*) The electric field in the gaps between the plates and the dielectric (see Fig. 24–17c) is the same as in the absence of the dielectric since the charge on the plates has not been altered. The result of Example 21–13 can be used here, which gives $E_0 = \sigma/\epsilon_0$. [Or we can note that, in the absence of the dielectric, $E_0 = V_0/d = Q/C_0 d$ (since $V_0 = Q/C_0$) $= Q/\epsilon_0 A$ (since $C_0 = \epsilon_0 A/d$) which is the same result.] Thus

$$E_0 = \frac{Q}{\epsilon_0 A} = \frac{1.66 \times 10^{-8}\,\text{C}}{\left(8.85 \times 10^{-12}\,\text{C}^2/\text{N} \cdot \text{m}^2\right)\left(2.50 \times 10^{-2}\,\text{m}^2\right)} = 7.50 \times 10^4\,\text{V/m}.$$

(*e*) In the dielectric the electric field is (Eq. 24–10)

$$E_{\text{D}} = \frac{E_0}{K} = \frac{7.50 \times 10^4\,\text{V/m}}{3.50} = 2.14 \times 10^4\,\text{V/m}.$$

(*f*) To obtain the potential difference in the presence of the dielectric we use Eq. 23–4a, and integrate from the surface of one plate to the other along a straight line parallel to the field lines:

$$V = -\int \vec{\mathbf{E}} \cdot d\vec{\boldsymbol{\ell}} = E_0(d - \ell) + E_{\text{D}}\ell,$$

which can be simplified to

$$V = E_0\left(d - \ell + \frac{\ell}{K}\right)$$

$$= \left(7.50 \times 10^4\,\text{V/m}\right)\left(1.00 \times 10^{-3}\,\text{m} + \frac{1.00 \times 10^{-3}\,\text{m}}{3.50}\right)$$

$$= 96.4\,\text{V}.$$

(*g*) In the presence of the dielectric, the capacitance is

$$C = \frac{Q}{V} = \frac{1.66 \times 10^{-8}\,\text{C}}{96.4\,\text{V}} = 172\,\text{pF}.$$

NOTE If the dielectric filled the space between the plates, the answers to (*f*) and (*g*) would be 42.9 V and 387 pF, respectively.

Summary

A **capacitor** is a device used to store charge (and electric energy), and consists of two nontouching conductors. The two conductors generally hold equal and opposite charges of magnitude Q. The ratio of this charge Q to the potential difference V between the conductors is called the **capacitance**, C.

$$C = \frac{Q}{V} \quad \text{or} \quad Q = CV. \qquad (24\text{--}1)$$

The capacitance of a parallel-plate capacitor is proportional to the area A of each plate and inversely proportional to their separation d:

$$C = \epsilon_0 \frac{A}{d}. \qquad (24\text{--}2)$$

When capacitors are connected in **parallel**, the equivalent capacitance is the sum of the individual capacitances:

$$C_{eq} = C_1 + C_2 + \cdots. \qquad (24\text{--}3)$$

When capacitors are connected in **series**, the reciprocal of the equivalent capacitance equals the sum of the reciprocals of the individual capacitances:

$$\frac{1}{C_{eq}} = \frac{1}{C_1} + \frac{1}{C_2} + \cdots. \qquad (24\text{--}4)$$

A charged capacitor stores an amount of electric energy given by

$$U = \tfrac{1}{2}QV = \tfrac{1}{2}CV^2 = \tfrac{1}{2}\frac{Q^2}{C}. \qquad (24\text{--}5)$$

This energy can be thought of as stored in the electric field between the plates. In any electric field $\vec{E}$ in free space the **energy density** u (energy per unit volume) is

$$u = \tfrac{1}{2}\epsilon_0 E^2. \qquad (24\text{--}6)$$

The space between the conductors contains a nonconducting material such as air, paper, or plastic. These materials are referred to as **dielectrics**, and the capacitance is proportional to a property of dielectrics called the **dielectric constant**, K (nearly equal to 1 for air). For a parallel-plate capacitor

$$C = K\epsilon_0 \frac{A}{d} = \epsilon \frac{A}{d} \qquad (24\text{--}8)$$

where $\epsilon = K\epsilon_0$ is called the **permittivity** of the dielectric material. When a dielectric is present, the energy density is

$$u = \tfrac{1}{2}K\epsilon_0 E^2 = \tfrac{1}{2}\epsilon E^2.$$

Questions

1. Suppose two nearby conductors carry the same negative charge. Can there be a potential difference between them? If so, can the definition of capacitance, $C = Q/V$, be used here?

2. Suppose the separation of plates d in a parallel-plate capacitor is not very small compared to the dimensions of the plates. Would you expect Eq. 24–2 to give an overestimate or underestimate of the true capacitance? Explain.

3. Suppose one of the plates of a parallel-plate capacitor was moved so that the area of overlap was reduced by half, but they are still parallel. How would this affect the capacitance?

4. When a battery is connected to a capacitor, why do the two plates acquire charges of the same magnitude? Will this be true if the two conductors are different sizes or shapes?

5. Describe a simple method of measuring ϵ_0 using a capacitor.

6. Suppose three identical capacitors are connected to a battery. Will they store more energy if connected in series or in parallel?

7. A large copper sheet of thickness ℓ is placed between the parallel plates of a capacitor, but does not touch the plates. How will this affect the capacitance?

8. The parallel plates of an isolated capacitor carry opposite charges, Q. If the separation of the plates is increased, is a force required to do so? Is the potential difference changed? What happens to the work done in the pulling process?

9. How does the energy in a capacitor change if (a) the potential difference is doubled, (b) the charge on each plate is doubled, and (c) the separation of the plates is doubled, as the capacitor remains connected to a battery in each case?

10. If the voltage across a capacitor is doubled, the amount of energy it can store (a) doubles; (b) is halved; (c) is quadrupled; (d) is unaffected; (e) none of these.

11. An isolated charged capacitor has horizontal plates. If a thin dielectric is inserted a short way between the plates, Fig. 24–19, will it move left or right when it is released?

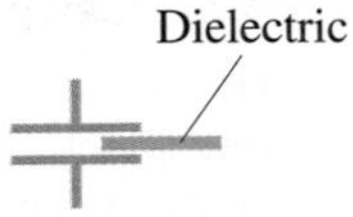

FIGURE 24–19
Question 11.

Dielectric

12. Suppose a battery remains connected to the capacitor in Question 11. What then will happen when the dielectric is released?

13. How does the energy stored in a capacitor change when a dielectric is inserted if (a) the capacitor is isolated so Q does not change; (b) the capacitor remains connected to a battery so V does not change?

14. For dielectrics consisting of polar molecules, how would you expect the dielectric constant to change with temperature?

15. A dielectric is pulled out from between the plates of a capacitor which remains connected to a battery. What changes occur to the capacitance, charge on the plates, potential difference, energy stored in the capacitor, and electric field?

16. We have seen that the capacitance C depends on the size, shape, and position of the two conductors, as well as on the dielectric constant K. What then did we mean when we said that C is a constant in Eq. 24–1?

17. What value might we assign to the dielectric constant for a good conductor? Explain.

Problems

24–1 Capacitors

1. (I) The two plates of a capacitor hold $+2800\,\mu C$ and $-2800\,\mu C$ of charge, respectively, when the potential difference is 930 V. What is the capacitance?

2. (I) How much charge flows from a 12.0-V battery when it is connected to a 12.6-μF capacitor?

3. (I) The potential difference between two short sections of parallel wire in air is 24.0 V. They carry equal and opposite charge of magnitude 75 pC. What is the capacitance of the two wires?

4. (I) The charge on a capacitor increases by $26\,\mu C$ when the voltage across it increases from 28 V to 78 V. What is the capacitance of the capacitor?

5. (II) A 7.7-μF capacitor is charged by a 125-V battery (Fig. 24–20a) and then is disconnected from the battery. When this capacitor (C_1) is then connected (Fig. 24–20b) to a second (initially uncharged) capacitor, C_2, the final voltage on each capacitor is 15 V. What is the value of C_2? [*Hint*: Charge is conserved.]

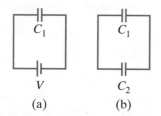

FIGURE 24–20 Problems 5 and 48.

6. (II) An isolated capacitor C_1 carries a charge Q_0. Its wires are then connected to those of a second capacitor C_2, previously uncharged. What charge will each carry now? What will be the potential difference across each?

7. (II) It takes 15 J of energy to move a 0.20-mC charge from one plate of a 15-μF capacitor to the other. How much charge is on each plate?

8. (II) A 2.70-μF capacitor is charged to 475 V and a 4.00-μF capacitor is charged to 525 V. (*a*) These capacitors are then disconnected from their batteries, and the positive plates are now connected to each other and the negative plates are connected to each other. What will be the potential difference across each capacitor and the charge on each? (*b*) What is the voltage and charge for each capacitor if plates of opposite sign are connected?

9. (II) Compact "ultracapacitors" with capacitance values up to several thousand farads are now commercially available. One application for ultracapacitors is in providing power for electrical circuits when other sources (such as a battery) are turned off. To get an idea of how much charge can be stored in such a component, assume a 1200-F ultracapacitor is initially charged to 12.0 V by a battery and is then disconnected from the battery. If charge is then drawn off the plates of this capacitor at a rate of 1.0 mC/s, say, to power the backup memory of some electrical gadget, how long (in days) will it take for the potential difference across this capacitor to drop to 6.0 V?

10. (II) In a **dynamic random access memory (DRAM)** computer chip, each memory cell contains a capacitor for charge storage. Each of these cells represents a single binary-bit value of 1 when its 35-fF capacitor $(1\,\text{fF} = 10^{-15}\,\text{F})$ is charged at 1.5 V, or 0 when uncharged at 0 V. (*a*) When it is fully charged, how many excess electrons are on a cell capacitor's negative plate? (*b*) After charge has been placed on a cell capacitor's plate, it slowly "leaks" off (through a variety of mechanisms) at a constant rate of 0.30 fC/s. How long does it take for the potential difference across this capacitor to decrease by 1.0% from its fully charged value? (Because of this leakage effect, the charge on a DRAM capacitor is "refreshed" many times per second.)

24–2 Determination of Capacitance

11. (I) To make a 0.40-μF capacitor, what area must the plates have if they are to be separated by a 2.8-mm air gap?

12. (I) What is the capacitance per unit length (F/m) of a coaxial cable whose inner conductor has a 1.0-mm diameter and the outer cylindrical sheath has a 5.0-mm diameter? Assume the space between is filled with air.

13. (I) Determine the capacitance of the Earth, assuming it to be a spherical conductor.

14. (II) Use Gauss's law to show that $\vec{\mathbf{E}} = 0$ inside the inner conductor of a cylindrical capacitor (see Fig. 24–6 and Example 24–2) as well as outside the outer cylinder.

15. (II) Dry air will break down if the electric field exceeds about 3.0×10^6 V/m. What amount of charge can be placed on a capacitor if the area of each plate is 6.8 cm²?

16. (II) An electric field of 4.80×10^5 V/m is desired between two parallel plates, each of area 21.0 cm² and separated by 0.250 cm of air. What charge must be on each plate?

17. (II) How strong is the electric field between the plates of a 0.80-μF air-gap capacitor if they are 2.0 mm apart and each has a charge of 92 μC?

18. (II) A large metal sheet of thickness ℓ is placed between, and parallel to, the plates of the parallel-plate capacitor of Fig. 24–4. It does not touch the plates, and extends beyond their edges. (*a*) What is now the net capacitance in terms of A, d, and ℓ? (*b*) If $\ell = 0.40\,d$, by what factor does the capacitance change when the sheet is inserted?

19. (III) Small distances are commonly measured capacitively. Consider an air-filled parallel-plate capacitor with fixed plate area $A = 25$ mm² and a variable plate-separation distance x. Assume this capacitor is attached to a capacitance-measuring instrument which can measure capacitance C in the range 1.0 pF to 1000.0 pF with an accuracy of $\Delta C = 0.1$ pF. (*a*) If C is measured while x is varied, over what range $(x_{\text{min}} \le x \le x_{\text{max}})$ can the plate-separation distance (in μm) be determined by this setup? (*b*) Define Δx to be the accuracy (magnitude) to which x can be determined, and determine a formula for Δx. (*c*) Determine the percent accuracy to which x_{min} and x_{max} can be measured.

20. (III) In an **electrostatic air cleaner** ("precipitator"), the strong nonuniform electric field in the central region of a cylindrical capacitor (with outer and inner cylindrical radii R_a and R_b) is used to create ionized air molecules for use in charging dust and soot particles (Fig. 24–21). Under standard atmospheric conditions, if air is subjected to an electric field magnitude that exceeds its dielectric strength $E_S = 2.7 \times 10^6$ N/C, air molecules will dissociate into positively charged ions and free electrons. In a precipitator, the region within which air is ionized (the *corona discharge* region) occupies a cylindrical volume of radius R that is typically five times that of the inner cylinder. Assume a particular precipitator is constructed with $R_b = 0.10$ mm and $R_a = 10.0$ cm. In order to create a corona discharge region with radius $R = 5.0\, R_b$, what potential difference V should be applied between the precipitator's inner and outer conducting cylinders? [Besides dissociating air, the charged inner cylinder repels the resulting positive ions from the corona discharge region, where they are put to use in charging dust particles, which are then "collected" on the negatively charged outer cylinder.]

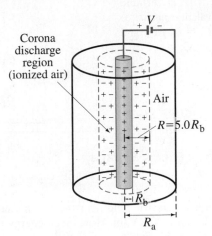

FIGURE 24–21 Problem 20.

24–3 Capacitors in Series and Parallel

21. (I) The capacitance of a portion of a circuit is to be reduced from 2900 pF to 1600 pF. What capacitance can be added to the circuit to produce this effect without removing existing circuit elements? Must any existing connections be broken to accomplish this?

22. (I) (a) Six 3.8-μF capacitors are connected in parallel. What is the equivalent capacitance? (b) What is their equivalent capacitance if connected in series?

23. (II) Given three capacitors, $C_1 = 2.0\,\mu$F, $C_2 = 1.5\,\mu$F, and $C_3 = 3.0\,\mu$F, what arrangement of parallel and series connections with a 12-V battery will give the minimum voltage drop across the 2.0-μF capacitor? What is the minimum voltage drop?

24. (II) Suppose three parallel-plate capacitors, whose plates have areas A_1, A_2, and A_3 and separations d_1, d_2, and d_3, are connected in parallel. Show, using only Eq. 24–2, that Eq. 24–3 is valid.

25. (II) An electric circuit was accidentally constructed using a 5.0-μF capacitor instead of the required 16-μF value. Without removing the 5.0-μF capacitor, what can a technician add to correct this circuit?

26. (II) Three conducting plates, each of area A, are connected as shown in Fig. 24–22. (a) Are the two capacitors thus formed connected in series or in parallel? (b) Determine C as a function of d_1, d_2, and A. Assume $d_1 + d_2$ is much less than the dimensions of the plates. (c) The middle plate can be moved (changing the values of d_1 and d_2), so as to vary the capacitance. What are the minimum and maximum values of the net capacitance?

FIGURE 24–22
Problem 26.

27. (II) Consider three capacitors, of capacitance 3600 pF, 5800 pF, and 0.0100 μF. What maximum and minimum capacitance can you form from these? How do you make the connection in each case?

28. (II) A 0.50-μF and a 0.80-μF capacitor are connected in series to a 9.0-V battery. Calculate (a) the potential difference across each capacitor and (b) the charge on each. (c) Repeat parts (a) and (b) assuming the two capacitors are in parallel.

29. (II) In Fig. 24–23, suppose $C_1 = C_2 = C_3 = C_4 = C$. (a) Determine the equivalent capacitance between points a and b. (b) Determine the charge on each capacitor and the potential difference across each in terms of V.

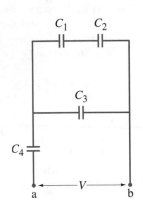

FIGURE 24–23
Problems 29 and 30.

30. (II) Suppose in Fig. 24–23 that $C_1 = C_2 = C_3 = 16.0\,\mu$F and $C_4 = 28.5\,\mu$F. If the charge on C_2 is $Q_2 = 12.4\,\mu$C, determine the charge on each of the other capacitors, the voltage across each capacitor, and the voltage V_{ab} across the entire combination.

31. (II) The switch S in Fig. 24–24 is connected downward so that capacitor C_2 becomes fully charged by the battery of voltage V_0. If the switch is then connected upward, determine the charge on each capacitor after the switching.

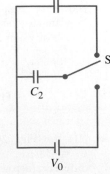

FIGURE 24–24
Problem 31.

32. (II) (a) Determine the equivalent capacitance between points a and b for the combination of capacitors shown in Fig. 24–25. (b) Determine the charge on each capacitor and the voltage across each if $V_{ba} = V$.

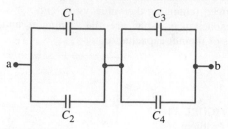

FIGURE 24–25 Problems 32 and 33.

33. (II) Suppose in Problem 32, Fig. 24–25, that $C_1 = C_3 = 8.0\,\mu\text{F}$, $C_2 = C_4 = 16\,\mu\text{F}$, and $Q_3 = 23\,\mu\text{C}$. Determine (a) the charge on each of the other capacitors, (b) the voltage across each capacitor, and (c) the voltage V_{ba} across the combination.

34. (II) Two capacitors connected in parallel produce an equivalent capacitance of $35.0\,\mu\text{F}$ but when connected in series the equivalent capacitance is only $5.5\,\mu\text{F}$. What is the individual capacitance of each capacitor?

35. (II) In the **capacitance bridge** shown in Fig. 24–26, a voltage V_0 is applied and the variable capacitor C_1 is adjusted until there is zero voltage between points a and b as measured on the voltmeter ($\!-\!\!\text{V}\!\!-\!$). Determine the unknown capacitance C_x if $C_1 = 8.9\,\mu\text{F}$ and the fixed capacitors have $C_2 = 18.0\,\mu\text{F}$ and $C_3 = 4.8\,\mu\text{F}$. Assume no charge flows through the voltmeter.

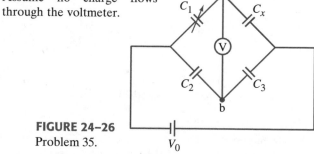

FIGURE 24–26
Problem 35.

36. (II) Two capacitors, $C_1 = 3200\,\text{pF}$ and $C_2 = 1800\,\text{pF}$, are connected in series to a 12.0-V battery. The capacitors are later disconnected from the battery and connected directly to each other, positive plate to positive plate, and negative plate to negative plate. What then will be the charge on each capacitor?

37. (II) (a) Determine the equivalent capacitance of the circuit shown in Fig. 24–27. (b) If $C_1 = C_2 = 2C_3 = 24.0\,\mu\text{F}$, how much charge is stored on each capacitor when $V = 35.0\,\text{V}$?

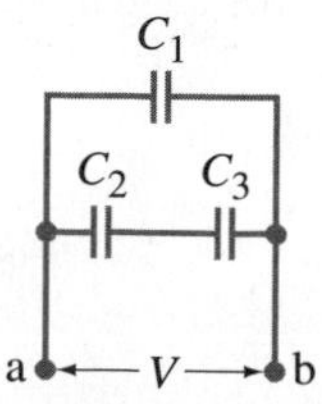

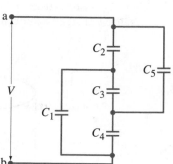

FIGURE 24–27
Problems 37, 38, and 45.

38. (II) In Fig. 24–27, let $C_1 = 2.00\,\mu\text{F}$, $C_2 = 3.00\,\mu\text{F}$, $C_3 = 4.00\,\mu\text{F}$, and $V = 24.0\,\text{V}$. What is the potential difference across each capacitor?

39. (III) Suppose one plate of a parallel-plate capacitor is tilted so it makes a small angle θ with the other plate, as shown in Fig. 24–28. Determine a formula for the capacitance C in terms of A, d, and θ, where A is the area of each plate and θ is small. Assume the plates are square. [Hint: Imagine the capacitor as many infinitesimal capacitors in parallel.]

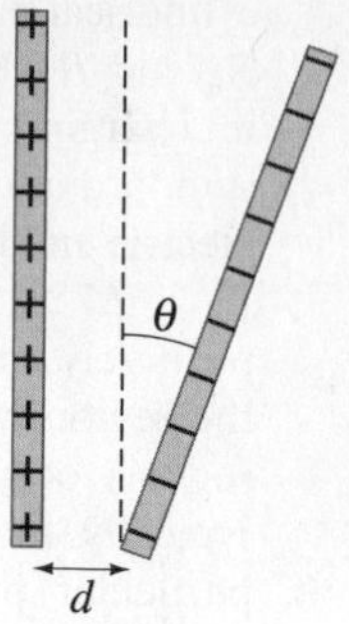

FIGURE 24–28
Problem 39.

40. (III) A voltage V is applied to the capacitor network shown in Fig. 24–29. (a) What is the equivalent capacitance? [Hint: Assume a potential difference V_{ab} exists across the network as shown; write potential differences for various pathways through the network from a to b in terms of the charges on the capacitors and the capacitances.] (b) Determine the equivalent capacitance if $C_2 = C_4 = 8.0\,\mu\text{F}$ and $C_1 = C_3 = C_5 = 4.5\,\mu\text{F}$.

FIGURE 24–29
Problem 40.

24–4 Electric Energy Storage

41. (I) 2200 V is applied to a 2800-pF capacitor. How much electric energy is stored?

42. (I) There is an electric field near the Earth's surface whose intensity is about 150 V/m. How much energy is stored per cubic meter in this field?

43. (I) How much energy is stored by the electric field between two square plates, 8.0 cm on a side, separated by a 1.3-mm air gap? The charges on the plates are equal and opposite and of magnitude $420\,\mu\text{C}$.

44. (II) A parallel-plate capacitor has fixed charges $+Q$ and $-Q$. The separation of the plates is then tripled. (a) By what factor does the energy stored in the electric field change? (b) How much work must be done to increase the separation of the plates from d to $3.0d$? The area of each plate is A.

45. (II) In Fig. 24–27, let $V = 10.0\,\text{V}$ and $C_1 = C_2 = C_3 = 22.6\,\mu\text{F}$. How much energy is stored in the capacitor network?

46. (II) How much energy must a 28-V battery expend to charge a $0.45\text{-}\mu\text{F}$ and a $0.20\text{-}\mu\text{F}$ capacitor fully when they are placed (a) in parallel, (b) in series? (c) How much charge flowed from the battery in each case?

47. (II) (a) Suppose the outer radius R_a of a cylindrical capacitor was tripled, but the charge was kept constant. By what factor would the stored energy change? Where would the energy come from? (b) Repeat part (a), assuming the voltage remains constant.

48. (II) A $2.20\text{-}\mu\text{F}$ capacitor is charged by a 12.0-V battery. It is disconnected from the battery and then connected to an uncharged $3.50\text{-}\mu\text{F}$ capacitor (Fig. 24–20). Determine the total stored energy (a) before the two capacitors are connected, and (b) after they are connected. (c) What is the change in energy?

49. (II) How much work would be required to remove a metal sheet from between the plates of a capacitor (as in Problem 18a), assuming: (a) the battery remains connected so the voltage remains constant; (b) the battery is disconnected so the charge remains constant?

50. (II) (a) Show that each plate of a parallel-plate capacitor exerts a force

$$F = \frac{1}{2}\frac{Q^2}{\epsilon_0 A}$$

on the other, by calculating dW/dx where dW is the work needed to increase the separation by dx. (b) Why does using $F = QE$, with E being the electric field between the plates, give the wrong answer?

51. (II) Show that the electrostatic energy stored in the electric field outside an isolated spherical conductor of radius r_0 carrying a net charge Q is

$$U = \frac{1}{8\pi\epsilon_0}\frac{Q^2}{r_0}.$$

Do this in three ways: (a) Use Eq. 24–6 for the energy density in an electric field [Hint: Consider spherical shells of thickness dr]; (b) use Eq. 24–5 together with the capacitance of an isolated sphere (Section 24–2); (c) by calculating the work needed to bring all the charge Q up from infinity in infinitesimal bits dq.

52. (II) When two capacitors are connected in parallel and then connected to a battery, the total stored energy is 5.0 times greater than when they are connected in series and then connected to the same battery. What is the ratio of the two capacitances? (Before the battery is connected in each case, the capacitors are fully discharged.)

53. (II) For commonly used **CMOS** (complementary metal oxide semiconductor) digital circuits, the charging of the component capacitors C to their working potential difference V accounts for the major contribution of its energy input requirements. Thus, if a given logical operation requires such circuitry to charge its capacitors N times, we can assume that the operation requires an energy of $N(\frac{1}{2}CV^2)$. In the past 20 years, the capacitance in digital circuits has been reduced by a factor of about 20 and the voltage to which these capacitors are charged has been reduced from 5.0 V to 1.5 V. Also, present-day alkaline batteries hold about five times the energy of older batteries. Two present-day AA alkaline cells, each of which measures 1 cm diameter by 4 cm long, can power the logic circuitry of a hand-held **personal digital assistant** (PDA) with its display turned off for about two months. If an attempt was made to construct a similar PDA (i.e., same digital capabilities so N remains constant) 20 years ago, how many (older) AA batteries would have been required to power its digital circuitry for two months? Would this PDA fit in a pocket or purse?

24–5 Dielectrics

54. (I) What is the capacitance of two square parallel plates 4.2 cm on a side that are separated by 1.8 mm of paraffin?

55. (II) Suppose the capacitor in Example 24–11 remains connected to the battery as the dielectric is removed. What will be the work required to remove the dielectric in this case?

56. (II) How much energy would be stored in the capacitor of Problem 43 if a mica dielectric is placed between the plates? Assume the mica is 1.3 mm thick (and therefore fills the space between the plates).

57. (II) In the DRAM computer chip of Problem 10, the cell capacitor's two conducting parallel plates are separated by a 2.0-nm thick insulating material with dielectric constant $K = 25$. (a) Determine the area A (in μm^2) of the cell capacitor's plates. (b) Assuming the plate area A accounts for half of the area of each cell, estimate how many megabytes of memory can be placed on such a 3.0-cm² silicon wafer. (1 byte = 8 bits.)

58. (II) A 3500-pF air-gap capacitor is connected to a 32-V battery. If a piece of mica fills the space between the plates, how much charge will flow from the battery?

59. (II) Two different dielectrics each fill half the space between the plates of a parallel-plate capacitor as shown in Fig. 24–30. Determine a formula for the capacitance in terms of K_1, K_2, the area A of the plates, and the separation d. [Hint: Can you consider this capacitor as two capacitors in series or in parallel?]

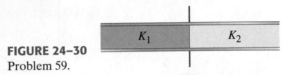

FIGURE 24–30
Problem 59.

60. (II) Two different dielectrics fill the space between the plates of a parallel-plate capacitor as shown in Fig. 24–31. Determine a formula for the capacitance in terms of K_1, K_2, the area A, of the plates, and the separation $d_1 = d_2 = d/2$. [Hint: Can you consider this capacitor as two capacitors in series or in parallel?]

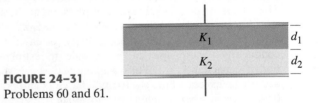

FIGURE 24–31
Problems 60 and 61.

61. (II) Repeat Problem 60 (Fig. 24–31) but assume the separation $d_1 \neq d_2$.

62. (II) Two identical capacitors are connected in parallel and each acquires a charge Q_0 when connected to a source of voltage V_0. The voltage source is disconnected and then a dielectric ($K = 3.2$) is inserted to fill the space between the plates of one of the capacitors. Determine (a) the charge now on each capacitor, and (b) the voltage now across each capacitor.

63. (III) A slab of width d and dielectric constant K is inserted a distance x into the space between the square parallel plates (of side ℓ) of a capacitor as shown in Fig. 24–32. Determine, as a function of x, (a) the capacitance, (b) the energy stored if the potential difference is V_0, and (c) the magnitude and direction of the force exerted on the slab (assume V_0 is constant).

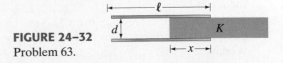

FIGURE 24–32
Problem 63.

64. (III) The quantity of liquid (such as cryogenic liquid nitrogen) available in its storage tank is often monitored by a capacitive level sensor. This sensor is a vertically aligned cylindrical capacitor with outer and inner conductor radii R_a and R_b, whose length ℓ spans the height of the tank. When a nonconducting liquid fills the tank to a height h ($\le \ell$) from the tank's bottom, the dielectric in the lower and upper region between the cylindrical conductors is the liquid (K_{liq}) and its vapor (K_V), respectively (Fig. 24–33). (a) Determine a formula for the fraction F of the tank filled by liquid in terms of the level-sensor capacitance C. [*Hint*: Consider the sensor as a combination of two capacitors.] (b) By connecting a capacitance-measuring instrument to the level sensor, F can be monitored. Assume the sensor dimensions are $\ell = 2.0$ m, $R_a = 5.0$ mm, and $R_b = 4.5$ mm. For liquid nitrogen ($K_{liq} = 1.4$, $K_V = 1.0$), what values of C (in pF) will correspond to the tank being completely full and completely empty?

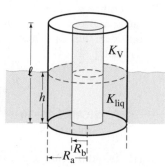

FIGURE 24–33
Problem 64.

*24–6 Molecular Description of Dielectrics

*65. (II) Show that the capacitor in Example 24–12 with dielectric inserted can be considered as equivalent to three capacitors in series, and using this assumption show that the same value for the capacitance is obtained as was obtained in part (g) of the Example.

*66. (II) Repeat Example 24–12 assuming the battery remains connected when the dielectric is inserted. Also, what is the free charge on the plates after the dielectric is added (let this be part (h) of this Problem)?

*67. (II) Using Example 24–12 as a model, derive a formula for the capacitance of a parallel-plate capacitor whose plates have area A, separation d, with a dielectric of dielectric constant K and thickness ℓ ($\ell < d$) placed between the plates.

*68. (II) In Example 24–12 what percent of the stored energy is stored in the electric field in the dielectric?

*69. (III) The capacitor shown in Fig. 24–34 is connected to a 90.0-V battery. Calculate (and sketch) the electric field everywhere between the capacitor plates. Find both the free charge on the capacitor plate and the induced charge on the faces of the glass dielectric plate.

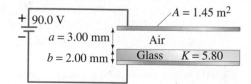

FIGURE 24–34
Problem 69.

General Problems

70. (a) A general rule for estimating the capacitance C of an isolated conducting sphere with radius r is C (in pF) $\approx r$ (in cm). That is, the numerical value of C in pF is about the same as the numerical value of the sphere's radius in cm. Justify this rule. (b) Modeling the human body as a 1-m-radius conducting sphere, use the given rule to estimate your body's capacitance. (c) While walking across a carpet, you acquire an excess "static electricity" charge Q and produce a 0.5-cm spark when reaching out to touch a metallic doorknob. The dielectric strength of air is 30 kV/cm. Use this information to estimate Q (in μC).

71. A *cardiac defibrillator* is used to shock a heart that is beating erratically. A capacitor in this device is charged to 7.5 kV and stores 1200 J of energy. What is its capacitance?

72. A homemade capacitor is assembled by placing two 9-in. pie pans 5.0 cm apart and connecting them to the opposite terminals of a 9-V battery. Estimate (a) the capacitance, (b) the charge on each plate, (c) the electric field halfway between the plates, and (d) the work done by the battery to charge the plates. (e) Which of the above values change if a dielectric is inserted?

73. An uncharged capacitor is connected to a 34.0-V battery until it is fully charged, after which it is disconnected from the battery. A slab of paraffin is then inserted between the plates. What will now be the voltage between the plates?

74. It takes 18.5 J of energy to move a 13.0-mC charge from one plate of a 17.0-μF capacitor to the other. How much charge is on each plate?

75. A huge 3.0-F capacitor has enough stored energy to heat 3.5 kg of water from 22°C to 95°C. What is the potential difference across the plates?

76. A coaxial cable, Fig. 24–35, consists of an inner cylindrical conducting wire of radius R_b surrounded by a dielectric insulator. Surrounding the dielectric insulator is an outer conducting sheath of radius R_a, which is usually "grounded." (a) Determine an expression for the capacitance per unit length of a cable whose insulator has dielectric constant K. (b) For a given cable, $R_b = 2.5$ mm and $R_a = 9.0$ mm. The dielectric constant of the dielectric insulator is $K = 2.6$. Suppose that there is a potential of 1.0 kV between the inner conducting wire and the outer conducting sheath. Find the capacitance per meter of the cable.

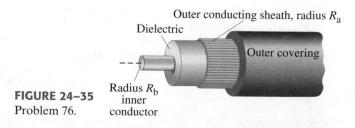

FIGURE 24–35
Problem 76.

77. The electric field between the plates of a paper-separated ($K = 3.75$) capacitor is 9.21×10^4 V/m. The plates are 1.95 mm apart and the charge on each plate is $0.675\ \mu$C. Determine the capacitance of this capacitor and the area of each plate.

78. Capacitors can be used as "electric charge counters." Consider an initially uncharged capacitor of capacitance C with its bottom plate grounded and its top plate connected to a source of electrons. (a) If N electrons flow onto the capacitor's top plate, show that the resulting potential difference V across the capacitor is directly proportional to N. (b) Assume the voltage-measuring device can accurately resolve voltage changes of about 1 mV. What value of C would be necessary to detect each new collected electron? (c) Using modern semiconductor technology, a micron-size capacitor can be constructed with parallel conducting plates separated by an insulating oxide of dielectric constant $K = 3$ and thickness $d = 100$ nm. To resolve the arrival of an individual electron on the plate of such a capacitor, determine the required value of ℓ (in μm) assuming square plates of side length ℓ.

79. A parallel-plate capacitor is isolated with a charge $\pm Q$ on each plate. If the separation of the plates is halved and a dielectric (constant K) is inserted in place of air, by what factor does the energy storage change? To what do you attribute the change in stored potential energy? How does the new value of the electric field between the plates compare with the original value?

80. In lightning storms, the potential difference between the Earth and the bottom of the thunderclouds can be as high as 35,000,000 V. The bottoms of thunderclouds are typically 1500 m above the Earth, and may have an area of 120 km². Modeling the Earth–cloud system as a huge capacitor, calculate (a) the capacitance of the Earth–cloud system, (b) the charge stored in the "capacitor," and (c) the energy stored in the "capacitor."

81. A multilayer film capacitor has a maximum voltage rating of 100 V and a capacitance of 1.0 μF. It is made from alternating sheets of metal foil connected together, separated by films of polyester dielectric. The sheets are 12.0 mm by 14.0 mm and the total thickness of the capacitor is 6.0 mm (not counting the thickness of the insulator on the outside). The metal foil is actually a very thin layer of metal deposited directly on the dielectric, so most of the thickness of the capacitor is due to the dielectric. The dielectric strength of the polyester is about 30×10^6 V/m. Estimate the dielectric constant of the polyester material in the capacitor.

82. A 3.5-μF capacitor is charged by a 12.4-V battery and then is disconnected from the battery. When this capacitor (C_1) is then connected to a second (initially uncharged) capacitor, C_2, the voltage on the first drops to 5.9 V. What is the value of C_2?

83. The power supply for a pulsed nitrogen laser has a 0.080-μF capacitor with a maximum voltage rating of 25 kV. (a) Estimate how much energy could be stored in this capacitor. (b) If 15% of this stored electrical energy is converted to light energy in a pulse that is 4.0-μs long, what is the power of the laser pulse?

84. A parallel-plate capacitor has square plates 12 cm on a side separated by 0.10 mm of plastic with a dielectric constant of $K = 3.1$. The plates are connected to a battery, causing them to become oppositely charged. Since the oppositely charged plates attract each other, they exert a pressure on the dielectric. If this pressure is 40.0 Pa, what is the battery voltage?

85. The variable capacitance of an old radio tuner consists of four plates connected together placed alternately between four other plates, also connected together (Fig. 24–36). Each plate is separated from its neighbor by 1.6 mm of air. One set of plates can move so that the area of overlap of each plate varies from 2.0 cm² to 9.0 cm². (a) Are these seven capacitors connected in series or in parallel? (b) Determine the range of capacitance values.

FIGURE 24–36
Problems 85 and 86.

86. A high-voltage supply can be constructed from a variable capacitor with interleaving plates which can be rotated as in Fig. 24–36. A version of this type of capacitor with more plates has a capacitance which can be varied from 10 pF to 1 pF. (a) Initially, this capacitor is charged by a 7500-V power supply when the capacitance is 8.0 pF. It is then disconnected from the power supply and the capacitance reduced to 1.0 pF by rotating the plates. What is the voltage across the capacitor now? (b) What is a major disadvantage of this as a high-voltage power supply?

87. A 175-pF capacitor is connected in series with an unknown capacitor, and as a series combination they are connected to a 25.0-V battery. If the 175-pF capacitor stores 125 pC of charge on its plates, what is the unknown capacitance?

88. A parallel-plate capacitor with plate area 2.0 cm² and air-gap separation 0.50 mm is connected to a 12-V battery, and fully charged. The battery is then disconnected. (a) What is the charge on the capacitor? (b) The plates are now pulled to a separation of 0.75 mm. What is the charge on the capacitor now? (c) What is the potential difference across the plates now? (d) How much work was required to pull the plates to their new separation?

89. In the circuit shown in Fig. 24–37, $C_1 = 1.0 \mu$F, $C_2 = 2.0 \mu$F, $C_3 = 2.4 \mu$F, and a voltage $V_{ab} = 24$ V is applied across points a and b. After C_1 is fully charged the switch is thrown to the right. What is the final charge and potential difference on each capacitor?

FIGURE 24–37
Problem 89.

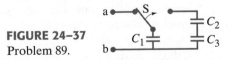

90. The long cylindrical capacitor shown in Fig. 24–38 consists of four concentric cylinders, with respective radii R_a, R_b, R_c, and R_d. The cylinders b and c are joined by metal strips. Determine the capacitance per unit length of this arrangement. (Assume equal and opposite charges are placed on the innermost and outermost cylinders.)

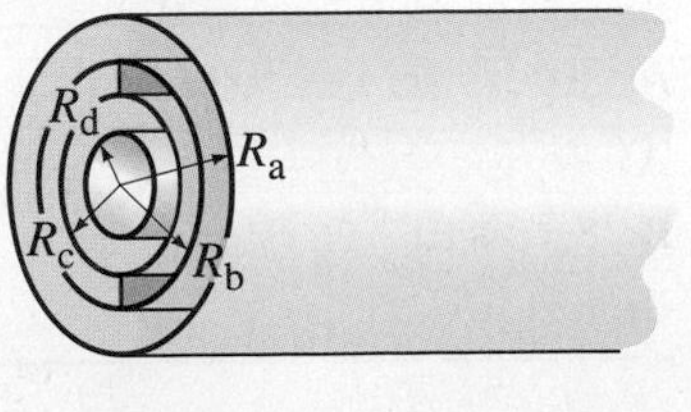

FIGURE 24–38
Problem 90.

91. A parallel-plate capacitor has plate area A, plate separation x, and has a charge Q stored on its plates (Fig. 24–39). Find the amount of work required to double the plate separation to $2x$, assuming the charge remains constant at Q. Show that your answer is consistent with the change in energy stored by the capacitor. (*Hint*: See Example 24–10.)

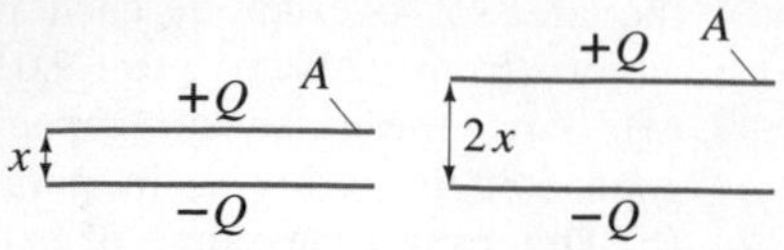

FIGURE 24–39
Problem 91.

x {$+Q$ A / $-Q$} $2x$ {$+Q$ A / $-Q$}

92. Consider the use of capacitors as memory cells. A charged capacitor would represent a one and an uncharged capacitor a zero. Suppose these capacitors were fabricated on a silicon chip and each has a capacitance of 30 femto-farads $(1\,\text{fF} = 10^{-15}\,\text{F}.)$ The dielectric filling the space between the parallel plates has dielectric constant $K = 25$ and a dielectric strength of $1.0 \times 10^9\,\text{V/m}$. (*a*) If the operating voltage is 1.5 V, how many electrons would be stored on one of these capacitors when charged? (*b*) If no safety factor is allowed, how thin a dielectric layer could we use for operation at 1.5 V? (*c*) Using the layer thickness from your answer to part (*b*), what would be the area of the capacitor plates?

93. To get an idea how big a farad is, suppose you want to make a 1-F air-filled parallel-plate capacitor for a circuit you are building. To make it a reasonable size, suppose you limit the plate area to $1.0\,\text{cm}^2$. What would the gap have to be between the plates? Is this practically achievable?

94. A student wearing shoes with thin insulating soles is standing on a grounded metal floor when he puts his hand flat against the screen of a CRT computer monitor. The voltage inside the monitor screen, 6.3 mm from his hand, is 25,000 V. The student's hand and the monitor form a capacitor; the student is a conductor, and there is another capacitor between the floor and his feet. Using reasonable numbers for hand and foot areas, estimate the student's voltage relative to the floor. Assume vinyl-soled shoes 1 cm thick.

95. A parallel-plate capacitor with plate area $A = 2.0\,\text{m}^2$ and plate separation $d = 3.0\,\text{mm}$ is connected to a 45-V battery (Fig. 24–40a). (*a*) Determine the charge on the capacitor, the electric field, the capacitance, and the energy stored in the capacitor. (*b*) With the capacitor still connected to the battery, a slab of plastic with dielectric strength $K = 3.2$ is placed between the plates of the capacitor, so that the gap is completely filled with the dielectric. What are the new values of charge, electric field, capacitance, and the energy U stored in the capacitor?

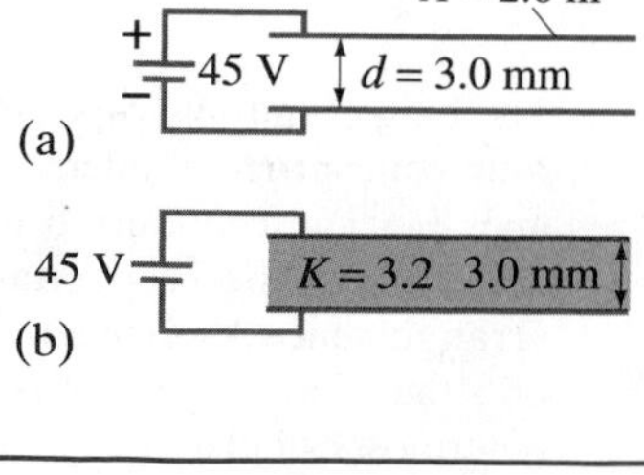

$A = 2.0\,\text{m}^2$
$+$ / $-$ 45 V $\quad d = 3.0\,\text{mm}$
(a)

FIGURE 24–40
Problem 95.
45 V $\quad$ $K = 3.2$ 3.0 mm
(b)

96. Let us try to estimate the maximum "static electricity" charge that might result during each walking step across an insulating floor. Assume the sole of a person's shoe has area $A \approx 150\,\text{cm}^2$, and when the foot is lifted from the ground during each step, the sole acquires an excess charge Q from rubbing contact with the floor. (*a*) Model the sole as a plane conducting surface with Q uniformly distributed across it as the foot is lifted from the ground. If the dielectric strength of the air between the sole and floor as the foot is lifted is $E_S = 3 \times 10^6\,\text{N/C}$, determine Q_{max}, the maximum possible excess charge that can be transferred to the sole during each step. (*b*) Modeling a person as an isolated conducting sphere of radius $r \approx 1\,\text{m}$, estimate a person's capacitance. (*c*) After lifting the foot from the floor, assume the excess charge Q quickly redistributes itself over the entire surface area of the person. Estimate the maximum potential difference that the person can develop with respect to the floor.

97. Paper has a dielectric constant $K = 3.7$ and a dielectric strength of $15 \times 10^6\,\text{V/m}$. Suppose that a typical sheet of paper has a thickness of 0.030 mm. You make a "homemade" capacitor by placing a sheet of $21 \times 14\,\text{cm}$ paper between two aluminum foil sheets (Fig. 24–41). The thickness of the aluminum foil is 0.040 mm. (*a*) What is the capacitance C_0 of your device? (*b*) About how much charge could you store on your capacitor before it would break down? (*c*) Show in a sketch how you could overlay sheets of paper and aluminum for a parallel combination. If you made 100 such capacitors, and connected the edges of the sheets in parallel so that you have a single large capacitor of capacitance $100\,C_0$, how thick would your new large capacitor be? (*d*) What is the maximum voltage you can apply to this $100\,C_0$ capacitor without breakdown?

FIGURE 24–41
Problem 97.

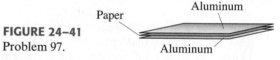
Paper   Aluminum
Aluminum

*Numerical/Computer

***98.** (II) Six physics students were each given an air filled capacitor. Although the areas were different, the spacing between the plates, d, was the same for all six capacitors, but was unknown. Each student made a measurement of the area A and capacitance C of their capacitor. Below is a Table for their data. Using the combined data and a graphing program or spreadsheet, determine the spacing d between the plates.

Area (m^2)	Capacitance (pF)
0.01	90
0.03	250
0.04	340
0.06	450
0.09	800
0.12	1050

Answers to Exercises

A: A.

B: $8.3 \times 10^{-9}\,\text{C}$.

C: (*a*).

D: (*e*).

E: (*b*).

The glow of the thin wire filament of a lightbulb is caused by the electric current passing through it. Electric energy is transformed to thermal energy (via collisions between moving electrons and atoms of the wire), which causes the wire's temperature to become so high that it glows. Electric current and electric power in electric circuits are of basic importance in everyday life. We examine both dc and ac in this Chapter, and include the microscopic analysis of electric current.

Electric Currents and Resistance

C H A P T E R

25

CHAPTER-OPENING QUESTION—Guess now!

The conductors shown are all made of copper and are at the same temperature. Which conductor would have the greatest resistance to the flow of charge entering from the left? Which would offer the least resistance?

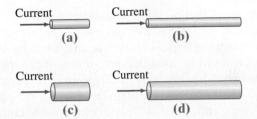

Current (a)   Current (b)

Current (c)   Current (d)

I n the previous four Chapters we have been studying static electricity: electric charges at rest. In this Chapter we begin our study of charges in motion, and we call a flow of charge an electric current.

In everyday life we are familiar with electric currents in wires and other conductors. Indeed, most practical electrical devices depend on electric current: current through a lightbulb, current in the heating element of a stove or electric heater, and currents in electronic devices. Electric currents can exist in conductors such as wires, and also in other devices such as the CRT of a television or computer monitor whose charged electrons flow through space (Section 23–9).

FIGURE 25–1 Alessandro Volta. In this portrait, Volta exhibits his battery to Napoleon in 1801.

In electrostatic situations, we saw in Sections 21–9 and 22–3 that the electric field must be zero inside a conductor (if it weren't, the charges would move). But when charges are *moving* in a conductor, there usually *is* an electric field in the conductor. Indeed, an electric field is needed to set charges into motion, and to keep them in motion in any normal conductor. We can control the flow of charge using electric fields and electric potential (voltage), concepts we have just been discussing. In order to have a current in a wire, a potential difference is needed, which can be provided by a battery.

We first look at electric current from a macroscopic point of view: that is, current as measured in a laboratory. Later in the Chapter we look at currents from a microscopic (theoretical) point of view as a flow of electrons in a wire.

Until the year 1800, the technical development of electricity consisted mainly of producing a static charge by friction. It all changed in 1800 when Alessandro Volta (1745–1827; Fig. 25–1) invented the electric battery, and with it produced the first steady flow of electric charge—that is, a steady electric current.

25–1 The Electric Battery

The events that led to the discovery of the battery are interesting. For not only was this an important discovery, but it also gave rise to a famous scientific debate.

In the 1780s, Luigi Galvani (1737–1798), professor at the University of Bologna, carried out a series of experiments on the contraction of a frog's leg muscle through electricity produced by static electricity. Galvani found that the muscle also contracted when dissimilar metals were inserted into the frog. Galvani believed that the source of the electric charge was in the frog muscle or nerve itself, and that the metal merely transmitted the charge to the proper points. When he published his work in 1791, he termed this charge "animal electricity." Many wondered, including Galvani himself, if he had discovered the long-sought "life-force."

Volta, at the University of Pavia 200 km away, was skeptical of Galvani's results, and came to believe that the source of the electricity was not in the animal itself, but rather in the *contact between the dissimilar metals*. Volta realized that a moist conductor, such as a frog muscle or moisture at the contact point of two dissimilar metals, was necessary in the circuit if it was to be effective. He also saw that the contracting frog muscle was a sensitive instrument for detecting electric "tension" or "electromotive force" (his words for what we now call potential), in fact more sensitive than the best available electroscopes (Section 21–4) that he and others had developed.[†]

FIGURE 25–2 A voltaic battery, from Volta's original publication.

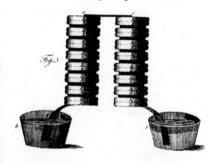

Volta's research found that certain combinations of metals produced a greater effect than others, and, using his measurements, he listed them in order of effectiveness. (This "electrochemical series" is still used by chemists today.) He also found that carbon could be used in place of one of the metals.

Volta then conceived his greatest contribution to science. Between a disc of zinc and one of silver, he placed a piece of cloth or paper soaked in salt solution or dilute acid and piled a "battery" of such couplings, one on top of another, as shown in Fig. 25–2. This "pile" or "battery" produced a much increased potential difference. Indeed, when strips of metal connected to the two ends of the pile were brought close, a spark was produced. Volta had designed and built the first electric battery; he published his discovery in 1800.

[†]Volta's most sensitive electroscope measured about 40 V per degree (angle of leaf separation). Nonetheless, he was able to estimate the potential differences produced by dissimilar metals in contact: for a silver–zinc contact he got about 0.7 V, remarkably close to today's value of 0.78 V.

Electric Cells and Batteries

A battery produces electricity by transforming chemical energy into electrical energy. Today a great variety of electric cells and batteries are available, from flashlight batteries to the storage battery of a car. The simplest batteries contain two plates or rods made of dissimilar metals (one can be carbon) called **electrodes**. The electrodes are immersed in a solution, such as a dilute acid, called the **electrolyte**. Such a device is properly called an **electric cell**, and several cells connected together is a **battery**, although today even a single cell is called a battery. The chemical reactions involved in most electric cells are quite complicated. Here we describe how one very simple cell works, emphasizing the physical aspects.

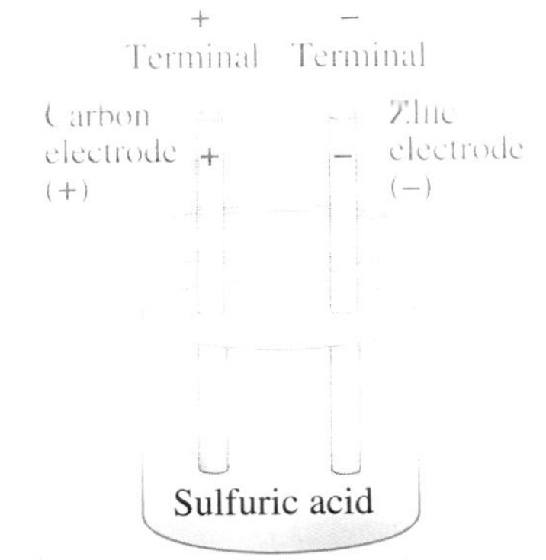

The cell shown in Fig. 25–3 uses dilute sulfuric acid as the electrolyte. One of the electrodes is made of carbon, the other of zinc. That part of each electrode outside the solution is called the **terminal**, and connections to wires and circuits are made here. The acid tends to dissolve the zinc electrode. Each zinc atom leaves two electrons behind on the electrode and enters the solution as a positive ion. The zinc electrode thus acquires a negative charge. As the electrolyte becomes positively charged, electrons are pulled off the carbon electrode by the electrolyte. Thus the carbon electrode becomes positively charged. Because there is an opposite charge on the two electrodes, there is a potential difference between the two terminals.

In a cell whose terminals are not connected, only a small amount of the zinc is dissolved, for as the zinc electrode becomes increasingly negative, any new positive zinc ions produced are attracted back to the electrode. Thus, a particular potential difference (or voltage) is maintained between the two terminals. If charge is allowed to flow between the terminals, say, through a wire (or a lightbulb), then more zinc can be dissolved. After a time, one or the other electrode is used up and the cell becomes "dead."

The voltage that exists between the terminals of a battery depends on what the electrodes are made of and their relative ability to be dissolved or give up electrons.

When two or more cells are connected so that the positive terminal of one is connected to the negative terminal of the next, they are said to be connected in *series* and their voltages add up. Thus, the voltage between the ends of two 1.5-V flashlight batteries connected in series is 3.0 V, whereas the six 2-V cells of an automobile storage battery give 12 V. Figure 25–4a shows a diagram of a common "dry cell" or "flashlight battery" used in portable radios and CD players, flashlights, etc., and Fig. 25–4b shows two smaller ones in series, connected to a flashlight bulb. A lightbulb consists of a thin, coiled wire (filament) inside an evacuated glass bulb, as shown in Fig. 25–5 and in the large photo opening this Chapter, page 651. The filament gets very hot (3000 K) and glows when charge passes through it.

FIGURE 25–3 Simple electric cell.

FIGURE 25–4 (a) Diagram of an ordinary dry cell (like a D-cell or AA). The cylindrical zinc cup is covered on the sides; its flat bottom is the negative terminal. (b) Two dry cells (AA type) connected in series. Note that the positive terminal of one cell pushes against the negative terminal of the other.

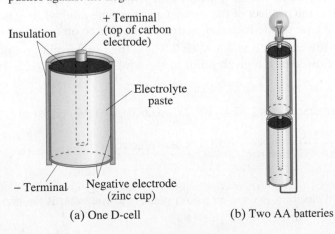

(a) One D-cell   (b) Two AA batteries

FIGURE 25–5 A lightbulb: the fine wire of the filament becomes so hot that it glows. This type of lightbulb is called an incandescent bulb (as compared, say, to a fluorescent bulb).

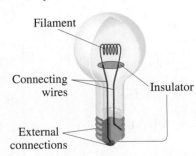

(a)

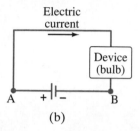

Electric
current

(b)

FIGURE 25–6 (a) A simple electric circuit. (b) Schematic drawing of the same circuit, consisting of a battery, connecting wires (thick gray lines), and a lightbulb or other device.

⚠ **CAUTION**

A battery does not create charge; a lightbulb does not destroy charge

25–2 Electric Current

The purpose of a battery is to produce a potential difference, which can then make charges move. When a continuous conducting path is connected between the terminals of a battery, we have an electric **circuit**, Fig. 25–6a. On any diagram of a circuit, as in Fig. 25–6b, we use the symbol

$$\dashv\vdash \qquad \text{[battery symbol]}$$

to represent a battery. The device connected to the battery could be a lightbulb, a heater, a radio, or whatever. When such a circuit is formed, charge can flow through the wires of the circuit, from one terminal of the battery to the other, as long as the conducting path is continuous. Any flow of charge such as this is called an **electric current**.

More precisely, the electric current in a wire is defined as the net amount of charge that passes through the wire's full cross section at any point per unit time. Thus, the average current $\bar{I}$ is defined as

$$\bar{I} = \frac{\Delta Q}{\Delta t}, \tag{25–1a}$$

where ΔQ is the amount of charge that passes through the conductor at any location during the time interval Δt. The instantaneous current is defined by the derivative limit

$$I = \frac{dQ}{dt}. \tag{25–1b}$$

Electric current is measured in coulombs per second; this is given a special name, the **ampere** (abbreviated amp or A), after the French physicist André Ampère (1775–1836). Thus, $1\,\text{A} = 1\,\text{C/s}$. Smaller units of current are often used, such as the milliampere $(1\,\text{mA} = 10^{-3}\,\text{A})$ and microampere $(1\,\mu\text{A} = 10^{-6}\,\text{A})$.

A current can flow in a circuit only if there is a *continuous* conducting path. We then have a **complete circuit**. If there is a break in the circuit, say, a cut wire, we call it an **open circuit** and no current flows. In any single circuit, with only a single path for current to follow such as in Fig. 25–6b, a steady current at any instant is the same at one point (say, point A) as at any other point (such as B). This follows from the conservation of electric charge: charge doesn't disappear. A battery does not create (or destroy) any net charge, nor does a lightbulb absorb or destroy charge.

EXAMPLE 25–1 **Current is flow of charge.** A steady current of 2.5 A exists in a wire for 4.0 min. (*a*) How much total charge passed by a given point in the circuit during those 4.0 min? (*b*) How many electrons would this be?

APPROACH Current is flow of charge per unit time, Eqs. 25–1, so the amount of charge passing a point is the product of the current and the time interval. To get the number of electrons (*b*), we divide the total charge by the charge on one electron.

SOLUTION (*a*) Since the current was 2.5 A, or 2.5 C/s, then in 4.0 min (= 240 s) the total charge that flowed past a given point in the wire was, from Eq. 25–1a,

$$\Delta Q = I\,\Delta t = (2.5\,\text{C/s})(240\,\text{s}) = 600\,\text{C}.$$

(*b*) The charge on one electron is $1.60 \times 10^{-19}\,\text{C}$, so 600 C would consist of

$$\frac{600\,\text{C}}{1.60 \times 10^{-19}\,\text{C/electron}} = 3.8 \times 10^{21}\,\text{electrons}.$$

EXERCISE A If 1 million electrons per second pass a point in a wire, what is the current in amps?

How to connect a battery. What is wrong with each of the schemes shown in Fig. 25–7 for lighting a flashlight bulb with a flashlight battery and a single wire?

RESPONSE (*a*) There is no closed path for charge to flow around. Charges might briefly start to flow from the battery toward the lightbulb, but there they run into a "dead end," and the flow would immediately come to a stop.

(*b*) Now there is a closed path passing to and from the lightbulb; but the wire touches only one battery terminal, so there is no potential difference in the circuit to make the charge move.

(*c*) Nothing is wrong here. This is a complete circuit: charge can flow out from one terminal of the battery, through the wire and the bulb, and into the other terminal. This scheme will light the bulb.

In many real circuits, wires are connected to a common conductor that provides continuity. This common conductor is called **ground**, usually represented as ⏚ or ⏚, and really is connected to the ground in a building or house. In a car, one terminal of the battery is called "ground," but is not connected to the ground—it is connected to the frame of the car, as is one connection to each lightbulb and other devices. Thus the car frame is a conductor in each circuit, ensuring a continuous path for charge flow.

We saw in Chapter 21 that conductors contain many free electrons. Thus, if a continuous conducting wire is connected to the terminals of a battery, negatively charged electrons flow in the wire. When the wire is first connected, the potential difference between the terminals of the battery sets up an electric field inside the wire[†] and parallel to it. Free electrons at one end of the wire are attracted into the positive terminal, and at the same time other electrons leave the negative terminal of the battery and enter the wire at the other end. There is a continuous flow of electrons throughout the wire that begins as soon as the wire is connected to *both* terminals. However, when the conventions of positive and negative charge were invented two centuries ago, it was assumed that positive charge flowed in a wire. For nearly all purposes, positive charge flowing in one direction is exactly equivalent to negative charge flowing in the opposite direction,[‡] as shown in Fig. 25–8. Today, we still use the historical convention of positive charge flow when discussing the direction of a current. So when we speak of the current direction in a circuit, we mean the direction positive charge would flow. This is sometimes referred to as **conventional current**. When we want to speak of the direction of electron flow, we will specifically state it is the electron current. In liquids and gases, both positive and negative charges (ions) can move.

25–3 Ohm's Law: Resistance and Resistors

To produce an electric current in a circuit, a difference in potential is required. One way of producing a potential difference along a wire is to connect its ends to the opposite terminals of a battery. It was Georg Simon Ohm (1787–1854) who established experimentally that the current in a metal wire is proportional to the potential difference V applied to its two ends:

$$I \propto V.$$

If, for example, we connect a wire to the two terminals of a 6-V battery, the current flow will be twice what it would be if the wire were connected to a 3-V battery. It is also found that reversing the sign of the voltage does not affect the magnitude of the current.

[†]This does not contradict what was said in Section 21–9 that in the *static* case, there can be no electric field within a conductor since otherwise the charges would move. Indeed, when there is an electric field in a conductor, charges do move, and we get an electric current.

[‡]An exception is discussed in Section 27–8.

(a)

(b)

(c)

FIGURE 25–7 Example 25–2.

FIGURE 25–8 Conventional current from + to − is equivalent to a negative electron flow from − to +.

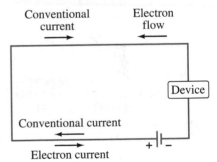

A useful analogy compares the flow of electric charge in a wire to the flow of water in a river, or in a pipe, acted on by gravity. If the river or pipe is nearly level, the flow rate is small. But if one end is somewhat higher than the other, the flow rate—or current—is greater. The greater the difference in height, the swifter the current. We saw in Chapter 23 that electric potential is analogous, in the gravitational case, to the height of a cliff. This applies in the present case to the height through which the fluid flows. Just as an increase in height can cause a greater flow of water, so a greater electric potential difference, or voltage, causes a greater electric current.

Exactly how large the current is in a wire depends not only on the voltage but also on the resistance the wire offers to the flow of electrons. The walls of a pipe, or the banks of a river and rocks in the middle, offer resistance to the water current. Similarly, electron flow is impeded because of interactions with the atoms of the wire. The higher this resistance, the less the current for a given voltage V. We then define electrical *resistance* so that the current is inversely proportional to the resistance: that is,

$$I = \frac{V}{R} \tag{25-2a}$$

where R is the **resistance** of a wire or other device, V is the potential difference applied across the wire or device, and I is the current through it. Equation 25–2a is often written as

$$V = IR. \tag{25-2b}$$

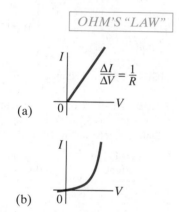

OHM'S "LAW"

(a)

(b)

FIGURE 25–9 Graphs of current vs. voltage for (a) a metal conductor which obeys Ohm's law, and (b) for a nonohmic device, in this case a semiconductor diode.

As mentioned above, Ohm found experimentally that in metal conductors R is a constant independent of V, a result known as **Ohm's law**. Equation 25–2b, $V = IR$, is itself sometimes called Ohm's law, but only when referring to materials or devices for which R is a constant independent of V. But R is not a constant for many substances other than metals, nor for devices such as diodes, vacuum tubes, transistors, and so on. Even for metals, R is not constant if the temperature changes much: for a lightbulb filament the measured resistance is low for small currents, but is much higher at its normal large operating current that puts it at the high temperature needed to make it glow (3000 K). Thus Ohm's "law" is not a fundamental law, but rather a description of a certain class of materials: metal conductors, whose temperature does not change much. Materials or devices that do not follow Ohm's law are said to be *nonohmic*. See Fig. 25–9.

The unit for resistance is called the **ohm** and is abbreviated Ω (Greek capital letter omega). Because $R = V/I$, we see that $1.0\,\Omega$ is equivalent to $1.0\,\text{V/A}$.

FIGURE 25–10 Example 25–3.

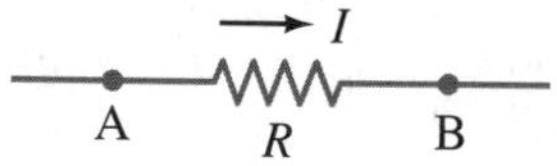

A R B

$\longrightarrow I$

| **CONCEPTUAL EXAMPLE 25–3** | **Current and potential.** Current I enters a resistor R as shown in Fig. 25–10. (a) Is the potential higher at point A or at point B? (b) Is the current greater at point A or at point B?

RESPONSE (a) Positive charge always flows from + to −, from high potential to low potential. Think again of the gravitational analogy: a mass will fall down from high gravitational potential to low. So for positive current I, point A is at a higher potential than point B.

(b) Conservation of charge requires that whatever charge flows into the resistor at point A, an equal amount of charge emerges at point B. Charge or current does not get "used up" by a resistor, just as an object that falls through a gravitational potential difference does not gain or lose mass. So the current is the same at A and B.

An electric potential decrease, as from point A to point B in Example 25–3, is often called a **potential drop** or a **voltage drop**.

EXAMPLE 25-4 **Flashlight bulb resistance.** A small flashlight bulb (Fig. 25–11) draws 300 mA from its 1.5-V battery. (a) What is the resistance of the bulb? (b) If the battery becomes weak and the voltage drops to 1.2 V, how would the current change?

APPROACH We can apply Ohm's law to the bulb, where the voltage applied across it is the battery voltage.

SOLUTION (a) We change 300 mA to 0.30 A and use Eq. 25–2a or b:

$$R = \frac{V}{I} = \frac{1.5 \text{ V}}{0.30 \text{ A}} = 5.0 \ \Omega.$$

(b) If the resistance stays the same, the current would be

$$I = \frac{V}{R} = \frac{1.2 \text{ V}}{5.0 \ \Omega} = 0.24 \text{ A} = 240 \text{ mA},$$

or a decrease of 60 mA.

NOTE With the smaller current in part (b), the bulb filament's temperature would be lower and the bulb less bright. Also, resistance does depend on temperature (Section 25–4), so our calculation is only a rough approximation.

FIGURE 25–11 Flashlight (Example 25–4). Note how the circuit is completed along the side strip.

EXERCISE B What resistance should be connected across a 9.0-V battery to make a 10-mA current? (a) 9 Ω, (b) 0.9 Ω, (c) 900 Ω, (d) 1.1 Ω, (e) 0.11 Ω.

All electric devices, from heaters to lightbulbs to stereo amplifiers, offer resistance to the flow of current. The filaments of lightbulbs (Fig. 25–5) and electric heaters are special types of wires whose resistance results in their becoming very hot. Generally, the connecting wires have very low resistance in comparison to the resistance of the wire filaments or coils, so the connecting wires usually have a minimal effect on the magnitude of the current. In many circuits, particularly in electronic devices, **resistors** are used to control the amount of current. Resistors have resistances ranging from less than an ohm to millions of ohms (see Figs. 25–12 and 25–13). The main types are "wire-wound" resistors which consist of a coil of fine wire, "composition" resistors which are usually made of carbon, and thin carbon or metal films.

When we draw a diagram of a circuit, we use the symbol

–/\/\/–   [resistor symbol]

to indicate a resistance. Wires whose resistance is negligible, however, are shown simply as straight lines.

FIGURE 25–12 Photo of resistors (striped), plus other devices on a circuit board.

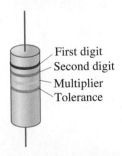

First digit
Second digit
Multiplier
Tolerance

FIGURE 25–13 The resistance value of a given resistor is written on the exterior, or may be given as a color code as shown above and in the Table: the first two colors represent the first two digits in the value of the resistance, the third color represents the power of ten that it must be multiplied by, and the fourth is the manufactured tolerance. For example, a resistor whose four colors are red, green, yellow, and silver has a resistance of $25 \times 10^4 \ \Omega = 250{,}000 \ \Omega = 250 \text{ k}\Omega$, plus or minus 10%. An alternate example of a simple code is a number such as 104, which means $R = 1.0 \times 10^4 \ \Omega$.

Resistor Color Code

Color	Number	Multiplier	Tolerance
Black	0	1	
Brown	1	10^1	1%
Red	2	10^2	2%
Orange	3	10^3	
Yellow	4	10^4	
Green	5	10^5	
Blue	6	10^6	
Violet	7	10^7	
Gray	8	10^8	
White	9	10^9	
Gold		10^{-1}	5%
Silver		10^{-2}	10%
No color			20%

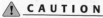

Some Helpful Clarifications

Here we briefly summarize some possible misunderstandings and clarifications. Batteries do not put out a constant current. Instead, batteries are intended to maintain a constant potential difference, or very nearly so. (Details in the next Chapter.) Thus a battery should be considered a source of voltage. The voltage is applied *across* a wire or device.

Electric current passes *through* a wire or device (connected to a battery), and its magnitude depends on that device's resistance. The resistance is a *property* of the wire or device. The voltage, on the other hand, is external to the wire or device, and is applied across the two ends of the wire or device. The current through the device might be called the "response": the current increases if the voltage increases or the resistance decreases, as $I = V/R$.

In a wire, the direction of the current is always parallel to the wire, no matter how the wire curves, just like water in a pipe. The direction of conventional (positive) current is from high potential $(+)$ toward lower potential $(-)$.

Current and charge do not increase or decrease or get "used up" when going through a wire or other device. The amount of charge that goes in at one end comes out at the other end.

25–4 Resistivity

It is found experimentally that the resistance R of any wire is directly proportional to its length ℓ and inversely proportional to its cross-sectional area A. That is,

$$R = \rho \frac{\ell}{A}, \tag{25–3}$$

where ρ, the constant of proportionality, is called the **resistivity** and depends on the material used. Typical values of ρ, whose units are $\Omega \cdot \text{m}$ (see Eq. 25–3), are given for various materials in the middle column of Table 25–1, which is divided into the categories *conductors*, *insulators*, and *semiconductors* (see Section 21–3). The values depend somewhat on purity, heat treatment, temperature, and other factors. Notice that silver has the lowest resistivity and is thus the best conductor (although it is expensive). Copper is close, and much less expensive, which is why most wires are made of copper. Aluminum, although it has a higher resistivity, is much less dense than copper; it is thus preferable to copper in some situations, such as for transmission lines, because its resistance for the same weight is less than that for copper.

TABLE 25–1 Resistivity and Temperature Coefficients (at 20°C)

Material	Resistivity, ρ ($\Omega \cdot$ m)	Temperature Coefficient, α (C°)$^{-1}$
Conductors		
Silver	1.59×10^{-8}	0.0061
Copper	1.68×10^{-8}	0.0068
Gold	2.44×10^{-8}	0.0034
Aluminum	2.65×10^{-8}	0.00429
Tungsten	5.6×10^{-8}	0.0045
Iron	9.71×10^{-8}	0.00651
Platinum	10.6×10^{-8}	0.003927
Mercury	98×10^{-8}	0.0009
Nichrome (Ni, Fe, Cr alloy)	100×10^{-8}	0.0004
Semiconductors†		
Carbon (graphite)	$(3-60) \times 10^{-5}$	−0.0005
Germanium	$(1-500) \times 10^{-3}$	−0.05
Silicon	0.1−60	−0.07
Insulators		
Glass	$10^{9} - 10^{12}$	
Hard rubber	$10^{13} - 10^{15}$	

†Values depend strongly on the presence of even slight amounts of impurities.

The reciprocal of the resistivity, called the **conductivity** σ, is

$$\sigma = \frac{1}{\rho} \tag{25-4}$$

and has units of $(\Omega \cdot m)^{-1}$.

EXERCISE C Return to the Chapter-Opening Question, page 651, and answer it again now. Try to explain why you may have answered differently the first time.

EXERCISE D A copper wire has a resistance of $10\,\Omega$. What will its resistance be if it is only half as long? (a) $20\,\Omega$, (b) $10\,\Omega$, (c) $5\,\Omega$, (d) $1\,\Omega$, (e) none of these.

EXAMPLE 25–5 **Speaker wires.** Suppose you want to connect your stereo to remote speakers (Fig. 25–14). (a) If each wire must be 20 m long, what diameter copper wire should you use to keep the resistance less than $0.10\,\Omega$ per wire? (b) If the current to each speaker is 4.0 A, what is the potential difference, or voltage drop, across each wire?

APPROACH We solve Eq. 25–3 to get the area A, from which we can calculate the wire's radius using $A = \pi r^2$. The diameter is $2r$. In (b) we can use Ohm's law, $V = IR$.

SOLUTION (a) We solve Eq. 25–3 for the area A and find ρ for copper in Table 25–1:

$$A = \rho \frac{\ell}{R} = \frac{(1.68 \times 10^{-8}\,\Omega \cdot m)(20\,m)}{(0.10\,\Omega)} = 3.4 \times 10^{-6}\,m^2.$$

The cross-sectional area A of a circular wire is $A = \pi r^2$. The radius must then be at least

$$r = \sqrt{\frac{A}{\pi}} = 1.04 \times 10^{-3}\,m = 1.04\,mm.$$

The diameter is twice the radius and so must be at least $2r = 2.1\,mm$.

(b) From $V = IR$ we find that the voltage drop across each wire is

$$V = IR = (4.0\,A)(0.10\,\Omega) = 0.40\,V.$$

NOTE The voltage drop across the wires reduces the voltage that reaches the speakers from the stereo amplifier, thus reducing the sound level a bit.

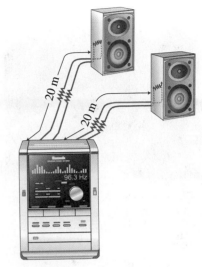

FIGURE 25–14 Example 25–5.

CONCEPTUAL EXAMPLE 25–6 **Stretching changes resistance.** Suppose a wire of resistance R could be stretched uniformly until it was twice its original length. What would happen to its resistance?

RESPONSE If the length ℓ doubles, then the cross-sectional area A is halved, because the volume $(V = A\ell)$ of the wire remains the same. From Eq. 25–3 we see that the resistance would increase by a factor of four $(2/\frac{1}{2} = 4)$.

EXERCISE E Copper wires in houses typically have a diameter of about 1.5 mm. How long a wire would have a 1.0-Ω resistance?

Temperature Dependence of Resistivity

The resistivity of a material depends somewhat on temperature. The resistance of metals generally increases with temperature. This is not surprising, for at higher temperatures, the atoms are moving more rapidly and are arranged in a less orderly fashion. So they might be expected to interfere more with the flow of electrons. If the temperature change is not too great, the resistivity of metals usually increases nearly linearly with temperature. That is,

$$\rho_T = \rho_0 \left[1 + \alpha (T - T_0) \right] \tag{25-5}$$

where ρ_0 is the resistivity at some reference temperature T_0 (such as 0°C or 20°C), ρ_T is the resistivity at a temperature T, and α is the *temperature coefficient of resistivity*. Values for α are given in Table 25–1. Note that the temperature coefficient for semiconductors can be negative. Why? It seems that at higher temperatures, some of the electrons that are normally not free in a semiconductor become free and can contribute to the current. Thus, the resistance of a semiconductor can decrease with an increase in temperature.

FIGURE 25–15 A thermistor shown next to a millimeter ruler for scale.

EXAMPLE 25–7 **Resistance thermometer.** The variation in electrical resistance with temperature can be used to make precise temperature measurements. Platinum is commonly used since it is relatively free from corrosive effects and has a high melting point. Suppose at 20.0°C the resistance of a platinum resistance thermometer is 164.2 Ω. When placed in a particular solution, the resistance is 187.4 Ω. What is the temperature of this solution?

APPROACH Since the resistance R is directly proportional to the resistivity ρ, we can combine Eq. 25–3 with Eq. 25–5 to find R as a function of temperature T, and then solve that equation for T.

SOLUTION We multiply Eq. 25–5 by (ℓ/A) to obtain (see also Eq. 25–3)

$$R = R_0\left[1 + \alpha(T - T_0)\right].$$

Here $R_0 = \rho_0 \ell / A$ is the resistance of the wire at $T_0 = 20.0°C$. We solve this equation for T and find (see Table 25–1 for α)

$$T = T_0 + \frac{R - R_0}{\alpha R_0} = 20.0°C + \frac{187.4\ \Omega - 164.2\ \Omega}{(3.927 \times 10^{-3}(C°)^{-1})(164.2\ \Omega)} = 56.0°C.$$

NOTE Resistance thermometers have the advantage that they can be used at very high or low temperatures where gas or liquid thermometers would be useless.

NOTE More convenient for some applications is a *thermistor* (Fig. 25–15), which consists of a metal oxide or semiconductor whose resistance also varies in a repeatable way with temperature. Thermistors can be made quite small and respond very quickly to temperature changes.

EXERCISE F The resistance of the tungsten filament of a common incandescent lightbulb is how many times greater at its operating temperature of 3000 K than its resistance at room temperature? (*a*) Less than 1% greater; (*b*) roughly 10% greater; (*c*) about 2 times greater; (*d*) roughly 10 times greater; (*e*) more than 100 times greater.

The value of α in Eq. 25–5 itself can depend on temperature, so it is important to check the temperature range of validity of any value (say, in a handbook of physical data). If the temperature range is wide, Eq. 25–5 is not adequate and terms proportional to the square and cube of the temperature are needed, but they are generally very small except when $T - T_0$ is large.

25–5 Electric Power

Electric energy is useful to us because it can be easily transformed into other forms of energy. Motors transform electric energy into mechanical energy, and are examined in Chapter 27.

In other devices such as electric heaters, stoves, toasters, and hair dryers, electric energy is transformed into thermal energy in a wire resistance known as a "heating element." And in an ordinary lightbulb, the tiny wire filament (Fig. 25–5 and Chapter-opening photo) becomes so hot it glows; only a few percent of the energy is transformed into visible light, and the rest, over 90%, into thermal energy. Lightbulb filaments and heating elements (Fig. 25–16) in household appliances have resistances typically of a few ohms to a few hundred ohms.

Electric energy is transformed into thermal energy or light in such devices, and there are many collisions between the moving electrons and the atoms of the wire. In each collision, part of the electron's kinetic energy is transferred to the atom with which it collides. As a result, the kinetic energy of the wire's atoms increases and hence the temperature of the wire element increases. The increased thermal energy can be transferred as heat by conduction and convection to the air in a heater or to food in a pan, by radiation to bread in a toaster, or radiated as light.

To find the power transformed by an electric device, recall that the energy transformed when an infinitesimal charge dq moves through a potential difference V is $dU = V\,dq$ (Eq. 23–3). Let dt be the time required for an amount of charge dq

FIGURE 25–16 Hot electric stove burner glows because of energy transformed by electric current.

to move through a potential difference V. Then the power P, which is the rate energy is transformed, is

$$P = \frac{dU}{dt} = \frac{dq}{dt}V$$

The charge that flows per second, dq/dt, is the electric current I. Thus we have

$$P = IV. \tag{25-6}$$

This general relation gives us the power transformed by any device, where I is the current passing through it and V is the potential difference across it. It also gives the power delivered by a source such as a battery. The SI unit of electric power is the same as for any kind of power, the **watt** $(1\,\text{W} = 1\,\text{J/s})$.

The rate of energy transformation in a resistance R can be written in two other ways, starting with the general relation $P = IV$ and substituting in $V = IR$:

$$P = IV = I(IR) = I^2R \tag{25-7a}$$

$$P = IV = \left(\frac{V}{R}\right)V = \frac{V^2}{R}. \tag{25-7b}$$

Equations 25–7a and b apply only to resistors, whereas Eq. 25–6, $P = IV$, is more general and applies to any device, including a resistor.

EXAMPLE 25–8 **Headlights.** Calculate the resistance of a 40-W automobile headlight designed for 12 V (Fig. 25–17).

APPROACH We solve Eq. 25–7b for R.

SOLUTION From Eq. 25–7b,

$$R = \frac{V^2}{P} = \frac{(12\,\text{V})^2}{(40\,\text{W})} = 3.6\,\Omega.$$

NOTE This is the resistance when the bulb is burning brightly at 40 W. When the bulb is cold, the resistance is much lower, as we saw in Eq. 25–5. Since the current is high when the resistance is low, lightbulbs burn out most often when first turned on.

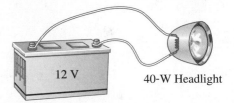

FIGURE 25–17 Example 25–8.

It is energy, not power, that you pay for on your electric bill. Since power is the *rate* energy is transformed, the total energy used by any device is simply its power consumption multiplied by the time it is on. If the power is in watts and the time is in seconds, the energy will be in joules since $1\,\text{W} = 1\,\text{J/s}$. Electric companies usually specify the energy with a much larger unit, the **kilowatt-hour** (kWh). One kWh $= (1000\,\text{W})(3600\,\text{s}) = 3.60 \times 10^6\,\text{J}$.

EXAMPLE 25–9 **Electric heater.** An electric heater draws a steady 15.0 A on a 120-V line. How much power does it require and how much does it cost per month (30 days) if it operates 3.0 h per day and the electric company charges 9.2 cents per kWh?

APPROACH We use Eq. 25–6, $P = IV$, to find the power. We multiply the power (in kW) by the time (h) used in a month and by the cost per energy unit, $0.092 per kWh, to get the cost per month.

SOLUTION The power is

$$P = IV = (15.0\,\text{A})(120\,\text{V}) = 1800\,\text{W}$$

or 1.80 kW. The time (in hours) the heater is used per month is $(3.0\,\text{h/d})(30\,\text{d}) = 90\,\text{h}$, which at 9.2¢/kWh would cost $(1.80\,\text{kW})(90\,\text{h})(\$0.092/\text{kWh}) = \$15$.

NOTE Household current is actually alternating (ac), but our solution is still valid assuming the given values for V and I are the proper averages (rms) as we discuss in Section 25–7.

FIGURE 25–18 Example 25–10. A lightning bolt.

EXAMPLE 25–10 **ESTIMATE** | **Lightning bolt.** Lightning is a spectacular example of electric current in a natural phenomenon (Fig. 25–18). There is much variability to lightning bolts, but a typical event can transfer 10^9 J of energy across a potential difference of perhaps 5×10^7 V during a time interval of about 0.2 s. Use this information to estimate (*a*) the total amount of charge transferred between cloud and ground, (*b*) the current in the lightning bolt, and (*c*) the average power delivered over the 0.2 s.

APPROACH We estimate the charge Q, recalling that potential energy change equals the potential difference ΔV times the charge Q, Eq. 23–3. We equate ΔU with the energy transferred, $\Delta U \approx 10^9$ J. Next, the current I is Q/t (Eq. 25–1a) and the power P is energy/time.

SOLUTION (*a*) From Eq. 23–3, the energy transformed is $\Delta U = Q\,\Delta V$. We solve for Q:

$$Q = \frac{\Delta U}{\Delta V} \approx \frac{10^9\,\text{J}}{5 \times 10^7\,\text{V}} = 20\,\text{coulombs.}$$

(*b*) The current during the 0.2 s is about

$$I = \frac{Q}{t} \approx \frac{20\,\text{C}}{0.2\,\text{s}} = 100\,\text{A.}$$

(*c*) The average power delivered is

$$P = \frac{\text{energy}}{\text{time}} = \frac{10^9\,\text{J}}{0.2\,\text{s}} = 5 \times 10^9\,\text{W} = 5\,\text{GW.}$$

We can also use Eq. 25–6:

$$P = IV = (100\,\text{A})(5 \times 10^7\,\text{V}) = 5\,\text{GW.}$$

NOTE Since most lightning bolts consist of several stages, it is possible that individual parts could carry currents much higher than the 100 A calculated above.

25–6 Power in Household Circuits

The electric wires that carry electricity to lights and other electric appliances have some resistance, although usually it is quite small. Nonetheless, if the current is large enough, the wires will heat up and produce thermal energy at a rate equal to I^2R, where R is the wire's resistance. One possible hazard is that the current-carrying wires in the wall of a building may become so hot as to start a fire. Thicker wires have less resistance (see Eq. 25–3) and thus can carry more current without becoming too hot. When a wire carries more current than is safe, it is said to be "overloaded." To prevent overloading, *fuses* or *circuit breakers* are installed in circuits. They are basically switches (Fig. 25–19)

FIGURE 25–19 (a) Fuses. When the current exceeds a certain value, the metallic ribbon melts and the circuit opens. Then the fuse must be replaced. (b) One type of circuit breaker. The electric current passes through a bimetallic strip. When the current exceeds a safe level, the heating of the bimetallic strip causes the strip to bend so far to the left that the notch in the spring-loaded metal strip drops down over the end of the bimetallic strip; (c) the circuit then opens at the contact points (one is attached to the metal strip) and the outside switch is also flipped. As soon as the bimetallic strip cools down, it can be reset using the outside switch. Magnetic-type circuit breakers are discussed in Chapters 28 and 29.

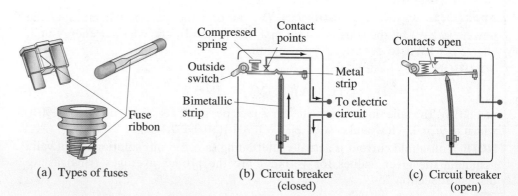

(a) Types of fuses

(b) Circuit breaker (closed)

(c) Circuit breaker (open)

that open the circuit when the current exceeds some particular value. A 20-A fuse or circuit breaker, for example, opens when the current passing through it exceeds 20 A. If a circuit repeatedly burns out a fuse or opens a circuit breaker, there are two possibilities: there may be too many devices drawing current in that circuit; or there is a fault somewhere, such as a "short." A short, or "short circuit," means that two wires have touched that should not have (perhaps because the insulation has worn through) so the resistance is much reduced and the current becomes very large. Short circuits should be remedied immediately.

Household circuits are designed with the various devices connected so that each receives the standard voltage (usually 120 V in the United States) from the electric company (Fig. 25–20). Circuits with the devices arranged as in Fig. 25–20 are called *parallel circuits*, as we will discuss in the next Chapter. When a fuse blows or circuit breaker opens, it is important to check the total current being drawn on that circuit, which is the sum of the currents in each device.

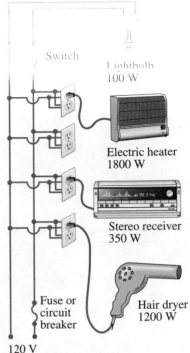

FIGURE 25–20 Connection of household appliances.

EXAMPLE 25–11 **Will a fuse blow?** Determine the total current drawn by all the devices in the circuit of Fig. 25–20.

APPROACH Each device has the same 120-V voltage across it. The current each draws from the source is found from $I = P/V$, Eq. 25–6.

SOLUTION The circuit in Fig. 25–20 draws the following currents: the lightbulb draws $I = P/V = 100 \text{ W}/120 \text{ V} = 0.8 \text{ A}$; the heater draws $1800 \text{ W}/120 \text{ V} = 15.0 \text{ A}$; the stereo draws a maximum of $350 \text{ W}/120 \text{ V} = 2.9 \text{ A}$; and the hair dryer draws $1200 \text{ W}/120 \text{ V} = 10.0 \text{ A}$. The total current drawn, if all devices are used at the same time, is

$$0.8 \text{ A} + 15.0 \text{ A} + 2.9 \text{ A} + 10.0 \text{ A} = 28.7 \text{ A}.$$

NOTE The heater draws as much current as 18 100-W lightbulbs. For safety, the heater should probably be on a circuit by itself.

If the circuit in Fig. 25–20 is designed for a 20-A fuse, the fuse should blow, and we hope it will, to prevent overloaded wires from getting hot enough to start a fire. Something will have to be turned off to get this circuit below 20 A. (Houses and apartments usually have several circuits, each with its own fuse or circuit breaker; try moving one of the devices to another circuit.) If the circuit is designed with heavier wire and a 30-A fuse, the fuse shouldn't blow—if it does, a short may be the problem. (The most likely place for a short is in the cord of one of the devices.) Proper fuse size is selected according to the wire used to supply the current. A properly rated fuse should *never* be replaced by a higher-rated one. A fuse blowing or a circuit breaker opening is acting like a switch, making an "open circuit." By an open circuit, we mean that there is no longer a complete conducting path, so no current can flow; it is as if $R = \infty$.

PHYSICS APPLIED
Proper fuses and shorts

CONCEPTUAL EXAMPLE 25–12 **A dangerous extension cord.** Your 1800-W portable electric heater is too far from your desk to warm your feet. Its cord is too short, so you plug it into an extension cord rated at 11 A. Why is this dangerous?

RESPONSE 1800 W at 120 V draws a 15-A current. The wires in the extension cord rated at 11 A could become hot enough to melt the insulation and cause a fire.

PHYSICS APPLIED
Extension cords and possible danger

EXERCISE G How many 60-W 120-V lightbulbs can operate on a 20-A line? (*a*) 2; (*b*) 3; (*c*) 6; (*d*) 20; (*e*) 40.

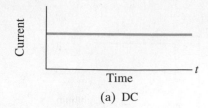

(a) DC

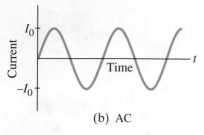

(b) AC

FIGURE 25–21 (a) Direct current. (b) Alternating current.

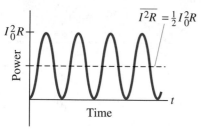

FIGURE 25–22 Power transformed in a resistor in an ac circuit.

25–7 Alternating Current

When a battery is connected to a circuit, the current moves steadily in one direction. This is called a **direct current**, or **dc**. Electric generators at electric power plants, however, produce **alternating current**, or **ac**. (Sometimes capital letters are used, DC and AC.) An alternating current reverses direction many times per second and is commonly sinusoidal, as shown in Fig. 25–21. The electrons in a wire first move in one direction and then in the other. The current supplied to homes and businesses by electric companies is ac throughout virtually the entire world. We will discuss and analyze ac circuits in detail in Chapter 30. But because ac circuits are so common in real life, we will discuss some of their basic aspects here.

The voltage produced by an ac electric generator is sinusoidal, as we shall see later. The current it produces is thus sinusoidal (Fig. 25–21b). We can write the voltage as a function of time as

$$V = V_0 \sin 2\pi f t = V_0 \sin \omega t.$$

The potential V oscillates between $+V_0$ and $-V_0$, and V_0 is referred to as the **peak voltage**. The frequency f is the number of complete oscillations made per second, and $\omega = 2\pi f$. In most areas of the United States and Canada, f is 60 Hz (the unit "hertz," as we saw in Chapters 10 and 14, means cycles per second). In many other countries, 50 Hz is used.

Equation 25–2, $V = IR$, works also for ac: if a voltage V exists across a resistance R, then the current I through the resistance is

$$I = \frac{V}{R} = \frac{V_0}{R} \sin \omega t = I_0 \sin \omega t. \tag{25–8}$$

The quantity $I_0 = V_0/R$ is the **peak current**. The current is considered positive when the electrons flow in one direction and negative when they flow in the opposite direction. It is clear from Fig. 25–21b that an alternating current is as often positive as it is negative. Thus, the average current is zero. This does not mean, however, that no power is needed or that no heat is produced in a resistor. Electrons do move back and forth, and do produce heat. Indeed, the power transformed in a resistance R at any instant is

$$P = I^2 R = I_0^2 R \sin^2 \omega t.$$

Because the current is squared, we see that the power is always positive, as graphed in Fig. 25–22. The quantity $\sin^2 \omega t$ varies between 0 and 1; and it is not too difficult to show[†] that its average value is $\frac{1}{2}$, as indicated in Fig. 25–22. Thus, the *average power* transformed, $\overline{P}$, is

$$\overline{P} = \tfrac{1}{2} I_0^2 R.$$

Since power can also be written $P = V^2/R = (V_0^2/R) \sin^2 \omega t$, we also have that the average power is

$$\overline{P} = \tfrac{1}{2} \frac{V_0^2}{R}.$$

The average or mean value of the *square* of the current or voltage is thus what is important for calculating average power: $\overline{I^2} = \tfrac{1}{2} I_0^2$ and $\overline{V^2} = \tfrac{1}{2} V_0^2$. The square root of each of these is the **rms** (root-mean-square) value of the current or voltage:

$$I_{\text{rms}} = \sqrt{\overline{I^2}} = \frac{I_0}{\sqrt{2}} = 0.707 I_0, \tag{25–9a}$$

$$V_{\text{rms}} = \sqrt{\overline{V^2}} = \frac{V_0}{\sqrt{2}} = 0.707 V_0. \tag{25–9b}$$

The rms values of V and I are sometimes called the *effective values*.

[†]A graph of $\cos^2 \omega t$ versus t is identical to that for $\sin^2 \omega t$ in Fig. 25–22, except that the points are shifted (by $\frac{1}{4}$ cycle) on the time axis. Hence the average value of $\sin^2$ and $\cos^2$, averaged over one or more full cycles, will be the same: $\overline{\sin^2 \omega t} = \overline{\cos^2 \omega t}$. From the trigonometric identity $\sin^2 \theta + \cos^2 \theta = 1$, we can write

$$\overline{(\sin^2 \omega t)} + \overline{(\cos^2 \omega t)} = 2\overline{(\sin^2 \omega t)} = 1.$$

Hence the average value of $\sin^2 \omega t$ is $\frac{1}{2}$.

They are useful because they can be substituted directly into the power formulas, Eqs. 25–6 and 25–7, to get the average power:

$$\overline{P} = I_{rms} V_{rms} \qquad (25\text{–}10a)$$

$$\overline{P} = \tfrac{1}{2} I_0^2 R = I_{rms}^2 R \qquad (25\text{–}10b)$$

$$\overline{P} = \tfrac{1}{2} \frac{V_0^2}{R} = \frac{V_{rms}^2}{R} \qquad (25\ 10c)$$

Thus, a direct current whose values of I and V equal the rms values of I and V for an alternating current will produce the same power. Hence it is usually the rms value of current and voltage that is specified or measured. For example, in the United States and Canada, standard line voltage[†] is 120-V ac. The 120 V is V_{rms}; the peak voltage V_0 is

$$V_0 = \sqrt{2}\, V_{rms} = 170\,\text{V}.$$

In much of the world (Europe, Australia, Asia) the rms voltage is 240 V, so the peak voltage is 340 V.

EXAMPLE 25–13 **Hair dryer.** (*a*) Calculate the resistance and the peak current in a 1000-W hair dryer (Fig. 25–23) connected to a 120-V line. (*b*) What happens if it is connected to a 240-V line in Britain?

APPROACH We are given $\overline{P}$ and V_{rms}, so $I_{rms} = \overline{P}/V_{rms}$ (Eq. 25–10a or 25–6), and $I_0 = \sqrt{2}\, I_{rms}$. Then we find R from $V = IR$.

SOLUTION (*a*) We solve Eq. 25–10a for the rms current:

$$I_{rms} = \frac{\overline{P}}{V_{rms}} = \frac{1000\,\text{W}}{120\,\text{V}} = 8.33\,\text{A}.$$

Then

$$I_0 = \sqrt{2}\, I_{rms} = 11.8\,\text{A}.$$

The resistance is

$$R = \frac{V_{rms}}{I_{rms}} = \frac{120\,\text{V}}{8.33\,\text{A}} = 14.4\,\Omega.$$

The resistance could equally well be calculated using peak values:

$$R = \frac{V_0}{I_0} = \frac{170\,\text{V}}{11.8\,\text{A}} = 14.4\,\Omega.$$

(*b*) When connected to a 240-V line, more current would flow and the resistance would change with the increased temperature (Section 25–4). But let us make an estimate of the power transformed based on the same 14.4-Ω resistance. The average power would be

$$\overline{P} = \frac{V_{rms}^2}{R} = \frac{(240\,\text{V})^2}{(14.4\,\Omega)} = 4000\,\text{W}.$$

This is four times the dryer's power rating and would undoubtedly melt the heating element or the wire coils of the motor.

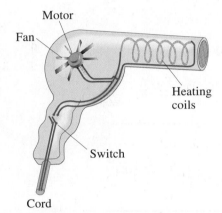

FIGURE 25–23 A hair dryer. Most of the current goes through the heating coils, a pure resistance; a small part goes to the motor to turn the fan. Example 25–13.

EXERCISE H Each channel of a stereo receiver is capable of an average power output of 100 W into an 8-Ω loudspeaker (see Fig. 25–14). What are the rms voltage and the rms current fed to the speaker (*a*) at the maximum power of 100 W, and (*b*) at 1.0 W when the volume is turned down?

[†]The line voltage can vary, depending on the total load; the frequency of 60 Hz or 50 Hz, however, remains extremely steady.

25–8 Microscopic View of Electric Current: Current Density and Drift Velocity

Up to now in this Chapter we have dealt mainly with a macroscopic view of electric current. We saw, however, that according to atomic theory, the electric current in metal wires is carried by negatively charged electrons, and that in liquid solutions current can also be carried by positive and/or negative ions. Let us now look at this microscopic picture in more detail.

When a potential difference is applied to the two ends of a wire of uniform cross section, the direction of the electric field $\vec{\mathbf{E}}$ is parallel to the walls of the wire (Fig. 25–24). The existence of $\vec{\mathbf{E}}$ within the conducting wire does not contradict our earlier result that $\vec{\mathbf{E}} = 0$ inside a conductor in the electrostatic case, as we are no longer dealing with the static case. Charges are free to move in a conductor, and hence can move under the action of the electric field. If all the charges are at rest, then $\vec{\mathbf{E}}$ must be zero (electrostatics).

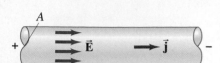

FIGURE 25–24 Electric field $\vec{\mathbf{E}}$ in a uniform wire of cross-sectional area A carrying a current I. The current density $j = I/A$.

We now define a new microscopic quantity, the **current density**, $\vec{\mathbf{j}}$. It is defined as the *electric current per unit cross-sectional area* at any point in space. If the current density $\vec{\mathbf{j}}$ in a wire of cross-sectional area A is uniform over the cross section, then j is related to the electric current by

$$j = \frac{I}{A} \quad \text{or} \quad I = jA. \tag{25–11}$$

If the current density is not uniform, then the general relation is

$$I = \int \vec{\mathbf{j}} \cdot d\vec{\mathbf{A}}, \tag{25–12}$$

where $d\vec{\mathbf{A}}$ is an element of surface and I is the current through the surface over which the integration is taken. The direction of the current density at any point is the direction that a positive charge would move when placed at that point—that is, the direction of $\vec{\mathbf{j}}$ at any point is generally the same as the direction of $\vec{\mathbf{E}}$, Fig. 25–24. The current density exists for any *point* in space. The current I, on the other hand, refers to a conductor as a whole, and hence is a macroscopic quantity.

The direction of $\vec{\mathbf{j}}$ is chosen to represent the direction of net flow of positive charge. In a conductor, it is negatively charged electrons that move, so they move in the direction of $-\vec{\mathbf{j}}$, or $-\vec{\mathbf{E}}$ (to the left in Fig. 25–24). We can imagine the free electrons as moving about randomly at high speeds, bouncing off the atoms of the wire (somewhat like the molecules of a gas—Chapter 18). When an electric field exists in the wire, Fig. 25–25, the electrons feel a force and initially begin to accelerate. But they soon reach a more or less steady average velocity in the direction of $\vec{\mathbf{E}}$, known as their **drift velocity**, $\vec{\boldsymbol{v}}_{\mathrm{d}}$ (collisions with atoms in the wire keep them from accelerating further). The drift velocity is normally very much smaller than the electrons' average random speed.

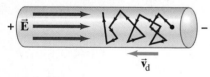

FIGURE 25–25 Electric field $\vec{\mathbf{E}}$ in a wire gives electrons in random motion a drift velocity v_{d}.

We can relate the drift velocity v_{d} to the macroscopic current I in the wire. In a time Δt, the electrons will travel a distance $\ell = v_{\mathrm{d}} \Delta t$ on average. Suppose the wire has cross-sectional area A. Then in time Δt, electrons in a volume $V = A\ell = Av_{\mathrm{d}} \Delta t$ will pass through the cross section A of wire, as shown in Fig. 25–26. If there are n free electrons (each of charge $-e$) per unit volume $(n = N/V)$, then the total charge ΔQ that passes through the area A in a time Δt is

$$\Delta Q = (\text{no. of charges, } N) \times (\text{charge per particle})$$
$$= (nV)(-e) = -(nAv_{\mathrm{d}} \Delta t)(e).$$

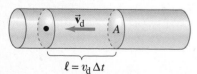

FIGURE 25–26 Electrons in the volume $A\ell$ will all pass through the cross section indicated in a time Δt, where $\ell = v_{\mathrm{d}} \Delta t$.

The current I in the wire is thus

$$I = \frac{\Delta Q}{\Delta t} = -neAv_{\mathrm{d}}. \tag{25–13}$$

The current density, $j = I/A$, is

$$j = -nev_{\mathrm{d}}. \tag{25–14}$$

In vector form, this is written

$$\vec{\mathbf{j}} = -ne\vec{\boldsymbol{v}}_{\mathrm{d}}, \tag{25–15}$$

where the minus sign indicates that the direction of (positive) current flow is opposite to the drift velocity of electrons.

We can generalize Eq. 25–15 to any type of charge flow, such as flow of ions in an electrolyte. If there are several types of ions (which can include free electrons), each of density n_i (number per unit volume), charge q_i ($q_i = -e$ for electrons) and drift velocity $\vec{v}_{di}$, then the net current density at any point is

$$\mathbf{j} = \sum_i n_i q_i \vec{v}_{di}. \tag{25–16}$$

The total current I passing through an area A perpendicular to a uniform $\mathbf{j}$ is then

$$I = \sum_i n_i q_i v_{di} A.$$

EXAMPLE 25–14 **Electron speeds in a wire.** A copper wire 3.2 mm in diameter carries a 5.0-A current. Determine (a) the current density in the wire, and (b) the drift velocity of the free electrons. (c) Estimate the rms speed of electrons assuming they behave like an ideal gas at 20°C. Assume that one electron per Cu atom is free to move (the others remain bound to the atom).

APPROACH For (a) $j = I/A = I/\pi r^2$. For (b) we can apply Eq. 25–14 to find v_d if we can determine the number n of free electrons per unit volume. Since we assume there is one free electron per atom, the density of free electrons, n, is the same as the density of Cu atoms. The atomic mass of Cu is 63.5 u (see Periodic Table inside the back cover), so 63.5 g of Cu contains one mole or 6.02×10^{23} free electrons. The mass density of copper (Table 13–1) is $\rho_D = 8.9 \times 10^3 \, \text{kg/m}^3$, where $\rho_D = m/V$. (We use ρ_D to distinguish it here from ρ for resistivity.) In (c) we use $K = \frac{3}{2}kT$, Eq. 18–4. (Do not confuse V for volume with V for voltage.)

SOLUTION (a) The current density is (with $r = \frac{1}{2}(3.2 \, \text{mm}) = 1.6 \times 10^{-3} \, \text{m}$)

$$j = \frac{I}{A} = \frac{I}{\pi r^2} = \frac{5.0 \, \text{A}}{\pi(1.6 \times 10^{-3} \, \text{m})^2} = 6.2 \times 10^5 \, \text{A/m}^2.$$

(b) The number of free electrons per unit volume, $n = N/V$ (where $V = m/\rho_D$), is

$$n = \frac{N}{V} = \frac{N}{m/\rho_D} = \frac{N(1 \, \text{mole})}{m(1 \, \text{mole})} \rho_D$$

$$n = \left(\frac{6.02 \times 10^{23} \, \text{electrons}}{63.5 \times 10^{-3} \, \text{kg}}\right)(8.9 \times 10^3 \, \text{kg/m}^3) = 8.4 \times 10^{28} \, \text{m}^{-3}.$$

Then, by Eq. 25–14, the drift velocity has magnitude

$$v_d = \frac{j}{ne} = \frac{6.2 \times 10^5 \, \text{A/m}^2}{(8.4 \times 10^{28} \, \text{m}^{-3})(1.6 \times 10^{-19} \, \text{C})} = 4.6 \times 10^{-5} \, \text{m/s} \approx 0.05 \, \text{mm/s}.$$

(c) If we model the free electrons as an ideal gas (a rather rough approximation), we use Eq. 18–5 to estimate the random rms speed of an electron as it darts around:

$$v_{rms} = \sqrt{\frac{3kT}{m}} = \sqrt{\frac{3(1.38 \times 10^{-23} \, \text{J/K})(293 \, \text{K})}{9.11 \times 10^{-31} \, \text{kg}}} = 1.2 \times 10^5 \, \text{m/s}.$$

The drift velocity (average speed in the direction of the current) is very much less than the rms thermal speed of the electrons, by a factor of about 10^9.

NOTE The result in (c) is an underestimate. Quantum theory calculations, and experiments, give the rms speed in copper to be about $1.6 \times 10^6 \, \text{m/s}$.

The drift velocity of electrons in a wire is very slow, only about 0.05 mm/s (Example 25–14 above), which means it takes an electron 20×10^3 s, or $5\frac{1}{2}$ h, to travel only 1 m. This is not, of course, how fast "electricity travels": when you flip a light switch, the light—even if many meters away—goes on nearly instantaneously. Why? Because electric fields travel essentially at the speed of light ($3 \times 10^8 \, \text{m/s}$). We can think of electrons in a wire as being like a pipe full of water: when a little water enters one end of the pipe, almost immediately some water comes out at the other end.

*Electric Field Inside a Wire

Equation 25–2b, $V = IR$, can be written in terms of microscopic quantities as follows. We write the resistance R in terms of the resistivity ρ:

$$R = \rho\frac{\ell}{A};$$

and we write V and I as

$$I = jA \qquad \text{and} \qquad V = E\ell.$$

The last relation follows from Eqs. 23–4, where we assume the electric field is uniform within the wire and ℓ is the length of the wire (or a portion of the wire) between whose ends the potential difference is V. Thus, from $V = IR$, we have

$$E\ell = (jA)\left(\rho\frac{\ell}{A}\right) = j\rho\ell$$

so

$$j = \frac{1}{\rho}E = \sigma E, \tag{25–17}$$

where $\sigma = 1/\rho$ is the *conductivity* (Eq. 25–4). For a metal conductor, ρ and σ do not depend on V (and hence not on E). Therefore the current density $\vec{\mathbf{j}}$ is proportional to the electrical field $\vec{\mathbf{E}}$ in the conductor. This is the "microscopic" statement of Ohm's law. Equation 25–17, which can be written in vector form as

$$\vec{\mathbf{j}} = \sigma\vec{\mathbf{E}} = \frac{1}{\rho}\vec{\mathbf{E}},$$

is sometimes taken as the definition of conductivity σ and resistivity ρ.

EXAMPLE 25–15 **Electric field inside a wire.** What is the electric field inside the wire of Example 25–14?

APPROACH We use Eq. 25–17 and $\rho = 1.68 \times 10^{-8}\ \Omega\cdot\text{m}$ for copper.

SOLUTION Example 25–14 gives $j = 6.2 \times 10^5\ \text{A/m}^2$, so

$$E = \rho j = (1.68 \times 10^{-8}\ \Omega\cdot\text{m})(6.2 \times 10^5\ \text{A/m}^2) = 1.0 \times 10^{-2}\ \text{V/m}.$$

NOTE For comparison, the electric field between the plates of a capacitor is often much larger; in Example 24–1, for example, E is on the order of $10^4\ \text{V/m}$. Thus we see that only a modest electric field is needed for current flow in practical cases.

*25–9 Superconductivity

At very low temperatures, well below 0°C, the resistivity (Section 25–4) of certain metals and certain compounds or alloys becomes zero as measured by the highest-precision techniques. Materials in such a state are said to be **superconducting**. It was first observed by H. K. Onnes (1853–1926) in 1911 when he cooled mercury below 4.2 K (-269°C) and found that the resistance of mercury suddenly dropped to zero. In general, superconductors become superconducting only below a certain *transition temperature* or *critical temperature*, T_C, which is usually within a few degrees of absolute zero. Current in a ring-shaped superconducting material has been observed to flow for years in the absence of a potential difference, with no measurable decrease. Measurements show that the resistivity ρ of superconductors is less than $4 \times 10^{-25}\ \Omega\cdot\text{m}$, which is over 10^{16} times smaller than that for copper, and is considered to be zero in practice. See Fig. 25–27.

Before 1986 the highest temperature at which a material was found to superconduct was 23 K, which required liquid helium to keep the material cold. In 1987, a compound of yttrium, barium, copper, and oxygen (YBCO) was developed that can be superconducting at 90 K. This was an important breakthrough since liquid nitrogen, which boils at 77 K (sufficiently cold to keep the material superconducting), is more easily and cheaply obtained than the liquid helium needed for conventional superconductors. Superconductivity at temperatures as high as 160 K has been reported, though in fragile compounds.

FIGURE 25–27 A superconducting material has zero resistivity when its temperature is below T_C, its "critical temperature." At T_C, the resistivity jumps to a "normal" nonzero value and increases with temperature as most materials do (Eq. 25–5).

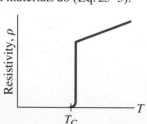

Most applications today use a bismuth-strontium-calcium-copper oxide, known (for short) as BSCCO. A major challenge is how to make a useable, bendable wire out of the BSCCO, which is very brittle. (One solution is to embed tiny filaments of the high-T_c superconductor in a metal alloy, which is not resistanceless, but the resistance is much less than that of a conventional copper cable.)

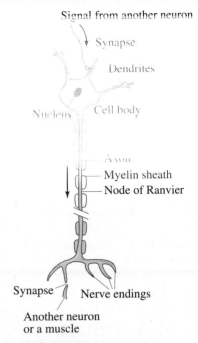

Signal from another neuron

Synapse

Dendrites

Nucleus — Cell body

Axon

Myelin sheath
Node of Ranvier

Synapse — Nerve endings

Another neuron
or a muscle

FIGURE 25–28 A simplified sketch of a typical neuron.

Electrical Conduction in the Nervous System

The flow of electric charge in the human nervous system provides us the means for being aware of the world. Although the detailed functioning is not well understood, we do have a reasonable understanding of how messages are transmitted within the nervous system: they are electrical signals passing along the basic element of the nervous system, the *neuron*.

Neurons are living cells of unusual shape (Fig. 25–28). Attached to the main cell body are several small appendages known as *dendrites* and a long tail called the *axon*. Signals are received by the dendrites and are propagated along the axon. When a signal reaches the nerve endings, it is transmitted to the next neuron or to a muscle at a connection called a *synapse*.

A neuron, before transmitting an electrical signal, is in the so-called "resting state." Like nearly all living cells, neurons have a net positive charge on the outer surface of the cell membrane and a negative charge on the inner surface. This difference in charge, or "dipole layer," means that a potential difference exists across the cell membrane. When a neuron is not transmitting a signal, this "resting potential," normally stated as

$$V_{\text{inside}} - V_{\text{outside}},$$

is typically $-60\,\text{mV}$ to $-90\,\text{mV}$, depending on the type of organism. The most common ions in a cell are K^+, Na^+, and Cl^-. There are large differences in the concentrations of these ions inside and outside a cell, as indicated by the typical values given in Table 25–2. Other ions are also present, so the fluids both inside and outside the axon are electrically neutral. Because of the differences in concentration, there is a tendency for ions to diffuse across the membrane (see Section 18–7 on diffusion). However, in the resting state the cell membrane prevents any net flow of Na^+ (through a mechanism of "active pumping" of Na^+ out of the cell). But it does allow the flow of Cl^- ions, and less so of K^+ ions, and it is these two ions that produce the dipole charge layer on the membrane. Because there is a greater concentration of K^+ inside the cell than outside, more K^+ ions tend to diffuse outward across the membrane than diffuse inward. A K^+ ion that passes through the membrane becomes attached to the outer surface of the membrane, and leaves behind an equal negative charge that lies on the inner surface of the membrane (Fig. 25–29). Independently, Cl^- ions tend to diffuse *into* the cell since their concentration outside is higher. Both K^+ and Cl^- diffusion tends to charge the interior surface of the membrane negatively and the outside positively. As charge accumulates on the membrane surface, it becomes increasingly difficult for more ions to diffuse: K^+ ions trying to move outward, for example, are repelled by the positive charge already there. Equilibrium is reached when the tendency to diffuse because of the concentration difference is just balanced by the electrical potential difference across the membrane. The greater the concentration difference, the greater the potential difference across the membrane ($-60\,\text{mV}$ to $-90\,\text{mV}$).

The most important aspect of a neuron is not that it has a resting potential (most cells do), but rather that it can respond to a stimulus and conduct an electrical signal along its length. The stimulus could be thermal (when you touch a hot stove) or chemical (as in taste buds); it could be pressure (as on the skin or at the eardrum), or light (as in the eye); or it could be the electric stimulus of a signal coming from the brain or another neuron. In the laboratory, the stimulus is usually electrical and is applied by a tiny probe at some point on the neuron. If the stimulus exceeds some threshold, a voltage pulse will travel down the axon. This voltage pulse can be detected at a point on the axon using a voltmeter or an oscilloscope connected as in Fig. 25–30.

TABLE 25–2
Concentrations of Ions Inside and Outside a Typical Axon

	Concentration inside axon (mol/m³)	Concentration outside axon (mol/m³)
K^+	140	5
Na^+	15	140
Cl^-	9	125

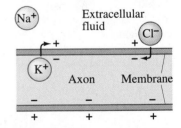

FIGURE 25–29 How a dipole layer of charge forms on a cell membrane.

FIGURE 25–30 Measuring the potential difference between the inside and outside of a nerve cell.

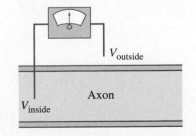

V_{outside}

Axon

V_{inside}

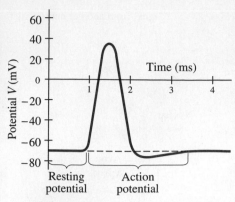

FIGURE 25–31 Action potential.

FIGURE 25–32 Propagation of an action potential along an axon membrane.

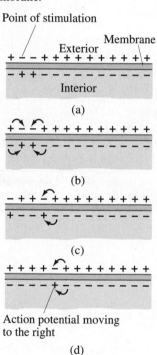

This voltage pulse has the shape shown in Fig. 25–31, and is called an **action potential**. As can be seen, the potential increases from a resting potential of about $-70\,mV$ and becomes a positive $30\,mV$ or $40\,mV$. The action potential lasts for about 1 ms and travels down an axon with a speed of 30 m/s to 150 m/s. When an action potential is stimulated, the nerve is said to have "fired."

What causes the action potential? Apparently, the cell membrane has the ability to alter its permeability properties. At the point where the stimulus occurs, the membrane suddenly becomes much more permeable to Na^+ than to K^+ and Cl^- ions. Thus, Na^+ ions rush into the cell and the inner surface of the wall becomes positively charged, and the potential difference quickly swings positive ($\approx +35\,mV$ in Fig. 25–31). Just as suddenly, the membrane returns to its original characteristics: it becomes impermeable to Na^+ and in fact pumps out Na^+ ions. The diffusion of Cl^- and K^+ ions again predominates and the original resting potential is restored ($-70\,mV$ in Fig. 25–31).

What causes the action potential to travel along the axon? The action potential occurs at the point of stimulation, as shown in Fig. 25–32a. The membrane momentarily is positive on the inside and negative on the outside at this point. Nearby charges are attracted toward this region, as shown in Fig. 25–32b. The potential in these adjacent regions then drops, causing an action potential there. Thus, as the membrane returns to normal at the original point, nearby it experiences an action potential, so the action potential moves down the axon (Figs. 25–32c and d).

You may wonder if the number of ions that pass through the membrane would significantly alter the concentrations. The answer is no; and we can show why by treating the axon as a capacitor in the following Example.

EXAMPLE 25–16 **ESTIMATE** **Capacitance of an axon.** (a) Do an order-of-magnitude estimate for the capacitance of an axon 10 cm long of radius $10\,\mu m$. The thickness of the membrane is about $10^{-8}\,m$, and the dielectric constant is about 3. (b) By what factor does the concentration (number of ions per volume) of Na^+ ions in the cell change as a result of one action potential?

APPROACH We model the membrane of an axon as a cylindrically shaped parallel-plate capacitor, with opposite charges on each side. The separation of the "plates" is the thickness of the membrane, $d \approx 10^{-8}\,m$. We first calculate the area of the cylinder and then can use Eq. 24–8, $C = K\epsilon_0 A/d$, to find the capacitance. In (b), we use the voltage change during one action potential to find the amount of charge moved across the membrane.

SOLUTION (a) The area A is the area of a cylinder of radius r and length ℓ:

$$A = 2\pi r\ell \approx (6.28)(10^{-5}\,m)(0.1\,m) \approx 6 \times 10^{-6}\,m^2.$$

From Eq. 24–8, we have

$$C = K\epsilon_0 \frac{A}{d} \approx (3)(8.85 \times 10^{-12}\,C^2/N\cdot m^2)\frac{6 \times 10^{-6}\,m^2}{10^{-8}\,m} \approx 10^{-8}\,F.$$

(b) Since the voltage changes from $-70\,mV$ to about $+30\,mV$, the total change is about 100 mV. The amount of charge that moves is then

$$Q = CV \approx (10^{-8}\,F)(0.1\,V) = 10^{-9}\,C.$$

Each ion carries a charge $e = 1.6 \times 10^{-19}\,C$, so the number of ions that flow per action potential is $Q/e = (10^{-9}\,C)/(1.6 \times 10^{-19}\,C) \approx 10^{10}$. The volume of our cylindrical axon is

$$V = \pi r^2\ell \approx (3)(10^{-5}\,m)^2(0.1\,m) = 3 \times 10^{-11}\,m^3,$$

and the concentration of Na^+ ions inside the cell (Table 25–2) is $15\,mol/m^3 = 15 \times 6.02 \times 10^{23}\,ions/m^3 \approx 10^{25}\,ions/m^3$. Thus, the cell contains $(10^{25}\,ions/m^3) \times (3 \times 10^{-11}\,m^3) \approx 3 \times 10^{14}\,Na^+$ ions. One action potential, then, will change the concentration of Na^+ ions by about $10^{10}/(3 \times 10^{14}) = \frac{1}{3} \times 10^{-4}$, or 1 part in 30,000. This tiny change would not be measurable.

Thus, even 1000 action potentials will not alter the concentration significantly. The sodium pump does not, therefore, have to remove Na^+ ions quickly after an action potential, but can operate slowly over time to maintain a relatively constant concentration.

Summary

An electric **battery** serves as a source of nearly constant potential difference by transforming chemical energy into electric energy. A simple battery consists of two electrodes made of different metals immersed in a solution or paste known as an electrolyte.

Electric current, I, refers to the rate of flow of electric charge and is measured in **amperes** (A): 1 A equals a flow of 1 C/s past a given point.

The direction of **conventional current** is that of positive charge flow. In a wire, it is actually negatively charged electrons that move, so they flow in a direction opposite to the conventional current. A positive charge flow in one direction is almost always equivalent to a negative charge flow in the opposite direction. Positive conventional current always flows from a high potential to a low potential.

The **resistance** R of a device is defined by the relation

$$V = IR, \qquad (25\text{--}2)$$

where I is the current in the device when a potential difference V is applied across it. For materials such as metals, R is a constant independent of V (thus $I \propto V$), a result known as **Ohm's law**. Thus, the current I coming from a battery of voltage V depends on the resistance R of the circuit connected to it.

Voltage is applied *across* a device or between the ends of a wire. Current passes *through* a wire or device. Resistance is a property *of* the wire or device.

The unit of resistance is the **ohm** (Ω), where $1\,\Omega = 1\,\text{V/A}$. See Table 25–3.

TABLE 25–3 Summary of Units

Current	$1\,\text{A} = 1\,\text{C/s}$
Potential difference	$1\,\text{V} = 1\,\text{J/C}$
Power	$1\,\text{W} = 1\,\text{J/s}$
Resistance	$1\,\Omega = 1\,\text{V/A}$

The resistance R of a wire is inversely proportional to its cross-sectional area A, and directly proportional to its length ℓ and to a property of the material called its resistivity:

$$R = \frac{\rho \ell}{A}. \qquad (25\text{--}3)$$

The **resistivity**, ρ, increases with temperature for metals, but for semiconductors it may decrease.

The rate at which energy is transformed in a resistance R from electric to other forms of energy (such as heat and light) is equal to the product of current and voltage. That is, the **power** transformed, measured in watts, is given by

$$P = IV, \qquad (25\text{--}6)$$

which for resistors can be written as

$$P = I^2 R = \frac{V^2}{R} \qquad (25\text{--}7)$$

The SI unit of power is the **watt** $(1\,\text{W} = 1\,\text{J/s})$.

The total electric energy transformed in any device equals the product of the power and the time during which the device is operated. In SI units, energy is given in joules $(1\,\text{J} = 1\,\text{W}\cdot\text{s})$, but electric companies use a larger unit, the **kilowatt-hour** $(1\,\text{kWh} = 3.6 \times 10^6\,\text{J})$.

Electric current can be **direct current** (**dc**), in which the current is steady in one direction; or it can be **alternating current** (**ac**), in which the current reverses direction at a particular frequency f, typically 60 Hz. Alternating currents are typically sinusoidal in time,

$$I = I_0 \sin \omega t, \qquad (25\text{--}8)$$

where $\omega = 2\pi f$, and are produced by an alternating voltage.

The **rms** values of sinusoidally alternating currents and voltages are given by

$$I_{\text{rms}} = \frac{I_0}{\sqrt{2}} \quad \text{and} \quad V_{\text{rms}} = \frac{V_0}{\sqrt{2}}, \qquad (25\text{--}9)$$

respectively, where I_0 and V_0 are the **peak** values. The power relationship, $P = IV = I^2 R = V^2/R$, is valid for the average power in alternating currents when the rms values of V and I are used.

Current density $\vec{\mathbf{j}}$ is the current per cross-sectional area. From a microscopic point of view, the current density is related to the number of charge carriers per unit volume, n, their charge, q, and their **drift velocity**, $\vec{\mathbf{v}}_d$, by

$$\vec{\mathbf{j}} = nq\vec{\mathbf{v}}_d. \qquad (25\text{--}16)$$

The electric field within a wire is related to $\vec{\mathbf{j}}$ by $\vec{\mathbf{j}} = \sigma \vec{\mathbf{E}}$ where $\sigma = 1/\rho$ is the **conductivity**.

[*At very low temperatures certain materials become **superconducting**, which means their electrical resistance becomes zero.]

[*The human nervous system operates via electrical conduction: when a nerve "fires," an electrical signal travels as a voltage pulse known as an **action potential**.]

Questions

1. What quantity is measured by a battery rating given in ampere-hours $(\text{A}\cdot\text{h})$?

2. When an electric cell is connected to a circuit, electrons flow away from the negative terminal in the circuit. But within the cell, electrons flow *to* the negative terminal. Explain.

3. When a flashlight is operated, what is being used up: battery current, battery voltage, battery energy, battery power, or battery resistance? Explain.

4. One terminal of a car battery is said to be connected to "ground." Since it is not really connected to the ground, what is meant by this expression?

5. When you turn on a water faucet, the water usually flows immediately. You don't have to wait for water to flow from the faucet valve to the spout. Why not? Is the same thing true when you connect a wire to the terminals of a battery?

6. Can a copper wire and an aluminum wire of the same length have the same resistance? Explain.

7. The equation $P = V^2/R$ indicates that the power dissipated in a resistor decreases if the resistance is increased, whereas the equation $P = I^2 R$ implies the opposite. Is there a contradiction here? Explain.

8. What happens when a lightbulb burns out?

9. If the resistance of a small immersion heater (to heat water for tea or soup, Fig. 25–33) was increased, would it speed up or slow down the heating process? Explain.

FIGURE 25–33
Question 9.

10. If a rectangular solid made of carbon has sides of lengths a, $2a$, and $3a$, how would you connect the wires from a battery so as to obtain (a) the least resistance, (b) the greatest resistance?

11. Explain why lightbulbs almost always burn out just as they are turned on and not after they have been on for some time.

12. Which draws more current, a 100-W lightbulb or a 75-W bulb? Which has the higher resistance?

13. Electric power is transferred over large distances at very high voltages. Explain how the high voltage reduces power losses in the transmission lines.

14. A 15-A fuse blows repeatedly. Why is it dangerous to replace this fuse with a 25-A fuse?

15. When electric lights are operated on low-frequency ac (say, 5 Hz), they flicker noticeably. Why?

16. Driven by ac power, the same electrons pass back and forth through your reading lamp over and over again. Explain why the light stays lit instead of going out after the first pass of electrons.

17. The heating element in a toaster is made of Nichrome wire. Immediately after the toaster is turned on, is the current (I_{rms}) in the wire increasing, decreasing, or staying constant? Explain.

18. Is current used up in a resistor? Explain.

19. Compare the drift velocities and electric currents in two wires that are geometrically identical and the density of atoms is similar, but the number of free electrons per atom in the material of one wire is twice that in the other.

20. A voltage V is connected across a wire of length ℓ and radius r. How is the electron drift velocity affected if (a) ℓ is doubled, (b) r is doubled, (c) V is doubled?

21. Why is it more dangerous to turn on an electric appliance when you are standing outside in bare feet than when you are inside wearing shoes with thick soles?

Problems

25–2 and 25–3 Electric Current, Resistance, Ohm's Law

(*Note*: The charge on one electron is 1.60×10^{-19} C.)

1. (I) A current of 1.30 A flows in a wire. How many electrons are flowing past any point in the wire per second?

2. (I) A service station charges a battery using a current of 6.7-A for 5.0 h. How much charge passes through the battery?

3. (I) What is the current in amperes if 1200 Na$^+$ ions flow across a cell membrane in 3.5 μs? The charge on the sodium is the same as on an electron, but positive.

4. (I) What is the resistance of a toaster if 120 V produces a current of 4.2 A?

5. (II) An electric clothes dryer has a heating element with a resistance of 8.6 Ω. (a) What is the current in the element when it is connected to 240 V? (b) How much charge passes through the element in 50 min? (Assume direct current.)

6. (II) A hair dryer draws 9.5 A when plugged into a 120-V line. (a) What is its resistance? (b) How much charge passes through it in 15 min? (Assume direct current.)

7. (II) A 4.5-V battery is connected to a bulb whose resistance is 1.6 Ω. How many electrons leave the battery per minute?

8. (II) A bird stands on a dc electric transmission line carrying 3100 A (Fig. 25–34). The line has 2.5×10^{-5} Ω resistance per meter, and the bird's feet are 4.0 cm apart. What is the potential difference between the bird's feet?

FIGURE 25–34
Problem 8.

9. (II) A 12-V battery causes a current of 0.60 A through a resistor. (a) What is its resistance, and (b) how many joules of energy does the battery lose in a minute?

10. (II) An electric device draws 6.50 A at 240 V. (a) If the voltage drops by 15%, what will be the current, assuming nothing else changes? (b) If the resistance of the device were reduced by 15%, what current would be drawn at 240 V?

25–4 Resistivity

11. (I) What is the diameter of a 1.00-m length of tungsten wire whose resistance is 0.32 Ω?

12. (I) What is the resistance of a 4.5-m length of copper wire 1.5 mm in diameter?

13. (II) Calculate the ratio of the resistance of 10.0 m of aluminum wire 2.0 mm in diameter, to 20.0 m of copper wire 1.8 mm in diameter.

14. (II) Can a 2.2-mm-diameter copper wire have the same resistance as a tungsten wire of the same length? Give numerical details.

15. (II) A sequence of potential differences V is applied across a wire (diameter $= 0.32$ mm, length $= 11$ cm) and the resulting currents I are measured as follows:

V (V)	0.100	0.200	0.300	0.400	0.500
I (mA)	72	144	216	288	360

(a) If this wire obeys Ohm's law, graphing I vs. V will result in a straight-line plot. Explain why this is so and determine the theoretical predictions for the straight line's slope and y-intercept. (b) Plot I vs. V. Based on this plot, can you conclude that the wire obeys Ohm's law (i.e., did you obtain a straight line with the expected y-intercept)? If so, determine the wire's resistance R. (c) Calculate the wire's resistivity and use Table 25–1 to identify the solid material from which it is composed.

16. (II) How much would you have to raise the temperature of a copper wire (originally at 20°C) to increase its resistance by 15%?

17. (II) A certain copper wire has a resistance of $10.0\,\Omega$. At what point along its length must the wire be cut so that the resistance of one piece is 4.0 times the resistance of the other? What is the resistance of each piece?

18. (II) Determine at what temperature aluminum will have the same resistivity as tungsten does at 20°C.

19. (II) A 100-W lightbulb has a resistance of about $12\,\Omega$ when cold (20°C) and $140\,\Omega$ when on (hot). Estimate the temperature of the filament when hot assuming an average temperature coefficient of resistivity $\alpha = 0.0045\,(\text{C}°)^{-1}$.

20. (II) Compute the voltage drop along a 26-m length of household no. 14 copper wire (used in 15-A circuits). The wire has diameter 1.628 mm and carries a 12-A current.

21. (II) Two aluminum wires have the same resistance. If one has twice the length of the other, what is the ratio of the diameter of the longer wire to the diameter of the shorter wire?

22. (II) A rectangular solid made of carbon has sides of lengths 1.0 cm, 2.0 cm, and 4.0 cm, lying along the x, y, and z axes, respectively (Fig. 25–35). Determine the resistance for current that passes through the solid in (a) the x direction, (b) the y direction, and (c) the z direction. Assume the resistivity is $\rho = 3.0 \times 10^{-5}\,\Omega\cdot\text{m}$.

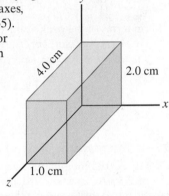

FIGURE 25–35
Problem 22.

23. (II) A length of aluminum wire is connected to a precision 10.00-V power supply, and a current of 0.4212 A is precisely measured at 20.0°C. The wire is placed in a new environment of unknown temperature where the measured current is 0.3818 A. What is the unknown temperature?

24. (II) Small changes in the length of an object can be measured using a **strain gauge** sensor, which is a wire with undeformed length ℓ_0, cross-sectional area A_0, and resistance R_0. This sensor is rigidly affixed to the object's surface, aligning its length in the direction in which length changes are to be measured. As the object deforms, the length of the wire sensor changes by $\Delta\ell$, and the resulting change ΔR in the sensor's resistance is measured. Assuming that as the solid wire is deformed to a length ℓ, its density (and volume) remains constant (only approximately valid), show that the strain $(= \Delta\ell/\ell_0)$ of the wire sensor, and thus of the object to which it is attached, is $\Delta R/2R_0$.

25. (II) A length of wire is cut in half and the two lengths are wrapped together side by side to make a thicker wire. How does the resistance of this new combination compare to the resistance of the original wire?

26. (III) For some applications, it is important that the value of a resistance not change with temperature. For example, suppose you made a 3.70-kΩ resistor from a carbon resistor and a Nichrome wire-wound resistor connected together so the total resistance is the sum of their separate resistances. What value should each of these resistors have (at 0°C) so that the combination is temperature independent?

27. (III) Determine a formula for the total resistance of a spherical shell made of material whose conductivity is σ and whose inner and outer radii are r_1 and r_2. Assume the current flows radially outward.

28. (III) The filament of a lightbulb has a resistance of $12\,\Omega$ at 20°C and $140\,\Omega$ when hot (as in Problem 19). (a) Calculate the temperature of the filament when it is hot, and take into account the change in length and area of the filament due to thermal expansion (assume tungsten for which the thermal expansion coefficient is $\approx 5.5 \times 10^{-6}\,\text{C}°^{-1}$). (b) In this temperature range, what is the percentage change in resistance due to thermal expansion, and what is the percentage change in resistance due solely to the change in ρ? Use Eq. 25–5.

29. (III) A 10.0-m length of wire consists of 5.0 m of copper followed by 5.0 m of aluminum, both of diameter 1.4 mm. A voltage difference of 85 mV is placed across the composite wire. (a) What is the total resistance (sum) of the two wires? (b) What is the current through the wire? (c) What are the voltages across the aluminum part and across the copper part?

30. (III) A hollow cylindrical resistor with inner radius r_1 and outer radius r_2, and length ℓ, is made of a material whose resistivity is ρ (Fig. 25–36). (a) Show that the resistance is given by

$$R = \frac{\rho}{2\pi\ell}\ln\frac{r_2}{r_1}$$

for current that flows radially outward. [*Hint*: Divide the resistor into concentric cylindrical shells and integrate.] (b) Evaluate the resistance R for such a resistor made of carbon whose inner and outer radii are 1.0 mm and 1.8 mm and whose length is 2.4 cm. (Choose $\rho = 15 \times 10^{-5}\,\Omega\cdot\text{m}$.) (c) What is the resistance in part (b) for current flowing *parallel* to the axis?

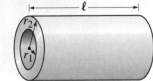

FIGURE 25–36
Problem 30.

25–5 and 25–6 Electric Power

31. (I) What is the maximum power consumption of a 3.0-V portable CD player that draws a maximum of 270 mA of current?

32. (I) The heating element of an electric oven is designed to produce 3.3 kW of heat when connected to a 240-V source. What must be the resistance of the element?

33. (I) What is the maximum voltage that can be applied across a 3.3-kΩ resistor rated at $\frac{1}{4}$ watt?

34. (I) (a) Determine the resistance of, and current through, a 75-W lightbulb connected to its proper source voltage of 110 V. (b) Repeat for a 440-W bulb.

35. (II) An electric power plant can produce electricity at a fixed power P, but the plant operator is free to choose the voltage V at which it is produced. This electricity is carried as an electric current I through a transmission line (resistance R) from the plant to the user, where it provides the user with electric power P'. (a) Show that the reduction in power $\Delta P = P - P'$ due to transmission losses is given by $\Delta P = P^2R/V^2$. (b) In order to reduce power losses during transmission, should the operator choose V to be as large or as small as possible?

36. (II) A 120-V hair dryer has two settings: 850 W and 1250 W. (a) At which setting do you expect the resistance to be higher? After making a guess, determine the resistance at (b) the lower setting; and (c) the higher setting.

37. (II) A 115-V fish-tank heater is rated at 95 W. Calculate (a) the current through the heater when it is operating, and (b) its resistance.

38. (II) You buy a 75-W lightbulb in Europe, where electricity is delivered to homes at 240 V. If you use the lightbulb in the United States at 120 V (assume its resistance does not change), how bright will it be relative to 75-W 120-V bulbs? [Hint: Assume roughly that brightness is proportional to power consumed.]

39. (II) How many kWh of energy does a 550-W toaster use in the morning if it is in operation for a total of 6.0 min? At a cost of 9.0 cents/kWh, estimate how much this would add to your monthly electric energy bill if you made toast four mornings per week.

40. (II) At $0.095/kWh, what does it cost to leave a 25-W porch light on day and night for a year?

41. (II) What is the total amount of energy stored in a 12-V, 75-A·h car battery when it is fully charged?

42. (II) An ordinary flashlight uses two D-cell 1.5-V batteries connected in series as in Fig. 25–4b (Fig. 25–37). The bulb draws 380 mA when turned on. (a) Calculate the resistance of the bulb and the power dissipated. (b) By what factor would the power increase if four D-cells in series were used with the same bulb? (Neglect heating effects of the filament.) Why shouldn't you try this?

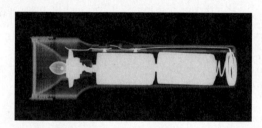

FIGURE 25–37 Problem 42.

43. (II) How many 75-W lightbulbs, connected to 120 V as in Fig. 25–20, can be used without blowing a 15-A fuse?

44. (II) An extension cord made of two wires of diameter 0.129 cm (no. 16 copper wire) and of length 2.7 m (9 ft) is connected to an electric heater which draws 15.0 A on a 120-V line. How much power is dissipated in the cord?

45. (II) A power station delivers 750 kW of power at 12,000 V to a factory through wires with total resistance 3.0 Ω. How much less power is wasted if the electricity is delivered at 50,000 V rather than 12,000 V?

46. (III) A small immersion heater can be used in a car to heat a cup of water for coffee or tea. If the heater can heat 120 mL of water from 25°C to 95°C in 8.0 min, (a) approximately how much current does it draw from the car's 12-V battery, and (b) what is its resistance? Assume the manufacturer's claim of 75% efficiency.

47. (III) The current in an electromagnet connected to a 240-V line is 17.5 A. At what rate must cooling water pass over the coils if the water temperature is to rise by no more than 6.50 C°?

48. (III) A 1.0-m-long round tungsten wire is to reach a temperature of 3100 K when a current of 15.0 A flows through it. What diameter should the wire be? Assume the wire loses energy only by radiation (emissivity ϵ = 1.0, Section 19–10) and the surrounding temperature is 20°C.

25–7 Alternating Current

49. (I) Calculate the peak current in a 2.7-kΩ resistor connected to a 220-V rms ac source.

50. (I) An ac voltage, whose peak value is 180 V, is across a 380-Ω resistor. What are the rms and peak currents in the resistor?

51. (II) Estimate the resistance of the 120-V_{rms} circuits in your house as seen by the power company, when (a) everything electrical is unplugged, and (b) there are two 75-W light-bulbs burning.

52. (II) The peak value of an alternating current in a 1500-W device is 5.4 A. What is the rms voltage across it?

53. (II) An 1800-W arc welder is connected to a 660-V_{rms} ac line. Calculate (a) the peak voltage and (b) the peak current.

54. (II) (a) What is the maximum instantaneous power dissipated by a 2.5-hp pump connected to a 240-V_{rms} ac power source? (b) What is the maximum current passing through the pump?

55. (II) A heater coil connected to a 240-V_{rms} ac line has a resistance of 44 Ω. (a) What is the average power used? (b) What are the maximum and minimum values of the instantaneous power?

56. (II) For a time-dependent voltage $V(t)$, which is periodic with period T, the rms voltage is defined to be $V_{rms} = [\frac{1}{T} \int_0^T V^2 \, dt]^{\frac{1}{2}}$. Use this definition to determine V_{rms} (in terms of the peak voltage V_0) for (a) a sinusoidal voltage, i.e., $V(t) = V_0 \sin(2\pi t/T)$ for $0 \le t \le T$; and (b) a positive square-wave voltage, i.e.,

$$V(t) = \begin{cases} V_0 & 0 \le t \le \dfrac{T}{2} \\ 0 & \dfrac{T}{2} \le t \le T \end{cases}.$$

25–8 Microscopic View of Electric Current

57. (II) A 0.65-mm-diameter copper wire carries a tiny current of 2.3 μA. Estimate (a) the electron drift velocity, (b) the current density, and (c) the electric field in the wire.

58. (II) A 5.80-m length of 2.0-mm-diameter wire carries a 750-mA current when 22.0 mV is applied to its ends. If the drift velocity is 1.7×10^{-5} m/s, determine (a) the resistance R of the wire, (b) the resistivity ρ, (c) the current density j, (d) the electric field inside the wire, and (e) the number n of free electrons per unit volume.

59. (II) At a point high in the Earth's atmosphere, He^{2+} ions in a concentration of 2.8×10^{12}/m^3 are moving due north at a speed of 2.0×10^6 m/s. Also, a 7.0×10^{11}/m^3 concentration of O$_2^-$ ions is moving due south at a speed of 6.2×10^6 m/s. Determine the magnitude and direction of the current density $\vec{j}$ at this point.

*25–10 Nerve Conduction

*60. (I) What is the magnitude of the electric field across an axon membrane 1.0×10^{-8} m thick if the resting potential is −70 mV?

*61. (II) A neuron is stimulated with an electric pulse. The action potential is detected at a point 3.40 cm down the axon 0.0052 s later. When the action potential is detected 7.20 cm from the point of stimulation, the time required is 0.0063 s. What is the speed of the electric pulse along the axon? (Why are two measurements needed instead of only one?)

*62. (III) During an action potential, Na$^+$ ions move into the cell at a rate of about 3×10^{-7} mol/m^2·s. How much power must be produced by the "active Na$^+$ pumping" system to produce this flow against a +30-mV potential difference? Assume that the axon is 10 cm long and 20 μm in diameter.

General Problems

63. A person accidentally leaves a car with the lights on. If each of the two headlights uses 40 W and each of the two tail-lights 6 W, for a total of 92 W, how long will a fresh 12-V battery last if it is rated at 85 A·h? Assume the full 12 V appears across each bulb.

64. How many coulombs are there in 1.00 ampere-hour?

65. You want to design a portable electric blanket that runs on a 1.5-V battery. If you use copper wire with a 0.50 mm diameter as the heating element, how long should the wire be if you want to generate 15 W of heating power? What happens if you accidentally connect the blanket to a 9.0-V battery?

66. What is the average current drawn by a 1.0-hp 120-V motor? (1 hp = 746 W.)

67. The *conductance* G of an object is defined as the reciprocal of the resistance R; that is, $G = 1/R$. The unit of conductance is a *mho* $(= \text{ohm}^{-1})$, which is also called the *siemens* (S). What is the conductance (in siemens) of an object that draws 480 mA of current at 3.0 V?

68. The heating element of a 110-V, 1500-W heater is 3.5 m long. If it is made of iron, what must its diameter be?

69. (a) A particular household uses a 1.8-kW heater 2.0 h/day ("on" time), four 100-W lightbulbs 6.0 h/day, a 3.0-kW electric stove element for a total of 1.0 h/day, and miscellaneous power amounting to 2.0 kWh/day. If electricity costs $0.105 per kWh, what will be their monthly bill (30 d)? (b) How much coal (which produces 7500 kcal/kg) must be burned by a 35%-efficient power plant to provide the yearly needs of this household?

70. A small city requires about 15 MW of power. Suppose that instead of using high-voltage lines to supply the power, the power is delivered at 120 V. Assuming a two-wire line of 0.50-cm-diameter copper wire, estimate the cost of the energy lost to heat per hour per meter. Assume the cost of electricity is about 9.0 cents per kWh.

71. A 1400-W hair dryer is designed for 117 V. (a) What will be the percentage change in power output if the voltage drops to 105 V? Assume no change in resistance. (b) How would the actual change in resistivity with temperature affect your answer?

72. The wiring in a house must be thick enough so it does not become so hot as to start a fire. What diameter must a copper wire be if it is to carry a maximum current of 35 A and produce no more than 1.5 W of heat per meter of length?

73. Determine the resistance of the tungsten filament in a 75-W 120-V incandescent lightbulb (a) at its operating temperature of about 3000 K, (b) at room temperature.

74. Suppose a current is given by the equation $I = 1.80 \sin 210t$, where I is in amperes and t in seconds. (a) What is the frequency? (b) What is the rms value of the current? (c) If this is the current through a 24.0-Ω resistor, write the equation that describes the voltage as a function of time.

75. A microwave oven running at 65% efficiency delivers 950 W to the interior. Find (a) the power drawn from the source, and (b) the current drawn. Assume a source voltage of 120 V.

76. A 1.00-Ω wire is stretched uniformly to 1.20 times its original length. What is its resistance now?

77. 220 V is applied to two different conductors made of the same material. One conductor is twice as long and twice the diameter of the second. What is the ratio of the power transformed in the first relative to the second?

78. An electric heater is used to heat a room of volume 54 m³. Air is brought into the room at 5°C and is completely replaced twice per hour. Heat loss through the walls amounts to approximately 850 kcal/h. If the air is to be maintained at 20°C, what minimum wattage must the heater have? (The specific heat of air is about 0.17 kcal/kg·C°.)

79. A 2800-W oven is connected to a 240-V source. (a) What is the resistance of the oven? (b) How long will it take to bring 120 mL of 15°C water to 100°C assuming 75% efficiency? (c) How much will this cost at 11 cents/kWh?

80. A proposed electric vehicle makes use of storage batteries as its source of energy. Its mass is 1560 kg and it is powered by 24 batteries, each 12 V, 95 A·h. Assume that the car is driven on level roads at an average speed of 45 km/h, and the average friction force is 240 N. Assume 100% efficiency and neglect energy used for acceleration. No energy is consumed when the vehicle is stopped, since the engine doesn't need to idle. (a) Determine the horsepower required. (b) After approximately how many kilometers must the batteries be recharged?

81. A 12.5-Ω resistor is made from a coil of copper wire whose total mass is 15.5 g. What is the diameter of the wire, and how long is it?

82. A fish-tank heater is rated at 95 W when connected to 120 V. The heating element is a coil of Nichrome wire. When uncoiled, the wire has a total length of 3.8 m. What is the diameter of the wire?

83. A 100-W, 120-V lightbulb has a resistance of 12 Ω when cold (20°C) and 140 Ω when on (hot). Calculate its power consumption (a) at the instant it is turned on, and (b) after a few moments when it is hot.

84. In an automobile, the system voltage varies from about 12 V when the car is off to about 13.8 V when the car is on and the charging system is in operation, a difference of 15%. By what percentage does the power delivered to the headlights vary as the voltage changes from 12 V to 13.8 V? Assume the headlight resistance remains constant.

85. The Tevatron accelerator at Fermilab (Illinois) is designed to carry an 11-mA beam of protons traveling at very nearly the speed of light $(3.0 \times 10^8 \text{ m/s})$ around a ring 6300 m in circumference. How many protons are in the beam?

86. Lightbulb A is rated at 120 V and 40 W for household applications. Lightbulb B is rated at 12 V and 40 W for automotive applications. (a) What is the current through each bulb? (b) What is the resistance of each bulb? (c) In one hour, how much charge passes through each bulb? (d) In one hour, how much energy does each bulb use? (e) Which bulb requires larger diameter wires to connect its power source and the bulb?

87. An air conditioner draws 14 A at 220-V ac. The connecting cord is copper wire with a diameter of 1.628 mm. (a) How much power does the air conditioner draw? (b) If the total length of wire is 15 m, how much power is dissipated in the wiring? (c) If no. 12 wire, with a diameter of 2.053 mm, was used instead, how much power would be dissipated in the wiring? (d) Assuming that the air conditioner is run 12 h per day, how much money per month (30 days) would be saved by using no. 12 wire? Assume that the cost of electricity is 12 cents per kWh.

88. Copper wire of diameter 0.259 cm is used to connect a set of appliances at 120 V, which draw 1750 W of power total. (a) What power is wasted in 25.0 m of this wire? (b) What is your answer if wire of diameter 0.412 cm is used?

89. Battery-powered electricity is very expensive compared with that available from a wall receptacle. Estimate the cost per kWh of (a) an alkaline D-cell (cost $1.70) and (b) an alkaline AA-cell (cost $1.25). These batteries can provide a continuous current of 25 mA for 820 h and 120 h, respectively, at 1.5 V. Compare to normal 120-V ac house current at $0.10/kWh.

90. How far does an average electron move along the wires of a 550-W toaster during an alternating current cycle? The power cord has copper wires of diameter 1.7 mm and is plugged into a standard 60-Hz 120-V ac outlet. [*Hint*: The maximum current in the cycle is related to the maximum drift velocity. The maximum velocity in an oscillation is related to the maximum displacement; see Chapter 14.]

91. A copper pipe has an inside diameter of 3.00 cm and an outside diameter of 5.00 cm (Fig. 25–38). What is the resistance of a 10.0-m length of this pipe?

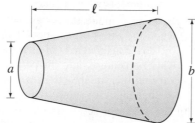

3.00 cm

5.00 cm

FIGURE 25–38
Problem 91.

92. For the wire in Fig. 25–39, whose diameter varies uniformly from a to b as shown, suppose a current $I = 2.0$ A enters at a. If $a = 2.5$ mm and $b = 4.0$ mm, what is the current density (assume uniform) at each end?

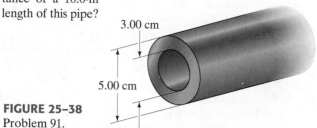

ℓ

a

b

FIGURE 25–39
Problems 92 and 93.

93. The cross section of a portion of wire increases uniformly as shown in Fig. 25–39 so it has the shape of a truncated cone. The diameter at one end is a and at the other it is b, and the total length along the axis is ℓ. If the material has resistivity ρ, determine the resistance R between the two ends in terms of $a, b, \ell,$ and ρ. Assume that the current flows uniformly through each section, and that the taper is small, i.e., $(b - a) \ll \ell$.

94. A tungsten filament used in a flashlight bulb operates at 0.20 A and 3.2 V. If its resistance at 20°C is 1.5 Ω, what is the temperature of the filament when the flashlight is on?

95. The level of liquid helium (temperature ≤ 4 K) in its storage tank can be monitored using a vertically aligned niobium–titanium (NbTi) wire, whose length ℓ spans the height of the tank. In this level-sensing setup, an electronic circuit maintains a constant electrical current I at all times in the NbTi wire and a voltmeter monitors the voltage difference V across this wire. Since the superconducting transition temperature for NbTi is 10 K, the portion of the wire immersed in the liquid helium is in the superconducting state, while the portion above the liquid (in helium vapor with temperature above 10 K) is in the normal state. Define $f = x/\ell$ to be the fraction of the tank filled with liquid helium (Fig. 25–40) and V_0 to be the value of V when the tank is empty ($f = 0$). Determine the relation between f and V (in terms of V_0).

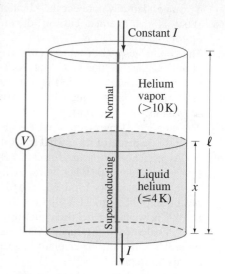

Constant I

Helium vapor ($>$10 K)

Normal

Superconducting

V

Liquid helium ($\leq$4 K)

ℓ

x

I

FIGURE 25–40
Problem 95.

*Numerical/Computer

*96. (II) The resistance, R, of a particular thermistor as a function of temperature T is shown in this Table:

T (°C)	R (Ω)	T (°C)	R (Ω)
20	126,740	36	60,743
22	115,190	38	55,658
24	104,800	40	51,048
26	95,447	42	46,863
28	87,022	44	43,602
30	79,422	46	39,605
32	72,560	48	36,458
34	66,356	50	33,591

Determine what type of best-fit equation (linear, quadratic, exponential, other) describes the variation of R with T. The resistance of the thermistor is 57,641 Ω when embedded in a substance whose temperature is unknown. Based on your equation, what is the unknown temperature?

Answers to Exercises

A: 1.6×10^{-13} A.

B: (c).

C: (b), (c).

D: (c).

E: 110 m.

F: (d).

G: (e).

H: (a) 28 V, 3.5 A; (b) 2.8 V, 0.35 A.

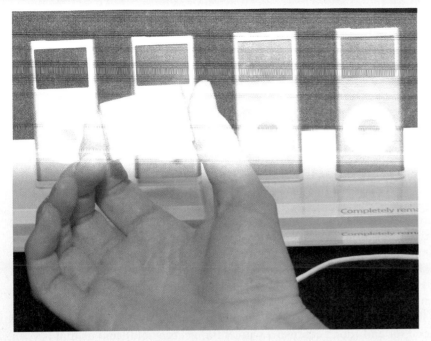

These MP3 players contain circuits that are dc, at least in part. (The audio signal is ac.) The circuit diagram below shows a possible amplifier circuit for each stereo channel. We have already met two of the circuit elements shown: resistors and capacitors, and we discuss them in circuits in this Chapter. (The large triangle is an amplifier chip containing transistors.) We also discuss voltmeters and ammeters, and how they are built and used to make measurements.

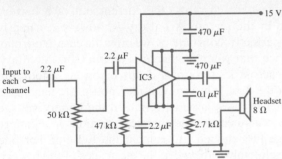

DC Circuits

CHAPTER-OPENING QUESTION—Guess now!

The automobile headlight bulbs shown in the circuits here are identical. The connection which produces more light is

(a) circuit 1.
(b) circuit 2.
(c) both the same.
(d) not enough information.

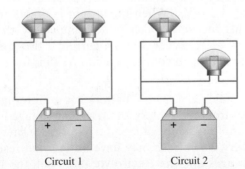

Circuit 1   Circuit 2

CONTENTS

Electric circuits are basic parts of all electronic devices from radio and TV sets to computers and automobiles. Scientific measurements, from physics to biology and medicine, make use of electric circuits. In Chapter 25, we discussed the basic principles of electric current. Now we will apply these principles to analyze dc circuits involving combinations of batteries, resistors, and capacitors. We also study the operation of some useful instruments.†

† AC circuits that contain only a voltage source and resistors can be analyzed like the dc circuits in this Chapter. However, ac circuits that contain capacitors and other circuit elements are more complicated, and we discuss them in Chapter 30.

TABLE 26–1 Symbols for Circuit Elements

Symbol	Device
⊣⊢	Battery
⊣⊢ or ⊣⊱	Capacitor
⌇	Resistor
——	Wire with negligible resistance
⌐	Switch
⏚ or ⏚	Ground

⚠ **CAUTION**

Why battery voltage isn't perfectly constant

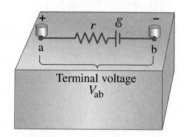

FIGURE 26–1 Diagram for an electric cell or battery.

When we draw a diagram for a circuit, we represent batteries, capacitors, and resistors by the symbols shown in Table 26–1. Wires whose resistance is negligible compared with other resistance in the circuit are drawn simply as straight lines. Some circuit diagrams show a ground symbol (⏚ or ⏚) which may mean a real connection to the ground, perhaps via a metal pipe, or it may simply mean a common connection, such as the frame of a car.

For the most part in this Chapter, except in Section 26–5 on *RC* circuits, we will be interested in circuits operating in their steady state. That is, we won't be looking at a circuit at the moment a change is made in it, such as when a battery or resistor is connected or disconnected, but rather later when the currents have reached their steady values.

26–1 EMF and Terminal Voltage

To have current in an electric circuit, we need a device such as a battery or an electric generator that transforms one type of energy (chemical, mechanical, or light, for example) into electric energy. Such a device is called a **source** of **electromotive force** or of **emf**. (The term "electromotive force" is a misnomer since it does not refer to a "force" that is measured in newtons. Hence, to avoid confusion, we prefer to use the abbreviation, emf.) The *potential difference* between the terminals of such a source, when no current flows to an external circuit, is called the **emf** of the source. The symbol $\mathscr{E}$ is usually used for emf (don't confuse it with E for electric field), and its unit is volts.

A battery is not a source of constant current—the current out of a battery varies according to the resistance in the circuit. A battery *is*, however, a nearly constant voltage source, but not perfectly constant as we now discuss. You may have noticed in your own experience that when a current is drawn from a battery, the potential difference (voltage) across its terminals drops below its rated emf. For example, if you start a car with the headlights on, you may notice the headlights dim. This happens because the starter draws a large current, and the battery voltage drops as a result. The voltage drop occurs because the chemical reactions in a battery (Section 25–1) cannot supply charge fast enough to maintain the full emf. For one thing, charge must move (within the electrolyte) between the electrodes of the battery, and there is always some hindrance to completely free flow. Thus, a battery itself has some resistance, which is called its **internal resistance**; it is usually designated r.

A real battery is modeled as if it were a perfect emf $\mathscr{E}$ in series with a resistor r, as shown in Fig. 26–1. Since this resistance r is inside the battery, we can never separate it from the battery. The two points a and b in the diagram represent the two terminals of the battery. What we measure is the **terminal voltage** $V_{ab} = V_a - V_b$. When no current is drawn from the battery, the terminal voltage equals the emf, which is determined by the chemical reactions in the battery: $V_{ab} = \mathscr{E}$. However, when a current I flows naturally from the battery there is an internal drop in voltage equal to Ir. Thus the terminal voltage (the actual voltage) is[†]

$$V_{ab} = \mathscr{E} - Ir. \qquad (26\text{–}1)$$

For example, if a 12-V battery has an internal resistance of 0.1 Ω, then when 10 A flows from the battery, the terminal voltage is $12\,\text{V} - (10\,\text{A})(0.1\,\Omega) = 11\,\text{V}$. The internal resistance of a battery is usually small. For example, an ordinary flashlight battery when fresh may have an internal resistance of perhaps 0.05 Ω. (However, as it ages and the electrolyte dries out, the internal resistance increases to many ohms.) Car batteries have lower internal resistance.

EXAMPLE 26–1 **Battery with internal resistance.** A 65.0-Ω resistor is connected to the terminals of a battery whose emf is 12.0 V and whose internal resistance is 0.5 Ω, Fig. 26–2. Calculate (*a*) the current in the circuit, (*b*) the terminal voltage of the battery, V_{ab}, and (*c*) the power dissipated in the resistor R and in the battery's internal resistance r.

APPROACH We first consider the battery as a whole, which is shown in Fig. 26–2 as an emf $\mathscr{E}$ and internal resistance r between points a and b. Then we apply $V = IR$ to the circuit itself.

[†]When a battery is being charged, a current is forced to pass through it; we then have to write
$$V_{ab} = \mathscr{E} + Ir.$$
See Section 26–4 or Problem 28 and Fig. 26–46.

FIGURE 26–2 Example 26–1.

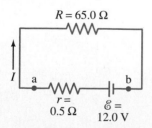

$R = 65.0\ \Omega$

I

$r = 0.5\ \Omega$

$\mathscr{E} = 12.0\ \text{V}$

SOLUTION (a) From Eq. 26–1, we have

$$V_{ab} = \mathscr{E} - Ir.$$

We apply Ohm's law (Eqs. 25–2) to this battery and the resistance R of the circuit: $V_{ab} = IR$. Hence $IR = \mathscr{E} - Ir$ or $\mathscr{E} = I(R + r)$, and so

$$I = \frac{\mathscr{E}}{R + r} = \frac{12.0\,\text{V}}{65.0\,\Omega + 0.5\,\Omega} = \frac{12.0\,\text{V}}{65.5\,\Omega} = 0.183\,\text{A}.$$

(b) The terminal voltage is

$$V_{ab} = \mathscr{E} - Ir = 12.0\,\text{V} - (0.183\,\text{A})(0.5\,\Omega) = 11.9\,\text{V}.$$

(c) The power dissipated (Eq. 25–7) in R is

$$P_R = I^2 R = (0.183\,\text{A})^2 (65.0\,\Omega) = 2.18\,\text{W},$$

and in r is

$$P_r = I^2 r = (0.183\,\text{A})^2 (0.5\,\Omega) = 0.02\,\text{W}.$$

EXERCISE A Repeat Example 26–1 assuming now that the resistance $R = 10.0\,\Omega$, whereas $\mathscr{E}$ and r remain as before.

In much of what follows, unless stated otherwise, we assume that the battery's internal resistance is negligible, and that the battery voltage given is its terminal voltage, which we will usually write simply as V rather than V_{ab}. Be careful not to confuse V (italic) for voltage and V (not italic) for the volt unit.

26–2 Resistors in Series and in Parallel

When two or more resistors are connected end to end along a single path as shown in Fig. 26–3a, they are said to be connected in **series**. The resistors could be simple resistors as were pictured in Fig. 25–12, or they could be lightbulbs (Fig. 26–3b), or heating elements, or other resistive devices. Any charge that passes through R_1 in Fig. 26–3a will also pass through R_2 and then R_3. Hence the same current I passes through each resistor. (If it did not, this would imply that either charge was not conserved, or that charge was accumulating at some point in the circuit, which does not happen in the steady state.)

We let V represent the potential difference (voltage) across all three resistors in Fig. 26–3a. We assume all other resistance in the circuit can be ignored, so V equals the terminal voltage supplied by the battery. We let V_1, V_2, and V_3 be the potential differences across each of the resistors, R_1, R_2, and R_3, respectively. From Ohm's law, $V = IR$, we can write $V_1 = IR_1$, $V_2 = IR_2$, and $V_3 = IR_3$. Because the resistors are connected end to end, energy conservation tells us that the total voltage V is equal to the sum of the voltages[†] across each resistor:

$$V = V_1 + V_2 + V_3 = IR_1 + IR_2 + IR_3. \qquad \text{[series]} \quad \textbf{(26–2)}$$

Now let us determine the equivalent single resistance R_{eq} that would draw the same current I as our combination of three resistors in series; see Fig. 26–3c. Such a single resistance R_{eq} would be related to V by

$$V = IR_{eq}.$$

We equate this expression with Eq. 26–2, $V = I(R_1 + R_2 + R_3)$, and find

$$R_{eq} = R_1 + R_2 + R_3. \qquad \text{[series]} \quad \textbf{(26–3)}$$

This is, in fact, what we expect. When we put several resistances in series, the total or equivalent resistance is the sum of the separate resistances. (Sometimes we may also call it the "net resistance.") This sum applies to any number of resistances in series. Note that when you add more resistance to the circuit, the current through the circuit will decrease. For example, if a 12-V battery is connected to a 4-Ω resistor, the current will be 3 A. But if the 12-V battery is connected to three 4-Ω resistors in series, the total resistance is 12 Ω and the current through the entire circuit will be only 1 A.

[†]To see in more detail why this is true, note that an electric charge q passing through R_1 loses an amount of potential energy equal to qV_1. In passing through R_2 and R_3, the potential energy U decreases by qV_2 and qV_3, for a total $\Delta U = qV_1 + qV_2 + qV_3$; this sum must equal the energy given to q by the battery, qV, so that energy is conserved. Hence $qV = q(V_1 + V_2 + V_3)$, and so $V = V_1 + V_2 + V_3$, which is Eq. 26–2.

FIGURE 26–3 (a) Resistances connected in series. (b) Resistances could be lightbulbs, or any other type of resistance. (c) Equivalent single resistance R_{eq} that draws the same current: $R_{eq} = R_1 + R_2 + R_3$.

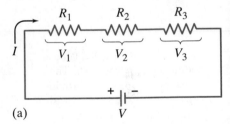

(a)

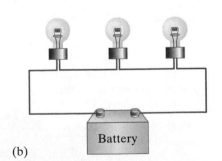

(b)

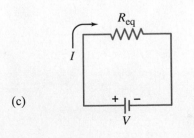

(c)

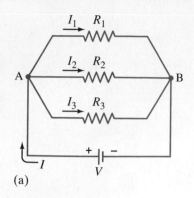

(a)

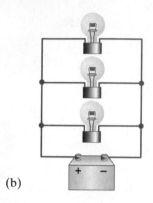

(b)

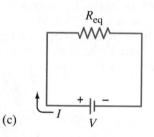

R_{eq}

(c)

FIGURE 26–4 (a) Resistances connected in parallel. (b) The resistances could be lightbulbs. (c) The equivalent circuit with R_{eq} obtained from Eq. 26–4:
$$\frac{1}{R_{eq}} = \frac{1}{R_1} + \frac{1}{R_2} + \frac{1}{R_3}.$$

FIGURE 26–5 Water pipes in parallel—analogy to electric currents in parallel.

Another simple way to connect resistors is in **parallel** so that the current from the source splits into separate branches or paths, as shown in Fig. 26–4a and b. The wiring in houses and buildings is arranged so all electric devices are in parallel, as we already saw in Chapter 25, Fig. 25–20. With parallel wiring, if you disconnect one device (say, R_1 in Fig. 26–4a), the current to the other devices is not interrupted. Compare to a series circuit, where if one device (say, R_1 in Fig. 26–3a) is disconnected, the current *is* stopped to all the others.

In a parallel circuit, Fig. 26–4a, the total current I that leaves the battery splits into three separate paths. We let I_1, I_2, and I_3 be the currents through each of the resistors, R_1, R_2, and R_3, respectively. Because *electric charge is conserved*, the current I flowing into junction A (where the different wires or conductors meet, Fig. 26–4a) must equal the current flowing out of the junction. Thus

$$I = I_1 + I_2 + I_3. \qquad \text{[parallel]}$$

When resistors are connected in parallel, each has the same voltage across it. (Indeed, any two points in a circuit connected by a wire of negligible resistance are at the same potential.) Hence the full voltage of the battery is applied to each resistor in Fig. 26–4a. Applying Ohm's law to each resistor, we have

$$I_1 = \frac{V}{R_1}, \qquad I_2 = \frac{V}{R_2}, \qquad \text{and} \qquad I_3 = \frac{V}{R_3}.$$

Let us now determine what single resistor R_{eq} (Fig. 26–4c) will draw the same current I as these three resistances in parallel. This equivalent resistance R_{eq} must satisfy Ohm's law too:

$$I = \frac{V}{R_{eq}}.$$

We now combine the equations above:

$$I = I_1 + I_2 + I_3,$$
$$\frac{V}{R_{eq}} = \frac{V}{R_1} + \frac{V}{R_2} + \frac{V}{R_3}.$$

When we divide out the V from each term, we have

$$\frac{1}{R_{eq}} = \frac{1}{R_1} + \frac{1}{R_2} + \frac{1}{R_3}. \qquad \text{[parallel]} \quad \textbf{(26–4)}$$

For example, suppose you connect two 4-Ω loudspeakers to a single set of output terminals of your stereo amplifier or receiver. (Ignore the other channel for a moment—our two speakers are both connected to the left channel, say.) The equivalent resistance of the two 4-Ω "resistors" in parallel is

$$\frac{1}{R_{eq}} = \frac{1}{4\,\Omega} + \frac{1}{4\,\Omega} = \frac{2}{4\,\Omega} = \frac{1}{2\,\Omega},$$

and so $R_{eq} = 2\,\Omega$. Thus the net (or equivalent) resistance is *less* than each single resistance. This may at first seem surprising. But remember that when you connect resistors in parallel, you are giving the current additional paths to follow. Hence the net resistance will be less.

Equations 26–3 and 26–4 make good sense. Recalling Eq. 25–3 for resistivity, $R = \rho \ell / A$, we see that placing resistors in series increases the length and therefore the resistance; putting resistors in parallel increases the area through which current flows, thus reducing the overall resistance.

An analogy may help here. Consider two identical pipes taking in water near the top of a dam and releasing it below as shown in Fig. 26–5. The gravitational potential difference, proportional to the height h, is the same for both pipes, just as the voltage is the same for parallel resistors. If both pipes are open, rather than only one, twice as much water will flow through. That is, with two equal pipes open, the net resistance to the flow of water will be reduced, by half, just as for electrical resistors in parallel. Note that if both pipes are closed, the dam offers infinite resistance to the flow of water. This corresponds in the electrical case to an open circuit—when the path is not continuous and no current flows—so the electrical resistance is infinite.

EXERCISE B You have a 10-Ω and a 15-Ω resistor. What is the smallest and largest equivalent resistance that you can make with these two resistors?

CONCEPTUAL EXAMPLE 26–2 **Series or parallel?** (*a*) The lightbulbs in Fig. 26–6 are identical. Which configuration produces more light? (*b*) Which way do you think the headlights of a car are wired? Ignore change of filament resistance R with current.

RESPONSE (*a*) The equivalent resistance of the parallel circuit is found from Eq. 26–4. $1/R_{eq} = 1/R + 1/R = 2/R$. Thus $R_{eq} = R/2$. The parallel combination then has lower resistance ($= R/2$) than the series combination ($R_{eq} = R + R = 2R$). There will be more total current in the parallel configuration (2), since $I = V/R_{eq}$ and V is the same for both circuits. The total power transformed, which is related to the light produced is $P = IV$ so the greater current in (2) means more light produced.

(*b*) Headlights are wired in parallel (2), because if one bulb goes out, the other bulb can stay lit. If they were in series (1), when one bulb burned out (the filament broke), the circuit would be open and no current would flow, so neither bulb would light.

NOTE When you answered the Chapter-Opening Question on page 677, was your answer circuit 2? Can you express any misconceptions you might have had?

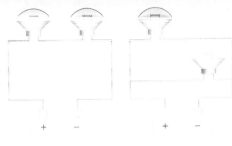

(1) Series   (2) Parallel

FIGURE 26–6 Example 26–2.

FIGURE 26–7 Example 26–3.

CONCEPTUAL EXAMPLE 26–3 **An illuminating surprise.** A 100-W, 120-V lightbulb and a 60-W, 120-V lightbulb are connected in two different ways as shown in Fig. 26–7. In each case, which bulb glows more brightly? Ignore change of filament resistance with current (and temperature).

RESPONSE (*a*) These are normal lightbulbs with their power rating given for 120 V. They both receive 120 V, so the 100-W bulb is naturally brighter.

(*b*) The resistance of the 100-W bulb is less than that of the 60-W bulb (calculated from $P = V^2/R$ at constant 120 V). Here they are connected in series and receive the same current. Hence, from $P = I^2R$ ($I =$ constant) the higher-resistance "60-W" bulb will transform more power and thus be brighter.

NOTE When connected in series as in (*b*), the two bulbs do *not* dissipate 60 W and 100 W because neither bulb receives 120 V.

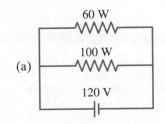

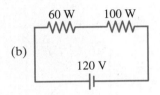

Note that whenever a group of resistors is replaced by the equivalent resistance, current and voltage and power in the rest of the circuit are unaffected.

EXAMPLE 26–4 **Circuit with series and parallel resistors.** How much current is drawn from the battery shown in Fig. 26–8a?

APPROACH The current I that flows out of the battery all passes through the 400-Ω resistor, but then it splits into I_1 and I_2 passing through the 500-Ω and 700-Ω resistors. The latter two resistors are in parallel with each other. We look for something that we already know how to treat. So let's start by finding the equivalent resistance, R_P, of the parallel resistors, 500 Ω and 700 Ω. Then we can consider this R_P to be in series with the 400-Ω resistor.

SOLUTION The equivalent resistance, R_P, of the 500-Ω and 700-Ω resistors in parallel is given by

$$\frac{1}{R_P} = \frac{1}{500\ \Omega} + \frac{1}{700\ \Omega} = 0.0020\ \Omega^{-1} + 0.0014\ \Omega^{-1} = 0.0034\ \Omega^{-1}.$$

This is $1/R_P$, so we take the reciprocal to find R_P. It is a common mistake to forget to take this reciprocal. Notice that the units of reciprocal ohms, Ω^{-1}, are a reminder. Thus

$$R_P = \frac{1}{0.0034\ \Omega^{-1}} = 290\ \Omega.$$

This 290 Ω is the equivalent resistance of the two parallel resistors, and is in series with the 400-Ω resistor as shown in the equivalent circuit of Fig. 26–8b. To find the total equivalent resistance R_{eq}, we add the 400-Ω and 290-Ω resistances together, since they are in series, and find

$$R_{eq} = 400\ \Omega + 290\ \Omega = 690\ \Omega.$$

The total current flowing from the battery is then

$$I = \frac{V}{R_{eq}} = \frac{12.0\ V}{690\ \Omega} = 0.0174\ A \approx 17\ mA.$$

NOTE This I is also the current flowing through the 400-Ω resistor, but not through the 500-Ω and 700-Ω resistors (both currents are less — see the next Example).

NOTE Complex resistor circuits can often be analyzed in this way, considering the circuit as a combination of series and parallel resistances.

FIGURE 26–8 (a) Circuit for Examples 26–4 and 26–5. (b) Equivalent circuit, showing the equivalent resistance of 290 Ω for the two parallel resistors in (a).

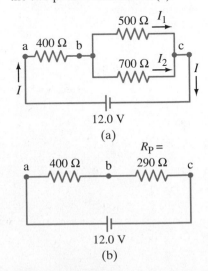

⚠ CAUTION

Remember to take the reciprocal

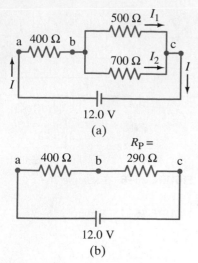

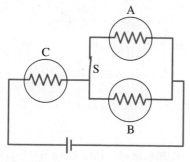

FIGURE 26-8 (repeated)
(a) Circuit for Examples 26–4 and 26–5. (b) Equivalent circuit, showing the equivalent resistance of 290 Ω for the two parallel resistors in (a).

EXAMPLE 26-5 Current in one branch. What is the current through the 500-Ω resistor in Fig. 26–8a?

APPROACH We need to find the voltage across the 500-Ω resistor, which is the voltage between points b and c in Fig. 26–8a, and we call it V_{bc}. Once V_{bc} is known, we can apply Ohm's law, $V = IR$, to get the current. First we find the voltage across the 400-Ω resistor, V_{ab}, since we know that 17.4 mA passes through it (Example 26–4).

SOLUTION V_{ab} can be found using $V = IR$:

$$V_{ab} = (0.0174\,A)(400\,\Omega) = 7.0\,V.$$

Since the total voltage across the network of resistors is $V_{ac} = 12.0\,V$, then V_{bc} must be $12.0\,V - 7.0\,V = 5.0\,V$. Then Ohm's law applied to the 500-Ω resistor tells us that the current I_1 through that resistor is

$$I_1 = \frac{5.0\,V}{500\,\Omega} = 1.0 \times 10^{-2}\,A = 10\,mA.$$

This is the answer we wanted. We can also calculate the current I_2 through the 700-Ω resistor since the voltage across it is also 5.0 V:

$$I_2 = \frac{5.0\,V}{700\,\Omega} = 7\,mA.$$

NOTE When I_1 combines with I_2 to form the total current I (at point c in Fig. 26–8a), their sum is $10\,mA + 7\,mA = 17\,mA$. This equals the total current I as calculated in Example 26–4, as it should.

CONCEPTUAL EXAMPLE 26-6 Bulb brightness in a circuit. The circuit shown in Fig. 26–9 has three identical lightbulbs, each of resistance R. (a) When switch S is closed, how will the brightness of bulbs A and B compare with that of bulb C? (b) What happens when switch S is opened? Use a minimum of mathematics in your answers.

RESPONSE (a) With switch S closed, the current that passes through bulb C must split into two equal parts when it reaches the junction leading to bulbs A and B. It splits into equal parts because the resistance of bulb A equals that of B. Thus, bulbs A and B each receive half of C's current; A and B will be equally bright, but they will be less bright than bulb C ($P = I^2R$). (b) When the switch S is open, no current can flow through bulb A, so it will be dark. We now have a simple one-loop series circuit, and we expect bulbs B and C to be equally bright. However, the equivalent resistance of this circuit ($= R + R$) is greater than that of the circuit with the switch closed. When we open the switch, we increase the resistance and reduce the current leaving the battery. Thus, bulb C will be dimmer when we open the switch. Bulb B gets more current when the switch is open (you may have to use some mathematics here), and so it will be brighter than with the switch closed; and B will be as bright as C.

FIGURE 26-9 Example 26–6, three identical lightbulbs. Each yellow circle with –W– inside represents a lightbulb and its resistance.

EXAMPLE 26-7 ESTIMATE A two-speed fan. One way a multiple-speed ventilation fan for a car can be designed is to put resistors in series with the fan motor. The resistors reduce the current through the motor and make it run more slowly. Suppose the current in the motor is 5.0 A when it is connected directly across a 12-V battery. (a) What series resistor should be used to reduce the current to 2.0 A for low-speed operation? (b) What power rating should the resistor have?

APPROACH An electric motor in series with a resistor can be treated as two resistors in series. The power comes from $P = IV$.

SOLUTION (a) When the motor is connected to 12 V and drawing 5.0 A, its resistance is $R = V/I = (12\,V)/(5.0\,A) = 2.4\,\Omega$. We will assume that this is the motor's resistance for all speeds. (This is an approximation because the current through the motor depends on its speed.) Then, when a current of 2.0 A is flowing, the voltage across the motor is $(2.0\,A)(2.4\,\Omega) = 4.8\,V$. The remaining $12.0\,V - 4.8\,V = 7.2\,V$ must appear across the series resistor. When 2.0 A flows through the resistor, its resistance must be $R = (7.2\,V)/(2.0\,A) = 3.6\,\Omega$. (b) The power dissipated by the resistor is $P = (7.2\,V)(2.0\,A) = 14.4\,W$. To be safe, a power rating of 20 W would be appropriate.

EXAMPLE 26–8 Analyzing a circuit. A 9.0-V battery whose internal resistance r is 0.50 Ω is connected in the circuit shown in Fig. 26–10a. (*a*) How much current is drawn from the battery? (*b*) What is the terminal voltage of the battery? (*c*) What is the current in the 6.0-Ω resistor?

APPROACH To find the current out of the battery, we first need to determine the equivalent resistance R_{eq} of the entire circuit, including r, which we do by identifying and isolating simple series or parallel combinations of resistors. Once we find I from Ohm's law, $I = \mathcal{E}/R_{eq}$, we get the terminal voltage using $V_{ab} = \mathcal{E} - Ir$. For (*c*) we apply Ohm's law to the 6.0-Ω resistor.

SOLUTION (*a*) We want to determine the equivalent resistance of the circuit. But where do we start? We note that the 4.0-Ω and 8.0-Ω resistors are in parallel, and so have an equivalent resistance R_{eq1} given by

$$\frac{1}{R_{eq1}} = \frac{1}{8.0\,\Omega} + \frac{1}{4.0\,\Omega} = \frac{3}{8.0\,\Omega};$$

so $R_{eq1} = 2.7\,\Omega$. This 2.7 Ω is in series with the 6.0-Ω resistor, as shown in the equivalent circuit of Fig. 26–10b. The net resistance of the lower arm of the circuit is then

$$R_{eq2} = 6.0\,\Omega + 2.7\,\Omega = 8.7\,\Omega,$$

as shown in Fig. 26–10c. The equivalent resistance R_{eq3} of the 8.7-Ω and 10.0-Ω resistances in parallel is given by

$$\frac{1}{R_{eq3}} = \frac{1}{10.0\,\Omega} + \frac{1}{8.7\,\Omega} = 0.21\,\Omega^{-1},$$

so $R_{eq3} = (1/0.21\,\Omega^{-1}) = 4.8\,\Omega$. This 4.8 Ω is in series with the 5.0-Ω resistor and the 0.50-Ω internal resistance of the battery (Fig. 26–10d), so the total equivalent resistance R_{eq} of the circuit is $R_{eq} = 4.8\,\Omega + 5.0\,\Omega + 0.50\,\Omega = 10.3\,\Omega$. Hence the current drawn is

$$I = \frac{\mathcal{E}}{R_{eq}} = \frac{9.0\,\text{V}}{10.3\,\Omega} = 0.87\,\text{A}.$$

(*b*) The terminal voltage of the battery is

$$V_{ab} = \mathcal{E} - Ir = 9.0\,\text{V} - (0.87\,\text{A})(0.50\,\Omega) = 8.6\,\text{V}.$$

(*c*) Now we can work back and get the current in the 6.0-Ω resistor. It must be the same as the current through the 8.7 Ω shown in Fig. 26–10c (why?). The voltage across that 8.7 Ω will be the emf of the battery minus the voltage drops across r and the 5.0-Ω resistor: $V_{8.7} = 9.0\,\text{V} - (0.87\,\text{A})(0.50\,\Omega + 5.0\,\Omega)$. Applying Ohm's law, we get the current (call it I')

$$I' = \frac{9.0\,\text{V} - (0.87\,\text{A})(0.50\,\Omega + 5.0\,\Omega)}{8.7\,\Omega} = 0.48\,\text{A}.$$

This is the current through the 6.0-Ω resistor.

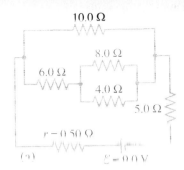

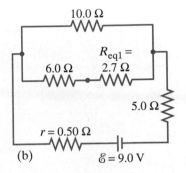

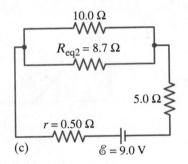

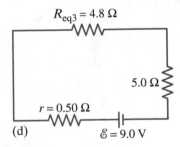

FIGURE 26–10 Circuit for Example 26–8, where r is the internal resistance of the battery.

26–3 Kirchhoff's Rules

In the last few Examples we have been able to find the currents in circuits by combining resistances in series and parallel, and using Ohm's law. This technique can be used for many circuits. However, some circuits are too complicated for that analysis. For example, we cannot find the currents in each part of the circuit shown in Fig. 26–11 simply by combining resistances as we did before.

To deal with such complicated circuits, we use Kirchhoff's rules, devised by G. R. Kirchhoff (1824–1887) in the mid-nineteenth century. There are two rules, and they are simply convenient applications of the laws of conservation of charge and energy.

FIGURE 26–11 Currents can be calculated using Kirchhoff's rules.

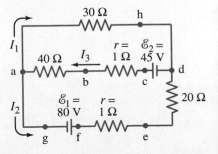

Kirchhoff's first rule or **junction rule** is based on the conservation of electric charge that we already used to derive the rule for parallel resistors. It states that

at any junction point, the sum of all currents entering the junction must equal the sum of all currents leaving the junction.

That is, whatever charge goes in must come out. We already saw an instance of this in the NOTE at the end of Example 26–5.

Kirchhoff's second rule or **loop rule** is based on the conservation of energy. It states that

the sum of the changes in potential around any closed loop of a circuit must be zero.

To see why this rule should hold, consider a rough analogy with the potential energy of a roller coaster on its track. When the roller coaster starts from the station, it has a particular potential energy. As it climbs the first hill, its potential energy increases and reaches a maximum at the top. As it descends the other side, its potential energy decreases and reaches a local minimum at the bottom of the hill. As the roller coaster continues on its path, its potential energy goes through more changes. But when it arrives back at the starting point, it has exactly as much potential energy as it had when it started at this point. Another way of saying this is that there was as much uphill as there was downhill.

Similar reasoning can be applied to an electric circuit. We will analyze the circuit of Fig. 26–11 shortly but first we consider the simpler circuit in Fig. 26–12. We have chosen it to be the same as the equivalent circuit of Fig. 26–8b already discussed. The current in this circuit is $I = (12.0\,\text{V})/(690\,\Omega) = 0.0174\,\text{A}$, as we calculated in Example 26–4. (We keep an extra digit in I to reduce rounding errors.) The positive side of the battery, point e in Fig. 26–12a, is at a high potential compared to point d at the negative side of the battery. That is, point e is like the top of a hill for a roller coaster. We follow the current around the circuit starting at any point. We choose to start at point d and follow a positive test charge completely around this circuit. As we go, we note all changes in potential. When the test charge returns to point d, the potential will be the same as when we started (total change in potential around the circuit is zero). We plot the changes in potential around the circuit in Fig. 26–12b; point d is arbitrarily taken as zero.

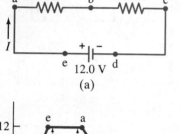

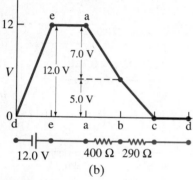

FIGURE 26–12 Changes in potential around the circuit in (a) are plotted in (b).

As our positive test charge goes from point d, which is the negative or low potential side of the battery, to point e, which is the positive terminal (high potential side) of the battery, the potential increases by 12.0 V. (This is like the roller coaster being pulled up the first hill.) That is,

$$V_{\text{ed}} = +12.0\,\text{V}.$$

When our test charge moves from point e to point a, there is no change in potential since there is no source of emf and we assume negligible resistance in the connecting wires. Next, as the charge passes through the 400-Ω resistor to get to point b, there is a decrease in potential of $V = IR = (0.0174\,\text{A})(400\,\Omega) = 7.0\,\text{V}$. The positive test charge is flowing "downhill" since it is heading toward the negative terminal of the battery, as indicated in the graph of Fig. 26–12b. Because this is a *decrease* in potential, we use a *negative* sign:

$$V_{\text{ba}} = V_{\text{b}} - V_{\text{a}} = -7.0\,\text{V}.$$

As the charge proceeds from b to c there is another potential decrease (a "voltage drop") of $(0.0174\,\text{A}) \times (290\,\Omega) = 5.0\,\text{V}$, and this too is a decrease in potential:

$$V_{\text{cb}} = -5.0\,\text{V}.$$

There is no change in potential as our test charge moves from c to d as we assume negligible resistance in the wires.

The sum of all the changes in potential around the circuit of Fig. 26–12 is

$$+12.0\,\text{V} - 7.0\,\text{V} - 5.0\,\text{V} = 0.$$

This is exactly what Kirchhoff's loop rule said it would be.

4. **Apply Kirchhoff's loop rule** for one or more loops; follow each loop in one direction only. Pay careful attention to subscripts, and to signs:

   (a) For a resistor, apply Ohm's law: the potential difference is negative (a decrease) if your chosen loop direction is the same as the chosen current direction through that resistor; the potential difference is positive (an increase) if your chosen loop direction is opposite to the chosen current direction.

   (b) For a battery, the potential difference is positive if your chosen loop direction is from the negative terminal toward the positive terminal; the potential difference is negative if the loop direction is from the positive terminal toward the negative terminal.

5. **Solve the equations** algebraically for the unknowns. Be careful when manipulating equations not to err with signs. At the end, check your answers by plugging them into the original equations, or even by using any additional loop or junction rule equations not used previously.

1. **Label the current** in each separate branch of the given circuit with a different subscript, such as I_1, I_2, I_3 (see Fig. 26–11 or 26–13). Each current refers to a segment between two junctions. Choose the direction of each current, using an arrow. The direction can be chosen arbitrarily: if the current is actually in the opposite direction, it will come out with a minus sign in the solution.

2. **Identify the unknowns.** You will need as many independent equations as there are unknowns. You may write down more equations than this, but you will find that some of the equations will be redundant (that is, not be independent in the sense of providing new information). You may use $V = IR$ for each resistor, which sometimes will reduce the number of unknowns.

3. **Apply Kirchhoff's junction rule** at one or more junctions.

EXAMPLE 26–9 Using Kirchhoff's rules.
Calculate the currents I_1, I_2, and I_3 in the three branches of the circuit in Fig. 26–13 (which is the same as Fig. 26–11).

APPROACH AND SOLUTION

1. **Label the currents** and their directions. Figure 26–13 uses the labels I_1, I_2, and I_3 for the current in the three separate branches. Since (positive) current tends to move away from the positive terminal of a battery, we choose I_2 and I_3 to have the directions shown in Fig. 26–13. The direction of I_1 is not obvious in advance, so we arbitrarily chose the direction indicated. If the current actually flows in the opposite direction, our answer will have a negative sign.

2. **Identify the unknowns.** We have three unknowns and therefore we need three equations, which we get by applying Kirchhoff's junction and loop rules.

3. **Junction rule:** We apply Kirchhoff's junction rule to the currents at point a, where I_3 enters and I_2 and I_1 leave:
$$I_3 = I_1 + I_2. \tag{a}$$
This same equation holds at point d, so we get no new information by writing an equation for point d.

4. **Loop rule:** We apply Kirchhoff's loop rule to two different closed loops. First we apply it to the upper loop ahdcba. We start (and end) at point a. From a to h we have a potential decrease $V_{ha} = -(I_1)(30\,\Omega)$. From h to d there is no change, but from d to c the potential increases by 45 V: that is, $V_{cd} = +45\,V$. From c to a the potential decreases through the two resistances by an amount $V_{ac} = -(I_3)(40\,\Omega + 1\,\Omega) = -(41\,\Omega)I_3$. Thus we have $V_{ha} + V_{cd} + V_{ac} = 0$, or
$$-30I_1 + 45 - 41I_3 = 0, \tag{b}$$
where we have omitted the units (volts and amps) so we can more easily do the algebra. For our second loop, we take the outer loop ahdefga. (We could have chosen the lower loop abcdefga instead.) Again we start at point a and have $V_{ha} = -(I_1)(30\,\Omega)$, and $V_{dh} = 0$. But when we take our positive test charge from d to e, it actually is going uphill, against the current—or at least against the *assumed* direction of the current, which is what counts in this calculation. Thus $V_{ed} = I_2(20\,\Omega)$ has a *positive* sign. Similarly, $V_{fe} = I_2(1\,\Omega)$. From f to g there is a decrease in potential of 80 V since we go from the high potential terminal of the battery to the low. Thus $V_{gf} = -80\,V$. Finally, $V_{ag} = 0$, and the sum of the potential changes around this loop is
$$-30I_1 + (20 + 1)I_2 - 80 = 0. \tag{c}$$
Our major work is done. The rest is algebra.

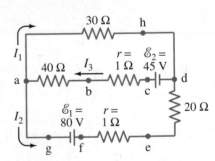

FIGURE 26–13 Currents can be calculated using Kirchhoff's rules. See Example 26–9.

5. Solve the equations. We have three equations—labeled (a), (b), and (c)—and three unknowns. From Eq. (c) we have

$$I_2 = \frac{80 + 30I_1}{21} = 3.8 + 1.4I_1. \qquad (d)$$

From Eq. (b) we have

$$I_3 = \frac{45 - 30I_1}{41} = 1.1 - 0.73I_1. \qquad (e)$$

We substitute Eqs. (d) and (e) into Eq. (a):

$$I_1 = I_3 - I_2 = 1.1 - 0.73I_1 - 3.8 - 1.4I_1.$$

We solve for I_1, collecting terms:

$$3.1I_1 = -2.7$$
$$I_1 = -0.87 \text{ A}.$$

The negative sign indicates that the direction of I_1 is actually opposite to that initially assumed and shown in Fig. 26–13. The answer automatically comes out in amperes because all values were in volts and ohms. From Eq. (d) we have

$$I_2 = 3.8 + 1.4I_1 = 3.8 + 1.4(-0.87) = 2.6 \text{ A},$$

and from Eq. (e)

$$I_3 = 1.1 - 0.73I_1 = 1.1 - 0.73(-0.87) = 1.7 \text{ A}.$$

This completes the solution.

NOTE The unknowns in different situations are not necessarily currents. It might be that the currents are given and we have to solve for unknown resistance or voltage. The variables are then different, but the technique is the same.

EXERCISE C Write the equation for the lower loop abcdefga of Example 26–9 and show, assuming the currents calculated in this Example, that the potentials add to zero for this lower loop.

26–4 Series and Parallel EMFs; Battery Charging

When two or more sources of emf, such as batteries, are arranged in series as in Fig. 26–14a, the total voltage is the algebraic sum of their respective voltages. On the other hand, when a 20-V and a 12-V battery are connected oppositely, as shown in Fig. 26–14b, the net voltage V_{ca} is 8 V (ignoring voltage drop across internal resistances). That is, a positive test charge moved from a to b gains in potential by 20 V, but when it passes from b to c it drops by 12 V. So the net change is $20 \text{ V} - 12 \text{ V} = 8 \text{ V}$. You might think that connecting batteries in reverse like this would be wasteful. For most purposes that would be true. But such a reverse arrangement is precisely how a battery charger works. In Fig. 26–14b, the 20-V source is charging up the 12-V battery. Because of its greater voltage, the 20-V source is forcing charge back into the 12-V battery: electrons are being forced into its negative terminal and removed from its positive terminal.

An automobile alternator keeps the car battery charged in the same way. A voltmeter placed across the terminals of a (12-V) car battery with the engine running fairly fast can tell you whether or not the alternator is charging the battery. If it is, the voltmeter reads 13 or 14 V. If the battery is not being charged, the voltage will be 12 V, or less if the battery is discharging. Car batteries can be recharged, but other batteries may not be rechargeable, since the chemical reactions in many cannot be reversed. In such cases, the arrangement of Fig. 26–14b would simply waste energy.

Sources of emf can also be arranged in parallel, Fig. 26–14c. With equal emfs, a parallel arrangement can provide more energy when large currents are needed. Each of the cells in parallel has to produce only a fraction of the total current, so the energy loss due to internal resistance is less than for a single cell; and the batteries will go dead less quickly.

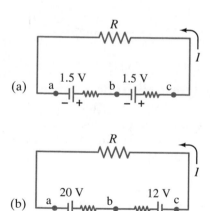

(a)

(b)

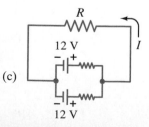

(c)

FIGURE 26–14 Batteries in series (a) and (b), and in parallel (c).

EXAMPLE 26–10 **Jump starting a car.** A good car battery is being used to jump start a car with a weak battery. The good battery has an emf of 12.5 V and internal resistance 0.020 Ω. Suppose the weak battery has an emf of 10.1 V and internal resistance 0.10 Ω. Each copper jumper cable is 3.0 m long and 0.50 cm in diameter, and can be attached as shown in Fig. 26–15. Assume the starter motor can be represented as a resistor $R_s = 0.15$ Ω. Determine the current through the starter motor (a) if only the weak battery is connected to it, and (b) if the good battery is also connected, as shown in Fig. 26–15.

APPROACH We apply Kirchhoff's rules, but in (b) we will first need to determine the resistance of the jumper cables using their dimensions and the resistivity ($\rho = 1.68 \times 10^{-8}$ Ω·m for copper) as discussed in Section 25–4.

SOLUTION (a) The circuit with only the weak battery and no jumper cables is simple: an emf of 10.1 V connected to two resistances in series, 0.10 Ω + 0.15 Ω = 0.25 Ω. Hence the current is $I = V/R = (10.1 \text{ V})/(0.25 \text{ Ω}) = 40$ A.

(b) We need to find the resistance of the jumper cables that connect the good battery. From Eq. 25–3, each has resistance $R_J = \rho\ell/A = (1.68 \times 10^{-8} \text{ Ω·m})(3.0 \text{ m})/(\pi)(0.25 \times 10^{-2} \text{ m})^2 = 0.0026$ Ω. Kirchhoff's loop rule for the full outside loop gives

$$12.5 \text{ V} - I_1(2R_J + r_1) - I_3 R_S = 0$$
$$12.5 \text{ V} - I_1(0.025 \text{ Ω}) - I_3(0.15 \text{ Ω}) = 0 \qquad \text{(i)}$$

since $(2R_J + r) = (0.0052 \text{ Ω} + 0.020 \text{ Ω}) = 0.025$ Ω.
The loop rule for the lower loop, including the weak battery and the starter, gives

$$10.1 \text{ V} - I_3(0.15 \text{ Ω}) - I_2(0.10 \text{ Ω}) = 0. \qquad \text{(ii)}$$

The junction rule at point B gives

$$I_1 + I_2 = I_3. \qquad \text{(iii)}$$

We have three equations in three unknowns. From Eq. (iii), $I_1 = I_3 - I_2$ and we substitute this into Eq. (i):

$$12.5 \text{ V} - (I_3 - I_2)(0.025 \text{ Ω}) - I_3(0.15 \text{ Ω}) = 0,$$
$$12.5 \text{ V} - I_3(0.175 \text{ Ω}) + I_2(0.025 \text{ Ω}) = 0.$$

Combining this last equation with Eq. (ii) gives $I_3 = 71$ A, quite a bit better than in part (a). The other currents are $I_2 = -5$ A and $I_1 = 76$ A. Note that $I_2 = -5$ A is in the opposite direction from that assumed in Fig. 26–15. The terminal voltage of the weak 10.1-V battery is now $V_{BA} = 10.1 \text{ V} - (-5 \text{ A})(0.10 \text{ Ω}) = 10.6$ V.

NOTE The circuit shown in Fig. 26–15, without the starter motor, is how a battery can be charged. The stronger battery pushes charge back into the weaker battery.

EXERCISE D If the jumper cables of Example 26–10 were mistakenly connected in reverse, the positive terminal of each battery would be connected to the negative terminal of the other battery (Fig. 26–16). What would be the current I even before the starter motor is engaged (the switch S in Fig. 26–16 is open)? Why could this cause the batteries to explode?

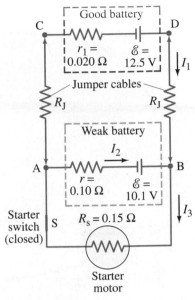

FIGURE 26–15 Example 26–10, a jump start.

FIGURE 26–16 Exercise D.

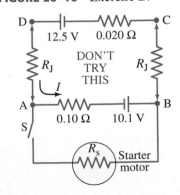

26–5 Circuits Containing Resistor and Capacitor (*RC* Circuits)

Our study of circuits in this Chapter has, until now, dealt with steady currents that do not change in time. Now we examine circuits that contain both resistance and capacitance. Such a circuit is called an ***RC* circuit**. *RC* circuits are common in everyday life: they are used to control the speed of a car's windshield wiper, and the timing of the change of traffic lights. They are used in camera flashes, in heart pacemakers, and in many other electronic devices. In *RC* circuits, we are not so interested in the final "steady state" voltage and charge on the capacitor, but rather in how these variables change in time.

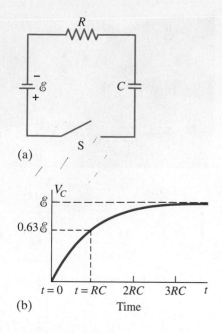

(a)

(b)

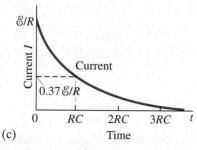

(c)

FIGURE 26–17 After the switch S closes in the RC circuit shown in (a), the voltage across the capacitor increases with time as shown in (b), and the current through the resistor decreases with time as shown in (c).

Let us now examine the RC circuit shown in Fig. 26–17a. When the switch S is closed, current immediately begins to flow through the circuit. Electrons will flow out from the negative terminal of the battery, through the resistor R, and accumulate on the upper plate of the capacitor. And electrons will flow into the positive terminal of the battery, leaving a positive charge on the other plate of the capacitor. As charge accumulates on the capacitor, the potential difference across it increases $(V_C = Q/C)$, and the current is reduced until eventually the voltage across the capacitor equals the emf of the battery, $\mathscr{E}$. There is then no potential difference across the resistor, and no further current flows. Potential difference V_C across the capacitor thus increases in time as shown in Fig. 26–17b. The mathematical form of this curve—that is, V_C as a function of time—can be derived using conservation of energy (or Kirchhoff's loop rule). The emf $\mathscr{E}$ of the battery will equal the sum of the voltage drops across the resistor (IR) and the capacitor (Q/C):

$$\mathscr{E} = IR + \frac{Q}{C}. \tag{26–5}$$

The resistance R includes all resistance in the circuit, including the internal resistance of the battery; I is the current in the circuit at any instant, and Q is the charge on the capacitor at that same instant. Although $\mathscr{E}$, R, and C are constants, both Q and I are functions of time. The rate at which charge flows through the resistor $(I = dQ/dt)$ is equal to the rate at which charge accumulates on the capacitor. Thus we can write

$$\mathscr{E} = R\frac{dQ}{dt} + \frac{1}{C}Q.$$

This equation can be solved by rearranging it:

$$\frac{dQ}{C\mathscr{E} - Q} = \frac{dt}{RC}.$$

We now integrate from $t = 0$, when $Q = 0$, to time t when a charge Q is on the capacitor:

$$\int_0^Q \frac{dQ}{C\mathscr{E} - Q} = \frac{1}{RC}\int_0^t dt$$

$$-\ln(C\mathscr{E} - Q)\Big|_0^Q = \frac{t}{RC}\Big|_0^t$$

$$-\ln(C\mathscr{E} - Q) - (-\ln C\mathscr{E}) = \frac{t}{RC}$$

$$\ln(C\mathscr{E} - Q) - \ln(C\mathscr{E}) = -\frac{t}{RC}$$

$$\ln\left(1 - \frac{Q}{C\mathscr{E}}\right) = -\frac{t}{RC}.$$

We take the exponential[†] of both sides

$$1 - \frac{Q}{C\mathscr{E}} = e^{-t/RC}$$

or

$$Q = C\mathscr{E}(1 - e^{-t/RC}). \tag{26–6a}$$

The potential difference across the capacitor is $V_C = Q/C$, so

$$V_C = \mathscr{E}(1 - e^{-t/RC}). \tag{26–6b}$$

From Eqs. 26–6 we see that the charge Q on the capacitor, and the voltage V_C across it, increase from zero at $t = 0$ to maximum values $Q_{max} = C\mathscr{E}$ and $V_C = \mathscr{E}$ after a very long time. The quantity RC that appears in the exponent is called the **time constant** τ of the circuit:

$$\tau = RC. \tag{26–7}$$

It represents the time[‡] required for the capacitor to reach $(1 - e^{-1}) = 0.63$ or 63% of its full charge and voltage. Thus the product RC is a measure of how quickly the

[†]The constant e, known as the base for natural logarithms, has the value $e = 2.718 \cdots$. Do not confuse this e with e for the charge on the electron.
[‡]The units of RC are $\Omega \cdot F = (V/A)(C/V) = C/(C/s) = s$.

capacitor gets charged. In a circuit, for example, where $R = 200\,\text{k}\Omega$ and $C = 3.0\,\mu\text{F}$, the time constant is $(2.0 \times 10^5\,\Omega)(3.0 \times 10^{-6}\,\text{F}) = 0.60\,\text{s}$. If the resistance is much lower, the time constant is much smaller. This makes sense, since a lower resistance will retard the flow of charge less. All circuits contain some resistance (if only in the connecting wires), so a capacitor never can be charged instantaneously when connected to a battery.

From Eq. 26–6 it appears that Q and V_C never quite reach their maximum values within a finite time. However, they reach 86% of maximum in $2RC$, 95% in $3RC$, 98% in $4RC$, and so on. Q and V_C approach their maximum values asymptotically. For example, if $R = 20\,\text{k}\Omega$ and $C = 0.30\,\mu\text{F}$, the time constant is $(2.0 \times 10^4\,\Omega)(3.0 \times 10^{-7}\,\text{F}) = 6.0 \times 10^{-3}\,\text{s}$. So the capacitor is more than 98% charged in less than $\frac{1}{40}$ of a second.

The current I through the circuit of Fig. 26–17a at any time t can be obtained by differentiating Eq. 26–6a:

$$I = \frac{dQ}{dt} = \frac{\mathscr{E}}{R} e^{-t/RC}. \qquad (26\text{–}8)$$

Thus, at $t = 0$, the current is $I = \mathscr{E}/R$, as expected for a circuit containing only a resistor (there is not yet a potential difference across the capacitor). The current then drops exponentially in time with a time constant equal to RC, as the voltage across the capacitor increases. This is shown in Fig. 26–17c. The time constant RC represents the time required for the current to drop to $1/e \approx 0.37$ of its initial value.

EXAMPLE 26–11 **RC circuit, with emf.** The capacitance in the circuit of Fig. 26–17a is $C = 0.30\,\mu\text{F}$, the total resistance is $20\,\text{k}\Omega$, and the battery emf is 12 V. Determine (a) the time constant, (b) the maximum charge the capacitor could acquire, (c) the time it takes for the charge to reach 99% of this value, (d) the current I when the charge Q is half its maximum value, (e) the maximum current, and (f) the charge Q when the current I is 0.20 its maximum value.

APPROACH We use Fig. 26–17 and Eqs. 26–5, 6, 7, and 8.

SOLUTION (a) The time constant is $RC = (2.0 \times 10^4\,\Omega)(3.0 \times 10^{-7}\,\text{F}) = 6.0 \times 10^{-3}\,\text{s}$.

(b) The maximum charge would be $Q = C\mathscr{E} = (3.0 \times 10^{-7}\,\text{F})(12\,\text{V}) = 3.6\,\mu\text{C}$.

(c) In Eq. 26–6a, we set $Q = 0.99C\mathscr{E}$:

$$0.99C\mathscr{E} = C\mathscr{E}(1 - e^{-t/RC}),$$

or

$$e^{-t/RC} = 1 - 0.99 = 0.01.$$

Then

$$\frac{t}{RC} = -\ln(0.01) = 4.6$$

so

$$t = 4.6RC = 28 \times 10^{-3}\,\text{s}$$

or 28 ms (less than $\frac{1}{30}\,\text{s}$).

(d) From part (b) the maximum charge is $3.6\,\mu\text{C}$. When the charge is half this value, $1.8\,\mu\text{C}$, the current I in the circuit can be found using the original differential equation, or Eq. 26–5:

$$I = \frac{1}{R}\left(\mathscr{E} - \frac{Q}{C}\right) = \frac{1}{2.0 \times 10^4\,\Omega}\left(12\,\text{V} - \frac{1.8 \times 10^{-6}\,\text{C}}{0.30 \times 10^{-6}\,\text{F}}\right) = 300\,\mu\text{A}.$$

(e) The current is a maximum when there is no charge on the capacitor ($Q = 0$):

$$I_{\text{max}} = \frac{\mathscr{E}}{R} = \frac{12\,\text{V}}{2.0 \times 10^4\,\Omega} = 600\,\mu\text{A}.$$

(f) Again using Eq. 26–5, now with $I = 0.20I_{\text{max}} = 120\,\mu\text{A}$, we have

$$Q = C(\mathscr{E} - IR)$$
$$= (3.0 \times 10^{-7}\,\text{F})[12\,\text{V} - (1.2 \times 10^{-4}\,\text{A})(2.0 \times 10^4\,\Omega)] = 2.9\,\mu\text{C}.$$

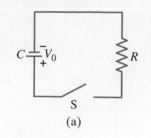

(a)

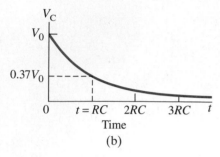

Time

(b)

FIGURE 26–18 For the RC circuit shown in (a), the voltage V_C across the capacitor decreases with time, as shown in (b), after the switch S is closed at $t = 0$. The charge on the capacitor follows the same curve since $V_C \propto Q$.

The circuit just discussed involved the *charging* of a capacitor by a battery through a resistance. Now let us look at another situation: when a capacitor is already charged (say to a voltage V_0), and it is then allowed to *discharge* through a resistance R as shown in Fig. 26–18a. (In this case there is no battery.) When the switch S is closed, charge begins to flow through resistor R from one side of the capacitor toward the other side, until the capacitor is fully discharged. The voltage across the resistor at any instant equals that across the capacitor:

$$IR = \frac{Q}{C}.$$

The rate at which charge leaves the capacitor equals the negative of the current in the resistor, $I = -dQ/dt$, because the capacitor is discharging (Q is decreasing). So we write the above equation as

$$-\frac{dQ}{dt} R = \frac{Q}{C}.$$

We rearrange this to

$$\frac{dQ}{Q} = -\frac{dt}{RC}$$

and integrate it from $t = 0$ when the charge on the capacitor is Q_0, to some time t later when the charge is Q:

$$\ln \frac{Q}{Q_0} = -\frac{t}{RC}$$

or

$$Q = Q_0 e^{-t/RC}. \tag{26–9a}$$

The voltage across the capacitor $(V_C = Q/C)$ as a function of time is

$$V_C = V_0 e^{-t/RC}, \tag{26–9b}$$

where the initial voltage $V_0 = Q_0/C$. Thus the charge on the capacitor, and the voltage across it, decrease exponentially in time with a time constant RC. This is shown in Fig. 26–18b. The current is

$$I = -\frac{dQ}{dt} = \frac{Q_0}{RC} e^{-t/RC} = I_0 e^{-t/RC}, \tag{26–10}$$

and it too is seen to decrease exponentially in time with the same time constant RC. The charge on the capacitor, the voltage across it, and the current in the resistor all decrease to 37% of their original value in one time constant $t = \tau = RC$.

EXERCISE E In 10 time constants, the charge on the capacitor in Fig. 26–18 will be about (a) $Q_0/20{,}000$, (b) $Q_0/5000$, (c) $Q_0/1000$, (d) $Q_0/10$, (e) $Q_0/3$?

FIGURE 26–19 Example 26–12.

EXAMPLE 26–12 Discharging RC circuit. In the RC circuit shown in Fig. 26–19, the battery has fully charged the capacitor, so $Q_0 = C\mathcal{E}$. Then at $t = 0$ the switch is thrown from position a to b. The battery emf is 20.0 V, and the capacitance $C = 1.02\,\mu\text{F}$. The current I is observed to decrease to 0.50 of its initial value in 40 μs. (a) What is the value of Q, the charge on the capacitor, at $t = 0$? (b) What is the value of R? (c) What is Q at $t = 60\,\mu$s?

APPROACH At $t = 0$, the capacitor has charge $Q_0 = C\mathcal{E}$, and then the battery is removed from the circuit and the capacitor begins discharging through the resistor, as in Fig. 26–18. At any time t later (Eq. 26–9a) we have

$$Q = Q_0 e^{-t/RC} = C\mathcal{E} e^{-t/RC}.$$

SOLUTION (a) At $t = 0$,

$$Q = Q_0 = C\mathcal{E} = (1.02 \times 10^{-6}\,\text{F})(20.0\,\text{V}) = 2.04 \times 10^{-5}\,\text{C} = 20.4\,\mu\text{C}.$$

(b) To find R, we are given that at $t = 40\,\mu\text{s}$, $I = 0.50I_0$. Hence

$$0.50I_0 = I_0 e^{-t/RC}.$$

Taking natural logs on both sides ($\ln 0.50 = -0.693$):

$$0.693 = \frac{t}{RC}$$

so

$$R = \frac{t}{(0.693)C} = \frac{(40 \times 10^{-6}\,\text{s})}{(0.693)(1.02 \times 10^{-6}\,\text{F})} = 57\,\Omega.$$

(c) At $t = 60\,\mu\text{s}$,

$$Q = Q_0 e^{-t/RC} = (20.4 \times 10^{-6}\,\text{C})e^{-\frac{60 \times 10^{-6}\,\text{s}}{(57\,\Omega)(1.02 \times 10^{-6}\,\text{F})}} = 7.3\,\mu\text{C}.$$

CONCEPTUAL EXAMPLE 26–13 | **Bulb in *RC* circuit.** In the circuit of Fig. 26–20, the capacitor is originally uncharged. Describe the behavior of the lightbulb from the instant switch S is closed until a long time later.

RESPONSE When the switch is first closed, the current in the circuit is high and the lightbulb burns brightly. As the capacitor charges, the voltage across the capacitor increases causing the current to be reduced, and the lightbulb dims. As the potential difference across the capacitor approaches the same voltage as the battery, the current decreases toward zero and the lightbulb goes out.

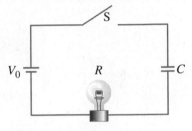

FIGURE 26–20 Example 26–13.

*Applications of *RC* Circuits

The charging and discharging in an *RC* circuit can be used to produce voltage pulses at a regular frequency. The charge on the capacitor increases to a particular voltage, and then discharges. One way of initiating the discharge of the capacitor is by the use of a gas-filled tube which has an electrical breakdown when the voltage across it reaches a certain value V_0. After the discharge is finished, the tube no longer conducts current and the recharging process repeats itself, starting at a lower voltage V_0'. Figure 26–21 shows a possible circuit, and the "sawtooth" voltage it produces.

A simple blinking light can be an application of a sawtooth oscillator circuit. Here the emf is supplied by a battery; the neon bulb flashes on at a rate of perhaps 1 cycle per second. The main component of a "flasher unit" is a moderately large capacitor.

The intermittent windshield wipers of a car can also use an *RC* circuit. The *RC* time constant, which can be changed using a multi-positioned switch for different values of R with fixed C, determines the rate at which the wipers come on.

(A) PHYSICS APPLIED
Sawtooth, blinkers, windshield wipers

FIGURE 26–21 (a) An *RC* circuit, coupled with a gas-filled tube as a switch, can produce a repeating "sawtooth" voltage, as shown in (b).

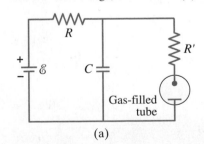

(a)

EXAMPLE 26–14 **ESTIMATE** | **Resistor in a turn signal.** Estimate the order of magnitude of the resistor in a turn-signal circuit.

APPROACH A typical turn signal flashes perhaps twice per second, so the time constant is on the order of 0.5 s. A moderate capacitor might have $C = 1\,\mu\text{F}$.
SOLUTION Setting $\tau = RC = 0.5$ s, we find

$$R = \frac{\tau}{C} = \frac{0.5\,\text{s}}{1 \times 10^{-6}\,\text{F}} \approx 500\,\text{k}\Omega.$$

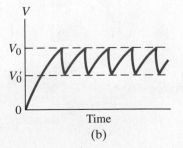

(b)

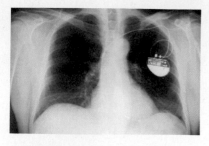

FIGURE 26–22 Electronic battery-powered pacemaker can be seen on the rib cage in this X-ray.

FIGURE 26–23 A person receives an electric shock when the circuit is completed.

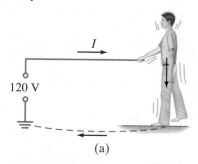

(a)

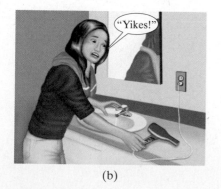

(b)

An interesting medical use of an *RC* circuit is the electronic heart pacemaker, which can make a stopped heart start beating again by applying an electric stimulus through electrodes attached to the chest. The stimulus can be repeated at the normal heartbeat rate if necessary. The heart itself contains *pacemaker* cells, which send out tiny electric pulses at a rate of 60 to 80 per minute. These signals induce the start of each heartbeat. In some forms of heart disease, the natural pacemaker fails to function properly, and the heart loses its beat. Such patients use *electronic pacemakers* which produce a regular voltage pulse that starts and controls the frequency of the heartbeat. The electrodes are implanted in or near the heart (Fig. 26–22), and the circuit contains a capacitor and a resistor. The charge on the capacitor increases to a certain point and then discharges a pulse to the heart. Then it starts charging again. The pulsing rate depends on the values of *R* and *C*.

26–6 Electric Hazards

Excess electric current can heat wires in buildings and cause fires, as discussed in Section 25–6. Electric current can also damage the human body or even be fatal. Electric current through the human body can cause damage in two ways: (1) Electric current heats tissue and can cause burns; (2) electric current stimulates nerves and muscles, and we feel a "shock." The severity of a shock depends on the magnitude of the current, how long it acts, and through what part of the body it passes. A current passing through vital organs such as the heart or brain is especially serious for it can interfere with their operation.

Most people can "feel" a current of about 1 mA. Currents of a few mA cause pain but rarely cause much damage in a healthy person. Currents above 10 mA cause severe contraction of the muscles, and a person may not be able to let go of the source of the current (say, a faulty appliance or wire). Death from paralysis of the respiratory system can occur. Artificial respiration, however, can sometimes revive a victim. If a current above about 80 to 100 mA passes across the torso, so that a portion passes through the heart for more than a second or two, the heart muscles will begin to contract irregularly and blood will not be properly pumped. This condition is called **ventricular fibrillation**. If it lasts for long, death results. Strangely enough, if the current is much larger, on the order of 1 A, death by heart failure may be less likely,[†] but such currents can cause serious burns, especially if concentrated through a small area of the body.

The seriousness of a shock depends on the applied voltage and on the effective resistance of the body. Living tissue has low resistance since the fluid of cells contains ions that can conduct quite well. However, the outer layer of skin, when dry, offers high resistance and is thus protective. The effective resistance between two points on opposite sides of the body when the skin is dry is in the range of 10^4 to $10^6\ \Omega$. But when the skin is wet, the resistance may be $10^3\ \Omega$ or less. A person who is barefoot or wearing thin-soled shoes will be in good contact with the ground, and touching a 120-V line with a wet hand can result in a current

$$I = \frac{120\,\text{V}}{1000\,\Omega} = 120\,\text{mA}.$$

As we saw, this could be lethal.

A person who has received a shock has become part of a complete circuit. Figure 26–23 shows two ways the circuit might be completed when a person

---

[†]Larger currents apparently bring the entire heart to a standstill. Upon release of the current, the heart returns to its normal rhythm. This may not happen when fibrillation occurs because, once started, it can be hard to stop. Fibrillation may also occur as a result of a heart attack or during heart surgery. A device known as a *defibrillator* (described in Section 24–4) can apply a brief high current to the heart, causing complete heart stoppage which is often followed by resumption of normal beating.

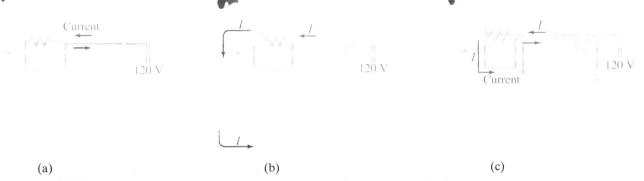

(a)   (b)   (c)

FIGURE 26–24 (a) An electric oven operating normally with a 2-prong plug. (b) Short to the case with ungrounded case: shock. (c) Short to the case with the case grounded by a 3-prong plug.

accidentally touches a "hot" electric wire—"hot" meaning a high potential such as 120 V (normal U.S. household voltage) relative to ground. The other wire of building wiring is connected to ground—either by a wire connected to a buried conductor, or via a metal water pipe into the ground. In Fig. 26–23a, the current passes from the high-voltage wire, through the person, to the ground through his bare feet, and back along the ground (a fair conductor) to the ground terminal of the source. If the person stands on a good insulator—thick rubber-soled shoes or a dry wood floor—there will be much more resistance in the circuit and consequently much less current through the person. If the person stands with bare feet on the ground, or is in a bathtub, there is lethal danger because the resistance is much less and the current greater. In a bathtub (or swimming pool), not only are you wet, which reduces your resistance, but the water is in contact with the drain pipe (typically metal) that leads to the ground. It is strongly recommended that you not touch anything electrical when wet or in bare feet. Building codes that require the use of non-metal pipes would be protective.

In Fig. 26–23b, a person touches a faulty "hot" wire with one hand, and the other hand touches a sink faucet (connected to ground via the pipe). The current is particularly dangerous because it passes across the chest, through the heart and lungs. A useful rule: if one hand is touching something electrical, keep your other hand in your pocket (don't use it!), and wear thick rubber-soled shoes. It is also a good idea to remove metal jewelry, especially rings (your finger is usually moist under a ring).

You can come into contact with a hot wire by touching a bare wire whose insulation has worn off, or from a bare wire inside an appliance when you're tinkering with it. (Always unplug an electrical device before investigating[†] its insides!) Another possibility is that a wire inside a device may break or lose its insulation and come in contact with the case. If the case is metal, it will conduct electricity. A person could then suffer a severe shock merely by touching the case, as shown in Fig. 26–24b. To prevent an accident, metal cases are supposed to be connected directly to ground by a separate ground wire. Then if a "hot" wire touches the grounded case, a short circuit to ground immediately occurs internally, as shown in Fig. 26–24c, and most of the current passes through the low-resistance ground wire rather than through the person. Furthermore, the high current should open the fuse or circuit breaker. Grounding a metal case is done by a separate ground wire connected to the third (round) prong of a 3-prong plug. Never cut off the third prong of a plug—it could save your life.

⚠ CAUTION

Keep one hand in your pocket when other touches electricity

🕴 PHYSICS APPLIED

Grounding and shocks

[†]Even then you can get a bad shock from a capacitor that hasn't been discharged until you touch it.

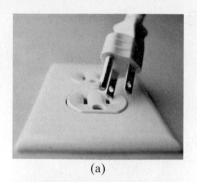

(a)

(b)

(c)

FIGURE 26–25 (a) A 3-prong plug, and (b) an adapter (gray) for old-fashioned 2-prong outlets—be sure to screw down the ground tab. (c) A polarized 2-prong plug.

FIGURE 26–26 Four wires entering a typical house. The color codes for wires are not always as shown here—be careful!

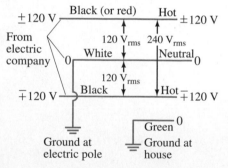

A three-prong plug, and an adapter, are shown in Figs. 26–25a and b.

Why is a third wire needed? The 120 V is carried by the other two wires—one **hot** (120 V ac), the other **neutral**, which is itself grounded. The third "dedicated" ground wire with the round prong may seem redundant. But it is protection for two reasons: (1) it protects against internal wiring that may have been done incorrectly; (2) the neutral wire carries normal current ("return" current from the 120 V) and it does have resistance; so there can be a voltage drop along it—normally small, but if connections are poor or corroded, or the plug is loose, the resistance could be large enough that you might feel that voltage if you touched the neutral wire some distance from its grounding point.

Some electrical devices come with only two wires, and the plug's two prongs are of different widths; the plug can be inserted only one way into the outlet so that the intended neutral (wider prong) in the device is connected to neutral in the wiring (Fig. 26–25c). For example, the screw threads of a lightbulb are meant to be connected to neutral (and the base contact to hot), to avoid shocks when changing a bulb in a possibly protruding socket. Devices with 2-prong plugs do *not* have their cases grounded; they are supposed to have double electric insulation. Take extra care anyway.

The insulation on a wire may be color coded. Hand-held meters may have red (hot) and black (ground) lead wires. But in a house, black is usually hot (or it may be red), whereas white is neutral and green is the dedicated ground, Fig. 26–26. But beware: these color codes cannot always be trusted. [In the U.S., three wires normally enter a house: two *hot* wires at 120 V each (which add together to 240 V for appliances or devices that run on 240 V) plus the grounded *neutral* (carrying return current for the two hots). See Fig. 26–26. The "dedicated" *ground* wire (non-current carrying) is a fourth wire that does not come from the electric company but enters the house from a nearby heavy stake in the ground or a buried metal pipe. The two hot wires can feed separate 120-V circuits in the house, so each 120-V circuit inside the house has only three wires, including ground.]

Normal circuit breakers (Sections 25–6 and 28–8) protect equipment and buildings from overload and fires. They protect humans only in some circumstances, such as the very high currents that result from a short, if they respond quickly enough. *Ground fault circuit interrupters* (GFCI), described in Section 29–8, are designed to protect people from the much lower currents (10 mA to 100 mA) that are lethal but would not throw a 15-A circuit breaker or blow a 20-A fuse.

It is current that harms, but it is voltage that drives the current. 30 volts is sometimes said to be the threshhold for danger. But even a 12-V car battery (which can supply large currents) can cause nasty burns and shock.

Another danger is **leakage current**, by which we mean a current along an unintended path. Leakage currents are often "capacitively coupled." For example, a wire in a lamp forms a capacitor with the metal case; charges moving in one conductor attract or repel charge in the other, so there is a current. Typical electrical codes limit leakage currents to 1 mA for any device. A 1-mA leakage current is usually harmless. It can be very dangerous, however, to a hospital patient with implanted electrodes connected to ground through the apparatus. This is due to the absence of the protective skin layer and because the current can pass directly through the heart as compared to the usual situation where the current enters at the hands and spreads out through the body. Although 100 mA may be needed to cause heart fibrillation when entering through the hands (very little of it actually passes through the heart), as little as 0.02 mA has been known to cause fibrillation when passing directly to the heart. Thus, a "wired" patient is in considerable danger from leakage current even from as simple an act as touching a lamp.

Finally, don't touch a downed power line (lethal!) or even get near it. A hot power line is at thousands of volts. A huge current can flow along the ground or pavement, from where the high-voltage wire touches the ground along its path to the grounding point of the neutral line, enough that the voltage between your two feet could be large. Tip: stand on one foot or run (only one foot touching the ground at a time).

Ammeters and Voltmeters

An **ammeter** is used to measure current, and a **voltmeter** measures potential difference or voltage. Measurements of current and voltage are made with meters that are of two types: (1) *analog* meters, which display numerical values by the position of a pointer that can move across a scale (Fig. 26–27a); and (2) *digital* meters, which display the numerical value in numbers (Fig. 26–27b). We now discuss the meters themselves and how they work, then how they are connected to circuits to make measurements. Finally we will discuss how using meters affects the circuit being measured, possibly causing erroneous results—and what to do about it.

*Analog Ammeters and Voltmeters

The crucial part of an analog ammeter or voltmeter, in which the reading is by a pointer on a scale (Fig. 26–27a), is a *galvanometer*. The galvanometer works on the principle of the force between a magnetic field and a current-carrying coil of wire, and will be discussed in Chapter 27. For now, we merely need to know that the deflection of the needle of a galvanometer is proportional to the current flowing through it. The *full-scale current sensitivity* of a galvanometer, I_m, is the electric current needed to make the needle deflect full scale.

A galvanometer can be used directly to measure small dc currents. For example, a galvanometer whose sensitivity I_m is 50 μA can measure currents from about 1 μA (currents smaller than this would be hard to read on the scale) up to 50 μA. To measure larger currents, a resistor is placed in parallel with the galvanometer. Thus, an analog **ammeter**, represented by the symbol •–Ⓐ–•, consists of a galvanometer (•–Ⓖ–•) in parallel with a resistor called the **shunt resistor**, as shown in Fig. 26–28. ("Shunt" is a synonym for "in parallel.") The shunt resistance is R_{sh}, and the resistance of the galvanometer coil, through which current passes, is r. The value of R_{sh} is chosen according to the full-scale deflection desired; R_{sh} is normally very small—giving an ammeter a very small net resistance—so most of the current passes through R_{sh} and very little ($\lesssim 50\,\mu$A) passes through the galvanometer to deflect the needle.

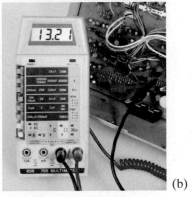

FIGURE 26–27 (a) An analog multimeter being used as a voltmeter. (b) An electronic digital meter.

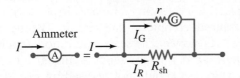

FIGURE 26–28 An ammeter is a galvanometer in parallel with a (shunt) resistor with low resistance, R_{sh}.

EXAMPLE 26–15 **Ammeter design.** Design an ammeter to read 1.0 A at full scale using a galvanometer with a full-scale sensitivity of 50 μA and a resistance $r = 30\,\Omega$. Check if the scale is linear.

APPROACH Only 50 μA $(= I_G = 0.000050\,\text{A})$ of the 1.0-A current must pass through the galvanometer to give full-scale deflection. The rest of the current $(I_R = 0.999950\,\text{A})$ passes through the small shunt resistor, R_{sh}, Fig. 26–28. The potential difference across the galvanometer equals that across the shunt resistor (they are in parallel). We apply Ohm's law to find R_{sh}.

SOLUTION Because $I = I_G + I_R$, when $I = 1.0$ A flows into the meter, we want I_R through the shunt resistor to be $I_R = 0.999950$ A. The potential difference across the shunt is the same as across the galvanometer, so Ohm's law tells us

$$I_R R_{sh} = I_G r;$$

then

$$R_{sh} = \frac{I_G r}{I_R} = \frac{(5.0 \times 10^{-5}\,\text{A})(30\,\Omega)}{(0.999950\,\text{A})} = 1.5 \times 10^{-3}\,\Omega,$$

or 0.0015 Ω. The shunt resistor must thus have a very low resistance and most of the current passes through it.

Because $I_G = I_R(R_{sh}/r)$ and (R_{sh}/r) is constant, we see that the scale is linear.

An analog **voltmeter** (•–Ⓥ–•) also consists of a galvanometer and a resistor. But the resistor R_{ser} is connected in series, Fig. 26–29, and it is usually large, giving a voltmeter a high internal resistance.

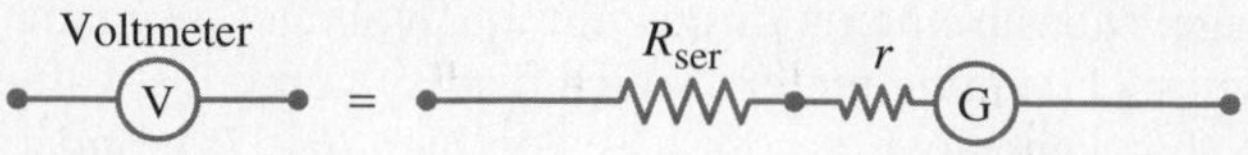

FIGURE 26–29 A voltmeter is a galvanometer in series with a resistor with high resistance, R_{ser}.

Voltmeter

•——Ⓥ——• = •——R_{ser} ⌇⌇⌇——•—⌇⌇—r—Ⓖ——•

EXAMPLE 26–16 Voltmeter design. Using a galvanometer with internal resistance $r = 30\,\Omega$ and full-scale current sensitivity of $50\,\mu A$, design a voltmeter that reads from 0 to 15 V. Is the scale linear?

APPROACH When a potential difference of 15 V exists across the terminals of our voltmeter, we want $50\,\mu A$ to be passing through it so as to give a full-scale deflection.

SOLUTION From Ohm's law, $V = IR$, we have (see Fig. 26–29)

$$15\,V = (50\,\mu A)(r + R_{ser}),$$

so

$$R_{ser} = \frac{15\,V}{5.0 \times 10^{-5}\,A} - r = 300\,k\Omega - 30\,\Omega = 300\,k\Omega.$$

Notice that $r = 30\,\Omega$ is so small compared to the value of R_{ser} that it doesn't influence the calculation significantly. The scale will again be linear: if the voltage to be measured is 6.0 V, the current passing through the voltmeter will be $(6.0\,V)/(3.0 \times 10^5\,\Omega) = 2.0 \times 10^{-5}\,A$, or $20\,\mu A$. This will produce two-fifths of full-scale deflection, as required $(6.0\,V/15.0\,V = 2/5)$.

The meters just described are for direct current. A dc meter can be modified to measure ac (alternating current, Section 25–7) with the addition of diodes (Chapter 40), which allow current to flow in one direction only. An ac meter can be calibrated to read rms or peak values.

Voltmeters and ammeters can have several series or shunt resistors to offer a choice of range. **Multimeters** can measure voltage, current, and resistance. Sometimes a multimeter is called a VOM (Volt-Ohm-Meter or Volt-Ohm-Milliammeter).

An **ohmmeter** measures resistance, and must contain a battery of known voltage connected in series to a resistor (R_{ser}) and to an ammeter (Fig. 26–30). The resistor whose resistance is to be measured completes the circuit. The needle deflection is inversely proportional to the resistance. The scale calibration depends on the value of the series resistor. Because an ohmmeter sends a current through the device whose resistance is to be measured, it should not be used on very delicate devices that could be damaged by the current.

The **sensitivity** of a meter is generally specified on the face. It may be given as so many ohms per volt, which indicates how many ohms of resistance there are in the meter per volt of full-scale reading. For example, if the sensitivity is 30,000 Ω/V, this means that on the 10-V scale the meter has a resistance of 300,000 Ω, whereas on a 100-V scale the meter resistance is 3 MΩ. The full-scale current sensitivity, I_m, discussed earlier, is just the reciprocal of the sensitivity in Ω/V.

*How to Connect Meters

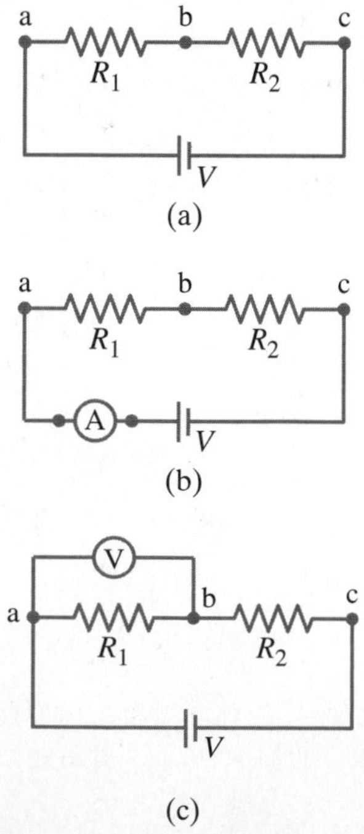

Suppose you wish to determine the current I in the circuit shown in Fig. 26–31a, and the voltage V across the resistor R_1. How exactly are ammeters and voltmeters connected to the circuit being measured?

Because an ammeter is used to measure the current flowing in the circuit, it must be inserted directly into the circuit, in series with the other elements, as shown in Fig. 26–31b. The smaller its internal resistance, the less it affects the circuit.

A voltmeter, on the other hand, is connected "externally," in parallel with the circuit element across which the voltage is to be measured. It is used to measure the potential difference between two points. Its two wire leads (connecting wires) are connected to the two points, as shown in Fig. 26–31c, where the voltage across R_1 is being measured. The larger its internal resistance, $(R_{ser} + r)$ in Fig. 26–29, the less it affects the circuit being measured.

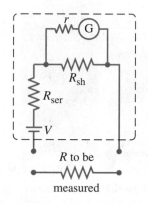

FIGURE 26–30 An ohmmeter.

FIGURE 26–31 Measuring current and voltage.

(a)

(b)

(c)

*Effects of Meter Resistance

It is important to know the sensitivity of a meter, for in many cases the resistance of the meter can seriously affect your results. Take the following Example.

EXAMPLE 26–17 **Voltage reading versus true voltage.** Suppose you are testing an electronic circuit which has two resistors, R_1 and R_2, each 15 kΩ, connected in series as shown in Fig. 26–32a. The battery maintains 8.0 V across them and has negligible internal resistance. A voltmeter whose sensitivity is 10,000 Ω/V is put on the 5.0-V scale. What voltage does the meter read when connected across R_1, Fig. 26–32b, and what error is caused by the finite resistance of the meter?

APPROACH The meter acts as a resistor in parallel with R_1. We use parallel and series resistor analyses and Ohm's law to find currents and voltages.

SOLUTION On the 5.0-V scale, the voltmeter has an internal resistance of $(5.0\ \text{V})(10{,}000\ \Omega/\text{V}) = 50{,}000\ \Omega$. When connected across R_1, as in Fig. 26–32b, we have this 50 kΩ in parallel with $R_1 = 15$ kΩ. The net resistance R_{eq} of these two is given by

$$\frac{1}{R_{eq}} = \frac{1}{50\ \text{k}\Omega} + \frac{1}{15\ \text{k}\Omega} = \frac{13}{150\ \text{k}\Omega};$$

so $R_{eq} = 11.5$ kΩ. This $R_{eq} = 11.5$ kΩ is in series with $R_2 = 15$ kΩ, so the total resistance of the circuit is now 26.5 kΩ (instead of the original 30 kΩ). Hence the current from the battery is

$$I = \frac{8.0\ \text{V}}{26.5\ \text{k}\Omega} = 3.0 \times 10^{-4}\ \text{A} = 0.30\ \text{mA}.$$

Then the voltage drop across R_1, which is the same as that across the voltmeter, is $(3.0 \times 10^{-4}\ \text{A})(11.5 \times 10^3\ \Omega) = 3.5$ V. [The voltage drop across R_2 is $(3.0 \times 10^{-4}\ \text{A})(15 \times 10^3\ \Omega) = 4.5$ V, for a total of 8.0 V.] If we assume the meter is precise, it will read 3.5 V. In the original circuit, without the meter, $R_1 = R_2$ so the voltage across R_1 is half that of the battery, or 4.0 V. Thus the voltmeter, because of its internal resistance, gives a low reading. In this case it is off by 0.5 V, or more than 10%.

Example 26–17 illustrates how seriously a meter can affect a circuit and give a misleading reading. If the resistance of a voltmeter is much higher than the resistance of the circuit, however, it will have little effect and its readings can be trusted, at least to the manufactured precision of the meter, which for ordinary analog meters is typically 3% to 4% of full-scale deflection. An ammeter also can interfere with a circuit, but the effect is minimal if its resistance is much less than that of the circuit as a whole. For both voltmeters and ammeters, the more sensitive the galvanometer, the less effect it will have. A 50,000-Ω/V meter is far better than a 1000-Ω/V meter.

*Digital Meters

Digital meters (see Fig. 26–27b) are used in the same way as analog meters: they are inserted directly into the circuit, in series, to measure current (Fig. 26–31b), and connected "outside," in parallel with the circuit, to measure voltage (Fig. 26–31c).

The internal construction of digital meters, however, is different from that of analog meters in that digital meters do not use a galvanometer. The electronic circuitry and digital readout are more sensitive than a galvanometer, and have less effect on the circuit to be measured. When we measure dc voltages, a digital meter's resistance is very high, commonly on the order of 10 to 100 MΩ $(10^7 – 10^8\ \Omega)$, and doesn't change significantly when different voltage scales are selected. A 100-MΩ digital meter draws off very little current when connected across even a 1-MΩ resistance.

The precision of digital meters is exceptional, often one part in 10^4 $(= 0.01\%)$ or better. This precision is not the same as accuracy, however. A precise meter of internal resistance $10^8\ \Omega$ will not give accurate results if used to measure a voltage across a 10^8-Ω resistor—in which case it is necessary to do a calculation like that in Example 26–17.

Whenever we make a measurement on a circuit, to some degree we affect that circuit (Example 26–17). This is true for other types of measurement as well: when we make a measurement on a system, we affect that system in some way. On a temperature measurement, for example, the thermometer can exchange heat with the system, thus altering its temperature. It is important to be able to make needed corrections, as we saw in Example 26–17.

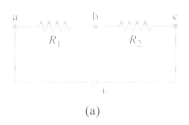

(a)

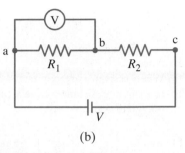

(b)

FIGURE 26–32 Example 26–17.

Summary

A device that transforms another type of energy into electrical energy is called a **source** of **emf**. A battery behaves like a source of emf in series with an **internal resistance**. The emf is the potential difference determined by the chemical reactions in the battery and equals the terminal voltage when no current is drawn. When a current is drawn, the voltage at the battery's terminals is less than its emf by an amount equal to the potential decrease Ir across the internal resistance.

When resistances are connected in **series** (end to end in a single linear path), the equivalent resistance is the sum of the individual resistances:

$$R_{eq} = R_1 + R_2 + \cdots. \tag{26–3}$$

In a series combination, R_{eq} is greater than any component resistance.

When resistors are connected in **parallel**, the reciprocal of the equivalent resistance equals the sum of the reciprocals of the individual resistances:

$$\frac{1}{R_{eq}} = \frac{1}{R_1} + \frac{1}{R_2} + \cdots. \tag{26–4}$$

In a parallel connection, the net resistance is less than any of the individual resistances.

Kirchhoff's rules are helpful in determining the currents and voltages in circuits. Kirchhoff's **junction rule** is based on conservation of electric charge and states that the sum of all currents entering any junction equals the sum of all currents leaving that junction. The second, or **loop rule**, is based on conservation of energy and states that the algebraic sum of the changes in potential around any closed path of the circuit must be zero.

When an **RC circuit** containing a resistor R in series with a capacitance C is connected to a dc source of emf, the voltage across the capacitor rises gradually in time characterized by an exponential of the form $(1 - e^{-t/RC})$, where the **time constant**,

$$\tau = RC, \tag{26–7}$$

is the time it takes for the voltage to reach 63 percent of its maximum value. The current through the resistor decreases as $e^{-t/RC}$.

A capacitor discharging through a resistor is characterized by the same time constant: in a time $\tau = RC$, the voltage across the capacitor drops to 37 percent of its initial value. The charge on the capacitor, and voltage across it, decreases as $e^{-t/RC}$, as does the current.

Electric shocks are caused by current passing through the body. To avoid shocks, the body must not become part of a complete circuit by allowing different parts of the body to touch objects at different potentials. Commonly, shocks are caused by one part of the body touching ground and another part touching a high electric potential.

[*An **ammeter** measures current. An analog ammeter consists of a galvanometer and a parallel **shunt resistor** that carries most of the current. An analog **voltmeter** consists of a galvanometer and a series resistor. An ammeter is inserted *into* the circuit whose current is to be measured. A voltmeter is external, being connected in parallel to the element whose voltage is to be measured. Digital voltmeters have greater internal resistance and affect the circuit to be measured less than do analog meters.]

Questions

1. Explain why birds can sit on power lines safely, whereas leaning a metal ladder up against a power line to fetch a stuck kite is extremely dangerous.

2. Discuss the advantages and disadvantages of Christmas tree lights connected in parallel versus those connected in series.

3. If all you have is a 120-V line, would it be possible to light several 6-V lamps without burning them out? How?

4. Two lightbulbs of resistance R_1 and R_2 $(R_2 > R_1)$ and a battery are all connected in series. Which bulb is brighter? What if they are connected in parallel? Explain.

5. Household outlets are often double outlets. Are these connected in series or parallel? How do you know?

6. With two identical lightbulbs and two identical batteries, how would you arrange the bulbs and batteries in a circuit to get the maximum possible total power to the lightbulbs? (Assume the batteries have negligible internal resistance.)

7. If two identical resistors are connected in series to a battery, does the battery have to supply more power or less power than when only one of the resistors is connected? Explain.

8. You have a single 60-W bulb on in your room. How does the overall resistance of your room's electric circuit change when you turn on an additional 100-W bulb?

9. When applying Kirchhoff's loop rule (such as in Fig. 26–33), does the sign (or direction) of a battery's emf depend on the direction of current through the battery? What about the terminal voltage?

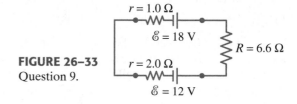

FIGURE 26–33
Question 9.

$r = 1.0\ \Omega$
$\mathscr{E} = 18\ V$
$R = 6.6\ \Omega$
$r = 2.0\ \Omega$
$\mathscr{E} = 12\ V$

10. Compare and discuss the formulas for resistors and for capacitors when connected in series and in parallel.

11. For what use are batteries connected in series? For what use are they connected in parallel? Does it matter if the batteries are nearly identical or not in either case?

12. Can the terminal voltage of a battery ever exceed its emf? Explain.

13. Explain in detail how you could measure the internal resistance of a battery.

14. In an *RC* circuit, current flows from the battery until the capacitor is completely charged. Is the total energy supplied by the battery equal to the total energy stored by the capacitor? If not, where does the extra energy go?

15. Given the circuit shown in Fig. 26–34, use the words "increases," "decreases," or "stays the same" to complete the following statements:

(a) If R_7 increases, the potential difference between A and E _____. Assume no resistance in $\textcircled{A}$ and $\mathscr{E}$.

(b) If R_7 increases, the potential difference between A and E _____. Assume $\textcircled{A}$ and $\mathscr{E}$ have resistance.

(c) If R_7 increases, the voltage drop across R_4 _____.

(d) If R_2 decreases, the current through R_1 _____.

(e) If R_2 decreases, the current through R_6 _____.

(f) If R_2 decreases, the current through R_3 _____.

(g) If R_5 increases, the voltage drop across R_2 _____.

(h) If R_5 increases, the voltage drop across R_4 _____.

(i) If R_2, R_5, and R_7 increase, $\mathscr{E} \, (r = 0)$ _____.

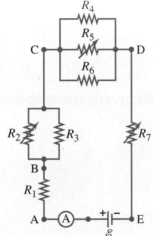

FIGURE 26–34
Question 15. R_2, R_5, and R_7 are *variable* resistors (you can change their resistance), given the symbol –⩛–.

16. Figure 26–35 is a diagram of a capacitor (or condenser) **microphone**. The changing air pressure in a sound wave causes one plate of the capacitor C to move back and forth. Explain how a current of the same frequency as the sound wave is produced.

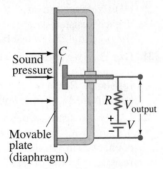

FIGURE 26–35 Diagram of a capacitor microphone. Question 16.

17. Design a circuit in which two different switches of the type shown in Fig. 26–36 can be used to operate the same lightbulb from opposite sides of a room.

FIGURE 26–36
Question 17.

*18. What is the main difference between an analog voltmeter and an analog ammeter?

*19. What would happen if you mistakenly used an ammeter where you needed to use a voltmeter?

*20. Explain why an ideal ammeter would have zero resistance and an ideal voltmeter infinite resistance.

*21. A voltmeter connected across a resistor always reads *less* than the actual voltage across the resistor when the meter is not present. Explain.

*22. A small battery-operated flashlight requires a single 1.5-V battery. The bulb is barely glowing, but when you take the battery out and check it with a voltmeter, it registers 1.5 V. How would you explain this?

23. Different lamps might have batteries connected in either of the two arrangements shown in Fig. 26–37. What would be the advantages of each scheme?

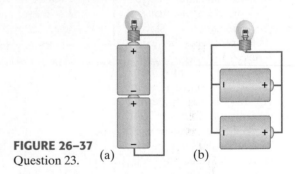

FIGURE 26–37
Question 23. (a)   (b)

Problems

26–1 Emf and Terminal Voltage

1. (I) Calculate the terminal voltage for a battery with an internal resistance of 0.900 Ω and an emf of 6.00 V when the battery is connected in series with (a) an 81.0-Ω resistor, and (b) an 810-Ω resistor.

2. (I) Four 1.50-V cells are connected in series to a 12-Ω lightbulb. If the resulting current is 0.45 A, what is the internal resistance of each cell, assuming they are identical and neglecting the resistance of the wires?

3. (II) A 1.5-V dry cell can be tested by connecting it to a low-resistance ammeter. It should be able to supply at least 25 A. What is the internal resistance of the cell in this case, assuming it is much greater than that of the ammeter?

4. (II) What is the internal resistance of a 12.0-V car battery whose terminal voltage drops to 8.4 V when the starter draws 95 A? What is the resistance of the starter?

26–2 Resistors in Series and Parallel

In these Problems neglect the internal resistance of a battery unless the Problem refers to it.

5. (I) A 650-Ω and a 2200-Ω resistor are connected in series with a 12-V battery. What is the voltage across the 2200-Ω resistor?

6. (I) Three 45-Ω lightbulbs and three 65-Ω lightbulbs are connected in series. (a) What is the total resistance of the circuit? (b) What is the total resistance if all six are wired in parallel?

7. (I) Suppose that you have a 680-Ω, a 720-Ω, and a 1.20-kΩ resistor. What is (a) the maximum, and (b) the minimum resistance you can obtain by combining these?

8. (I) How many 10-Ω resistors must be connected in series to give an equivalent resistance to five 100-Ω resistors connected in parallel?

9. (II) Suppose that you have a 9.0-V battery and you wish to apply a voltage of only 4.0 V. Given an unlimited supply of 1.0-Ω resistors, how could you connect them so as to make a "voltage divider" that produces a 4.0-V output for a 9.0-V input?

10. (II) Three 1.70-kΩ resistors can be connected together in four different ways, making combinations of series and/or parallel circuits. What are these four ways, and what is the net resistance in each case?

11. (II) A battery with an emf of 12.0 V shows a terminal voltage of 11.8 V when operating in a circuit with two light-bulbs, each rated at 4.0 W (at 12.0 V), which are connected in parallel. What is the battery's internal resistance?

12. (II) Eight identical bulbs are connected in series across a 110-V line. (a) What is the voltage across each bulb? (b) If the current is 0.42 A, what is the resistance of each bulb, and what is the power dissipated in each?

13. (II) Eight bulbs are connected in parallel to a 110-V source by two long leads of total resistance 1.4 Ω. If 240 mA flows through each bulb, what is the resistance of each, and what fraction of the total power is wasted in the leads?

14. (II) The performance of the starter circuit in an automobile can be significantly degraded by a small amount of corrosion on a battery terminal. Figure 26–38a depicts a properly functioning circuit with a battery (12.5-V emf, 0.02-Ω internal resistance) attached via corrosion-free cables to a starter motor of resistance $R_S = 0.15 \, \Omega$. Suppose that later, corrosion between a battery terminal and a starter cable introduces an extra series resistance of just $R_C = 0.10 \, \Omega$ into the circuit as suggested in Fig. 26–38b. Let P_0 be the power delivered to the starter in the circuit free of corrosion, and let P be the power delivered to the circuit with corrosion. Determine the ratio P/P_0.

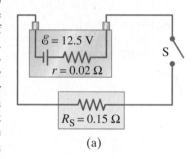

(a)

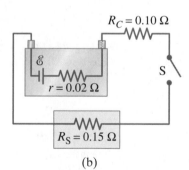

(b)

FIGURE 26–38
Problem 14.

15. (II) A close inspection of an electric circuit reveals that a 480-Ω resistor was inadvertently soldered in the place where a 370-Ω resistor is needed. How can this be fixed without removing anything from the existing circuit?

16. (II) Determine (a) the equivalent resistance of the circuit shown in Fig. 26–39, and (b) the voltage across each resistor.

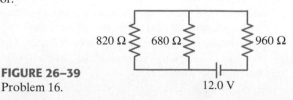

FIGURE 26–39
Problem 16.

820 Ω   680 Ω   960 Ω

12.0 V

17. (II) A 75-W, 110-V bulb is connected in parallel with a 25-W, 110-V bulb. What is the net resistance?

18. (II) (a) Determine the equivalent resistance of the "ladder" of equal 125-Ω resistors shown in Fig. 26–40. In other words, what resistance would an ohmmeter read if connected between points A and B? (b) What is the current through each of the three resistors on the left if a 50.0-V battery is connected between points A and B?

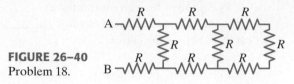

FIGURE 26–40
Problem 18.

19. (II) What is the net resistance of the circuit connected to the battery in Fig. 26–41?

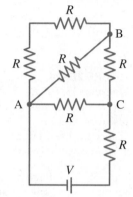

FIGURE 26–41
Problems 19 and 20.

20. (II) Calculate the current through each resistor in Fig. 26–41 if each resistance $R = 1.20 \, \text{k}\Omega$ and $V = 12.0 \, \text{V}$. What is the potential difference between points A and B?

21. (II) The two terminals of a voltage source with emf $\mathscr{E}$ and internal resistance r are connected to the two sides of a load resistance R. For what value of R will the maximum power be delivered from the source to the load?

22. (II) Two resistors when connected in series to a 110-V line use one-fourth the power that is used when they are connected in parallel. If one resistor is 3.8 kΩ, what is the resistance of the other?

23. (III) Three equal resistors (R) are connected to a battery as shown in Fig. 26–42. Qualitatively, what happens to (a) the voltage drop across each of these resistors, (b) the current flow through each, and (c) the terminal voltage of the battery, when the switch S is opened, after having been closed for a long time? (d) If the emf of the battery is 9.0 V, what is its terminal voltage when the switch is closed if the internal resistance r is 0.50 Ω and $R = 5.50 \, \Omega$? (e) What is the terminal voltage when the switch is open?

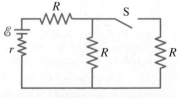

FIGURE 26–42
Problem 23.

24. (III) A 2.8-kΩ and a 3.7-kΩ resistor are connected in parallel; this combination is connected in series with a 1.8-kΩ resistor. If each resistor is rated at $\frac{1}{2}$ W (maximum without overheating), what is the maximum voltage that can be applied across the whole network?

25. (III) Consider the network of resistors shown in Fig. 26–43. Answer qualitatively: (*a*) What happens to the voltage across each resistor when the switch S is closed? (*b*) What happens to the current through each when the switch is closed? (*c*) What happens to the power output of the battery when the switch is closed? (*d*) Let $R_1 = R_2 = R_3 = R_4 = 125\ \Omega$ and $V = 22.0\ \text{V}$. Determine the current through each resistor before and after closing the switch. Are your qualitative predictions confirmed?

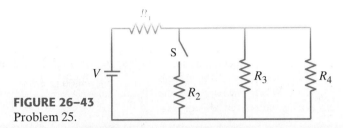

FIGURE 26–43
Problem 25.

26. (III) You are designing a wire resistance heater to heat an enclosed volume of gas. For the apparatus to function properly, this heater must transfer heat to the gas at a very constant rate. While in operation, the resistance of the heater will always be close to the value $R = R_0$, but may fluctuate slightly causing its resistance to vary a small amount $\Delta R\ (\ll R_0)$. To maintain the heater at constant power, you design the circuit shown in Fig. 26–44, which includes two resistors, each of resistance r. Determine the value for r so that the heater power will remain constant even if its resistance R fluctuates by a small amount. [*Hint:* If $\Delta R \ll R_0$, then $\Delta P \approx \Delta R \dfrac{dP}{dR}\Big|_{R=R_0}$.]

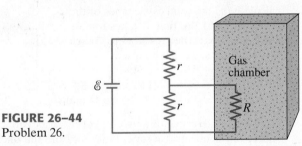

FIGURE 26–44
Problem 26.

26–3 Kirchhoff's Rules

27. (I) Calculate the current in the circuit of Fig. 26–45, and show that the sum of all the voltage changes around the circuit is zero.

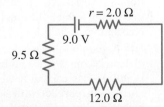

FIGURE 26–45
Problem 27.

28. (II) Determine the terminal voltage of each battery in Fig. 26–46.

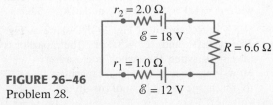

FIGURE 26–46
Problem 28.

29. (II) For the circuit shown in Fig. 26–47, find the potential difference between points a and b. Each resistor has $R = 1.30\ \Omega$ and each battery is 1.5 V.

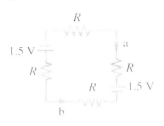

FIGURE 26–47
Problem 29.

30. (II) (*a*) A network of five equal resistors R is connected to a battery $\mathscr{E}$ as shown in Fig. 26–48. Determine the current I that flows out of the battery. (*b*) Use the value determined for I to find the single resistor R_{eq} that is equivalent to the five-resistor network.

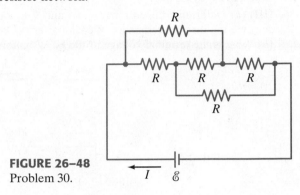

FIGURE 26–48
Problem 30.

31. (II) (*a*) What is the potential difference between points a and d in Fig. 26–49 (similar to Fig. 26–13, Example 26–9), and (*b*) what is the terminal voltage of each battery?

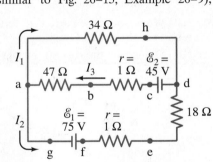

FIGURE 26–49
Problem 31.

32. (II) Calculate the currents in each resistor of Fig. 26–50.

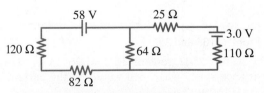

FIGURE 26–50   Problem 32.

33. (II) Determine the magnitudes and directions of the currents through R_1 and R_2 in Fig. 26–51.

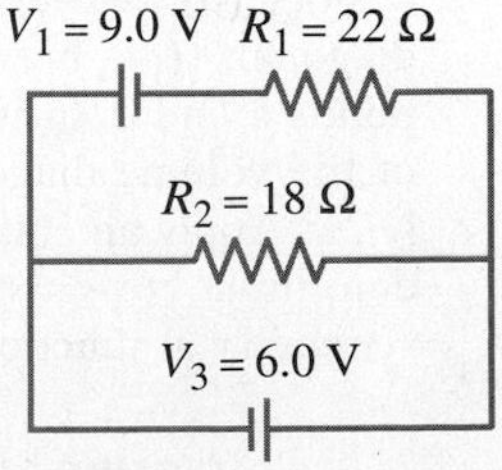

$V_1 = 9.0\ \text{V}\quad R_1 = 22\ \Omega$

$R_2 = 18\ \Omega$

$V_3 = 6.0\ \text{V}$

FIGURE 26–51
Problem 33.

34. (II) Determine the magnitudes and directions of the currents in each resistor shown in Fig. 26–52. The batteries have emfs of $\mathscr{E}_1 = 9.0\text{ V}$ and $\mathscr{E}_2 = 12.0\text{ V}$ and the resistors have values of $R_1 = 25\ \Omega$, $R_2 = 48\ \Omega$, and $R_3 = 35\ \Omega$. (*a*) Ignore internal resistance of the batteries. (*b*) Assume each battery has internal resistance $r = 1.0\ \Omega$.

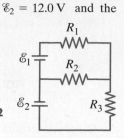

FIGURE 26–52
Problem 34.

35. (II) A voltage V is applied to n identical resistors connected in parallel. If the resistors are instead all connected in series with the applied voltage, show that the power transformed is decreased by a factor n^2.

36. (III) (*a*) Determine the currents I_1, I_2, and I_3 in Fig. 26–53. Assume the internal resistance of each battery is $r = 1.0\ \Omega$. (*b*) What is the terminal voltage of the 6.0-V battery?

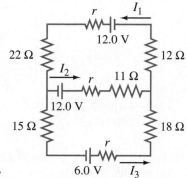

FIGURE 26–53
Problems 36 and 37.

37. (III) What would the current I_1 be in Fig. 26–53 if the 12-Ω resistor is shorted out (resistance = 0)? Let $r = 1.0\ \Omega$.

38. (III) Determine the current through each of the resistors in Fig. 26–54.

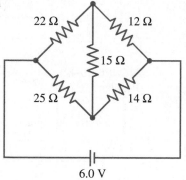

FIGURE 26–54
Problems 38 and 39.

39. (III) If the 25-Ω resistor in Fig. 26–54 is shorted out (resistance = 0), what then would be the current through the 15-Ω resistor?

40. (III) Twelve resistors, each of resistance R, are connected as the edges of a cube as shown in Fig. 26–55. Determine the equivalent resistance (*a*) between points a and b, the ends of a side; (*b*) between points a and c, the ends of a face diagonal; (*c*) between points a and d, the ends of the volume diagonal. [*Hint*: Apply an emf and determine currents; use symmetry at junctions.]

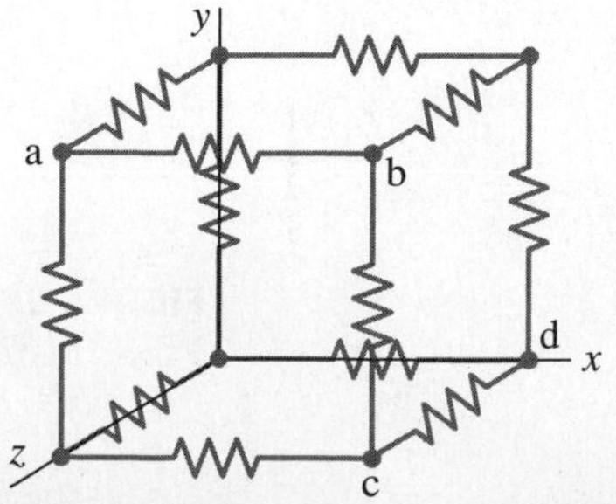

FIGURE 26–55
Problem 40.

41. (III) Determine the net resistance in Fig. 26–56 (*a*) between points a and c, and (*b*) between points a and b. Assume $R' \neq R$. [*Hint*: Apply an emf and determine currents; use symmetry at junctions.]

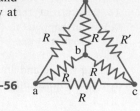

FIGURE 26–56
Problem 41.

26–4 Emfs Combined, Battery Charging

42. (II) Suppose two batteries, with unequal emfs of 2.00 V and 3.00 V, are connected as shown in Fig. 26–57. If each internal resistance is $r = 0.450\ \Omega$, and $R = 4.00\ \Omega$, what is the voltage across the resistor R?

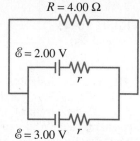

FIGURE 26–57
Problem 42.

26–5 *RC* Circuits

43. (I) Estimate the range of resistance needed to make a variable timer for typical intermittent windshield wipers if the capacitor used is on the order of $1\ \mu\text{F}$.

44. (II) In Fig. 26–58 (same as Fig. 26–17a), the total resistance is 15.0 kΩ, and the battery's emf is 24.0 V. If the time constant is measured to be 24.0 μs, calculate (*a*) the total capacitance of the circuit and (*b*) the time it takes for the voltage across the resistor to reach 16.0 V after the switch is closed.

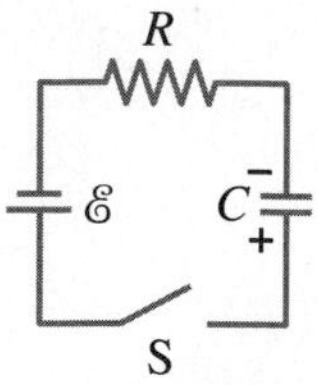

FIGURE 26–58
Problems 44 and 46.

45. (II) Two 3.8-μF capacitors, two 2.2-kΩ resistors, and a 12.0-V source are connected in series. Starting from the uncharged state, how long does it take for the current to drop from its initial value to 1.50 mA?

46. (II) How long does it take for the energy stored in a capacitor in a series RC circuit (Fig. 26–58) to reach 75% of its maximum value? Express answer in terms of the time constant $\tau = RC$.

47. (II) A parallel-plate capacitor is filled with a dielectric of dielectric constant K and high resistivity ρ (it conducts very slightly). This capacitor can be modeled as a pure capacitance C in parallel with a resistance R. Assume a battery places a charge $+Q$ and $-Q$ on the capacitor's opposing plates and is then disconnected. Show that the capacitor discharges with a time constant $\tau = K\varepsilon_0\rho$ (known as the *dielectric relaxation time*). Evaluate τ if the dielectric is glass with $\rho = 1.0 \times 10^{12}\ \Omega\cdot\text{m}$ and $K = 5.0$.

48. (II) The RC circuit of Fig. 26–59 (same as Fig. 26–18a) has $R = 8.7$ kΩ and $C = 3.0\ \mu$F. The capacitor is at voltage V_0 at $t = 0$, when the switch is closed. How long does it take the capacitor to discharge to 0.10% of its initial voltage?

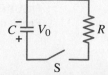

FIGURE 26–59
Problem 48.

49. (II) Consider the circuit shown in Fig. 26–60, where all resistors have the same resistance R. At $t = 0$, with the capacitor C uncharged, the switch is closed. (a) At $t = 0$, the three currents can be determined by analyzing a simpler, but equivalent, circuit. Identify this simpler circuit and use it to find the values of I_1, I_2, and I_3 at $t = 0$. (b) At $t = \infty$, the currents can be determined by analyzing a simpler, equivalent circuit. Identify this simpler circuit and implement it in finding the values of I_1, I_2, and I_3 at $t = \infty$. (c) At $t = \infty$, what is the potential difference across the capacitor?

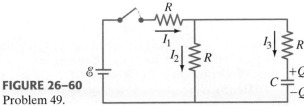

FIGURE 26–60
Problem 49.

50. (III) Determine the time constant for charging the capacitor in the circuit of Fig. 26–61. [*Hint:* Use Kirchhoff's rules.] (b) What is the maximum charge on the capacitor?

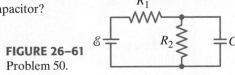

FIGURE 26–61
Problem 50.

51. (III) Two resistors and two uncharged capacitors are arranged as shown in Fig. 26–62. Then a potential difference of 24 V is applied across the combination as shown. (a) What is the potential at point a with switch S open? (Let $V = 0$ at the negative terminal of the source.) (b) What is the potential at point b with the switch open? (c) When the switch is closed, what is the final potential of point b? (d) How much charge flows through the switch S after it is closed?

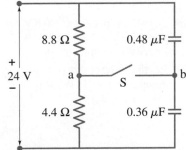

FIGURE 26–62
Problems 51 and 52.

52. (III) Suppose the switch S in Fig. 26–62 is closed. What is the time constant (or time constants) for charging the capacitors after the 24 V is applied?

*26–7 Ammeters and Voltmeters

*53. (I) An ammeter has a sensitivity of $35{,}000\,\Omega/V$. What current in the galvanometer produces full-scale deflection?

*54. (I) What is the resistance of a voltmeter on the 250-V scale if the meter sensitivity is $35{,}000\,\Omega/V$?

*55. (II) A galvanometer has a sensitivity of $45\,k\Omega/V$ and internal resistance $20.0\,\Omega$. How could you make this into (a) an ammeter that reads 2.0 A full scale, or (b) a voltmeter reading 1.00 V full scale?

*56. (II) A galvanometer has an internal resistance of $32\,\Omega$ and deflects full scale for a 55-μA current. Describe how to use this galvanometer to make (a) an ammeter to read currents up to 25 A, and (b) a voltmeter to give a full scale deflection of 250 V.

*57. (II) A particular digital meter is based on an electronic module that has an internal resistance of $100\,M\Omega$ and a full-scale sensitivity of 400 mV. Two resistors connected as shown in Fig. 26–63 can be used to change the voltage range. Assume $R_1 = 10\,M\Omega$. Find the value of R_2 that will result in a voltmeter with a full-scale range of 40 V.

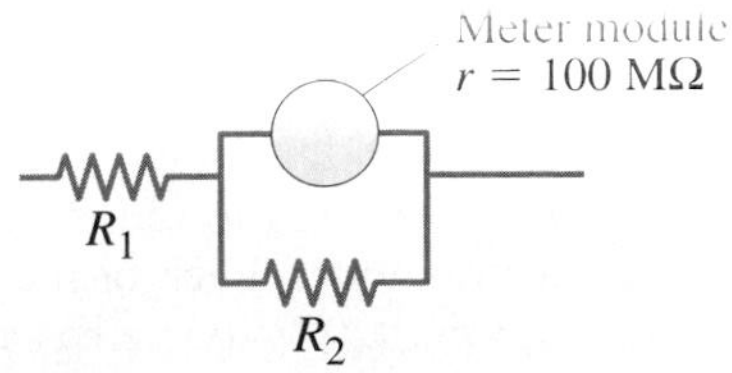

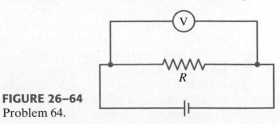

FIGURE 26–63
Problem 57.

*58. (II) A milliammeter reads 25 mA full scale. It consists of a 0.20-Ω resistor in parallel with a 33-Ω galvanometer. How can you change this ammeter to a voltmeter giving a full scale reading of 25 V without taking the ammeter apart? What will be the sensitivity (Ω/V) of your voltmeter?

*59. (II) A 45-V battery of negligible internal resistance is connected to a 44-kΩ and a 27-kΩ resistor in series. What reading will a voltmeter, of internal resistance 95 kΩ, give when used to measure the voltage across each resistor? What is the percent inaccuracy due to meter resistance for each case?

*60. (II) An ammeter whose internal resistance is $53\,\Omega$ reads 5.25 mA when connected in a circuit containing a battery and two resistors in series whose values are 650 Ω and 480 Ω. What is the actual current when the ammeter is absent?

*61. (II) A battery with $\mathscr{E} = 12.0\,V$ and internal resistance $r = 1.0\,\Omega$ is connected to two 7.5-kΩ resistors in series. An ammeter of internal resistance $0.50\,\Omega$ measures the current, and at the same time a voltmeter with internal resistance 15 kΩ measures the voltage across one of the 7.5-kΩ resistors in the circuit. What do the ammeter and voltmeter read?

*62. (II) A 12.0-V battery (assume the internal resistance = 0) is connected to two resistors in series. A voltmeter whose internal resistance is $18.0\,k\Omega$ measures 5.5 V and 4.0 V, respectively, when connected across each of the resistors. What is the resistance of each resistor?

*63. (III) Two 9.4-kΩ resistors are placed in series and connected to a battery. A voltmeter of sensitivity $1000\,\Omega/V$ is on the 3.0-V scale and reads 2.3 V when placed across either resistor. What is the emf of the battery? (Ignore its internal resistance.)

*64. (III) When the resistor R in Fig. 26–64 is $35\,\Omega$, the high-resistance voltmeter reads 9.7 V. When R is replaced by a 14.0-Ω resistor, the voltmeter reading drops to 8.1 V. What are the emf and internal resistance of the battery?

FIGURE 26–64
Problem 64.

General Problems

65. Suppose that you wish to apply a 0.25-V potential difference between two points on the human body. The resistance is about 1800 Ω, and you only have a 1.5-V battery. How can you connect up one or more resistors to produce the desired voltage?

66. A **three-way lightbulb** can produce 50 W, 100 W, or 150 W, at 120 V. Such a bulb contains two filaments that can be connected to the 120 V individually or in parallel. (a) Describe how the connections to the two filaments are made to give each of the three wattages. (b) What must be the resistance of each filament?

67. Suppose you want to run some apparatus that is 65 m from an electric outlet. Each of the wires connecting your apparatus to the 120-V source has a resistance per unit length of 0.0065 Ω/m. If your apparatus draws 3.0 A, what will be the voltage drop across the connecting wires and what voltage will be applied to your apparatus?

68. For the circuit shown in Fig. 26–18a, show that the decrease in energy stored in the capacitor from $t = 0$ until one time constant has elapsed equals the energy dissipated as heat in the resistor.

69. A heart pacemaker is designed to operate at 72 beats/min using a 6.5-μF capacitor in a simple RC circuit. What value of resistance should be used if the pacemaker is to fire (capacitor discharge) when the voltage reaches 75% of maximum?

70. Suppose that a person's body resistance is 950 Ω. (a) What current passes through the body when the person accidentally is connected to 110 V? (b) If there is an alternative path to ground whose resistance is 35 Ω, what current passes through the person? (c) If the voltage source can produce at most 1.5 A, how much current passes through the person in case (b)?

71. A **Wheatstone bridge** is a type of "bridge circuit" used to make measurements of resistance. The unknown resistance to be measured, R_x, is placed in the circuit with accurately known resistances R_1, R_2, and R_3 (Fig. 26–65). One of these, R_3, is a variable resistor which is adjusted so that when the switch is closed momentarily, the ammeter Ⓐ shows zero current flow. (a) Determine R_x in terms of R_1, R_2, and R_3. (b) If a Wheatstone bridge is "balanced" when $R_1 = 630\,\Omega$, $R_2 = 972\,\Omega$, and $R_3 = 78.6\,\Omega$, what is the value of the unknown resistance?

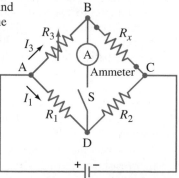

FIGURE 26–65
Problems 71 and 72.
Wheatstone bridge.

72. An unknown length of platinum wire 1.22 mm in diameter is placed as the unknown resistance in a Wheatstone bridge (see Problem 71, Fig. 26–65). Arms 1 and 2 have resistance of 38.0 Ω and 29.2 Ω, respectively. Balance is achieved when R_3 is 3.48 Ω. How long is the platinum wire?

73. The internal resistance of a 1.35-V mercury cell is 0.030 Ω, whereas that of a 1.5-V dry cell is 0.35 Ω. Explain why three mercury cells can more effectively power a 2.5-W hearing aid that requires 4.0 V than can three dry cells.

74. How many $\frac{1}{2}$-W resistors, each of the same resistance, must be used to produce an equivalent 3.2-kΩ, 3.5-W resistor? What is the resistance of each, and how must they be connected? Do not exceed $P = \frac{1}{2}$ W in each resistor.

75. A solar cell, 3.0 cm square, has an output of 350 mA at 0.80 V when exposed to full sunlight. A solar panel that delivers close to 1.3 A of current at an emf of 120 V to an external load is needed. How many cells will you need to create the panel? How big a panel will you need, and how should you connect the cells to one another? How can you optimize the output of your solar panel?

76. A power supply has a fixed output voltage of 12.0 V, but you need $V_T = 3.0$ V output for an experiment. (a) Using the voltage divider shown in Fig. 26–66, what should R_2 be if R_1 is 14.5 Ω? (b) What will the terminal voltage V_T be if you connect a load to the 3.0-V output, assuming the load has a resistance of 7.0 Ω?

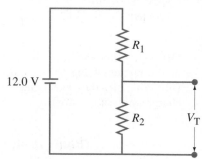

FIGURE 26–66
Problem 76.

77. The current through the 4.0-kΩ resistor in Fig. 26–67 is 3.10 mA. What is the terminal voltage V_{ba} of the "unknown" battery? (There are two answers. Why?)

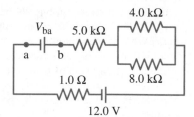

FIGURE 26–67
Problem 77.

78. A battery produces 40.8 V when 7.40 A is drawn from it, and 47.3 V when 2.80 A is drawn. What are the emf and internal resistance of the battery?

79. In the circuit shown in Fig. 26–68, the 33-Ω resistor dissipates 0.80 W. What is the battery voltage?

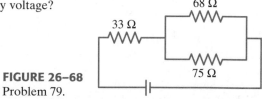

FIGURE 26–68
Problem 79.

80. The current through the 20-Ω resistor in Fig. 26–69 does not change whether the two switches S_1 and S_2 are both open or both closed. Use this clue to determine the value of the unknown resistance R.

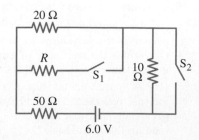

FIGURE 26–69
Problem 80.

*81. (*a*) A voltmeter and an ammeter can be connected as shown in Fig. 26–70a to measure a resistance R. If V is the voltmeter reading, and I is the ammeter reading, the value of R will not quite be V/I (as in Ohm's law) because some of the current actually goes through the voltmeter. Show that the actual value of R is given by

$$\frac{1}{R} = \frac{I}{V} - \frac{1}{R_V},$$

where R_V is the voltmeter resistance. Note that $R \approx V/I$ if $R_V \gg R$. (*b*) A voltmeter and an ammeter can also be connected as shown in Fig. 26–70b to measure a resistance R. Show in this case that

$$R = \frac{V}{I} - R_A,$$

where V and I are the voltmeter and ammeter readings and R_A is the resistance of the ammeter. Note that $R \approx V/I$ if $R_A \ll R$.

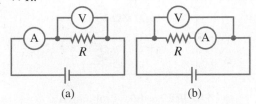

(a)   (b)

FIGURE 26–70 Problem 81.

82. (*a*) What is the equivalent resistance of the circuit shown in Fig. 26–71? (*b*) What is the current in the 18-Ω resistor? (*c*) What is the current in the 12-Ω resistor? (*d*) What is the power dissipation in the 4.5-Ω resistor?

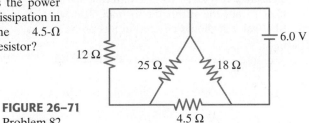

FIGURE 26–71
Problem 82.

83. A flashlight bulb rated at 2.0 W and 3.0 V is operated by a 9.0-V battery. To light the bulb at its rated voltage and power, a resistor R is connected in series as shown in Fig. 26–72. What value should the resistor have?

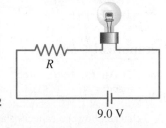

FIGURE 26–72
Problem 83.

84. Some light-dimmer switches use a variable resistor as shown in Fig. 26–73. The slide moves from position $x = 0$ to $x = 1$, and the resistance up to slide position x is proportional to x (the total resistance is $R_{var} = 150\,\Omega$ at $x = 1$). What is the power expended in the lightbulb if (*a*) $x = 1.00$, (*b*) $x = 0.65$, (*c*) $x = 0.35$?

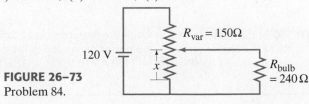

FIGURE 26–73
Problem 84.

85. A **potentiometer** is a device to precisely measure potential differences or emf, using a "null" technique. In the simple potentiometer circuit shown in Fig. 26–74, R' represents the total resistance of the resistor from A to B (which could be a long uniform "slide" wire), whereas R represents the resistance of only the part from A to the movable contact at C. When the unknown emf to be measured, $\mathscr{E}_x$, is placed into the circuit as shown, the movable contact C is moved until the galvanometer G gives a null reading (i.e., zero) when the switch S is closed. The resistance between A and C for this situation we call R_x. Next, a standard emf, $\mathscr{E}_s$, which is known precisely, is inserted into the circuit in place of $\mathscr{E}_x$ and again the contact C is moved until zero current flows through the galvanometer when the switch S is closed. The resistance between A and C now is called R_s. (*a*) Show that the unknown emf is given by

$$\mathscr{E}_x = \left(\frac{R_x}{R_s}\right)\mathscr{E}_s$$

where R_x, R_s, and $\mathscr{E}_s$ are all precisely known. The working battery is assumed to be fresh and to give a constant voltage. (*b*) A slide-wire potentiometer is balanced against a 1.0182-V standard cell when the slide wire is set at 33.6 cm out of a total length of 100.0 cm. For an unknown source, the setting is 45.8 cm. What is the emf of the unknown? (*c*) The galvanometer of a potentiometer has an internal resistance of 35 Ω and can detect a current as small as 0.012 mA. What is the minimum uncertainty possible in measuring an unknown voltage? (*d*) Explain the advantage of using this "null" method of measuring emf.

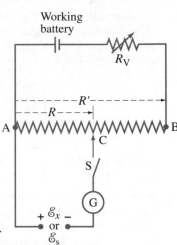

FIGURE 26–74
Potentiometer circuit.
Problem 85.

86. Electronic devices often use an RC circuit to protect against power outages as shown in Fig. 26–75. (*a*) If the protector circuit is supposed to keep the supply voltage at least 75% of full voltage for as long as 0.20 s, how big a resistance R is needed? The capacitor is 8.5 μF. Assume the attached "electronics" draws negligible current. (*b*) Between which two terminals should the device be connected, a and b, b and c, or a and c?

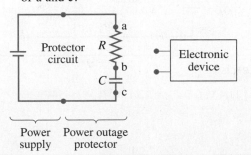

FIGURE 26–75
Problem 86.

87. The circuit shown in Fig. 26–76 is a primitive 4-bit **digital-to-analog converter** (DAC). In this circuit, to represent each digit (2^n) of a binary number, a "1" has the nth switch closed whereas zero ("0") has the switch open. For example, 0010 is represented by closing switch $n = 1$, while all other switches are open. Show that the voltage V across the 1.0-Ω resistor for the binary numbers 0001, 0010, 0100, and 1010 (which represent 1, 2, 4, 10) follows the pattern that you expect for a 4-bit DAC.

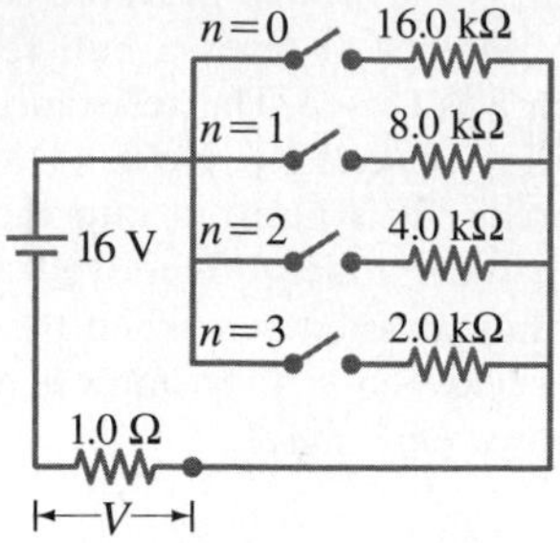

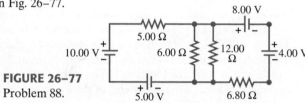

FIGURE 26–76
Problem 87.

88. Determine the current in each resistor of the circuit shown in Fig. 26–77.

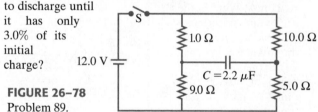

FIGURE 26–77
Problem 88.

89. In the circuit shown in Fig. 26–78, switch S is closed at time $t = 0$. (a) After the capacitor is fully charged, what is the voltage across it? How much charge is on it? (b) Switch S is now opened. How long does it now take for the capacitor to discharge until it has only 3.0% of its initial charge?

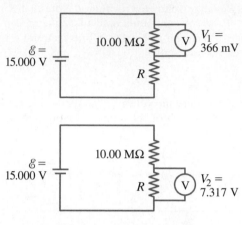

FIGURE 26–78
Problem 89.

90. Figure 26–79 shows the circuit for a simple **sawtooth oscillator**. At time $t = 0$, its switch S is closed. The neon bulb has initially infinite resistance until the voltage across it reaches 90.0 V, and then it begins to conduct with very little resistance (essentially zero). It stops conducting (its resistance becomes essentially infinite) when the voltage drops down to 65.0 V. (a) At what time t_1 does the neon bulb reach 90.0 V and start conducting? (b) At what time t_2 does the bulb reach 90.0 V for a second time and again become conducting? (c) Sketch the sawtooth waveform between $t = 0$ and $t = 0.70$ s.

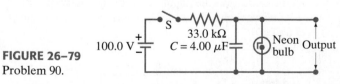

FIGURE 26–79
Problem 90.

*91. Measurements made on circuits that contain large resistances can be confusing. Consider a circuit powered by a battery $\mathcal{E} = 15.000$ V with a 10.00-MΩ resistor in series with an unknown resistor R. As shown in Fig. 26–80, a particular voltmeter reads $V_1 = 366$ mV when connected across the 10.00-MΩ resistor, and this meter reads $V_2 = 7.317$ V when connected across R. Determine the value of R. [*Hint*: Define R_V as the voltmeter's internal resistance.]

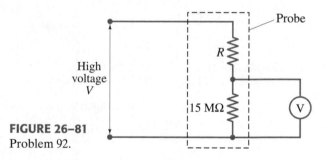

FIGURE 26–80   Problem 91.

*92. A typical voltmeter has an internal resistance of 10 MΩ and can only measure voltage differences of up to several hundred volts. Figure 26–81 shows the design of a probe to measure a very large voltage difference V using a voltmeter. If you want the voltmeter to read 50 V when $V = 50$ kV, what value R should be used in this probe?

FIGURE 26–81
Problem 92.

Numerical/Computer

*93. (II) An *RC* series circuit contains a resistor $R = 15$ kΩ, a capacitor $C = 0.30$ μF, and a battery of emf $\mathcal{E} = 9.0$ V. Starting at $t = 0$, when the battery is connected, determine the charge Q on the capacitor and the current I in the circuit from $t = 0$ to $t = 10.0$ ms (at 0.1-ms intervals). Make graphs showing how the charge Q and the current I change with time within this time interval. From the graphs find the time at which the charge attains 63% of its final value, $C\mathcal{E}$, and the current drops to 37% of its initial value, $\mathcal{E}/R$.

Answers to Exercises

A: (a) 1.14 A; (b) 11.4 V; (c) $P_R = 13.1$ W, $P_r = 0.65$ W.

B: 6 Ω and 25 Ω.

C: $41I_3 - 45 + 21I_2 - 80 = 0$.

D: 180 A; this high current through the batteries could cause them to become very hot: the power dissipated in the weak battery would be $P = I^2r = (180$ A$)^2(0.10 \ \Omega) = 3200$ W!

E: (a).

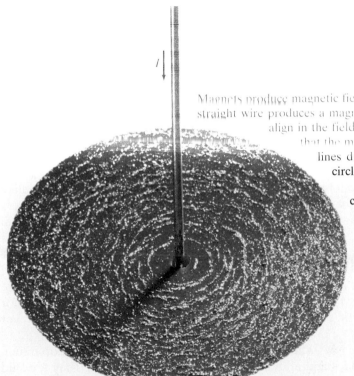

Magnets produce magnetic fields, but so do electric currents. An electric current flowing in this straight wire produces a magnetic field which causes the tiny pieces of iron (iron "filings") to align in the field. We shall see in this Chapter how magnetic field is defined, and that the magnetic field direction is along the iron filings. The magnetic field lines due to the electric current in this long wire are in the shape of circles around the wire.

We also discuss how magnetic fields exert forces on electric currents and on charged particles, as well as useful applications of the interaction between magnetic fields and electric currents and moving electric charges.

Magnetism

CHAPTER 27

CHAPTER-OPENING QUESTION—Guess now!
Which of the following can experience a force when placed in the magnetic field of a magnet?

(a) An electric charge at rest.
(b) An electric charge moving.
(c) An electric current in a wire.
(d) Another magnet.

The history of magnetism begins thousands of years ago. In a region of Asia Minor known as Magnesia, rocks were found that could attract each other. These rocks were called "magnets" after their place of discovery. Not until the nineteenth century, however, was it seen that magnetism and electricity are closely related. A crucial discovery was that electric currents produce magnetic effects (we will say "magnetic fields") like magnets do. All kinds of practical devices depend on magnetism, from compasses to motors, loudspeakers, computer memory, and electric generators.

CONTENTS

27–1 Magnets and Magnetic Fields

We have all observed a magnet attract paper clips, nails, and other objects made of iron, Fig. 27–1. Any magnet, whether it is in the shape of a bar or a horseshoe, has two ends or faces, called **poles**, which is where the magnetic effect is strongest. If a bar magnet is suspended from a fine thread, it is found that one pole of the magnet will always point toward the north. It is not known for sure when this fact was discovered, but it is known that the Chinese were making use of it as an aid to navigation by the eleventh century and perhaps earlier. This is the principle of a compass.

FIGURE 27–1 A horseshoe magnet attracts iron tacks and paper clips.

707

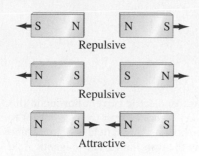

FIGURE 27–2 Like poles of a magnet repel; unlike poles attract. Red arrows indicate force direction.

FIGURE 27–3 If you split a magnet, you won't get isolated north and south poles; instead, two new magnets are produced, each with a north and a south pole.

⚠ **CAUTION**

Magnets do not attract all metals

FIGURE 27–4 (a) Visualizing magnetic field lines around a bar magnet, using iron filings and compass needles. The red end of the bar magnet is its north pole. The N pole of a nearby compass needle points away from the north pole of the magnet. (b) Magnetic field lines for a bar magnet.

⚠ **CAUTION**

Magnetic field lines form closed loops, unlike electric field lines

A compass needle is simply a bar magnet which is supported at its center of gravity so that it can rotate freely. The pole of a freely suspended magnet that points toward geographic north is called the **north pole** of the magnet. The other pole points toward the south and is called the **south pole**.

It is a familiar observation that when two magnets are brought near one another, each exerts a force on the other. The force can be either attractive or repulsive and can be felt even when the magnets don't touch. If the north pole of one bar magnet is brought near the north pole of a second magnet, the force is repulsive. Similarly, if the south poles of two magnets are brought close, the force is repulsive. But when a north pole is brought near the south pole of another magnet, the force is attractive. These results are shown in Fig. 27–2, and are reminiscent of the forces between electric charges: like poles repel, and unlike poles attract. *But do not confuse magnetic poles with electric charge.* They are very different. One important difference is that a positive or negative electric charge can easily be isolated. But an isolated single magnetic pole has never been observed. If a bar magnet is cut in half, you do not obtain isolated north and south poles. Instead, two new magnets are produced, Fig. 27–3, each with north (N) and south (S) poles. If the cutting operation is repeated, more magnets are produced, each with a north and a south pole. Physicists have searched for isolated single magnetic poles (monopoles), but no **magnetic monopole** has ever been observed.

Only iron and a few other materials, such as cobalt, nickel, gadolinium, and some of their oxides and alloys, show strong magnetic effects. They are said to be **ferromagnetic** (from the Latin word *ferrum* for iron). Other materials show some slight magnetic effect, but it is very weak and can be detected only with delicate instruments. We will look in more detail at ferromagnetism in Section 28–7.

In Chapter 21, we used the concept of an electric field surrounding an electric charge. In a similar way, we can picture a **magnetic field** surrounding a magnet. The force one magnet exerts on another can then be described as the interaction between one magnet and the magnetic field of the other. Just as we drew electric field lines, we can also draw **magnetic field lines**. They can be drawn, as for electric field lines, so that (1) the direction of the magnetic field is tangent to a field line at any point, and (2) the number of lines per unit area is proportional to the strength of the magnetic field.

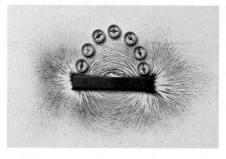
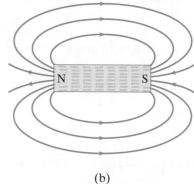

(a)   (b)

The *direction* of the magnetic field at a given point can be defined as the direction that the north pole of a compass needle would point if placed at that point. (A more precise definition will be given in Section 27–3.) Figure 27–4a shows how thin iron filings (acting like tiny magnets) reveal the magnetic field lines by lining up like the compass needles. The magnetic field determined in this way for the field surrounding a bar magnet is shown in Fig. 27–4b. Notice that because of our definition, the lines always point out from the north pole and in toward the south pole of a magnet (the north pole of a magnetic compass needle is attracted to the south pole of the magnet).

Magnetic field lines continue inside a magnet, as indicated in Fig. 27–4b. Indeed, given the lack of single magnetic poles, magnetic field lines always form closed loops, unlike electric field lines that begin on positive charges and end on negative charges.

Earth's Magnetic Field

The Earth's magnetic field is shown in Fig. 27–5. The pattern of field lines is as if there were an imaginary bar magnet inside the Earth. Since the north pole (N) of a compass needle points north, the Earth's magnetic pole which is in the geographic north is magnetically a south pole, as indicated in Fig. 27–5 by the S on the schematic bar magnet inside the Earth. Remember that the north pole of one magnet is attracted to the south pole of another magnet. Nonetheless, Earth's pole in the north is still often called the "north magnetic pole," or "geomagnetic north," simply because it is in the north. Similarly, the Earth's southern magnetic pole, which is near the geographic south pole, is magnetically a north pole (N). The Earth's magnetic poles do not coincide with the *geographic* poles, which are on the Earth's axis of rotation. The north magnetic pole, for example, is in the Canadian Arctic,[†] about 900 km from the geographic north pole, or "true north." This difference must be taken into account when you use a compass (Fig. 27–6). The angular difference between magnetic north and true (geographical) north is called the **magnetic declination**. In the U.S. it varies from 0° to about 20°, depending on location.

Notice in Fig. 27–5 that the Earth's magnetic field at most locations is not tangent to the Earth's surface. The angle that the Earth's magnetic field makes with the horizontal at any point is referred to as the **angle of dip**.

EXERCISE A Does the Earth's magnetic field have a greater magnitude near the poles or near the equator? [*Hint*: Note the field lines in Fig. 27–5.]

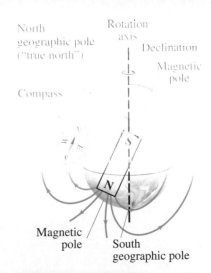

FIGURE 27–5 The Earth acts like a huge magnet; but its magnetic poles are not at the geographic poles, which are on the Earth's rotation axis.

PHYSICS APPLIED
Use of a compass

FIGURE 27–6 Using a map and compass in the wilderness. First you align the compass case so the needle points away from true north (N) exactly the number of degrees of declination as stated on the map (15° for the place shown on this topographic map of a part of California). Then align the map with true north, as shown, *not* with the compass needle.

Uniform Magnetic Field

The simplest magnetic field is one that is uniform—it doesn't change in magnitude or direction from one point to another. A perfectly uniform field over a large area is not easy to produce. But the field between two flat parallel pole pieces of a magnet is nearly uniform if the area of the pole faces is large compared to their separation, as shown in Fig. 27–7. At the edges, the field "fringes" out somewhat: the magnetic field lines are no longer quite parallel and uniform. The parallel evenly spaced field lines in the central region of the gap indicate that the field is uniform at points not too near the edges, much like the electric field between two parallel plates (Fig. 23–16).

FIGURE 27–7 Magnetic field between two wide poles of a magnet is nearly uniform, except near the edges.

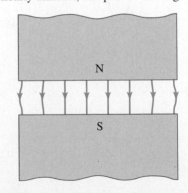

[†]Magnetic north is moving many kilometers a year at present. Magnetism in rocks suggests that the Earth's poles have not only moved significantly over geologic time, but have also reversed direction 400 times over the last 330 million years.

FIGURE 27–8 (a) Deflection of compass needles near a current-carrying wire, showing the presence and direction of the magnetic field. (b) Magnetic field lines around an electric current in a straight wire. See also the Chapter-Opening photo. (c) Right-hand rule for remembering the direction of the magnetic field: when the thumb points in the direction of the conventional current, the fingers wrapped around the wire point in the direction of the magnetic field.

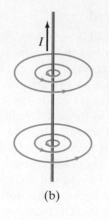

(a)　　　　　　　(b)　　　　　　　(c)

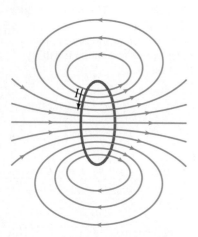

FIGURE 27–9 Magnetic field lines due to a circular loop of wire.

Right-hand-rule 1: magnetic field direction produced by electric current

FIGURE 27–10 Right-hand rule for determining the direction of the magnetic field relative to the current.

I　　Magnetic field

27–2 Electric Currents Produce Magnetic Fields

During the eighteenth century, many scientists sought to find a connection between electricity and magnetism. A stationary electric charge and a magnet were shown to have no influence on each other. But in 1820, Hans Christian Oersted (1777–1851) found that when a compass needle is placed near a wire, the needle deflects as soon as the two ends of the wire are connected to the terminals of a battery and the wire carries an electric current. As we have seen, a compass needle is deflected by a magnetic field. So Oersted's experiment showed that **an electric current produces a magnetic field**. He had found a connection between electricity and magnetism.

A compass needle placed near a straight section of current-carrying wire experiences a force, causing the needle to align tangent to a circle around the wire, Fig. 27–8a. Thus, the magnetic field lines produced by a current in a straight wire are in the form of circles with the wire at their center, Fig. 27–8b. The direction of these lines is indicated by the north pole of the compasses in Fig. 27–8a. There is a simple way to remember the direction of the magnetic field lines in this case. It is called a **right-hand rule**: grasp the wire with your right hand so that your thumb points in the direction of the conventional (positive) current; then your fingers will encircle the wire in the direction of the magnetic field, Fig. 27–8c.

The magnetic field lines due to a circular loop of current-carrying wire can be determined in a similar way using a compass. The result is shown in Fig. 27–9. Again the right-hand rule can be used, as shown in Fig. 27–10. Unlike the uniform field shown in Fig. 27–7, the magnetic fields shown in Figs. 27–8 and 27–9 are *not* uniform—the fields are different in magnitude and direction at different points.

EXERCISE B A straight wire carries a current directly toward you. In what direction are the magnetic field lines surrounding the wire?

27–3 Force on an Electric Current in a Magnetic Field; Definition of $\vec{B}$

In Section 27–2 we saw that an electric current exerts a force on a magnet, such as a compass needle. By Newton's third law, we might expect the reverse to be true as well: we should expect that *a magnet exerts a force on a current-carrying wire*. Experiments indeed confirm this effect, and it too was first observed by Oersted.

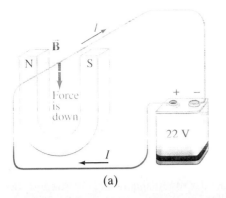

(a)

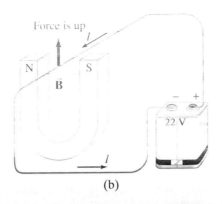

Force is up

(b)

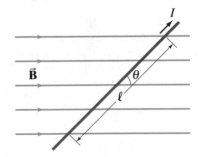

(c)   Right-hand rule

Suppose a straight wire is placed in the magnetic field between the poles of a horseshoe magnet as shown in Fig. 27–11. When a current flows in the wire, experiment shows that a force is exerted on the wire. But this force is *not* toward one or the other pole of the magnet. Instead, the force is directed at right angles to the magnetic field direction, downward in Fig. 27–11a. If the current is reversed in direction, the force is in the opposite direction, upward as shown in Fig. 27–11b. Experiments show that *the direction of the force is always perpendicular to the direction of the current and also perpendicular to the direction of the magnetic field,* $\vec{\mathbf{B}}$.

The direction of the force is given by another **right-hand rule**, as illustrated in Fig. 27–11c. Orient your right hand until your outstretched fingers can point in the direction of the conventional current I, and when you bend your fingers they point in the direction of the magnetic field lines, $\vec{\mathbf{B}}$. Then your outstretched thumb will point in the direction of the force $\vec{\mathbf{F}}$ on the wire.

This right-hand rule describes the direction of the force. What about the magnitude of the force on the wire? It is found experimentally that the magnitude of the force is directly proportional to the current I in the wire, and to the length ℓ of wire exposed to the magnetic field (assumed uniform). Furthermore, if the magnetic field B is made stronger, the force is found to be proportionally greater. The force also depends on the angle θ between the current direction and the magnetic field (Fig. 27–12), being proportional to $\sin\theta$. Thus, the force on a wire carrying a current I with length ℓ in a uniform magnetic field B is given by

$$F \propto I\ell B \sin\theta.$$

When the current is perpendicular to the field lines ($\theta = 90°$), the force is strongest. When the wire is parallel to the magnetic field lines ($\theta = 0°$), there is no force at all.

Up to now we have not defined the magnetic field strength precisely. In fact, the magnetic field B can be conveniently defined in terms of the above proportion so that the proportionality constant is precisely 1. Thus we have

$$F = I\ell B \sin\theta. \tag{27–1}$$

If the direction of the current is perpendicular to the field $\vec{\mathbf{B}}$ ($\theta = 90°$), then the force is

$$F_{max} = I\ell B. \qquad [\text{current} \perp \vec{\mathbf{B}}] \tag{27–2}$$

If the current is parallel to the field ($\theta = 0°$), the force is zero. The magnitude of $\vec{\mathbf{B}}$ can be defined using Eq. 27–2 as $B = F_{max}/I\ell$, where F_{max} is the magnitude of the force on a straight length ℓ of wire carrying a current I when the wire is perpendicular to $\vec{\mathbf{B}}$.

The relation between the force $\vec{\mathbf{F}}$ on a wire carrying current I, and the magnetic field $\vec{\mathbf{B}}$ that causes the force, can be written as a vector equation. To do so, we recall that the direction of $\vec{\mathbf{F}}$ is given by the right-hand rule (Fig. 27–11c), and the magnitude by Eq. 27–1. This is consistent with the definition of the vector cross product (see Section 11–2), so we can write

$$\vec{\mathbf{F}} = I\vec{\boldsymbol{\ell}} \times \vec{\mathbf{B}}; \tag{27–3}$$

here, $\vec{\boldsymbol{\ell}}$ is a vector whose magnitude is the length of the wire and its direction is along the wire (assumed straight) in the direction of the conventional (positive) current.

FIGURE 27–11 (a) Force on a current-carrying wire placed in a magnetic field $\vec{\mathbf{B}}$; (b) same, but current reversed; (c) right-hand rule for setup in (b).

Right-hand-rule 2:
force on current exerted by $\vec{\mathbf{B}}$

FIGURE 27–12 Current-carrying wire in a magnetic field. Force on the wire is directed into the page.

Equation 27–3 applies if the magnetic field is uniform and the wire is straight. If $\vec{\mathbf{B}}$ is not uniform, or if the wire does not everywhere make the same angle θ with $\vec{\mathbf{B}}$, then Eq. 27–3 can be written more generally as

$$d\vec{\mathbf{F}} = I\,d\vec{\boldsymbol{\ell}} \times \vec{\mathbf{B}}, \qquad\qquad (27\text{–}4)$$

where $d\vec{\mathbf{F}}$ is the infinitesimal force acting on a differential length $d\vec{\boldsymbol{\ell}}$ of the wire. The total force on the wire is then found by integrating.

Equation 27–4 can serve (just as well as Eq. 27–2 or 27–3) as a practical definition of $\vec{\mathbf{B}}$. An equivalent way to define $\vec{\mathbf{B}}$, in terms of the force on a moving electric charge, is discussed in the next Section.

EXERCISE C A wire carrying current I is perpendicular to a magnetic field of strength B. Assuming a fixed length of wire, which of the following changes will result in decreasing the force on the wire by a factor of 2? (*a*) Decrease the angle from 90° to 45°; (*b*) decrease the angle from 90° to 30°; (*c*) decrease the current in the wire to $I/2$; (*d*) decrease the magnetic field strength to $B/2$; (*e*) none of these will do it.

The SI unit for magnetic field B is the **tesla** (T). From Eqs. 27–1, 2, 3, or 4, we see that $1\,\text{T} = 1\,\text{N/A}\cdot\text{m}$. An older name for the tesla is the "weber per meter squared" $(1\,\text{Wb/m}^2 = 1\,\text{T})$. Another unit sometimes used to specify magnetic field is a cgs unit, the **gauss** (G): $1\,\text{G} = 10^{-4}\,\text{T}$. A field given in gauss should always be changed to teslas before using with other SI units. To get a "feel" for these units, we note that the magnetic field of the Earth at its surface is about $\frac{1}{2}\,\text{G}$ or $0.5 \times 10^{-4}\,\text{T}$. On the other hand, the field near a small magnet attached to your refrigerator may be $100\,\text{G}$ $(0.01\,\text{T})$ whereas strong electromagnets can produce fields on the order of $2\,\text{T}$ and superconducting magnets can produce over $10\,\text{T}$.

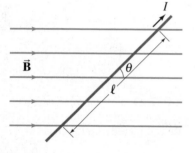

FIGURE 27–12 (Repeated for Example 27–1.) Current-carrying wire in a magnetic field. Force on the wire is directed into the page.

EXAMPLE 27–1 **Magnetic force on a current-carrying wire.** A wire carrying a 30-A current has a length $\ell = 12\,\text{cm}$ between the pole faces of a magnet at an angle $\theta = 60°$ (Fig. 27–12). The magnetic field is approximately uniform at $0.90\,\text{T}$. We ignore the field beyond the pole pieces. What is the magnitude of the force on the wire?

APPROACH We use Eq. 27–1, $F = I\ell B \sin\theta$.

SOLUTION The force F on the 12-cm length of wire within the uniform field B is

$$F = I\ell B \sin\theta = (30\,\text{A})(0.12\,\text{m})(0.90\,\text{T})(0.866) = 2.8\,\text{N}.$$

EXERCISE D A straight power line carries 30 A and is perpendicular to the Earth's magnetic field of $0.50 \times 10^{-4}\,\text{T}$. What magnitude force is exerted on 100 m of this power line?

On a diagram, when we want to represent an electric current or a magnetic field that is pointing out of the page (toward us) or into the page, we use $\odot$ or $\times$, respectively. The $\odot$ is meant to resemble the tip of an arrow pointing directly toward the reader, whereas the $\times$ or $\otimes$ resembles the tail of an arrow moving away. (See Figs. 27–13 and 27–14.)

EXAMPLE 27–2 **Measuring a magnetic field.** A rectangular loop of wire hangs vertically as shown in Fig. 27–13. A magnetic field $\vec{\mathbf{B}}$ is directed horizontally, perpendicular to the wire, and points out of the page at all points as represented by the symbol $\odot$. The magnetic field $\vec{\mathbf{B}}$ is very nearly uniform along the horizontal portion of wire ab (length $\ell = 10.0\,\text{cm}$) which is near the center of the gap of a large magnet producing the field. The top portion of the wire loop is free of the field. The loop hangs from a balance which measures a downward magnetic force (in addition to the gravitational force) of $F = 3.48 \times 10^{-2}\,\text{N}$ when the wire carries a current $I = 0.245\,\text{A}$. What is the magnitude of the magnetic field B?

APPROACH Three straight sections of the wire loop are in the magnetic field: a horizontal section and two vertical sections. We apply Eq. 27–1 to each section and use the right-hand rule.

SOLUTION The magnetic force on the left vertical section of wire points to the left; the force on the vertical section on the right points to the right. These two forces are equal and in opposite directions and so add up to zero. Hence, the net magnetic force on the loop is that on the horizontal section ab, whose length is $\ell = 0.100$ m. The angle θ between $\vec{B}$ and the wire is $\theta = 90°$, so $\sin\theta = 1$. Thus Eq. 27–1 gives

$$B = \frac{F}{I\ell} = \frac{3.48 \times 10^{-2}\,\text{N}}{(0.245\,\text{A})(0.100\,\text{m})} = 1.42\,\text{T}.$$

NOTE This technique can be a precise means of determining magnetic field strength.

EXAMPLE 27–3 **Magnetic force on a semicircular wire.** A rigid wire, carrying a current I, consists of a semicircle of radius R and two straight portions as shown in Fig. 27–14. The wire lies in a plane perpendicular to a uniform magnetic field $\vec{B}_0$. Note choice of x and y axis. The straight portions each have length ℓ within the field. Determine the net force on the wire due to the magnetic field $\vec{B}_0$.

APPROACH The forces on the two straight sections are equal $(= I\ell B_0)$ and in opposite directions, so they cancel. Hence the net force is that on the semicircular portion.

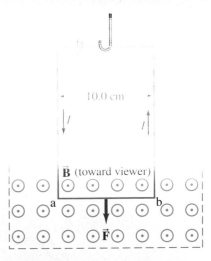

FIGURE 27–13 Measuring a magnetic field $\vec{B}$. Example 27–2.

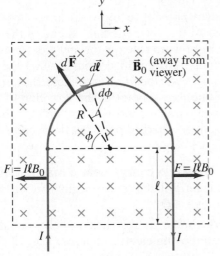

FIGURE 27–14 Example 27–3.

SOLUTION We divide the semicircle into short lengths $d\ell = R\,d\phi$ as indicated in Fig. 27–14, and use Eq. 27–4, $d\vec{F} = I\,d\vec{\ell} \times \vec{B}$, to find

$$dF = IB_0 R\,d\phi,$$

where dF is the force on the length $d\ell = R\,d\phi$, and the angle between $d\vec{\ell}$ and $\vec{B}_0$ is $90°$ (so $\sin\theta = 1$ in the cross product). The x component of the force $d\vec{F}$ on the segment $d\vec{\ell}$ shown, and the x component of $d\vec{F}$ for a symmetrically located $d\vec{\ell}$ on the other side of the semicircle, will cancel each other. Thus for the entire semicircle there will be no x component of force. Hence we need be concerned only with the y components, each equal to $dF\sin\phi$, and the total force will have magnitude

$$F = \int_0^\pi dF\sin\phi = IB_0 R \int_0^\pi \sin\phi\,d\phi = -IB_0 R\cos\phi \Big|_0^\pi = 2IB_0 R,$$

with direction vertically upward along the y axis in Fig. 27–14.

27–4 Force on an Electric Charge Moving in a Magnetic Field

We have seen that a current-carrying wire experiences a force when placed in a magnetic field. Since a current in a wire consists of moving electric charges, we might expect that freely moving charged particles (not in a wire) would also experience a force when passing through a magnetic field. Indeed, this is the case.

From what we already know we can predict the force on a single moving electric charge. If N such particles of charge q pass by a given point in time t, they constitute a current $I = Nq/t$. We let t be the time for a charge q to travel a distance ℓ in a magnetic field $\vec{\mathbf{B}}$; then $\vec{\boldsymbol{\ell}} = \vec{\mathbf{v}}t$ where $\vec{\mathbf{v}}$ is the velocity of the particle. Thus, the force on these N particles is, by Eq. 27–3, $\vec{\mathbf{F}} = I\vec{\boldsymbol{\ell}} \times \vec{\mathbf{B}} = (Nq/t)(\vec{\mathbf{v}}t) \times \vec{\mathbf{B}} = Nq\vec{\mathbf{v}} \times \vec{\mathbf{B}}$. The force on *one* of the N particles is then

$$\vec{\mathbf{F}} = q\vec{\mathbf{v}} \times \vec{\mathbf{B}}. \tag{27–5a}$$

This basic and important result can be considered as an alternative way of defining the magnetic field $\vec{\mathbf{B}}$, in place of Eq. 27–4 or 27–3. The magnitude of the force in Eq. 27–5a is

$$F = qvB \sin\theta. \tag{27–5b}$$

This gives the magnitude of the force on a particle of charge q moving with velocity $\vec{\mathbf{v}}$ at a point where the magnetic field has magnitude B. The angle between $\vec{\mathbf{v}}$ and $\vec{\mathbf{B}}$ is θ. The force is greatest when the particle moves perpendicular to $\vec{\mathbf{B}}$ ($\theta = 90°$):

$$F_{\text{max}} = qvB. \qquad [\vec{\mathbf{v}} \perp \vec{\mathbf{B}}]$$

The force is *zero* if the particle moves *parallel* to the field lines ($\theta = 0°$). The *direction* of the force is perpendicular to the magnetic field $\vec{\mathbf{B}}$ and to the velocity $\vec{\mathbf{v}}$ of the particle. It is given again by a **right-hand rule** (as for any cross product): you orient your right hand so that your outstretched fingers point along the direction of the particle's velocity ($\vec{\mathbf{v}}$), and when you bend your fingers they must point along the direction of $\vec{\mathbf{B}}$. Then your thumb will point in the direction of the force. This is true only for *positively* charged particles, and will be "up" for the positive particle shown in Fig. 27–15. For negatively charged particles, the force is in exactly the opposite direction, "down" in Fig. 27–15.

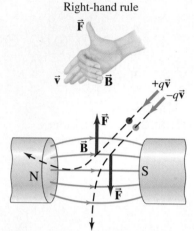

Right-hand rule

FIGURE 27–15 Force on charged particles due to a magnetic field is perpendicular to the magnetic field direction. If $\vec{\mathbf{v}}$ is horizontal, then $\vec{\mathbf{F}}$ is vertical.

Right-hand-rule 3: force on moving charge exerted by $\vec{\mathbf{B}}$

CONCEPTUAL EXAMPLE 27–4 **Negative charge near a magnet.** A negative charge $-Q$ is placed at rest near a magnet. Will the charge begin to move? Will it feel a force? What if the charge were positive, $+Q$?

RESPONSE No to all questions. A charge at rest has velocity equal to zero. Magnetic fields exert a force only on moving electric charges (Eqs. 27–5).

EXERCISE E Return to the Chapter-Opening Question, page 707, and answer it again now. Try to explain why you may have answered differently the first time.

EXAMPLE 27–5 **Magnetic force on a proton.** A magnetic field exerts a force of 8.0×10^{-14} N toward the west on a proton moving vertically upward at a speed of 5.0×10^6 m/s (Fig. 27–16a). When moving horizontally in a northerly direction, the force on the proton is zero (Fig. 27–16b). Determine the magnitude and direction of the magnetic field in this region. (The charge on a proton is $q = +e = 1.6 \times 10^{-19}$ C.)

APPROACH Since the force on the proton is zero when moving north, the field must be in a north–south direction. In order to produce a force to the west when the proton moves upward, the right-hand rule tells us that $\vec{\mathbf{B}}$ must point toward the north. (Your thumb points west and the outstretched fingers of your right hand point upward only when your bent fingers point north.) The magnitude of $\vec{\mathbf{B}}$ is found using Eq. 27–5b.

SOLUTION Equation 27–5b with $\theta = 90°$ gives

$$B = \frac{F}{qv} = \frac{8.0 \times 10^{-14}\,\text{N}}{(1.6 \times 10^{-19}\,\text{C})(5.0 \times 10^6\,\text{m/s})} = 0.10\,\text{T}.$$

FIGURE 27–16 Example 27–5.

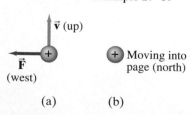

(a)   (b)

EXAMPLE 27–6 ESTIMATE Magnetic force on ions during a nerve pulse. Estimate the magnetic force due to the Earth's magnetic field on ions crossing a cell membrane during an action potential. Assume the speed of the ions is 10^{-2} m/s (Section 25–10).

APPROACH Using $F = qvB$, set the magnetic field of the Earth to be roughly $B \approx 10^{-4}$ T, and the charge $q \approx e \approx 10^{-19}$ C.

SOLUTION $F \approx (10^{-19} \text{C})(10^{-2} \text{m/s})(10^{-4} \text{T}) = 10^{-25}$ N

NOTE This is an extremely small force. Yet it is thought migrating animals do somehow detect the Earth's magnetic field, and this is an area of active research.

The path of a charged particle moving in a plane perpendicular to a uniform magnetic field is a circle as we shall now show. In Fig. 27–17 the magnetic field is directed *into* the paper, as represented by ×'s. An electron at point P is moving to the right, and the force on it at this point is downward as shown (use the right-hand rule and reverse the direction for negative charge). The electron is thus deflected toward the page bottom. A moment later, say, when it reaches point Q, the force is still perpendicular to the velocity and is in the direction shown. Because the force is always perpendicular to $\vec{v}$, the magnitude of $\vec{v}$ does not change—the electron moves at constant speed. We saw in Chapter 5 that if the force on a particle is always perpendicular to its velocity $\vec{v}$, the particle moves in a circle and has a centripetal acceleration $a = v^2/r$ (Eq. 5–1). Thus a charged particle moves in a circular path with constant centripetal acceleration in a uniform magnetic field (see Example 27–7). The electron moves clockwise in Fig. 27–17. A positive particle in this field would feel a force in the opposite direction and would thus move counterclockwise.

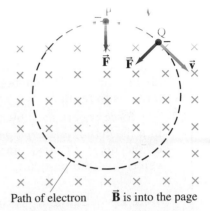

Path of electron $\vec{B}$ is into the page

FIGURE 27–17 Force exerted by a uniform magnetic field on a moving charged particle (in this case, an electron) produces a circular path.

EXAMPLE 27–7 Electron's path in a uniform magnetic field. An electron travels at 2.0×10^7 m/s in a plane perpendicular to a uniform 0.010-T magnetic field. Describe its path quantitatively.

APPROACH The electron moves at speed v in a curved path and so must have a centripetal acceleration $a = v^2/r$ (Eq. 5–1). We find the radius of curvature using Newton's second law. The force is given by Eq. 27–5b with $\sin\theta = 1$: $F = qvB$.

SOLUTION We insert F and a into Newton's second law:

$$\Sigma F = ma$$
$$qvB = \frac{mv^2}{r}.$$

We solve for r and find

$$r = \frac{mv}{qB}.$$

Since $\vec{F}$ is perpendicular to $\vec{v}$, the magnitude of $\vec{v}$ doesn't change. From this equation we see that if $\vec{B}$ = constant, then r = constant, and the curve must be a circle as we claimed above. To get r we put in the numbers:

$$r = \frac{(9.1 \times 10^{-31} \text{kg})(2.0 \times 10^7 \text{m/s})}{(1.6 \times 10^{-19} \text{C})(0.010 \text{T})} = 1.1 \times 10^{-2} \text{m} = 1.1 \text{cm}.$$

NOTE See Fig. 27–18.

FIGURE 27–18 The blue ring inside the glass tube is the glow of a beam of electrons that ionize the gas molecules. The red coils of current-carrying wire produce a nearly uniform magnetic field, illustrating the circular path of charged particles in a uniform magnetic field.

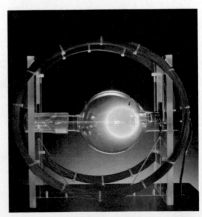

The time T required for a particle of charge q moving with constant speed v to make one circular revolution in a uniform magnetic field $\vec{B}$ ($\perp \vec{v}$) is $T = 2\pi r/v$, where $2\pi r$ is the circumference of its circular path. From Example 27–7, $r = mv/qB$, so

$$T = \frac{2\pi m}{qB}.$$

Since T is the period of rotation, the frequency of rotation is

$$f = \frac{1}{T} = \frac{qB}{2\pi m}. \tag{27–6}$$

This is often called the **cyclotron frequency** of a particle in a field because this is the frequency at which particles revolve in a cyclotron (see Problem 66).

CONCEPTUAL EXAMPLE 27–8 **Stopping charged particles.** Can a magnetic field be used to stop a single charged particle, as an electric field can?

RESPONSE No, because the force is always *perpendicular* to the velocity of the particle and thus cannot change the magnitude of its velocity. It also means the magnetic force cannot do work on the particle and so cannot change the kinetic energy of the particle.

PROBLEM SOLVING

Magnetic Fields

Magnetic fields are somewhat analogous to the electric fields of Chapter 21, but there are several important differences to recall:

1. The force experienced by a charged particle moving in a magnetic field is *perpendicular* to the direction of the magnetic field (and to the direction of the velocity of the particle), whereas the force exerted by an electric field is *parallel* to the direction of the field (and unaffected by the velocity of the particle).

2. The *right-hand rule*, in its different forms, is intended to help you determine the directions of magnetic field, and the forces they exert, and/or the directions of electric current or charged particle velocity. The right-hand rules (Table 27–1) are designed to deal with the "perpendicular" nature of these quantities.

TABLE 27–1 Summary of Right-hand Rules (= RHR)

Physical Situation	Example	How to Orient Right Hand	Result
1. Magnetic field produced by current (RHR-1)	I $\vec{B}$ Fig. 27–8c	Wrap fingers around wire with thumb pointing in direction of current I	Fingers point in direction of $\vec{B}$
2. Force on electric current I due to magnetic field (RHR-2)	$\vec{F}$ $\vec{I}$ $\vec{B}$ Fig. 27–11c	Fingers point straight along current I, then bend along magnetic field $\vec{B}$	Thumb points in direction of the force $\vec{F}$
3. Force on electric charge $+q$ due to magnetic field (RHR-3)	$\vec{F}$ $\vec{v}$ $\vec{B}$ Fig. 27–15	Fingers point along particle's velocity $\vec{v}$, then along $\vec{B}$	Thumb points in direction of the force $\vec{F}$

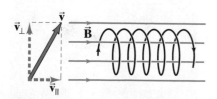

FIGURE 27–19 Example 27–9.

CONCEPTUAL EXAMPLE 27–9 **A helical path.** What is the path of a charged particle in a uniform magnetic field if its velocity is *not* perpendicular to the magnetic field?

RESPONSE The velocity vector can be broken down into components parallel and perpendicular to the field. The velocity component parallel to the field lines experiences no force ($\theta = 0$), so this component remains constant. The velocity component perpendicular to the field results in circular motion about the field lines. Putting these two motions together produces a helical (spiral) motion around the field lines as shown in Fig. 27–19.

EXERCISE F What is the sign of the charge in Fig. 27–19? How would you modify the drawing if the sign were reversed?

*Aurora Borealis

Charged ions approach the Earth from the Sun (the "solar wind") and enter the atmosphere mainly near the poles, sometimes causing a phenomenon called the **aurora borealis** or "northern lights" in northern latitudes. To see why, consider Example 27–9 and Fig. 27–20 (see also Fig. 27–19). In Fig. 27–20 we imagine a stream of charged particles approaching the Earth. The velocity component *perpendicular* to the field for each particle becomes a circular orbit around the field lines. Whereas the velocity component *parallel* to the field carries the particle along the field lines toward the poles. As a particle approaches the N pole, the magnetic field is stronger and the radius of the helical path becomes smaller.

A high concentration of charged particles ionizes the air, and as the electrons recombine with atoms, light is emitted (Chapter 37) which is the aurora. Auroras are especially spectacular during periods of high sunspot activity when the solar wind brings more charged particles toward Earth.

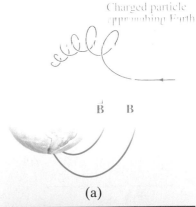

(a)

Lorentz Equation

If a particle of charge q moves with velocity $\vec{v}$ in the presence of both a magnetic field $\vec{B}$ and an electric field $\vec{E}$, it will feel a force

$$\vec{F} = q(\vec{E} + \vec{v} \times \vec{B}) \qquad (27\text{–}7)$$

where we have made use of Eqs. 21–3 and 27–5a. Equation 27–7 is often called the **Lorentz equation** and is considered one of the basic equations in physics.

(b)

FIGURE 27–20 (a) Diagram showing a negatively charged particle that approaches the Earth and is "captured" by the magnetic field of the Earth. Such particles follow the field lines toward the poles as shown. (b) Photo of aurora borealis (here, in Kansas, where it is a rare sight).

| CONCEPTUAL EXAMPLE 27–10 | **Velocity selector, or filter: Crossed $\vec{E}$ and $\vec{B}$ fields.**

Some electronic devices and experiments need a beam of charged particles all moving at nearly the same velocity. This can be achieved using both a uniform electric field and a uniform magnetic field, arranged so they are at right angles to each other. As shown in Fig. 27–21a, particles of charge q pass through slit S_1 and enter the region where $\vec{B}$ points into the page and $\vec{E}$ points down from the positive plate toward the negative plate. If the particles enter with different velocities, show how this device "selects" a particular velocity, and determine what this velocity is.

RESPONSE After passing through slit S_1, each particle is subject to two forces as shown in Fig. 27–21b. If q is positive, the magnetic force is upwards and the electric force downwards. (Vice versa if q is negative.) The exit slit, S_2, is assumed to be directly in line with S_1 and the particles' velocity $\vec{v}$. Depending on the magnitude of $\vec{v}$, some particles will be bent upwards and some downwards. The only ones to make it through the slit S_2 will be those for which the net force is zero: $\Sigma F = qvB - qE = 0$. Hence this device selects particles whose velocity is

$$v = \frac{E}{B}. \qquad (27\text{–}8)$$

This result does not depend on the sign of the charge q.

FIGURE 27–21 A velocity selector: if $v = E/B$, the particles passing through S_1 make it through S_2.

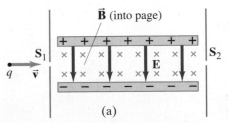

(a)

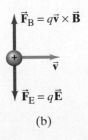

(b)

EXERCISE G A particle in a velocity selector as diagrammed in Fig. 27–21 hits below the exit hole, S_2. This means that the particle (*a*) is going faster than the selected speed; (*b*) is going slower than the selected speed; (*c*) answer *a* is true if $q > 0$, *b* is true if $q < 0$; (*d*) answer *a* is true if $q < 0$, *b* is true if $q > 0$.

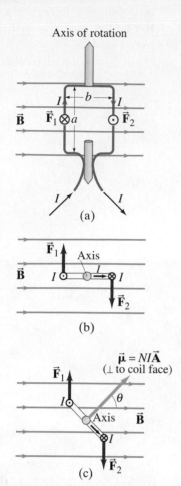

FIGURE 27–22 Calculating the torque on a current loop in a magnetic field $\vec{\mathbf{B}}$. (a) Loop face parallel to $\vec{\mathbf{B}}$ field lines; (b) top view; (c) loop makes an angle to $\vec{\mathbf{B}}$, reducing the torque since the lever arm is reduced.

27–5 Torque on a Current Loop; Magnetic Dipole Moment

When an electric current flows in a closed loop of wire placed in an external magnetic field, as shown in Fig. 27–22, the magnetic force on the current can produce a torque. This is the principle behind a number of important practical devices, including motors and analog voltmeters and ammeters, which we discuss in the next Section. The interaction between a current and a magnetic field is important in other areas as well, including atomic physics.

Current flows through the rectangular loop in Fig. 27–22a, whose face we assume is parallel to $\vec{\mathbf{B}}$. $\vec{\mathbf{B}}$ exerts no force and no torque on the horizontal segments of wire because they are parallel to the field and $\sin\theta = 0$ in Eq. 27–1. But the magnetic field does exert a force on each of the vertical sections of wire as shown, $\vec{\mathbf{F}}_1$ and $\vec{\mathbf{F}}_2$ (see also top view, Fig. 27–22b). By right-hand-rule 2 (Fig. 27–11c or Table 27–1) the direction of the force on the upward current on the left is in the opposite direction from the equal magnitude force $\vec{\mathbf{F}}_2$ on the downward current on the right. These forces give rise to a net torque that acts to rotate the coil about its vertical axis.

Let us calculate the magnitude of this torque. From Eq. 27–2 (current $\perp \vec{\mathbf{B}}$), the force $F = IaB$, where a is the length of the vertical arm of the coil. The lever arm for each force is $b/2$, where b is the width of the coil and the "axis" is at the midpoint. The torques produced by $\vec{\mathbf{F}}_1$ and $\vec{\mathbf{F}}_2$ act in the same direction, so the total torque is the sum of the two torques:

$$\tau = IaB\frac{b}{2} + IaB\frac{b}{2} = IabB = IAB,$$

where $A = ab$ is the area of the coil. If the coil consists of N loops of wire, the current is then NI, so the torque becomes

$$\tau = NIAB.$$

If the coil makes an angle θ with the magnetic field, as shown in Fig. 27–22c, the forces are unchanged, but each lever arm is reduced from $\frac{1}{2}b$ to $\frac{1}{2}b\sin\theta$. Note that the angle θ is taken to be the angle between $\vec{\mathbf{B}}$ and the perpendicular to the face of the coil, Fig. 27–22c. So the torque becomes

$$\tau = NIAB\sin\theta. \qquad \textbf{(27–9)}$$

This formula, derived here for a rectangular coil, is valid for any shape of flat coil.

The quantity NIA is called the **magnetic dipole moment** of the coil and is considered a vector:

$$\vec{\boldsymbol{\mu}} = NI\vec{\mathbf{A}}, \qquad \textbf{(27–10)}$$

where the direction of $\vec{\mathbf{A}}$ (and therefore of $\vec{\boldsymbol{\mu}}$) is *perpendicular* to the plane of the coil (the green arrow in Fig. 27–22c) consistent with the right-hand rule (cup your right hand so your fingers wrap around the loop in the direction of current flow, then your thumb points in the direction of $\vec{\boldsymbol{\mu}}$ and $\vec{\mathbf{A}}$). With this definition of $\vec{\boldsymbol{\mu}}$, we can rewrite Eq. 27–9 in vector form:

$$\vec{\boldsymbol{\tau}} = NI\vec{\mathbf{A}} \times \vec{\mathbf{B}}$$

or

$$\vec{\boldsymbol{\tau}} = \vec{\boldsymbol{\mu}} \times \vec{\mathbf{B}}, \qquad \textbf{(27–11)}$$

which gives the correct magnitude and direction for the torque $\vec{\boldsymbol{\tau}}$.

Equation 27–11 has the same form as Eq. 21–9b for an electric dipole (with electric dipole moment $\vec{\mathbf{p}}$) in an electric field $\vec{\mathbf{E}}$, which is $\vec{\boldsymbol{\tau}} = \vec{\mathbf{p}} \times \vec{\mathbf{E}}$. And just as an electric dipole has potential energy given by $U = -\vec{\mathbf{p}} \cdot \vec{\mathbf{E}}$ when in an electric field, we expect a similar form for a magnetic dipole in a magnetic field. In order to rotate a current loop (Fig. 27–22) so as to increase θ, we must do work against the torque due to the magnetic field.

Hence the potential energy depends on angle (see Eq. 10–22, the work-energy principle for rotational motion) as

$$U = \int \tau \, d\theta = \int NIAB \sin\theta \, d\theta = -\mu B \cos\theta + C.$$

If we choose $U = 0$ at $\theta = \pi/2$, then the arbitrary constant C is zero and the potential energy is

$$U = -\mu B \cos\theta = -\vec{\mu} \cdot \vec{B}. \tag{27–12}$$

as expected (compare Eq. 21–10). Bar magnets and compass needles as well as current loops, can be considered as magnetic dipoles. Note the striking similarities of the fields produced by a bar magnet and a current loop, Figs. 27–4b and 27–9.

EXAMPLE 27–11 Torque on a coil. A circular coil of wire has a diameter of 20.0 cm and contains 10 loops. The current in each loop is 3.00 A, and the coil is placed in a 2.00-T external magnetic field. Determine the maximum and minimum torque exerted on the coil by the field.

APPROACH Equation 27–9 is valid for any shape of coil, including circular loops. Maximum and minimum torque are determined by the angle θ the coil makes with the magnetic field.

SOLUTION The area of one loop of the coil is

$$A = \pi r^2 = \pi(0.100\,\text{m})^2 = 3.14 \times 10^{-2}\,\text{m}^2.$$

The maximum torque occurs when the coil's face is parallel to the magnetic field, so $\theta = 90°$ in Fig. 27–22c, and $\sin\theta = 1$ in Eq. 27–9:

$$\tau = NIAB \sin\theta = (10)(3.00\,\text{A})(3.14 \times 10^{-2}\,\text{m}^2)(2.00\,\text{T})(1) = 1.88\,\text{N·m}.$$

The minimum torque occurs if $\sin\theta = 0$, for which $\theta = 0°$, and then $\tau = 0$ from Eq. 27–9.

NOTE If the coil is free to turn, it will rotate toward the orientation with $\theta = 0°$.

EXAMPLE 27–12 Magnetic moment of a hydrogen atom. Determine the magnetic dipole moment of the electron orbiting the proton of a hydrogen atom at a given instant, assuming (in the Bohr model) it is in its ground state with a circular orbit of radius $0.529 \times 10^{-10}\,\text{m}$. [This is a very rough picture of atomic structure, but nonetheless gives an accurate result.]

APPROACH We start by setting the electrostatic force on the electron due to the proton equal to $ma = mv^2/r$ since the electron's acceleration is centripetal.

SOLUTION The electron is held in its orbit by the coulomb force, so Newton's second law, $F = ma$, gives

$$\frac{e^2}{4\pi\epsilon_0 r^2} = \frac{mv^2}{r};$$

so

$$v = \sqrt{\frac{e^2}{4\pi\epsilon_0 mr}}$$

$$= \sqrt{\frac{(8.99 \times 10^9\,\text{N·m}^2/\text{C}^2)(1.60 \times 10^{-19}\,\text{C})^2}{(9.11 \times 10^{-31}\,\text{kg})(0.529 \times 10^{-10}\,\text{m})}} = 2.19 \times 10^6\,\text{m/s}.$$

Since current is the electric charge that passes a given point per unit time, the revolving electron is equivalent to a current

$$I = \frac{e}{T} = \frac{ev}{2\pi r},$$

where $T = 2\pi r/v$ is the time required for one orbit. Since the area of the orbit is $A = \pi r^2$, the magnetic dipole moment is

$$\mu = IA = \frac{ev}{2\pi r}(\pi r^2) = \tfrac{1}{2}evr$$

$$= \tfrac{1}{2}(1.60 \times 10^{-19}\,\text{C})(2.19 \times 10^6\,\text{m/s})(0.529 \times 10^{-10}\,\text{m}) = 9.27 \times 10^{-24}\,\text{A·m}^2,$$

or $9.27 \times 10^{-24}\,\text{J/T}$.

*27–6 Applications: Motors, Loudspeakers, Galvanometers

*Electric Motors

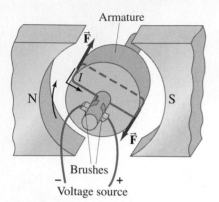

FIGURE 27–23 Diagram of a simple dc motor.

An **electric motor** changes electric energy into (rotational) mechanical energy. A motor works on the principle that a torque is exerted on a coil of current-carrying wire suspended in the magnetic field of a magnet, described in Section 27–5. The coil is mounted on a large cylinder called the **rotor** or **armature**, Fig. 27–23. The curved pole pieces and iron armature tend to concentrate the magnetic field lines so the field lines are parallel to the face of the wire (see Fig. 27–28 which shows the field lines). Actually, there are several coils, although only one is indicated in Fig. 27–23. The armature is mounted on a shaft or axle. When the armature is in the position shown in Fig. 27–23, the magnetic field exerts forces on the current in the loop as shown (perpendicular to $\vec{B}$ and to the current direction). However, when the coil, which is rotating clockwise in Fig. 27–23, passes beyond the vertical position, the forces would then act to return the coil back to vertical if the current remained the same. But if the current could somehow be reversed at that critical moment, the forces would reverse, and the coil would continue rotating in the same direction. Thus, alternation of the current is necessary if a motor is to turn continuously in one direction. This can be achieved in a **dc motor** with the use of **commutators** and **brushes**: as shown in Fig. 27–24, input current passes through stationary brushes that rub against the conducting commutators mounted on the motor shaft. At every half revolution, each commutator changes its connection over to the other brush. Thus the current in the coil reverses every half revolution as required for continuous rotation.

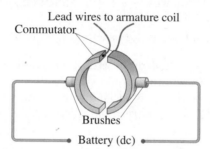

FIGURE 27–24 The commutator-brush arrangement in a dc motor ensures alternation of the current in the armature to keep rotation continuous. The commutators are attached to the motor shaft and turn with it, whereas the brushes remain stationary.

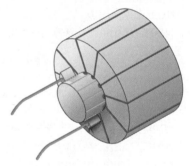

FIGURE 27–25 Motor with many windings.

Most motors contain several coils, called *windings*, each located in a different place on the armature, Fig. 27–25. Current flows through each coil only during a small part of a revolution, at the time when its orientation results in the maximum torque. In this way, a motor produces a much steadier torque than can be obtained from a single coil.

An **ac motor**, with ac current as input, can work without commutators since the current itself alternates. Many motors use wire coils to produce the magnetic field (electromagnets) instead of a permanent magnet. Indeed the design of most motors is more complex than described here, but the general principles remain the same.

*Loudspeakers

FIGURE 27–26 Loudspeaker.

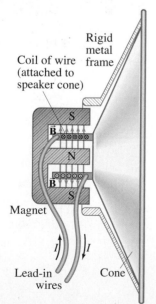

A **loudspeaker** also works on the principle that a magnet exerts a force on a current-carrying wire. The electrical output of a stereo or TV set is connected to the wire leads of the speaker. The speaker leads are connected internally to a coil of wire, which is itself attached to the speaker cone, Fig. 27–26. The speaker cone is usually made of stiffened cardboard and is mounted so that it can move back and forth freely. A permanent magnet is mounted directly in line with the coil of wire. When the alternating current of an audio signal flows through the wire coil, which is free to move within the magnet, the coil experiences a force due to the magnetic field of the magnet. (The force is to the right at the instant shown in Fig. 27–26.)

As the current alternates at the frequency of the audio signal, the coil and attached speaker cone move back and forth at the same frequency, causing alternate compressions and rarefactions of the adjacent air, and sound waves are produced. A speaker thus changes electrical energy into sound energy, and the frequencies and intensities of the emitted sound waves can be an accurate reproduction of the electrical input.

*Galvanometer

The basic component of analog meters (those with pointer and dial), including analog ammeters, voltmeters, and ohmmeters, is a galvanometer. We have already seen how these meters are designed (Section 26–7), and now we can examine how the crucial element, a galvanometer, works. As shown in Fig. 27–27, a **galvanometer** consists of a coil of wire (with attached pointer) suspended in the magnetic field of a permanent magnet. When current flows through the loop of wire, the magnetic field exerts a torque on the loop, as given by Eq. 27–9,

$$\tau = NIAB \sin \theta.$$

This torque is opposed by a spring which exerts a torque τ_s approximately proportional to the angle ϕ through which it is turned (Hooke's law). That is,

$$\tau_s = k\phi,$$

where k is the stiffness constant of the spring. The coil and attached pointer rotate to the angle where the torques balance. When the needle is in equilibrium at rest, the torques are equal: $k\phi = NIAB \sin \theta$, or

$$\phi = \frac{NIAB_r \sin \theta}{k}.$$

The deflection of the pointer, ϕ, is directly proportional to the current I flowing in the coil, but also depends on the angle θ the coil makes with $\vec{B}$. For a useful meter we need ϕ to depend only on the current I, independent of θ. To solve this problem, magnets with curved pole pieces are used and the galvanometer coil is wrapped around a cylindrical iron core as shown in Fig. 27–28. The iron tends to concentrate the magnetic field lines so that $\vec{B}$ always points parallel to the face of the coil at the wire outside the core. The force is then always perpendicular to the face of the coil, and the torque will not vary with angle. Thus ϕ will be proportional to I as required.

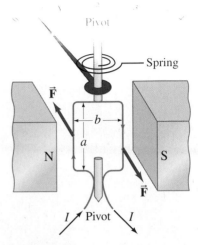

FIGURE 27–27 Galvanometer.

FIGURE 27–28 Galvanometer coil wrapped on an iron core.

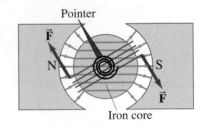

27–7 Discovery and Properties of the Electron

The electron plays a basic role in our understanding of electricity and magnetism today. But its existence was not suggested until the 1890s. We discuss it here because magnetic fields were crucial for measuring its properties.

Toward the end of the nineteenth century, studies were being done on the discharge of electricity through rarefied gases. One apparatus, diagrammed in Fig. 27–29, was a glass tube fitted with electrodes and evacuated so only a small amount of gas remained inside. When a very high voltage was applied to the electrodes, a dark space seemed to extend outward from the cathode (negative electrode) toward the opposite end of the tube; and that far end of the tube would glow. If one or more screens containing a small hole was inserted as shown, the glow was restricted to a tiny spot on the end of the tube. It seemed as though something being emitted by the cathode traveled to the opposite end of the tube. These "somethings" were named **cathode rays**.

There was much discussion at the time about what these rays might be. Some scientists thought they might resemble light. But the observation that the bright spot at the end of the tube could be deflected to one side by an electric or magnetic field suggested that cathode rays could be charged particles; and the direction of the deflection was consistent with a negative charge. Furthermore, if the tube contained certain types of rarefied gas, the path of the cathode rays was made visible by a slight glow.

FIGURE 27–29 Discharge tube. In some models, one of the screens is the anode (positive plate).

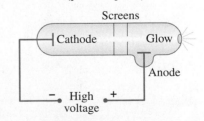

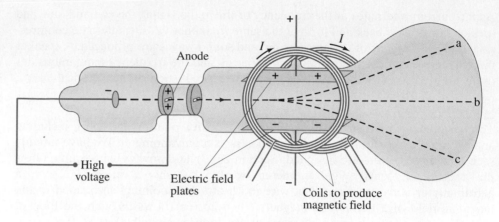

FIGURE 27–30 Cathode rays deflected by electric and magnetic fields.

Estimates of the charge e of the (assumed) cathode-ray particles, as well as of their charge-to-mass ratio e/m, had been made by 1897. But in that year, J. J. Thomson (1856–1940) was able to measure e/m directly, using the apparatus shown in Fig. 27–30. Cathode rays are accelerated by a high voltage and then pass between a pair of parallel plates built into the tube. The voltage applied to the plates produces an electric field, and a pair of coils produces a magnetic field. When only the electric field is present, say with the upper plate positive, the cathode rays are deflected upward as in path a in Fig. 27–30. If only a magnetic field exists, say inward, the rays are deflected downward along path c. These observations are just what is expected for a negatively charged particle. The force on the rays due to the magnetic field is $F = evB$, where e is the charge and v is the velocity of the cathode rays. In the absence of an electric field, the rays are bent into a curved path, so we have, from $F = ma$,

$$evB = m\frac{v^2}{r},$$

and thus

$$\frac{e}{m} = \frac{v}{Br}.$$

The radius of curvature r can be measured and so can B. The velocity v can be found by applying an electric field in addition to the magnetic field. The electric field E is adjusted so that the cathode rays are undeflected and follow path b in Fig. 27–30. This is just like the velocity selector of Example 27–10 where the force due to the electric field, $F = eE$, is balanced by the force due to the magnetic field, $F = evB$. Thus $eE = evB$ and $v = E/B$. Combining this with the above equation we have

$$\frac{e}{m} = \frac{E}{B^2 r}. \tag{27–13}$$

The quantities on the right side can all be measured so that although e and m could not be determined separately, the ratio e/m could be determined. The accepted value today is $e/m = 1.76 \times 10^{11}$ C/kg. Cathode rays soon came to be called **electrons**.

It is worth noting that the "discovery" of the electron, like many others in science, is not quite so obvious as discovering gold or oil. Should the discovery of the electron be credited to the person who first saw a glow in the tube? Or to the person who first called them cathode rays? Perhaps neither one, for they had no conception of the electron as we know it today. In fact, the credit for the discovery is generally given to Thomson, but not because he was the first to see the glow in the tube. Rather it is because he believed that this phenomenon was due to tiny negatively charged particles and made careful measurements on them. Furthermore he argued that these particles were constituents of atoms, and not ions or atoms themselves as many thought, and he developed an electron theory of matter. His view is close to what we accept today, and this is why Thomson is credited with the "discovery." Note, however, that neither he nor anyone else ever actually saw an electron itself. We discuss this briefly, for it illustrates the fact that discovery in science is not always a clear-cut matter. In fact some philosophers of science think the word "discovery" is often not appropriate, such as in this case.

Thomson believed that an electron was not an atom, but rather a constituent, or part, of an atom. Convincing evidence for this came soon with the determination of the charge and the mass of the cathode rays. Thomson's student J. S. Townsend made the first direct (but rough) measurement of e in 1897. But it was the more refined **oil-drop experiment** of Robert A. Millikan (1868–1953) that yielded a precise value for the charge on the electron and showed that charge comes in discrete amounts. In this experiment, tiny droplets of mineral oil carrying an electric charge were allowed to fall under gravity between two parallel plates, Fig. 27–31. The electric field E between the plates was adjusted until the drop was suspended in midair. The downward pull of gravity, mg, was then just balanced by the upward force due to the electric field. Thus $qE = mg$, so the charge $q = mg/E$. The mass of the droplet was determined by measuring its terminal velocity in the absence of the electric field. Sometimes the drop was charged negatively, and sometimes positively, suggesting that the drop had acquired or lost electrons (by friction, leaving the atomizer). Millikan's painstaking observations and analysis presented convincing evidence that any charge was an integral multiple of a smallest charge, e, that was ascribed to the electron, and that the value of e was 1.6×10^{-19} C. This value of e, combined with the measurement of e/m, gives the mass of the electron to be $(1.6 \times 10^{-19}\,\text{C})/(1.76 \times 10^{11}\,\text{C/kg}) = 9.1 \times 10^{-31}$ kg. This mass is less than a thousandth the mass of the smallest atom, and thus confirmed the idea that the electron is only a part of an atom. The accepted value today for the mass of the electron is $m_e = 9.11 \times 10^{-31}$ kg.

FIGURE 27–31 Millikan's oil-drop experiment.

CRT, Revisited

The cathode ray tube (CRT), which can serve as the picture tube of TV sets, oscilloscopes, and computer monitors, was discussed in Chapter 23. There, in Fig. 23–22, we saw a design using electric deflection plates to maneuver the electron beam. Many CRTs, however, make use of the magnetic field produced by coils to maneuver the electron beam. They operate much like the coils shown in Fig. 27–30.

27–8 The Hall Effect

When a current-carrying conductor is held fixed in a magnetic field, the field exerts a sideways force on the charges moving in the conductor. For example, if electrons move to the right in the rectangular conductor shown in Fig. 27–32a, the inward magnetic field will exert a downward force on the electrons $\vec{\mathbf{F}}_B = -e\vec{\mathbf{v}}_d \times \vec{\mathbf{B}}$, where $\vec{\mathbf{v}}_d$ is the drift velocity of the electrons (Section 25–8). Thus the electrons will tend to move nearer to face D than face C. There will thus be a potential difference between faces C and D of the conductor. This potential difference builds up until the electric field $\vec{\mathbf{E}}_H$ that it produces exerts a force, $e\vec{\mathbf{E}}_H$, on the moving charges that is equal and opposite to the magnetic force. This effect is called the **Hall effect** after E. H. Hall, who discovered it in 1879. The difference of potential produced is called the **Hall emf**.

The electric field due to the separation of charge is called the *Hall field*, $\vec{\mathbf{E}}_H$, and points downward in Fig. 27–32a, as shown. In equilibrium, the force due to this electric field is balanced by the magnetic force $ev_d B$, so

$$eE_H = ev_d B.$$

Hence $E_H = v_d B$. The Hall emf is then (Eq. 23–4b, assuming the conductor is long and thin so E_H is uniform)

$$\mathscr{E}_H = E_H d = v_d B d, \tag{27–14}$$

where d is the width of the conductor.

A current of negative charges moving to the right is equivalent to positive charges moving to the left, at least for most purposes. But the Hall effect can distinguish these two. As can be seen in Fig. 27–32b, positive particles moving to the left are deflected downward, so that the bottom surface is positive relative to the top surface. This is the reverse of part (a). Indeed, the direction of the emf in the Hall effect first revealed that it is negative particles that move in metal conductors.

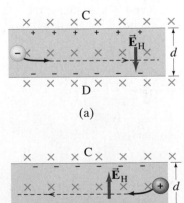

FIGURE 27–32 The Hall effect. (a) Negative charges moving to the right as the current. (b) Positive charges moving to the left as the current.

(a)

(b)

The magnitude of the Hall emf is proportional to the strength of the magnetic field. The Hall effect can thus be used to measure magnetic field strengths. First the conductor, called a *Hall probe*, is calibrated with known magnetic fields. Then, for the same current, its emf output will be a measure of B. Hall probes can be made very small and are convenient and accurate to use.

The Hall effect can also be used to measure the drift velocity of charge carriers when the external magnetic field B is known. Such a measurement also allows us to determine the density of charge carriers in the material.

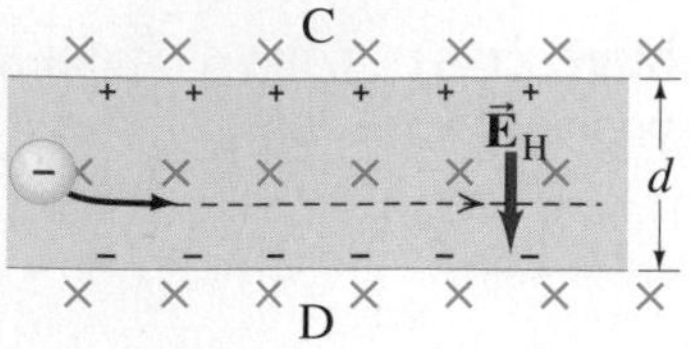

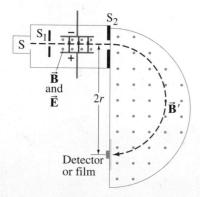

FIGURE 27–32a (Repeated here for Example 27–13.)

EXAMPLE 27–13 **Drift velocity using the Hall effect.** A long copper strip 1.8 cm wide and 1.0 mm thick is placed in a 1.2-T magnetic field as in Fig. 27–32a. When a steady current of 15 A passes through it, the Hall emf is measured to be $1.02\,\mu\text{V}$. Determine the drift velocity of the electrons and the density of free (conducting) electrons (number per unit volume) in the copper.

APPROACH We use Eq. 27–14 to obtain the drift velocity, and Eq. 25–13 of Chapter 25 to find the density of conducting electrons.

SOLUTION The drift velocity (Eq. 27–14) is

$$v_{\text{d}} = \frac{\mathscr{E}_{\text{H}}}{Bd} = \frac{1.02 \times 10^{-6}\,\text{V}}{(1.2\,\text{T})(1.8 \times 10^{-2}\,\text{m})} = 4.7 \times 10^{-5}\,\text{m/s}.$$

The density of charge carriers n is obtained from Eq. 25–13, $I = nev_{\text{d}}A$, where A is the cross-sectional area through which the current I flows. Then

$$n = \frac{I}{ev_{\text{d}}A} = \frac{15\,\text{A}}{(1.6 \times 10^{-19}\,\text{C})(4.7 \times 10^{-5}\,\text{m/s})(1.8 \times 10^{-2}\,\text{m})(1.0 \times 10^{-3}\,\text{m})}$$

$$= 11 \times 10^{28}\,\text{m}^{-3}.$$

This value for the density of free electrons in copper, $n = 11 \times 10^{28}$ per m^3, is the experimentally measured value. It represents *more* than one free electron per atom, which as we saw in Example 25–14 is $8.4 \times 10^{28}\,\text{m}^{-3}$.

*27–9 Mass Spectrometer

FIGURE 27–33 Bainbridge-type mass spectrometer. The magnetic fields B and B' point out of the paper (indicated by the dots), for positive ions.

A **mass spectrometer** is a device to measure masses of atoms. It is used today not only in physics but also in chemistry, geology, and medicine, often to identify atoms (and their concentration) in given samples. As shown in Fig. 27–33, ions are produced by heating, or by an electric current, in the source or sample S. The particles, of mass m and electric charge q, pass through slit S_1 and enter crossed electric and magnetic fields. Ions follow a straight-line path in this "velocity selector" (as in Example 27–10) if the electric force qE is balanced by the magnetic force qvB: that is, if $qE = qvB$, or $v = E/B$. Thus only those ions whose speed is $v = E/B$ will pass through undeflected and emerge through slit S_2. In the semicircular region, after S_2, there is only a magnetic field, B', so the ions follow a circular path. The radius of the circular path is found from their mark on film (or detectors) if B' is fixed; or else r is fixed by the position of a detector and B' is varied until detection occurs. Newton's second law, $\Sigma F = ma$, applied to an ion moving in a circle under the influence only of the magnetic field B' gives $qvB' = mv^2/r$. Since $v = E/B$, we have

$$m = \frac{qB'r}{v} = \frac{qBB'r}{E}.$$

All the quantities on the right side are known or can be measured, and thus m can be determined.

Historically, the masses of many atoms were measured this way. When a pure substance was used, it was sometimes found that two or more closely spaced marks would appear on the film. For example, neon produced two marks whose radii corresponded to atoms of mass 20 and 22 atomic mass units (u). Impurities were ruled out and it was concluded that there must be two types of neon with different masses. These different forms were called **isotopes**. It was soon found that most elements are mixtures of isotopes, and the difference in mass is due to different numbers of neutrons (discussed in Chapter 41).

EXAMPLE 27–14 **Mass spectrometry.** Carbon atoms of atomic mass 12.0 u are found to be mixed with another, unknown, element. In a mass spectrometer with fixed B', the carbon traverses a path of radius 22.4 cm and the unknown's path has a 26.2-cm radius. What is the unknown element? Assume the ions of both elements have the same charge.

APPROACH The carbon and unknown atoms pass through the same electric and magnetic fields. Hence their masses are proportional to the radius of their respective paths (see equation on previous page).

SOLUTION We write a ratio for the masses, using the equation at the bottom of the previous page:

$$\frac{m_x}{m_C} = \frac{qBB'r_x/E}{qBB'r_C/E} = \frac{26.2 \text{ cm}}{22.4 \text{ cm}} = 1.17.$$

Thus $m_x = 1.17 \times 12.0 \text{ u} = 14.0 \text{ u}$. The other element is probably nitrogen (see the Periodic Table, inside the back cover).

NOTE The unknown could also be an isotope such as carbon-14 ($^{14}_{6}\text{C}$). See Appendix F. Further physical or chemical analysis would be needed.

Summary

A magnet has two **poles**, north and south. The north pole is that end which points toward geographic north when the magnet is freely suspended. Like poles of two magnets repel each other, whereas unlike poles attract.

We can picture that a **magnetic field** surrounds every magnet. The SI unit for magnetic field is the **tesla** (T).

Electric currents produce magnetic fields. For example, the lines of magnetic field due to a current in a straight wire form circles around the wire, and the field exerts a force on magnets (or currents) near it.

A magnetic field exerts a force on an electric current. The force on an infinitesimal length of wire $d\vec{\boldsymbol{\ell}}$ carrying a current I in a magnetic field $\vec{\mathbf{B}}$ is

$$d\vec{\mathbf{F}} = I \, d\vec{\boldsymbol{\ell}} \times \vec{\mathbf{B}}. \tag{27-4}$$

If the field $\vec{\mathbf{B}}$ is uniform over a straight length $\vec{\boldsymbol{\ell}}$ of wire, then the force is

$$\vec{\mathbf{F}} = I\vec{\boldsymbol{\ell}} \times \vec{\mathbf{B}} \tag{27-3}$$

which has magnitude

$$F = I\ell B \sin\theta \tag{27-1}$$

where θ is the angle between magnetic field $\vec{\mathbf{B}}$ and the current direction. The direction of the force is perpendicular to the wire and to the magnetic field, and is given by the right-hand rule. This relation serves as the definition of magnetic field $\vec{\mathbf{B}}$.

Similarly, a magnetic field $\vec{\mathbf{B}}$ exerts a force on a charge q moving with velocity $\vec{\mathbf{v}}$ given by

$$\vec{\mathbf{F}} = q\vec{\mathbf{v}} \times \vec{\mathbf{B}}. \tag{27-5a}$$

The magnitude of the force is

$$F = qvB \sin\theta, \tag{27-5b}$$

where θ is the angle between $\vec{\mathbf{v}}$ and $\vec{\mathbf{B}}$.

The path of a charged particle moving perpendicular to a uniform magnetic field is a circle.

If both electric and magnetic fields ($\vec{\mathbf{E}}$ and $\vec{\mathbf{B}}$) are present, the force on a charge q moving with velocity $\vec{\mathbf{v}}$ is

$$\vec{\mathbf{F}} = q\vec{\mathbf{E}} + q\vec{\mathbf{v}} \times \vec{\mathbf{B}}. \tag{27-7}$$

The torque on a current loop in a magnetic field $\vec{\mathbf{B}}$ is

$$\vec{\boldsymbol{\tau}} = \vec{\boldsymbol{\mu}} \times \vec{\mathbf{B}}, \tag{27-11}$$

where $\vec{\boldsymbol{\mu}}$ is the **magnetic dipole moment** of the loop:

$$\vec{\boldsymbol{\mu}} = NI\vec{\mathbf{A}}. \tag{27-10}$$

Here N is the number of coils carrying current I in the loop and $\vec{\mathbf{A}}$ is a vector perpendicular to the plane of the loop (use right-hand rule, fingers along current in loop) and has magnitude equal to the area of the loop.

The measurement of the charge-to-mass ratio (e/m) of the electron was done using magnetic and electric fields. The charge e on the electron was first measured in the Millikan oil-drop experiment and then its mass was obtained from the measured value of the e/m ratio.

In the **Hall effect**, moving charges in a conductor placed in a magnetic field are forced to one side, producing an emf between the two sides of the conductor.

[*A **mass spectrometer** uses magnetic and electric fields to measure the mass of ions.]

Questions

1. A compass needle is not always balanced parallel to the Earth's surface, but one end may dip downward. Explain.

2. Draw the magnetic field lines around a straight section of wire carrying a current horizontally to the left.

3. A horseshoe magnet is held vertically with the north pole on the left and south pole on the right. A wire passing between the poles, equidistant from them, carries a current directly away from you. In what direction is the force on the wire?

4. In the relation $\vec{F} = I\vec{\ell} \times \vec{B}$, which pairs of the vectors $(\vec{F}, \vec{\ell}, \vec{B})$ are always at 90°? Which can be at other angles?

5. The magnetic field due to current in wires in your home can affect a compass. Discuss the effect in terms of currents, including if they are ac or dc.

6. If a negatively charged particle enters a region of uniform magnetic field which is perpendicular to the particle's velocity, will the kinetic energy of the particle increase, decrease, or stay the same? Explain your answer. (Neglect gravity and assume there is no electric field.)

7. In Fig. 27–34, charged particles move in the vicinity of a current-carrying wire. For each charged particle, the arrow indicates the direction of motion of the particle, and the + or − indicates the sign of the charge. For each of the particles, indicate the direction of the magnetic force due to the magnetic field produced by the wire.

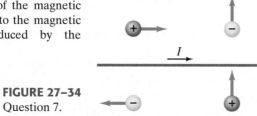

FIGURE 27–34
Question 7.

8. A positively charged particle in a nonuniform magnetic field follows the trajectory shown in Fig. 27–35. Indicate the direction of the magnetic field at points near the path, assuming the path is always in the plane of the page, and indicate the relative magnitudes of the field in each region.

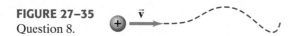

FIGURE 27–35
Question 8.

9. Note that the pattern of magnetic field lines surrounding a bar magnet is similar to that of the electric field around an electric dipole. From this fact, predict how the magnetic field will change with distance (a) when near one pole of a very long bar magnet, and (b) when far from a magnet as a whole.

10. Explain why a strong magnet held near a CRT television screen causes the picture to become distorted. Also, explain why the picture sometimes goes completely black where the field is the strongest. [But don't risk damage to your TV by trying this.]

11. Describe the trajectory of a negatively charged particle in the velocity selector of Fig. 27–21 if its speed exceeds E/B. What is its trajectory if $v < E/B$? Would it make any difference if the particle were positively charged?

12. Can you set a resting electron into motion with a steady magnetic field? With an electric field? Explain.

13. A charged particle is moving in a circle under the influence of a uniform magnetic field. If an electric field that points in the same direction as the magnetic field is turned on, describe the path the charged particle will take.

14. The force on a particle in a magnetic field is the idea behind **electromagnetic pumping**. It is used to pump metallic fluids (such as sodium) and to pump blood in artificial heart machines. The basic design is shown in Fig. 27–36. An electric field is applied perpendicular to a blood vessel and to a magnetic field. Explain how ions are caused to move. Do positive and negative ions feel a force in the same direction?

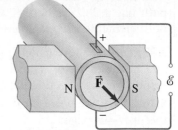

FIGURE 27–36
Electromagnetic pumping
in a blood vessel.
Question 14.

15. A beam of electrons is directed toward a horizontal wire carrying a current from left to right (Fig. 27–37). In what direction is the beam deflected?

FIGURE 27–37
Question 15.

16. A charged particle moves in a straight line through a particular region of space. Could there be a nonzero magnetic field in this region? If so, give two possible situations.

17. If a moving charged particle is deflected sideways in some region of space, can we conclude, for certain, that $\vec{B} \neq 0$ in that region? Explain.

18. How could you tell whether moving electrons in a certain region of space are being deflected by an electric field or by a magnetic field (or by both)?

19. How can you make a compass without using iron or other ferromagnetic material?

20. Describe how you could determine the dipole moment of a bar magnet or compass needle.

21. In what positions (if any) will a current loop placed in a uniform magnetic field be in (a) stable equilibrium, and (b) unstable equilibrium?

*22. A rectangular piece of semiconductor is inserted in a magnetic field and a battery is connected to its ends as shown in Fig. 27–38. When a sensitive voltmeter is connected between points a and b, it is found that point a is at a higher potential than b. What is the sign of the charge carriers in this semiconductor material?

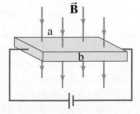

FIGURE 27–38
Question 22.

*23. Two ions have the same mass, but one is singly ionized and the other is doubly ionized. How will their positions on the film of the mass spectrometer of Fig. 27–33 differ?

Problems

1. (I) (a) What is the force per meter of length on a straight wire carrying a 9.40-A current when perpendicular to a 0.90-T uniform magnetic field? (b) What if the angle between the wire and field is 35.0°?

2. (I) Calculate the magnitude of the magnetic force on a 240-m length of wire stretched between two towers and carrying a 150-A current. The Earth's magnetic field of 5.0×10^{-5} T makes an angle of 68° with the wire.

3. (I) A 1.6-m length of wire carrying 4.5 A of current toward the south is oriented horizontally. At that point on the Earth's surface, the dip angle of the Earth's magnetic field makes an angle of 41° to the wire. Estimate the magnitude of the magnetic force on the wire due to the Earth's magnetic field of 5.5×10^{-5} T at this point.

4. (II) The magnetic force per meter on a wire is measured to be only 25 percent of its maximum possible value. Sketch the relationship of the wire and the field if the force had been a maximum, and sketch the relationship as it actually is, calculating the angle between the wire and the magnetic field.

5. (II) The force on a wire is a maximum of 7.50×10^{-2} N when placed between the pole faces of a magnet. The current flows horizontally to the right and the magnetic field is vertical. The wire is observed to "jump" toward the observer when the current is turned on. (a) What type of magnetic pole is the top pole face? (b) If the pole faces have a diameter of 10.0 cm, estimate the current in the wire if the field is 0.220 T. (c) If the wire is tipped so that it makes an angle of 10.0° with the horizontal, what force will it now feel? [Hint: What length of wire will now be in the field?]

6. (II) Suppose a straight 1.00-mm-diameter copper wire could just "float" horizontally in air because of the force due to the Earth's magnetic field $\vec{\mathbf{B}}$, which is horizontal, perpendicular to the wire, and of magnitude 5.0×10^{-5} T. What current would the wire carry? Does the answer seem feasible? Explain briefly.

7. (II) A stiff wire 50.0 cm long is bent at a right angle in the middle. One section lies along the z axis and the other is along the line $y = 2x$ in the xy plane. A current of 20.0 A flows in the wire—down the z axis and out the line in the xy plane. The wire passes through a uniform magnetic field given by $\vec{\mathbf{B}} = (0.318\hat{\mathbf{i}})$T. Determine the magnitude and direction of the total force on the wire.

8. (II) A long wire stretches along the x axis and carries a 3.0-A current to the right $(+x)$. The wire is in a uniform magnetic field $\vec{\mathbf{B}} = (0.20\hat{\mathbf{i}} - 0.36\hat{\mathbf{j}} + 0.25\hat{\mathbf{k}})$ T. Determine the components of the force on the wire per cm of length.

9. (II) A current-carrying circular loop of wire (radius r, current I) is partially immersed in a magnetic field of constant magnitude B_0 directed out of the page as shown in Fig. 27–39. Determine the net force on the loop due to the field in terms of θ_0. (Note that θ_0 points to the dashed line, above which $B = 0$.)

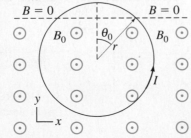

FIGURE 27–39
Problem 9.

10. (II) A 2.0-m-long wire carries a current of 8.2 A and is immersed within a uniform magnetic field $\vec{\mathbf{B}}$. When this wire lies along the $+x$ axis, a magnetic force $\vec{\mathbf{F}} = (-2.5\hat{\mathbf{j}})$ N acts on the wire, and when it lies on the $+y$ axis, the force is $\vec{\mathbf{F}} = (2.5\hat{\mathbf{i}} - 5.0\hat{\mathbf{k}})$ N. Find $\vec{\mathbf{B}}$.

11. (III) A curved wire connecting two points a and b, lies in a plane perpendicular to a uniform magnetic field $\vec{\mathbf{B}}$ and carries a current I. Show that the resultant magnetic force on the wire, no matter what its shape, is the same as that on a straight wire connecting the two points carrying the same current I. See Fig. 27–40.

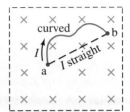

FIGURE 27–40
Problem 11.

12. (III) A circular loop of wire, of radius r, carries current I. It is placed in a magnetic field whose straight lines seem to diverge from a point a distance d below the loop on its axis. (That is, the field makes an angle θ with the loop at all points, Fig. 27–41, where $\tan\theta = r/d$.) Determine the force on the loop.

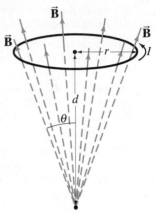

FIGURE 27–41
Problem 12.

27–4 Force on Charge Moving in Magnetic Field

13. (I) Determine the magnitude and direction of the force on an electron traveling 8.75×10^5 m/s horizontally to the east in a vertically upward magnetic field of strength 0.45 T.

14. (I) An electron is projected vertically upward with a speed of 1.70×10^6 m/s into a uniform magnetic field of 0.480 T that is directed horizontally away from the observer. Describe the electron's path in this field.

15. (I) Alpha particles of charge $q = +2e$ and mass $m = 6.6 \times 10^{-27}$ kg are emitted from a radioactive source at a speed of 1.6×10^7 m/s. What magnetic field strength would be required to bend them into a circular path of radius $r = 0.18$ m?

16. (I) Find the direction of the force on a negative charge for each diagram shown in Fig. 27–42, where $\vec{\mathbf{v}}$ (green) is the velocity of the charge and $\vec{\mathbf{B}}$ (blue) is the direction of the magnetic field. ($\otimes$ means the vector points inward. $\odot$ means it points outward, toward you.)

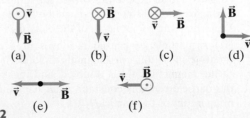

FIGURE 27–42
Problem 16.

17. (I) Determine the direction of $\vec{\mathbf{B}}$ for each case in Fig. 27–43, where $\vec{\mathbf{F}}$ represents the maximum magnetic force on a positively charged particle moving with velocity $\vec{\mathbf{v}}$.

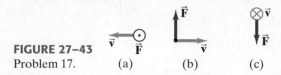

FIGURE 27–43
Problem 17.   (a)   (b)   (c)

18. (II) What is the velocity of a beam of electrons that goes undeflected when passing through perpendicular electric and magnetic fields of magnitude 8.8×10^3 V/m and 7.5×10^{-3} T, respectively? What is the radius of the electron orbit if the electric field is turned off?

19. (II) A doubly charged helium atom whose mass is 6.6×10^{-27} kg is accelerated by a voltage of 2700 V. (*a*) What will be its radius of curvature if it moves in a plane perpendicular to a uniform 0.340-T field? (*b*) What is its period of revolution?

20. (II) A proton (mass m_p), a deuteron $(m = 2m_p, Q = e)$, and an alpha particle $(m = 4m_p, Q = 2e)$ are accelerated by the same potential difference V and then enter a uniform magnetic field $\vec{\mathbf{B}}$, where they move in circular paths perpendicular to $\vec{\mathbf{B}}$. Determine the radius of the paths for the deuteron and alpha particle in terms of that for the proton.

21. (II) For a particle of mass m and charge q moving in a circular path in a magnetic field B, (*a*) show that its kinetic energy is proportional to r^2, the square of the radius of curvature of its path, and (*b*) show that its angular momentum is $L = qBr^2$, about the center of the circle.

22. (II) An electron moves with velocity $\vec{\mathbf{v}} = (7.0\hat{\mathbf{i}} - 6.0\hat{\mathbf{j}}) \times 10^4$ m/s in a magnetic field $\vec{\mathbf{B}} = (-0.80\hat{\mathbf{i}} + 0.60\hat{\mathbf{j}})$ T. Determine the magnitude and direction of the force on the electron.

23. (II) A 6.0-MeV (kinetic energy) proton enters a 0.20-T field, in a plane perpendicular to the field. What is the radius of its path? See Section 23–8.

24. (II) An electron experiences the greatest force as it travels 2.8×10^6 m/s in a magnetic field when it is moving northward. The force is vertically upward and of magnitude 8.2×10^{-13} N. What is the magnitude and direction of the magnetic field?

25. (II) A proton moves through a region of space where there is a magnetic field $\vec{\mathbf{B}} = (0.45\hat{\mathbf{i}} + 0.38\hat{\mathbf{j}})$ T and an electric field $\vec{\mathbf{E}} = (3.0\hat{\mathbf{i}} - 4.2\hat{\mathbf{j}}) \times 10^3$ V/m. At a given instant, the proton's velocity is $\vec{\mathbf{v}} = (6.0\hat{\mathbf{i}} + 3.0\hat{\mathbf{j}} - 5.0\hat{\mathbf{k}}) \times 10^3$ m/s. Determine the components of the total force on the proton.

26. (II) An electron experiences a force $\vec{\mathbf{F}} = (3.8\hat{\mathbf{i}} - 2.7\hat{\mathbf{j}}) \times 10^{-13}$ N when passing through a magnetic field $\vec{\mathbf{B}} = (0.85 \text{ T})\hat{\mathbf{k}}$. Determine the components of the electron's velocity.

27. (II) A particle of charge q moves in a circular path of radius r in a uniform magnetic field $\vec{\mathbf{B}}$. If the magnitude of the magnetic field is doubled, and the kinetic energy of the particle remains constant, what happens to the angular momentum of the particle?

28. (II) An electron enters a uniform magnetic field $B = 0.28$ T at a 45° angle to $\vec{\mathbf{B}}$. Determine the radius r and pitch p (distance between loops) of the electron's helical path assuming its speed is 3.0×10^6 m/s. See Fig. 27–44.

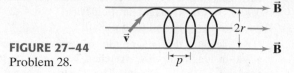

FIGURE 27–44
Problem 28.

29. (II) A particle with charge q and momentum p, initially moving along the x axis, enters a region where a uniform magnetic field $\vec{\mathbf{B}} = B_0\hat{\mathbf{k}}$ extends over a width $x = \ell$ as shown in Fig. 27–45. The particle is deflected a distance d in the $+y$ direction as it traverses the field. Determine (*a*) whether q is positive or negative, and (*b*) the magnitude of its momentum p.

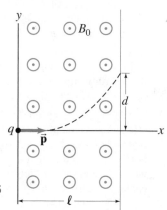

FIGURE 27–45
Problem 29.

30. (II) The path of protons emerging from an accelerator must be bent by 90° by a "bending magnet" so as not to strike a barrier in their path a distance d from their exit hole in the accelerator. Show that the field $\vec{\mathbf{B}}$ in the bending magnet, which we assume is uniform and can extend over an area $d \times d$, must have magnitude $B \geq (2mK/e^2d^2)^{\frac{1}{2}}$, where m is the mass of a proton and K is its kinetic energy.

31. (III) Suppose the Earth's magnetic field at the equator has magnitude 0.50×10^{-4} T and a northerly direction at all points. Estimate the speed a singly ionized uranium ion $(m = 238 \text{ u}, q = e)$ would need to circle the Earth 5.0 km above the equator. Can you ignore gravity? [Ignore relativity.]

32. (III) A 3.40-g bullet moves with a speed of 155 m/s perpendicular to the Earth's magnetic field of 5.00×10^{-5} T. If the bullet possesses a net charge of 18.5×10^{-9} C, by what distance will it be deflected from its path due to the Earth's magnetic field after it has traveled 1.00 km?

33. (III) A proton moving with speed $v = 1.3 \times 10^5$ m/s in a field-free region abruptly enters an essentially uniform magnetic field $B = 0.850$ T $(\vec{\mathbf{B}} \perp \vec{\mathbf{v}})$. If the proton enters the magnetic field region at a 45° angle as shown in Fig. 27–46, (*a*) at what angle does it leave, and (*b*) at what distance x does it exit the field?

FIGURE 27–46
Problem 33.

34. (III) A particle with charge $+q$ and mass m travels in a uniform magnetic field $\vec{\mathbf{B}} = B_0\hat{\mathbf{k}}$. At time $t = 0$, the particle's speed is v_0 and its velocity vector lies in the xy plane directed at an angle of 30° with respect to the y axis as shown in Fig. 27–47. At a later time $t = t_a$, the particle will cross the x axis at $x = a$. In terms of q, m, v_0, and B_0, determine (a) a, and (b) t_a.

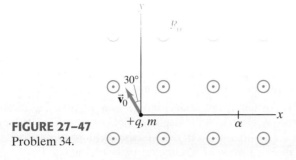

FIGURE 27–47
Problem 34.

27–5 Torque on a Current Loop; Magnetic Moment

35. (I) How much work is required to rotate the current loop (Fig. 27–22) in a uniform magnetic field $\vec{\mathbf{B}}$ from (a) $\theta = 0°\,(\vec{\boldsymbol{\mu}} \| \vec{\mathbf{B}})$ to $\theta = 180°$, (b) $\theta = 90°$ to $\theta = -90°$?

36. (I) A 13.0-cm-diameter circular loop of wire is placed with the plane of the loop parallel to the uniform magnetic field between the pole pieces of a large magnet. When 4.20 A flows in the coil, the torque on it is 0.185 m·N. What is the magnetic field strength?

37. (II) A circular coil 18.0 cm in diameter and containing twelve loops lies flat on the ground. The Earth's magnetic field at this location has magnitude 5.50×10^{-5} T and points into the Earth at an angle of 66.0° below a line pointing due north. If a 7.10-A clockwise current passes through the coil, determine (a) the torque on the coil, and (b) which edge of the coil rises up, north, east, south, or west.

38. (II) Show that the magnetic dipole moment μ of an electron orbiting the proton nucleus of a hydrogen atom is related to the orbital momentum L of the electron by

$$ \mu = \frac{e}{2m}L. $$

39. (II) A 15-loop circular coil 22 cm in diameter lies in the xy plane. The current in each loop of the coil is 7.6 A clockwise, and an external magnetic field $\vec{\mathbf{B}} = (0.55\hat{\mathbf{i}} + 0.60\hat{\mathbf{j}} - 0.65\hat{\mathbf{k}})$ T passes through the coil. Determine (a) the magnetic moment of the coil, $\vec{\boldsymbol{\mu}}$; (b) the torque on the coil due to the external magnetic field; (c) the potential energy U of the coil in the field (take the same zero for U as we did in our discussion of Fig. 27–22).

40. (III) Suppose a nonconducting rod of length d carries a uniformly distributed charge Q. It is rotated with angular velocity ω about an axis perpendicular to the rod at one end, Fig. 27–48. Show that the magnetic dipole moment of this rod is $\frac{1}{6}Q\omega d^2$. [Hint: Consider the motion of each infinitesimal length of the rod.]

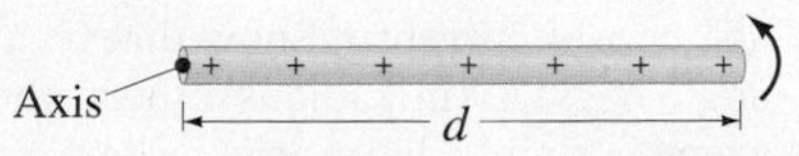

FIGURE 27–48
Problem 40.   Axis d

*41. (I) If the current to a motor drops by 12%, by what factor does the output torque change?

42. (I) A galvanometer needle deflects full scale for a 63.0-μA current. What current will give full-scale deflection if the magnetic field weakens to 0.800 of its original value?

*43. (I) If the restoring spring of a galvanometer weakens by 15% over the years, what current will give full-scale deflection if it originally required 46 μA?

27–7 Discovery of Electron

44. (I) What is the value of q/m for a particle that moves in a circle of radius 8.0 mm in a 0.46-T magnetic field if a crossed 260-V/m electric field will make the path straight?

45. (II) An oil drop whose mass is determined to be 3.3×10^{-15} kg is held at rest between two large plates separated by 1.0 cm as in Fig. 27–31. If the potential difference between the plates is 340 V, how many excess electrons does this drop have?

27–8 Hall Effect

46. (II) A Hall probe, consisting of a rectangular slab of current-carrying material, is calibrated by placing it in a known magnetic field of magnitude 0.10 T. When the field is oriented normal to the slab's rectangular face, a Hall emf of 12 mV is measured across the slab's width. The probe is then placed in a magnetic field of unknown magnitude B, and a Hall emf of 63 mV is measured. Determine B assuming that the angle θ between the unknown field and the plane of the slab's rectangular face is (a) $\theta = 90°$, and (b) $\theta = 60°$.

47. (II) A Hall probe used to measure magnetic field strengths consists of a rectangular slab of material (free-electron density n) with width d and thickness t, carrying a current I along its length ℓ. The slab is immersed in a magnetic field of magnitude B oriented perpendicular to its rectangular face (of area ℓd), so that a Hall emf $\mathscr{E}_H$ is produced across its width d. The probe's magnetic sensitivity, defined as $K_H = \mathscr{E}_H/IB$, indicates the magnitude of the Hall emf achieved for a given applied magnetic field and current. A slab with a large K_H is a good candidate for use as a Hall probe. (a) Show that $K_H = 1/ent$. Thus, a good Hall probe has small values for both n and t. (b) As possible candidates for the material used in a Hall probe, consider (i) a typical metal ($n \approx 1 \times 10^{29}/\text{m}^3$) and (ii) a (doped) semiconductor ($n \approx 3 \times 10^{22}/\text{m}^3$). Given that a semiconductor slab can be manufactured with a thickness of 0.15 mm, how thin (nm) should a metal slab be to yield a K_H value equal to that of the semiconductor slab? Compare this metal slab thickness with the 0.3-nm size of a typical metal atom. (c) For the typical semiconductor slab described in part (b), what is the expected value for $\mathscr{E}_H$ when $I = 100$ mA and $B = 0.1$ T?

48. (II) A rectangular sample of a metal is 3.0 cm wide and 680 μm thick. When it carries a 42-A current and is placed in a 0.80-T magnetic field it produces a 6.5-μV Hall emf. Determine: (a) the Hall field in the conductor; (b) the drift speed of the conduction electrons; (c) the density of free electrons in the metal.

49. (II) In a probe that uses the Hall effect to measure magnetic fields, a 12.0-A current passes through a 1.50-cm-wide 1.30-mm-thick strip of sodium metal. If the Hall emf is 1.86 μV, what is the magnitude of the magnetic field (take it perpendicular to the flat face of the strip)? Assume one free electron per atom of Na, and take its specific gravity to be 0.971.

50. (II) The Hall effect can be used to measure blood flow rate because the blood contains ions that constitute an electric current. (a) Does the sign of the ions influence the emf? (b) Determine the flow velocity in an artery 3.3 mm in diameter if the measured emf is 0.13 mV and B is 0.070 T. (In actual practice, an alternating magnetic field is used.)

*27–9 Mass Spectrometer

*51. (I) In a mass spectrometer, germanium atoms have radii of curvature equal to 21.0, 21.6, 21.9, 22.2, and 22.8 cm. The largest radius corresponds to an atomic mass of 76 u. What are the atomic masses of the other isotopes?

*52. (II) One form of mass spectrometer accelerates ions by a voltage V before they enter a magnetic field B. The ions are assumed to start from rest. Show that the mass of an ion is $m = qB^2R^2/2V$, where R is the radius of the ions' path in the magnetic field and q is their charge.

*53. (II) Suppose the electric field between the electric plates in the mass spectrometer of Fig. 27–33 is 2.48×10^4 V/m and the magnetic fields are $B = B' = 0.58$ T. The source contains carbon isotopes of mass numbers 12, 13, and 14 from a long dead piece of a tree. (To estimate atomic masses, multiply by 1.66×10^{-27} kg.) How far apart are the lines formed by the singly charged ions of each type on the photographic film? What if the ions were doubly charged?

*54. (II) A mass spectrometer is being used to monitor air pollutants. It is difficult, however, to separate molecules with nearly equal mass such as CO (28.0106 u) and N_2 (28.0134 u). How large a radius of curvature must a spectrometer have if these two molecules are to be separated at the film or detectors by 0.65 mm?

*55. (II) An unknown particle moves in a straight line through crossed electric and magnetic fields with $E = 1.5$ kV/m and $B = 0.034$ T. If the electric field is turned off, the particle moves in a circular path of radius $r = 2.7$ cm. What might the particle be?

General Problems

56. Protons move in a circle of radius 5.10 cm in a 0.625-T magnetic field. What value of electric field could make their paths straight? In what direction must the electric field point?

57. Protons with momentum 3.8×10^{-16} kg·m/s are magnetically steered clockwise in a circular path 2.0 km in diameter at Fermi National Accelerator Laboratory in Illinois. Determine the magnitude and direction of the field in the magnets surrounding the beam pipe.

58. A proton and an electron have the same kinetic energy upon entering a region of constant magnetic field. What is the ratio of the radii of their circular paths?

59. Two stiff parallel wires a distance d apart in a horizontal plane act as rails to support a light metal rod of mass m (perpendicular to each rail), Fig. 27–49. A magnetic field $\vec{B}$, directed vertically upward (outward in diagram), acts throughout. At $t = 0$, a constant current I begins to flow through the system. Determine the speed of the rod, which starts from rest at $t = 0$, as a function of time (a) assuming no friction between the rod and the rails, and (b) if the coefficient of friction is μ_k. (c) In which direction does the rod move, east or west, if the current through it heads north?

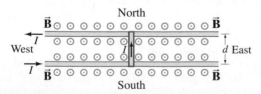

FIGURE 27–49 Looking down on a rod sliding on rails. Problems 59 and 60.

60. Suppose the rod in Fig. 27–49 (Problem 59) has mass $m = 0.40$ kg and length 22 cm and the current through it is $I = 36$ A. If the coefficient of static friction is $\mu_s = 0.50$, determine the minimum magnetic field $\vec{B}$ (not necessarily vertical) that will just cause the rod to slide. Give the magnitude of $\vec{B}$ and its direction relative to the vertical (outwards towards us).

61. Near the equator, the Earth's magnetic field points almost horizontally to the north and has magnitude $B = 0.50 \times 10^{-4}$ T. What should be the magnitude and direction for the velocity of an electron if its weight is to be exactly balanced by the magnetic force?

62. Calculate the magnetic force on an airplane which has acquired a net charge of 1850 μC and moves with a speed of 120 m/s perpendicular to the Earth's magnetic field of 5.0×10^{-5} T.

63. A motor run by a 9.0-V battery has a 20 turn square coil with sides of length 5.0 cm and total resistance 24 Ω. When spinning, the magnetic field felt by the wire in the coil is 0.020 T. What is the maximum torque on the motor?

64. Estimate the approximate maximum deflection of the electron beam near the center of a CRT television screen due to the Earth's 5.0×10^{-5} T field. Assume the screen is 18 cm from the electron gun, where the electrons are accelerated (a) by 2.0 kV, or (b) by 28 kV. Note that in color TV sets, the beam must be directed accurately to within less than 1 mm in order to strike the correct phosphor. Because the Earth's field is significant here, mu-metal shields are used to reduce the Earth's field in the CRT. (See Section 23–9.)

65. The rectangular loop of wire shown in Fig. 27–22 has mass m and carries current I. Show that if the loop is oriented at an angle $\theta \ll 1$ (in radians), then when it is released it will execute simple harmonic motion about $\theta = 0$. Calculate the period of the motion.

66. The **cyclotron** (Fig. 27–50) is a device used to accelerate elementary particles such as protons to high speeds. Particles starting at point A with some initial velocity travel in circular orbits in the magnetic field B. The particles are accelerated to higher speeds each time they pass in the gap between the metal "dees," where there is an electric field E. (There is no electric field within the hollow metal dees.) The electric field changes direction each half-cycle, due to an ac voltage $V = V_0 \sin 2\pi f t$, so that the particles are increased in speed at each passage through the gap. (a) Show that the frequency f of the voltage must be $f = Bq/2\pi m$, where q is the charge on the particles and m their mass. (b) Show that the kinetic energy of the particles increases by $2qV_0$ each revolution, assuming that the gap is small. (c) If the radius of the cyclotron is 0.50 m and the magnetic field strength is 0.60 T, what will be the maximum kinetic energy of accelerated protons in MeV?

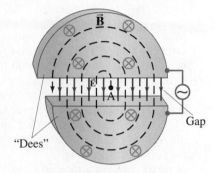

FIGURE 27–50
A cyclotron. "Dees"
Problem 66.

67. Magnetic fields are very useful in particle accelerators for "beam steering"; that is, magnetic fields can be used to change the beam's direction without altering its speed (Fig. 27–51). Show how this could work with a beam of protons. What happens to protons that are not moving with the speed that the magnetic field is designed for? If the field extends over a region 5.0 cm wide and has a magnitude of 0.38 T, by approximately what angle will a beam of protons traveling at 0.85×10^7 m/s be bent?

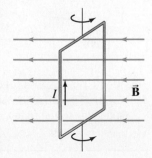

FIGURE 27–51
Problem 67.

Evacuated tubes, inside of which the protons move with velocity indicated by the green arrows

68. A square loop of aluminum wire is 20.0 cm on a side. It is to carry 15.0 A and rotate in a uniform 1.35-T magnetic field as shown in Fig. 27–52. (a) Determine the minimum diameter of the wire so that it will not fracture from tension or shear. Assume a safety factor of 10. (See Table 12–2.) (b) What is the resistance of a single loop of this wire?

FIGURE 27–52
Problem 68.

69. A sort of "projectile launcher" is shown in Fig. 27–53. A large current moves in a closed loop composed of fixed rails, a power supply, and a very light, almost frictionless bar touching the rails. A 1.8 T magnetic field is perpendicular to the plane of the circuit. If the rails are a distance $d = 24$ cm apart, and the bar has a mass of 1.5 g, what constant current flow is needed to accelerate the bar from rest to 25 m/s in a distance of 1.0 m? In what direction must the field point?

FIGURE 27–53   Problem 69.

70. (a) What value of magnetic field would make a beam of electrons, traveling to the right at a speed of 4.8×10^6 m/s, go undeflected through a region where there is a uniform electric field of 8400 V/m pointing vertically up? (b) What is the direction of the magnetic field if it is known to be perpendicular to the electric field? (c) What is the frequency of the circular orbit of the electrons if the electric field is turned off?

71. In a certain cathode ray tube, electrons are accelerated horizontally by 25 kV. They then pass through a uniform magnetic field B for a distance of 3.5 cm, which deflects them upward so they reach the top of the screen 22 cm away, 11 cm above the center. Estimate the value of B.

72. **Zeeman effect.** In the Bohr model of the hydrogen atom, the electron is held in its circular orbit of radius r about its proton nucleus by electrostatic attraction. If the atoms are placed in a weak magnetic field $\vec{\mathbf{B}}$, the rotation frequency of electrons rotating in a plane perpendicular to $\vec{\mathbf{B}}$ is changed by an amount

$$\Delta f = \pm \frac{eB}{4\pi m}$$

where e and m are the charge and mass of an electron. (a) Derive this result, assuming the force due to $\vec{\mathbf{B}}$ is much less than that due to electrostatic attraction of the nucleus. (b) What does the $\pm$ sign indicate?

73. A proton follows a spiral path through a gas in a magnetic field of 0.018 T, perpendicular to the plane of the spiral, as shown in Fig. 27–54. In two successive loops, at points P and Q, the radii are 10.0 mm and 8.5 mm, respectively. Calculate the change in the kinetic energy of the proton as it travels from P to Q.

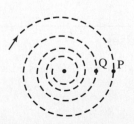

FIGURE 27–54   Problem 73.

74. The net force on a current loop whose face is perpendicular to a uniform magnetic field is zero, since contributions to the net force from opposite sides of the loop cancel. However, if the field varies in magnitude from one side of the loop to the other, then there can be a net force on the loop. Consider a square loop with sides whose length is a, located with one side at $x = b$ in the xy plane (Fig. 27–55). A magnetic field is directed along z, with a magnitude that varies with x according to

$$B = B_0\left(1 - \frac{x}{b}\right).$$

If the current in the loop circulates counterclockwise (that is, the magnetic dipole moment of the loop is along the z axis), find an expression for the net force on the loop.

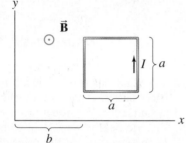

FIGURE 27–55
Problem 74.

75. The power cable for an electric trolley (Fig. 27–56) carries a horizontal current of 330 A toward the east. The Earth's magnetic field has a strength 5.0×10^{-5} T and makes an angle of dip of 22° at this location. Calculate the magnitude and direction of the magnetic force on a 5.0-m length of this cable.

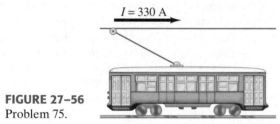

FIGURE 27–56
Problem 75.

76. A uniform conducting rod of length d and mass m sits atop a fulcrum, which is placed a distance $d/4$ from the rod's left-hand end and is immersed in a uniform magnetic field of magnitude B directed into the page (Fig. 27–57). An object whose mass M is 8.0 times greater than the rod's mass is hung from the rod's left-hand end. What current (direction and magnitude) should flow through the rod in order for it to be "balanced" (i.e., be at rest horizontally) on the fulcrum? (Flexible connecting wires which exert negligible force on the rod are not shown.)

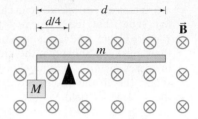

FIGURE 27–57
Problem 76.

77. In a simple device for measuring the magnitude B of a magnetic field, a conducting rod (length $d = 1.0$ m, mass $m = 150$ g) hangs from a friction-free pivot and is oriented so that its axis of rotation is aligned with the direction of the magnetic field to be measured. Thin flexible wires (which exert negligible force on the rod) carry a current $I = 12$ A, which causes the rod to deflect an angle θ with respect to the vertical, where it remains at rest (Fig. 27–58). (a) Is the current flowing upward (toward the pivot) or downward in Fig. 27–58? (b) If $\theta = 13°$, determine B. (c) What is the largest magnetic field magnitude that can be measured using this device?

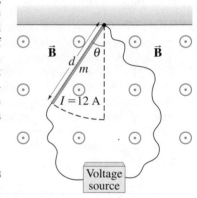

FIGURE 27–58
Problem 77.

Answers to Exercises

A: Near the poles, where the field lines are closer together.

B: Counterclockwise.

C: (b), (c), (d).

D: 0.15 N.

E: (b), (c), (d).

F: Negative; the direction of the helical path would be reversed (still going to the right).

G: (d).

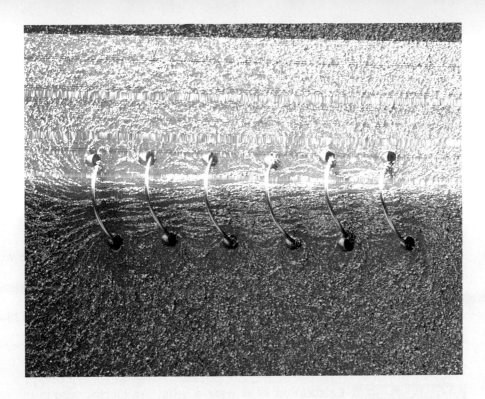

A long coil of wire with many closely spaced loops is called a solenoid. When a long solenoid carries an electric current, a nearly uniform magnetic field is produced within the loops as suggested by the alignment of the iron filings in this photo. The magnitude of the field inside a solenoid is readily found using Ampère's law, one of the great general laws of electromagnetism, relating magnetic fields and electric currents. We examine these connections in detail in this Chapter, as well as other means for producing magnetic fields.

Sources of Magnetic Field

CHAPTER-OPENING QUESTION—Guess now!

Which of the following will produce a magnetic field?

(a) An electric charge at rest.
(b) A moving electric charge.
(c) An electric current.
(d) The voltage of a battery not connected to anything.
(e) Any piece of iron.
(f) A piece of any metal.

I n the previous Chapter, we discussed the effects (forces and torques) that a magnetic field has on electric currents and on moving electric charges. We also saw that magnetic fields are produced not only by magnets but also by electric currents (Oersted's great discovery). It is this aspect of magnetism, the production of magnetic fields, that we discuss in this Chapter. We will now see how magnetic field strengths are determined for some simple situations, and discuss some general relations between magnetic fields and their sources, electric current. Most elegant is Ampère's law. We also study the Biot-Savart Law, which can be very helpful for solving practical problems.

CONTENTS

28–1 Magnetic Field Due to a Straight Wire

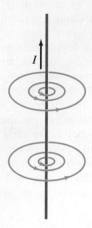

FIGURE 28–1 Same as Fig. 27–8b. Magnetic field lines around a long straight wire carrying an electric current I.

FIGURE 28–2 Example 28–1.

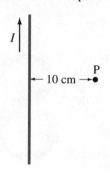

⚠ **CAUTION**

A compass, near a current, may not point north

FIGURE 28–3 Example 28–2. Wire 1 carrying current I_1 out towards us, and wire 2 carrying current I_2 into the page, produce magnetic fields whose lines are circles around their respective wires.

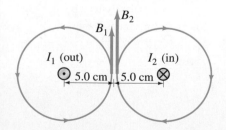

We saw in Section 27–2 that the magnetic field due to the electric current in a long straight wire is such that the field lines are circles with the wire at the center (Fig. 28–1). You might expect that the field strength at a given point would be greater if the current flowing in the wire were greater; and that the field would be less at points farther from the wire. This is indeed the case. Careful experiments show that the magnetic field B due to a long straight wire at a point near it is directly proportional to the current I in the wire and inversely proportional to the distance r from the wire:

$$B \propto \frac{I}{r}.$$

This relation $B \propto I/r$ is valid as long as r, the perpendicular distance to the wire, is much less than the distance to the ends of the wire (i.e., the wire is long).

The proportionality constant is written[†] as $\mu_0/2\pi$; thus,

$$B = \frac{\mu_0}{2\pi} \frac{I}{r}. \qquad \text{[near a long straight wire]} \quad \textbf{(28–1)}$$

The value of the constant μ_0, which is called the **permeability of free space**, is $\mu_0 = 4\pi \times 10^{-7}\,\text{T}\cdot\text{m/A}$.

EXAMPLE 28–1 **Calculation of $\vec{B}$ near a wire.** An electric wire in the wall of a building carries a dc current of 25 A vertically upward. What is the magnetic field due to this current at a point P, 10 cm due north of the wire (Fig. 28–2)?

APPROACH We assume the wire is much longer than the 10-cm distance to the point P so we can apply Eq. 28–1.

SOLUTION According to Eq. 28–1:

$$B = \frac{\mu_0 I}{2\pi r} = \frac{(4\pi \times 10^{-7}\,\text{T}\cdot\text{m/A})(25\,\text{A})}{(2\pi)(0.10\,\text{m})} = 5.0 \times 10^{-5}\,\text{T},$$

or 0.50 G. By the right-hand rule (Table 27–1, page 716), the field due to the current points to the west (into the page in Fig. 28–2) at point P.

NOTE The wire's field has about the same magnitude as Earth's magnetic field, so a compass at P would not point north but in a northwesterly direction.

NOTE Most electrical wiring in buildings consists of cables with two wires in each cable. Since the two wires carry current in opposite directions, their magnetic fields cancel to a large extent, but may still affect sensitive electronic devices.

EXERCISE A In Example 25–10 we saw that a typical lightning bolt produces a 100-A current for 0.2 s. Estimate the magnetic field 10 m from a lightning bolt. Would it have a significant effect on a compass?

EXAMPLE 28–2 **Magnetic field midway between two currents.** Two parallel straight wires 10.0 cm apart carry currents in opposite directions (Fig. 28–3). Current $I_1 = 5.0\,\text{A}$ is out of the page, and $I_2 = 7.0\,\text{A}$ is into the page. Determine the magnitude and direction of the magnetic field halfway between the two wires.

APPROACH The magnitude of the field produced by each wire is calculated from Eq. 28–1. The direction of *each* wire's field is determined with the right-hand rule. The total field is the vector sum of the two fields at the midway point.

SOLUTION The magnetic field lines due to current I_1 form circles around the wire of I_1, and right-hand-rule-1 (Fig. 27–8c) tells us they point counterclockwise around the wire. The field lines due to I_2 form circles around the wire of I_2 and point clockwise, Fig. 28–3. At the midpoint, both fields point upward as shown, and so add together.

[†]The constant is chosen in this complicated way so that Ampère's law (Section 28–4), which is considered more fundamental, will have a simple and elegant form.

The midpoint is 0.050 m from each wire, and from Eq. 28–1 the magnitudes of B_1 and B_2 are

$$B_1 = \frac{\mu_0 I_1}{2\pi r} = \frac{(4\pi \times 10^{-7}\,\text{T·m/A})(5.0\,\text{A})}{2\pi(0.050\,\text{m})} = 2.0 \times 10^{-5}\,\text{T};$$

$$B_2 = \frac{\mu_0 I_2}{2\pi r} = \frac{(4\pi \times 10^{-7}\,\text{T·m/A})(7.0\,\text{A})}{2\pi(0.050\,\text{m})} = 2.8 \times 10^{-5}\,\text{T}.$$

The total field is up with a magnitude of

$$B = B_1 + B_2 = 4.8 \times 10^{-5}\,\text{T}.$$

EXERCISE B Suppose both I_1 and I_2 point into the page in Fig. 28–3. What then is the field midway between the two wires?

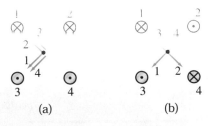

(a)   (b)

FIGURE 28–4 Example 28–3.

CONCEPTUAL EXAMPLE 28–3 | **Magnetic field due to four wires.** Figure 28–4 shows four long parallel wires which carry equal currents into or out of the page as shown. In which configuration, (a) or (b), is the magnetic field greater at the center of the square?

RESPONSE It is greater in (a). The arrows illustrate the directions of the field produced by each wire; check it out, using the right-hand rule to confirm these results. The net field at the center is the superposition of the four fields, which will point to the left in (a) and is zero in (b).

28–2 Force between Two Parallel Wires

We have seen that a wire carrying a current produces a magnetic field (magnitude given by Eq. 28–1 for a long straight wire). Also, a current-carrying wire feels a force when placed in a magnetic field (Section 27–3, Eq. 27–1). Thus, we expect that two current-carrying wires will exert a force on each other.

Consider two long parallel wires separated by a distance d, as in Fig. 28–5a. They carry currents I_1 and I_2, respectively. Each current produces a magnetic field that is "felt" by the other, so each must exert a force on the other. For example, the magnetic field B_1 produced by I_1 in Fig. 28–5 is given by Eq. 28–1, which at the location of wire 2 is

$$B_1 = \frac{\mu_0}{2\pi} \frac{I_1}{d}.$$

See Fig. 28–5b, where the field due *only* to I_1 is shown. According to Eq. 27–2, the force F_2 exerted by B_1 on a length ℓ_2 of wire 2, carrying current I_2, is

$$F_2 = I_2 B_1 \ell_2.$$

Note that the force on I_2 is due only to the field produced by I_1. Of course, I_2 also produces a field, but it does not exert a force on itself. We substitute B_1 into the formula for F_2 and find that the force on a length ℓ_2 of wire 2 is

$$F_2 = \frac{\mu_0}{2\pi} \frac{I_1 I_2}{d} \ell_2. \qquad \text{[parallel wires]} \quad \textbf{(28–2)}$$

If we use right-hand-rule-1 of Fig. 27–8c, we see that the lines of B_1 are as shown in Fig. 28–5b. Then using right-hand-rule-2 of Fig. 27–11c, we see that the force exerted on I_2 will be to the left in Fig. 28–5b. That is, I_1 exerts an attractive force on I_2 (Fig. 28–6a). This is true as long as the currents are in the same direction. If I_2 is in the opposite direction, the right-hand rule indicates that the force is in the opposite direction. That is, I_1 exerts a repulsive force on I_2 (Fig. 28–6b).

Reasoning similar to that above shows that the magnetic field produced by I_2 exerts an equal but opposite force on I_1. We expect this to be true also from Newton's third law. Thus, as shown in Fig. 28–6, parallel currents in the same direction attract each other, whereas parallel currents in opposite directions repel.

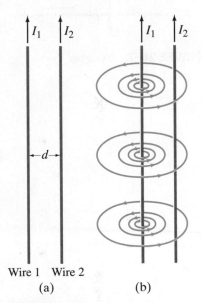

FIGURE 28–5 (a) Two parallel conductors carrying currents I_1 and I_2. (b) Magnetic field $\vec{B}_1$ produced by I_1. (Field produced by I_2 is not shown.) $\vec{B}_1$ points into page at position of I_2.

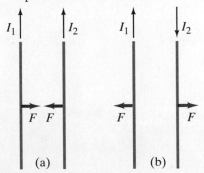

FIGURE 28–6 (a) Parallel currents in the same direction exert an attractive force on each other. (b) Antiparallel currents (in opposite directions) exert a repulsive force on each other.

EXAMPLE 28–4 **Force between two current-carrying wires.** The two wires of a 2.0-m-long appliance cord are 3.0 mm apart and carry a current of 8.0 A dc. Calculate the force one wire exerts on the other.

APPROACH Each wire is in the magnetic field of the other when the current is on, so we can apply Eq. 28–2.

SOLUTION Equation 28–2 gives

$$F = \frac{(4\pi \times 10^{-7}\,\text{T·m/A})(8.0\,\text{A})^2(2.0\,\text{m})}{(2\pi)(3.0 \times 10^{-3}\,\text{m})} = 8.5 \times 10^{-3}\,\text{N}.$$

The currents are in opposite directions (one toward the appliance, the other away from it), so the force would be repulsive and tend to spread the wires apart.

EXAMPLE 28–5 **Suspending a wire with a current.** A horizontal wire carries a current $I_1 = 80$ A dc. A second parallel wire 20 cm below it (Fig. 28–7) must carry how much current I_2 so that it doesn't fall due to gravity? The lower wire has a mass of 0.12 g per meter of length.

APPROACH If wire 2 is not to fall under gravity, which acts downward, the magnetic force on it must be upward. This means that the current in the two wires must be in the same direction (Fig. 28–6). We can find the current I_2 by equating the magnitudes of the magnetic force and the gravitational force on the wire.

SOLUTION The force of gravity on wire 2 is downward. For each 1.0 m of wire length, the gravitational force has magnitude

$$F = mg = (0.12 \times 10^{-3}\,\text{kg/m})(1.0\,\text{m})(9.8\,\text{m/s}^2) = 1.18 \times 10^{-3}\,\text{N}.$$

The magnetic force on wire 2 must be upward, and Eq. 28–2 gives

$$F = \frac{\mu_0}{2\pi} \frac{I_1 I_2}{d} \ell$$

where $d = 0.20$ m and $I_1 = 80$ A. We solve this for I_2 and set the two force magnitudes equal (letting $\ell = 1.0$ m):

$$I_2 = \frac{2\pi d}{\mu_0 I_1}\left(\frac{F}{\ell}\right) = \frac{2\pi(0.20\,\text{m})}{(4\pi \times 10^{-7}\,\text{T·m/A})(80\,\text{A})}\frac{(1.18 \times 10^{-3}\,\text{N/m})}{(1.0\,\text{m})} = 15\,\text{A}.$$

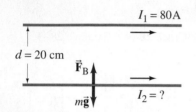

$I_1 = 80\,\text{A}$

$d = 20\,\text{cm}$

$\vec{\mathbf{F}}_B$

$m\vec{\mathbf{g}}$

$I_2 = ?$

FIGURE 28–7 Example 28–5.

28–3 Definitions of the Ampere and the Coulomb

You may have wondered how the constant μ_0 in Eq. 28–1 could be exactly $4\pi \times 10^{-7}$ T·m/A. Here is how it happened. With an older definition of the ampere, μ_0 was measured experimentally to be very close to this value. Today, μ_0 is *defined* to be exactly $4\pi \times 10^{-7}$ T·m/A. This could not be done if the ampere were defined independently. The ampere, the unit of current, is now defined in terms of the magnetic field B it produces using the defined value of μ_0.

In particular, we use the force between two parallel current-carrying wires, Eq. 28–2, to define the ampere precisely. If $I_1 = I_2 = 1$ A exactly, and the two wires are exactly 1 m apart, then

$$\frac{F}{\ell} = \frac{\mu_0}{2\pi} \frac{I_1 I_2}{d} = \frac{(4\pi \times 10^{-7}\,\text{T·m/A})}{(2\pi)}\frac{(1\,\text{A})(1\,\text{A})}{(1\,\text{m})} = 2 \times 10^{-7}\,\text{N/m}.$$

Thus, *one* **ampere** *is defined as that current flowing in each of two long parallel wires 1 m apart, which results in a force of exactly 2×10^{-7} N per meter of length of each wire.*

This is the precise definition of the ampere. The **coulomb** is then defined as being *exactly* one ampere-second: $1 \text{ C} = 1 \text{ A} \cdot \text{s}$. The value of k or ϵ_0 in Coulomb's law (Section 21–5) is obtained from experiment.

This may seem a rather roundabout way of defining quantities. The reason behind it is the desire for **operational definitions** of quantities—that is, definitions of quantities that can actually be measured given a definite set of operations to carry out. For example, the unit of charge, the coulomb, could be defined in terms of the force between two equal charges after defining a value for ϵ_0 or k in Eqs. 21–1 or 21–2. However, to carry out an actual experiment to measure the force between two charges is very difficult. For one thing, any desired amount of charge is not easily obtained precisely; and charge tends to leak from objects into the air. The amount of current in a wire, on the other hand, can be varied accurately and continuously (by putting a variable resistor in a circuit). Thus the force between two current-carrying conductors is far easier to measure precisely. This is why we first define the ampere, and then define the coulomb in terms of the ampere. At the National Institute of Standards and Technology in Maryland, precise measurement of current is made using circular coils of wire rather than straight lengths because it is more convenient and accurate.

Electric and magnetic field strengths are also defined operationally: the electric field in terms of the measurable force on a charge, via Eq. 21–3; and the magnetic field in terms of the force per unit length on a current-carrying wire, via Eq. 27–2.

28–4 Ampère's Law

In Section 28–1 we saw that Eq. 28–1 gives the relation between the current in a long straight wire and the magnetic field it produces. This equation is valid *only* for a long straight wire. Is there a general relation between a current in a wire of any shape and the magnetic field around it? The answer is yes: the French scientist André Marie Ampère (1775–1836) proposed such a relation shortly after Oersted's discovery. Consider an arbitrary closed path around a current as shown in Fig. 28–8, and imagine this path as being made up of short segments each of length $\Delta\ell$. First, we take the product of the length of each segment times the component of $\vec{B}$ parallel to that segment (call this component $B_\parallel$). If we now sum all these terms, according to Ampère, the result will be equal to μ_0 times the net current I_{encl} that passes through the surface enclosed by the path:

$$\sum B_\parallel \, \Delta\ell = \mu_0 I_{\text{encl}}.$$

The lengths $\Delta\ell$ are chosen so that $B_\parallel$ is essentially constant along each length. The sum must be made over a *closed path*; and I_{encl} is the net current passing through the surface bounded by this closed path (orange in Fig. 28–8). In the limit $\Delta\ell \to 0$, this relation becomes

$$\oint \vec{B} \cdot d\vec{\ell} = \mu_0 I_{\text{encl}}, \qquad (28\text{–}3)$$

where $d\vec{\ell}$ is an infinitesimal length vector and the vector dot product assures that the parallel component of $\vec{B}$ is taken. Equation 28–3 is known as **Ampère's law**. The integrand in Eq. 28–3 is taken around a closed path, and I_{encl} is the current passing through the space enclosed by the chosen path or loop.

FIGURE 28–8 Arbitrary path enclosing a current, for Ampère's law. The path is broken down into segments of equal length $\Delta\ell$.

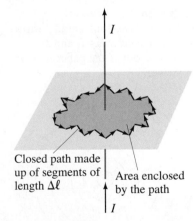

Closed path made up of segments of length $\Delta\ell$

Area enclosed by the path

AMPÈRE'S LAW

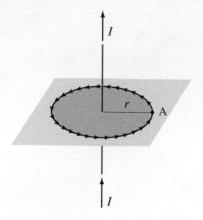

FIGURE 28–9 Circular path of radius r.

To understand Ampère's law better, let us apply it to the simple case of a single long straight wire carrying a current I which we've already examined, and which served as an inspiration for Ampère himself. Suppose we want to find the magnitude of $\vec{B}$ at some point A which is a distance r from the wire (Fig. 28–9). We know the magnetic field lines are circles with the wire at their center. So to apply Eq. 28–3 we choose as our path of integration a circle of radius r. The choice of path is ours, so we choose one that will be convenient: at any point on this circular path, $\vec{B}$ will be tangent to the circle. Furthermore, since all points on the path are the same distance from the wire, by symmetry we expect B to have the same magnitude at each point. Thus for any short segment of the circle (Fig. 28–9), $\vec{B}$ will be parallel to that segment, and (setting $I_{\text{encl}} = I$)

$$\mu_0 I = \oint \vec{B} \cdot d\vec{\ell}$$

$$= \oint B \, d\ell = B \oint d\ell = B(2\pi r),$$

where $\oint d\ell = 2\pi r$, the circumference of the circle. We solve for B and obtain

$$B = \frac{\mu_0 I}{2\pi r}.$$

This is just Eq. 28–1 for the field near a long straight wire as discussed earlier.

Ampère's law thus works for this simple case. A great many experiments indicate that Ampère's law is valid in general. However, as with Gauss's law for the electric field, its practical value as a means to calculate the magnetic field is limited mainly to simple or symmetric situations. Its importance is that it relates the magnetic field to the current in a direct and mathematically elegant way. Ampère's law is thus considered one of the basic laws of electricity and magnetism. It is valid for any situation where the currents and fields are steady and not changing in time, and no magnetic materials are present.

We now can see why the constant in Eq. 28–1 is written $\mu_0 / 2\pi$. This is done so that only μ_0 appears in Eq. 28–3, rather than, say, $2\pi k$ if we had used k in Eq. 28–1. In this way, the more fundamental equation, Ampère's law, has the simpler form.

It should be noted that the $\vec{B}$ in Ampère's law is not necessarily due only to the current I_{encl}. Ampère's law, like Gauss's law for the electric field, is valid in general. $\vec{B}$ is the field at each point in space along the chosen path due to all sources—including the current I enclosed by the path, but also due to any other sources. For example, the field surrounding two parallel current-carrying wires is the vector sum of the fields produced by each, and the field lines are shown in Fig. 28–10. If the path chosen for the integral (Eq. 28–3) is a circle centered on one of the wires with radius less than the distance between the wires (the dashed line in Fig. 28–10), only the current (I_1) in the encircled wire is included on the right side of Eq. 28–3. $\vec{B}$ on the left side of the equation must be the total $\vec{B}$ at each point due to both wires. Note also that $\oint \vec{B} \cdot d\vec{\ell}$ for the path shown in Fig. 28–10 is the same whether the second wire is present or not (in both cases, it equals $\mu_0 I_1$ according to Ampère's law). How can this be? It can be so because the fields due to the two wires partially cancel one another at some points between them, such as point C in the diagram ($\vec{B} = 0$ at a point midway between the wires if $I_1 = I_2$); at other points, such as D in Fig. 28–10, the fields add together to produce a larger field. In the *sum*, $\oint \vec{B} \cdot d\vec{\ell}$, these effects just balance so that $\oint \vec{B} \cdot d\vec{\ell} = \mu_0 I_1$, whether the second wire is there or not. The integral $\oint \vec{B} \cdot d\vec{\ell}$ will be the same in each case, even though $\vec{B}$ will not be the same at every point for each of the two cases.

FIGURE 28–10 Magnetic field lines around two long parallel wires whose equal currents, I_1 and I_2, are coming out of the paper toward the viewer.

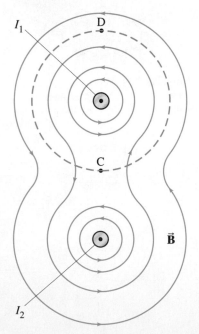

EXAMPLE 28–6 **Field inside and outside a wire.** A long straight cylindrical wire conductor of radius R carries a current I of uniform current density in the conductor. Determine the magnetic field due to this current at (a) points outside the conductor ($r > R$), and (b) points inside the conductor ($r < R$). See Fig. 28–11. Assume that r, the radial distance from the axis, is much less than the length of the wire. (c) If $R = 2.0$ mm and $I = 60$ A, what is B at $r = 1.0$ mm, $r = 2.0$ mm, and $r = 3.0$ mm?

APPROACH We can use *symmetry*: Because the wire is long, straight, and cylindrical, we expect from symmetry that the magnetic field must be the same at all points that are the same distance from the center of the conductor. There is no reason why any such point should have preference over others at the same distance from the wire (they are physically equivalent). So B must have the same value at all points the same distance from the center. We also expect $\vec{\mathbf{B}}$ to be tangent to circles around the wire (Fig. 28–1), so we choose a circular path of integration as we did in Fig. 28–9.

SOLUTION (a) We apply Ampère's law, integrating around a circle ($r > R$) centered on the wire (Fig. 28–11a), and then $I_{encl} = I$:

$$\oint \vec{\mathbf{B}} \cdot d\vec{\boldsymbol{\ell}} = B(2\pi r) = \mu_0 I_{encl}$$

or

$$B = \frac{\mu_0 I}{2\pi r}, \qquad\qquad [r > R]$$

which is the same result as for a thin wire.

(b) Inside the wire ($r < R$), we again choose a circular path concentric with the cylinder; we expect $\vec{\mathbf{B}}$ to be tangential to this path, and again, because of the symmetry, it will have the same magnitude at all points on the circle. The current enclosed in this case is less than I by a factor of the ratio of the areas:

$$I_{encl} = I \frac{\pi r^2}{\pi R^2}.$$

So Ampère's law gives

$$\oint \vec{\mathbf{B}} \cdot d\vec{\boldsymbol{\ell}} = \mu_0 I_{encl}$$

$$B(2\pi r) = \mu_0 I\left(\frac{\pi r^2}{\pi R^2}\right)$$

so

$$B = \frac{\mu_0 I r}{2\pi R^2}. \qquad\qquad [r < R]$$

The field is zero at the center of the conductor and increases linearly with r until $r = R$; beyond $r = R$, B decreases as $1/r$. This is shown in Fig. 28–11b. Note that these results are valid only for points close to the center of the conductor as compared to its length. For a current to flow, there must be connecting wires (to a battery, say), and the field due to these conducting wires, if not very far away, will destroy the assumed symmetry.

(c) At $r = 2.0$ mm, the surface of the wire, $r = R$, so

$$B = \frac{\mu_0 I}{2\pi R} = \frac{(4\pi \times 10^{-7} \text{ T·m/A})(60 \text{ A})}{(2\pi)(2.0 \times 10^{-3} \text{ m})} = 6.0 \times 10^{-3} \text{ T}.$$

We saw in (b) that inside the wire B is linear in r. So at $r = 1.0$ mm, B will be half what it is at $r = 2.0$ mm or 3.0×10^{-3} T. Outside the wire, B falls off as $1/r$, so at $r = 3.0$ mm it will be two-thirds as great as at $r = 2.0$ mm, or $B = 4.0 \times 10^{-3}$ T. To check, we use our result in (a), $B = \mu_0 I/2\pi r$, which gives the same result.

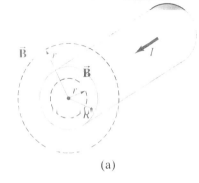
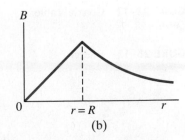

(a)

(b)

FIGURE 28–11 Magnetic field inside and outside a cylindrical conductor (Example 28–6).

⚠ **CAUTION**

Connecting wires can destroy assumed symmetry

Coaxial cable
(shielding)

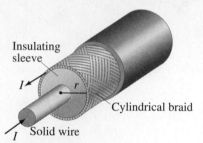

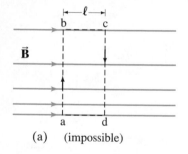

FIGURE 28–12 Coaxial cable.
Example 28–7.

FIGURE 28–13 Exercise C.

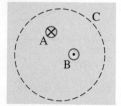

FIGURE 28–14 Example 28–8.

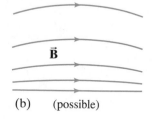

(a)   (impossible)

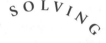

$\vec{B}$

(b)   (possible)

CONCEPTUAL EXAMPLE 28–7 Coaxial cable. A *coaxial cable* is a single wire surrounded by a cylindrical metallic braid, as shown in Fig. 28–12. The two conductors are separated by an insulator. The central wire carries current to the other end of the cable, and the outer braid carries the return current and is usually considered ground. Describe the magnetic field (*a*) in the space between the conductors, and (*b*) outside the cable.

RESPONSE (*a*) In the space between the conductors, we can apply Ampère's law for a circular path around the center wire, just as we did for the case shown in Figs. 28–9 and 28–11. The magnetic field lines will be concentric circles centered on the center of the wire, and the magnitude is given by Eq. 28–1. The current in the outer conductor has no bearing on this result. (Ampère's law uses only the current enclosed *inside* the path; as long as the currents outside the path don't affect the *symmetry* of the field, they do not contribute to the field along the path at all). (*b*) Outside the cable, we can draw a similar circular path, for we expect the field to have the same cylindrical symmetry. Now, however, there are two currents enclosed by the path, and they add up to zero. The field outside the cable is zero.

The nice feature of coaxial cables is that they are self-shielding: no stray magnetic fields exist outside the cable. The outer cylindrical conductor also shields external electric fields from coming in (see also Example 21–14). This makes them ideal for carrying signals near sensitive equipment. Audiophiles use coaxial cables between stereo equipment components and even to the loudspeakers.

EXERCISE C In Fig. 28–13, A and B are wires each carrying a 3.0-A current but in opposite directions. On the circle C, which statement is true? (*a*) $B = 0$; (*b*) $\oint \vec{B} \cdot d\vec{\ell} = 0$; (*c*) $B = 3\mu_0$; (*d*) $B = -3\mu_0$; (*e*) $\oint \vec{B} \cdot d\vec{\ell} = 6\mu_0$.

EXAMPLE 28–8 A nice use for Ampère's law. Use Ampère's law to show that in any region of space where there are no currents the magnetic field cannot be both unidirectional and nonuniform as shown in Fig. 28–14a.

APPROACH The wider spacing of lines near the top of Fig. 28–14a indicates the field $\vec{B}$ has a smaller magnitude at the top than it does lower down. We apply Ampère's law to the rectangular path abcd shown dashed in Fig. 28–14a.

SOLUTION Because no current is enclosed by the chosen path, Ampère's law gives

$$\oint \vec{B} \cdot d\vec{\ell} = 0.$$

The integral along sections ab and cd is zero, since $\vec{B} \perp d\vec{\ell}$. Thus

$$\oint \vec{B} \cdot d\vec{\ell} = B_{bc}\ell - B_{da}\ell = (B_{bc} - B_{da})\ell,$$

which is not zero since the field B_{bc} along the path bc is less than the field B_{da} along path da. Hence we have a contradiction: $\oint \vec{B} \cdot d\vec{\ell}$ cannot be both zero (since $I = 0$) and nonzero. Thus we have shown that a nonuniform unidirectional field is not consistent with Ampère's law. A nonuniform field whose direction also changes, as in Fig. 28–14b, is consistent with Ampère's law (convince yourself this is so), and possible. The fringing of a permanent magnet's field (Fig. 27–7) has this shape.

Ampère's Law

1. Ampère's law, like Gauss's law, is always a valid statement. But as a calculation tool it is limited mainly to systems with a high degree of symmetry. The first step in applying Ampère's law is to identify useful **symmetry**.

2. Choose an integration path that reflects the symmetry (see the Examples). Search for paths where B has constant magnitude along the entire path or along segments of the path. Make sure your integration path passes through the point where you wish to evaluate the magnetic field.

3. Use symmetry to determine the direction of $\vec{B}$ along the integration path. With a smart choice of path, $\vec{B}$ will be either parallel or perpendicular to the path.

4. Determine the enclosed current, I_{encl}. Be careful with signs. Let the fingers of your right hand curl along the direction of $\vec{B}$ so that your thumb shows the direction of positive current. If you have a solid conductor and your integration path does not enclose the full current, you can use the current density (current per unit area) multiplied by the enclosed area (as in Example 28–6).

Magnetic Field of a Solenoid and a Toroid

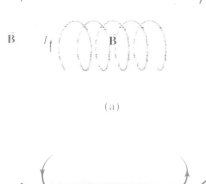

A long coil of wire consisting of many loops is called a **solenoid**. Each loop produces a magnetic field as was shown in Fig. 27–9. In Fig. 28–15a, we see the field due to a solenoid when the coils are far apart. Near each wire, the field lines are very nearly circles as for a straight wire (that is, at distances that are small compared to the curvature of the wire). Between any two wires, the fields due to each loop tend to cancel. Toward the center of the solenoid, the fields add up to give a field that can be fairly large and fairly uniform. For a long solenoid with closely packed coils, the field is nearly uniform and parallel to the solenoid axis within the entire cross section, as shown in Fig. 28–15b. The field outside the solenoid is very small compared to the field inside, except near the ends. Note that the same number of field lines that are concentrated inside the solenoid, spread out into the vast open space outside.

We now use Ampère's law to determine the magnetic field inside a very long (ideally, infinitely long) closely packed solenoid. We choose the path abcd shown in Fig. 28–16, far from either end, for applying Ampère's law. We will consider this path as made up of four segments, the sides of the rectangle: ab, bc, cd, da. Then the left side of Eq. 28–3, Ampère's law, becomes

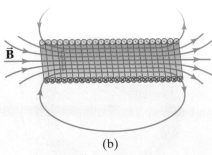

(a)

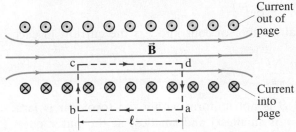

(b)

FIGURE 28–15 Magnetic field due to a solenoid: (a) loosely spaced turns, (b) closely spaced turns.

$$\oint \vec{B} \cdot d\vec{\ell} = \int_a^b \vec{B} \cdot d\vec{\ell} + \int_b^c \vec{B} \cdot d\vec{\ell} + \int_c^d \vec{B} \cdot d\vec{\ell} + \int_d^a \vec{B} \cdot d\vec{\ell}.$$

The field outside the solenoid is so small as to be negligible compared to the field inside. Thus the first term in this sum will be zero. Furthermore, $\vec{B}$ is perpendicular to the segments bc and da inside the solenoid, and is nearly zero between and outside the coils,

FIGURE 28–16 Cross-sectional view into a solenoid. The magnetic field inside is straight except at the ends. Red dashed lines indicate the path chosen for use in Ampère's law. $\odot$ and $\otimes$ are electric current direction (in the wire loops) out of the page and into the page.

so these terms too are zero. Therefore we have reduced the integral to the segment cd where $\vec{B}$ is the nearly uniform field inside the solenoid, and is parallel to $d\vec{\ell}$, so

$$\oint \vec{B} \cdot d\vec{\ell} = \int_c^d \vec{B} \cdot d\vec{\ell} = B\ell,$$

where ℓ is the length cd. Now we determine the current enclosed by this loop for the right side of Ampère's law, Eq. 28–3. If a current I flows in the wire of the solenoid, the total current enclosed by our path abcd is NI where N is the number of loops our path encircles (five in Fig. 28–16). Thus Ampère's law gives us

$$B\ell = \mu_0 NI.$$

If we let $n = N/\ell$ be the *number of loops per unit length*, then

$$B = \mu_0 nI. \qquad \text{[solenoid]} \quad (28\text{–}4)$$

This is the magnitude of the magnetic field within a solenoid. Note that B depends only on the number of loops per unit length, n, and the current I. The field does not depend on position within the solenoid, so B is uniform. This is strictly true only for an infinite solenoid, but it is a good approximation for real ones for points not close to the ends.

EXAMPLE 28–9 **Field inside a solenoid.** A thin 10-cm-long solenoid used for fast electromechanical switching has a total of 400 turns of wire and carries a current of 2.0 A. Calculate the field inside near the center.

APPROACH We use Eq. 28–4, where the number of turns per unit length is $n = 400/0.10 \text{ m} = 4.0 \times 10^3 \text{ m}^{-1}$.

SOLUTION $B = \mu_0 nI = (4\pi \times 10^{-7} \text{ T} \cdot \text{m/A})(4.0 \times 10^3 \text{ m}^{-1})(2.0 \text{ A}) = 1.0 \times 10^{-2} \text{ T}.$

A close look at Fig. 28–15 shows that the field outside of a solenoid is much like that of a bar magnet (Fig. 27–4). Indeed, a solenoid acts like a magnet, with one end acting as a north pole and the other as south pole, depending on the direction of the current in the loops. Since magnetic field lines leave the north pole of a magnet, the north poles of the solenoids in Fig. 28–15 are on the right.

Solenoids have many practical applications, and we discuss some of them later in the Chapter, in Section 28–8.

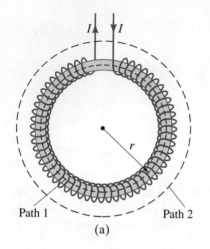

(b)

FIGURE 28–17 (a) A toroid. (b) A section of the toroid showing direction of the current for three loops: ⊙ means current toward you, ⊗ means current away from you.

EXAMPLE 28–10 **Toroid.** Use Ampère's law to determine the magnetic field (a) inside and (b) outside a toroid, which is like a solenoid bent into the shape of a circle as shown in Fig. 28–17a.

APPROACH The magnetic field lines inside the toroid will be circles concentric with the toroid. (If you think of the toroid as a solenoid bent into a circle, the field lines bend along with the solenoid.) The direction of $\vec{\mathbf{B}}$ is clockwise. We choose as our path of integration one of these field lines of radius r inside the toroid as shown by the dashed line labeled "path 1" in Fig. 28–17a. We make this choice to use the *symmetry* of the situation, so B will be tangent to the path and will have the same magnitude at all points along the path (although it is not necessarily the same across the whole cross section of the toroid). This chosen path encloses *all* the coils; if there are N coils, each carrying current I, then $I_{\text{encl}} = NI$.

SOLUTION (a) Ampère's law applied along this path gives

$$\oint \vec{\mathbf{B}} \cdot d\vec{\boldsymbol{\ell}} = \mu_0 I_{\text{encl}}$$

$$B(2\pi r) = \mu_0 NI,$$

where N is the total number of coils and I is the current in each of the coils. Thus

$$B = \frac{\mu_0 NI}{2\pi r}.$$

The magnetic field B is not uniform within the toroid: it is largest along the inner edge (where r is smallest) and smallest at the outer edge. However, if the toroid is large, but thin (so that the difference between the inner and outer radii is small compared to the average radius), the field will be essentially uniform within the toroid. In this case, the formula for B reduces to that for a straight solenoid $B = \mu_0 nI$ where $n = N/(2\pi r)$ is the number of coils per unit length. (b) Outside the toroid, we choose as our path of integration a circle concentric with the toroid, "path 2" in Fig. 28–17a. This path encloses N loops carrying current I in one direction and N loops carrying the same current in the opposite direction. (Figure 28–17b shows the directions of the current for the parts of the loop on the inside and outside of the toroid.) Thus the net current enclosed by path 2 is zero. For a very tightly packed toroid, all points on path 2 are equidistant from the toroid and equivalent, so we expect B to be the same at all points along the path. Hence, Ampère's law gives

$$\oint \vec{\mathbf{B}} \cdot d\vec{\boldsymbol{\ell}} = \mu_0 I_{\text{encl}}$$

$$B(2\pi r) = 0$$

or

$$B = 0.$$

The same is true for a path taken at a radius smaller than that of the toroid. So there is no field exterior to a very tightly wound toroid. It is all inside the loops.

Biot-Savart Law

The usefulness of Ampère's law for determining the magnetic field $\vec{B}$ due to particular electric currents is restricted to situations where the symmetry of the given currents allows us to evaluate $\oint \vec{B} \cdot d\vec{\ell}$ readily. This does not, of course, invalidate Ampère's law nor does it reduce its fundamental importance. Recall the electric case, where Gauss's law is considered fundamental but is limited in its use for actually calculating $\vec{E}$. We must often determine the electric field $\vec{E}$ by another method summing over contributions due to infinitesimal charge elements dq via Coulomb's law: $dE = (1/4\pi\epsilon_0)(dq/r^2)$. A magnetic equivalent to this infinitesimal form of Coulomb's law would be helpful for currents that do not have great symmetry. Such a law was developed by Jean Baptiste Biot (1774–1862) and Felix Savart (1791–1841) shortly after Oersted's discovery in 1820 that a current produces a magnetic field.

According to Biot and Savart, a current I flowing in any path can be considered as many tiny (infinitesimal) current elements, such as in the wire of Fig. 28–18. If $d\vec{\ell}$ represents any infinitesimal length along which the current is flowing, then the magnetic field, $d\vec{B}$, at any point P in space, due to this element of current, is given by

$$d\vec{B} = \frac{\mu_0 I}{4\pi} \frac{d\vec{\ell} \times \hat{r}}{r^2}, \qquad (28\text{--}5)$$

Biot-Savart law

where $\vec{r}$ is the displacement vector from the element $d\vec{\ell}$ to the point P, and $\hat{r} = \vec{r}/r$ is the unit vector (magnitude = 1) in the direction of $\vec{r}$ (see Fig. 28–18).

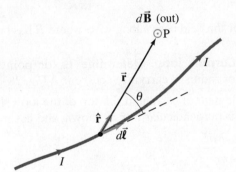

FIGURE 28–18 Biot-Savart law: the field at P due to current element $I d\vec{\ell}$ is $d\vec{B} = (\mu_0 I/4\pi)(d\vec{\ell} \times \hat{r}/r^2)$.

Equation 28–5 is known as the **Biot-Savart law**. The magnitude of $d\vec{B}$ is

$$dB = \frac{\mu_0 I d\ell \sin\theta}{4\pi r^2}, \qquad (28\text{--}6)$$

where θ is the angle between $d\vec{\ell}$ and $\vec{r}$ (Fig. 28–18). The total magnetic field at point P is then found by summing (integrating) over all current elements:

$$\vec{B} = \int d\vec{B} = \frac{\mu_0 I}{4\pi} \int \frac{d\vec{\ell} \times \hat{r}}{r^2}.$$

Note that this is a *vector* sum. The Biot-Savart law is the magnetic equivalent of Coulomb's law in its infinitesimal form. It is even an inverse square law, like Coulomb's law.

An important difference between the Biot-Savart law and Ampère's law (Eq. 28–3) is that in Ampère's law $[\oint \vec{B} \cdot d\vec{\ell} = \mu_0 I_{encl}]$, $\vec{B}$ is not necessarily due only to the current enclosed by the path of integration. But in the Biot-Savart law the field $d\vec{B}$ in Eq. 28–5 is due only, and entirely, to the current element $I d\vec{\ell}$. To find the total $\vec{B}$ at any point in space, it is necessary to include *all* currents.

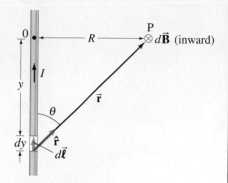

FIGURE 28–19 Determining $\vec{\mathbf{B}}$ due to a long straight wire using the Biot-Savart law.

EXAMPLE 28–11 **$\vec{\mathbf{B}}$ due to current I in straight wire.** For the field near a long straight wire carrying a current I, show that the Biot-Savart law gives the same result as Eq. 28–1, $B = \mu_0 I/2\pi r$.

APPROACH We calculate the magnetic field in Fig. 28–19 at point P, which is a perpendicular distance R from an infinitely long wire. The current is moving upwards, and both $d\vec{\boldsymbol{\ell}}$ and $\hat{\mathbf{r}}$, which appear in the cross product of Eq. 28–5, are in the plane of the page. Hence the direction of the field $d\vec{\mathbf{B}}$ due to each element of current must be directed into the plane of the page as shown (right-hand rule for the cross product $d\vec{\boldsymbol{\ell}} \times \hat{\mathbf{r}}$). Thus all the $d\vec{\mathbf{B}}$ have the same direction at point P, and add up to give $\vec{\mathbf{B}}$ the same direction consistent with our previous results (Figs. 28–1 and 28–11).

SOLUTION The magnitude of $\vec{\mathbf{B}}$ will be

$$ B = \frac{\mu_0 I}{4\pi} \int_{y=-\infty}^{+\infty} \frac{dy \sin\theta}{r^2}, $$

where $dy = d\ell$ and $r^2 = R^2 + y^2$. Note that we are integrating over y (the length of the wire) so R is considered constant. Both y and θ are variables, but they are not independent. In fact, $y = -R/\tan\theta$. Note that we measure y as positive upward from point 0, so for the current element we are considering $y < 0$. Then

$$ dy = +R \csc^2\theta\, d\theta = \frac{R\, d\theta}{\sin^2\theta} = \frac{R\, d\theta}{(R/r)^2} = \frac{r^2\, d\theta}{R}. $$

From Fig. 28–19 we can see that $y = -\infty$ corresponds to $\theta = 0$ and that $y = +\infty$ corresponds to $\theta = \pi$ radians. So our integral becomes

$$ B = \frac{\mu_0 I}{4\pi} \frac{1}{R} \int_{\theta=0}^{\pi} \sin\theta\, d\theta = -\left.\frac{\mu_0 I}{4\pi R} \cos\theta\right|_0^\pi = \frac{\mu_0 I}{2\pi R}. $$

This is just Eq. 28–1 for the field near a long wire, where R has been used instead of r.

EXAMPLE 28–12 **Current loop.** Determine $\vec{\mathbf{B}}$ for points on the axis of a circular loop of wire of radius R carrying a current I, Fig. 28–20.

APPROACH For an element of current at the top of the loop, the magnetic field $d\vec{\mathbf{B}}$ at point P on the axis is perpendicular to $\vec{\mathbf{r}}$ as shown, and has magnitude (Eq. 28–5)

$$ dB = \frac{\mu_0 I\, d\ell}{4\pi r^2} $$

since $d\vec{\boldsymbol{\ell}}$ is perpendicular to $\vec{\mathbf{r}}$ so $|d\vec{\boldsymbol{\ell}} \times \hat{\mathbf{r}}| = d\ell$. We can break $d\vec{\mathbf{B}}$ down into components $dB_\parallel$ and $dB_\perp$, which are parallel and perpendicular to the axis as shown.

SOLUTION When we sum over all the elements of the loop, *symmetry* tells us that the perpendicular components will cancel on opposite sides, so $B_\perp = 0$. Hence, the total $\vec{\mathbf{B}}$ will point along the axis, and will have magnitude

$$ B = B_\parallel = \int dB \cos\phi = \int dB\, \frac{R}{r} = \int dB\, \frac{R}{(R^2 + x^2)^{\frac{1}{2}}}, $$

where x is the distance of P from the center of the ring, and $r^2 = R^2 + x^2$. Now we put in dB from the equation above and integrate around the current loop, noting that all segments $d\vec{\boldsymbol{\ell}}$ of current are the same distance, $(R^2 + x^2)^{\frac{1}{2}}$, from point P:

$$ B = \frac{\mu_0 I}{4\pi} \frac{R}{(R^2 + x^2)^{\frac{3}{2}}} \int d\ell = \frac{\mu_0 I R^2}{2(R^2 + x^2)^{\frac{3}{2}}} $$

since $\int d\ell = 2\pi R$, the circumference of the loop.

NOTE At the very center of the loop (where $x = 0$) the field has its maximum value

$$ B = \frac{\mu_0 I}{2R}. \qquad \text{[at center of current loop]} $$

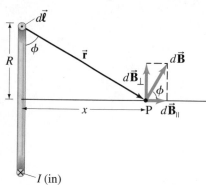

FIGURE 28–20 Determining $\vec{\mathbf{B}}$ due to a current loop.

Recall from Section 27–5 that a current loop, such as that just discussed (Fig. 28–20), is considered a **magnetic dipole**. We saw there that a current loop has a magnetic dipole moment

$$\mu = NIA,$$

where A is the area of the loop and N is the number of coils in the loop, each carrying current I. We also saw in Chapter 27 that a magnetic dipole placed in an external magnetic field experiences a torque and possesses potential energy, just like an electric dipole. In Example 28–12 we looked at another aspect of a magnetic dipole: the magnetic field *produced by* a magnetic dipole has magnitude, along the dipole axis, of

$$B = \frac{\mu_0 I R^2}{2(R^2 + x^2)^{\frac{3}{2}}}.$$

We can write this in terms of the magnetic dipole moment $\mu = IA = I\pi R^2$ (for a single loop $N = 1$):

$$B = \frac{\mu_0}{2\pi} \frac{\mu}{(R^2 + x^2)^{\frac{3}{2}}}. \qquad \text{[magnetic dipole]} \quad \textbf{(28–7a)}$$

(Be careful to distinguish μ for dipole moment from μ_0, the magnetic permeability constant.) For distances far from the loop, $x \gg R$, this becomes

$$B \approx \frac{\mu_0}{2\pi} \frac{\mu}{x^3}. \qquad \left[\begin{array}{c}\text{on axis,} \\ \text{magnetic dipole, } x \gg R\end{array}\right] \quad \textbf{(28–7b)}$$

The magnetic field on the axis of a magnetic dipole decreases with the cube of the distance, just as the electric field does for an electric dipole. B decreases as the cube of the distance also for points not on the axis, although the multiplying factor is not the same. The magnetic field due to a current loop can be determined at various points using the Biot-Savart law and the results are in accord with experiment. The field lines around a current loop are shown in Fig. 28–21.

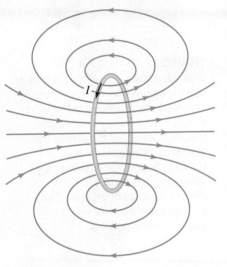

FIGURE 28–21 Magnetic field due to a circular loop of wire. (Same as Fig. 27–9.)

EXAMPLE 28–13 **$\vec{B}$ due to a wire segment.** One quarter of a circular loop of wire carries a current I as shown in Fig. 28–22. The current I enters and leaves on straight segments of wire, as shown; the straight wires are along the radial direction from the center C of the circular portion. Find the magnetic field at point C.

APPROACH The current in the straight sections produces no magnetic field at point C because $d\vec{\ell}$ and $\hat{r}$ in the Biot-Savart law (Eq. 28–5) are parallel and therefore $d\vec{\ell} \times \hat{r} = 0$. Each piece $d\vec{\ell}$ of the curved section of wire produces a field $d\vec{B}$ that points into the page at C (right-hand rule).

SOLUTION The magnitude of each $d\vec{B}$ due to each $d\ell$ of the circular portion of wire is (Eq. 28–6)

$$dB = \frac{\mu_0 I \, d\ell}{4\pi R^2}$$

where $r = R$ is the radius of the curved section, and $\sin\theta$ in Eq. 28–6 is $\sin 90° = 1$. With $r = R$ for all pieces $d\vec{\ell}$, we integrate over a quarter of a circle.

$$B = \int dB = \frac{\mu_0 I}{4\pi R^2} \int d\ell = \frac{\mu_0 I}{4\pi R^2} \left(\frac{1}{4} 2\pi R\right) = \frac{\mu_0 I}{8R}.$$

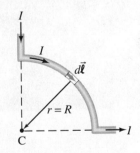

FIGURE 28–22 Example 28–13.

28-7 Magnetic Materials—Ferromagnetism

Magnetic fields can be produced (1) by magnetic materials (magnets) and (2) by electric currents. Common magnetic materials include ordinary magnets, iron cores in motors and electromagnets, recording tape, computer hard drives and magnetic stripes on credit cards. We saw in Section 27–1 that iron (and a few other materials) can be made into strong magnets. These materials are said to be **ferromagnetic**. We now look into the sources of ferromagnetism.

A bar magnet, with its two opposite poles near either end, resembles an electric dipole (equal-magnitude positive and negative charges separated by a distance). Indeed, a bar magnet is sometimes referred to as a "magnetic dipole." There are opposite "poles" separated by a distance. And the magnetic field lines of a bar magnet form a pattern much like that for the electric field of an electric dipole: compare Fig. 21–34a with Fig. 27–4 (or 28–24).

Microscopic examination reveals that a piece of iron is made up of tiny regions known as **domains**, less than 1 mm in length or width. Each domain behaves like a tiny magnet with a north and a south pole. In an unmagnetized piece of iron, these domains are arranged randomly, as shown in Fig. 28–23a. The magnetic effects of the domains cancel each other out, so this piece of iron is not a magnet. In a magnet, the domains are preferentially aligned in one direction as shown in Fig. 28–23b (downward in this case). A magnet can be made from an unmagnetized piece of iron by placing it in a strong magnetic field. (You can make a needle magnetic, for example, by stroking it with one pole of a strong magnet.) The magnetization direction of domains may actually rotate slightly to be more nearly parallel to the external field, and the borders of domains may move so domains with magnetic orientation parallel to the external field grow larger (compare Figs. 28–23a and b).

We can now explain how a magnet can pick up unmagnetized pieces of iron like paper clips. The field of the magnet's south pole (say) causes a slight realignment of the domains in the unmagnetized object, which then becomes a temporary magnet with its north pole facing the south pole of the permanent magnet; thus, attraction results. Similarly, elongated iron filings in a magnetic field acquire aligned domains and align themselves to reveal the shape of the magnetic field, Fig. 28–24.

An iron magnet can remain magnetized for a long time, and is referred to as a "permanent magnet." But if you drop a magnet on the floor or strike it with a hammer, you can jar the domains into randomness and the magnet loses some or all of its magnetism. Heating a permanent magnet can also cause loss of magnetism, for raising the temperature increases the random thermal motion of atoms, which tends to randomize the domains. Above a certain temperature known as the **Curie temperature** (1043 K for iron), a magnet cannot be made at all. Iron, nickel, cobalt, gadolinium, and certain alloys are ferromagnetic at room temperature; several other elements and alloys have low Curie temperature and thus are ferromagnetic only at low temperatures. Most other metals, such as aluminum and copper, do not show any noticeable magnetic effect (but see Section 28–10).

The striking similarity between the fields produced by a bar magnet and by a loop of electric current (Figs. 27–4b and 28–21) offers a clue that perhaps magnetic fields produced by electric currents may have something to do with ferromagnetism. According to modern atomic theory, atoms can be visualized as having electrons that orbit around a central nucleus. The electrons are charged, and so constitute an electric current and therefore produce a magnetic field; but the fields due to orbiting electrons generally all add up to zero. Electrons themselves produce an additional magnetic field, as if they and their electric charge were spinning about their own axes. It is the magnetic field due to electron **spin**[†] that is believed to produce ferromagnetism in most ferromagnetic materials.

It is believed today that *all* magnetic fields are caused by electric currents. This means that magnetic field lines always form closed loops, unlike electric field lines which begin on positive charges and end on negative charges.

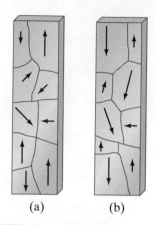

FIGURE 28–23 (a) An unmagnetized piece of iron is made up of domains that are randomly arranged. Each domain is like a tiny magnet; the arrows represent the magnetization direction, with the arrowhead being the N pole. (b) In a magnet, the domains are preferentially aligned in one direction (down in this case), and may be altered in size by the magnetization process.

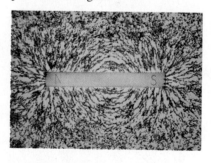

FIGURE 28–24 Iron filings line up along magnetic field lines due to a permanent magnet.

⚠ **CAUTION**

$\vec{B}$ *lines form closed loops,* $\vec{E}$ *start on* ⊕ *and end on* ⊖

[†]The name "spin" comes from an early suggestion that this intrinsic magnetic moment arises from the electron "spinning" on its axis (as well as "orbiting" the nucleus) to produce the extra field. However this view of a spinning electron is oversimplified and not valid.

EXERCISE D Return to the Chapter-Opening Question, page 733, and answer it again now. Try to explain why you may have answered differently the first time.

Electromagnets and Solenoids—Applications

A long coil of wire consisting of many loops of wire, as discussed in Section 28–5, is called a solenoid. The magnetic field within a solenoid can be fairly large since it will be the sum of the fields due to the current in each loop (see Fig. 28–25). The solenoid acts like a magnet; one end can be considered the north pole and the other the south pole, depending on the direction of the current in the loops (use the right-hand rule). Since the magnetic field lines leave the north pole of a magnet, the north pole of the solenoid in Fig. 28–25 is on the right.

If a piece of iron is placed inside a solenoid, the magnetic field is increased greatly because the domains of the iron are aligned by the magnetic field produced by the current. The resulting magnetic field is the sum of that due to the current and that due to the iron, and can be hundreds or thousands of times larger than that due to the current alone (see Section 28–9). This arrangement is called an **electromagnet**. The alloys of iron used in electromagnets acquire and lose their magnetism quite readily when the current is turned on or off, and so are referred to as "soft iron." (It is "soft" only in a magnetic sense.) Iron that holds its magnetism even when there is no externally applied field is called "hard iron." Hard iron is used in permanent magnets. Soft iron is usually used in electromagnets so that the field can be turned on and off readily. Whether iron is hard or soft depends on heat treatment, type of alloy, and other factors.

Electromagnets have many practical applications, from use in motors and generators to producing large magnetic fields for research. Sometimes an iron core is not present—the magnetic field comes only from the current in the wire coils. When the current flows continuously in a normal electromagnet, a great deal of waste heat (I^2R power) can be produced. Cooling coils, which are tubes carrying water, are needed to absorb the heat in larger installations.

For some applications, the current-carrying wires are made of superconducting material kept below the transition temperature (Section 25–9). Very high fields can be produced with superconducting wire without an iron core. No electric power is needed to maintain large current in the superconducting coils, which means large savings of electricity; nor must huge amounts of heat be dissipated. It is not a free ride, though, because energy is needed to keep the superconducting coils at the necessary low temperature.

Another useful device consists of a solenoid into which a rod of iron is partially inserted. This combination is also referred to as a solenoid. One simple use is as a doorbell (Fig. 28–26). When the circuit is closed by pushing the button, the coil effectively becomes a magnet and exerts a force on the iron rod. The rod is pulled into the coil and strikes the bell. A large solenoid is used in the starters of cars; when you engage the starter, you are closing a circuit that not only turns the starter motor, but activates a solenoid that first moves the starter into direct contact with the gears on the engine's flywheel. Solenoids are used as switches in many devices. They have the advantage of moving mechanical parts quickly and accurately.

*Magnetic Circuit Breakers

Modern circuit breakers that protect houses and buildings from overload and fire contain not only a "thermal" part (bimetallic strip as described in Section 25–6, Fig. 25–19) but also a magnetic sensor. If the current is above a certain level, the magnetic field it produces pulls an iron plate that breaks the same contact points as in Fig. 25–19b and c. In more sophisticated circuit breakers, including ground fault circuit interrupters (GFCIs—discussed in Section 29–8), a solenoid is used. The iron rod of Fig. 28–26, instead of striking a bell, strikes one side of a pair of points, opening them and opening the circuit. Magnetic circuit breakers react quickly (<10 msec), and for buildings are designed to react to the high currents of shorts (but not shut off for the start-up surges of motors).

FIGURE 28–25 Magnetic field of a solenoid. The north pole of this solenoid, thought of as a magnet, is on the right, and the south pole is on the left.

PHYSICS APPLIED
Electromagnets and solenoids

PHYSICS APPLIED
Doorbell, car starter

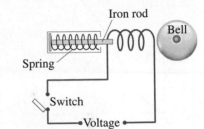

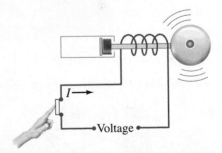

FIGURE 28–26 Solenoid used as a doorbell.

PHYSICS APPLIED
Magnetic circuit breakers

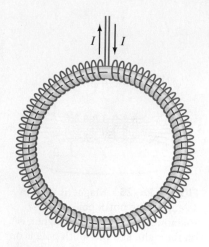

FIGURE 28–27 Iron-core toroid.

FIGURE 28–28 Total magnetic field B in an iron-core toroid as a function of the external field B_0 (B_0 is caused by the current I in the coil).

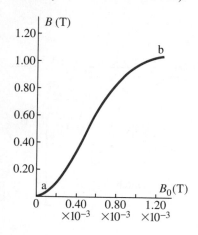

FIGURE 28–29 Hysteresis curve.

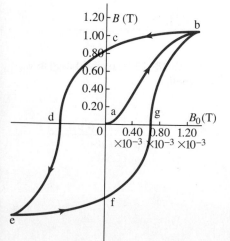

*28–9 Magnetic Fields in Magnetic Materials; Hysteresis

The field of a long solenoid is directly proportional to the current. Indeed, Eq. 28–4 tells us that the field B_0 inside a solenoid is given by

$$B_0 = \mu_0 n I.$$

This is valid if there is only air inside the coil. If we put a piece of iron or other ferromagnetic material inside the solenoid, the field will be greatly increased, often by hundreds or thousands of times. This occurs because the domains in the iron become preferentially aligned by the external field. The resulting magnetic field is the sum of that due to the current and that due to the iron. It is sometimes convenient to write the total field in this case as a sum of two terms:

$$\vec{\mathbf{B}} = \vec{\mathbf{B}}_0 + \vec{\mathbf{B}}_M. \tag{28–8}$$

Here, $\vec{\mathbf{B}}_0$ refers to the field due only to the current in the wire (the "external field"). It is the field that would be present in the absence of a ferromagnetic material. Then $\vec{\mathbf{B}}_M$ represents the additional field due to the ferromagnetic material itself; often $\vec{\mathbf{B}}_M \gg \vec{\mathbf{B}}_0$.

The total field inside a solenoid in such a case can also be written by replacing the constant μ_0 in Eq. 28–4 by another constant, μ, characteristic of the material inside the coil:

$$B = \mu n I; \tag{28–9}$$

μ is called the **magnetic permeability** of the material (do not confuse it with $\vec{\boldsymbol{\mu}}$ for magnetic dipole moment). For ferromagnetic materials, μ is much greater than μ_0. For all other materials, its value is very close to μ_0 (Section 28–10). The value of μ, however, is not constant for ferromagnetic materials; it depends on the value of the external field B_0, as the following experiment shows.

Measurements on magnetic materials are generally done using a toroid, which is essentially a long solenoid bent into the shape of a circle (Fig. 28–27), so that practically all the lines of $\vec{\mathbf{B}}$ remain within the toroid. Suppose the toroid has an iron core that is initially unmagnetized and there is no current in the windings of the toroid. Then the current I is slowly increased, and B_0 increases linearly with I. The total field B also increases, but follows the curved line shown in the graph of Fig. 28–28. (Note the different scales: $B \gg B_0$.) Initially, point a, the domains (Section 28–7) are randomly oriented. As B_0 increases, the domains become more and more aligned until at point b, nearly all are aligned. The iron is said to be approaching **saturation**. Point b is typically 70% of full saturation. (If B_0 is increased further, the curve continues to rise very slowly, and reaches 98% saturation only when B_0 reaches a value about a thousandfold above that at point b; the last few domains are very difficult to align.) Next, suppose the external field B_0 is reduced by decreasing the current in the toroid coils. As the current is reduced to zero, shown as point c in Fig. 28–29, the domains do not become completely random. Some permanent magnetism remains. If the current is then reversed in direction, enough domains can be turned around so $B = 0$ (point d). As the reverse current is increased further, the iron approaches saturation in the opposite direction (point e). Finally, if the current is again reduced to zero and then increased in the original direction, the total field follows the path efgb, again approaching saturation at point b.

Notice that the field did not pass through the origin (point a) in this cycle. The fact that the curves do not retrace themselves on the same path is called **hysteresis**. The curve bcdefgb is called a **hysteresis loop**. In such a cycle, much energy is transformed to thermal energy (friction) due to realigning of the domains. It can be shown that the energy dissipated in this way is proportional to the area of the hysteresis loop.

At points c and f, the iron core is magnetized even though there is no current in the coils. These points correspond to a permanent magnet. For a permanent magnet, it is desired that ac and af be as large as possible. Materials for which this is true are said to have high **retentivity**.

Materials with a broad hysteresis curve as in Fig. 28–29 are said to be magnetically "hard" and make good permanent magnets. On the other hand, a hysteresis curve such as that in Fig. 28–30 occurs for "soft" iron, which is preferred for electromagnets and transformers (Section 29–6) since the field can be more readily switched off, and the field can be reversed with less loss of energy.

A ferromagnetic material can be demagnetized—that is, made unmagnetized. This can be done by reversing the magnetizing current repeatedly while decreasing its magnitude. This results in the curve of Fig. 28–31. The heads of a tape recorder are demagnetized in this way. The alternating magnetic field acting at the heads due to a handheld demagnetizer is strong when the demagnetizer is placed near the heads and decreases as it is moved slowly away. Video and audio tapes themselves can be erased and ruined by a magnetic field, as can computer hard disks, other magnetic storage devices, and the magnetic stripes on credit cards.

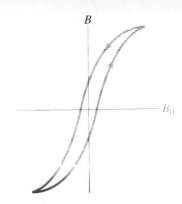

FIGURE 28–30 Hysteresis curve for soft iron.

FIGURE 28–31 Successive hysteresis loops during demagnetization.

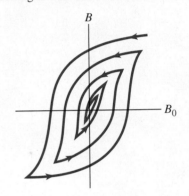

*28–10 Paramagnetism and Diamagnetism

All materials are magnetic to at least a tiny extent. Nonferromagnetic materials fall into two principal classes: *paramagnetic*, in which the magnetic permeability μ is very slightly greater than μ_0; and *diamagnetic*, in which μ is very slightly less than μ_0. The ratio of μ to μ_0 for any material is called the **relative permeability** K_m:

$$K_m = \frac{\mu}{\mu_0}.$$

Another useful parameter is the **magnetic susceptibility** χ_m defined as

$$\chi_m = K_m - 1.$$

Paramagnetic substances have $K_m > 1$ and $\chi_m > 0$, whereas diamagnetic substances have $K_m < 1$ and $\chi_m < 0$. See Table 28–1, and note how small the effect is.

TABLE 28–1 Paramagnetism and Diamagnetism: Magnetic Susceptibilities			
Paramagnetic substance	χ_m	**Diamagnetic substance**	χ_m
Aluminum	2.3×10^{-5}	Copper	-9.8×10^{-6}
Calcium	1.9×10^{-5}	Diamond	-2.2×10^{-5}
Magnesium	1.2×10^{-5}	Gold	-3.6×10^{-5}
Oxygen (STP)	2.1×10^{-6}	Lead	-1.7×10^{-5}
Platinum	2.9×10^{-4}	Nitrogen (STP)	-5.0×10^{-9}
Tungsten	6.8×10^{-5}	Silicon	-4.2×10^{-6}

The difference between paramagnetic and diamagnetic materials can be understood theoretically at the molecular level on the basis of whether or not the molecules have a permanent magnetic dipole moment. One type of **paramagnetism** occurs in materials whose molecules (or ions) have a permanent magnetic dipole moment.[†] In the absence of an external field, the molecules are randomly oriented and no magnetic effects are observed. However, when an external magnetic field is applied, say, by putting the material in a solenoid, the applied field exerts a torque on the magnetic dipoles (Section 27–5), tending to align them parallel to the field. The total magnetic field (external plus that due to aligned magnetic dipoles) will be slightly greater than B_0. The thermal motion of the molecules reduces the alignment, however.

[†]Other types of paramagnetism also occur whose origin is different from that described here, such as in metals where free electrons can contribute.

A useful quantity is the **magnetization vector**, $\vec{M}$, defined as the magnetic dipole moment per unit volume,

$$\vec{M} = \frac{\vec{\mu}}{V},$$

where $\vec{\mu}$ is the magnetic dipole moment of the sample and V its volume. It is found experimentally that M is directly proportional to the external magnetic field (tending to align the dipoles) and inversely proportional to the kelvin temperature T (tending to randomize dipole directions). This is called *Curie's law*, after Pierre Curie (1859–1906), who first noted it:

$$M = C\frac{B}{T},$$

where C is a constant. If the ratio B/T is very large (B very large or T very small) Curie's law is no longer accurate; as B is increased (or T decreased), the magnetization approaches some maximum value, M_{max}. This makes sense, of course, since M_{max} corresponds to complete alignment of all the permanent magnetic dipoles. However, even for very large magnetic fields, $\approx 2.0\,\text{T}$, deviations from Curie's law are normally noted only at very low temperatures, on the order of a few kelvins.

Ferromagnetic materials, as mentioned in Section 28–7, are no longer ferromagnetic above a characteristic temperature called the Curie temperature (1043 K for iron). Above this Curie temperature, they generally are paramagnetic.

Diamagnetic materials (for which μ_m is slightly less than μ_0) are made up of molecules that have no permanent magnetic dipole moment. When an external magnetic field is applied, magnetic dipoles are induced, but the induced magnetic dipole moment is in the direction opposite to that of the field. Hence the total field will be slightly less than the external field. The effect of the external field—in the crude model of electrons orbiting nuclei—is to increase the "orbital" speed of electrons revolving in one direction, and to decrease the speed of electrons revolving in the other direction; the net result is a net dipole moment opposing the external field. Diamagnetism is present in all materials, but is weaker even than paramagnetism and so is overwhelmed by paramagnetic and ferromagnetic effects in materials that display these other forms of magnetism.

Summary

The magnetic field B at a distance r from a long straight wire is directly proportional to the current I in the wire and inversely proportional to r:

$$B = \frac{\mu_0}{2\pi}\frac{I}{r}. \tag{28–1}$$

The magnetic field lines are circles centered at the wire.

The force that one long current-carrying wire exerts on a second parallel current-carrying wire 1 m away serves as the definition of the ampere unit, and ultimately of the coulomb as well.

Ampère's law states that the line integral of the magnetic field $\vec{B}$ around any closed loop is equal to μ_0 times the total net current I_{encl} enclosed by the loop:

$$\oint \vec{B} \cdot d\vec{\ell} = \mu_0 I_{encl}. \tag{28–3}$$

The magnetic field inside a long tightly wound solenoid is

$$B = \mu_0 n I \tag{28–4}$$

where n is the number of coils per unit length and I is the current in each coil.

The **Biot-Savart law** is useful for determining the magnetic field due to a known arrangement of currents. It states that

$$d\vec{B} = \frac{\mu_0 I}{4\pi}\frac{d\vec{\ell} \times \hat{r}}{r^2}, \tag{28–5}$$

where $d\vec{B}$ is the contribution to the total field at some point P due to a current I along an infinitesimal length $d\vec{\ell}$ of its path, and $\hat{r}$ is the unit vector along the direction of the displacement vector $\vec{r}$ from $d\vec{\ell}$ to P. The total field $\vec{B}$ will be the integral over all $d\vec{B}$.

Iron and a few other materials can be made into strong permanent magnets. They are said to be **ferromagnetic**. Ferromagnetic materials are made up of tiny **domains**—each a tiny magnet—which are preferentially aligned in a permanent magnet, but randomly aligned in a nonmagnetized sample.

[*When a ferromagnetic material is placed in a magnetic field B_0 due to a current, say inside a solenoid or toroid, the material becomes magnetized. When the current is turned off, however, the material remains magnetized, and when the current is increased in the opposite direction (and then again reversed), a graph of the total field B versus B_0 is a **hysteresis loop**, and the fact that the curves do not retrace themselves is called **hysteresis**.]

[*All materials exhibit some magnetic effects. Nonferromagnetic materials have much smaller paramagnetic or diamagnetic properties.]

Questions

1. The magnetic field due to current in wires in your home can affect a compass. Discuss the problem in terms of currents, depending on whether they are ac or dc, and their distance away.

2. Compare and contrast the magnetic field due to a long straight current and the electric field due to a long straight line of electric charge at rest (Section 21–7).

3. Two insulated long wires carrying equal currents I cross at right angles to each other. Describe the magnetic force one exerts on the other.

4. A horizontal wire carries a large current. A second wire carrying a current in the same direction is suspended below it. Can the current in the upper wire hold the lower wire in suspension against gravity? Under what conditions will the lower wire be in equilibrium?

5. A horizontal current-carrying wire, free to move in Earth's gravitational field, is suspended directly above a second, parallel, current-carrying wire. (a) In what direction is the current in the lower wire? (b) Can the upper wire be held in stable equilibrium due to the magnetic force of the lower wire? Explain.

6. (a) Write Ampère's law for a path that surrounds both conductors in Fig. 28–10. (b) Repeat, assuming the lower current, I_2, is in the opposite direction $(I_2 = -I_1)$.

7. Suppose the cylindrical conductor of Fig. 28–11a has a concentric cylindrical hollow cavity inside it (so it looks like a pipe). What can you say about $\vec{B}$ in the cavity?

8. Explain why a field such as that shown in Fig. 28–14b is consistent with Ampère's law. Could the lines curve upward instead of downward?

9. What would be the effect on B inside a long solenoid if (a) the diameter of all the loops was doubled, or (b) the spacing between loops was doubled, or (c) the solenoid's length was doubled along with a doubling in the total number of loops.

10. Use the Biot-Savart law to show that the field of the current loop in Fig. 28–21 is correct as shown for points off the axis.

11. Do you think $\vec{B}$ will be the same for all points in the plane of the current loop of Fig. 28–21? Explain.

12. Why does twisting the lead-in wires to electrical devices reduce the magnetic effects of the leads?

13. Compare the Biot-Savart law with Coulomb's law. What are the similarities and differences?

14. How might you define or determine the magnetic pole strength (the magnetic equivalent of a single electric charge) for (a) a bar magnet, (b) a current loop?

15. How might you measure the magnetic dipole moment of the Earth?

16. A type of magnetic switch similar to a solenoid is a **relay** (Fig. 28–32). A relay is an electromagnet (the iron rod inside the coil does not move) which, when activated, attracts a piece of iron on a pivot. Design a relay to close an electrical switch. A relay is used when you need to switch on a circuit carrying a very large current but you do not want that large current flowing through the main switch. For example, the starter switch of a car is connected to a relay so that the large current needed for the starter doesn't pass to the dashboard switch.

FIGURE 28–32
Question 16.

17. A heavy magnet attracts, from rest, a heavy block of iron. Before striking the magnet the block has acquired considerable kinetic energy. (a) What is the source of this kinetic energy? (b) When the block strikes the magnet, some of the latter's domains may be jarred into randomness; describe the energy transformations.

18. Will a magnet attract any metallic object, such as those made of aluminum, or only those made of iron? (Try it and see.) Why is this so?

19. An unmagnetized nail will not attract an unmagnetized paper clip. However, if one end of the nail is in contact with a magnet, the other end *will* attract a paper clip. Explain.

20. Can an iron rod attract a magnet? Can a magnet attract an iron rod? What must you consider to answer these questions?

21. How do you suppose the first magnets found in Magnesia were formed?

22. Why will either pole of a magnet attract an unmagnetized piece of iron?

23. Suppose you have three iron rods, two of which are magnetized but the third is not. How would you determine which two are the magnets without using any additional objects?

24. Two iron bars attract each other no matter which ends are placed close together. Are both magnets? Explain.

*25. Describe the magnetization curve for (a) a paramagnetic substance and (b) a diamagnetic substance, and compare to that for a ferromagnetic substance (Fig. 28–29).

*26. Can all materials be considered (a) diamagnetic, (b) paramagnetic, (c) ferromagnetic? Explain.

Problems

28–1 and 28–2 Straight Wires, Magnetic Field, and Force

1. (I) Jumper cables used to start a stalled vehicle often carry a 65-A current. How strong is the magnetic field 3.5 cm from one cable? Compare to the Earth's magnetic field (5.0×10^{-5} T).

2. (I) If an electric wire is allowed to produce a magnetic field no larger than that of the Earth (0.50×10^{-4} T) at a distance of 15 cm from the wire, what is the maximum current the wire can carry?

3. (I) Determine the magnitude and direction of the force between two parallel wires 25 m long and 4.0 cm apart, each carrying 35 A in the same direction.

4. (I) A vertical straight wire carrying an upward 28-A current exerts an attractive force per unit length of 7.8×10^{-4} N/m on a second parallel wire 7.0 cm away. What current (magnitude and direction) flows in the second wire?

5. (I) In Fig. 28–33, a long straight wire carries current I out of the page toward the viewer. Indicate, with appropriate arrows, the direction of $\vec{\mathbf{B}}$ at each of the points C, D, and E in the plane of the page.

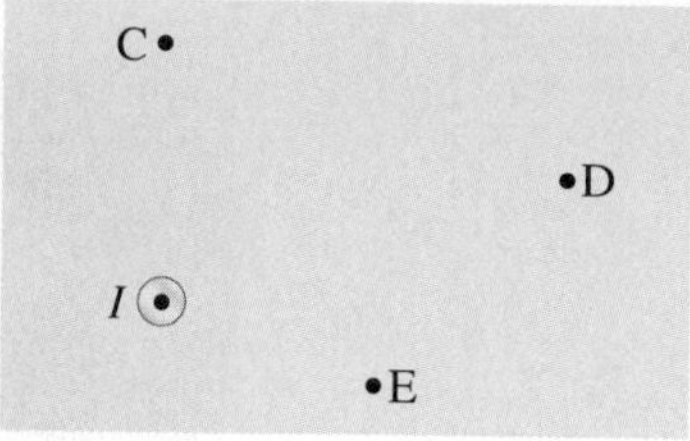

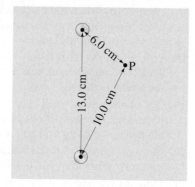

FIGURE 28–33
Problem 5.

6. (II) An experiment on the Earth's magnetic field is being carried out 1.00 m from an electric cable. What is the maximum allowable current in the cable if the experiment is to be accurate to ± 2.0%?

7. (II) Two long thin parallel wires 13.0 cm apart carry 35-A currents in the same direction. Determine the magnetic field vector at a point 10.0 cm from one wire and 6.0 cm from the other (Fig. 28–34).

FIGURE 28–34
Problem 7.

8. (II) A horizontal compass is placed 18 cm due south from a straight vertical wire carrying a 43-A current downward. In what direction does the compass needle point at this location? Assume the horizontal component of the Earth's field at this point is 0.45×10^{-4} T and the magnetic declination is 0°.

9. (II) A long horizontal wire carries 24.0 A of current due north. What is the net magnetic field 20.0 cm due west of the wire if the Earth's field there points downward, 44° below the horizontal, and has magnitude 5.0×10^{-5} T?

10. (II) A straight stream of protons passes a given point in space at a rate of 2.5×10^9 protons/s. What magnetic field do they produce 2.0 m from the beam?

11. (II) Determine the magnetic field midway between two long straight wires 2.0 cm apart in terms of the current I in one when the other carries 25 A. Assume these currents are (a) in the same direction, and (b) in opposite directions.

12. (II) Two straight parallel wires are separated by 6.0 cm. There is a 2.0-A current flowing in the first wire. If the magnetic field strength is found to be zero between the two wires at a distance of 2.2 cm from the first wire, what is the magnitude and direction of the current in the second wire?

13. (II) Two long straight wires each carry a current I out of the page toward the viewer, Fig. 28–35. Indicate, with appropriate arrows, the direction of $\vec{\mathbf{B}}$ at each of the points 1 to 6 in the plane of the page. State if the field is zero at any of the points.

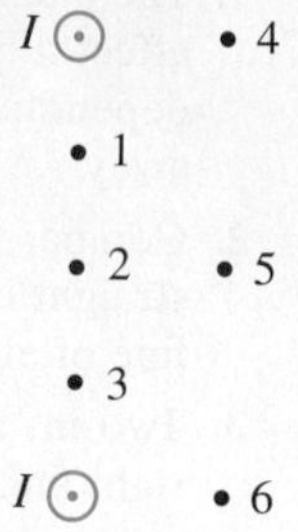

FIGURE 28–35
Problem 13.

14. (II) A long pair of insulated wires serves to conduct 28.0 A of dc current to and from an instrument. If the wires are of negligible diameter but are 2.8 mm apart, what is the magnetic field 10.0 cm from their midpoint, in their plane (Fig. 28–36)? Compare to the magnetic field of the Earth.

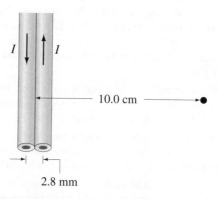

FIGURE 28–36   Problems 14 and 15.

15. (II) A third wire is placed in the plane of the two wires shown in Fig. 28–36 parallel and just to the right. If it carries 25.0 A upward, what force per meter of length does it exert on each of the other two wires? Assume it is 2.8 mm from the nearest wire, center to center.

16. (II) A power line carries a current of 95 A west along the tops of 8.5-m-high poles. (a) What is the magnitude and direction of the magnetic field produced by this wire at the ground directly below? How does this compare with the Earth's field of about $\frac{1}{2}$ G? (b) Where would the line's field cancel the Earth's?

17. (II) A compass needle points 28° E of N outdoors. However, when it is placed 12.0 cm to the east of a vertical wire inside a building, it points 55° E of N. What is the magnitude and direction of the current in the wire? The Earth's field there is 0.50×10^{-4} T and is horizontal.

18. (II) A rectangular loop of wire is placed next to a straight wire, as shown in Fig. 28–37. There is a current of 3.5 A in both wires. Determine the magnitude and direction of the net force on the loop.

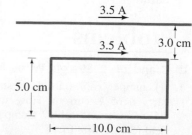

FIGURE 28–37
Problem 18.

19. (II) Let two long parallel wires, a distance d apart, carry equal currents I in the same direction. One wire is at $x = 0$, the other at $x = d$, Fig. 28–38. Determine $\vec{\mathbf{B}}$ along the x axis between the wires as a function of x.

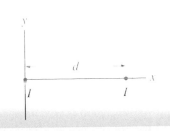

FIGURE 28–38
Problems 19 and 20.

20. (II) Repeat Problem 19 if the wire at $x = 0$ carries twice the current ($2I$) as the other wire, and in the opposite direction.

21. (II) Two long wires are oriented so that they are perpendicular to each other. At their closest, they are 20.0 cm apart (Fig. 28–39). What is the magnitude of the magnetic field at a point midway between them if the top one carries a current of 20.0 A and the bottom one carries 12.0 A?

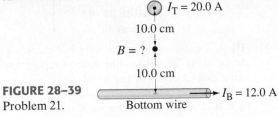

FIGURE 28–39
Problem 21.

22. (II) Two long parallel wires 8.20 cm apart carry 16.5-A currents in the same direction. Determine the magnetic field vector at a point P, 12.0 cm from one wire and 13.0 cm from the other. See Fig. 28–40.
[*Hint*: Use the law of cosines.]

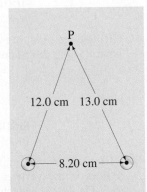

FIGURE 28–40
Problem 22.

23. (III) A very long flat conducting strip of width d and negligible thickness lies in a horizontal plane and carries a uniform current I across its cross section. (*a*) Show that at points a distance y directly above its center, the field is given by

$$B = \frac{\mu_0 I}{\pi d} \tan^{-1} \frac{d}{2y},$$

assuming the strip is infinitely long. [*Hint*: Divide the strip into many thin "wires," and sum (integrate) over these.] (*b*) What value does B approach for $y \gg d$? Does this make sense? Explain.

24. (III) A triangular loop of side length a carries a current I (Fig. 28–41). If this loop is placed a distance d away from a very long straight wire carrying a current I', determine the force on the loop.

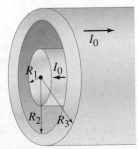

Problem 24.

28–4 and 28–5 Ampère's Law, Solenoids and Toroids

25. (I) A 40.0-cm-long solenoid 1.35 cm in diameter is to produce a field of 0.385 mT at its center. How much current should the solenoid carry if it has 765 turns of wire?

26. (I) A 32-cm-long solenoid, 1.8 cm in diameter, is to produce a 0.30-T magnetic field at its center. If the maximum current is 4.5 A, how many turns must the solenoid have?

27. (I) A 2.5-mm-diameter copper wire carries a 33-A current (uniform across its cross section). Determine the magnetic field: (*a*) at the surface of the wire; (*b*) inside the wire, 0.50 mm below the surface; (*c*) outside the wire 2.5 mm from the surface.

28. (II) A toroid (Fig. 28–17) has a 50.0-cm inner diameter and a 54.0-cm outer diameter. It carries a 25.0 A current in its 687 coils. Determine the range of values for B inside the toroid.

29. (II) A 20.0-m-long copper wire, 2.00 mm in diameter including insulation, is tightly wrapped in a single layer with adjacent coils touching, to form a solenoid of diameter 2.50 cm (outer edge). What is (*a*) the length of the solenoid and (*b*) the field at the center when the current in the wire is 16.7 A?

30. (II) (*a*) Use Eq. 28–1, and the vector nature of $\vec{\mathbf{B}}$, to show that the magnetic field lines around two long parallel wires carrying equal currents $I_1 = I_2$ are as shown in Fig. 28–10. (*b*) Draw the equipotential lines around two stationary positive electric charges. (*c*) Are these two diagrams similar? Identical? Why or why not?

31. (II) A coaxial cable consists of a solid inner conductor of radius R_1, surrounded by a concentric cylindrical tube of inner radius R_2 and outer radius R_3 (Fig. 28–42). The conductors carry equal and opposite currents I_0 distributed uniformly across their cross sections. Determine the magnetic field at a distance R from the axis for: (*a*) $R < R_1$; (*b*) $R_1 < R < R_2$; (*c*) $R_2 < R < R_3$; (*d*) $R > R_3$. (*e*) Let $I_0 = 1.50$ A, $R_1 = 1.00$ cm, $R_2 = 2.00$ cm, and $R_3 = 2.50$ cm. Graph B from $R = 0$ to $R = 3.00$ cm.

FIGURE 28–42
Problems 31 and 32.

32. (III) Suppose the current in the coaxial cable of Problem 31, Fig. 28–42, is not uniformly distributed, but instead the current density j varies linearly with distance from the center: $j_1 = C_1 R$ for the inner conductor and $j_2 = C_2 R$ for the outer conductor. Each conductor still carries the same total current I_0, in opposite directions. Determine the magnetic field in terms of I_0 in the same four regions of space as in Problem 31.

33. (I) The Earth's magnetic field is essentially that of a magnetic dipole. If the field near the North Pole is about $1.0 \times 10^{-4}\,\text{T}$, what will it be (approximately) 13,000 km above the surface at the North Pole?

34. (II) A wire, in a plane, has the shape shown in Fig. 28–43, two arcs of a circle connected by radial lengths of wire. Determine $\vec{\mathbf{B}}$ at point C in terms of R_1, R_2, θ, and the current I.

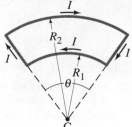

FIGURE 28–43
Problem 34.

35. (II) A circular conducting ring of radius R is connected to two exterior straight wires at two ends of a diameter (Fig. 28–44). The current I splits into unequal portions (as shown) while passing through the ring. What is $\vec{\mathbf{B}}$ at the center of the ring?

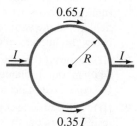

FIGURE 28–44
Problem 35.

36. (II) A small loop of wire of radius 1.8 cm is placed at the center of a wire loop with radius 25.0 cm. The planes of the loops are perpendicular to each other, and a 7.0-A current flows in each. Estimate the torque the large loop exerts on the smaller one. What simplifying assumption did you make?

37. (II) A wire is formed into the shape of two half circles connected by equal-length straight sections as shown in Fig. 28–45. A current I flows in the circuit clockwise as shown. Determine (*a*) the magnitude and direction of the magnetic field at the center, C, and (*b*) the magnetic dipole moment of the circuit.

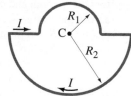

FIGURE 28–45
Problem 37.

38. (II) A single point charge q is moving with velocity $\vec{\mathbf{v}}$. Use the Biot-Savart law to show that the magnetic field $\vec{\mathbf{B}}$ it produces at a point P, whose position vector relative to the charge is $\vec{\mathbf{r}}$ (Fig. 28–46), is given by

$$\vec{\mathbf{B}} = \frac{\mu_0}{4\pi}\frac{q\vec{\mathbf{v}} \times \vec{\mathbf{r}}}{r^3}.$$

(Assume v is much less than the speed of light.)

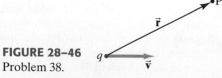

FIGURE 28–46
Problem 38.

39. (II) A nonconducting circular disk, of radius R, carries a uniformly distributed electric charge Q. The plate is set spinning with angular velocity ω about an axis perpendicular to the plate through its center (Fig. 28–47). Determine (*a*) its magnetic dipole moment and (*b*) the magnetic field at points on its axis a distance x from its center; (*c*) does Eq. 28–7b apply in this case for $x \gg R$?

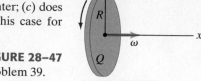

FIGURE 28–47
Problem 39.

40. (II) Consider a straight section of wire of length d, as in Fig. 28–48, which carries a current I. (*a*) Show that the magnetic field at a point P a distance R from the wire along its perpendicular bisector is

$$B = \frac{\mu_0 I}{2\pi R}\frac{d}{(d^2 + 4R^2)^{\frac{1}{2}}}.$$

(*b*) Show that this is consistent with Example 28–11 for an infinite wire.

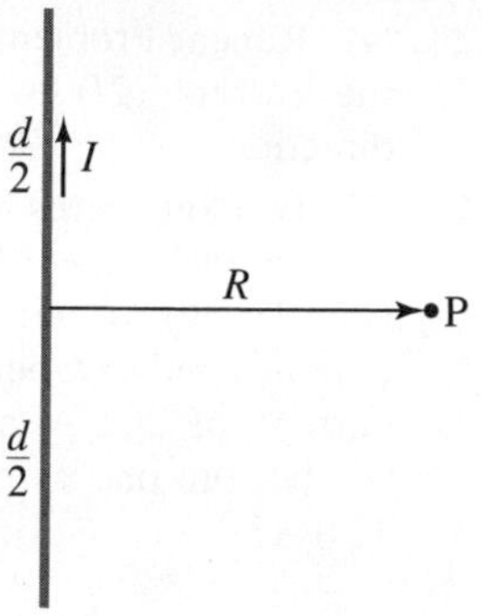

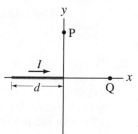

FIGURE 28–48
Problem 40.

41. (II) A segment of wire of length d carries a current I as shown in Fig. 28–49. (*a*) Show that for points along the positive x axis (the axis of the wire), such as point Q, the magnetic field $\vec{\mathbf{B}}$ is zero. (*b*) Determine a formula for the field at points along the y axis, such as point P.

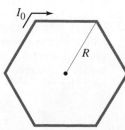

FIGURE 28–49
Problem 41.

42. (III) Use the result of Problem 41 to find the magnetic field at point P in Fig. 28–50 due to the current in the square loop.

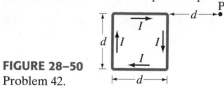

FIGURE 28–50
Problem 42.

43. (III) A wire is bent into the shape of a regular polygon with n sides whose vertices are a distance R from the center. (See Fig. 28–51, which shows the special case of $n = 6$.) If the wire carries a current I_0, (*a*) determine the magnetic field at the center; (*b*) if n is allowed to become very large $(n \to \infty)$, show that the formula in part (*a*) reduces to that for a circular loop (Example 28–12).

FIGURE 28–51
Problem 43.

44. (III) Start with the result of Example 28–12 for the magnetic field along the axis of a single loop to obtain the field inside a very long solenoid with n turns per meter (Eq. 28–4) that stretches from $+\infty$ to $-\infty$.

45. (III) A single rectangular loop of wire, with sides a and b, carries a current I. An xy coordinate system has its origin at the lower left corner of the rectangle with the x axis parallel to side b (Fig. 28–52) and the y axis parallel to side a. Determine the magnetic field B at all points (x, y) within the loop.

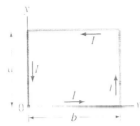

FIGURE 28–52
Problem 45.

46. (III) A square loop of wire, of side d, carries a current I. (a) Determine the magnetic field B at points on a line perpendicular to the plane of the square which passes through the center of the square (Fig. 28–53). Express B as a function of x, the distance along the line from the center of the square. (b) For $x \gg d$, does the square appear to be a magnetic dipole? If so, what is its dipole moment?

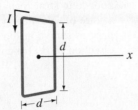

FIGURE 28–53
Problem 46.

28–7 Magnetic Materials—Ferromagnetism

47. (II) An iron atom has a magnetic dipole moment of about $1.8 \times 10^{-23}\,\text{A}\cdot\text{m}^2$. (a) Determine the dipole moment of an iron bar 9.0 cm long, 1.2 cm wide, and 1.0 cm thick, if it is 100 percent saturated. (b) What torque would be exerted on this bar when placed in a 0.80-T field acting at right angles to the bar?

*48. (I) The following are some values of B and B_0 for a piece of annealed iron as it is being magnetized:

$B_0 (10^{-4}\,\text{T})$	0.0	0.13	0.25	0.50	0.63	0.78	1.0	1.3
$B(\text{T})$	0.0	0.0042	0.010	0.028	0.043	0.095	0.45	0.67
$B_0 (10^{-4}\,\text{T})$	1.9	2.5	6.3	13.0	130	1300	10,000	
$B(\text{T})$	1.01	1.18	1.44	1.58	1.72	2.26	3.15	

Determine the magnetic permeability μ for each value and plot a graph of μ versus B_0.

*49. (I) A large thin toroid has 285 loops of wire per meter, and a 3.0-A current flows through the wire. If the relative permeability of the iron is $\mu/\mu_0 = 2200$, what is the total field B inside the toroid?

*50. (II) An iron-core solenoid is 38 cm long and 1.8 cm in diameter, and has 640 turns of wire. The magnetic field inside the solenoid is 2.2 T when 48 A flows in the wire. What is the permeability μ at this high field strength?

General Problems

51. Three long parallel wires are 3.5 cm from one another. (Looking along them, they are at three corners of an equilateral triangle.) The current in each wire is 8.00 A, but its direction in wire M is opposite to that in wires N and P (Fig. 28–54). Determine the magnetic force per unit length on each wire due to the other two.

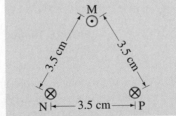

FIGURE 28–54
Problems 51, 52, and 53.

52. In Fig. 28–54, determine the magnitude and direction of the magnetic field midway between points M and N.

53. In Fig. 28–54 the top wire is 1.00-mm-diameter copper wire and is suspended in air due to the two magnetic forces from the bottom two wires. The current is 40.0 A in each of the two bottom wires. Calculate the required current flow in the suspended wire.

54. An electron enters a large solenoid at a 7.0° angle to the axis. If the field is a uniform $3.3 \times 10^{-2}\,\text{T}$, determine the radius and pitch (distance between loops) of the electron's helical path if its speed is $1.3 \times 10^7\,\text{m/s}$.

55. Two long straight parallel wires are 15 cm apart. Wire A carries 2.0-A current. Wire B's current is 4.0 A in the same direction. (a) Determine the magnetic field due to wire A at the position of wire B. (b) Determine the magnetic field due to wire B at the position of wire A. (c) Are these two magnetic fields equal and opposite? Why or why not? (d) Determine the force per unit length on wire A due to wire B, and that on wire B due to wire A. Are these two forces equal and opposite? Why or why not?

56. A rectangular loop of wire carries a 2.0-A current and lies in a plane which also contains a very long straight wire carrying a 10.0-A current as shown in Fig. 28–55. Determine (a) the net force and (b) the net torque on the loop due to the straight wire.

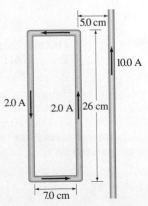

FIGURE 28–55
Problem 56.

57. A very large flat conducting sheet of thickness t carries a uniform current density $\vec{j}$ throughout (Fig. 28–56). Determine the magnetic field (magnitude and direction) at a distance y above the plane. (Assume the plane is infinitely long and wide.)

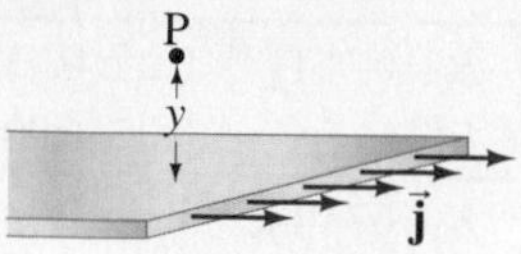

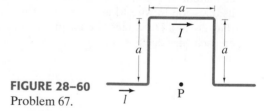

FIGURE 28–56
Problem 57.

58. A long horizontal wire carries a current of 48 A. A second wire, made of 1.00-mm-diameter copper wire and parallel to the first, is kept in suspension magnetically 5.0 cm below (Fig. 28–57). (*a*) Determine the magnitude and direction of the current in the lower wire. (*b*) Is the lower wire in stable equilibrium? (*c*) Repeat parts (*a*) and (*b*) if the second wire is suspended 5.0 cm *above* the first due to the first's magnetic field.

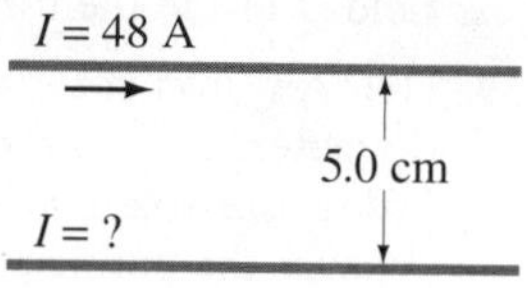

$I = 48$ A

5.0 cm

$I = ?$

FIGURE 28–57
Problem 58.

59. A square loop of wire, of side d, carries a current I. Show that the magnetic field at the center of the square is

$$B = \frac{2\sqrt{2}\,\mu_0 I}{\pi d}.$$

[*Hint*: Determine $\vec{B}$ for each segment of length d.]

60. In Problem 59, if you reshaped the square wire into a circle, would B increase or decrease at the center? Explain.

61. Helmholtz coils are two identical circular coils having the same radius R and the same number of turns N, separated by a distance equal to the radius R and carrying the same current I in the same direction. (See Fig. 28–58.) They are used in scientific instruments to generate nearly uniform magnetic fields.(They can be seen in the photo, Fig. 27–18.) (*a*) Determine the magnetic field B at points x along the line joining their centers. Let $x = 0$ at the center of one coil, and $x = R$ at the center of the other. (*b*) Show that the field midway between the coils is particularly uniform by showing that $\dfrac{dB}{dx} = 0$ and $\dfrac{d^2B}{dx^2} = 0$ at the midpoint between the coils. (*c*) If $R = 10.0$ cm, $N = 250$ turns and $I = 2.0$ A, what is the field at the midpoint between the coils, $x = R/2$?

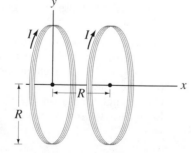

FIGURE 28–58
Problem 61.

62. For two long parallel wires separated by a distance d, carrying currents I_1 and I_2 as in Fig. 28–10, show directly (Eq. 28–1) that Ampère's law is valid (but do not use Ampère's law) for a circular path of radius r ($r < d$) centered on I_1:

$$\oint \vec{B} \cdot d\vec{\ell} = \mu_0 I_1.$$

63. Near the Earth's poles the magnetic field is about 1 G $(1 \times 10^{-4}\,\text{T})$. Imagine a simple model in which the Earth's field is produced by a single current loop around the equator. Estimate roughly the current this loop would carry.

64. A 175-g model airplane charged to 18.0 mC and traveling at 2.8 m/s passes within 8.6 cm of a wire, nearly parallel to its path, carrying a 25-A current. What acceleration (in g's) does this interaction give the airplane?

65. Suppose that an electromagnet uses a coil 2.0 m in diameter made from square copper wire 2.0 mm on a side; the power supply produces 35 V at a maximum power output of 1.0 kW. (*a*) How many turns are needed to run the power supply at maximum power? (*b*) What is the magnetic field strength at the center of the coil? (*c*) If you use a greater number of turns and this same power supply, will a greater magnetic field result? Explain.

66. Four long straight parallel wires located at the corners of a square of side d carry equal currents I_0 perpendicular to the page as shown in Fig. 28–59. Determine the magnitude and direction of $\vec{B}$ at the center C of the square.

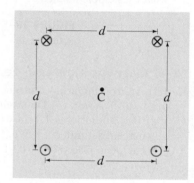

FIGURE 28–59
Problem 66.

67. Determine the magnetic field at the point P due to a very long wire with a square bend as shown in Fig. 28–60. The point P is halfway between the two corners. [*Hint*: You can use the results of Problems 40 and 41.]

FIGURE 28–60
Problem 67.

68. A thin 12-cm-long solenoid has a total of 420 turns of wire and carries a current of 2.0 A. Calculate the field inside the solenoid near the center.

69. A 550-turn solenoid is 15 cm long. The current into it is 33 A. A 3.0-cm-long straight wire cuts through the center of the solenoid, along a diameter. This wire carries a 22-A current downward (and is connected by other wires that don't concern us). What is the force on this wire assuming the solenoid's field points due east?

70. You have 1.0 kg of copper and want to make a practical solenoid that produces the greatest possible magnetic field for a given voltage. Should you make your copper wire long and thin, short and fat, or something else? Consider other variables, such as solenoid diameter, length, and so on.

71. A small solenoid (radius r_a) is inside a larger solenoid (radius $r_b > r_a$). They are coaxial with n_a and n_b turns per unit length, respectively. The solenoids carry the same current, but in opposite directions. Let r be the radial distance from the common axis of the solenoids. If the magnetic field inside the inner solenoid ($r < r_a$) is to be in the opposite direction as the field between the solenoids ($r_a < r < r_b$), but have half the magnitude, determine the required ratio n_b/n_a.

72. Find B at the center of the 10 cm radius semicircle in Fig. 28–61. The straight wires extend a great distance outward to the left and carry a current $I = 6.0 A$.

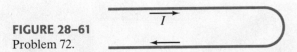

FIGURE 28–61
Problem 72.

73. The design of a magneto-optical atom trap requires a magnetic field B that is directly proportional to position x along an axis. Such a field perturbs the absorption of laser light by atoms in the manner needed to spatially confine atoms in the trap. Let us demonstrate that "anti-Helmholtz" coils will provide the required field $B = Cx$, where C is a constant. Anti-Helmholtz coils consist of two identical circular wire coils, each with radius R and N turns, carrying current I in opposite directions (Fig. 28–62). The coils share a common axis (defined as the x axis with $x = 0$ at the midpoint (0) between the coils). Assume that the centers of the coils are separated by a distance equal to the radius R of the coils. (a) Show that the magnetic field at position x along the x axis is given by

$$B(x) = \frac{4\,\mu_0 NI}{R}\left\{\left[4 + \left(1 - \frac{2x}{R}\right)^2\right]^{-\frac{3}{2}} - \left[4 + \left(1 + \frac{2x}{R}\right)^2\right]^{-\frac{3}{2}}\right\}.$$

(b) For small excursions from the origin where $|x| \ll R$, show that the magnetic field is given by $B \approx Cx$, where the constant $C = 48\,\mu_0 NI/25\sqrt{5}\,R^2$. (c) For optimal atom trapping, dB/dx should be about 0.15 T/m. Assume an atom trap uses anti-Helmholtz coils with $R = 4.0$ cm and $N = 150$. What current should flow through the coils? [Coil separation equal to coil radius, as assumed in this problem, is not a strict requirement for anti-Helmholtz coils.]

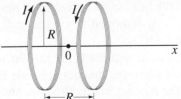

FIGURE 28–62
Problem 75.

74. You want to get an idea of the magnitude of magnetic fields produced by overhead power lines. You estimate that a transmission wire is about 12 m above the ground. The local power company tells you that the line operates at 15 kV and provide a maximum of 45 MW to the local area. Estimate the maximum magnetic field you might experience walking under such a power line, and compare to the Earth's field. [For an ac current, values are rms, and the magnetic field will be changing.]

*Numerical/Computer

* 75. (II) A circular current loop of radius 15 cm containing 250 turns carries a current of 2.0 A. Its center is at the origin and its axis lies along the x axis. Calculate the magnetic field B at a point x on the x axis for $x = -40$ cm to $+40$ cm in steps of 2 cm and make a graph of B as a function of x.

* 76. (III) A set of Helmholtz coils (see Problem 61, Fig. 28–58) have a radius $R = 10.0$ cm and are separated by a distance $R = 10.0$ cm. Each coil has 250 loops carrying a current $I = 2.0$ A. (a) Determine the total magnetic field B along the x axis (the center line for the two coils) in steps of 0.2 cm from the center of one coil ($x = 0$) to the center of the other ($x = R$). (b) Graph B as a function of x. (c) By what % does B vary from $x = 5.0$ cm to $x = 6.0$ cm?

Answers to Exercises

A: 2×10^{-6} T; not at this distance, and then only briefly.

B: 0.8×10^{-5} T, up.

C: (b).

D: (b), (c).

One of the great laws of physics is Faraday's law of induction, which says that a changing magnetic flux produces an induced emf. This photo shows a bar magnet moving inside a coil of wire, and the galvanometer registers an induced current. This phenomenon of electromagnetic induction is the basis for many practical devices, including generators, alternators, transformers, tape recording, and computer memory.

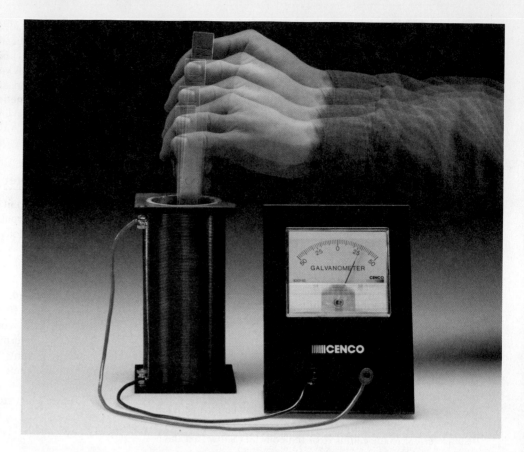

^{C H A P T E R}

29
Electromagnetic Induction and Faraday's Law

CONTENTS

CHAPTER-OPENING QUESTION—Guess now!

In the photograph above, the bar magnet is inserted into the coil of wire, and is left there for 1 minute; then it is removed from the coil. What would an observer watching the galvanometer see?

(a) No change; without a battery there is no current to detect.

(b) A small current flows while the magnet is inside the coil of wire.

(c) A current spike as the magnet enters the coil, and then nothing.

(d) A current spike as the magnet enters the coil, and then a steady small current.

(e) A current spike as the magnet enters the coil, then nothing, and then a current spike in the opposite direction as the magnet leaves the coil.

I n Chapter 27, we discussed two ways in which electricity and magnetism are related: (1) an electric current produces a magnetic field; and (2) a magnetic field exerts a force on an electric current or moving electric charge. These discoveries were made in 1820–1821. Scientists then began to wonder: if electric currents produce a magnetic field, is it possible that a magnetic field can produce an electric current? Ten years later the American Joseph Henry (1797–1878) and the Englishman Michael Faraday (1791–1867) independently found that it was possible. Henry actually made the discovery first. But Faraday published his results earlier and investigated the subject in more detail. We now discuss this phenomenon and some of its world-changing applications including the electric generator.

Induced EMF

In his attempt to produce an electric current from a magnetic field, Faraday used an
apparatus like that shown in Fig. 29–1. A coil of wire X was connected to a battery.
The current that flowed through X produced a magnetic field that was intensified
by the ring-shaped iron core around which the wire was wrapped. Faraday hoped
that a strong steady current in X would produce a great enough magnetic field to
produce a current in a second coil Y wrapped on the same iron ring. This second
circuit, Y, contained a galvanometer to detect any current but contained no battery.

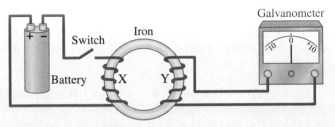

FIGURE 29–1   Faraday's experiment
to induce an emf.

He met no success with constant currents. But the long-sought effect was finally
observed when Faraday noticed the galvanometer in circuit Y deflect strongly at the
moment he closed the switch in circuit X. And the galvanometer deflected strongly
in the opposite direction when he opened the switch in X. A constant current in X
produced a constant magnetic field which produced *no* current in Y. Only when the
current in X was starting or stopping was a current produced in Y.

Faraday concluded that although a constant magnetic field produces no current in
a conductor, a *changing* magnetic field can produce an electric current. Such a current
is called an **induced current**. When the magnetic field through coil Y changes, a
current occurs in Y as if there were a source of emf in circuit Y. We therefore say that

a changing magnetic field induces an emf.

Faraday did further experiments on **electromagnetic induction**, as this
phenomenon is called. For example, Fig. 29–2 shows that if a magnet is moved
quickly into a coil of wire, a current is induced in the wire. If the magnet is quickly
removed, a current is induced in the opposite direction ($\vec{\mathbf{B}}$ through the coil
decreases). Furthermore, if the magnet is held steady and the coil of wire is moved
toward or away from the magnet, again an emf is induced and a current flows.
Motion or change is required to induce an emf. It doesn't matter whether the
magnet or the coil moves. It is their *relative motion* that counts.

> ⚠ **CAUTION**
> *Changing* $\vec{\mathbf{B}}$*, not* $\vec{\mathbf{B}}$ *itself,*
> *induces current*

> ⚠ **CAUTION**
> *Relative motion—magnet*
> *or coil moving induces current*

FIGURE 29–2   (a) A current is induced when a magnet is moved toward a coil, momentarily increasing the magnetic
field through the coil. (b) The induced current is opposite when the magnet is moved away from the coil ($\vec{\mathbf{B}}$ decreases).
Note that the galvanometer zero is at the center of the scale and the needle deflects left or right, depending on the
direction of the current. In (c), no current is induced if the magnet does not move relative to the coil. It is the
relative motion that counts here: the magnet can be held steady and the coil moved, which also induces an emf.

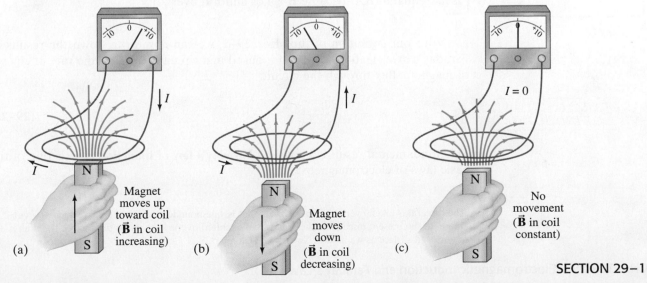

29–2 Faraday's Law of Induction; Lenz's Law

Faraday investigated quantitatively what factors influence the magnitude of the emf induced. He found first of all that the more rapidly the magnetic field changes, the greater the induced emf. He also found that the induced emf depends on the area of the circuit loop. Thus we say that the emf is proportional to the rate of change of the **magnetic flux**, Φ_B, passing through the circuit or loop of area A. Magnetic flux for a uniform magnetic field is defined in the same way we did for electric flux in Chapter 22, namely as

$$\Phi_B = B_\perp A = BA\cos\theta = \vec{B}\cdot\vec{A}. \qquad [\vec{B}\text{ uniform}] \quad (29\text{–}1a)$$

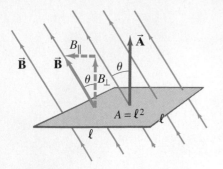

Here $B_\perp$ is the component of the magnetic field $\vec{B}$ perpendicular to the face of the loop, and θ is the angle between $\vec{B}$ and the vector $\vec{A}$ (representing the area) whose direction is perpendicular to the face of the loop. These quantities are shown in Fig. 29–3 for a square loop of side ℓ whose area is $A = \ell^2$. If the area is of some other shape, or $\vec{B}$ is not uniform, the magnetic flux can be written[†]

$$\Phi_B = \int \vec{B}\cdot d\vec{A}. \qquad (29\text{–}1b)$$

FIGURE 29–3 Determining the flux through a flat loop of wire. This loop is square, of side ℓ and area $A = \ell^2$.

FIGURE 29–4 Magnetic flux Φ_B is proportional to the number of lines of $\vec{B}$ that pass through the loop.

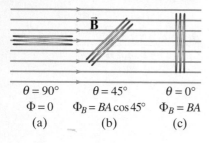

$\theta = 90°$ $\theta = 45°$ $\theta = 0°$
$\Phi = 0$ $\Phi_B = BA\cos45°$ $\Phi_B = BA$
(a)   (b)   (c)

As we saw in Chapter 27, the lines of $\vec{B}$ (like lines of $\vec{E}$) can be drawn such that the number of lines per unit area is proportional to the field strength. Then the flux Φ_B can be thought of as being proportional to the *total number of lines passing through the area enclosed by the loop*. This is illustrated in Fig. 29–4, where the loop is viewed from the side (on edge). For $\theta = 90°$, no magnetic field lines pass through the loop and $\Phi_B = 0$, whereas Φ_B is a maximum when $\theta = 0°$. The unit of magnetic flux is the tesla-meter²; this is called a **weber**: $1\text{ Wb} = 1\text{ T·m}^2$.

FIGURE 29–5 Example 29–1.

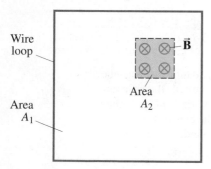

Wire loop

Area A_2

$\vec{B}$

Area A_1

CONCEPTUAL EXAMPLE 29–1 **Determining flux.** A square loop of wire encloses area A_1 as shown in Fig. 29–5. A uniform magnetic field $\vec{B}$ perpendicular to the loop extends over the area A_2. What is the magnetic flux through the loop A_1?

RESPONSE We assume that the magnetic field is zero outside the area A_2. The total magnetic flux through area A_1 is the flux through area A_2, which by Eq. 29–1a for a uniform field is BA_2, plus the flux through the remaining area $(= A_1 - A_2)$, which is zero because $B = 0$. So the total flux is $\Phi_B = BA_2 + 0(A_1 - A_2) = BA_2$. It is *not* equal to BA_1 because $\vec{B}$ is not uniform over A_1.

With our definition of flux, Eqs. 29–1, we can now write down the results of Faraday's investigations: The emf induced in a circuit is equal to the rate of change of magnetic flux through the circuit:

FARADAY'S LAW OF INDUCTION

$$\mathscr{E} = -\frac{d\Phi_B}{dt}. \qquad (29\text{–}2a)$$

This fundamental result is known as **Faraday's law of induction**, and is one of the basic laws of electromagnetism.

[†]The integral is taken over an open surface—that is, one bounded by a closed curve such as a circle or square. In the present discussion, the area is that enclosed by the loop under discussion. The area is not an enclosed surface as we used in Gauss's law, Chapter 22.

If the circuit contains N loops that are closely wrapped so the same flux passes through each, the emfs induced in each loop add together, so

$$\mathcal{E} = -N \frac{d\Phi_B}{dt} \qquad \text{[N loops]} \quad (29\text{--}2\text{b})$$

EXAMPLE 29-2 **A loop of wire in a magnetic field.** A square loop of wire of side $\ell = 5.0\,cm$ is in a uniform magnetic field $B = 0.16\,T$. What is the magnetic flux in the loop (*a*) when $\vec{B}$ is perpendicular to the face of the loop and (*b*) when $\vec{B}$ is at an angle of 30° to the area $\vec{A}$ of the loop? (*c*) What is the magnitude of the average current in the loop if it has a resistance of $0.012\,\Omega$ and it is rotated from position (*b*) to position (*a*) in 0.14 s?

APPROACH We use the definition $\Phi_B = \vec{B} \cdot \vec{A}$ to calculate the magnetic flux. Then we use Faraday's law of induction to find the induced emf in the coil, and from that the induced current ($I = \mathcal{E}/R$).

SOLUTION The area of the coil is $A = \ell^2 = (5.0 \times 10^{-2}\,m)^2 = 2.5 \times 10^{-3}\,m^2$, and the direction of $\vec{A}$ is perpendicular to the face of the loop (Fig. 29–3).

(*a*) $\vec{B}$ is perpendicular to the coil's face, and thus parallel to $\vec{A}$ (Fig. 29–3), so

$$\Phi_B = \vec{B} \cdot \vec{A}$$
$$= BA \cos 0° = (0.16\,T)(2.5 \times 10^{-3}\,m^2)(1) = 4.0 \times 10^{-4}\,Wb.$$

(*b*) The angle between $\vec{B}$ and $\vec{A}$ is 30°, so

$$\Phi_B = \vec{B} \cdot \vec{A}$$
$$= BA \cos \theta = (0.16\,T)(2.5 \times 10^{-3}\,m^2) \cos 30° = 3.5 \times 10^{-4}\,Wb.$$

(*c*) The magnitude of the induced emf is

$$\mathcal{E} = \frac{\Delta \Phi_B}{\Delta t} = \frac{(4.0 \times 10^{-4}\,Wb) - (3.5 \times 10^{-4}\,Wb)}{0.14\,s} = 3.6 \times 10^{-4}\,V.$$

The current is then

$$I = \frac{\mathcal{E}}{R} = \frac{3.6 \times 10^{-4}\,V}{0.012\,\Omega} = 0.030\,A = 30\,mA.$$

The minus signs in Eqs. 29–2a and b are there to remind us in which direction the induced emf acts. Experiments show that

a current produced by an induced emf moves in a direction so that the magnetic field created by that current opposes the original change in flux.

This is known as **Lenz's law**. Be aware that we are now discussing two distinct magnetic fields: (1) the changing magnetic field or flux that induces the current, and (2) the magnetic field produced by the induced current (all currents produce a field). The second field opposes the change in the first.

Lenz's law can be said another way, valid even if no current can flow (as when a circuit is not complete):

An induced emf is always in a direction that opposes the original change in flux that caused it.

⚠ **CAUTION**

Distinguish two different magnetic fields

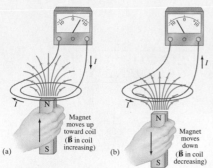

FIGURE 29–2 (repeated).

Let us apply Lenz's law to the relative motion between a magnet and a coil, Fig. 29–2. The changing flux through the coil induces an emf in the coil, producing a current. This induced current produces its own magnetic field. In Fig. 29–2a the distance between the coil and the magnet decreases. The magnet's magnetic field (and number of field lines) through the coil increases, and therefore the flux increases. The magnetic field of the magnet points upward. To oppose the upward increase, the magnetic field inside the coil produced by the induced current needs to point *downward*. Thus, Lenz's law tells us that the current moves as shown (use the right-hand rule). In Fig. 29–2b, the flux *decreases* (because the magnet is moved away and B decreases), so the induced current in the coil produces an *upward* magnetic field through the coil that is "trying" to maintain the status quo. Thus the current in Fig. 29–2b is in the opposite direction from Fig. 29–2a.

It is important to note that an emf is induced whenever there is a change in *flux* through the coil, and we now consider some more possibilities.

FIGURE 29–6 A current can be induced by changing the area of the coil, even though B doesn't change. Here the area is reduced by pulling on its sides: the *flux* through the coil is reduced as we go from (a) to (b). Here the brief induced current acts in the direction shown so as to try to maintain the original flux ($\Phi = BA$) by producing its own magnetic field into the page. That is, as the area A decreases, the current acts to increase B in the original (inward) direction.

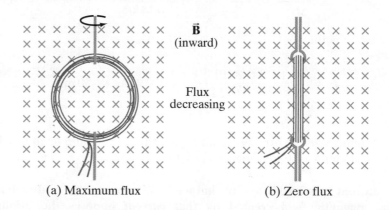

(a)   (b)

Since magnetic flux $\Phi_B = \int \vec{B} \cdot d\vec{A} = \int B \cos\theta \, dA$, we see that an emf can be induced in three ways: (1) by a changing magnetic field B; (2) by changing the area A of the loop in the field; or (3) by changing the loop's orientation θ with respect to the field. Figures 29–1 and 29–2 illustrated case 1. Examples of cases 2 and 3 are illustrated in Figs. 29–6 and 29–7, respectively.

FIGURE 29–7 A current can be induced by rotating a coil in a magnetic field. The flux through the coil changes from (a) to (b) because θ (in Eq. 29–1a, $\Phi = BA \cos\theta$) went from $0°$ ($\cos\theta = 1$) to $90°$ ($\cos\theta = 0$).

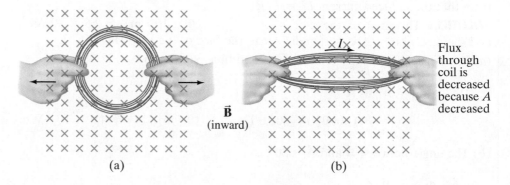

(a) Maximum flux   (b) Zero flux

FIGURE 29–8 Example 29–3: An induction stove.

CONCEPTUAL EXAMPLE 29–3 **Induction stove.** In an induction stove (Fig. 29–8), an ac current exists in a coil that is the "burner" (a burner that never gets hot). Why will it heat a metal pan but not a glass container?

RESPONSE The ac current sets up a changing magnetic field that passes through the pan bottom. This changing magnetic field induces a current in the pan bottom, and since the pan offers resistance, electric energy is transformed to thermal energy which heats the pot and its contents. A glass container offers such high resistance that little current is induced and little energy is transferred ($P = V^2/R$).

Lenz's law is used to determine the direction of the (conventional) electric current induced in a loop due to a change in magnetic flux inside the loop. To produce an induced current you need

(a) a closed conducting loop, and

(b) an external magnetic flux through the loop that is changing in time.

1. Determine whether the magnetic flux ($\Phi_B = BA \cos\theta$) inside the loop is decreasing, increasing, or unchanged.

2. The magnetic field due to the induced current: (a) points in the same direction as the external field if the flux is decreasing; (b) points in the opposite direction from the external field if the flux is increasing; or (c) is zero if the flux is not changing.

3. Once you know the direction of the induced magnetic field, use the right-hand rule to find the direction of the induced current.

4. Always keep in mind that there are two magnetic fields: (1) an external field whose flux must be changing if it is to induce an electric current, and (2) a magnetic field produced by the induced current.

FIGURE 29–9 Example 29–4.

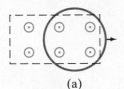

(a)

Pulling a round loop to the right out of a magnetic field which points out of the page

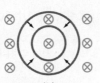

(b)

Shrinking a loop in a magnetic field pointing into the page

(c)

S magnetic pole moving from below, up toward the loop

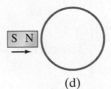

(d)

N magnetic pole moving toward loop in the plane of the page

(e)

Rotating the loop by pulling the left side toward us and pushing the right side in; the magnetic field points from right to left

CONCEPTUAL EXAMPLE 29–4 **Practice with Lenz's law.** In which direction is the current induced in the circular loop for each situation in Fig. 29–9?

RESPONSE (*a*) Initially, the magnetic field pointing out of the page passes through the loop. If you pull the loop out of the field, magnetic flux through the loop decreases; so the induced current will be in a direction to maintain the decreasing flux through the loop: the current will be counterclockwise to produce a magnetic field outward (toward the reader).

(*b*) The external field is into the page. The coil area gets smaller, so the flux will decrease; hence the induced current will be clockwise, producing its own field into the page to make up for the flux decrease.

(*c*) Magnetic field lines point into the S pole of a magnet, so as the magnet moves toward us and the loop, the magnet's field points into the page and is getting stronger. The current in the loop will be induced in the counterclockwise direction in order to produce a field $\vec{B}$ *out* of the page.

(*d*) The field is in the plane of the loop, so no magnetic field lines pass through the loop and the flux through the loop is zero throughout the process; hence there is no change in external magnetic flux with time, and there will be no induced emf or current in the loop.

(*e*) Initially there is no flux through the loop. When you start to rotate the loop, the external field through the loop begins increasing to the left. To counteract this change in flux, the loop will have current induced in a counterclockwise direction so as to produce its own field to the right.

⚠ **C A U T I O N**

Magnetic field created by induced current opposes change in external flux, not necessarily opposing the external field

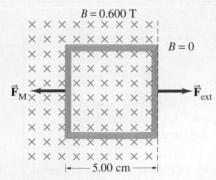

$B = 0.600 \, \text{T}$

$B = 0$

$\vec{F}_{M}$

$\vec{F}_{ext}$

5.00 cm

FIGURE 29–10 Example 29–5. The square coil in a magnetic field $B = 0.600 \, \text{T}$ is pulled abruptly to the right to a region where $B = 0$.

FIGURE 29–11 Exercise B.

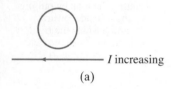

I increasing

(a)

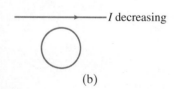

I decreasing

(b)

I constant

(c)

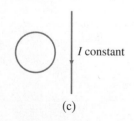

I increasing

(d)

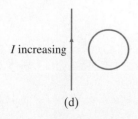

EXAMPLE 29–5 **Pulling a coil from a magnetic field.** A 100-loop square coil of wire, with side $\ell = 5.00 \, \text{cm}$ and total resistance $100 \, \Omega$, is positioned perpendicular to a uniform 0.600-T magnetic field, as shown in Fig. 29–10. It is quickly pulled from the field at constant speed (moving perpendicular to $\vec{B}$) to a region where B drops abruptly to zero. At $t = 0$, the right edge of the coil is at the edge of the field. It takes 0.100 s for the whole coil to reach the field-free region. Find (a) the rate of change in flux through the coil, and (b) the emf and current induced. (c) How much energy is dissipated in the coil? (d) What was the average force required (F_{ext})?

APPROACH We start by finding how the magnetic flux, $\Phi_B = BA$, changes during the time interval $\Delta t = 0.100 \, \text{s}$. Faraday's law then gives the induced emf and Ohm's law gives the current.

SOLUTION (a) The area of the coil is $A = \ell^2 = (5.00 \times 10^{-2} \, \text{m})^2 = 2.50 \times 10^{-3} \, \text{m}^2$. The flux through one loop is initially $\Phi_B = BA = (0.600 \, \text{T})(2.50 \times 10^{-3} \, \text{m}^2) = 1.50 \times 10^{-3} \, \text{Wb}$. After 0.100 s, the flux is zero. The rate of change in flux is constant (because the coil is square), equal to

$$\frac{\Delta \Phi_B}{\Delta t} = \frac{0 - (1.50 \times 10^{-3} \, \text{Wb})}{0.100 \, \text{s}} = -1.50 \times 10^{-2} \, \text{Wb/s}.$$

(b) The emf induced (Eq. 29–2) in the 100-loop coil during this 0.100-s interval is

$$\mathcal{E} = -N \frac{\Delta \Phi_B}{\Delta t} = -(100)(-1.50 \times 10^{-2} \, \text{Wb/s}) = 1.50 \, \text{V}.$$

The current is found by applying Ohm's law to the 100-Ω coil:

$$I = \frac{\mathcal{E}}{R} = \frac{1.50 \, \text{V}}{100 \, \Omega} = 1.50 \times 10^{-2} \, \text{A} = 15.0 \, \text{mA}.$$

By Lenz's law, the current must be clockwise to produce more $\vec{B}$ into the page and thus oppose the decreasing flux into the page.

(c) The total energy dissipated in the coil is the product of the power ($= I^2 R$) and the time:

$$E = Pt = I^2 R t = (1.50 \times 10^{-2} \, \text{A})^2 (100 \, \Omega)(0.100 \, \text{s}) = 2.25 \times 10^{-3} \, \text{J}.$$

(d) We can use the result of part (c) and apply the work-energy principle: the energy dissipated E is equal to the work W needed to pull the coil out of the field (Chapters 7 and 8). Because $W = \bar{F}d$ where $d = 5.00 \, \text{cm}$, then

$$\bar{F} = \frac{W}{d} = \frac{2.25 \times 10^{-3} \, \text{J}}{5.00 \times 10^{-2} \, \text{m}} = 0.0450 \, \text{N}.$$

Alternate Solution (d) We can also calculate the force directly using $\vec{F} = I\vec{\ell} \times \vec{B}$, Eq. 27–3, which here for constant $\vec{B}$ is $F = I\ell B$. The force the magnetic field exerts on the top and bottom sections of the square coil of Fig. 29–10 are in opposite directions and cancel each other. The magnetic force $\vec{F}_M$ exerted on the left vertical section of the square coil acts to the left as shown because the current is up (clockwise). The right side of the loop is in the region where $\vec{B} = 0$. Hence the external force, to the right, needed to just overcome the magnetic force to the left (on $N = 100$ loops) is

$$F_{ext} = NI\ell B = (100)(0.0150 \, \text{A})(0.0500 \, \text{m})(0.600 \, \text{T}) = 0.0450 \, \text{N},$$

which is the same answer, confirming our use of energy conservation above.

EXERCISE B What is the direction of the induced current in the circular loop due to the current shown in each part of Fig. 29–11?

EMF Induced in a Moving Conductor

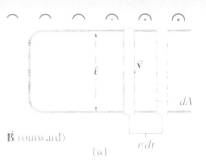

Another way to induce an emf is shown in Fig. 29–12a, and this situation helps illuminate the nature of the induced emf. Assume that a uniform magnetic field $\vec{B}$ is perpendicular to the area bounded by the U-shaped conductor and the movable rod resting on it. If the rod is made to move at a speed v, it travels a distance $dx = v\,dt$ in a time dt. Therefore, the area of the loop increases by an amount $dA = \ell\,dx = \ell v\,dt$ in a time dt. By Faraday's law there is an induced emf $\mathscr{E}$ whose magnitude is given by

$$\mathscr{E} = \frac{d\Phi_B}{dt} = \frac{B\,dA}{dt} = \frac{B\ell v\,dt}{dt} = B\ell v. \qquad (29\text{–}3)$$

Equation 29–3 is valid as long as B, ℓ, and v are mutually perpendicular. (If they are not, we use only the components of each that are mutually perpendicular.) An emf induced on a conductor moving in a magnetic field is sometimes called *motional emf*.

We can also obtain Eq. 29–3 without using Faraday's law. We saw in Chapter 27 that a charged particle moving perpendicular to a magnetic field B with speed v experiences a force $\vec{F} = q\vec{v} \times \vec{B}$ (Eq. 27–5a). When the rod of Fig. 29–12a moves to the right with speed v, the electrons in the rod also move with this speed. Therefore, since $\vec{v} \perp \vec{B}$, each electron feels a force $F = qvB$, which acts up the page as shown in Fig. 29–12b. If the rod was not in contact with the U-shaped conductor, electrons would collect at the upper end of the rod, leaving the lower end positive (see signs in Fig. 29–12b). There must thus be an induced emf. If the rod is in contact with the U-shaped conductor (Fig. 29–12a), the electrons will flow into the U. There will then be a clockwise (conventional) current in the loop. To calculate the emf, we determine the work W needed to move a charge q from one end of the rod to the other against this potential difference: $W = \text{force} \times \text{distance} = (qvB)(\ell)$. The emf equals the work done per unit charge, so $\mathscr{E} = W/q = qvB\ell/q = B\ell v$, the same result[†] as from Faraday's law above, Eq. 29–3.

EXERCISE C In what direction will the electrons flow in Fig. 29–12 if the rod moves to the left, decreasing the area of the current loop?

EXAMPLE 29–6 **ESTIMATE** **Does a moving airplane develop a large emf?** An airplane travels 1000 km/h in a region where the Earth's magnetic field is about $5 \times 10^{-5}\,\text{T}$ and is nearly vertical (Fig. 29–13). What is the potential difference induced between the wing tips that are 70 m apart?

APPROACH We consider the wings to be a 70-m-long conductor moving through the Earth's magnetic field. We use Eq. 29–3 to get the emf.

SOLUTION Since $v = 1000\,\text{km/h} = 280\,\text{m/s}$, and $\vec{v} \perp \vec{B}$, we have

$$\mathscr{E} = B\ell v = (5 \times 10^{-5}\,\text{T})(70\,\text{m})(280\,\text{m/s}) \approx 1\,\text{V}.$$

NOTE Not much to worry about.

EXAMPLE 29–7 **Electromagnetic blood-flow measurement.** The rate of blood flow in our body's vessels can be measured using the apparatus shown in Fig. 29–14, since blood contains charged ions. Suppose that the blood vessel is 2.0 mm in diameter, the magnetic field is 0.080 T, and the measured emf is 0.10 mV. What is the flow velocity v of the blood?

APPROACH The magnetic field $\vec{B}$ points horizontally from left to right (N pole toward S pole). The induced emf acts over the width $\ell = 2.0\,\text{mm}$ of the blood vessel, perpendicular to $\vec{B}$ and $\vec{v}$ (Fig. 29–14), just as in Fig. 29–12. We can then use Eq. 29–3 to get v. ($\vec{v}$ in Fig. 29–14 corresponds to $\vec{v}$ in Fig. 29–12.)

SOLUTION We solve for v in Eq. 29–3:

$$v = \frac{\mathscr{E}}{B\ell} = \frac{(1.0 \times 10^{-4}\,\text{V})}{(0.080\,\text{T})(2.0 \times 10^{-3}\,\text{m})} = 0.63\,\text{m/s}.$$

NOTE In actual practice, an alternating current is used to produce an alternating magnetic field. The induced emf is then alternating.

[†]This force argument, which is basically the same as for the Hall effect (Section 27–8), explains this one way of inducing an emf. It does not explain the general case of electromagnetic induction.

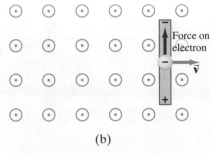

(b)

FIGURE 29–12 (a) A conducting rod is moved to the right on a U-shaped conductor in a uniform magnetic field $\vec{B}$ that points out of the page. The induced current is clockwise. (b) Upward force on an electron in the metal rod (moving to the right) due to $\vec{B}$ pointing out of the page; hence electrons can collect at top of rod, leaving + charge at bottom.

FIGURE 29–13 Example 29–6.

PHYSICS APPLIED
Blood-flow measurement

FIGURE 29–14 Measurement of blood velocity from the induced emf. Example 29–7.

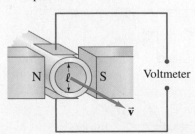

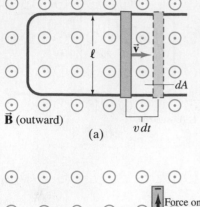

$\vec{B}$ (outward)

(a)

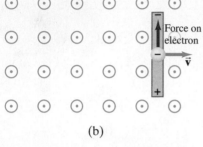

(b)

FIGURE 29–12 (repeated)
(a) A conducting rod is moved to the right on a U-shaped conductor in a uniform magnetic field $\vec{B}$ that points out of the page. The induced current is clockwise. (b) Upward force on an electron in the metal rod (moving to the right) due to $\vec{B}$ pointing out of the page; hence electrons can collect at top of rod, leaving + charge at bottom.

EXAMPLE 29–8 **Force on the rod.** To make the rod of Fig. 29–12a move to the right at constant speed v, you need to apply an external force on the rod to the right. (a) Explain and determine the magnitude of the required force. (b) What external power is needed to move the rod? (Do not confuse this external force on the rod with the upward force on the electrons shown in Fig. 29–12b.)

APPROACH When the rod moves to the right, electrons flow upward in the rod according to the right-hand rule. So the conventional current is downward in the rod. We can see this also from Lenz's law: the outward magnetic flux through the loop is increasing, so the induced current must oppose the increase. Thus the current is clockwise so as to produce a magnetic into the page (right-hand rule). The magnetic force on the moving rod is $\vec{F} = I\vec{\ell} \times \vec{B}$ for a constant $\vec{B}$ (Eq. 27–3). The right-hand rule tells us this magnetic force is to the left, and is thus a "drag force" opposing our effort to move the rod to the right.

SOLUTION (a) The magnitude of the external force, to the right, needs to balance the magnetic force $F = I\ell B$, to the left. The current $I = \mathcal{E}/R = B\ell v/R$ (see Eq. 29–3), and the resistance R is that of the whole circuit: the rod and the U-shaped conductor. The force F required to move the rod is thus

$$F = I\ell B = \left(\frac{B\ell v}{R}\right)\ell B = \frac{B^2\ell^2}{R}v.$$

If B, ℓ, and R are constant, then a constant speed v is produced by a constant external force. (Constant R implies that the parallel rails have negligible resistance.)

(b) The external power needed to move the rod for constant R is

$$P_{\text{ext}} = Fv = \frac{B^2\ell^2v^2}{R}.$$

The power dissipated in the resistance is $P = I^2R$. With $I = \mathcal{E}/R = B\ell v/R$,

$$P_{\text{R}} = I^2R = \frac{B^2\ell^2v^2}{R},$$

so the power input equals the power dissipated in the resistance at any moment.

29–4 Electric Generators

We discussed alternating currents (ac) briefly in Section 25–7. Now we examine how ac is generated: by an **electric generator** or **dynamo**, one of the most important practical results of Faraday's great discovery. A generator transforms mechanical energy into electric energy, just the opposite of what a motor does. A simplified diagram of an **ac generator** is shown in Fig. 29–15. A generator consists of many loops of wire (only one is shown) wound on an *armature* that can rotate in a magnetic field. The axle is turned by some mechanical means (falling water, steam turbine, car motor belt), and an emf is induced in the rotating coil. An electric current is thus the *output* of a generator. Suppose in Fig. 29–15 that the armature is rotating clockwise; then $\vec{F} = q\vec{v} \times \vec{B}$ applied to charged particles in the wire (or Lenz's law) tells us that the (conventional) current in the wire labeled b on the armature is outward, toward us; therefore the current is outward from brush b. (Each brush is fixed and presses against a continuous slip ring that rotates with the armature.) After one-half revolution, wire b will be where wire a is now in the drawing, and the current then at brush b will be inward. Thus the current produced is alternating.

FIGURE 29–15 An ac generator.

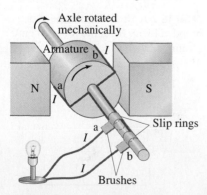

Let us assume the loop is being made to rotate in a uniform magnetic field $\vec{\mathbf{B}}$ with constant angular velocity ω. From Faraday's law (Eq. 29–2a), the induced emf is

$$\mathscr{E} = -\frac{d\Phi_B}{dt} = -\frac{d}{dt}\int \vec{\mathbf{B}}\cdot d\vec{\mathbf{A}} = -\frac{d}{dt}[BA\cos\theta]$$

where A is the area of the loop and θ is the angle between $\vec{\mathbf{B}}$ and $\vec{\mathbf{A}}$. Since $\omega = d\theta/dt$, then $\theta = \theta_0 + \omega t$. We arbitrarily take $\theta_0 = 0$, so

$$\mathscr{E} = -BA\frac{d}{dt}(\cos\omega t) = BA\omega\sin\omega t.$$

If the rotating coil contains N loops,

$$\mathscr{E} = NBA\omega\sin\omega t$$
$$= \mathscr{E}_0\sin\omega t. \qquad (29\text{–}4)$$

Thus the output emf is sinusoidal (Fig. 29–16) with amplitude $\mathscr{E}_0 = NBA\omega$. Such a rotating coil in a magnetic field is the basic operating principle of an ac generator.

The frequency $f\,(=\omega/2\pi)$ is 60 Hz for general use in the United States and Canada, whereas 50 Hz is used in many countries. Most of the power generated in the United States is done at steam plants, where the burning of fossil fuels (coal, oil, natural gas) boils water to produce high-pressure steam that turns a turbine connected to the generator axle. Falling water from the top of a dam (hydroelectric) is also common (Fig. 29–17). At nuclear power plants, the nuclear energy released is used to produce steam to turn turbines. Indeed, a heat engine (Chapter 20) connected to a generator is the principal means of generating electric power. The frequency of 60 Hz or 50 Hz is maintained very precisely by power companies, and in doing Problems, we will assume it is at least as precise as other numbers given.

EXAMPLE 29–9 **An ac generator.** The armature of a 60-Hz ac generator rotates in a 0.15-T magnetic field. If the area of the coil is $2.0\times10^{-2}\,\text{m}^2$, how many loops must the coil contain if the peak output is to be $\mathscr{E}_0 = 170\,\text{V}$?

APPROACH From Eq. 29–4 we see that the maximum emf is $\mathscr{E}_0 = NBA\omega$.

SOLUTION We solve Eq. 29–4 for N with $\omega = 2\pi f = (6.28)(60\,\text{s}^{-1}) = 377\,\text{s}^{-1}$:

$$N = \frac{\mathscr{E}_0}{BA\omega} = \frac{170\,\text{V}}{(0.15\,\text{T})(2.0\times10^{-2}\,\text{m}^2)(377\,\text{s}^{-1})} = 150\text{ turns}.$$

A **dc generator** is much like an ac generator, except the slip rings are replaced by split-ring commutators, Fig. 29–18a, just as in a dc motor (Section 27–6). The output of such a generator is as shown and can be smoothed out by placing a capacitor in parallel with the output (Section 26–5). More common is the use of many armature windings, as in Fig. 29–18b, which produces a smoother output.

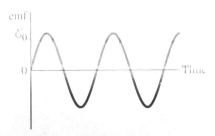

FIGURE 29–16 An ac generator produces an alternating current. The output emf $\mathscr{E} = \mathscr{E}_0\sin\omega t$, where $\mathscr{E}_0 = NAB\omega$ (Eq. 29–4).

PHYSICS APPLIED
Power plants

FIGURE 29–17 Water-driven generators at the base of Bonneville Dam, Oregon.

PHYSICS APPLIED
DC generator

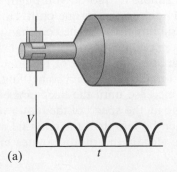 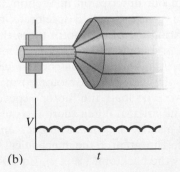

FIGURE 29–18 (a) A dc generator with one set of commutators, and (b) a dc generator with many sets of commutators and windings.

FIGURE 29–19 (a) Simplified schematic diagram of an alternator. The input current to the rotor from the battery is connected through continuous slip rings. Sometimes the rotor electromagnet is replaced by a permanent magnet. (b) Actual shape of an alternator. The rotor is made to turn by a belt from the engine. The current in the wire coil of the rotor produces a magnetic field inside it on its axis that points horizontally from left to right, thus making north and south poles of the plates attached at either end. These end plates are made with triangular fingers that are bent over the coil—hence there are alternating N and S poles quite close to one another, with magnetic field lines between them as shown by the blue lines. As the rotor turns, these field lines pass through the fixed stator coils (shown on the right for clarity, but in operation the rotor rotates within the stator), inducing a current in them, which is the output.

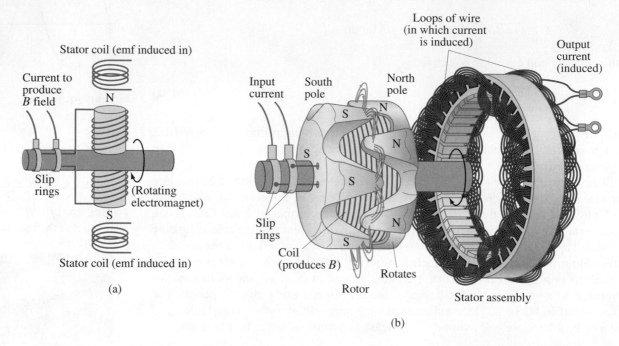

(a)

(b)

Automobiles used to use dc generators. Today they mainly use **alternators**, which avoid the problems of wear and electrical arcing (sparks) across the split-ring commutators of dc generators. Alternators differ from generators in that an electromagnet, called the *rotor*, is fed by current from the battery and is made to rotate by a belt from the engine. The magnetic field of the turning rotor passes through a surrounding set of stationary coils called the *stator* (Fig. 29–19), inducing an alternating current in the stator coils, which is the output. This ac output is changed to dc for charging the battery by the use of semiconductor diodes, which allow current flow in one direction only.

*29–5 Back EMF and Counter Torque; Eddy Currents

*Back EMF, in a Motor

A motor turns and produces mechanical energy when a current is made to flow in it. From our description in Section 27–6 of a simple dc motor, you might expect that the armature would accelerate indefinitely due to the torque on it. However, as the armature of the motor turns, the magnetic flux through the coil changes and an emf is generated. This induced emf acts to oppose the motion (Lenz's law) and is called the **back emf** or **counter emf**. The greater the speed of the motor, the greater the back emf. A motor normally turns and does work on something, but if there were no load, the motor's speed would increase until the back emf equaled the input voltage. When there is a mechanical load, the speed of the motor may be limited also by the load. The back emf will then be less than the external applied voltage. The greater the mechanical load, the slower the motor rotates and the lower is the back emf ($\mathscr{E} \propto \omega$, Eq. 29–4).

EXAMPLE 29–10 **Back emf in a motor.** The armature windings of a dc motor have a resistance of 5.0 Ω. The motor is connected to a 120-V line, and when the motor reaches full speed against its normal load, the back emf is 108 V. Calculate (a) the current into the motor when it is just starting up, and (b) the current when the motor reaches full speed.

APPROACH As the motor is just starting up, it is turning very slowly, so there is no induced back emf. The only voltage is the 120-V line. The current is given by Ohm's law with $R = 5.0$ Ω. At full speed, we must include as emfs both the 120-V applied emf and the opposing back emf.

SOLUTION (a) At start up, the current is controlled by the 120 V applied to the coil's 5.0-Ω resistance. By Ohm's law,

$$I = \frac{V}{R} = \frac{120\,\text{V}}{5.0\,\Omega} = 24\,\text{A}.$$

(b) When the motor is at full speed, the back emf must be included in the equivalent circuit shown in Fig. 29–20. In this case, Ohm's law (or Kirchhoff's rule) gives

$$120\,\text{V} - 108\,\text{V} = I(5.0\,\Omega).$$

Therefore

$$I = \frac{12\,\text{V}}{5.0\,\Omega} = 2.4\,\text{A}.$$

NOTE This result shows that the current can be very high when a motor first starts up. This is why the lights in your house may dim when the motor of the refrigerator (or other large motor) starts up. The large initial current causes the voltage to the lights and at the outlets to drop, since the house wiring has resistance and there is some voltage drop across it when large currents are drawn.

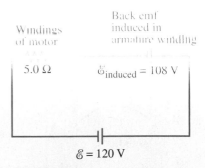

Windings of motor   Back emf induced in armature winding

5.0 Ω $\mathscr{E}_{\text{induced}} = 108$ V

$\mathscr{E} = 120$ V

FIGURE 29–20 Circuit of a motor showing induced back emf. Example 29–10.

CONCEPTUAL EXAMPLE 29–11 | **Motor overload.** When using an appliance such as a blender, electric drill, or sewing machine, if the appliance is overloaded or jammed so that the motor slows appreciably or stops while the power is still connected, the device can burn out and be ruined. Explain why this happens.

RESPONSE The motors are designed to run at a certain speed for a given applied voltage, and the designer must take the expected back emf into account. If the rotation speed is reduced, the back emf will not be as high as expected ($\mathscr{E} \propto \omega$, Eq. 29–4), and the current will increase, and may become large enough that the windings of the motor heat up to the point of ruining the motor.

*Counter Torque

In a generator, the situation is the reverse of that for a motor. As we saw, the mechanical turning of the armature induces an emf in the loops, which is the output. If the generator is not connected to an external circuit, the emf exists at the terminals but there is no current. In this case, it takes little effort to turn the armature. But if the generator *is* connected to a device that draws current, then a current flows in the coils of the armature. Because this current-carrying coil is in an external magnetic field, there will be a torque exerted on it (as in a motor), and this torque opposes the motion (use the right-hand rule for the force on a wire in Fig. 29–15). This is called a **counter torque**. The greater the electrical load—that is, the more current that is drawn—the greater will be the counter torque. Hence the external applied torque will have to be greater to keep the generator turning. This makes sense from the conservation of energy principle. More mechanical-energy input is needed to produce more electrical-energy output.

EXERCISE D A bicycle headlight is powered by a generator that is turned by the bicycle wheel. (a) If you pedal faster, how does the power to the light change? (b) Does the generator resist being turned as the bicycle's speed increases, and if so how?

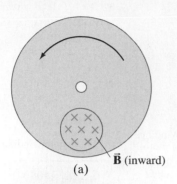

(a)

(b)

FIGURE 29–21 Production of eddy currents in a rotating wheel. The grey lines in (b) indicate induced current.

FIGURE 29–22 Airport metal detector.

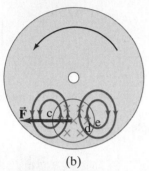

PHYSICS APPLIED
Airport metal detector

FIGURE 29–23 Repairing a step-down transformer on a utility pole.

*Eddy Currents

Induced currents are not always confined to well-defined paths such as in wires. Consider, for example, the rotating metal wheel in Fig. 29–21a. An external magnetic field is applied to a limited area of the wheel as shown and points into the page. The section of wheel in the magnetic field has an emf induced in it because the conductor is moving, carrying electrons with it. The flow of induced (conventional) current in the wheel is upward in the region of the magnetic field (Fig. 29–21b), and the current follows a downward return path outside that region. Why? According to Lenz's law, the induced currents oppose the change that causes them. Consider the part of the wheel labeled c in Fig. 29–21b, where the magnetic field is zero but is just about to enter a region where $\vec{B}$ points into the page. To oppose this inward increase in magnetic field, the induced current is counterclockwise to produce a field pointing out of the page (right-hand-rule 1). Similarly, region d is about to move to e, where $\vec{B}$ is zero; hence the current is clockwise to produce an inward field opposed to this decreasing flux inward. These currents are referred to as **eddy currents**. They can be present in any conductor that is moving across a magnetic field or through which the magnetic flux is changing.

In Fig. 29–21b, the magnetic field exerts a force $\vec{F}$ on the induced currents it has created, and that force opposes the rotational motion. Eddy currents can be used in this way as a smooth braking device on, say, a rapid-transit car. In order to stop the car, an electromagnet can be turned on that applies its field either to the wheels or to the moving steel rail below. Eddy currents can also be used to dampen (reduce) the oscillation of a vibrating system. Eddy currents, however, can be a problem. For example, eddy currents induced in the armature of a motor or generator produce heat $(P = I\mathscr{E})$ and waste energy. To reduce the eddy currents, the armatures are *laminated*; that is, they are made of very thin sheets of iron that are well insulated from one another. The total path length of the eddy currents is confined to each slab, which increases the total resistance; hence the current is less and there is less wasted energy.

Walk-through metal detectors at airports (Fig. 29–22) detect metal objects using electromagnetic induction and eddy currents. Several coils are situated in the walls of the walk-through at different heights. In a technique called "pulse induction," the coils are given repeated brief pulses of current (on the order of microseconds), hundreds or thousands of times a second. Each pulse in a coil produces a magnetic field for a very brief period of time. When a passenger passes through the walk-through, any metal object being carried will have eddy currents induced in it. The eddy currents persist briefly after each input pulse, and the small magnetic field produced by the persisting eddy current (before the next external pulse) can be detected, setting off an alert or alarm. Stores and libraries sometimes use similar systems to discourage theft.

29–6 Transformers and Transmission of Power

A transformer is a device for increasing or decreasing an ac voltage. Transformers are found everywhere: on utility poles (Fig. 29–23) to reduce the high voltage from the electric company to a usable voltage in houses (120 V or 240 V), in chargers for cell phones, laptops, and other electronic devices, in CRT monitors and in your car to give the needed high voltage (to the spark plugs), and in many other applications. A **transformer** consists of two coils of wire known as the **primary** and **secondary** coils. The two coils can be interwoven (with insulated wire); or they can be linked by an iron core which is laminated to minimize eddy-current losses (Section 29–5), as shown in Fig. 29–24. Transformers are designed so that (nearly) all the magnetic flux produced by the current in the primary coil also passes through the secondary coil, and we assume this is true in what follows. We also assume that energy losses (in resistance and hysteresis) can be ignored—a good approximation for real transformers, which are often better than 99% efficient.

When an ac voltage is applied to the primary coil, the changing magnetic field it produces will induce an ac voltage of the same frequency in the secondary coil. However, the voltage will be different according to the number of loops in each coil. From Faraday's law, the voltage or emf induced in the secondary coil is

$$V_S = N_S \frac{d\Phi_B}{dt},$$

where N_S is the number of turns in the secondary coil, and $d\Phi_B/dt$ is the rate at which the magnetic flux changes.

The input primary voltage, V_P, is related to the rate at which the flux changes through it,

$$V_P = N_P \frac{d\Phi_B}{dt},$$

where N_P is the number of turns in the primary coil. This follows because the changing flux produces a back emf, $N_P d\Phi_B/dt$, in the primary that exactly balances the applied voltage V_P if the resistance of the primary can be ignored (Kirchhoff's rules). We divide these two equations, assuming little or no flux is lost, to find

$$\frac{V_S}{V_P} = \frac{N_S}{N_P}. \tag{29–5}$$

This **transformer equation** tells how the secondary (output) voltage is related to the primary (input) voltage; V_S and V_P in Eq. 29–5 can be the rms values (Section 25–7) for both, or peak values for both. DC voltages don't work in a transformer because there would be no changing magnetic flux.

If the secondary coil contains more loops than the primary coil $(N_S > N_P)$, we have a **step-up transformer**. The secondary voltage is greater than the primary voltage. For example, if the secondary coil has twice as many turns as the primary coil, then the secondary voltage will be twice that of the primary voltage. If N_S is less than N_P, we have a **step-down transformer**.

Although ac voltage can be increased (or decreased) with a transformer, we don't get something for nothing. Energy conservation tells us that the power output can be no greater than the power input. A well-designed transformer can be greater than 99% efficient, so little energy is lost to heat. The power output thus essentially equals the power input. Since power $P = IV$ (Eq. 25–6), we have

$$I_P V_P = I_S V_S,$$

or

$$\frac{I_S}{I_P} = \frac{N_P}{N_S}. \tag{29–6}$$

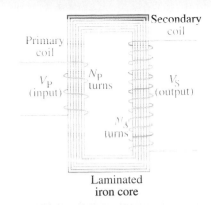

FIGURE 29–24 Step-up transformer $(N_P = 4, N_S = 12)$.

EXAMPLE 29–12 **Cell phone charger.** The charger for a cell phone contains a transformer that reduces 120-V (or 240-V) ac to 5.0-V ac to charge the 3.7-V battery (Section 26–4). (It also contains diodes to change the 5.0-V ac to 5.0-V dc.) Suppose the secondary coil contains 30 turns and the charger supplies 700 mA. Calculate (a) the number of turns in the primary coil, (b) the current in the primary, and (c) the power transformed.

APPROACH We assume the transformer is ideal, with no flux loss, so we can use Eq. 29–5 and then Eq. 29–6.

SOLUTION (a) This is a step-down transformer, and from Eq. 29–5 we have

$$N_P = N_S \frac{V_P}{V_S} = \frac{(30)(120 \text{ V})}{(5.0 \text{ V})} = 720 \text{ turns}.$$

(b) From Eq. 29–6

$$I_P = I_S \frac{N_S}{N_P} = (0.70 \text{ A})\left(\frac{30}{720}\right) = 29 \text{ mA}.$$

(c) The power transformed is

$$P = I_S V_S = (0.70 \text{ A})(5.0 \text{ V}) = 3.5 \text{ W}.$$

NOTE The power in the primary coil, $P = (0.029 \text{ A})(120 \text{ V}) = 3.5 \text{ W}$, is the same as the power in the secondary coil. There is 100% efficiency in power transfer for our ideal transformer.

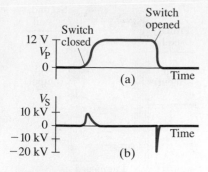

12 V Switch closed | Switch opened
V_P
0
(a) Time

V_S
10 kV
0
−10 kV
−20 kV
(b) Time

FIGURE 29–25 A dc voltage turned on and off as shown in (a) produces voltage pulses in the secondary (b). Voltage scales in (a) and (b) are not the same.

A transformer operates only on ac. A dc current in the primary coil does not produce a changing flux and therefore induces no emf in the secondary. However, if a dc voltage is applied to the primary through a switch, at the instant the switch is opened or closed there will be an induced voltage in the secondary. For example, if the dc is turned on and off as shown in Fig. 29–25a, the voltage induced in the secondary is as shown in Fig. 29–25b. Notice that the secondary voltage drops to zero when the dc voltage is steady. This is basically how, in the **ignition system** of an automobile, the high voltage is created to produce the spark across the gap of a spark plug that ignites the gas-air mixture. The transformer is referred to simply as an "ignition coil," and transforms the 12 V of the battery (when switched off in the primary) into a spike of as much as 30 kV in the secondary.

Transformers play an important role in the transmission of electricity. Power plants are often situated some distance from metropolitan areas, so electricity must then be transmitted over long distances (Fig. 29–26). There is always some power loss in the transmission lines, and this loss can be minimized if the power is transmitted at high voltage, using transformers, as the following Example shows.

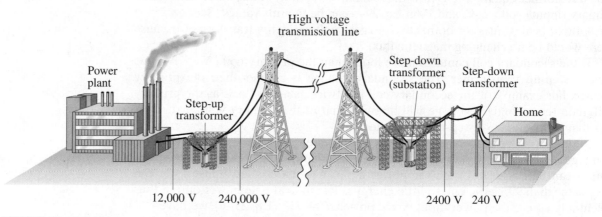

FIGURE 29–26 The transmission of electric power from power plants to homes makes use of transformers at various stages.

EXAMPLE 29–13 **Transmission lines.** An average of 120 kW of electric power is sent to a small town from a power plant 10 km away. The transmission lines have a total resistance of 0.40 Ω. Calculate the power loss if the power is transmitted at (a) 240 V and (b) 24,000 V.

APPROACH We cannot use $P = V^2/R$ because if R is the resistance of the transmission lines, we don't know the voltage drop along them; the given voltages are applied across the lines plus the load (the town). But we can determine the current I in the lines $(= P/V)$, and then find the power loss from $P_L = I^2R$, for both cases (a) and (b).

SOLUTION (a) If 120 kW is sent at 240 V, the total current will be

$$I = \frac{P}{V} = \frac{1.2 \times 10^5 \,\text{W}}{2.4 \times 10^2 \,\text{V}} = 500 \,\text{A}.$$

The power loss in the lines, P_L, is then

$$P_L = I^2R = (500 \,\text{A})^2(0.40 \,\Omega) = 100 \,\text{kW}.$$

Thus, over 80% of all the power would be wasted as heat in the power lines!

(*b*) If 120 kW is sent at 24,000 V, the total current will be

$$I = \frac{P}{V} = \frac{1.2 \times 10^5\,\text{W}}{2.4 \times 10^4\,\text{V}} = 5.0\,\text{A}.$$

The power loss in the lines is then

$$P_1 = I^2 R = (5.0\,\text{A})^2 (0.40\,\Omega) = 10\,\text{W}.$$

Which is less than $\frac{1}{100}$ of 1%, a far better efficiency!

NOTE We see that the higher voltage results in less current, and thus less power is wasted as heat in the transmission lines. It is for this reason that power is usually transmitted at very high voltages, as high as 700 kV.

The great advantage of ac, and a major reason it is in nearly universal use, is that the voltage can easily be stepped up or down by a transformer. The output voltage of an electric generating plant is stepped up prior to transmission. Upon arrival in a city, it is stepped down in stages at electric substations prior to distribution. The voltage in lines along city streets is typically 2400 V or 7200 V (but sometimes less), and is stepped down to 240 V or 120 V for home use by transformers (Figs. 29–23 and 29–26).

Fluorescent lights require a very high voltage initially to ionize the gas inside the bulb. The high voltage is obtained using a step-up transformer, called a ballast, and can be replaced independently of the bulb in many fluorescent light fixtures. When the ballast starts to fail, the tube is slow to light. Replacing the bulb will not solve the problem. In newer compact fluorescent bulbs designed to replace incandescent bulbs, the ballast (transformer) is part of the bulb, and is very small.

29–7 A Changing Magnetic Flux Produces an Electric Field

We have seen in earlier Chapters (especially Chapter 25, Section 25–8) that when an electric current flows in a wire, there is an electric field in the wire that does the work of moving the electrons in the wire. In this Chapter we have seen that a changing magnetic flux induces a current in the wire, which implies that there is an electric field in the wire induced by the changing magnetic flux. Thus we come to the important conclusion that

a changing magnetic flux produces an electric field.

This result applies not only to wires and other conductors, but is actually a general result that applies to any region in space. Indeed, an electric field will be produced at any point in space where there is a changing magnetic field.

Faraday's Law—General Form

We can put these ideas into mathematical form by generalizing our relation between an electric field and the potential difference between two points a and b: $V_{\text{ab}} = \int_a^b \vec{E} \cdot d\vec{\ell}$ (Eq. 23–4a) where $d\vec{\ell}$ is an element of displacement along the path of integration. The emf $\mathscr{E}$ induced in a circuit is equal to the work done per unit charge by the electric field, which equals the integral of $\vec{E} \cdot d\vec{\ell}$ around the closed path:

$$\mathscr{E} = \oint \vec{E} \cdot d\vec{\ell}. \qquad (29\text{--}7)$$

We combine this with Eq. 29–2a, to obtain a more elegant and general form of Faraday's law

$$\oint \vec{E} \cdot d\vec{\ell} = -\frac{d\Phi_B}{dt} \qquad (29\text{--}8)$$

FARADAY'S LAW (general form)

which relates the changing magnetic flux to the electric field it produces. The integral on the left is taken around a path enclosing the area through which the magnetic flux Φ_B is changing. This more elegant statement of Faraday's law (Eq. 29–8) is valid not only in conductors, but in any region of space. To illustrate this, let us take an Example.

FIGURE 29–27 Example 29–14.
(a) Side view of nearly constant $\vec{\mathbf{B}}$.
(b) Top view, for determining the
electric field $\vec{\mathbf{E}}$ at point P. (c) Lines of
$\vec{\mathbf{E}}$ produced by increasing $\vec{\mathbf{B}}$ (pointing
outward). (d) Graph of E vs. r.

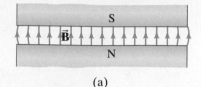

(a)

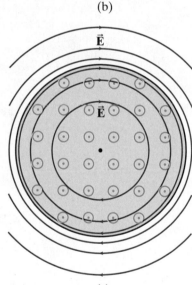

(b)

(c)

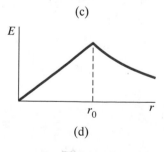

(d)

EXAMPLE 29–14 $\vec{\mathbf{E}}$ **produced by changing** $\vec{\mathbf{B}}$**.** A magnetic field $\vec{\mathbf{B}}$ between
the pole faces of an electromagnet is nearly uniform at any instant over a circular
area of radius r_0 as shown in Figs. 29–27a and b. The current in the windings of
the electromagnet is increasing in time so that $\vec{\mathbf{B}}$ changes in time at a constant
rate $d\vec{\mathbf{B}}/dt$ at each point. Beyond the circular region $(r > r_0)$, we assume $\vec{\mathbf{B}} = 0$
at all times. Determine the electric field $\vec{\mathbf{E}}$ at any point P a distance r from the
center of the circular area due to the changing $\vec{\mathbf{B}}$.

APPROACH The changing magnetic flux through a circle of radius r, shown
dashed in Fig. 29–27b, will produce an emf around this circle. Because all points
on the dashed circle are equivalent physically, the electric field too will show this
symmetry and will be in the plane perpendicular to $\vec{\mathbf{B}}$. Thus we can expect $\vec{\mathbf{E}}$ to
be perpendicular to $\vec{\mathbf{B}}$ and to be tangent to the circle of radius r. The direction of
$\vec{\mathbf{E}}$ will be as shown in Fig. 29–27b and c, since by Lenz's law the induced $\vec{\mathbf{E}}$ needs
to be capable of producing a current that generates a magnetic field opposing the
original change in $\vec{\mathbf{B}}$. By *symmetry*, we also expect $\vec{\mathbf{E}}$ to have the same magnitude
at all points on the circle of radius r.

SOLUTION We take the circle shown in Fig 29–27b as our path of integration in
Eq. 29–8. We ignore the minus sign so we can concentrate on magnitude since we
already found the direction of $\vec{\mathbf{E}}$ from Lenz's law, and obtain

$$E(2\pi r) = (\pi r^2)\frac{dB}{dt}, \qquad [r < r_0]$$

since $\Phi_B = BA = B(\pi r^2)$ at any instant. We solve for E and obtain

$$E = \frac{r}{2}\frac{dB}{dt}. \qquad [r < r_0]$$

This expression is valid up to the edge of the circle $(r \le r_0)$, beyond which
$\vec{\mathbf{B}} = 0$. If we now consider a point where $r > r_0$, the flux through a circle of
radius r is $\Phi_B = \pi r_0^2 B$. Then Eq. 29–8 gives

$$E(2\pi r) = \pi r_0^2\frac{dB}{dt} \qquad [r > r_0]$$

or

$$E = \frac{r_0^2}{2r}\frac{dB}{dt}. \qquad [r > r_0]$$

Thus the magnitude of the induced electric field increases linearly from zero at the
center of the magnet to $E = (dB/dt)(r_0/2)$ at the edge, and then decreases inversely
with distance in the region beyond the edge of the magnetic field. The electric field
lines are circles as shown in Fig. 29–27c. A graph of E vs. r is shown in Fig. 29–27d.

EXERCISE E Consider the magnet shown in Fig. 29–27 with a radius $r_0 = 6.0$ cm. If the
magnetic field changes uniformly from 0.040 T to 0.090 T in 0.18 s, what is the magnitude
of the resulting electric field at (a) $r = 3.0$ cm and (b) $r = 9.0$ cm?

*Forces Due to Changing $\vec{\mathbf{B}}$ are Nonconservative

Example 29–14 illustrates an important difference between electric fields
produced by changing magnetic fields and electric fields produced by electric
charges at rest (electrostatic fields). Electric field lines produced in the electrostatic case (Chapters 21 to 24) start and stop on electric charges. But the electric
field lines produced by a changing magnetic field are continuous; they form closed
loops. This distinction goes even further and is an important one. In the electrostatic case, the potential difference between two points is given by (Eq. 23–4a)

$$V_{ba} = V_b - V_a = -\int_a^b \vec{\mathbf{E}}\cdot d\vec{\boldsymbol{\ell}}.$$

If the integral is around a closed loop, so points a and b are the same, then $V_{ba} = 0$.

Hence the integral of $\vec{\mathbf{E}} \cdot d\vec{\boldsymbol{\ell}}$ around a closed path is zero:

$$\oint \vec{\mathbf{E}} \cdot d\vec{\boldsymbol{\ell}} = 0. \qquad\qquad \text{[electrostatic field]}$$

This followed from the fact that the electrostatic force (Coulomb's law) is a conservative force, and so a potential energy function could be defined. Indeed, the relation above, $\oint \vec{\mathbf{E}} \cdot d\vec{\boldsymbol{\ell}} = 0$, tells us that the work done per unit charge around any closed path is zero (or the work done between any two points is independent of path—see Chapter 8), which is a property only of a conservative force. But in the nonelectrostatic case, when the electric field is produced by a changing magnetic field, the integral around a closed path is *not* zero, but is given by Eq. 29–8:

$$\oint \vec{\mathbf{E}} \cdot d\vec{\boldsymbol{\ell}} = -\frac{d\Phi_B}{dt}.$$

We thus come to the conclusion that the forces due to changing magnetic fields are *nonconservative*. We are not able therefore to define a potential energy, or potential function, at a given point in space for the nonelectrostatic case. Although static electric fields are *conservative fields*, the electric field produced by a changing magnetic field is a **nonconservative field**.

*29–8 Applications of Induction: Sound Systems, Computer Memory, Seismograph, GFCI

*Microphone

There are various types of *microphones*, and many operate on the principle of induction. In one form, a microphone is just the inverse of a loudspeaker (Section 27–6). A small coil connected to a membrane is suspended close to a small permanent magnet, as shown in Fig. 29–28. The coil moves in the magnetic field when sound waves strike the membrane and this motion induces an emf. The frequency of the induced emf will be just that of the impinging sound waves, and this emf is the "signal" that can be amplified and sent to loudspeakers, or sent to a recorder.

*Read/Write on Tape and Disks

Recording and playback on tape or disks is done by magnetic *heads*. Recording tapes for use in audio and video tape recorders contain a thin layer of magnetic oxide on a thin plastic tape. During recording, the audio and/or video signal voltage is sent to the recording head, which acts as a tiny electromagnet (Fig. 29–29) that magnetizes the tiny section of tape passing over the narrow gap in the head at each instant. In playback, the changing magnetism of the moving tape at the gap causes corresponding changes in the magnetic field within the soft-iron head, which in turn induces an emf in the coil (Faraday's law). This induced emf is the output signal that can be amplified and sent to a loudspeaker (audio) or to the picture tube (video). In audio and video recorders, the signals may be *analog*—they vary continuously in amplitude over time. The variation in degree of magnetization of the tape at any point reflects the variation in amplitude and frequency of the audio or video signal.

Digital information, such as used on computer hard drives or on magnetic computer tape and some types of digital tape recorders, is read and written using heads that are basically the same as just described (Fig. 29–29). The essential difference is in the signals, which are not analog, but are digital, and in particular binary, meaning that only two values are possible for each of the extremely high number of predetermined spaces on the tape or disk. The two possible values are usually referred to as 1 and 0. The signal voltage does not vary continuously but rather takes on only two values, $+5$ V and 0 V, for example, corresponding to the 1 or 0. Thus, information is carried as a series of **bits**, each of which can have only one of two values, 1 or 0.

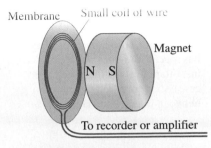

FIGURE 29–28 Diagram of a microphone that works by induction.

FIGURE 29–29 (a) Read/Write (playback/recording) head for tape or disk. In writing or recording, the electric input signal to the head, which acts as an electromagnet, magnetizes the passing tape or disk. In reading or playback, the changing magnetic field of the passing tape or disk induces a changing magnetic field in the head, which in turn induces in the coil an emf that is the output signal. (b) Photo of a hard drive showing several platters and read/write heads that can quickly move from the edge of the disk to the center.

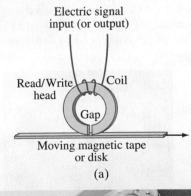

(a)

(b)

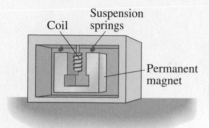

FIGURE 29–30 One type of seismograph, in which the coil is fixed to the case and moves with the Earth. The magnet, suspended by springs, has inertia and does not move instantaneously with the coil (and case), so there is relative motion between magnet and coil.

*Credit Card Swipe

When you swipe your credit card at a store or gas station, the magnetic stripe on the back of the card passes over a read head just as in a tape recorder or computer. The magnetic stripe contains personal information about your account and connects by telephone line for approval if your account is in order.

*Seismograph

In geophysics, a **seismograph** measures the intensity of earthquake waves using a magnet and a coil of wire. Either the magnet or the coil is fixed to the case, and the other is inertial (suspended by a spring; Fig. 29–30). The relative motion of magnet and coil when the Earth shakes induces an emf output.

*Ground Fault Circuit Interrupter (GFCI)

Fuses and circuit breakers (Sections 25–6 and 28–8) protect buildings from fire, and apparatus from damage, due to undesired high currents. But they do not turn off the current until it is very much greater than that which causes permanent damage to humans or death (≈ 100 mA). If fast enough, they may protect in case of a short. A *ground fault circuit interrupter* (GFCI) is meant to protect humans; GFCIs can react to currents as small as 5 mA.

FIGURE 29–31 A ground fault circuit interrupter (GFCI).

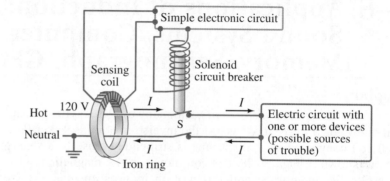

FIGURE 29–32 (a) A GFCI wall outlet. GFCIs can be recognized because they have "test" and "reset" buttons. (b) Add-on GFCI that plugs into outlet.

(a)

(b)

Electromagnetic induction is the physical basis of a GFCI. As shown in Fig. 29–31, the two conductors of a power line leading to an electrical circuit or device (red) pass through a small iron ring. Around the ring are many loops of thin wire that serve as a sensing coil. Under normal conditions (no ground fault), the current moving in the hot wire is exactly balanced by the returning current in the neutral wire. If something goes wrong and the hot wire touches the ungrounded metal case of the device or appliance, some of the entering current can pass through a person who touches the case and then to ground (a *ground fault*). Then the return current in the neutral wire will be less than the entering current in the hot wire, so there is a *net current* passing through the GFCI's iron ring. Because the current is ac, it is changing and the current difference produces a changing magnetic field in the iron, thus inducing an emf in the sensing coil wrapped around the iron. For example, if a device draws 8.0 A, and there is a ground fault through a person of 100 mA (= 0.1 A), then 7.9 A will appear in the neutral wire. The emf induced in the sensing coil by this 100-mA difference is amplified by a simple transistor circuit and sent to its own solenoid circuit breaker that opens the circuit at the switch S, thus protecting your life.

If the case of the faulty device is grounded, the current difference is even higher when there is a fault, and the GFCI trips immediately.

GFCIs can sense currents as low as 5 mA and react in 1 msec, saving lives. They can be small enough to fit as a wall outlet (Fig. 29–32a), or as a plug-in unit into which you plug a hair dryer or toaster (Fig. 29–32b). It is especially important to have GFCIs installed in kitchens, bathrooms, outdoors, and near swimming pools, where people are most in danger of touching ground. GFCIs always have a "test" button (to be sure it works) and a "reset" button (after it goes off).

Summary

The **magnetic flux** passing through a loop is equal to the product of the area of the loop times the perpendicular component of the (uniform) magnetic field: $\Phi_B = B_\perp A = BA \cos\theta$. If **B** is not uniform, then

$$\Phi_B = \int \vec{B} \cdot d\vec{A}. \qquad (29\text{-}1\text{b})$$

If the magnetic flux through a coil of wire changes in time, an emf is induced in the coil. The magnitude of the induced emf equals the time rate of change of the magnetic flux through the loop times the number N of loops in the coil:

$$\mathscr{E} = -N\frac{d\Phi_B}{dt}. \qquad (29\text{-}2\text{b})$$

This is **Faraday's law of induction**.

The induced emf can produce a current whose magnetic field opposes the original change in flux (**Lenz's law**).

We can also see from Faraday's law that a straight wire of length ℓ moving with speed v perpendicular to a magnetic field of strength B has an emf induced between its ends equal to:

$$\mathscr{E} = B\ell v. \qquad (29\text{-}3)$$

Faraday's law also tells us that *a changing magnetic field produces an electric field*. The mathematical relation is

$$\oint \vec{E} \cdot d\vec{\ell} = -\frac{d\Phi_B}{dt} \qquad (29\text{-}8)$$

and is the general form of Faraday's law. The integral on the left is taken around the loop through which the magnetic flux Φ_B is changing.

An electric **generator** changes mechanical energy into electrical energy. Its operation is based on Faraday's law: a coil of wire is made to rotate uniformly by mechanical means in a magnetic field, and the changing flux through the coil induces a sinusoidal current, which is the output of the generator.

[*A motor, which operates in the reverse of a generator, acts like a generator in that a **back emf** is induced in its rotating coil; since this counter emf opposes the input voltage, it can act to limit the current in a motor coil. Similarly, a generator acts somewhat like a motor in that a **counter torque** acts on its rotating coil.]

A **transformer**, which is a device to change the magnitude of an ac voltage, consists of a primary coil and a secondary coil. The changing flux due to an ac voltage in the primary coil induces an ac voltage in the secondary coil. In a 100% efficient transformer, the ratio of output to input voltages (V_S/V_P) equals the ratio of the number of turns N_S in the secondary to the number N_P in the primary:

$$\frac{V_S}{V_P} = \frac{N_S}{N_P}. \qquad (29\text{-}5)$$

The ratio of secondary to primary current is in the inverse ratio of turns:

$$\frac{I_S}{I_P} = \frac{N_P}{N_S}. \qquad (29\text{-}6)$$

[*Microphones, ground fault circuit interrupters, seismographs, and read/write heads for computer drives and tape recorders are applications of electromagnetic induction.]

Questions

1. What would be the advantage, in Faraday's experiments (Fig. 29–1), of using coils with many turns?

2. What is the difference between magnetic flux and magnetic field?

3. Suppose you are holding a circular ring of wire and suddenly thrust a magnet, south pole first, away from you toward the center of the circle. Is a current induced in the wire? Is a current induced when the magnet is held steady within the ring? Is a current induced when you withdraw the magnet? In each case, if your answer is yes, specify the direction.

4. Two loops of wire are moving in the vicinity of a very long straight wire carrying a steady current as shown in Fig. 29–33. Find the direction of the induced current in each loop.

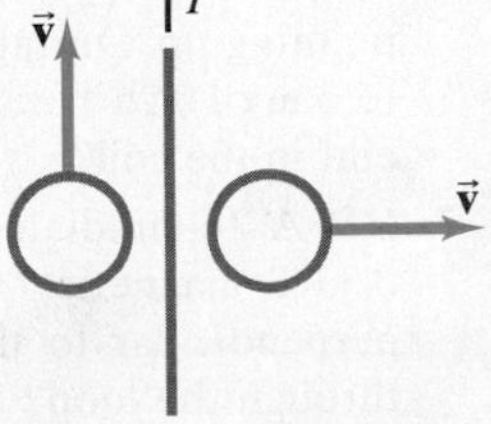

FIGURE 29–33
Questions 4 and 5.

5. Is there a force between the two loops discussed in Question 4? If so, in what direction?

6. Suppose you are looking along a line through the centers of two circular (but separate) wire loops, one behind the other. A battery is suddenly connected to the front loop, establishing a clockwise current. (a) Will a current be induced in the second loop? (b) If so, when does this current start? (c) When does it stop? (d) In what direction is this current? (e) Is there a force between the two loops? (f) If so, in what direction?

7. The battery mentioned in Question 6 is disconnected. Will a current be induced in the second loop? If so, when does it start and stop? In what direction is this current?

8. In what direction will the current flow in Fig. 29–12a if the rod moves to the left, which decreases the area of the loop to the left?

9. In Fig. 29–34, determine the direction of the induced current in resistor R_A (a) when coil B is moved toward coil A, (b) when coil B is moved away from A, (c) when the resistance R_B is increased.

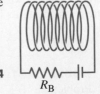

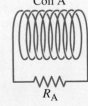

FIGURE 29–34
Question 9.

10. In situations where a small signal must travel over a distance, a *shielded cable* is used in which the signal wire is surrounded by an insulator and then enclosed by a cylindrical conductor carrying the return current (Fig. 28–12). Why is a "shield" necessary?

11. What is the advantage of placing the two insulated electric wires carrying ac close together or even twisted about each other?

12. Which object will fall faster in a nonuniform magnetic field, a conducting loop with radius ℓ or a straight wire of length $\ell/2$?

13. A region where no magnetic field is desired is surrounded by a sheet of low-resistivity metal. (*a*) Will this sheet shield the interior from a rapidly changing magnetic field outside? Explain. (*b*) Will it act as a shield to a static magnetic field? (*c*) What if the sheet is superconducting (resistivity = 0)?

14. A cell phone charger contains a transformer. Why can't you just buy one universal charger to charge your old cell phone, your new cell phone, your drill, and your toy electric train?

15. An enclosed transformer has four wire leads coming from it. How could you determine the ratio of turns on the two coils without taking the transformer apart? How would you know which wires paired with which?

16. The use of higher-voltage lines in homes—say, 600 V or 1200 V—would reduce energy waste. Why are they not used?

17. A transformer designed for a 120-V ac input will often "burn out" if connected to a 120-V dc source. Explain. [*Hint*: The resistance of the primary coil is usually very low.]

*18. Explain why, exactly, the lights may dim briefly when a refrigerator motor starts up. When an electric heater is turned on, the lights may stay dimmed as long as the heater is on. Explain the difference.

*19. Use Fig. 29–15 plus the right-hand rules to show why the counter torque in a generator *opposes* the motion.

*20. Will an eddy current brake (Fig. 29–21) work on a copper or aluminum wheel, or must the wheel be ferromagnetic? Explain.

*21. It has been proposed that eddy currents be used to help sort solid waste for recycling. The waste is first ground into tiny pieces and iron removed with a dc magnet. The waste then is allowed to slide down an incline over permanent magnets. How will this aid in the separation of nonferrous metals (Al, Cu, Pb, brass) from nonmetallic materials?

*22. The pivoted metal bar with slots in Fig. 29–35 falls much more quickly through a magnetic field than does a solid bar. Explain.

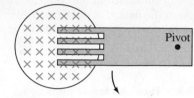

FIGURE 29–35
Question 22.

*23. If an aluminum sheet is held between the poles of a large bar magnet, it requires some force to pull it out of the magnetic field even though the sheet is not ferromagnetic and does not touch the pole faces. Explain.

*24. A bar magnet falling inside a vertical metal tube reaches a terminal velocity even if the tube is evacuated so that there is no air resistance. Explain.

*25. A metal bar, pivoted at one end, oscillates freely in the absence of a magnetic field; but in a magnetic field, its oscillations are quickly damped out. Explain. (This *magnetic damping* is used in a number of practical devices.)

*26. Since a magnetic microphone is basically like a loudspeaker, could a loudspeaker (Section 27–6) actually serve as a microphone? That is, could you speak into a loudspeaker and obtain an output signal that could be amplified? Explain. Discuss, in light of your response, how a microphone and loudspeaker differ in construction.

Problems

29–1 and 29–2 Faraday's Law of Induction

1. (I) The magnetic flux through a coil of wire containing two loops changes at a constant rate from −58 Wb to +38 Wb in 0.42 s. What is the emf induced in the coil?

2. (I) The north pole of the magnet in Fig. 29–36 is being inserted into the coil. In which direction is the induced current flowing through the resistor R?

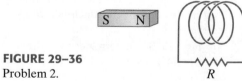

FIGURE 29–36
Problem 2.

3. (I) The rectangular loop shown in Fig. 29–37 is pushed into the magnetic field which points inward. In what direction is the induced current?

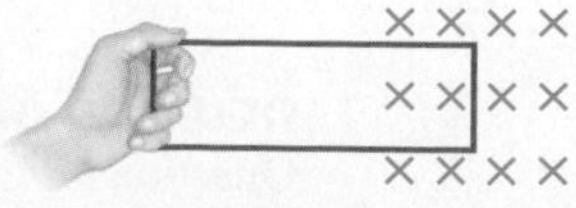

FIGURE 29–37
Problem 3.

4. (I) A 22.0-cm-diameter loop of wire is initially oriented perpendicular to a 1.5-T magnetic field. The loop is rotated so that its plane is parallel to the field direction in 0.20 s. What is the average induced emf in the loop?

5. (II) A circular wire loop of radius $r = 12$ cm is immersed in a uniform magnetic field $B = 0.500$ T with its plane normal to the direction of the field. If the field magnitude then decreases at a constant rate of −0.010 T/s, at what rate should r increase so that the induced emf within the loop is zero?

6. (II) A 10.8-cm-diameter wire coil is initially oriented so that its plane is perpendicular to a magnetic field of 0.68 T pointing up. During the course of 0.16 s, the field is changed to one of 0.25 T pointing down. What is the average induced emf in the coil?

7. (II) A 16-cm-diameter circular loop of wire is placed in a 0.50-T magnetic field. (*a*) When the plane of the loop is perpendicular to the field lines, what is the magnetic flux through the loop? (*b*) The plane of the loop is rotated until it makes a 35° angle with the field lines. What is the angle θ in Eq. 29–1a for this situation? (*c*) What is the magnetic flux through the loop at this angle?

8. (II) (a) If the resistance of the resistor in Fig. 29–38 is slowly increased, what is the direction of the current induced in the small circular loop inside the larger loop? (b) What would it be if the small loop were placed outside the larger one, to the left?

FIGURE 29–38
Problem 8.

9. (II) If the solenoid in Fig. 29–39 is being pulled away from the loop shown, in what direction is the induced current in the loop?

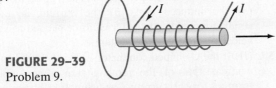

FIGURE 29–39
Problem 9.

10. (II) The magnetic field perpendicular to a circular wire loop 8.0 cm in diameter is changed from +0.52 T to −0.45 T in 180 ms, where + means the field points away from an observer and − toward the observer. (a) Calculate the induced emf. (b) In what direction does the induced current flow?

11. (II) A circular loop in the plane of the paper lies in a 0.75-T magnetic field pointing into the paper. If the loop's diameter changes from 20.0 cm to 6.0 cm in 0.50 s, (a) what is the direction of the induced current, (b) what is the magnitude of the average induced emf, and (c) if the coil resistance is 2.5 Ω, what is the average induced current?

12. (II) Part of a single rectangular loop of wire with dimensions shown in Fig. 29–40 is situated inside a region of uniform magnetic field of 0.650 T. The total resistance of the loop is 0.280 Ω. Calculate the force required to pull the loop from the field (to the right) at a constant velocity of 3.40 m/s. Neglect gravity.

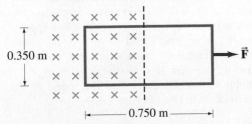

FIGURE 29–40 Problem 12.

13. (II) While demonstrating Faraday's law to her class, a physics professor inadvertently moves the gold ring on her finger from a location where a 0.80-T magnetic field points along her finger to a zero-field location in 45 ms. The 1.5-cm-diameter ring has a resistance and mass of 55 μΩ and 15 g, respectively. (a) Estimate the thermal energy produced in the ring due to the flow of induced current. (b) Find the temperature rise of the ring, assuming all of the thermal energy produced goes into increasing the ring's temperature. The specific heat of gold is 129 J/kg·C°.

14. (II) A 420-turn solenoid, 25 cm long, has a diameter of 2.5 cm. A 15-turn coil is wound tightly around the center of the solenoid. If the current in the solenoid increases uniformly from 0 to 5.0 A in 0.60 s, what will be the induced emf in the short coil during this time?

15. (II) A 22.0-cm-diameter coil consists of 28 turns of circular copper wire 2.6 mm in diameter. A uniform magnetic field, perpendicular to the plane of the coil, changes at a rate of 0.65×10^{-2} T/s. Determine (a) the current in the loop, and (b) the rate at which thermal energy is produced.

16. (II) A power line carrying a sinusoidally varying current with frequency $f = 60$ Hz and peak value $I_0 = 55$ kA runs at a height of 7.0 m across a farmer's land (Fig. 29–41). The farmer constructs a vertically oriented 2.0-m-high 10-turn rectangular wire coil below the power line. The farmer hopes to use the induced voltage in this coil to power 120-Volt electrical equipment, which requires a sinusoidally varying voltage with frequency $f = 60$ Hz and peak value $V_0 = 170$ V. What should the length ℓ of the coil be? Would this be unethical?

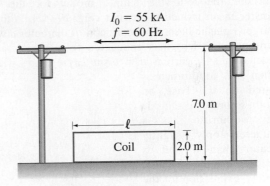

FIGURE 29–41 Problem 16.

17. (II) The magnetic field perpendicular to a single 18.2-cm-diameter circular loop of copper wire decreases uniformly from 0.750 T to zero. If the wire is 2.35 mm in diameter, how much charge moves past a point in the coil during this operation?

18. (II) The magnetic flux through each loop of a 75-loop coil is given by $(8.8t - 0.51t^3) \times 10^{-2}$ T·m², where the time t is in seconds. (a) Determine the emf $\mathscr{E}$ as a function of time. (b) What is $\mathscr{E}$ at $t = 1.0$ s and $t = 4.0$ s?

19. (II) A 25-cm-diameter circular loop of wire has a resistance of 150 Ω. It is initially in a 0.40-T magnetic field, with its plane perpendicular to $\vec{B}$, but is removed from the field in 120 ms. Calculate the electric energy dissipated in this process.

20. (II) The area of an elastic circular loop decreases at a constant rate, $dA/dt = -3.50 \times 10^{-2}$ m²/s. The loop is in a magnetic field $B = 0.28$ T whose direction is perpendicular to the plane of the loop. At $t = 0$, the loop has area $A = 0.285$ m². Determine the induced emf at $t = 0$, and at $t = 2.00$ s.

21. (II) Suppose the radius of the elastic loop in Problem 20 increases at a constant rate, $dr/dt = 4.30$ cm/s. Determine the emf induced in the loop at $t = 0$ and at $t = 1.00$ s.

22. (II) A single circular loop of wire is placed inside a long solenoid with its plane perpendicular to the axis of the solenoid. The area of the loop is A_1 and that of the solenoid, which has n turns per unit length, is A_2. A current $I = I_0 \cos \omega t$ flows in the solenoid turns. What is the induced emf in the small loop?

23. (II) We are looking down on an elastic conducting loop with resistance $R = 2.0\,\Omega$, immersed in a magnetic field. The field's magnitude is uniform spatially, but varies with time t according to $B(t) = \alpha t$, where $\alpha = 0.60\,\text{T/s}$. The area A of the loop also increases at a constant rate, according to $A(t) = A_0 + \beta t$, where $A_0 = 0.50\,\text{m}^2$ and $\beta = 0.70\,\text{m}^2/\text{s}$. Find the magnitude and direction (clockwise or counterclockwise, when viewed from above the page) of the induced current within the loop at time $t = 2.0\,\text{s}$ if the magnetic field (a) is parallel to the plane of the loop to the right; (b) is perpendicular to the plane of the loop, down.

24. (II) Inductive battery chargers, which allow transfer of electrical power without the need for exposed electrical contacts, are commonly used in appliances that need to be safely immersed in water, such as electric toothbrushes. Consider the following simple model for the power transfer in an inductive charger (Fig. 29–42). Within the charger's plastic base, a primary coil of diameter d with n_P turns per unit length is connected to a home's ac wall outlet so that a current $I = I_0 \sin(2\pi f t)$ flows within it. When the toothbrush is seated on the base, an N-turn secondary coil inside the toothbrush has a diameter only slightly greater than d and is centered on the primary. Find an expression for the emf induced in the secondary coil. [This induced emf recharges the battery.]

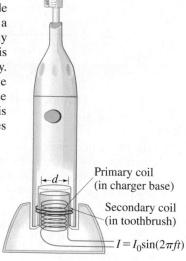

Primary coil (in charger base)

Secondary coil (in toothbrush)

$I = I_0 \sin(2\pi f t)$

FIGURE 29–42
Problem 24.

25. (III) (a) Determine the magnetic flux through a square loop of side a (Fig. 29–43) if one side is parallel to, and a distance b from, a straight wire that carries a current I. (b) If the loop is pulled away from the wire at speed v, what emf is induced in it? (c) Does the induced current flow clockwise or counterclockwise? (d) Determine the force F required to pull the loop away.

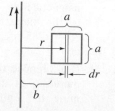

FIGURE 29–43
Problems 25 and 26.

26. (III) Determine the emf induced in the square loop in Fig. 29–43 if the loop stays at rest and the current in the straight wire is given by $I(t) = (15.0\,\text{A}) \sin(2500 t)$ where t is in seconds. The distance a is 12.0 cm, and b is 15.0 cm.

29–3 Motional EMF

27. (I) The moving rod in Fig. 29–12b is 13.2 cm long and generates an emf of 120 mV while moving in a 0.90-T magnetic field. What is its speed?

28. (I) The moving rod in Fig. 29–12b is 12.0 cm long and is pulled at a speed of 15.0 cm/s. If the magnetic field is 0.800 T, calculate the emf developed.

29. (II) In Fig. 29–12a, the rod moves to the right with a speed of 1.3 m/s and has a resistance of 2.5 Ω. The rail separation is $\ell = 25.0$ cm. The magnetic field is 0.35 T, and the resistance of the U-shaped conductor is 25.0 Ω at a given instant. Calculate (a) the induced emf, (b) the current in the U-shaped conductor, and (c) the external force needed to keep the rod's velocity constant at that instant.

30. (II) If the U-shaped conductor in Fig. 29–12a has resistivity ρ, whereas that of the moving rod is negligible, derive a formula for the current I as a function of time. Assume the rod starts at the bottom of the U at $t = 0$, and moves with uniform speed v in the magnetic field B. The cross-sectional area of the rod and all parts of the U is A.

31. (II) Suppose that the U-shaped conductor and connecting rod in Fig. 29–12a are oriented vertically (but still in contact) so that the rod is falling due to the gravitational force. Find the terminal speed of the rod if it has mass $m = 3.6$ grams, length $\ell = 18$ cm, and resistance $R = 0.0013\,\Omega$. It is falling in a uniform horizontal field $B = 0.060$ T. Neglect the resistance of the U-shaped conductor.

32. (II) When a car drives through the Earth's magnetic field, an emf is induced in its vertical 75.0-cm-long radio antenna. If the Earth's field $(5.0 \times 10^{-5}\,\text{T})$ points north with a dip angle of $45°$, what is the maximum emf induced in the antenna and which direction(s) will the car be moving to produce this maximum value? The car's speed is 30.0 m/s on a horizontal road.

33. (II) A conducting rod rests on two long frictionless parallel rails in a magnetic field $\vec{\mathbf{B}}$ ($\perp$ to the rails and rod) as in Fig. 29–44. (a) If the rails are horizontal and the rod is given an initial push, will the rod travel at constant speed even though a magnetic field is present? (b) Suppose at $t = 0$, when the rod has speed $v = v_0$, the two rails are connected electrically by a wire from point a to point b. Assuming the rod has resistance R and the rails have negligible resistance, determine the speed of the rod as a function of time. Discuss your answer.

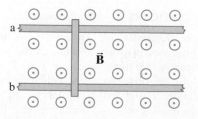

FIGURE 29–44   Problems 33 and 34.

34. (III) Suppose a conducting rod (mass m, resistance R) rests on two frictionless and resistanceless parallel rails a distance ℓ apart in a uniform magnetic field $\vec{\mathbf{B}}$ ($\perp$ to the rails and to the rod) as in Fig. 29–44. At $t = 0$, the rod is at rest and a source of emf is connected to the points a and b. Determine the speed of the rod as a function of time if (a) the source puts out a constant current I, (b) the source puts out a constant emf $\mathcal{E}_0$, (c) Does the rod reach a terminal speed in either case? If so, what is it?

35. (III) A short section of wire of length a is moving with velocity $\vec{\mathbf{v}}$, parallel to a very long wire carrying a current I as shown in Fig. 29–45. The near end of the wire section is a distance b from the long wire. Assuming the vertical wire is very long compared to $a + b$, determine the emf between the ends of the short section. Assume $\vec{\mathbf{v}}$ is (a) in the same direction as I, (b) in the opposite direction to I.

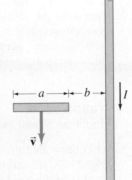

FIGURE 29–45
Problem 35.

29–4 Generators

36. (I) The generator of a car idling at 875-rpm produces 12.4 V. What will the output be at a rotation speed of 1550 rpm assuming nothing else changes?

37. (I) A simple generator is used to generate a peak output voltage of 24.0 V. The square armature consists of windings that are 5.15 cm on a side and rotates in a field of 0.420 T at a rate of 60.0 rev/s. How many loops of wire should be wound on the square armature?

38. (II) A simple generator has a 480-loop square coil 22.0 cm on a side. How fast must it turn in a 0.550-T field to produce a 120-V peak output?

39. (II) Show that the rms output of an ac generator is $V_{\text{rms}} = NAB\omega/\sqrt{2}$ where $\omega = 2\pi f$.

40. (II) A 250-loop circular armature coil with a diameter of 10.0 cm rotates at 120 rev/s in a uniform magnetic field of strength 0.45 T. What is the rms voltage output of the generator? What would you do to the rotation frequency in order to double the rms voltage output?

*29–5 Back EMF, Counter Torque; Eddy Current

***41.** (I) The back emf in a motor is 72 V when operating at 1200 rpm. What would be the back emf at 2500 rpm if the magnetic field is unchanged?

***42.** (I) A motor has an armature resistance of 3.05 Ω. If it draws 7.20 A when running at full speed and connected to a 120-V line, how large is the back emf?

***43.** (II) What will be the current in the motor of Example 29–10 if the load causes it to run at half speed?

***44.** (II) The back emf in a motor is 85 V when the motor is operating at 1100 rpm. How would you change the motor's magnetic field if you wanted to reduce the back emf to 75 V when the motor was running at 2300 rpm?

***45.** (II) A dc generator is rated at 16 kW, 250 V, and 64 A when it rotates at 1000 rpm. The resistance of the armature windings is 0.40 Ω. (a) Calculate the "no-load" voltage at 1000 rpm (when there is no circuit hooked up to the generator). (b) Calculate the full-load voltage (i.e. at 64 A) when the generator is run at 750 rpm. Assume that the magnitude of the magnetic field remains constant.

[Assume 100% efficiency, unless stated otherwise.]

46. (I) A transformer has 620 turns in the primary coil and 85 in the secondary coil. What kind of transformer is this, and by what factor does it change the voltage? By what factor does it change the current?

47. (I) Neon signs require 12 kV for their operation. To operate from a 240-V line, what must be the ratio of secondary to primary turns of the transformer? What would the voltage output be if the transformer were connected backward?

48. (II) A model-train transformer plugs into 120-V ac and draws 0.35 A while supplying 7.5 A to the train. (a) What voltage is present across the tracks? (b) Is the transformer step-up or step-down?

49. (II) The output voltage of a 75-W transformer is 12 V, and the input current is 22 A. (a) Is this a step-up or a step-down transformer? (b) By what factor is the voltage multiplied?

50. (II) If 65 MW of power at 45 kV (rms) arrives at a town from a generator via 3.0-Ω transmission lines, calculate (a) the emf at the generator end of the lines, and (b) the fraction of the power generated that is wasted in the lines.

51. (II) Assume a voltage source supplies an ac voltage of amplitude V_0 between its output terminals. If the output terminals are connected to an external circuit, and an ac current of amplitude I_0 flows out of the terminals, then the equivalent resistance of the external circuit is $R_{\text{eq}} = V_0/I_0$. (a) If a resistor R is connected directly to the output terminals, what is R_{eq}? (b) If a transformer with N_{P} and N_{S} turns in its primary and secondary, respectively, is placed between the source and the resistor as shown in Fig. 29–46, what is R_{eq}? [Transformers can be used in ac circuits to alter the apparent resistance of circuit elements, such as loud speakers, in order to maximize transfer of power.]

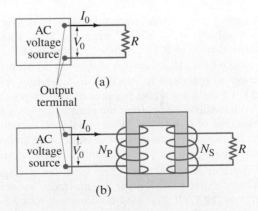

FIGURE 29–46   Problem 51.

52. (III) Design a dc transmission line that can transmit 225 MW of electricity 185 km with only a 2.0% loss. The wires are to be made of aluminum and the voltage is 660 kV.

53. (III) Suppose 85 kW is to be transmitted over two 0.100-Ω lines. Estimate how much power is saved if the voltage is stepped up from 120 V to 1200 V and then down again, rather than simply transmitting at 120 V. Assume the transformers are each 99% efficient.

29–7 Changing Φ_B Produces $\vec{E}$

54. (II) In a circular region, there is a uniform magnetic field $\vec{B}$ pointing into the page (Fig. 29–47). An xy coordinate system has its origin at the circular region's center. A free positive point charge $+Q = 1.0 \, \mu C$ is initially at rest at a position $x = +10$ cm on the x axis. If the magnitude of the magnetic field is now decreased at a rate of -0.10 T/s, what force (magnitude and direction) will act on $+Q$?

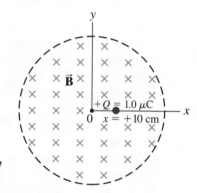

FIGURE 29–47
Problem 54.

55. (II) The **betatron**, a device used to accelerate electrons to high energy, consists of a circular vacuum tube placed in a magnetic field (Fig. 29–48), into which electrons are injected. The electromagnet produces a field that (1) keeps the electrons in their circular orbit inside the tube, and (2) increases the speed of the electrons when B changes. (a) Explain how the electrons are accelerated. (See Fig. 29–48.) (b) In what direction are the electrons moving in Fig. 29–48 (give directions as if looking down from above)? (c) Should B increase or decrease to accelerate the electrons? (d) The magnetic field is actually 60 Hz ac; show that the electrons can be accelerated only during $\frac{1}{4}$ of a cycle $\left(\frac{1}{240} \text{s}\right)$. (During this time they make hundreds of thousands of revolutions and acquire very high energy.)

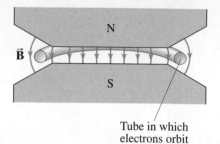

FIGURE 29–48
Problems 55 and 56.

Tube in which electrons orbit

56. (III) Show that the electrons in a betatron, Problem 55 and Fig. 29–48, are accelerated at constant radius if the magnetic field B_0 at the position of the electron orbit in the tube is equal to half the average value of the magnetic field (B_{avg}) over the area of the circular orbit at each moment: $B_0 = \frac{1}{2} B_{avg}$. (This is the reason the pole faces have a rather odd shape, as indicated in Fig. 29–48.)

57. (III) Find a formula for the net electric field in the moving rod of Problem 34 as a function of time for each case, (a) and (b).

General Problems

58. Suppose you are looking at two current loops in the plane of the page as shown in Fig. 29–49. When the switch S is closed in the left-hand coil, (a) what is the direction of the induced current in the other loop? (b) What is the situation after a "long" time? (c) What is the direction of the induced current in the right-hand loop if that loop is quickly pulled horizontally to the right (S having been closed for a long time)?

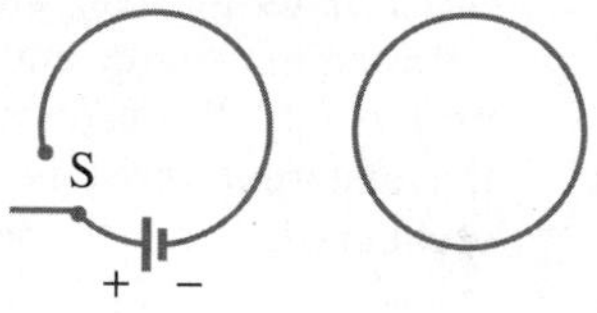

FIGURE 29–49
Problem 58.

59. A square loop 27.0 cm on a side has a resistance of 7.50 Ω. It is initially in a 0.755-T magnetic field, with its plane perpendicular to $\vec{B}$, but is removed from the field in 40.0 ms. Calculate the electric energy dissipated in this process.

60. Power is generated at 24 kV at a generating plant located 85 km from a town that requires 65 MW of power at 12 kV. Two transmission lines from the plant to the town each have a resistance of 0.10 Ω/km. What should the output voltage of the transformer at the generating plant be for an overall transmission efficiency of 98.5%, assuming a perfect transformer?

61. A circular loop of area 12 m² encloses a magnetic field perpendicular to the plane of the loop; its magnitude is $B(t) = (10 \text{ T/s})t$. The loop is connected to a 7.5-Ω resistor and a 5.0-pF capacitor in series. When fully charged, how much charge is stored on the capacitor?

62. The primary windings of a transformer which has an 85% efficiency are connected to 110-V ac. The secondary windings are connected across a 2.4-Ω, 75-W lightbulb. (a) Calculate the current through the primary windings of the transformer. (b) Calculate the ratio of the number of primary windings of the transformer to the number of secondary windings of the transformer.

63. A pair of power transmission lines each have a 0.80-Ω resistance and carry 740 A over 9.0 km. If the rms input voltage is 42 kV, calculate (a) the voltage at the other end, (b) the power input, (c) power loss in the lines, and (d) the power output.

64. Show that the power loss in transmission lines, P_L, is given by $P_L = (P_T)^2 R_L / V^2$, where P_T is the power transmitted to the user, V is the delivered voltage, and R_L is the resistance of the power lines.

65. A high-intensity desk lamp is rated at 35 W but requires only 12 V. It contains a transformer that converts 120-V household voltage. (a) Is the transformer step-up or step-down? (b) What is the current in the secondary coil when the lamp is on? (c) What is the current in the primary coil? (d) What is the resistance of the bulb when on?

66. Two resistanceless rails rest 32 cm apart on a 6.0° ramp. They are joined at the bottom by a 0.60-Ω resistor. At the top a copper bar of mass 0.040 kg (ignore its resistance) is laid across the rails. The whole apparatus is immersed in a vertical 0.55-T field. What is the terminal (steady) velocity of the bar as it slides frictionlessly down the rails?

67. A coil with 150 turns, a radius of 5.0 cm, and a resistance of 12 Ω surrounds a solenoid with 230 turns/cm and a radius of 4.5 cm; see Fig. 29–50. The current in the solenoid changes at a constant rate from 0 to 2.0 A in 0.10 s. Calculate the magnitude and direction of the induced current in the 150-turn coil.

FIGURE 29–50
Problem 67.

I

68. A **search coil** for measuring B (also called a **flip coil**) is a small coil with N turns, each of cross-sectional area A. It is connected to a so-called **ballistic galvanometer**, which is a device to measure the total charge Q that passes through it in a short time. The flip coil is placed in the magnetic field to be measured with its face perpendicular to the field. It is then quickly rotated 180° about a diameter. Show that the total charge Q that flows in the induced current during this short "flip" time is proportional to the magnetic field B. In particular, show that B is given by

$$B = \frac{QR}{2NA}$$

where R is the total resistance of the circuit, including that of the coil and that of the ballistic galvanometer which measures the charge Q.

69. A ring with a radius of 3.0 cm and a resistance of 0.025 Ω is rotated about an axis through its diameter by 90° in a magnetic field of 0.23 T perpendicular to that axis. What is the largest number of electrons that would flow past a fixed point in the ring as this process is accomplished?

70. A flashlight can be made that is powered by the induced current from a magnet moving through a coil of wire. The coil and magnet are inside a plastic tube that can be shaken causing the magnet to move back and forth through the coil. Assume the magnet has a maximum field strength of 0.05 T. Make reasonable assumptions and specify the size of the coil and the number of turns necessary to light a standard 1-watt, 3-V flashlight bulb.

*71. A small electric car overcomes a 250-N friction force when traveling 35 km/h. The electric motor is powered by ten 12-V batteries connected in series and is coupled directly to the wheels whose diameters are 58 cm. The 270 armature coils are rectangular, 12 cm by 15 cm, and rotate in a 0.60-T magnetic field. (a) How much current does the motor draw to produce the required torque? (b) What is the back emf? (c) How much power is dissipated in the coils? (d) What percent of the input power is used to drive the car?

72. What is the energy dissipated as a function of time in a circular loop of 18 turns of wire having a radius of 10.0 cm and a resistance of 2.0 Ω if the plane of the loop is perpendicular to a magnetic field given by

$$B(t) = B_0 e^{-t/\tau}$$

with $B_0 = 0.50$ T and $\tau = 0.10$ s?

73. A thin metal rod of length ℓ rotates with angular velocity ω about an axis through one end (Fig. 29–51). The rotation axis is perpendicular to the rod and is parallel to a uniform magnetic field $\vec{B}$. Determine the emf developed between the ends of the rod.

FIGURE 29–51
Problem 75.

$\vec{B}$

*74. The magnetic field of a "shunt-wound" dc motor is produced by field coils placed in parallel with the armature coils. Suppose that the field coils have a resistance of 36.0 Ω and the armature coils 3.00 Ω. The back emf at full speed is 105 V when the motor is connected to 115 V dc. (a) Draw the equivalent circuit for the situations when the motor is just starting and when it is running full speed. (b) What is the total current drawn by the motor at start up? (c) What is the total current drawn when the motor runs at full speed?

75. Apply Faraday's law, in the form of Eq. 29–8, to show that the static electric field between the plates of a parallel-plate capacitor cannot drop abruptly to zero at the edges, but must, in fact, fringe. Use the path shown dashed in Fig. 29–52.

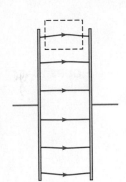

FIGURE 29–52
Problem 75.

76. A circular metal disk of radius R rotates with angular velocity ω about an axis through its center perpendicular to its face. The disk rotates in a uniform magnetic field B whose direction is parallel to the rotation axis. Determine the emf induced between the center and the edges.

77. What is the magnitude and direction of the electric field at each point in the rotating disk of Problem 76?

78. A circular-shaped circuit of radius r, containing a resistance R and capacitance C, is situated with its plane perpendicular to a spatially uniform magnetic field $\vec{B}$ directed into the page (Fig. 29–53). Starting at time $t = 0$, the voltage difference $V_{ba} = V_b - V_a$ across the capacitor plates is observed to increase with time t according to $V_{ba} = V_0(1 - e^{-t/\tau})$, where V_0 and τ are positive constants. Determine dB/dt, the rate at which the magnetic field magnitude changes with time. Is B becoming larger or smaller as time increases?

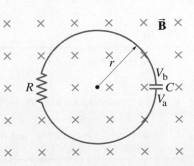

FIGURE 29–53
Problem 78.

79. In a certain region of space near Earth's surface, a uniform horizontal magnetic field of magnitude B exists above a level defined to be $y = 0$. Below $y = 0$, the field abruptly becomes zero (Fig. 29–54). A vertical square wire loop has resistivity ρ, mass density ρ_m, diameter d, and side length ℓ. It is initially at rest with its lower horizontal side at $y = 0$ and is then allowed to fall under gravity, with its plane perpendicular to the direction of the magnetic field. (*a*) While the loop is still partially immersed in the magnetic field (as it falls into the zero-field region), determine the magnetic "drag" force that acts on it at the moment when its speed is v. (*b*) Assume that the loop achieves a terminal velocity v_T before its upper horizontal side exits the field. Determine a formula for v_T. (*c*) If the loop is made of copper and $B = 0.80\,T$, find v_T.

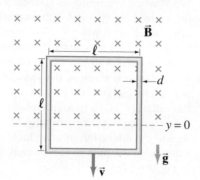

FIGURE 29–54
Problem 79.

*80. (III) In an experiment, a coil was mounted on a low-friction cart that moved through the magnetic field B of a permanent magnet. The speed of the cart v and the induced voltage V were simultaneously measured, as the cart moved through the magnetic field, using a computer-interfaced motion sensor and a voltmeter. The Table below shows the collected data:

Speed, v (m/s)	0.367	0.379	0.465	0.623	0.630
Induced voltage, V (V)	0.128	0.135	0.164	0.221	0.222

(*a*) Make a graph of the induced voltage, V, vs. the speed, v. Determine a best-fit linear equation for the data. Theoretically, the relationship between V and v is given by $V = BN\ell v$ where N is the number of turns of the coil, B is the magnetic field, and ℓ is the average of the inside and outside widths of the coil. In the experiment, $B = 0.126\,T$, $N = 50$, and $\ell = 0.0561\,m$. (*b*) Find the % error between the slope of the experimental graph and the theoretical value for the slope. (*c*) For each of the measured speeds v, determine the theoretical value of V and find the % error of each.

Answers to Exercises

A: (*e*).

B: (*a*) Counterclockwise; (*b*) clockwise; (*c*) zero; (*d*) counterclockwise.

C: Electrons flow clockwise (conventional current counterclockwise).

D: (*a*) increases; (*b*) yes; increases (counter torque).

E: (*a*) $4.2 \times 10^{-3}\,V/m$; (*b*) $5.6 \times 10^{-3}\,V/m$.

A spark plug in a car receives a high voltage, which produces a high enough electric field in the air across its gap to pull electrons off the atoms in the air–gasoline mixture and form a spark. The high voltage is produced, from the basic 12 V of the car battery, by an induction coil which is basically a transformer or mutual inductance. Any coil of wire has a self-inductance, and a changing current in it causes an emf to be induced. Such inductors are useful in many circuits.

C H A P T E R

30

Inductance, Electromagnetic Oscillations, and AC Circuits

CHAPTER-OPENING QUESTION—Guess now!

Consider a circuit with only a capacitor C and a coil of many loops of wire (called an inductor, L) as shown. If the capacitor is initially charged $(Q = Q_0)$, what will happen when the switch S is closed?

(a) Nothing will happen—the capacitor will remain charged with charge $Q = Q_0$.

(b) The capacitor will quickly discharge and remain discharged $(Q = 0)$.

(c) Current will flow until the positive charge is on the opposite plate of the capacitor, and then will reverse—back and forth.

(d) The energy initially in the capacitor $\left(U_E = \frac{1}{2}Q_0^2/C\right)$ will all transfer to the coil and then remain that way.

(e) The system will quickly transfer half of the capacitor energy to the coil and then remain that way.

We discussed in the last Chapter how a changing magnetic flux through a circuit induces an emf in that circuit. Before that we saw that an electric current produces a magnetic field. Combining these two ideas, we could predict that a changing current in one circuit ought to induce an emf and a current in a second nearby circuit, and even induce an emf in itself. We already saw an example in the previous Chapter (transformers), but now we will treat this effect in a more general way in terms of what we will call mutual inductance and self-inductance. The concept of inductance also gives us a springboard to treat energy storage in a magnetic field. This Chapter concludes with an analysis of circuits that contain inductance as well as resistance and/or capacitance.

785

30–1 Mutual Inductance

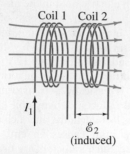

FIGURE 30–1 A changing current in one coil will induce a current in the second coil.

If two coils of wire are placed near each other, as in Fig. 30–1, a changing current in one will induce an emf in the other. According to Faraday's law, the emf $\mathscr{E}_2$ induced in coil 2 is proportional to the rate of change of magnetic flux passing through it. This flux is due to the current I_1 in coil 1, and it is often convenient to express the emf in coil 2 in terms of the current in coil 1.

We let Φ_{21} be the magnetic flux in each loop of coil 2 created by the current in coil 1. If coil 2 contains N_2 closely wrapped loops, then $N_2 \Phi_{21}$ is the total flux passing through coil 2. If the two coils are fixed in space, $N_2 \Phi_{21}$ is proportional to the current I_1 in coil 1; the proportionality constant is called the **mutual inductance**, M_{21}, defined by

$$M_{21} = \frac{N_2 \Phi_{21}}{I_1}. \tag{30–1}$$

The emf $\mathscr{E}_2$ induced in coil 2 due to a changing current in coil 1 is, by Faraday's law,

$$\mathscr{E}_2 = -N_2 \frac{d\Phi_{21}}{dt}.$$

We combine this with Eq. 30–1 rewritten as $\Phi_{21} = M_{21} I_1 / N_2$ (and take its derivative) and obtain

$$\mathscr{E}_2 = -M_{21} \frac{dI_1}{dt}. \tag{30–2}$$

This relates the change in current in coil 1 to the emf it induces in coil 2. The mutual inductance of coil 2 with respect to coil 1, M_{21}, is a "constant" in that it does not depend on I_1; M_{21} depends on "geometric" factors such as the size, shape, number of turns, and relative positions of the two coils, and also on whether iron (or some other ferromagnetic material) is present. For example, if the two coils in Fig. 30–1 are farther apart, fewer lines of flux can pass through coil 2, so M_{21} will be less. For some arrangements, the mutual inductance can be calculated (see Example 30–1). More often it is determined experimentally.

Suppose, now, we consider the reverse situation: when a changing current in coil 2 induces an emf in coil 1. In this case,

$$\mathscr{E}_1 = -M_{12} \frac{dI_2}{dt}$$

where M_{12} is the mutual inductance of coil 1 with respect to coil 2. It is possible to show, although we will not prove it here, that $M_{12} = M_{21}$. Hence, for a given arrangement we do not need the subscripts and we can let

$$M = M_{12} = M_{21},$$

so that

$$\mathscr{E}_1 = -M \frac{dI_2}{dt} \tag{30–3a}$$

and

$$\mathscr{E}_2 = -M \frac{dI_1}{dt}. \tag{30–3b}$$

The SI unit for mutual inductance is the henry (H), where $1\,\text{H} = 1\,\text{V}\cdot\text{s/A} = 1\,\Omega\cdot\text{s}$.

EXERCISE A Two coils which are close together have a mutual inductance of 330 mH. (*a*) If the emf in coil 1 is 120 V, what is the rate of change of the current in coil 2? (*b*) If the rate of change of current in coil 1 is 36 A/s, what is the emf in coil 2?

EXAMPLE 30–1 **Solenoid and coil.** A long thin solenoid of length ℓ and cross-sectional area A contains N_1 closely packed turns of wire. Wrapped around it is an insulated coil of N_2 turns, Fig. 30–2. Assume all the flux from coil 1 (the solenoid) passes through coil 2, and calculate the mutual inductance.

APPROACH We first determine the flux produced by the solenoid, all of which passes uniformly through coil N_2, using Eq. 28–4 for the magnetic field inside the solenoid:

$$B = \mu_0 \frac{N_1}{\ell} I_1,$$

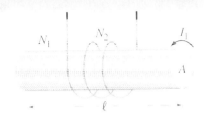

FIGURE 30–2 Example 30–1.

where $n = N_1/\ell$ is the number of loops in the solenoid per unit length, and I_1 is the current in the solenoid.

SOLUTION The solenoid is closely packed, so we assume that all the flux in the solenoid stays inside the secondary coil. Then the flux Φ_{21} through coil 2 is

$$\Phi_{21} = BA = \mu_0 \frac{N_1}{\ell} I_1 A.$$

Then the mutual inductance is

$$M = \frac{N_2 \Phi_{21}}{I_1} = \frac{\mu_0 N_1 N_2 A}{\ell}.$$

NOTE We calculated M_{21}; if we had tried to calculate M_{12}, it would have been difficult. Given $M_{12} = M_{21} = M$, we did the simpler calculation to obtain M. Note again that M depends only on geometric factors, and not on the currents.

CONCEPTUAL EXAMPLE 30–2 **Reversing the coils.** How would Example 30–1 change if the coil with N_2 turns was inside the solenoid rather than outside the solenoid?

RESPONSE The magnetic field inside the solenoid would be unchanged. The flux through the coil would be BA where A is the area of the coil, not of the solenoid as in Example 30–1. Solving for M would give the same formula except that A would refer to the coil, and would be smaller.

EXERCISE B Which solenoid and coil combination shown in Fig. 30–3 has the largest mutual inductance? Assume each solenoid is the same.

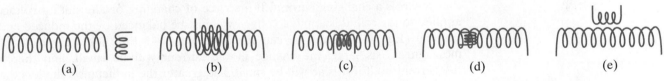

(a) (b) (c) (d) (e)

FIGURE 30–3 Exercise B.

A transformer is an example of mutual inductance in which the coupling is maximized so that nearly all flux lines pass through both coils. Mutual inductance has other uses as well, including some types of *pacemakers* used to maintain blood flow in heart patients (Section 26–5). Power in an external coil is transmitted via mutual inductance to a second coil in the pacemaker at the heart. This type has the advantage over battery-powered pacemakers in that surgery is not needed to replace a battery when it wears out.

Mutual inductance can sometimes be a problem, however. Any changing current in a circuit can induce an emf in another part of the same circuit or in a different circuit even though the conductors are not in the shape of a coil. The mutual inductance M is usually small unless coils with many turns and/or iron cores are involved. However, in situations where small voltages are being used, problems due to mutual inductance often arise. Shielded cable, in which an inner conductor is surrounded by a cylindrical grounded conductor (Fig. 28–12), is often used to reduce the problem.

PHYSICS APPLIED
Pacemaker

30–2 Self-Inductance

The concept of inductance applies also to a single isolated coil of N turns. When a changing current passes through a coil (or solenoid), a changing magnetic flux is produced inside the coil, and this in turn induces an emf in that same coil. This induced emf opposes the change in flux (Lenz's law). For example, if the current through the coil is increasing, the increasing magnetic flux induces an emf that opposes the original current and tends to retard its increase. If the current is decreasing in the coil, the decreasing flux induces an emf in the same direction as the current, thus tending to maintain the original current.

The magnetic flux Φ_B passing through the N turns of a coil is proportional to the current I in the coil, so we define the **self-inductance** L (in analogy to mutual inductance, Eq. 30–1) as

$$L = \frac{N\Phi_B}{I}. \tag{30–4}$$

Then the emf $\mathscr{E}$ induced in a coil of self-inductance L is, from Faraday's law,

$$\mathscr{E} = -N\frac{d\Phi_B}{dt} = -L\frac{dI}{dt}. \tag{30–5}$$

Like mutual inductance, self-inductance is measured in henrys. The magnitude of L depends on the geometry and on the presence of a ferromagnetic material. Self-inductance (inductance, for short) can be defined, as above, for any circuit or part of a circuit.

Circuits always contain some inductance, but often it is quite small unless the circuit contains a coil of many turns. A coil that has significant self-inductance L is called an **inductor**. Inductance is shown on circuit diagrams by the symbol

—ᴍᴍ— ;   [inductor symbol]

any resistance an inductor has should also be shown separately. Inductance can serve a useful purpose in certain circuits. Often, however, inductance is to be avoided in a circuit. Precision resistors are normally wire wound and thus would have inductance as well as resistance. The inductance can be minimized by winding the insulated wire back on itself in the opposite sense so that the current going in opposite directions produces little net magnetic flux; this is called a **noninductive winding**.

If an inductor has negligible resistance, it is the inductance (or induced emf) that controls a changing current. If a source of changing or alternating voltage is applied to the coil, this applied voltage will just be balanced by the induced emf of the coil (Eq. 30–5). Thus we can see from Eq. 30–5 that, for a given $\mathscr{E}$, if the inductance L is large, the change in the current will be small, and therefore the current itself if it is ac will be small. The greater the inductance, the less the ac current. An inductance thus acts something like a resistance to impede the flow of alternating current. We use the term *reactance* or *impedance* for this quality of an inductor. We will discuss reactance and impedance more fully in Sections 30–7 and 30–8. We shall see that reactance depends not only on the inductance L, but also on the frequency. Here we mention one example of its importance. The resistance of the primary in a transformer is usually quite small, perhaps less than $1\,\Omega$. If resistance alone limited the current in a transformer, tremendous currents would flow when a high voltage was applied. Indeed, a dc voltage applied to a transformer can burn it out. It is the induced emf (or reactance) of the coil that limits the current to a reasonable value.

Common inductors have inductances in the range from about $1\,\mu\text{H}$ to about $1\,\text{H}$ (where $1\,\text{H} = 1\ \text{henry} = 1\,\Omega\cdot\text{s}$).

EXAMPLE 30–3 **Solenoid inductance.** (*a*) Determine a formula for the self-inductance L of a tightly wrapped and long solenoid containing N turns of wire in its length ℓ and whose cross-sectional area is A. (*b*) Calculate the value of L if $N = 100$, $\ell = 5.0\,\text{cm}$, $A = 0.30\,\text{cm}^2$ and the solenoid is air filled.

APPROACH To determine the inductance L, it is usually simplest to start with Eq. 30–4, so we need to first determine the flux.

SOLUTION (a) According to Eq. 28–4, the magnetic field inside a solenoid (ignoring end effects) is constant: $B = \mu_0 nI$ where $n = N/\ell$. The flux is $\Phi_B = BA = \mu_0 NIA/\ell$, so

$$L = \frac{N\Phi_B}{I} = \frac{\mu_0 N^2 A}{\ell} ,$$

(b) Since $\mu_0 = 4\pi \times 10^{-7}\,\text{T}\cdot\text{m/A}$, then

$$L = \frac{(4\pi \times 10^{-7}\,\text{T}\cdot\text{m/A})(100)^2(5.0 \times 10^{-5}\,\text{m}^2)}{(5.0 \times 10^{-2}\,\text{m})} = 1.5\,\mu\text{H}.$$

NOTE Magnetic field lines "stray" out of the solenoid (see Fig. 28–15), especially near the ends, so our formula is only an approximation.

CONCEPTUAL EXAMPLE 30–4 | **Direction of emf in inductor.** Current passes through the coil in Fig. 30–4 from left to right as shown. (a) If the current is increasing with time, in which direction is the induced emf? (b) If the current is decreasing in time, what then is the direction of the induced emf?

RESPONSE (a) From Lenz's law we know that the induced emf must oppose the change in magnetic flux. If the current is increasing, so is the magnetic flux. The induced emf acts to oppose the increasing flux, which means it acts like a source of emf that opposes the outside source of emf driving the current. So the induced emf in the coil acts to oppose I in Fig. 30–4a. In other words, the inductor might be thought of as a battery with a positive terminal at point A (tending to block the current entering at A), and negative at point B. (b) If the current is decreasing, then by Lenz's law the induced emf acts to bolster the flux—like a source of emf reinforcing the external emf. The induced emf acts to increase I in Fig. 30–4b, so in this situation you can think of the induced emf as a battery with its negative terminal at point A to attract more (+) current to move to the right.

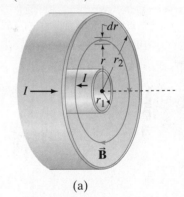

FIGURE 30–4 Example 30–4. The + and − signs refer to the induced emf due to the changing current, as if points A and B were the terminals of a battery (and the coiled loops were the inside of the battery).

EXAMPLE 30–5 **Coaxial cable inductance.** Determine the inductance per unit length of a coaxial cable whose inner conductor has a radius r_1 and the outer conductor has a radius r_2, Fig. 30–5. Assume the conductors are thin hollow tubes so there is no magnetic field within the inner conductor, and the magnetic field inside both thin conductors can be ignored. The conductors carry equal currents I in opposite directions.

APPROACH We need to find the magnetic flux, $\Phi_B = \int \vec{B} \cdot d\vec{A}$, between the conductors. The lines of $\vec{B}$ are circles surrounding the inner conductor (only one is shown in Fig. 30–5a). From Ampère's law, $\oint \vec{B} \cdot d\vec{\ell} = \mu_0 I$, the magnitude of the field along the circle at a distance r from the center, when the inner conductor carries a current I, is (Example 28–6):

$$B = \frac{\mu_0 I}{2\pi r}.$$

The magnetic flux through a rectangle of width dr and length ℓ (along the cable, Fig. 30–5b), a distance r from the center, is

$$d\Phi_B = B(\ell\, dr) = \frac{\mu_0 I}{2\pi r}\ell\, dr.$$

SOLUTION The total flux in a length ℓ of cable is

$$\Phi_B = \int d\Phi_B = \frac{\mu_0 I\ell}{2\pi}\int_{r_1}^{r_2}\frac{dr}{r} = \frac{\mu_0 I\ell}{2\pi}\ln\frac{r_2}{r_1}.$$

Since the current I all flows in one direction in the inner conductor, and the same current I all flows in the opposite direction in the outer conductor, we have only one turn, so $N = 1$ in Eq. 30–4. Hence the self-inductance for a length ℓ is

$$L = \frac{\Phi_B}{I} = \frac{\mu_0 \ell}{2\pi}\ln\frac{r_2}{r_1}.$$

The inductance per unit length is

$$\frac{L}{\ell} = \frac{\mu_0}{2\pi}\ln\frac{r_2}{r_1}.$$

Note that L depends only on geometric factors and not on the current I.

FIGURE 30–5 Example 30–5. Coaxial cable: (a) end view, (b) side view (cross section).

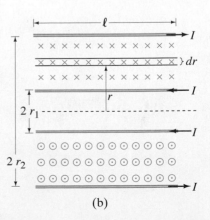

(a)

(b)

30–3 Energy Stored in a Magnetic Field

When an inductor of inductance L is carrying a current I which is changing at a rate dI/dt, energy is being supplied to the inductor at a rate

$$P = I\mathscr{E} = LI\frac{dI}{dt}$$

where P stands for power and we used[†] Eq. 30–5. Let us calculate the work needed to increase the current in an inductor from zero to some value I. Using this last equation, the work dW done in a time dt is

$$dW = P\,dt = LI\,dI.$$

Then the total work done to increase the current from zero to I is

$$W = \int dW = \int_0^I LI\,dI = \tfrac{1}{2}LI^2.$$

This work done is equal to the energy U stored in the inductor when it is carrying a current I (and we take $U = 0$ when $I = 0$):

$$U = \tfrac{1}{2}LI^2. \tag{30–6}$$

This can be compared to the energy stored in a capacitor, C, when the potential difference across it is V (see Section 24–4):

$$U = \tfrac{1}{2}CV^2.$$

EXERCISE C What is the inductance of an inductor if it has a stored energy of 1.5 J when there is a current of 2.5 A in it? (*a*) 0.48 H, (*b*) 1.2 H, (*c*) 2.1 H, (*d*) 4.7 H, (*e*) 19 H.

Just as the energy stored in a capacitor can be considered to reside in the electric field between its plates, so the energy in an inductor can be considered to be stored in its magnetic field. To write the energy in terms of the magnetic field, let us use the result of Example 30–3, that the inductance of an ideal solenoid (end effects ignored) is $L = \mu_0 N^2 A/\ell$. Because the magnetic field B in a solenoid is related to the current I by $B = \mu_0 NI/\ell$, we have

$$U = \tfrac{1}{2}LI^2 = \frac{1}{2}\left(\frac{\mu_0 N^2 A}{\ell}\right)\left(\frac{B\ell}{\mu_0 N}\right)^2$$

$$= \frac{1}{2}\frac{B^2}{\mu_0}A\ell.$$

We can think of this energy as residing in the volume enclosed by the windings, which is $A\ell$. Then the energy per unit volume or **energy density** is

$$u = \text{energy density} = \frac{1}{2}\frac{B^2}{\mu_0}. \tag{30–7}$$

This formula, which was derived for the special case of a solenoid, can be shown to be valid for any region of space where a magnetic field exists. If a ferromagnetic material is present, μ_0 is replaced by μ. This equation is analogous to that for an electric field, $\tfrac{1}{2}\epsilon_0 E^2$, Eq. 24–6.

30–4 *LR* Circuits

Any inductor will have some resistance. We represent this situation by drawing its inductance L and its resistance R separately, as in Fig. 30–6a. The resistance R could also include any other resistance present in the circuit. Now we ask, what happens when a battery or other source of dc voltage V_0 is connected in series to such an LR circuit?

[†]No minus sign here because we are supplying power to oppose the emf of the inductor.

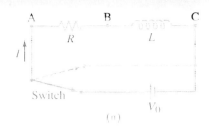

At the instant the switch connecting the battery is closed, the current starts to flow. It is opposed by the induced emf in the inductor which means point B in Fig. 30–6a is positive relative to point C. However, as soon as current starts to flow, there is also a voltage drop of magnitude IR across the resistance. Hence the voltage applied across the inductance is reduced and the current increases less rapidly. The current thus rises gradually as shown in Fig. 30–6b, and approaches the steady value $I_{max} = V_0/R$, for which all the voltage drop is across the resistance.

We can show this analytically by applying Kirchhoff's loop rule to the circuit of Fig. 30–6a. The emfs in the circuit are the battery voltage V_0 and the emf $\mathscr{E} = -L(dI/dt)$ in the inductor opposing the increasing current. Hence the sum of the potential changes around the loop is

$$V_0 - IR - L\frac{dI}{dt} = 0,$$

where I is the current in the circuit at any instant. We rearrange this to obtain

$$L\frac{dI}{dt} + RI = V_0. \tag{30–8}$$

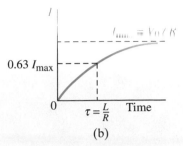

FIGURE 30–6 (a) LR circuit; (b) growth of current when connected to battery.

This is a linear differential equation and can be integrated in the same way we did in Section 26–5 for an RC circuit. We rewrite Eq. 30–8 and then integrate:

$$\int_{I=0}^{I} \frac{dI}{V_0 - IR} = \int_0^t \frac{dt}{L}.$$

Then

$$-\frac{1}{R}\ln\left(\frac{V_0 - IR}{V_0}\right) = \frac{t}{L}$$

or

$$I = \frac{V_0}{R}\left(1 - e^{-t/\tau}\right) \tag{30–9}$$

where

$$\tau = \frac{L}{R} \tag{30–10}$$

is the **time constant** of the LR circuit. The symbol τ represents the time required for the current I to reach $(1 - 1/e) = 0.63$ or 63% of its maximum value (V_0/R). Equation 30–9 is plotted in Fig. 30–6b. (Compare to the RC circuit, Section 26–5.)

| **EXERCISE D** Show that L/R does have dimensions of time. (See Section 1–7.)

Now let us flip the switch in Fig. 30–6a so that the battery is taken out of the circuit, and points A and C are connected together as shown in Fig. 30–7 at the moment when the switching occurs (call it $t = 0$) and the current is I_0. Then the differential equation (Eq. 30–8) becomes (since $V_0 = 0$):

$$L\frac{dI}{dt} + RI = 0.$$

We rearrange this equation and integrate:

$$\int_{I_0}^{I} \frac{dI}{I} = -\int_0^t \frac{R}{L}\,dt$$

where $I = I_0$ at $t = 0$, and $I = I$ at time t.

We integrate this last equation to obtain

$$\ln\frac{I}{I_0} = -\frac{R}{L}t$$

or

$$I = I_0 e^{-t/\tau} \tag{30–11}$$

where again the time constant is $\tau = L/R$. The current thus decays exponentially to zero as shown in Fig. 30–8.

This analysis shows that there is always some "reaction time" when an electromagnet, for example, is turned on or off. We also see that an LR circuit has properties similar to an RC circuit (Section 26–5). Unlike the capacitor case, however, the time constant here is *inversely* proportional to R.

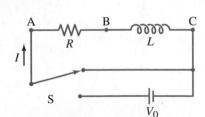

FIGURE 30–7 The switch is flipped quickly so the battery is removed but we still have a circuit. The current at this moment (call it $t = 0$) is I_0.

FIGURE 30–8 Decay of the current in Fig. 30–7 in time after the battery is switched out of the circuit.

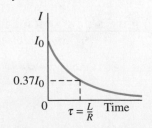

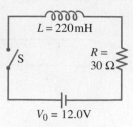

$L = 220\,\text{mH}$

S

$R = 30\,\Omega$

$V_0 = 12.0\,\text{V}$

FIGURE 30–9 Example 30–6.

EXAMPLE 30–6 **An _LR_ circuit.** At $t = 0$, a 12.0-V battery is connected in series with a 220-mH inductor and a total of 30-Ω resistance, as shown in Fig. 30–9. (a) What is the current at $t = 0$? (b) What is the time constant? (c) What is the maximum current? (d) How long will it take the current to reach half its maximum possible value? (e) At this instant, at what rate is energy being delivered by the battery, and (f) at what rate is energy being stored in the inductor's magnetic field?

APPROACH We have the situation shown in Figs. 30–6a and b, and we can apply the equations we just developed.

SOLUTION (a) The current cannot instantaneously jump from zero to some other value when the switch is closed because the inductor opposes the change $\left(\mathscr{E}_L = -L(dI/dt)\right)$. Hence just after the switch is closed, I is still zero at $t = 0$ and then begins to increase.

(b) The time constant is, from Eq. 30–10, $\tau = L/R = (0.22\,\text{H})/(30\,\Omega) = 7.3\,\text{ms}$.

(c) The current reaches its maximum steady value after a long time, when $dI/dt = 0$ so $I_{\max} = V_0/R = 12.0\,\text{V}/30\,\Omega = 0.40\,\text{A}$.

(d) We set $I = \frac{1}{2}I_{\max} = V_0/2R$ in Eq. 30–9, which gives us

$$1 - e^{-t/\tau} = \tfrac{1}{2}$$

or

$$e^{-t/\tau} = \tfrac{1}{2}.$$

We solve for t:

$$t = \tau \ln 2 = (7.3 \times 10^{-3}\,\text{s})(0.69) = 5.0\,\text{ms}.$$

(e) At this instant, $I = I_{\max}/2 = 200\,\text{mA}$, so the power being delivered by the battery is

$$P = IV = (0.20\,\text{A})(12\,\text{V}) = 2.4\,\text{W}.$$

(f) From Eq. 30–6, the energy stored in an inductor L at any instant is

$$U = \tfrac{1}{2}LI^2$$

where I is the current in the inductor at that instant. The _rate_ at which the energy changes is

$$\frac{dU}{dt} = LI\frac{dI}{dt}.$$

We can differentiate Eq. 30–9 to obtain dI/dt, or use the differential equation, Eq. 30–8, directly:

$$\frac{dU}{dt} = I\left(L\frac{dI}{dt}\right) = I(V_0 - RI)$$

$$= (0.20\,\text{A})[12\,\text{V} - (30\,\Omega)(0.20\,\text{A})] = 1.2\,\text{W}.$$

Since only part of the battery's power is feeding the inductor at this instant, where is the rest going?

EXERCISE E A resistor in series with an inductor has a time constant of 10 ms. When the same resistor is placed in series with a 5-μF capacitor, the time constant is 5×10^{-6} s. What is the value of the inductor? (a) 5 μH; (b) 10 μH; (c) 5 mH; (d) 10 mH; (e) not enough information to determine it.

⊛ **PHYSICS APPLIED**
Surge protection

An inductor can act as a "surge protector" for sensitive electronic equipment that can be damaged by high currents. If equipment is plugged into a standard wall plug, a sudden "surge," or increase, in voltage will normally cause a corresponding large change in current and damage the electronics. However, if there is an inductor in series with the voltage to the device, the sudden change in current produces an opposing emf preventing the current from reaching dangerous levels.

LC Circuits and Electromagnetic Oscillations

In any electric circuit, there can be three basic components: resistance, capacitance, and inductance, in addition to a source of emf. (There can also be more complex components, such as diodes or transistors.) We have previously discussed both RC and LR circuits. Now we look at an LC circuit, one that contains only a capacitance C and an inductance L, Fig. 30–10. This is an idealized circuit in which we assume there is no resistance; in the next Section we will include resistance. Let us suppose the capacitor in Fig. 30–10 is initially charged so that one plate has charge Q_0 and the other plate has charge $-Q_0$, and the potential difference across it is $V = Q/C$ (Eq. 24–1). Suppose that at $t = 0$, the switch is closed. The capacitor immediately begins to discharge. As it does so, the current I through the inductor increases. We now apply Kirchhoff's loop rule (sum of potential changes around a loop is zero):

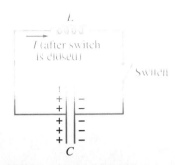

FIGURE 30–10 An LC circuit.

$$-L\frac{dI}{dt} + \frac{Q}{C} = 0.$$

Because charge leaves the positive plate on the capacitor to produce the current I as shown in Fig. 30–10, the charge Q on the (positive) plate of the capacitor is decreasing, so $I = -dQ/dt$. We can then rewrite the above equation as

$$\frac{d^2Q}{dt^2} + \frac{Q}{LC} = 0. \tag{30–12}$$

This is a familiar differential equation. It has the same form as the equation for simple harmonic motion (Chapter 14, Eq. 14–3). The solution of Eq. 30–12 can be written as

$$Q = Q_0 \cos(\omega t + \phi) \tag{30–13}$$

where Q_0 and ϕ are constants that depend on the initial conditions. We insert Eq. 30–13 into Eq. 30–12, noting that $d^2Q/dt^2 = -\omega^2 Q_0 \cos(\omega t + \phi)$; thus

$$-\omega^2 Q_0 \cos(\omega t + \phi) + \frac{1}{LC} Q_0 \cos(\omega t + \phi) = 0$$

or

$$\left(-\omega^2 + \frac{1}{LC}\right) \cos(\omega t + \phi) = 0.$$

This relation can be true for all times t only if $(-\omega^2 + 1/LC) = 0$, which tells us that

$$\omega = 2\pi f = \sqrt{\frac{1}{LC}}. \tag{30–14}$$

Equation 30–13 shows that the charge on the capacitor in an LC circuit oscillates sinusoidally. The current in the inductor is

$$I = -\frac{dQ}{dt} = \omega Q_0 \sin(\omega t + \phi)$$

$$= I_0 \sin(\omega t + \phi); \tag{30–15}$$

so the current too is sinusoidal. The maximum value of I is $I_0 = \omega Q_0 = Q_0/\sqrt{LC}$. Equations 30–13 and 30–15 for Q and I when $\phi = 0$ are plotted in Fig. 30–11.

FIGURE 30–11 Charge Q and current I in an LC circuit. The period $T = \frac{1}{f} = \frac{2\pi}{\omega} = 2\pi\sqrt{LC}$.

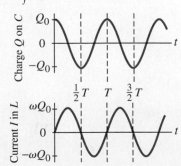

Now let us look at LC oscillations from the point of view of energy. The energy stored in the electric field of the capacitor at any time t is (see Eq. 24–5):

$$U_E = \frac{1}{2}\frac{Q^2}{C} = \frac{Q_0^2}{2C}\cos^2(\omega t + \phi).$$

The energy stored in the magnetic field of the inductor at the same instant is (Eq. 30–6)

$$U_B = \frac{1}{2}LI^2 = \frac{L\omega^2 Q_0^2}{2}\sin^2(\omega t + \phi) = \frac{Q_0^2}{2C}\sin^2(\omega t + \phi)$$

where we used Eq. 30–14. If we let $\phi = 0$, then at times $t = 0$, $t = \frac{1}{2}T$, $t = T$, and so on (where T is the period $= 1/f = 2\pi/\omega$), we have $U_E = Q_0^2/2C$ and $U_B = 0$. That is, all the energy is stored in the electric field of the capacitor. But at $t = \frac{1}{4}T, \frac{3}{4}T$, and so on, $U_E = 0$ and $U_B = Q_0^2/2C$, and so all the energy is stored in the magnetic field of the inductor. At any time t, the total energy is

$$
\begin{aligned}
U &= U_E + U_B = \frac{1}{2}\frac{Q^2}{C} + \frac{1}{2}LI^2 \\
&= \frac{Q_0^2}{2C}\left[\cos^2(\omega t + \phi) + \sin^2(\omega t + \phi)\right] = \frac{Q_0^2}{2C}. \quad \textbf{(30–16)}
\end{aligned}
$$

Hence the total energy is constant, and energy is conserved.

What we have in this LC circuit is an **LC oscillator** or **electromagnetic oscillation**. The charge Q oscillates back and forth, from one plate of the capacitor to the other, and repeats this continuously. Likewise, the current oscillates back and forth as well. They are also energy oscillations: when Q is a maximum, the energy is all stored in the electric field of the capacitor; but when Q reaches zero, the current I is a maximum and all the energy is stored in the magnetic field of the inductor. Thus the energy oscillates between being stored in the electric field of the capacitor and in the magnetic field of the inductor. See Fig. 30–12.

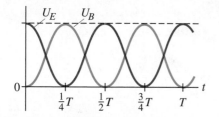

FIGURE 30–12 Energy U_E (red line) and U_B (blue line) stored in the capacitor and the inductor as a function of time. Note how the energy oscillates between electric and magnetic. The dashed line at the top is the (constant) total energy $U = U_E + U_B$.

EXERCISE F Return to the Chapter-Opening Question, page 785, and answer it again now. Try to explain why you may have answered differently the first time.

EXAMPLE 30–7 *LC circuit.* A 1200-pF capacitor is fully charged by a 500-V dc power supply. It is disconnected from the power supply and is connected, at $t = 0$, to a 75-mH inductor. Determine: (*a*) the initial charge on the capacitor; (*b*) the maximum current; (*c*) the frequency f and period T of oscillation; and (*d*) the total energy oscillating in the system.

APPROACH We use the analysis above, and the definition of capacitance $Q = CV$ (Chapter 24).

SOLUTION (*a*) The 500-V power supply, before being disconnected, charged the capacitor to a charge of

$$Q_0 = CV = (1.2 \times 10^{-9}\,\text{F})(500\,\text{V}) = 6.0 \times 10^{-7}\,\text{C}.$$

(*b*) The maximum current, I_{max}, is (see Eqs. 30-14 and 30–15)

$$I_{max} = \omega Q_0 = \frac{Q_0}{\sqrt{LC}} = \frac{(6.0 \times 10^{-7}\,\text{C})}{\sqrt{(0.075\,\text{H})(1.2 \times 10^{-9}\,\text{F})}} = 63\,\text{mA}.$$

(*c*) Equation 30–14 gives us the frequency:

$$f = \frac{\omega}{2\pi} = \frac{1}{(2\pi\sqrt{LC})} = 17\,\text{kHz},$$

and the period T is

$$T = \frac{1}{f} = 6.0 \times 10^{-5}\,\text{s}.$$

(*d*) Finally the total energy (Eq. 30–16) is

$$U = \frac{Q_0^2}{2C} = \frac{(6.0 \times 10^{-7}\,\text{C})^2}{2(1.2 \times 10^{-9}\,\text{F})} = 1.5 \times 10^{-4}\,\text{J}.$$

LC Oscillations with Resistance (*LRC* Circuit)

The *LC* circuit discussed in the previous Section is an idealization. There is always some resistance R in any circuit, and so we now discuss such a simple *LRC* circuit, Fig. 30–13.

Suppose again that the capacitor is initially given a charge Q_0 and the battery or other source is then removed from the circuit. The switch is closed at $t = 0$. Since there is now a resistance in the circuit, we expect some of the energy to be converted to thermal energy, and so we don't expect undamped oscillations as in a pure *LC* circuit. Indeed, if we use Kirchhoff's loop rule around this circuit, we obtain

$$-L\frac{dI}{dt} - IR + \frac{Q}{C} = 0,$$

which is the same equation we had in Section 30–5 with the addition of the voltage drop IR across the resistor. Since $I = -dQ/dt$, as we saw in Section 30–5, this equation becomes

$$L\frac{d^2Q}{dt^2} + R\frac{dQ}{dt} + \frac{1}{C}Q = 0. \qquad (30\text{–}17)$$

This second-order differential equation in the variable Q has precisely the same form as that for the damped harmonic oscillator, Eq. 14–15:

$$m\frac{d^2x}{dt^2} + b\frac{dx}{dt} + kx = 0.$$

Hence we can analyze our *LRC* circuit in the same way as for damped harmonic motion, Section 14–7. Our system may undergo damped oscillations, curve A in Fig. 30–14 (underdamped system), or it may be critically damped (curve B), or overdamped (curve C), depending on the relative values of R, L, and C. Using the results of Section 14–7, with m replaced by L, b by R, and k by C^{-1}, we find that the system will be underdamped when

$$R^2 < \frac{4L}{C},$$

and overdamped for $R^2 > 4L/C$. Critical damping (curve B in Fig. 30–14) occurs when $R^2 = 4L/C$. If R is smaller than $\sqrt{4L/C}$, the angular frequency, ω', will be

$$\omega' = \sqrt{\frac{1}{LC} - \frac{R^2}{4L^2}} \qquad (30\text{–}18)$$

(compare to Eq. 14–18). And the charge Q as a function of time will be

$$Q = Q_0 e^{-\frac{R}{2L}t} \cos(\omega't + \phi) \qquad (30\text{–}19)$$

where ϕ is a phase constant (compare to Eq. 14–19).

Oscillators are an important element in many electronic devices: radios and television sets use them for tuning, tape recorders use them (the "bias frequency") when recording, and so on. Because some resistance is always present, electrical oscillators generally need a periodic input of power to compensate for the energy converted to thermal energy in the resistance.

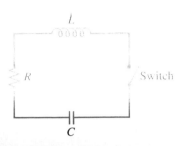

FIGURE 30–13   An *LRC* circuit.

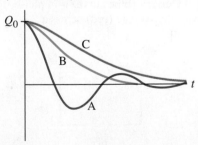

FIGURE 30–14   Charge Q on the capacitor in an *LRC* circuit as a function of time: curve A is for underdamped oscillation $(R^2 < 4L/C)$, curve B is for critically damped $(R^2 = 4L/C)$, and curve C is for overdamped $(R^2 > 4L/C)$.

EXAMPLE 30-8 **Damped oscillations.** At $t = 0$, a 40-mH inductor is placed in series with a resistance $R = 3.0\,\Omega$ and a charged capacitor $C = 4.8\,\mu\text{F}$. (a) Show that this circuit will oscillate. (b) Determine the frequency. (c) What is the time required for the charge amplitude to drop to half its starting value? (d) What value of R will make the circuit nonoscillating?

APPROACH We first check R^2 vs. $4L/C$; then use Eqs. 30–18 and 30–19.

SOLUTION (a) In order to oscillate, the circuit must be underdamped, so we must have $R^2 < 4L/C$. Since $R^2 = 9.0\,\Omega^2$ and $4L/C = 4(0.040\,\text{H})/(4.8 \times 10^{-6}\,\text{F}) = 3.3 \times 10^4\,\Omega^2$, this relation is satisfied, so the circuit will oscillate.

(b) We use Eq. 30–18:

$$f' = \frac{\omega'}{2\pi} = \frac{1}{2\pi}\sqrt{\frac{1}{LC} - \frac{R^2}{4L^2}} = 3.6 \times 10^2\,\text{Hz}.$$

(c) From Eq. 30–19, the amplitude will be half when

$$e^{-\frac{R}{2L}t} = \tfrac{1}{2}$$

or

$$t = \frac{2L}{R}\ln 2 = 18\,\text{ms}.$$

(d) To make the circuit critically damped or overdamped, we must use the criterion $R^2 \geq 4L/C = 3.3 \times 10^4\,\Omega^2$. Hence we must have $R \geq 180\,\Omega$.

30–7 AC Circuits with AC Source

We have previously discussed circuits that contain combinations of resistor, capacitor, and inductor, but only when they are connected to a dc source of emf or to no source. Now we discuss these circuit elements when they are connected to a source of alternating voltage that produces an alternating current (ac).

First we examine, one at a time, how a resistor, a capacitor, and an inductor behave when connected to a source of alternating voltage, represented by the symbol

[alternating voltage]

which produces a sinusoidal voltage of frequency f. We assume in each case that the emf gives rise to a current

$$I = I_0 \cos 2\pi f t = I_0 \cos \omega t \tag{30-20}$$

where t is time and I_0 is the peak current. Remember (Section 25–7) that $V_{\text{rms}} = V_0/\sqrt{2}$ and $I_{\text{rms}} = I_0/\sqrt{2}$ (Eqs. 25–9).

Resistor

When an ac source is connected to a resistor as in Fig. 30–15a, the current increases and decreases with the alternating voltage according to Ohm's law

$$V = IR = I_0 R \cos \omega t = V_0 \cos \omega t$$

where $V_0 = I_0 R$ is the peak voltage. Figure 30–15b shows the voltage (red curve) and the current (blue curve) as a function of time. Because the current is zero when the voltage is zero and the current reaches a peak when the voltage does, we say that the current and voltage are **in phase**. Energy is transformed into heat (Section 25–7), at an average rate

$$\overline{P} = \overline{IV} = I_{\text{rms}}^2 R = V_{\text{rms}}^2/R.$$

Inductor

In Fig. 30–16a an inductor of inductance L (symbol $-\!\text{mmm}\!-$) is connected to the ac source. We ignore any resistance it might have (it is usually small). The voltage applied to the inductor will be equal to the "back" emf generated in the inductor by the changing current as given by Eq. 30–5. This is because the sum of the electric potential changes around any closed circuit must add up to zero, according to Kirchhoff's rule.

FIGURE 30–15 (a) Resistor connected to an ac source. (b) Current (blue curve) is in phase with the voltage (red) across a resistor.

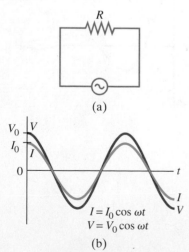

Thus

$$V - L\frac{dI}{dt} = 0$$

or (inserting Eq. 30–20)

$$V = L\frac{dI}{dt} = -\omega L I_0 \sin \omega t. \tag{30–21}$$

Using the identity $\sin \theta = -\cos(\theta + 90°)$ we can write

$$V = \omega L I_0 \cos(\omega t + 90°) = V_0 \cos(\omega t + 90°) \tag{30–22a}$$

where

$$V_0 = I_0 \omega L \tag{30–22b}$$

is the peak voltage. The current I and voltage V as a function of time are graphed for the inductor in Fig. 30–16b. It is clear from this graph, as well as from Eqs. 30–22, that the current and voltage are out of phase by a quarter cycle, which is equivalent to $\pi/2$ radians or 90°. We see from the graph that

the current lags the voltage by 90° in an inductor.

That is, the current in an inductor reaches its peaks a quarter cycle later than the voltage does. Alternatively, we can say that the voltage leads the current by 90°.

Because the current and voltage in an inductor are out of phase by 90°, the product IV (=power) is as often positive as it is negative (Fig. 30–16b). So no energy is transformed in an inductor on the average; and no energy is dissipated as thermal energy.

Just as a resistor impedes the flow of charge, so too an inductor impedes the flow of charge in an alternating current due to the back emf produced. For a resistor R, the peak current and peak voltage are related by $V_0 = I_0 R$. We can write a similar relation for an inductor:

$$V_0 = I_0 X_L \qquad \begin{bmatrix} \text{maximum or rms values,} \\ \text{not at any instant} \end{bmatrix} \tag{30–23a}$$

where, from Eq. 30–22b (and using $\omega = 2\pi f$ where f is the frequency of the ac),

$$X_L = \omega L = 2\pi f L. \tag{30–23b}$$

The term X_L is called the **inductive reactance** of the inductor, and has units of ohms. The greater X_L is, the more it impedes the flow of charge and the smaller the current. X_L is larger for higher frequencies f and larger inductance L.

Equation 30–23a is valid for peak values I_0 and V_0; it is also valid for rms values, $V_{rms} = I_{rms} X_L$. Because the peak values of current and voltage are not reached at the same time, Eq. 30–23a is *not valid at a particular instant*, as is the case for a resistor $(V = IR)$.

Note from Eq. 30–23b that if $\omega = 2\pi f = 0$ (so the current is dc), there is no back emf and no impedance to the flow of charge.

EXAMPLE 30–9 **Reactance of a coil.** A coil has a resistance $R = 1.00 \, \Omega$ and an inductance of 0.300 H. Determine the current in the coil if (*a*) 120-V dc is applied to it, (*b*) 120-V ac (rms) at 60.0 Hz is applied.

APPROACH When the voltage is dc, there is no inductive reactance $(X_L = 2\pi f L = 0$ since $f = 0)$, so we apply Ohm's law for the resistance. When the voltage is ac, we calculate the reactance X_L and then use Eq. 30–23a.

SOLUTION (*a*) With dc, we have no X_L so we simply apply Ohm's law:

$$I = \frac{V}{R} = \frac{120 \, \text{V}}{1.00 \, \Omega} = 120 \, \text{A}.$$

(*b*) The inductive reactance is

$$X_L = 2\pi f L = (6.283)(60.0 \, \text{s}^{-1})(0.300 \, \text{H}) = 113 \, \Omega.$$

In comparison to this, the resistance can be ignored. Thus,

$$I_{rms} = \frac{V_{rms}}{X_L} = \frac{120 \, \text{V}}{113 \, \Omega} = 1.06 \, \text{A}.$$

NOTE It might be tempting to say that the total impedance is $113 \, \Omega + 1 \, \Omega = 114 \, \Omega$. This might imply that about 1% of the voltage drop is across the resistor, or about 1 V; and that across the inductance is 119 V. Although the 1 V across the resistor is correct, the other statements are not true because of the alteration in phase in an inductor. This will be discussed in the next Section.

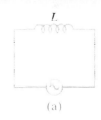

(a)

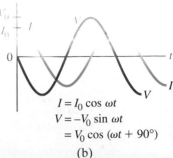

$I = I_0 \cos \omega t$
$V = -V_0 \sin \omega t$
$= V_0 \cos (\omega t + 90°)$

(b)

FIGURE 30–16 (a) Inductor connected to an ac source. (b) Current (blue curve) lags voltage (red curve) by a quarter cycle or 90°.

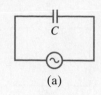

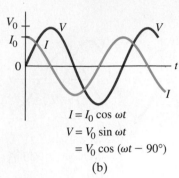

$I = I_0 \cos \omega t$

$V = V_0 \sin \omega t$

$\quad = V_0 \cos (\omega t - 90°)$

(b)

FIGURE 30–17 (a) Capacitor connected to an ac source. (b) Current leads voltage by a quarter cycle, or 90°.

Capacitor

When a capacitor is connected to a battery, the capacitor plates quickly acquire equal and opposite charges; but no steady current flows in the circuit. A capacitor prevents the flow of a dc current. But if a capacitor is connected to an alternating source of voltage, as in Fig. 30–17a, an alternating current will flow continuously. This can happen because when the ac voltage is first turned on, charge begins to flow and one plate acquires a negative charge and the other a positive charge. But when the voltage reverses itself, the charges flow in the opposite direction. Thus, for an alternating applied voltage, an ac current is present in the circuit continuously.

Let us look at this in more detail. By Kirchhoff's loop rule, the applied source voltage must equal the voltage V across the capacitor at any moment:

$$V = \frac{Q}{C}$$

where C is the capacitance and Q is the charge on the capacitor plates. The current I at any instant (given as $I = I_0 \cos \omega t$, Eq. 30–20) is

$$I = \frac{dQ}{dt} = I_0 \cos \omega t.$$

Hence the charge Q on the plates at any instant is given by

$$Q = \int_0^t dQ = \int_0^t I_0 \cos \omega t \, dt = \frac{I_0}{\omega} \sin \omega t.$$

Then the voltage across the capacitor is

$$V = \frac{Q}{C} = I_0 \left(\frac{1}{\omega C} \right) \sin \omega t.$$

Using the trigonometric identity $\sin \theta = \cos(90° - \theta) = \cos(\theta - 90°)$, we can rewrite this as

$$V = I_0 \left(\frac{1}{\omega C} \right) \cos(\omega t - 90°) = V_0 \cos(\omega t - 90°) \qquad \textbf{(30–24a)}$$

where

$$V_0 = I_0 \left(\frac{1}{\omega C} \right) \qquad \textbf{(30–24b)}$$

is the peak voltage. The current $I \left(= I_0 \cos \omega t \right)$ and voltage V (Eq. 30–24a) across the capacitor are graphed in Fig. 30–17b. It is clear from this graph, as well as a comparison of Eq. 30–24a with Eq. 30–20, that the current and voltage are out of phase by a quarter cycle or 90° ($\pi/2$ radians):

The current leads the voltage across a capacitor by 90°.

Alternatively we can say that the voltage lags the current by 90°. This is the opposite of what happens for an inductor.

Because the current and voltage are out of phase by 90°, the average power dissipated is zero, just as for an inductor. Energy from the source is fed to the capacitor, where it is stored in the electric field between its plates. As the field decreases, the energy returns to the source. Thus *only a resistance will dissipate energy* as thermal energy in an ac circuit.

A relationship between the applied voltage and the current in a capacitor can be written just as for an inductance:

$$V_0 = I_0 X_C \qquad \begin{bmatrix} \text{maximum or rms values,} \\ \text{not at any instant} \end{bmatrix} \qquad \textbf{(30–25a)}$$

where X_C is the **capacitive reactance** of the capacitor, and has units of ohms; X_C is given by (see Eq. 30–24b):

$$X_C = \frac{1}{\omega C} = \frac{1}{2\pi f C}. \qquad \textbf{(30–25b)}$$

When frequency f and/or capacitance C are smaller, X_C is larger and thus impedes the flow of charge more. That is, when X_C is larger, the current is smaller (Eq. 30–25a). In the next Section we use the term **impedance** to represent reactances and resistance.

Equation 30–25a relates the peak values of V and I, or the rms values $\left(V_{\text{rms}} = I_{\text{rms}} X_C \right)$. But it is not valid at a particular instant because I and V are not in phase.

Note from Eq. 30–25b that for dc conditions, $\omega = 2\pi f = 0$ and X_c becomes infinite. This is as it should be, since a pure capacitor does not pass dc current. Also, note that the reactance of an inductor increases with frequency, but that of a capacitor decreases with frequency.

EXAMPLE 30–10 **Capacitor reactance.** What is the rms current in the circuit of Fig. 30–17a if $C = 1.0\,\mu\text{F}$ and $V_{\text{rms}} = 120\,\text{V}$? Calculate (a) for $f = 60\,\text{Hz}$, and then (b) for $f = 6.0 \times 10^5\,\text{Hz}$.

APPROACH We find the reactance using Eq. 30–25b, and solve for current in the equivalent form of Ohm's law, Eq. 30–25a.

SOLUTION (a) $X_C = 1/2\pi fC = 1/(6.28)(60\,\text{s}^{-1})(1.0 \times 10^{-6}\,\text{F}) = 2.7\,\text{k}\Omega$. The rms current is (Eq. 30–25a):

$$I_{\text{rms}} = \frac{V_{\text{rms}}}{X_C} = \frac{120\,\text{V}}{2.7 \times 10^3\,\Omega} = 44\,\text{mA}.$$

(b) For $f = 6.0 \times 10^5\,\text{Hz}$, X_C will be $0.27\,\Omega$ and $I_{\text{rms}} = 440\,\text{A}$, vastly larger!

NOTE The dependence on f is dramatic. For high frequencies, the capacitive reactance is very small, and the current can be large.

Two common applications of capacitors are illustrated in Fig. 30–18a and b. In Fig. 30–18a, circuit A is said to be capacitively coupled to circuit B. The purpose of the capacitor is to prevent a dc voltage from passing from A to B but allowing an ac signal to pass relatively unimpeded (if C is sufficiently large, Eq. 30–25b). The capacitor in Fig. 30–18a is called a **high-pass filter** because it allows high-frequency ac to pass easily, but not dc.

In Fig. 30–18b, the capacitor passes ac to ground. In this case, a dc voltage can be maintained between circuits A and B, but an ac signal leaving A passes to ground instead of into B. Thus the capacitor in Fig. 30–18b acts like a **low-pass filter** when a constant dc voltage is required; any high-frequency variation in voltage will pass to ground instead of into circuit B. (Very low-frequency ac will also be able to reach circuit B, at least in part.)

Loudspeakers having separate "woofer" (low-frequency speaker) and "tweeter" (high-frequency speaker) may use a simple "cross-over" that consists of a capacitor in the tweeter circuit to impede low-frequency signals, and an inductor in the woofer circuit to impede high-frequency signals $(X_L = 2\pi fL)$. Hence mainly low-frequency sounds reach and are emitted by the woofer. See Fig. 30–18c.

EXERCISE G At what frequency is the reactance of a 1.0-μF capacitor equal to 500 Ω? (a) 320 Hz, (b) 500 Hz, (c) 640 Hz, (d) 2000 Hz, (e) 4000 Hz.

EXERCISE H At what frequency is the reactance of a 1.0-μH inductor equal to 500 Ω? (a) 80 Hz, (b) 500 Hz, (c) 80 MHz, (d) 160 MHz, (e) 500 MHz.

30–8 *LRC* Series AC Circuit

Let us examine a circuit containing all three elements in series: a resistor R, an inductor L, and a capacitor C, Fig. 30–19. If a given circuit contains only two of these elements, we can still use the results of this Section by setting $R = 0$, $X_L = 0$, or $X_C = 0$, as needed. We let V_R, V_L, and V_C represent the voltage across each element at a *given instant* in time; and V_{R0}, V_{L0}, and V_{C0} represent the *maximum* (peak) values of these voltages. The voltage across each of the elements will follow the phase relations we discussed in the previous Section. At any instant the voltage V supplied by the source will be, by Kirchhoff's loop rule,

$$V = V_R + V_L + V_C. \tag{30–26}$$

Because the various voltages are not in phase, they do not reach their peak values at the same time, so the peak voltage of the source V_0 will *not* equal $V_{R0} + V_{L0} + V_{C0}$.

(a) High-pass filter

(b) Low-pass filter

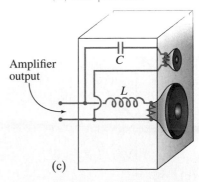

(c)

FIGURE 30–18 (a) and (b) Two common uses for a capacitor as a filter. (c) Simple loudspeaker cross-over.

 **PHYSICS APPLIED**
Capacitors as filters

PHYSICS APPLIED
Loudspeaker cross-over

FIGURE 30–19 An *LRC* circuit.

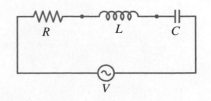

! CAUTION
Peak voltages do not add to yield source voltage

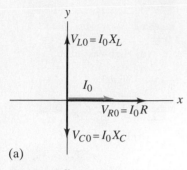

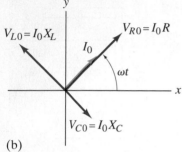

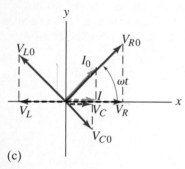

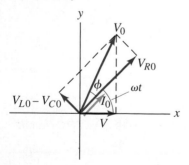

FIGURE 30–20 Phasor diagram for a series *LRC* circuit at (a) $t = 0$, (b) a time t later. (c) Projections on x axis reflect Eqs. 30–20, 30–22a, and 30–24a.

FIGURE 30–21 Phasor diagram for a series *LRC* circuit showing the sum vector, V_0.

Let us now find the impedance of an *LRC* circuit as a whole (the effect of R, X_C, and X_L), as well as the peak current I_0, and the phase relation between V and I. The current at any instant must be the same at all points in the circuit. Thus the *currents in each element are in phase with each other, even though the voltages are not.* We choose our origin in time $(t = 0)$ so that the current I at any time t is (as in Eq. 30–20)

$$I = I_0 \cos \omega t.$$

We analyze an *LRC* circuit using[†] a **phasor diagram**. Arrows (treated like vectors) are drawn in an xy coordinate system to represent each voltage. The *length of each arrow represents the magnitude of the peak voltage across each element:*

$$V_{R0} = I_0 R, \qquad V_{L0} = I_0 X_L, \qquad \text{and} \qquad V_{C0} = I_0 X_C.$$

V_{R0} is in phase with the current and is initially $(t = 0)$ drawn along the positive x axis, as is the current (Fig. 30–20a). V_{L0} leads the current by 90°, so it leads V_{R0} by 90° and is initially drawn along the positive y axis. V_{C0} lags the current by 90°, so V_{C0} is drawn initially along the negative y axis, Fig. 30–20a.

If we let the vector diagram rotate counterclockwise at frequency f, we get the diagram shown in Fig. 30–20b; after a time, t, each arrow has rotated through an angle ωt. Then the *projections of each arrow on the x axis represent the voltages across each element* at the instant t (Fig. 30–20c). For example $I = I_0 \cos \omega t$. Compare Eqs. 30–22a and 30–24a with Fig. 30–20c to confirm the validity of the phasor diagram.

The sum of the projections of the three voltage vectors represents the instantaneous voltage across the whole circuit, V. Therefore, the vector sum of these vectors will be the vector that represents the peak source voltage, V_0, as shown in Fig. 30–21 where it is seen that V_0 makes an angle ϕ with I_0 and V_{R0}. As time passes, V_0 rotates with the other vectors, so the instantaneous voltage V (projection of V_0 on the x axis) is (see Fig. 30–21)

$$V = V_0 \cos(\omega t + \phi).$$

The voltage V across the whole circuit must equal the source voltage (Fig. 30–19). Thus the voltage from the source is out of phase[‡] with the current by an angle ϕ.

From this analysis we can now determine the total **impedance** Z of the circuit, which is defined in analogy to resistance and reactance as

$$V_{\text{rms}} = I_{\text{rms}} Z, \qquad \text{or} \qquad V_0 = I_0 Z. \qquad \textbf{(30–27)}$$

From Fig. 30–21 we see, using the Pythagorean theorem (V_0 is the hypotenuse of a right triangle), that

$$V_0 = \sqrt{V_{R0}^2 + (V_{L0} - V_{C0})^2}$$

$$= I_0 \sqrt{R^2 + (X_L - X_C)^2}.$$

Thus, from Eq. 30–27, the total impedance Z is

$$Z = \sqrt{R^2 + (X_L - X_C)^2} \qquad \textbf{(30–28a)}$$

$$= \sqrt{R^2 + \left(\omega L - \frac{1}{\omega C}\right)^2}. \qquad \textbf{(30–28b)}$$

Also from Fig. 30–21, we can find the phase angle ϕ between voltage and current:

$$\tan \phi = \frac{V_{L0} - V_{C0}}{V_{R0}} = \frac{I_0(X_L - X_C)}{I_0 R} = \frac{X_L - X_C}{R}. \qquad \textbf{(30–29a)}$$

[†]We could instead do our analysis by rewriting Eq. 30–26 as a differential equation (setting $V_C = Q/C$, $V_R = IR = (dQ/dt)R$, and $V_L = L\, dI/dt$) and trying to solve the differential equation. The differential equation we would get would look like Eq. 14–21 in Section 14–8 (on forced vibrations), and would be solved in the same way. Phasor diagrams are easier, and at the same time give us some physical insight.

[‡]As a check, note that if $R = X_C = 0$, then $\phi = 90°$, and V_0 would lead the current by 90°, as it must for an inductor alone. Similarly, if $R = L = 0$, $\phi = -90°$ and V_0 would lag the current by 90°, as it must for a capacitor alone.

We can also write

$$\cos \phi = \frac{V_{R0}}{V_0} = \frac{I_0 R}{I_0 Z} = \frac{R}{Z}. \qquad (30\text{-}29b)$$

Figure 30-21 was drawn for the case $X_L > X_C$, and the current lags the source voltage by ϕ. When the reverse is true, $X_L < X_C$, then ϕ in Eqs. 30-29 is less than zero, and the current leads the source voltage.

We saw earlier that power is dissipated only by a resistance; none is dissipated by inductance or capacitance. Therefore, the average power $\overline{P} = I_{rms}^2 R$. But from Eq. 30-29b, $R = Z \cos \phi$. Therefore

$$\overline{P} = I_{rms}^2 Z \cos \phi = I_{rms} V_{rms} \cos \phi. \qquad (30\text{-}30)$$

The factor $\cos \phi$ is referred to as the **power factor** of the circuit. For a pure resistor, $\cos \phi = 1$ and $\overline{P} = I_{rms} V_{rms}$. For a capacitor or inductor alone, $\phi = -90°$ or $+90°$, respectively, so $\cos \phi = 0$ and no power is dissipated.

EXAMPLE 30-11 *LRC circuit.* Suppose $R = 25.0 \, \Omega$, $L = 30.0 \, \text{mH}$, and $C = 12.0 \, \mu\text{F}$ in Fig. 30-19, and they are connected in series to a 90.0-V ac (rms) 500-Hz source. Calculate (*a*) the current in the circuit, (*b*) the voltmeter readings (rms) across each element, (*c*) the phase angle ϕ, and (*d*) the power dissipated in the circuit.

APPROACH To obtain the current we need to determine the impedance (Eq. 30-28 plus Eqs. 30-23b and 30-25b), and then use $I_{rms} = V_{rms}/Z$. Voltage drops across each element are found using Ohm's law or equivalent for each element: $V_R = IR$, $V_L = IX_L$, and $V_C = IX_C$.

SOLUTION (*a*) First, we find the reactance of the inductor and capacitor at $f = 500 \, \text{Hz} = 500 \, \text{s}^{-1}$:

$$X_L = 2\pi f L = 94.2 \, \Omega, \qquad X_C = \frac{1}{2\pi f C} = 26.5 \, \Omega.$$

Then the total impedance is

$$Z = \sqrt{R^2 + (X_L - X_C)^2} = \sqrt{(25.0 \, \Omega)^2 + (94.2 \, \Omega - 26.5 \, \Omega)^2} = 72.2 \, \Omega.$$

From the impedance version of Ohm's law, Eq. 30-27,

$$I_{rms} = \frac{V_{rms}}{Z} = \frac{90.0 \, \text{V}}{72.2 \, \Omega} = 1.25 \, \text{A}.$$

(*b*) The rms voltage across each element is

$$(V_R)_{rms} = I_{rms} R = (1.25 \, \text{A})(25.0 \, \Omega) = 31.2 \, \text{V}$$
$$(V_L)_{rms} = I_{rms} X_L = (1.25 \, \text{A})(94.2 \, \Omega) = 118 \, \text{V}$$
$$(V_C)_{rms} = I_{rms} X_C = (1.25 \, \text{A})(26.5 \, \Omega) = 33.1 \, \text{V}.$$

NOTE These voltages do *not* add up to the source voltage, 90.0 V (rms). Indeed, the rms voltage across the inductance *exceeds* the source voltage. This can happen because the different voltages are out of phase with each other, and at any instant one voltage can be negative, to compensate for a large positive voltage of another. The rms voltages, however, are always positive by definition. Although the rms voltages need not add up to the source voltage, the instantaneous voltages at any time must add up to the source voltage at that instant.

(*c*) The phase angle ϕ is given by Eq. 30-29b,

$$\cos \phi = \frac{R}{Z} = \frac{25.0 \, \Omega}{72.2 \, \Omega} = 0.346,$$

so $\phi = 69.7°$. Note that ϕ is positive because $X_L > X_C$ in this case, so $V_{L0} > V_{C0}$ in Fig. 30-21.

(*d*) $\overline{P} = I_{rms} V_{rms} \cos \phi = (1.25 \, \text{A})(90.0 \, \text{V})(25.0 \, \Omega/72.2 \, \Omega) = 39.0 \, \text{W}.$

⚠ CAUTION

Individual peak or rms voltages do NOT add up to source voltage (due to phase differences)

30–9 Resonance in AC Circuits

The rms current in an LRC series circuit is given by (see Eqs. 30–27 and 30–28b):

$$I_{rms} = \frac{V_{rms}}{Z} = \frac{V_{rms}}{\sqrt{R^2 + \left(\omega L - \frac{1}{\omega C}\right)^2}}. \qquad (30\text{–}31)$$

Because the reactance of inductors and capacitors depends on the frequency f $(= \omega/2\pi)$ of the source, the current in an LRC circuit will depend on frequency. From Eq. 30–31 we can see that the current will be maximum at a frequency that satisfies

$$\left(\omega L - \frac{1}{\omega C}\right) = 0.$$

We solve this for ω and call the solution ω_0:

$$\omega_0 = \sqrt{\frac{1}{LC}}. \qquad \text{[resonance]} \quad (30\text{–}32)$$

When $\omega = \omega_0$, the circuit is in **resonance**, and $f_0 = \omega_0/2\pi$ is the **resonant frequency** of the circuit. At this frequency, $X_C = X_L$, so the impedance is purely resistive and $\cos\phi = 1$. A graph of I_{rms} versus ω is shown in Fig. 30–22 for particular values of R, L, and C. For small R compared to X_L and X_C, the resonance peak will be higher and sharper. When R is very small, the circuit approaches the pure LC circuit we discussed in Section 30–5. When R is large compared to X_L and X_C, the resonance curve is relatively flat—there is little frequency dependence.

This electrical resonance is analogous to mechanical resonance, which we discussed in Chapter 14. The energy transferred to the system by the source is a maximum at resonance whether it is electrical resonance, the oscillation of a spring, or pushing a child on a swing (Section 14–8). That this is true in the electrical case can be seen from Eq. 30–30; at resonance, $\cos\phi = 1$, and power $\overline{P}$ is a maximum. A graph of power versus frequency peaks like that for the current, Fig. 30–22.

Electric resonance is used in many circuits. Radio and TV sets, for example, use resonant circuits for tuning in a station. Many frequencies reach the circuit, but a significant current flows only for those at or near the resonant frequency. Either L or C is variable so that different stations can be tuned in.

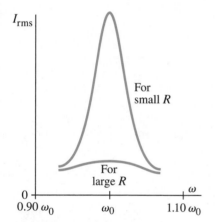

FIGURE 30–22 Current in LRC circuit as a function of angular frequency, ω, showing resonance peak at $\omega = \omega_0 = \sqrt{1/LC}$.

*30–10 Impedance Matching

It is common to connect one electric circuit to a second circuit. For example, a TV antenna is connected to a TV receiver, an amplifier is connected to a loudspeaker; electrodes for an electrocardiogram are connected to a recorder. Maximum power is transferred from one to the other, with a minimum of loss, when the output impedance of the one device matches the input impedance of the second.

To show why, we consider simple circuits that contain only resistance. In Fig. 30–23 the source in circuit 1 could represent the signal from an antenna or a laboratory probe, and R_1 represents its resistance including internal resistance of the source. R_1 is called the output impedance (or resistance) of circuit 1. The output of circuit 1 is across the terminals a and b which are connected to the input of circuit 2 which may be very complicated. We let R_2 be the equivalent "input resistance" of circuit 2.

FIGURE 30–23 Output of the circuit on the left is input to the circuit on the right.

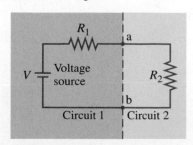

The power delivered to circuit 2 is $P = I^2 R_2$, where $I = V/(R_1 + R_2)$. Thus

$$P = I^2 R_1 = \frac{V^2 R_2}{(R_1 + R_2)^2}.$$

If the resistance of the source is R_1, what value should R_2 have so that the maximum power is transferred to circuit 2? To determine this, we take the derivative of P with respect to R_2 and set it equal to zero, which gives

$$V^2 \left[\frac{1}{(R_1 + R_2)^2} - \frac{2R_2}{(R_1 + R_2)^3} \right] = 0$$

or

$$R_2 = R_1.$$

Thus, the maximum power is transmitted when the *output impedance* of one device *equals the input impedance* of the second. This is called **impedance matching**.

In an ac circuit that contains capacitors and inductors, the different phases are important and the analysis is more complicated. However, the same result holds: to maximize power transfer it is important to match impedances $(Z_2 = Z_1)$.

In addition, it is possible to seriously distort a signal if impedances do not match, and this can lead to meaningless or erroneous experimental results.

*30–11 Three-Phase AC

Transmission lines typically consist of four wires, rather than two. One of these wires is the ground; the remaining three are used to transmit three-phase ac power which is a superposition of three ac voltages 120° out of phase with each other:

$$V_1 = V_0 \sin \omega t$$
$$V_2 = V_0 \sin(\omega t + 2\pi/3)$$
$$V_3 = V_0 \sin(\omega t + 4\pi/3).$$

(See Fig. 30–24.) Why is three-phase power used? We saw in Fig. 25–22 that single-phase ac (i.e., the voltage V_1 by itself) delivers power to the load in pulses. A much smoother flow of power can be delivered using three-phase power. Suppose that each of the three voltages making up the three-phase source is hooked up to a resistor R. Then the power delivered is:

$$P = \frac{1}{R}(V_1^2 + V_2^2 + V_3^2).$$

You can show that this power is a constant equal to $3V_0^2/2R$, which is three times the rms power delivered by a single-phase source. This smooth flow of power makes electrical equipment run smoothly. Although houses use single-phase ac power, most industrial-grade machinery is wired for three-phase power.

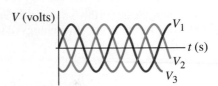

V (volts)

V_1

t (s)

V_2

V_3

FIGURE 30–24 The three voltages, out of phase by 120° $(=\frac{2}{3}\pi$ radians), in a three-phase power line.

EXAMPLE 30–12 **Three-phase circuit.** In a three-phase circuit, 266 V rms exists between line 1 and ground. What is the rms voltage between lines 2 and 3?

SOLUTION We are given $V_{rms} = V_0/\sqrt{2} = 266$ V. Hence $V_0 = 376$ V. Now
$$V_3 - V_2 = V_0[\sin(\omega t + 4\pi/3) - \sin(\omega t + 2\pi/3)] = 2V_0 \sin\tfrac{1}{2}(\tfrac{2\pi}{3}) \cos\tfrac{1}{2}(2\omega t)$$
where we used the identity: $\sin A - \sin B = 2 \sin\tfrac{1}{2}(A - B) \cos\tfrac{1}{2}(A + B)$. The rms voltage is

$$(V_3 - V_2)_{rms} = \frac{1}{\sqrt{2}} 2V_0 \sin\frac{\pi}{3} = \sqrt{2}(376 \text{ V})(0.866) = 460 \text{ V (rms)}.$$

Summary

A changing current in a coil of wire will induce an emf in a second coil placed nearby. The **mutual inductance**, M, is defined as the proportionality constant between the induced emf $\mathcal{E}_2$ in the second coil and the time rate of change of current in the first:

$$\mathcal{E}_2 = -M\, dI_1/dt. \qquad (30\text{-}3b)$$

We can also write M as

$$M = \frac{N_2 \Phi_{21}}{I_1} \qquad (30\text{-}1)$$

where Φ_{21} is the magnetic flux through coil 2 with N_2 loops, produced by the current I_1 in another coil (coil 1).

Within a single coil, a changing current induces an opposing emf, $\mathcal{E}$, so a coil has a **self-inductance** L defined by

$$\mathcal{E} = -L\, dI/dt. \qquad (30\text{-}5)$$

This induced emf acts as an *impedance* to the flow of an alternating current. We can also write L as

$$L = N\frac{\Phi_B}{I} \qquad (30\text{-}4)$$

where Φ_B is the flux through the inductance when a current I flows in its N loops.

When the current in an inductance L is I, the energy stored in the inductance is given by

$$U = \tfrac{1}{2}LI^2. \qquad (30\text{-}6)$$

This energy can be thought of as being stored in the magnetic field of the inductor. The energy density u in any magnetic field B is given by

$$u = \frac{1}{2}\frac{B^2}{\mu_0}, \qquad (30\text{-}7)$$

where μ_0 is replaced by μ if a ferromagnetic material is present.

When an inductance L and resistor R are connected in series to a constant source of emf, V_0, the current rises according to an exponential of the form

$$I = \frac{V_0}{R}\left(1 - e^{-t/\tau}\right), \qquad (30\text{-}9)$$

where

$$\tau = L/R \qquad (30\text{-}10)$$

is the **time constant**. The current eventually levels out at $I = V_0/R$. If the battery is suddenly switched out of the **LR circuit**, and the circuit remains complete, the current drops exponentially, $I = I_0 e^{-t/\tau}$, with the same time constant τ.

The current in a pure **LC circuit** (or charge on the capacitor) would oscillate sinusoidally. The energy too would oscillate back and forth between electric and magnetic, from the capacitor to the inductor, and back again. If such a circuit has resistance (LRC), and the capacitor at some instant is charged, it can undergo damped oscillations or exhibit critically damped or overdamped behavior.

Capacitance and inductance offer *impedance* to the flow of alternating current just as resistance does. This impedance is referred to as **reactance**, X, and is defined (as for resistors) as the proportionality constant between voltage and current (either the rms or peak values). Across an inductor,

$$V_0 = I_0 X_L, \qquad (30\text{-}23a)$$

and across a capacitor,

$$V_0 = I_0 X_C. \qquad (30\text{-}25a)$$

The reactance of an inductor increases with frequency:

$$X_L = \omega L. \qquad (30\text{-}23b)$$

where $\omega = 2\pi f$ and f is the frequency of the ac. The reactance of a capacitor decreases with frequency:

$$X_C = \frac{1}{\omega C}. \qquad (30\text{-}25b)$$

Whereas the current through a resistor is always in phase with the voltage across it, this is not true for inductors and capacitors: in an inductor, the current lags the voltage by 90°, and in a capacitor the current leads the voltage by 90°.

In an ac **LRC series circuit**, the total **impedance** Z is defined by the equivalent of $V = IR$ for resistance: namely $V_0 = I_0 Z$ or $V_{\text{rms}} = I_{\text{rms}} Z$. The impedance Z is related to R, C, and L by

$$Z = \sqrt{R^2 + (X_L - X_C)^2}. \qquad (30\text{-}28a)$$

The current in the circuit lags (or leads) the source voltage by an angle ϕ given by $\cos\phi = R/Z$. Only the resistor in an ac LRC circuit dissipates energy, and at a rate

$$\overline{P} = I_{\text{rms}}^2 Z \cos\phi \qquad (30\text{-}30)$$

where the factor $\cos\phi$ is referred to as the **power factor**.

An LRC series circuit **resonates** at a frequency given by

$$\omega_0 = \sqrt{\frac{1}{LC}}. \qquad (30\text{-}32)$$

The rms current in the circuit is largest when the applied voltage has a frequency equal to $f_0 (= \omega_0/2\pi)$. The lower the resistance R, the higher and sharper the resonance peak.

Questions

1. How would you arrange two flat circular coils so that their mutual inductance was (a) greatest, (b) least (without separating them by a great distance)?

2. Suppose the second coil of N_2 turns in Fig. 30-2 were moved so it was near the end of the solenoid. How would this affect the mutual inductance?

3. Would two coils with mutual inductance also have self-inductance? Explain.

4. Is the energy density inside a solenoid greatest near the ends of the solenoid or near its center?

5. If you are given a fixed length of wire, how would you shape it to obtain the greatest self-inductance? The least?

6. Does the emf of the battery in Fig. 30-6a affect the time needed for the LR circuit to reach (a) a given fraction of its maximum possible current, (b) a given value of current?

7. A circuit with large inductive time constant carries a steady current. If a switch is opened, there can be a very large (and sometimes dangerous) spark or "arcing over." Explain.

8. At the instant the battery is connected into the LR circuit of Fig. 30–6a, the emf in the inductor has its maximum value even though the current is zero. Explain.

9. What keeps an LC circuit oscillating even after the capacitor has discharged completely?

10. Is the ac current in the inductor always the same as the current in the resistor of the LRC circuit of Fig. 30–13?

11. When an ac generator is connected to an LRC circuit, where does the energy come from ultimately? Where does it go? How do the values of L, C, and R affect the energy supplied by the generator?

12. In an ac LRC circuit, if $X_L > X_C$, the circuit is said to be predominantly "inductive." And if $X_C > X_L$, the circuit is said to be predominantly "capacitive." Discuss the reasons for these terms. In particular, do they say anything about the relative values of L and C at a given frequency?

13. Do the results of Section 30–8 approach the proper expected results when ω approaches zero? What are the expected results?

14. Under what conditions is the impedance in an LRC circuit a minimum?

15. Is it possible for the instantaneous power output of an ac generator connected to an LRC circuit ever to be negative? Explain.

16. In an ac LRC circuit, does the power factor, $\cos\phi$, depend on frequency? Does the power dissipated depend on frequency?

17. Describe briefly how the frequency of the source emf affects the impedance of (a) a pure resistance, (b) a pure capacitance, (c) a pure inductance, (d) an LRC circuit near resonance (R small), (e) an LRC circuit far from resonance (R small).

18. Discuss the response of an LRC circuit as $R \to 0$ when the frequency is (a) at resonance, (b) near resonance, (c) far from resonance. Is there energy dissipation in each case? Discuss the transformations of energy that occur in each case.

19. An LRC resonant circuit is often called an *oscillator* circuit. What is it that oscillates?

20. Compare the oscillations of an LRC circuit to the vibration of a mass m on a spring. What do L and C correspond to in the mechanical system?

Problems

30–1 Mutual Inductance

1. (II) A 2.44-m-long coil containing 225 loops is wound on an iron core (average $\mu = 1850\mu_0$) along with a second coil of 115 loops. The loops of each coil have a radius of 2.00 cm. If the current in the first coil drops uniformly from 12.0 A to zero in 98.0 ms, determine: (a) the mutual inductance M; (b) the emf induced in the second coil.

2. (II) Determine the mutual inductance per unit length between two long solenoids, one inside the other, whose radii are r_1 and r_2 ($r_2 < r_1$) and whose turns per unit length are n_1 and n_2.

3. (II) A small thin coil with N_2 loops, each of area A_2, is placed inside a long solenoid, near its center. The solenoid has N_1 loops in its length ℓ and has area A_1. Determine the mutual inductance as a function of θ, the angle between the plane of the small coil and the axis of the solenoid.

4. (III) A long straight wire and a small rectangular wire loop lie in the same plane, Fig. 30–25. Determine the mutual inductance in terms of ℓ_1, ℓ_2, and w. Assume the wire is very long compared to ℓ_1, ℓ_2, and w, and that the rest of its circuit is very far away compared to ℓ_1, ℓ_2, and w.

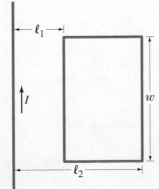

FIGURE 30–25
Problem 4.

30–2 Self-Inductance

5. (I) If the current in a 280-mH coil changes steadily from 25.0 A to 10.0 A in 360 ms, what is the magnitude of the induced emf?

6. (I) How many turns of wire would be required to make a 130-mH inductance out of a 30.0-cm-long air-filled coil with a diameter of 4.2 cm?

7. (I) What is the inductance of a coil if the coil produces an emf of 2.50 V when the current in it changes from −28.0 mA to +25.0 mA in 12.0 ms?

8. (II) An air-filled cylindrical inductor has 2800 turns, and it is 2.5 cm in diameter and 21.7 cm long. (a) What is its inductance? (b) How many turns would you need to generate the same inductance if the core were filled with iron of magnetic permeability 1200 times that of free space?

9. (II) A coil has 3.25-Ω resistance and 440-mH inductance. If the current is 3.00 A and is increasing at a rate of 3.60 A/s, what is the potential difference across the coil at this moment?

10. (II) If the outer conductor of a coaxial cable has radius 3.0 mm, what should be the radius of the inner conductor so that the inductance per unit length does not exceed 55 nH per meter?

11. (II) To demonstrate the large size of the henry unit, a physics professor wants to wind an air-filled solenoid with self-inductance of 1.0 H on the outside of a 12-cm diameter plastic hollow tube using copper wire with a 0.81-mm diameter. The solenoid is to be tightly wound with each turn touching its neighbor (the wire has a thin insulating layer on its surface so the neighboring turns are not in electrical contact). How long will the plastic tube need to be and how many kilometers of copper wire will be required? What will be the resistance of this solenoid?

12. (II) The wire of a tightly wound solenoid is unwound and used to make another tightly wound solenoid of 2.5 times the diameter. By what factor does the inductance change?

13. (II) A toroid has a rectangular cross section as shown in Fig. 30–26. Show that the self-inductance is

$$L = \frac{\mu_0 N^2 h}{2\pi} \ln \frac{r_2}{r_1}$$

where N is the total number of turns and r_1, r_2, and h are the dimensions shown in Fig. 30–26. [*Hint*: Use Ampère's law to get B as a function of r inside the toroid, and integrate.]

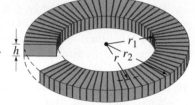

FIGURE 30–26
Problems 13 and 19.
A toroid of rectangular cross section, with N turns carrying a current I.

14. (II) Ignoring any mutual inductance, what is the equivalent inductance of two inductors connected (*a*) in series, (*b*) in parallel?

30–3 Magnetic Energy Storage

15. (I) The magnetic field inside an air-filled solenoid 38.0 cm long and 2.10 cm in diameter is 0.600 T. Approximately how much energy is stored in this field?

16. (I) Typical large values for electric and magnetic fields attained in laboratories are about 1.0×10^4 V/m and 2.0 T. (*a*) Determine the energy density for each field and compare. (*b*) What magnitude electric field would be needed to produce the same energy density as the 2.0-T magnetic field?

17. (II) What is the energy density at the center of a circular loop of wire carrying a 23.0-A current if the radius of the loop is 28.0 cm?

18. (II) Calculate the magnetic and electric energy densities at the surface of a 3.0-mm-diameter copper wire carrying a 15-A current.

19. (II) For the toroid of Fig. 30–26, determine the energy density in the magnetic field as a function of r ($r_1 < r < r_2$) and integrate this over the volume to obtain the total energy stored in the toroid, which carries a current I in each of its N loops.

20. (II) Determine the total energy stored per unit length in the magnetic field between the coaxial cylinders of a coaxial cable (Fig. 30–5) by using Eq. 30–7 for the energy density and integrating over the volume.

21. (II) A long straight wire of radius R carries current I uniformly distributed across its cross-sectional area. Find the magnetic energy stored per unit length in the interior of this wire.

30–4 LR Circuits

22. (II) After how many time constants does the current in Fig. 30–6 reach within (*a*) 5.0%, (*b*) 1.0%, and (*c*) 0.10% of its maximum value?

23. (II) How many time constants does it take for the potential difference across the resistor in an LR circuit like that in Fig. 30–7 to drop to 3.0% of its original value?

24. (II) It takes 2.56 ms for the current in an LR circuit to increase from zero to 0.75 its maximum value. Determine (*a*) the time constant of the circuit, (*b*) the resistance of the circuit if $L = 31.0$ mH.

25. (II) (*a*) Determine the energy stored in the inductor L as a function of time for the LR circuit of Fig. 30–6a. (*b*) After how many time constants does the stored energy reach 99.9% of its maximum value?

26. (II) In the circuit of Fig. 30–27, determine the current in each resistor (I_1, I_2, I_3) at the moment (*a*) the switch is closed, (*b*) a long time after the switch is closed. After the switch has been closed for a long time, and then reopened, what is each current (*c*) just after it is opened, (*d*) after a long time?

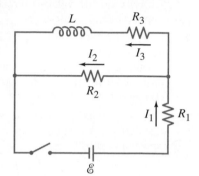

FIGURE 30–27
Problem 26.

27. (II) (*a*) In Fig. 30–28a, assume that the switch S has been in position A for sufficient time so that a steady current $I_0 = V_0/R$ flows through the resistor R. At time $t = 0$, the switch is quickly switched to position B and the current through R decays according to $I = I_0 e^{-t/\tau}$. Show that the maximum emf $\mathcal{E}_{max}$ induced in the inductor during this time period equals the battery voltage V_0. (*b*) In Fig. 30–28b, assume that the switch has been in position A for sufficient time so that a steady current $I_0 = V_0/R$ flows through the resistor R. At time $t = 0$, the switch is quickly switched to position B and the current decays through resistor R' (which is much greater than R) according to $I = I_0 e^{-t/\tau'}$. Show that the maximum emf $\mathcal{E}_{max}$ induced in the inductor during this time period is $(R'/R)V_0$. If $R' = 55R$ and $V_0 = 120$ V, determine $\mathcal{E}_{max}$. [When a mechanical switch is opened, a high-resistance air gap is created, which is modeled as R' here. This Problem illustrates why high-voltage sparking can occur if a current-carrying inductor is suddenly cut off from its power source.]

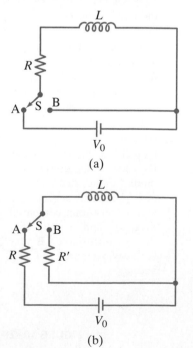

FIGURE 30–28
Problem 27.
(a)
(b)

28. (II) You want to turn on the current through a coil of self-inductance L in a controlled manner, so you place it in series with a resistor $R = 2200\,\Omega$, a switch, and a dc voltage source $V_0 = 240\,\text{V}$. After closing the switch, you find that the current through the coil builds up to its steady-state value with a time constant τ. You are pleased with the current's steady-state value, but want τ to be half as long. What new values should you use for R and V_0?

29. (II) A 12-V battery has been connected to an LR circuit for sufficient time so that a steady current flows through the resistor $R = 2.2\,\text{k}\Omega$ and inductor $L = 18\,\text{mH}$. At $t = 0$, the battery is removed from the circuit and the current decays exponentially through R. Determine the emf $\mathscr{E}$ across the inductor as time t increases. At what time is $\mathscr{E}$ greatest and what is this maximum value (V)?

30. (III) Two tightly wound solenoids have the same length and circular cross-sectional area. But solenoid 1 uses wire that is 1.5 times as thick as solenoid 2. (a) What is the ratio of their inductances? (b) What is the ratio of their inductive time constants (assuming no other resistance in the circuits)?

30–5 LC Circuits and Oscillations

31. (I) The variable capacitor in the tuner of an AM radio has a capacitance of 1350 pF when the radio is tuned to a station at 550 kHz. (a) What must be the capacitance for a station at 1600 kHz? (b) What is the inductance (assumed constant)? Ignore resistance.

32. (I) (a) If the initial conditions of an LC circuit were $I = I_0$ and $Q = 0$ at $t = 0$, write Q as a function of time. (b) Practically, how could you set up these initial conditions?

33. (II) In some experiments, short distances are measured by using capacitance. Consider forming an LC circuit using a parallel-plate capacitor with plate area A, and a known inductance L. (a) If charge is found to oscillate in this circuit at frequency $f = \omega/2\pi$ when the capacitor plates are separated by distance x, show that $x = 4\pi^2 A\epsilon_0 f^2 L$. (b) When the plate separation is changed by Δx, the circuit's oscillation frequency will change by Δf. Show that $\Delta x/x \approx 2(\Delta f/f)$. (c) If f is on the order of 1 MHz and can be measured to a precision of $\Delta f = 1\,\text{Hz}$, with what percent accuracy can x be determined? Assume fringing effects at the capacitor's edges can be neglected.

34. (II) A 425-pF capacitor is charged to 135 V and then quickly connected to a 175-mH inductor. Determine (a) the frequency of oscillation, (b) the peak value of the current, and (c) the maximum energy stored in the magnetic field of the inductor.

35. (II) At $t = 0$, let $Q = Q_0$, and $I = 0$ in an LC circuit. (a) At the first moment when the energy is shared equally by the inductor and the capacitor, what is the charge on the capacitor? (b) How much time has elapsed (in terms of the period T)?

30–6 LC Oscillations with Resistance

36. (II) A damped LC circuit loses 3.5% of its electromagnetic energy per cycle to thermal energy. If $L = 65\,\text{mH}$ and $C = 1.00\,\mu\text{F}$, what is the value of R?

37. (II) In an oscillating LRC circuit, how much time does it take for the energy stored in the fields of the capacitor and inductor to fall to 75% of the initial value? (See Fig. 30–13; assume $R \ll \sqrt{4L/C}$.)

38. (III) How much resistance must be added to a pure LC circuit ($L = 350\,\text{mH}$, $C = 1800\,\text{pF}$) to change the oscillator's frequency by 0.25%? Will it be increased or decreased?

30–7 AC Circuits: Reactance

39. (I) At what frequency will a 32.0-mH inductor have a reactance of 660 Ω?

40. (I) What is the reactance of a 9.2-μF capacitor at a frequency of (a) 60.0 Hz, (b) 1.00 MHz?

41. (I) Plot a graph of the reactance of a 1.0-μF capacitor as a function of frequency from 10 Hz to 1000 Hz.

42. (I) Calculate the reactance of, and rms current in, a 260-mH radio coil connected to a 250-V (rms) 33.3-kHz ac line. Ignore resistance.

43. (II) A resistor R is in parallel with a capacitor C, and this parallel combination is in series with a resistor R'. If connected to an ac voltage source of frequency ω, what is the equivalent impedance of this circuit at the two extremes in frequency (a) $\omega = 0$, and (b) $\omega = \infty$?

44. (II) What is the inductance L of the primary of a transformer whose input is 110 V at 60 Hz and the current drawn is 3.1 A? Assume no current in the secondary.

45. (II) (a) What is the reactance of a 0.086-μF capacitor connected to a 22-kV (rms), 660-Hz line? (b) Determine the frequency and the peak value of the current.

46. (II) A capacitor is placed in parallel with some device, B, as in Fig. 30–18b, to filter out stray high-frequency signals, but to allow ordinary 60-Hz ac to pass through with little loss. Suppose that circuit B in Fig. 30–18b is a resistance $R = 490\,\Omega$ connected to ground, and that $C = 0.35\,\mu\text{F}$. Calculate the ratio of the capacitor's current amplitude to the incoming current's amplitude if the incoming current has a frequency of (a) 60 Hz; (b) 60,000 Hz.

47. (II) A current $I = 1.80 \cos 377t$ (I in amps, t in seconds, and the "angle" is in radians) flows in a series LR circuit in which $L = 3.85\,\text{mH}$ and $R = 1.35\,\text{k}\Omega$. What is the average power dissipation?

30–8 LRC Series AC Circuit

48. (I) A 10.0-kΩ resistor is in series with a 26.0-mH inductor and an ac source. Calculate the impedance of the circuit if the source frequency is (a) 55.0 Hz; (b) 55,000 Hz.

49. (I) A 75-Ω resistor and a 6.8-μF capacitor are connected in series to an ac source. Calculate the impedance of the circuit if the source frequency is (a) 60 Hz; (b) 6.0 MHz.

50. (I) For a 120-V, 60-Hz voltage, a current of 70 mA passing through the body for 1.0 s could be lethal. What must be the impedance of the body for this to occur?

51. (II) A 2.5-kΩ resistor in series with a 420-mH inductor is driven by an ac power supply. At what frequency is the impedance double that of the impedance at 60 Hz?

52. (II) (a) What is the rms current in a series RC circuit if $R = 3.8\,\text{k}\Omega$, $C = 0.80\,\mu\text{F}$, and the rms applied voltage is 120 V at 60.0 Hz? (b) What is the phase angle between voltage and current? (c) What is the power dissipated by the circuit? (d) What are the voltmeter readings across R and C?

53. (II) An ac voltage source is connected in series with a 1.0-μF capacitor and a 750-Ω resistor. Using a digital ac voltmeter, the amplitude of the voltage source is measured to be 4.0 V rms, while the voltages across the resistor and across the capacitor are found to be 3.0 V rms and 2.7 V rms, respectively. Determine the frequency of the ac voltage source. Why is the voltage measured across the voltage source not equal to the sum of the voltages measured across the resistor and across the capacitor?

54. (II) Determine the total impedance, phase angle, and rms current in an *LRC* circuit connected to a 10.0-kHz, 725-V (rms) source if $L = 32.0\,\text{mH}$, $R = 8.70\,\text{k}\Omega$, and $C = 6250\,\text{pF}$.

55. (II) (*a*) What is the rms current in a series *LR* circuit when a 60.0-Hz, 120-V rms ac voltage is applied, where $R = 965\,\Omega$ and $L = 225\,\text{mH}$? (*b*) What is the phase angle between voltage and current? (*c*) How much power is dissipated? (*d*) What are the rms voltage readings across R and L?

56. (II) A 35-mH inductor with 2.0-Ω resistance is connected in series to a 26-μF capacitor and a 60-Hz, 45-V (rms) source. Calculate (*a*) the rms current, (*b*) the phase angle, and (*c*) the power dissipated in this circuit.

57. (II) A 25-mH coil whose resistance is $0.80\,\Omega$ is connected to a capacitor C and a 360-Hz source voltage. If the current and voltage are to be in phase, what value must C have?

58. (II) A 75-W lightbulb is designed to operate with an applied ac voltage of 120 V rms. The bulb is placed in series with an inductor L, and this series combination is then connected to a 60-Hz 240-V rms voltage source. For the bulb to operate properly, determine the required value for L. Assume the bulb has resistance R and negligible inductance.

59. (II) In the *LRC* circuit of Fig. 30–19, suppose $I = I_0 \sin \omega t$ and $V = V_0 \sin(\omega t + \phi)$. Determine the instantaneous power dissipated in the circuit from $P = IV$ using these equations and show that on the average, $\overline{P} = \frac{1}{2} V_0 I_0 \cos \phi$, which confirms Eq. 30–30.

60. (II) An *LRC* series circuit with $R = 150\,\Omega$, $L = 25\,\text{mH}$, and $C = 2.0\,\mu$F is powered by an ac voltage source of peak voltage $V_0 = 340\,\text{V}$ and frequency $f = 660\,\text{Hz}$. (*a*) Determine the peak current that flows in this circuit. (*b*) Determine the phase angle of the source voltage relative to the current. (*c*) Determine the peak voltage across R and its phase angle relative to the source voltage. (*d*) Determine the peak voltage across L and its phase angle relative to the source voltage. (*e*) Determine the peak voltage across C and its phase angle relative to the source voltage.

61. (II) An *LR* circuit can be used as a "phase shifter." Assume that an "input" source voltage $V = V_0 \sin(2\pi f t + \phi)$ is connected across a series combination of an inductor $L = 55\,\text{mH}$ and resistor R. The "output" of this circuit is taken across the resistor. If $V_0 = 24\,\text{V}$ and $f = 175\,\text{Hz}$, determine the value of R so that the output voltage V_R lags the input voltage V by 25°. Compare (as a ratio) the peak output voltage with V_0.

30–9 Resonance in AC Circuits

62. (I) A 3800-pF capacitor is connected in series to a 26.0-μH coil of resistance $2.00\,\Omega$. What is the resonant frequency of this circuit?

63. (I) What is the resonant frequency of the *LRC* circuit of Example 30–11? At what rate is energy taken from the generator, on the average, at this frequency?

64. (II) An *LRC* circuit has $L = 4.15\,\text{mH}$ and $R = 3.80\,\text{k}\Omega$. (*a*) What value must C have to produce resonance at 33.0 kHz? (*b*) What will be the maximum current at resonance if the peak external voltage is 136 V?

65. (II) The frequency of the ac voltage source (peak voltage V_0) in an *LRC* circuit is tuned to the circuit's resonant frequency $f_0 = 1/(2\pi\sqrt{LC})$. (*a*) Show that the peak voltage across the capacitor is $V_{C0} = V_0 T_0/2\pi\,\tau)$, where $T_0(= 1/f_0)$ is the period of the resonant frequency and $\tau = RC$ is the time constant for charging the capacitor C through a resistor R. (*b*) Define $\beta = T_0/(2\pi\tau)$ so that $V_{C0} = \beta V_0$. Then β is the "amplification" of the source voltage across the capacitor. If a particular *LRC* circuit contains a 2.0-nF capacitor and has a resonant frequency of 5.0 kHz, what value of R will yield $\beta = 125$?

66. (II) Capacitors made from piezoelectric materials are commonly used as sound transducers ("speakers"). They often require a large operating voltage. One method for providing the required voltage is to include the speaker as part of an *LRC* circuit as shown in Fig. 30–29, where the speaker is modeled electrically as the capacitance $C = 1.0\,\text{nF}$. Take $R = 35\,\Omega$ and $L = 55\,\text{mH}$. (*a*) What is the resonant frequency f_0 for this circuit? (*b*) If the voltage source has peak amplitude $V_0 = 2.0\,\text{V}$ at frequency $f = f_0$, find the peak voltage V_{C0} across the speaker (i.e., the capacitor C). (*c*) Determine the ratio V_{C0}/V_0.

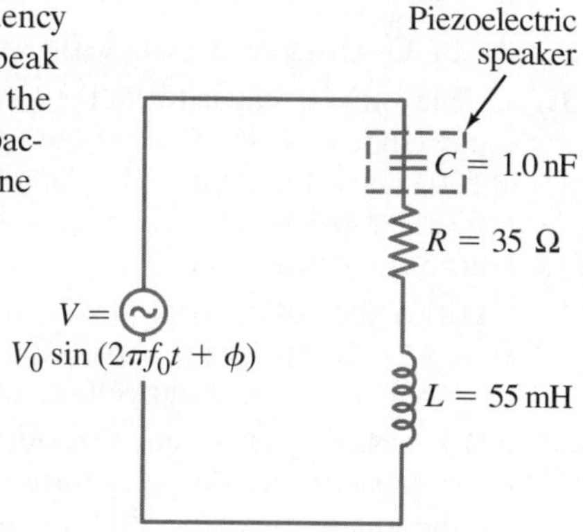

FIGURE 30–29
Problem 66.

67. (II) (*a*) Determine a formula for the average power $\overline{P}$ dissipated in an *LRC* circuit in terms of L, R, C, ω, and V_0. (*b*) At what frequency is the power a maximum? (*c*) Find an approximate formula for the width of the resonance peak in average power, $\Delta\omega$, which is the difference in the two (angular) frequencies where $\overline{P}$ has half its maximum value. Assume a sharp peak.

68. (II) (*a*) Show that oscillation of charge Q on the capacitor of an *LRC* circuit has amplitude

$$Q_0 = \frac{V_0}{\sqrt{(\omega R)^2 + \left(\omega^2 L - \dfrac{1}{C}\right)^2}}.$$

(*b*) At what angular frequency, ω', will Q_0 be a maximum? (*c*) Compare to a forced damped harmonic oscillator (Chapter 14), and discuss. (See also Question 20 in this Chapter.)

69. (II) A resonant circuit using a 220-nF capacitor is to resonate at 18.0 kHz. The air-core inductor is to be a solenoid with closely packed coils made from 12.0 m of insulated wire 1.1 mm in diameter. How many loops will the inductor contain?

*30–10 Impedance Matching

***70.** (II) The output of an electrocardiogram amplifier has an impedance of 45 kΩ. It is to be connected to an 8.0-Ω loudspeaker through a transformer. What should be the turns ratio of the transformer?

General Problems

71. A 2200-pF capacitor is charged to 120 V and then quickly connected to an inductor. The frequency of oscillation is observed to be 17 kHz. Determine (a) the inductance, (b) the peak value of the current, and (c) the maximum energy stored in the magnetic field of the inductor.

72. At $t = 0$, the current through a 60.0-mH inductor is 50.0 mA and is increasing at the rate of 78.0 mA/s. What is the initial energy stored in the inductor, and how long does it take for the energy to increase by a factor of 5.0 from the initial value?

73. At time $t = 0$, the switch in the circuit shown in Fig. 30–30 is closed. After a sufficiently long time, steady currents I_1, I_2, and I_3 flow through resistors R_1, R_2, and R_3, respectively. Determine these three currents.

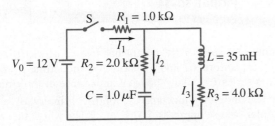

FIGURE 30–30   Problem 73.

74. (a) Show that the self-inductance L of a toroid (Fig. 30–31) of radius r_0 containing N loops each of diameter d is

$$L \approx \frac{\mu_0 N^2 d^2}{8r_0}$$

if $r_0 \gg d$. Assume the field is uniform inside the toroid; is this actually true? Is this result consistent with L for a solenoid? Should it be? (b) Calculate the inductance L of a large toroid if the diameter of the coils is 2.0 cm and the diameter of the whole ring is 66 cm. Assume the field inside the toroid is uniform. There are a total of 550 loops of wire.

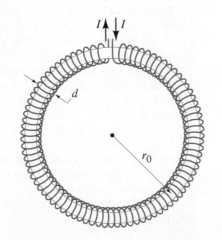

FIGURE 30–31
A toroid.
Problem 74.

75. A pair of straight parallel thin wires, such as a lamp cord, each of radius r, are a distance ℓ apart and carry current to a circuit some distance away. Ignoring the field within each wire, show that the inductance per unit length is $(\mu_0/\pi) \ln[(\ell - r)/r]$.

76. Assuming the Earth's magnetic field averages about 0.50×10^{-4} T near the surface of the Earth, estimate the total energy stored in this field in the first 5.0 km above the Earth's surface.

77. (a) For an underdamped LRC circuit, determine a formula for the energy $U = U_L + U_R$ stored in the electric and magnetic fields as a function of time. Give answer in terms of the initial charge Q_0 on the capacitor. (b) Show how dU/dt is related to the rate energy is transformed in the resistor, I^2R.

78. An electronic device needs to be protected against sudden surges in current. In particular, after the power is turned on the current should rise to no more than 7.5 mA in the first 75 μs. The device has resistance 150 Ω and is designed to operate at 33 mA. How would you protect this device?

79. The circuit shown in Fig. 30–32a can integrate (in the calculus sense) the input voltage V_{in}, if the time constant L/R is large compared with the time during which V_{in} varies. Explain how this integrator works and sketch its output for the square wave signal input shown in Fig. 30–32b. [*Hint:* Write Kirchhoff's loop rule for the circuit. Multiply each term in this differential equation (in I) by a factor $e^{Rt/L}$ to make it easier to integrate.]

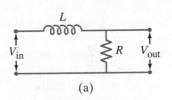

(a)

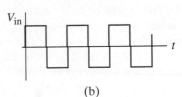

FIGURE 30–32
Problem 79.
(b)

80. Suppose circuit B in Fig. 30–18a consists of a resistance $R = 550\ \Omega$. The filter capacitor has capacitance $C = 1.2\ \mu$F. Will this capacitor act to eliminate 60-Hz ac but pass a high-frequency signal of frequency 6.0 kHz? To check this, determine the voltage drop across R for a 130-mV signal of frequency (a) 60 Hz; (b) 6.0 kHz.

81. An ac voltage source $V = V_0 \sin(\omega t + 90°)$ is connected across an inductor L and current $I = I_0 \sin(\omega t)$ flows in this circuit. Note that the current and source voltage are 90° out of phase. (a) Directly calculate the average power delivered by the source over one period T of its sinusoidal cycle via the integral $\overline{P} = \int_0^T VI\, dt/T$. (b) Apply the relation $\overline{P} = I_{rms}V_{rms} \cos\phi$ to this circuit and show that the answer you obtain is consistent with that found in part (a). Comment on your results.

82. A circuit contains two elements, but it is not known if they are L, R, or C. The current in this circuit when connected to a 120-V 60-Hz source is 5.6 A and lags the voltage by 65°. What are the two elements and what are their values?

83. A 3.5-kΩ resistor in series with a 440-mH inductor is driven by an ac power supply. At what frequency is the impedance double that of the impedance at 60 Hz?

84. (a) What is the rms current in an RC circuit if $R = 5.70\,k\Omega$, $C = 1.80\,\mu F$, and the rms applied voltage is 120 V at 60.0 Hz? (b) What is the phase angle between voltage and current? (c) What is the power dissipated by the circuit? (d) What are the voltmeter readings across R and C?

85. An inductance coil draws 2.5 A dc when connected to a 45-V battery. When connected to a 60-Hz 120-V (rms) source, the current drawn is 3.8 A (rms). Determine the inductance and resistance of the coil.

86. The **Q-value** of a resonance circuit can be defined as the ratio of the voltage across the capacitor (or inductor) to the voltage across the resistor, at resonance. The larger the Q factor, the sharper the resonance curve will be and the sharper the tuning. (a) Show that the Q factor is given by the equation $Q = (1/R)\sqrt{L/C}$. (b) At a resonant frequency $f_0 = 1.0\,MHz$, what must be the value of L and R to produce a Q factor of 350? Assume that $C = 0.010\,\mu F$.

87. Show that the fraction of electromagnetic energy lost (to thermal energy) per cycle in a lightly damped $(R^2 \ll 4L/C)$ LRC circuit is approximately

$$\frac{\Delta U}{U} = \frac{2\pi R}{L\omega} = \frac{2\pi}{Q}.$$

The quantity Q can be defined as $Q = L\omega/R$, and is called the Q-value, or *quality factor*, of the circuit and is a measure of the damping present. A high Q-value means smaller damping and less energy input required to maintain oscillations.

88. In a series LRC circuit, the inductance is 33 mH, the capacitance is 55 nF, and the resistance is 1.50 kΩ. At what frequencies is the power factor equal to 0.17?

89. In our analysis of a series LRC circuit, Fig. 30–19, suppose we chose $V = V_0 \sin \omega t$. (a) Construct a phasor diagram, like that of Fig. 30–21, for this case. (b) Write a formula for the current I, defining all terms.

90. A voltage $V = 0.95 \sin 754t$ is applied to an LRC circuit (I is in amperes, t is in seconds, V is in volts, and the "angle" is in radians) which has $L = 22.0\,mH$, $R = 23.2\,k\Omega$, and $C = 0.42\,\mu F$. (a) What is the impedance and phase angle? (b) How much power is dissipated in the circuit? (c) What is the rms current and voltage across each element?

91. *Filter circuit.* Figure 30–33 shows a simple filter circuit designed to pass dc voltages with minimal attenuation and to remove, as much as possible, any ac components (such as 60-Hz line voltage that could cause hum in a stereo receiver, for example). Assume $V_{in} = V_1 + V_2$ where V_1 is dc and $V_2 = V_{20} \sin \omega t$, and that any resistance is very small. (a) Determine the current through the capacitor: give amplitude and phase (assume $R = 0$ and $X_L > X_C$). (b) Show that the ac component of the output voltage, $V_{2\,out}$, equals $(Q/C) - V_1$, where Q is the charge on the capacitor at any instant, and determine the amplitude and phase of $V_{2\,out}$. (c) Show that the attenuation of the ac voltage is greatest when $X_C \ll X_L$, and calculate the ratio of the output to input ac voltage in this case. (d) Compare the dc output voltage to input voltage.

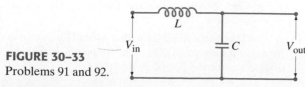

FIGURE 30–33
Problems 91 and 92.

92. Show that if the inductor L in the filter circuit of Fig. 30–33 (Problem 91) is replaced by a large resistor R, there will still be significant attenuation of the ac voltage and little attenuation of the dc voltage if the input dc voltage is high and the current (and power) are low.

93. A resistor R, capacitor C, and inductor L are connected in parallel across an ac generator as shown in Fig. 30–34. The source emf is $V = V_0 \sin \omega t$. Determine the current as a function of time (including amplitude and phase): (a) in the resistor, (b) in the inductor, (c) in the capacitor. (d) What is the total current leaving the source? (Give amplitude I_0 and phase.) (e) Determine the impedance Z defined as $Z = V_0/I_0$. (f) What is the power factor?

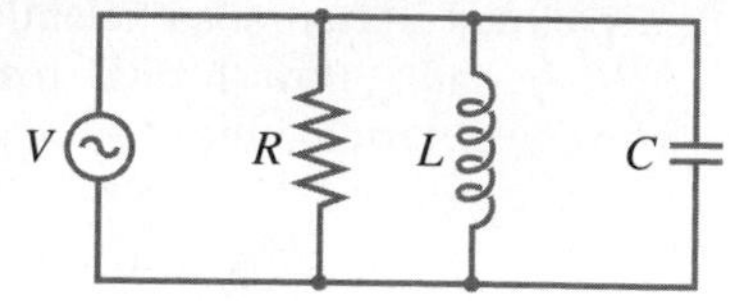

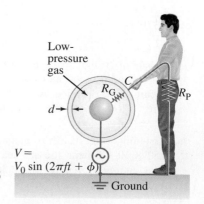

FIGURE 30–34
Problem 93.

94. Suppose a series LRC circuit has two resistors, R_1 and R_2, two capacitors, C_1 and C_2, and two inductors, L_1 and L_2, all in series. Calculate the total impedance of the circuit.

95. Determine the inductance L of the primary of a transformer whose input is 220 V at 60 Hz when the current drawn is 4.3 A. Assume no current in the secondary.

96. In a *plasma globe*, a hollow glass sphere is filled with low-pressure gas and a small spherical metal electrode is located at its center. Assume an ac voltage source of peak voltage V_0 and frequency f is applied between the metal sphere and the ground, and that a person is touching the outer surface of the globe with a fingertip, whose approximate area is 1.0 cm^2. The equivalent circuit for this situation is shown in Fig. 30–35, where R_G and R_P are the resistances of the gas and the person, respectively, and C is the capacitance formed by the gas, glass, and finger. (a) Determine C assuming it is a parallel-plate capacitor. The conductive gas and the person's fingertip form the opposing plates of area $A = 1.0\,cm^2$. The plates are separated by glass (dielectric constant $K = 5.0$) of thickness $d = 2.0\,mm$. (b) In a typical plasma globe, $f = 12\,kHz$. Determine the reactance X_C of C at this frequency in MΩ. (c) The voltage may be $V_0 = 2500\,V$. With this high voltage, the dielectric strength of the gas is exceeded and the gas becomes ionized. In this "plasma" state, the gas emits light ("sparks") and is highly conductive so that $R_G \ll X_C$. Assuming also that $R_P \ll X_C$, estimate the peak current that flows in the given circuit. Is this level of current dangerous? (d) If the plasma globe operated at $f = 1.0\,MHz$, estimate the peak current that would flow in the given circuit. Is this level of current dangerous?

FIGURE 30–35
Problem 96.

97. You have a small electromagnet that consumes 350 W from a residential circuit operating at 120 V at 60 Hz. Using your ac multimeter, you determine that the unit draws 4.0 A rms. What are the values of the inductance and the internal resistance?

98. An inductor L in series with a resistor R, driven by a sinusoidal voltage source, responds as described by the following differential equation:

$$V_0 \sin \omega t = L \frac{dI}{dt} + RI.$$

Show that a current of the form $I = I_0 \sin(\omega t - \phi)$ flows through the circuit by direct substitution into the differential equation. Determine the amplitude of the current (I_0) and the phase difference ϕ between the current and the voltage source.

99. In a certain LRC series circuit, when the ac voltage source has a particular frequency f, the peak voltage across the inductor is 6.0 times greater than the peak voltage across the capacitor. Determine f in terms of the resonant frequency f_0 of this circuit.

100. For the circuit shown in Fig. 30–36, $V = V_0 \sin \omega t$. Calculate the current in each element of the circuit, as well as the total impedance. [*Hint*: Try a trial solution of the form $I = I_0 \sin(\omega t + \phi)$ for the current leaving the source.]

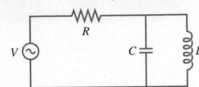

FIGURE 30–36
Problem 100.

101. To detect vehicles at traffic lights, wire loops with dimensions on the order of 2 m are often buried horizontally under roadways. Assume the self-inductance of such a loop is $L = 5.0$ mH and that it is part of an LRC circuit as shown in Fig. 30–37 with $C = 0.10\,\mu\text{F}$ and $R = 45\,\Omega$. The ac voltage has frequency f and rms voltage V_{rms}. (*a*) The frequency f is chosen to match the resonant frequency f_0 of the circuit. Find f_0 and determine what the rms voltage $(V_R)_{\text{rms}}$ across the resistor will be when $f = f_0$. (*b*) Assume that f, C, and R never change, but that, when a car is located above the buried loop, the loop's self-inductance decreases by 10% (due to induced eddy currents in the car's metal parts). Determine by what factor the voltage $(V_R)_{\text{rms}}$ decreases in this situation in comparison to no car above the loop. [Monitoring $(V_R)_{\text{rms}}$ detects the presence of a car.]

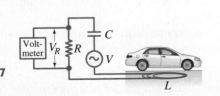

FIGURE 30–37
Problem 101.

102. For the circuit shown in Fig. 30–38, show that if the condition $R_1 R_2 = L/C$ is satisfied then the potential difference between points a and b is zero for all frequencies.

FIGURE 30–38
Problem 102.

*Numerical/Computer

*103. (II) The RC circuit shown in Fig. 30–39 is called a **low-pass filter** because it passes low-frequency ac signals with less attenuation than high-frequency ac signals. (*a*) Show that the voltage gain is $A = V_{\text{out}}/V_{\text{in}} = 1/(4\pi^2 f^2 R^2 C^2 + 1)^{\frac{1}{2}}$. (*b*) Discuss the behavior of the gain A for $f \to 0$ and $f \to \infty$. (*c*) Choose $R = 850\,\Omega$ and $C = 1.0 \times 10^{-6}\,\text{F}$, and graph log A versus log f with suitable scales to show the behavior of the circuit at low and high frequencies.

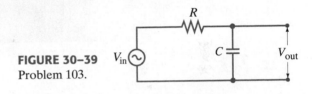

FIGURE 30–39
Problem 103.

*104. (II) The RC circuit shown in Fig. 30–40 is called a **high-pass filter** because it passes high-frequency ac signals with less attenuation than low-frequency ac signals. (*a*) Show that the voltage gain is $A = V_{\text{out}}/V_{\text{in}} = 2\pi f RC/(4\pi^2 f^2 R^2 C^2 + 1)^{\frac{1}{2}}$. (*b*) Discuss the behavior of the gain A for $f \to 0$ and $f \to \infty$. (*c*) Choose $R = 850\,\Omega$ and $C = 1.0 \times 10^{-6}\,\text{F}$, and then graph log A versus log f with suitable scales to show the behavior of the circuit at high and low frequencies.

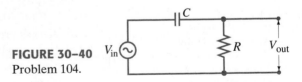

FIGURE 30–40
Problem 104.

*105. (III) Write a computer program or use a spreadsheet program to plot I_{rms} for an ac LRC circuit with a sinusoidal voltage source (Fig. 30–19) with $V_{\text{rms}} = 0.100$ V. For $L = 50\,\mu\text{H}$ and $C = 50\,\mu\text{F}$, plot the I_{rms} graph for (*a*) $R = 0.10\,\Omega$, and (*b*) $R = 1.0\,\Omega$ from $\omega = 0.1\omega_0$ to $\omega = 3.0\omega_0$ on the same graph.

Answers to Exercises

A: (*a*) 360 A/s; (*b*) 12 V.

B: (*b*).

C: (*a*).

D: From Eq. 30–5, L has dimensions VT/A so (L/R) has dimensions $(VT/A)/(V/A) = T$.

E: (*d*).

F: (*c*)

G: (*a*).

H: (*c*).

Wireless technology is all around us: in this photo we see a Bluetooth earpiece for wireless telephone communication and a wi-fi computer. The wi-fi antenna is just visible at the lower left. All these devices work by electromagnetic waves traveling through space, based on the great work of Maxwell which we investigate in this Chapter. Modern wireless devices are applications of Marconi's development of long distance transmission of information a century ago.

We will see in this Chapter that Maxwell predicted the existence of EM waves from his famous equations. Maxwell's equations themselves are a magnificent summary of electromagnetism. We will also examine how EM waves carry energy and momentum.

CHAPTER 31

Maxwell's Equations and Electromagnetic Waves

CONTENTS

CHAPTER-OPENING QUESTION—Guess now!

Which of the following best describes the difference between radio waves and X-rays?

(a) X-rays are radiation while radio waves are electromagnetic waves.

(b) Both can be thought of as electromagnetic waves. They differ only in wavelength and frequency.

(c) X-rays are pure energy. Radio waves are made of fields, not energy.

(d) Radio waves come from electric currents in an antenna. X-rays are not related to electric charge.

(e) The fact that X-rays can expose film, and radio waves cannot, means they are fundamentally different.

The culmination of electromagnetic theory in the nineteenth century was the prediction, and the experimental verification, that waves of electromagnetic fields could travel through space. This achievement opened a whole new world of communication: first the wireless telegraph, then radio and television, and more recently cell phones, remote-control devices, wi-fi, and Bluetooth. Most important was the spectacular prediction that visible light is an electromagnetic wave.

The theoretical prediction of electromagnetic waves was the work of the Scottish physicist James Clerk Maxwell (1831–1879; Fig. 31–1), who unified, in one magnificent theory, all the phenomena of electricity and magnetism.

The development of electromagnetic theory in the early part of the nineteenth century by Oersted, Ampère, and others was not actually done in terms of electric and magnetic fields. The idea of the field was introduced somewhat later by Faraday and was not generally used until Maxwell showed that all electric and magnetic phenomena could be described using only four equations involving electric and magnetic fields. These equations, known as **Maxwell's equations**, are the basic equations for all electromagnetism. They are fundamental in the same sense that Newton's three laws of motion and the law of universal gravitation are for mechanics. In a sense, they are even more fundamental, since they are consistent with the theory of relativity (Chapter 36), whereas Newton's laws are not. Because all of electromagnetism is contained in this set of four equations, Maxwell's equations are considered one of the great triumphs of human intellect.

Before we discuss Maxwell's equations and electromagnetic waves, we first need to discuss a major new prediction of Maxwell's, and, in addition, Gauss's law for magnetism.

FIGURE 31–1 James Clerk Maxwell (1831–1879).

31–1 Changing Electric Fields Produce Magnetic Fields; Ampère's Law and Displacement Current

Ampère's Law

That a magnetic field is produced by an electric current was discovered by Oersted, and the mathematic relation is given by Ampère's law (Eq. 28–3):

$$\oint \vec{\mathbf{B}} \cdot d\vec{\boldsymbol{\ell}} = \mu_0 I_{encl}.$$

Is it possible that magnetic fields could be produced in another way as well? For if a changing magnetic field produces an electric field, as discussed in Section 29–7, then perhaps the reverse might be true as well: that *a changing electric field will produce a magnetic field*. If this were true, it would signify a beautiful *symmetry* in nature.

To back up this idea that a changing electric field might produce a magnetic field, we use an indirect argument that goes something like this. According to Ampère's law, we divide any chosen closed path into short segments $d\vec{\boldsymbol{\ell}}$, take the dot product of each $d\vec{\boldsymbol{\ell}}$ with the magnetic field $\vec{\mathbf{B}}$ at that segment, and sum (integrate) all these products over the chosen closed path. That sum will equal μ_0 times the total current I that passes through a surface bounded by the path of the line integral. When we applied Ampère's law to the field around a straight wire (Section 28–4), we imagined the current as passing through the circular area enclosed by our circular loop, and that area is the flat surface 1 shown in Fig. 31–2. However, we could just as well use the sackshaped surface 2 in Fig. 31–2 as the surface for Ampère's law, since the same current I passes through it.

Now consider the closed circular path for the situation of Fig. 31–3, where a capacitor is being discharged. Ampère's law works for surface 1 (current I passes through surface 1), but it does not work for surface 2, since no current passes through surface 2. There is a magnetic field around the wire, so the left side of Ampère's law ($\int \vec{\mathbf{B}} \cdot d\vec{\boldsymbol{\ell}}$) is not zero; yet no current flows through surface 2, so the right side of Ampère's law *is* zero. We seem to have a contradiction of Ampère's law.

There is a magnetic field present in Fig. 31–3, however, only if charge is flowing to or away from the capacitor plates. The changing charge on the plates means that the electric field between the plates is changing in time. Maxwell resolved the problem of no current through surface 2 in Fig. 31–3 by proposing that there needs to be an extra term on the right in Ampère's law involving the changing electric field.

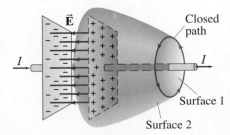

FIGURE 31–2 Ampère's law applied to two different surfaces bounded by the same closed path.

FIGURE 31–3 A capacitor discharging. A conduction current passes through surface 1, but no conduction current passes through surface 2. An extra term is needed in Ampère's law.

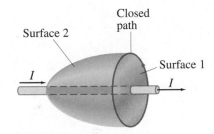

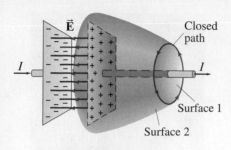

FIGURE 31–3 (repeated) See text.

Ampère's law (general form)

Let us see what this term should be by determining it for the changing electric field between the capacitor plates in Fig. 31–3. The charge Q on a capacitor of capacitance C is $Q = CV$ where V is the potential difference between the plates (Eq. 24–1). Also recall that $V = Ed$ (Eq. 23–4) where d is the (small) separation of the plates and E is the (uniform) electric field strength between them, if we ignore any fringing of the field. Also, for a parallel-plate capacitor, $C = \epsilon_0 A/d$, where A is the area of each plate (Eq. 24–2). We combine these to obtain:

$$Q = CV = \left(\epsilon_0 \frac{A}{d}\right)(Ed) = \epsilon_0 AE.$$

If the charge on each plate changes at a rate dQ/dt, the electric field changes at a proportional rate. That is, by differentiating this expression for Q, we have:

$$\frac{dQ}{dt} = \epsilon_0 A \frac{dE}{dt}.$$

Now dQ/dt is also the current I flowing into or out of the capacitor:

$$I = \frac{dQ}{dt} = \epsilon_0 A \frac{dE}{dt} = \epsilon_0 \frac{d\Phi_E}{dt}$$

where $\Phi_E = EA$ is the **electric flux** through the closed path (surface 2 in Fig. 31–3). In order to make Ampère's law work for surface 2 in Fig. 31–3, as well as for surface 1 (where current I flows), we therefore write:

$$\oint \vec{\mathbf{B}} \cdot d\vec{\boldsymbol{\ell}} = \mu_0 I_{encl} + \mu_0 \epsilon_0 \frac{d\Phi_E}{dt}. \qquad (31\text{–}1)$$

This equation represents the general form of **Ampère's law**,[†] and embodies Maxwell's idea that a magnetic field can be caused not only by an ordinary electric current, but also by a changing electric field or changing electric flux. Although we arrived at it for a special case, Eq. 31–1 has proved valid in general. The last term on the right in Eq. 31–1 is usually very small, and not easy to measure experimentally.

EXAMPLE 31–1 **Charging capacitor.** A 30-pF air-gap capacitor has circular plates of area $A = 100\ \text{cm}^2$. It is charged by a 70-V battery through a 2.0-Ω resistor. At the instant the battery is connected, the electric field between the plates is changing most rapidly. At this instant, calculate (*a*) the current into the plates, and (*b*) the rate of change of electric field between the plates. (*c*) Determine the magnetic field induced between the plates. Assume $\vec{\mathbf{E}}$ is uniform between the plates at any instant and is zero at all points beyond the edges of the plates.

APPROACH In Section 26–5 we discussed *RC* circuits and saw that the charge on a capacitor being charged, as a function of time, is

$$Q = CV_0\left(1 - e^{-t/RC}\right),$$

where V_0 is the voltage of the battery. To find the current at $t = 0$, we differentiate this and substitute the values $V_0 = 70\ \text{V}$, $C = 30\ \text{pF}$, $R = 2.0\ \Omega$.

SOLUTION (*a*) We take the derivative of Q and evaluate it at $t = 0$:

$$\left.\frac{dQ}{dt}\right|_{t=0} = \left.\frac{CV_0}{RC} e^{-t/RC}\right|_{t=0} = \frac{V_0}{R} = \frac{70\ \text{V}}{2.0\ \Omega} = 35\ \text{A}.$$

This is the rate at which charge accumulates on the capacitor and equals the current flowing in the circuit at $t = 0$.

(*b*) The electric field between two closely spaced conductors is given by (Eq. 21–8)

$$E = \frac{\sigma}{\epsilon_0} = \frac{Q/A}{\epsilon_0}$$

as we saw in Chapter 21 (see Example 21–13).

[†] Actually, there is a third term on the right for the case when a magnetic field is produced by magnetized materials. This can be accounted for by changing μ_0 to μ, but we will mainly be interested in cases where no magnetic material is present. In the presence of a dielectric, ϵ_0 is replaced by $\epsilon = K\epsilon_0$ (see Section 24–5).

Hence

$$\frac{dE}{dt} = \frac{dQ/dt}{\epsilon_0 A} = \frac{35\ \text{A}}{(8.85 \times 10^{-12}\ \text{C}^2/\text{N}\cdot\text{m}^2)(1.0 \times 10^{-2}\ \text{m}^2)} = 4.0 \times 10^{14}\ \text{V/m}\cdot\text{s}.$$

(c) Although we will not prove it, we might expect the lines of $\vec{B}$, because of symmetry, to be circles, and to be perpendicular to $\vec{E}$, as shown in Fig. 31-4; this is the same symmetry we saw for the inverse situation of a changing magnetic field producing an electric field (Section 29-7, see Fig. 29-27). To determine the magnitude of B between the plates we apply Ampère's law, Eq. 31-1, with the current $I_{\text{encl}} = 0$:

$$\oint \vec{B}\cdot d\vec{\ell} = \mu_0 \epsilon_0 \frac{d\Phi_E}{dt}.$$

We choose our path to be a circle of radius r, centered at the center of the plate, and thus following a magnetic field line such as the one shown in Fig. 31-4. For $r \le r_0$ (the radius of plate) the flux through a circle of radius r is $E(\pi r^2)$ since E is assumed uniform between the plates at any moment. So from Ampère's law we have

$$B(2\pi r) = \mu_0 \epsilon_0 \frac{d}{dt}(\pi r^2 E)$$

$$= \mu_0 \epsilon_0 \pi r^2 \frac{dE}{dt}.$$

Hence

$$B = \frac{\mu_0 \epsilon_0}{2} r \frac{dE}{dt}. \qquad [r \le r_0]$$

We assume $\vec{E} = 0$ for $r > r_0$, so for points beyond the edge of the plates all the flux is contained within the plates (area $= \pi r_0^2$) and $\Phi_E = E\pi r_0^2$. Thus Ampère's law gives

$$B(2\pi r) = \mu_0 \epsilon_0 \frac{d}{dt}(\pi r_0^2 E)$$

$$= \mu_0 \epsilon_0 \pi r_0^2 \frac{dE}{dt}$$

or

$$B = \frac{\mu_0 \epsilon_0 r_0^2}{2r} \frac{dE}{dt}. \qquad [r \ge r_0]$$

B has its maximum value at $r = r_0$ which, from either relation above (using $r_0 = \sqrt{A/\pi} = 5.6\ \text{cm}$), is

$$B_{\text{max}} = \frac{\mu_0 \epsilon_0 r_0}{2} \frac{dE}{dt}$$

$$= \tfrac{1}{2}(4\pi \times 10^{-7}\ \text{T}\cdot\text{m/A})(8.85 \times 10^{-12}\ \text{C}^2/\text{N}\cdot\text{m}^2)(5.6 \times 10^{-2}\ \text{m})(4.0 \times 10^{14}\ \text{V/m}\cdot\text{s})$$

$$= 1.2 \times 10^{-4}\ \text{T}.$$

This is a very small field and lasts only briefly (the time constant $RC = 6.0 \times 10^{-11}\ \text{s}$) and so would be very difficult to measure.

Let us write the magnetic field B outside the capacitor plates of Example 31-1 in terms of the current I that leaves the plates. The electric field between the plates is $E = \sigma/\epsilon_0 = Q/(\epsilon_0 A)$, as we saw in part b, so $dE/dt = I/(\epsilon_0 A)$. Hence B for $r > r_0$ is,

$$B = \frac{\mu_0 \epsilon_0 r_0^2}{2r} \frac{dE}{dt} = \frac{\mu_0 \epsilon_0 r_0^2}{2r} \frac{I}{\epsilon_0 \pi r_0^2} = \frac{\mu_0 I}{2\pi r}.$$

This is the same formula for the field that surrounds a wire (Eq. 28-1). Thus the B field outside the capacitor is the same as that outside the wire. In other words, the magnetic field produced by the changing electric field between the plates is the same as that produced by the current in the wire.

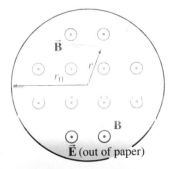

FIGURE 31-4 Frontal view of a circular plate of a parallel-plate capacitor. $\vec{E}$ between plates points out toward viewer; lines of $\vec{B}$ are circles. (Example 31-1.)

Displacement Current

Maxwell interpreted the second term on the right in Eq. 31–1 as being *equivalent* to an electric current. He called it a **displacement current**, I_D. An ordinary current I is then called a **conduction current**. Ampère's law can then be written

$$\oint \vec{\mathbf{B}} \cdot d\vec{\boldsymbol{\ell}} = \mu_0 (I + I_D)_{\text{encl}} \tag{31–2}$$

where

Displacement current

$$I_D = \epsilon_0 \frac{d\Phi_E}{dt}. \tag{31–3}$$

The term "displacement current" was based on an old discarded theory. Don't let it confuse you: I_D does not represent a flow of electric charge[†], nor is there a displacement.

31–2 Gauss's Law for Magnetism

We are almost in a position to state Maxwell's equations, but first we need to discuss the magnetic equivalent of Gauss's law. As we saw in Chapter 29, for a magnetic field $\vec{\mathbf{B}}$ the *magnetic flux* Φ_B through a surface is defined as

$$\Phi_B = \int \vec{\mathbf{B}} \cdot d\vec{\mathbf{A}}$$

where the integral is over the area of either an open or a closed surface. The magnetic flux through a closed surface—that is, a surface which completely encloses a volume—is written

$$\Phi_B = \oint \vec{\mathbf{B}} \cdot d\vec{\mathbf{A}}.$$

In the electric case, we saw in Section 22–2 that the electric flux Φ_E through a closed surface is equal to the total net charge Q enclosed by the surface, divided by ϵ_0 (Eq. 22–4):

$$\oint \vec{\mathbf{E}} \cdot d\vec{\mathbf{A}} = \frac{Q}{\epsilon_0}.$$

This relation is Gauss's law for electricity.

We can write a similar relation for the magnetic flux. We have seen, however, that in spite of intense searches, no isolated magnetic poles (monopoles)—the magnetic equivalent of single electric charges—have ever been observed. Hence, **Gauss's law for magnetism** is

$$\oint \vec{\mathbf{B}} \cdot d\vec{\mathbf{A}} = 0. \tag{31–4}$$

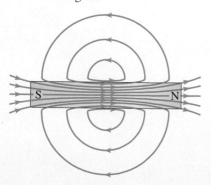

FIGURE 31–5 Magnetic field lines for a bar magnet.

In terms of magnetic field lines, this relation tells us that as many lines enter the enclosed volume as leave it. If, indeed, magnetic monopoles do not exist, then there are no "sources" or "sinks" for magnetic field lines to start or stop on, corresponding to electric field lines starting on positive charges and ending on negative charges. Magnetic field lines must then be continuous. Even for a bar magnet, a magnetic field $\vec{\mathbf{B}}$ exists inside as well as outside the magnetic material, and the lines of $\vec{\mathbf{B}}$ are closed loops as shown in Fig. 31–5.

[†]The interpretation of the changing electric field as a current does fit in well with our discussion in Chapter 30 where we saw that an alternating current can be said to pass through a capacitor (although charge doesn't). It also means that Kirchhoff's junction rule will be valid even at a capacitor plate: conduction current flows into the plate, but no conduction current flows out of the plate—instead a "displacement current" flows out of one plate (toward the other plate).

Maxwell's Equations

With the extension of Ampère's law given by Eq. 31–1, plus Gauss's law for magnetism (Eq. 31–4), we are now ready to state all four of Maxwell's equations. We have seen them all before in the past ten Chapters. In the absence of dielectric or magnetic materials, **Maxwell's equations** are:

$$\oint \vec{E} \cdot d\vec{A} = \frac{Q}{\epsilon_0} \qquad (31\text{–}5a)$$

$$\oint \vec{B} \cdot d\vec{A} = 0 \qquad (31\text{–}5b)$$

$$\oint \vec{E} \cdot d\vec{\ell} = -\frac{d\Phi_B}{dt} \qquad (31\text{–}5c)$$

$$\oint \vec{B} \cdot d\vec{\ell} = \mu_0 I + \mu_0 \epsilon_0 \frac{d\Phi_E}{dt}. \qquad (31\text{–}5d)$$

MAXWELL'S EQUATIONS

The first two of Maxwell's equations are the same as Gauss's law for electricity (Chapter 22, Eq. 22–4) and Gauss's law for magnetism (Section 31–2, Eq. 31–4). The third is Faraday's law (Chapter 29, Eq. 29–8) and the fourth is Ampère's law as modified by Maxwell (Eq. 31–1). (We dropped the subscripts on Q_{encl} and I_{encl} for simplicity.)

They can be summarized in words: (1) a generalized form of Coulomb's law relating electric field to its sources, electric charges; (2) the same for the magnetic field, except that if there are no magnetic monopoles, magnetic field lines are continuous—they do not begin or end (as electric field lines do on charges); (3) an electric field is produced by a changing magnetic field; (4) a magnetic field is produced by an electric current or by a changing electric field.

Maxwell's equations are the basic equations for all electromagnetism, and are as fundamental as Newton's three laws of motion and the law of universal gravitation. Maxwell's equations can also be written in differential form; see Appendix E.

In earlier Chapters, we have seen that we can treat electric and magnetic fields separately if they do not vary in time. But we cannot treat them independently if they do change in time. For a changing magnetic field produces an electric field; and a changing electric field produces a magnetic field. An important outcome of these relations is the production of electromagnetic waves.

31–4 Production of Electromagnetic Waves

A magnetic field will be produced in empty space if there is a changing electric field. A changing magnetic field produces an electric field that is itself changing. This changing electric field will, in turn, produce a magnetic field, which will be changing, and so it too will produce a changing electric field; and so on. Maxwell found that the net result of these interacting changing fields was a *wave* of electric and magnetic fields that can propagate (travel) through space! We now examine, in a simplified way, how such **electromagnetic waves** can be produced.

Consider two conducting rods that will serve as an "antenna" (Fig. 31–6a). Suppose these two rods are connected by a switch to the opposite terminals of a battery. When the switch is closed, the upper rod quickly becomes positively charged and the lower one negatively charged. Electric field lines are formed as indicated by the lines in Fig. 31–6b. While the charges are flowing, a current exists whose direction is indicated by the black arrows. A magnetic field is therefore produced near the antenna. The magnetic field lines encircle the rod-like antenna and therefore, in Fig. 31–6, $\vec{B}$ points into the page ($\otimes$) on the right and out of the page ($\odot$) on the left. How far out do these electric and magnetic fields extend? In the static case, the fields extend outward indefinitely far. However, when the switch in Fig. 31–6 is closed, the fields quickly appear nearby, but it takes time for them to reach distant points. Both electric and magnetic fields store energy, and this energy cannot be transferred to distant points at infinite speed.

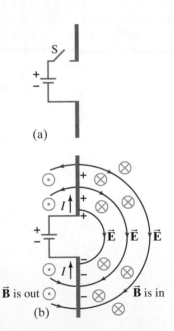

(a)

(b)

$\vec{B}$ is out $\odot$ $\vec{B}$ is in

FIGURE 31–6 Fields produced by charge flowing into conductors. It takes time for the $\vec{E}$ and $\vec{B}$ fields to travel outward to distant points. The fields are shown to the right of the antenna, but they move out in all directions, symmetrically about the (vertical) antenna.

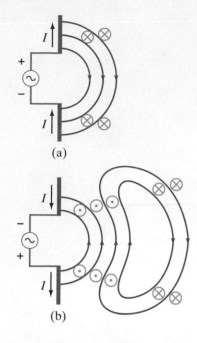

(a)

(b)

FIGURE 31–7 Sequence showing electric and magnetic fields that spread outward from oscillating charges on two conductors (the antenna) connected to an ac source (see the text).

FIGURE 31–8 (a) The radiation fields (far from the antenna) produced by a sinusoidal signal on the antenna. The red closed loops represent electric field lines. The magnetic field lines, perpendicular to the page and represented by blue ⊗ and ⊙, also form closed loops. (b) Very far from the antenna the wave fronts (field lines) are essentially flat over a fairly large area, and are referred to as *plane waves*.

Now we look at the situation of Fig. 31–7 where our antenna is connected to an ac generator. In Fig. 31–7a, the connection has just been completed. Charge starts building up and fields form just as in Fig. 31–6. The + and − signs in Fig. 31–7a indicate the net charge on each rod at a given instant. The black arrows indicate the direction of the current. The electric field is represented by the red lines in the plane of the page; and the magnetic field, according to the right-hand rule, is into (⊗) or out of (⊙) the page, in blue. In Fig. 31–7b, the voltage of the ac generator has reversed in direction; the current is reversed and the new magnetic field is in the opposite direction. Because the new fields have changed direction, the old lines fold back to connect up to some of the new lines and form closed loops as shown.[†] The old fields, however, don't suddenly disappear; they are on their way to distant points. Indeed, because a changing magnetic field produces an electric field, and a changing electric field produces a magnetic field, this combination of changing electric and magnetic fields moving outward is self-supporting, no longer depending on the antenna charges.

The fields not far from the antenna, referred to as the *near field*, become quite complicated, but we are not so interested in them. We are instead mainly interested in the fields far from the antenna (they are generally what we detect), which we refer to as the **radiation field**, or *far field*. The electric field lines form loops, as shown in Fig. 31–8, and continue moving outward. The magnetic field lines also form closed loops, but are not shown since they are perpendicular to the page. Although the lines are shown only on the right of the source, fields also travel in other directions. The field strengths are greatest in directions perpendicular to the oscillating charges; and they drop to zero along the direction of oscillation—above and below the antenna in Fig. 31–8.

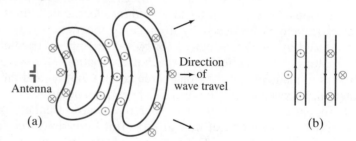

Antenna

Direction of wave travel

(a)

(b)

The magnitudes of both $\vec{E}$ and $\vec{B}$ in the radiation field are found to decrease with distance as $1/r$. (Compare this to the static electric field given by Coulomb's law where $\vec{E}$ decreases as $1/r^2$.) The energy carried by the electromagnetic wave is proportional (as for any wave, Chapter 15) to the square of the amplitude, E^2 or B^2, as will be discussed further in Section 31–8. So the intensity of the wave decreases as $1/r^2$, which we call the **inverse square law**.

Several things about the radiation field can be noted from Fig. 31–8. First, *the electric and magnetic fields at any point are perpendicular to each other, and to the direction of wave travel*. Second, we can see that the fields alternate in direction ($\vec{B}$ is into the page at some points and out of the page at others; $\vec{E}$ points up at some points and down at others). Thus, the field strengths vary from a maximum in one direction, to zero, to a maximum in the other direction. The electric and magnetic fields are "in phase": that is, they each are zero at the same points and reach their maxima at the same points in space. Finally, very far from the antenna (Fig. 31–8b) the field lines are quite flat over a reasonably large area, and the waves are referred to as **plane waves**.

If the source voltage varies sinusoidally, then the electric and magnetic field strengths in the radiation field will also vary sinusoidally. The sinusoidal character of the waves is diagrammed in Fig. 31–9, which shows the field directions and magnitudes plotted as a function of position. Notice that $\vec{B}$ and $\vec{E}$ are perpendicular to each other and to the direction of travel (= the direction of the wave velocity $\vec{v}$). The direction of $\vec{v}$ can be had from a right-hand rule using $\vec{E} \times \vec{B}$: fingers along $\vec{E}$, then along $\vec{B}$, gives $\vec{v}$ along thumb.

[†]We are considering waves traveling through empty space. There are no charges for lines of $\vec{E}$ to start or stop on, so they form closed loops. Magnetic field lines always form closed loops.

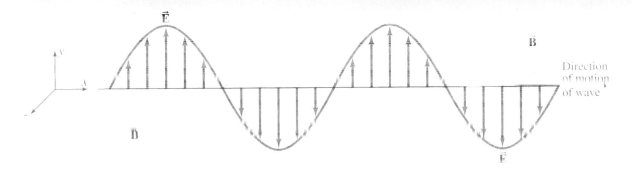

We call these waves electromagnetic (EM) waves. They are *transverse* waves because the amplitude is perpendicular to the direction of wave travel. However, EM waves are always waves of *fields*, not of matter (like waves on water or a rope). Because they are fields, EM waves can propagate in empty space.

As we have seen, EM waves are produced by electric charges that are oscillating, and hence are undergoing acceleration. In fact, we can say in general that

accelerating electric charges give rise to electromagnetic waves.

Electromagnetic waves can be produced in other ways as well, requiring description at the atomic and nuclear levels, as we will discuss later.

EXERCISE A At a particular instant in time, a wave has its electric field pointing north and its magnetic field pointing up. In which direction is the wave traveling? (*a*) South, (*b*) west, (*c*) east, (*d*) down, (*e*) not enough information.

31–5 Electromagnetic Waves, and Their Speed, Derived from Maxwell's Equations

Let us now examine how the existence of EM waves follows from Maxwell's equations. We will see that Maxwell's prediction of the existence of EM waves was startling. Equally startling was the speed at which they were predicted to travel.

We begin by considering a region of free space, where there are *no charges or conduction currents*—that is, far from the source so that the wave fronts (the field lines in Fig. 31–8) are essentially flat over a reasonable area. We call them **plane waves**, as we saw, because at any instant $\vec{E}$ and $\vec{B}$ are uniform over a reasonably large plane perpendicular to the direction of propagation. We choose a coordinate system, so that the wave is traveling in the x direction with velocity $\vec{v} = v\hat{i}$, with $\vec{E}$ parallel to the y axis and $\vec{B}$ parallel to the z axis, as in Fig. 31–9.

Maxwell's equations, with $Q = I = 0$, become

$$\oint \vec{E} \cdot d\vec{A} = 0 \tag{31–6a}$$

$$\oint \vec{B} \cdot d\vec{A} = 0 \tag{31–6b}$$

$$\oint \vec{E} \cdot d\vec{\ell} = -\frac{d\Phi_B}{dt} \tag{31–6c}$$

$$\oint \vec{B} \cdot d\vec{\ell} = \mu_0 \epsilon_0 \frac{d\Phi_E}{dt}. \tag{31–6d}$$

Notice the beautiful *symmetry* of these equations. The term on the right in the last equation, conceived by Maxwell, is essential for this symmetry. It is also essential if electromagnetic waves are to be produced, as we will now see.

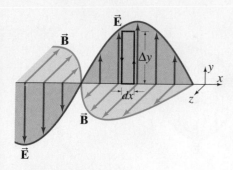

FIGURE 31–10 Applying Faraday's law to the rectangle $(\Delta y)(dx)$.

If the wave is sinusoidal with wavelength λ and frequency f, then, as we saw in Chapter 15, Section 15–4, such a traveling wave can be written as

$$E = E_y = E_0 \sin(kx - \omega t)$$
$$B = B_z = B_0 \sin(kx - \omega t) \qquad \textbf{(31–7)}$$

where

$$k = \frac{2\pi}{\lambda}, \qquad \omega = 2\pi f, \qquad \text{and} \qquad f\lambda = \frac{\omega}{k} = v, \qquad \textbf{(31–8)}$$

with v being the speed of the wave. Although visualizing the wave as sinusoidal is helpful, we will not have to assume so in most of what follows.

Consider now a small rectangle in the plane of the electric field as shown in Fig. 31–10. This rectangle has a finite height Δy, and a very thin width which we take to be the infinitesimal distance dx. First we show that $\vec{\mathbf{E}}$, $\vec{\mathbf{B}}$, and $\vec{\mathbf{v}}$ are in the orientation shown by applying Lenz's law to this rectangular loop. The changing magnetic flux through this loop is related to the electric field around the loop by Faraday's law (Maxwell's third equation, Eq. 31–6c). For the case shown, B through the loop is decreasing in time (the wave is moving to the right). So the electric field must be in a direction to oppose this change, meaning E must be greater on the right side of the loop than on the left, as shown (so it could produce a counterclockwise current whose magnetic field would act to oppose the change in Φ_B—but of course there is no current). This brief argument shows that the orientation of $\vec{\mathbf{E}}$, $\vec{\mathbf{B}}$, and $\vec{\mathbf{v}}$ are in the correct relation as shown. That is, $\vec{\mathbf{v}}$ is in the direction of $\vec{\mathbf{E}} \times \vec{\mathbf{B}}$. Now let us apply Faraday's law, which is Maxwell's third equation (Eq. 31–6c),

$$\oint \vec{\mathbf{E}} \cdot d\vec{\boldsymbol{\ell}} = -\frac{d\Phi_B}{dt}$$

to the rectangle of height Δy and width dx shown in Fig. 31–10. First we consider $\oint \vec{\mathbf{E}} \cdot d\vec{\boldsymbol{\ell}}$. Along the short top and bottom sections of length dx, $\vec{\mathbf{E}}$ is perpendicular to $d\vec{\boldsymbol{\ell}}$, so $\vec{\mathbf{E}} \cdot d\vec{\boldsymbol{\ell}} = 0$. Along the vertical sides, we let E be the electric field along the left side, and on the right side where it will be slightly larger, it is $E + dE$. Thus, if we take our loop counterclockwise,

$$\oint \vec{\mathbf{E}} \cdot d\vec{\boldsymbol{\ell}} = (E + dE)\,\Delta y - E\,\Delta y = dE\,\Delta y.$$

For the right side of Faraday's law, the magnetic flux through the loop changes as

$$\frac{d\Phi_B}{dt} = \frac{dB}{dt}\,dx\,\Delta y,$$

since the area of the loop, $(dx)(\Delta y)$, is not changing. Thus, Faraday's law gives us

$$dE\,\Delta y = -\frac{dB}{dt}\,dx\,\Delta y$$

or

$$\frac{dE}{dx} = -\frac{dB}{dt}.$$

Actually, both E and B are functions of position x and time t. We should therefore use partial derivatives:

$$\frac{\partial E}{\partial x} = -\frac{\partial B}{\partial t} \qquad \textbf{(31–9)}$$

where $\partial E/\partial x$ means the derivative of E with respect to x while t is held fixed, and $\partial B/\partial t$ is the derivative of B with respect to t while x is kept fixed.

We can obtain another important relation between E and B in addition to Eq. 31–9. To do so, we consider now a small rectangle in the plane of $\vec{\mathbf{B}}$, whose length and width are Δz and dx as shown in Fig. 31–11. To this rectangular loop we apply Maxwell's fourth equation (the extension of Ampère's law), Eq. 31–6d:

$$\oint \vec{\mathbf{B}} \cdot d\vec{\boldsymbol{\ell}} = \mu_0 \epsilon_0 \frac{d\Phi_E}{dt}$$

where we have taken $I = 0$ since we assume the absence of conduction currents.

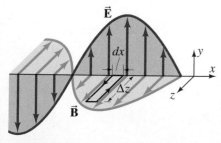

FIGURE 31–11 Applying Maxwell's fourth equation to the rectangle $(\Delta z)(dx)$.

Along the short sides (dx), $\vec{B} \cdot d\vec{\ell}$ is zero since $\vec{B}$ is perpendicular to $d\vec{\ell}$. Along the longer sides (Δz), we let B be the magnetic field along the left side of length Δz, and $B + dB$ be the field along the right side. We again integrate counterclockwise, so

$$\oint \vec{B} \cdot d\vec{\ell} = B\,\Delta z - (B + dB)\,\Delta z = -dB\,\Delta z.$$

The right side of Maxwell's fourth equation is

$$\mu_0 \epsilon_0 \frac{d\Phi_E}{dt} = \mu_0 \epsilon_0 \frac{dE}{dt}\,dx\,\Delta z.$$

Equating the two expressions, we obtain

$$-dB\,\Delta z = \mu_0 \epsilon_0 \frac{dE}{dt}\,dx\,\Delta z$$

or

$$\frac{\partial B}{\partial x} = -\mu_0 \epsilon_0 \frac{\partial E}{\partial t} \tag{31-10}$$

where we have replaced dB/dx and dE/dt by the proper partial derivatives as before.

We can use Eqs. 31–9 and 31–10 to obtain a relation between the magnitudes of $\vec{E}$ and $\vec{B}$, and the speed v. Let E and B be given by Eqs. 31–7 as a function of x and t. When we apply Eq. 31–9, taking the derivatives of E and B as given by Eqs. 31–7, we obtain

$$kE_0 \cos(kx - \omega t) = \omega B_0 \cos(kx - \omega t)$$

or

$$\frac{E_0}{B_0} = \frac{\omega}{k} = v,$$

since $v = \omega/k$ (see Eq. 31–8 or 15–12). Since E and B are in phase, we see that E and B are related by

$$\frac{E}{B} = v \tag{31-11}$$

at any point in space, where v is the velocity of the wave.

Now we apply Eq. 31–10 to the sinusoidal fields (Eqs. 31–7) and we obtain

$$kB_0 \cos(kx - \omega t) = \mu_0 \epsilon_0 \omega E_0 \cos(kx - \omega t)$$

or

$$\frac{B_0}{E_0} = \frac{\mu_0 \epsilon_0 \omega}{k} = \mu_0 \epsilon_0 v.$$

We just saw that $B_0/E_0 = 1/v$, so

$$\mu_0 \epsilon_0 v = \frac{1}{v}.$$

Solving for v we find

$$v = c = \frac{1}{\sqrt{\epsilon_0 \mu_0}}, \tag{31-12}$$

where c is the special symbol for the speed of electromagnetic waves in free space. We see that c is a constant, independent of the wavelength or frequency. If we put in values for ϵ_0 and μ_0 we find

$$c = \frac{1}{\sqrt{\epsilon_0 \mu_0}} = \frac{1}{\sqrt{(8.85 \times 10^{-12}\,\mathrm{C^2/N \cdot m^2})(4\pi \times 10^{-7}\,\mathrm{T \cdot m/A})}}$$

$$= 3.00 \times 10^8\,\mathrm{m/s}.$$

This is a remarkable result. For this is precisely equal to the measured speed of light!

EXAMPLE 31–2 Determining $\vec{E}$ and $\vec{B}$ in EM waves. Assume a 60.0-Hz EM wave is a sinusoidal wave propagating in the z direction with $\vec{E}$ pointing in the x direction, and $E_0 = 2.00\,\text{V/m}$. Write vector expressions for $\vec{E}$ and $\vec{B}$ as functions of position and time.

APPROACH We find λ from $\lambda f = v = c$. Then we use Fig. 31–9 and Eqs. 31–7 and 31–8 for the mathematical form of traveling electric and magnetic fields of an EM wave.

SOLUTION The wavelength is

$$\lambda = \frac{c}{f} = \frac{3.00 \times 10^8\,\text{m/s}}{60.0\,\text{s}^{-1}} = 5.00 \times 10^6\,\text{m}.$$

From Eq. 31–8 we have

$$k = \frac{2\pi}{\lambda} = \frac{2\pi}{5.00 \times 10^6\,\text{m}} = 1.26 \times 10^{-6}\,\text{m}^{-1}$$

$$\omega = 2\pi f = 2\pi(60.0\,\text{Hz}) = 377\,\text{rad/s}.$$

From Eq. 31–11 with $v = c$, we find that

$$B_0 = \frac{E_0}{c} = \frac{2.00\,\text{V/m}}{3.00 \times 10^8\,\text{m/s}} = 6.67 \times 10^{-9}\,\text{T}.$$

The direction of propagation is that of $\vec{E} \times \vec{B}$, as in Fig. 31–9. With $\vec{E}$ pointing in the x direction, and the wave propagating in the z direction, $\vec{B}$ must point in the y direction. Using Eqs. 31–7 we find:

$$\vec{E} = \hat{i}(2.00\,\text{V/m}) \sin\big[(1.26 \times 10^{-6}\,\text{m}^{-1})z - (377\,\text{rad/s})t\big]$$

$$\vec{B} = \hat{j}(6.67 \times 10^{-9}\,\text{T}) \sin\big[(1.26 \times 10^{-6}\,\text{m}^{-1})z - (377\,\text{rad/s})t\big]$$

*Derivation of Speed of Light (General)

We can derive the speed of EM waves without having to assume sinusoidal waves by combining Eqs. 31–9 and 31–10 as follows. We take the derivative, with respect to t of Eq. 31–10

$$\frac{\partial^2 B}{\partial t\,\partial x} = -\mu_0 \epsilon_0 \frac{\partial^2 E}{\partial t^2}.$$

We next take the derivative of Eq. 31–9 with respect to x:

$$\frac{\partial^2 E}{\partial x^2} = -\frac{\partial^2 B}{\partial t\,\partial x}.$$

Since $\partial^2 B/\partial t\,\partial x$ appears in both relations, we obtain

$$\frac{\partial^2 E}{\partial t^2} = \frac{1}{\mu_0 \epsilon_0} \frac{\partial^2 E}{\partial x^2}. \tag{31–13a}$$

By taking other derivatives of Eqs. 31–9 and 31–10 we obtain the same relation for B:

$$\frac{\partial^2 B}{\partial t^2} = \frac{1}{\mu_0 \epsilon_0} \frac{\partial^2 B}{\partial x^2}. \tag{31–13b}$$

Both of Eqs. 31–13 have the form of the **wave equation** for a plane wave traveling in the x direction, as discussed in Section 15–5 (Eq. 15–16):

$$\frac{\partial^2 D}{\partial t^2} = v^2 \frac{\partial^2 D}{\partial x^2},$$

where D stands for any type of displacement. We see that the velocity v for Eqs. 31–13 is given by

$$v^2 = \frac{1}{\mu_0 \epsilon_0}$$

in agreement with Eq. 31–12. Thus we see that a natural outcome of Maxwell's equations is that E and B obey the wave equation for waves traveling with speed $v = 1/\sqrt{\mu_0 \epsilon_0}$. It was on this basis that Maxwell predicted the existence of electromagnetic waves and predicted their speed.

31–6 Light as an Electromagnetic Wave and the Electromagnetic Spectrum

The calculations in Section 31–5 gave the result that Maxwell himself determined: that the speed of EM waves in empty space is given by

$$c = \frac{E}{B} = \frac{1}{\sqrt{\epsilon_0 \mu_0}} - 3.00 \times 10^8 \,\text{m/s},$$

the same as the measured speed of light in vacuum.

Light had been shown some 60 years previously to behave like a wave (we'll discuss this in Chapter 34). But nobody knew what kind of wave it was. What is it that is oscillating in a light wave? Maxwell, on the basis of the calculated speed of EM waves, argued that light must be an electromagnetic wave. This idea soon came to be generally accepted by scientists, but not fully until after EM waves were experimentally detected. EM waves were first generated and detected experimentally by Heinrich Hertz (1857–1894) in 1887, eight years after Maxwell's death. Hertz used a spark-gap apparatus in which charge was made to rush back and forth for a short time, generating waves whose frequency was about 10^9 Hz. He detected them some distance away using a loop of wire in which an emf was produced when a changing magnetic field passed through. These waves were later shown to travel at the speed of light, 3.00×10^8 m/s, and to exhibit all the characteristics of light such as reflection, refraction, and interference. The only difference was that they were not visible. Hertz's experiment was a strong confirmation of Maxwell's theory.

The wavelengths of visible light were measured in the first decade of the nineteenth century, long before anyone imagined that light was an electromagnetic wave. The wavelengths were found to lie between 4.0×10^{-7} m and 7.5×10^{-7} m, or 400 nm to 750 nm ($1\,\text{nm} = 10^{-9}$ m). The frequencies of visible light can be found using Eq. 15–1 or 31–8, which we rewrite here:

$$c = \lambda f, \tag{31–14}$$

where f and λ are the frequency and wavelength, respectively, of the wave. Here, c is the speed of light, 3.00×10^8 m/s; it gets the special symbol c because of its universality for all EM waves in free space. Equation 31–14 tells us that the frequencies of visible light are between 4.0×10^{14} Hz and 7.5×10^{14} Hz. (Recall that $1\,\text{Hz} = 1$ cycle per second $= 1\,\text{s}^{-1}$.)

But visible light is only one kind of EM wave. As we have seen, Hertz produced EM waves of much lower frequency, about 10^9 Hz. These are now called **radio waves**, because frequencies in this range are used to transmit radio and TV signals. Electromagnetic waves, or EM radiation as we sometimes call it, have been produced or detected over a wide range of frequencies. They are usually categorized as shown in Fig. 31–12, which is known as the **electromagnetic spectrum**.

FIGURE 31–12
Electromagnetic spectrum.

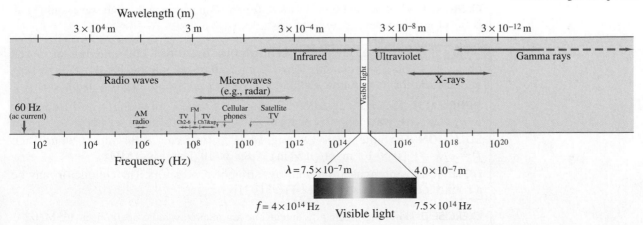

Radio waves and microwaves can be produced in the laboratory using electronic equipment (Fig. 31–7). Higher-frequency waves are very difficult to produce electronically. These and other types of EM waves are produced in natural processes, as emission from atoms, molecules, and nuclei (more on this later). EM waves can be produced by the acceleration of electrons or other charged particles, such as electrons in the antenna of Fig. 31–7. Another example is X-rays, which are produced (Chapter 35) when fast-moving electrons are rapidly decelerated upon striking a metal target. Even the visible light emitted by an ordinary incandescent light is due to electrons undergoing acceleration within the hot filament.

We will meet various types of EM waves later. However, it is worth mentioning here that infrared (IR) radiation (EM waves whose frequency is just less than that of visible light) is mainly responsible for the heating effect of the Sun. The Sun emits not only visible light but substantial amounts of IR and UV (ultraviolet) as well. The molecules of our skin tend to "resonate" at infrared frequencies, so it is these that are preferentially absorbed and thus warm us up. We humans experience EM waves differently, depending on their wavelengths: Our eyes detect wavelengths between about 4×10^{-7} m and 7.5×10^{-7} m (visible light), whereas our skin detects longer wavelengths (IR). Many EM wavelengths we don't detect directly at all.

EXERCISE B Return to the Chapter-Opening Question, page 812, and answer it again now. Try to explain why you may have answered differently the first time.

CAUTION

Sound and EM waves are different

Light and other electromagnetic waves travel at a speed of 3×10^8 m/s. Compare this to sound, which travels (see Chapter 16) at a speed of about 300 m/s in air, a million times slower; or to typical freeway speeds of a car, 30 m/s (100 km/h, or 60 mi/h), 10 million times slower than light. EM waves differ from sound waves in another big way: sound waves travel in a medium such as air, and involve motion of air molecules; EM waves do not involve any material—only fields, and they can travel in empty space.

EXAMPLE 31–3 **Wavelengths of EM waves.** Calculate the wavelength (a) of a 60-Hz EM wave, (b) of a 93.3-MHz FM radio wave, and (c) of a beam of visible red light from a laser at frequency 4.74×10^{14} Hz.

APPROACH All of these waves are electromagnetic waves, so their speed is $c = 3.00 \times 10^8$ m/s. We solve for λ in Eq. 31–14: $\lambda = c/f$.

SOLUTION (a) $\lambda = \dfrac{c}{f} = \dfrac{3.00 \times 10^8 \text{ m/s}}{60 \text{ s}^{-1}} = 5.0 \times 10^6$ m,

or 5000 km. 60 Hz is the frequency of ac current in the United States, and, as we see here, one wavelength stretches all the way across the continental USA.

(b) $\lambda = \dfrac{3.00 \times 10^8 \text{ m/s}}{93.3 \times 10^6 \text{ s}^{-1}} = 3.22$ m.

The length of an FM antenna is about half this $\left(\tfrac{1}{2}\lambda\right)$, or $1\tfrac{1}{2}$ m.

(c) $\lambda = \dfrac{3.00 \times 10^8 \text{ m/s}}{4.74 \times 10^{14} \text{ s}^{-1}} = 6.33 \times 10^{-7}$ m ($= 633$ nm).

EXERCISE C What are the frequencies of (a) an 80-m-wavelength radio wave, and (b) an X-ray of wavelength 5.5×10^{-11} m?

EXAMPLE 31–4 **ESTIMATE** **Cell phone antenna.** The antenna of a cell phone is often $\tfrac{1}{4}$ wavelength long. A particular cell phone has an 8.5-cm-long straight rod for its antenna. Estimate the operating frequency of this phone.

APPROACH The basic equation relating wave speed, wavelength, and frequency is $c = \lambda f$; the wavelength λ equals four times the antenna's length.

SOLUTION The antenna is $\tfrac{1}{4}\lambda$ long, so $\lambda = 4(8.5 \text{ cm}) = 34 \text{ cm} = 0.34$ m. Then $f = c/\lambda = (3.0 \times 10^8 \text{ m/s})/(0.34 \text{ m}) = 8.8 \times 10^8$ Hz $= 880$ MHz.

NOTE Radio antennas are not always straight conductors. The conductor may be a round loop to save space. See Fig. 31–21b.

EXERCISE D How long should a $\tfrac{1}{4}\lambda$ antenna be for an aircraft radio operating at 165 MHz?

Electromagnetic waves can travel along transmission lines as well as in empty space. When a source of emf is connected to a transmission line—be it two parallel wires or a coaxial cable (Fig. 31–13)—the electric field within the wire is not set up immediately at all points along the wires. This is based on the same argument we used in Section 31–4 with reference to Fig. 31–7. Indeed, it can be shown that if the wires are separated by empty space or air, the electrical signal travels along the wires at the speed $c = 3.0 \times 10^8$ m/s. For example, when you flip a light switch, the light actually goes on a tiny fraction of a second later. If the wires are in a medium whose electric permittivity is ϵ and magnetic permeability is μ (Sections 24–5 and 28–9, respectively), the speed is not given by Eq. 31–12, but by

$$v = \frac{1}{\sqrt{\epsilon \mu}}.$$

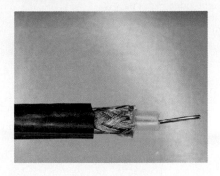

FIGURE 31–13 Coaxial cable.

EXAMPLE 31–5 ESTIMATE Phone call time lag. You make a telephone call from New York to a friend in London. Estimate how long it will take the electrical signal generated by your voice to reach London, assuming the signal is (a) carried on a telephone cable under the Atlantic Ocean, and (b) sent via satellite 36,000 km above the ocean. Would this cause a noticeable delay in either case?

APPROACH The signal is carried on a telephone wire or in the air via satellite. In either case it is an electromagnetic wave. Electronics as well as the wire or cable slow things down, but as a rough estimate we take the speed to be $c = 3.0 \times 10^8$ m/s.

SOLUTION The distance from New York to London is about 5000 km.
(a) The time delay via the cable is $t = d/c \approx (5 \times 10^6 \text{ m})/(3.0 \times 10^8 \text{ m/s}) = 0.017$ s.
(b) Via satellite the time would be longer because communications satellites, which are usually geosynchronous (Example 6–6), move at a height of 36,000 km. The signal would have to go up to the satellite and back down, or about 72,000 km. The actual distance the signal would travel would be a little more than this as the signal would go up and down on a diagonal. Thus $t = d/c \approx 7.2 \times 10^7 \text{ m}/(3 \times 10^8 \text{ m/s}) = 0.24$ s.

NOTE When the signal travels via the underwater cable, there is only a hint of a delay and conversations are fairly normal. When the signal is sent via satellite, the delay *is* noticeable. The length of time between the end of when you speak and your friend receives it and replies, and then you hear the reply, is about a half second beyond the normal time in a conversation. This is enough to be noticeable, and you have to adjust for it so you don't start talking again while your friend's reply is on the way back to you.

EXERCISE E If you are on the phone via satellite to someone only 100 km away, would you hear the same effect?

EXERCISE F If your voice traveled as a sound wave, how long would it take to go from New York to London?

31–7 Measuring the Speed of Light

Galileo attempted to measure the speed of light by trying to measure the time required for light to travel a known distance between two hilltops. He stationed an assistant on one hilltop and himself on another, and ordered the assistant to lift the cover from a lamp the instant he saw a flash from Galileo's lamp. Galileo measured the time between the flash of his lamp and when he received the light from his assistant's lamp. The time was so short that Galileo concluded it merely represented human reaction time, and that the speed of light must be extremely high.

The first successful determination that the speed of light is finite was made by the Danish astronomer Ole Roemer (1644–1710). Roemer had noted that the carefully measured orbital period of Io, a moon of Jupiter with an average period of 42.5 h, varied slightly, depending on the relative position of Earth and Jupiter. He attributed this variation in the apparent period to the change in distance between the Earth and Jupiter during one of Io's periods, and the time it took light to travel the extra distance. Roemer concluded that the speed of light—though great—is finite.

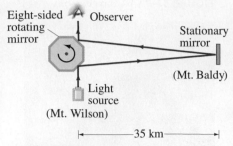

Eight-sided rotating mirror

Observer

Stationary mirror

(Mt. Baldy)

Light source

(Mt. Wilson)

|←———35 km———→|

FIGURE 31–14 Michelson's speed-of-light apparatus (not to scale).

Since then a number of techniques have been used to measure the speed of light. Among the most important were those carried out by the American Albert A. Michelson (1852–1931). Michelson used the rotating mirror apparatus diagrammed in Fig. 31–14 for a series of high-precision experiments carried out from 1880 to the 1920s. Light from a source would hit one face of a rotating eight-sided mirror. The reflected light traveled to a stationary mirror a large distance away and back again as shown. If the rotating mirror was turning at just the right rate, the returning beam of light would reflect from one of the eight mirrors into a small telescope through which the observer looked. If the speed of rotation was only slightly different, the beam would be deflected to one side and would not be seen by the observer. From the required speed of the rotating mirror and the known distance to the stationary mirror, the speed of light could be calculated. In the 1920s, Michelson set up the rotating mirror on the top of Mt. Wilson in southern California and the stationary mirror on Mt. Baldy (Mt. San Antonio) 35 km away. He later measured the speed of light in vacuum using a long evacuated tube.

Today the speed of light, c, in vacuum is taken as

$$c = 2.99792458 \times 10^8 \text{ m/s},$$

and is defined to be this value. This means that the standard for length, the meter, is no longer defined separately. Instead, as we noted in Section 1–4, the meter is now formally defined as the distance light travels in vacuum in 1/299,792,458 of a second. We usually round off c to

$$c = 3.00 \times 10^8 \text{ m/s}$$

when extremely precise results are not required. In air, the speed is only slightly less.

31–8 Energy in EM Waves; the Poynting Vector

Electromagnetic waves carry energy from one region of space to another. This energy is associated with the moving electric and magnetic fields. In Section 24–4, we saw that the energy density u_E (J/m^3) stored in an electric field E is $u_E = \frac{1}{2}\epsilon_0 E^2$ (Eq. 24–6). The energy density stored in a magnetic field B, as we discussed in Section 30–3, is given by $u_B = \frac{1}{2}B^2/\mu_0$ (Eq. 30–7). Thus, the total energy stored per unit volume in a region of space where there is an electromagnetic wave is

$$u = u_E + u_B = \frac{1}{2}\epsilon_0 E^2 + \frac{1}{2}\frac{B^2}{\mu_0}. \qquad \textbf{(31–15)}$$

In this equation, E and B represent the electric and magnetic field strengths of the wave at any instant in a small region of space. We can write Eq. 31–15 in terms of the E field alone, using Eqs. 31–11 $(B = E/c)$ and 31–12 $(c = 1/\sqrt{\epsilon_0\mu_0})$ to obtain

$$u = \frac{1}{2}\epsilon_0 E^2 + \frac{1}{2}\frac{\epsilon_0\mu_0 E^2}{\mu_0} = \epsilon_0 E^2. \qquad \textbf{(31–16a)}$$

Note here that the energy density associated with the B field equals that due to the E field, and each contributes half to the total energy. We can also write the energy density in terms of the B field only:

$$u = \epsilon_0 E^2 = \epsilon_0 c^2 B^2 = \frac{B^2}{\mu_0}, \qquad \textbf{(31–16b)}$$

or in one term containing both E and B,

$$u = \epsilon_0 E^2 = \epsilon_0 EcB = \frac{\epsilon_0 EB}{\sqrt{\epsilon_0\mu_0}} = \sqrt{\frac{\epsilon_0}{\mu_0}}\, EB. \qquad \textbf{(31–16c)}$$

Equations 31–16 give the energy density in any region of space at any instant.

Now let us determine the energy the wave transports per unit time per unit area. This is given by a vector $\vec{S}$, which is called the **Poynting vector**.[†] The units of $\vec{S}$ are W/m^2. The direction of $\vec{S}$ is the direction in which the energy is transported, which is the direction in which the wave is moving.

[†]After J. H. Poynting (1852–1914).

Let us imagine the wave is passing through an area A perpendicular to the x axis as shown in Fig. 31–15. In a short time dt, the wave moves to the right a distance $dx = c\,dt$ where c is the wave speed. The energy that passes through A in the time dt is the energy that occupies the volume $dV = A\,dx = Ac\,dt$. The energy density u is $u = \epsilon_0 E^2$ where E is the electric field in this volume at the given instant. So the total energy dU contained in this volume dV is the energy density u times the volume: $dU = u\,dV = (\epsilon_0 E^2)(Ac\,dt)$. Therefore the energy crossing the area A per time dt is

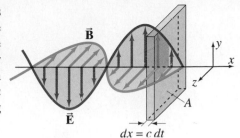

FIGURE 31–15 Electromagnetic wave carrying energy through area A.

$$S = \frac{1}{A}\frac{dU}{dt} = \epsilon_0 cE^2. \tag{31–17}$$

Since $E = cB$ and $c = 1/\sqrt{\epsilon_0 \mu_0}$, this can also be written:

$$S = \epsilon_0 cE^2 = \frac{cB^2}{\mu_0} = \frac{EB}{\mu_0}.$$

The direction of $\vec{S}$ is along $\vec{v}$, perpendicular to $\vec{E}$ and $\vec{B}$, so the Poynting vector $\vec{S}$ can be written

$$\vec{S} = \frac{1}{\mu_0}(\vec{E} \times \vec{B}). \tag{31–18}$$

Equation 31–17 or 31–18 gives the energy transported per unit area per unit time at any *instant*. We often want to know the *average* over an extended period of time since the frequencies are usually so high we don't detect the rapid time variation. If E and B are sinusoidal, then $\overline{E^2} = E_0^2/2$, just as for electric currents and voltages (Section 25–7), where E_0 is the *maximum* value of E. Thus we can write for the magnitude of the Poynting vector, on the average,

$$\overline{S} = \frac{1}{2}\epsilon_0 cE_0^2 = \frac{1}{2}\frac{c}{\mu_0}B_0^2 = \frac{E_0 B_0}{2\mu_0}, \tag{31–19a}$$

where B_0 is the maximum value of B. This time averaged value of $\vec{S}$ is the **intensity**, defined as the average power transferred across unit area (Section 15–3). We can also write for the average value of S:

$$\overline{S} = \frac{E_{rms}B_{rms}}{\mu_0} \tag{31–19b}$$

where E_{rms} and B_{rms} are the rms values $\left(E_{rms} = \sqrt{\overline{E^2}},\ B_{rms} = \sqrt{\overline{B^2}}\right)$.

EXAMPLE 31–6 **E and B from the Sun.** Radiation from the Sun reaches the Earth (above the atmosphere) at a rate of about $1350\,\text{J/s}\cdot\text{m}^2\ (= 1350\,\text{W/m}^2)$. Assume that this is a single EM wave, and calculate the maximum values of E and B.

APPROACH We solve Eq. 31–19a $\left(\overline{S} = \frac{1}{2}\epsilon_0 cE_0^2\right)$ for E_0 in terms of $\overline{S}$ using $\overline{S} = 1350\,\text{J/s}\cdot\text{m}^2$.

SOLUTION
$$E_0 = \sqrt{\frac{2\overline{S}}{\epsilon_0 c}} = \sqrt{\frac{2(1350\,\text{J/s}\cdot\text{m}^2)}{(8.85 \times 10^{-12}\,\text{C}^2/\text{N}\cdot\text{m}^2)(3.00 \times 10^8\,\text{m/s})}}$$
$$= 1.01 \times 10^3\,\text{V/m}.$$

From Eq. 31–11, $B = E/c$, so

$$B_0 = \frac{E_0}{c} = \frac{1.01 \times 10^3\,\text{V/m}}{3.00 \times 10^8\,\text{m/s}} = 3.37 \times 10^{-6}\,\text{T}.$$

NOTE Although B has a small numerical value compared to E (because of the way the different units for E and B are defined), B contributes the same energy to the wave as E does, as we saw earlier (Eqs. 31–15 and 16).

31–9 Radiation Pressure

If electromagnetic waves carry energy, then we might expect them to also carry linear momentum. When an electromagnetic wave encounters the surface of an object, a force will be exerted on the surface as a result of the momentum transfer ($F = dp/dt$), just as when a moving object strikes a surface. The force per unit area exerted by the waves is called **radiation pressure**, and its existence was predicted by Maxwell. He showed that if a beam of EM radiation (light, for example) is completely absorbed by an object, then the momentum transferred is

$$\Delta p = \frac{\Delta U}{c} \qquad \begin{bmatrix} \text{radiation} \\ \text{fully} \\ \text{absorbed} \end{bmatrix} \quad \textbf{(31–20a)}$$

where ΔU is the energy absorbed by the object in a time Δt, and c is the speed of light.[†] If instead, the radiation is fully reflected (suppose the object is a mirror), then the momentum transferred is twice as great, just as when a ball bounces elastically off a surface:

$$\Delta p = \frac{2\,\Delta U}{c}. \qquad \begin{bmatrix} \text{radiation} \\ \text{fully} \\ \text{reflected} \end{bmatrix} \quad \textbf{(31–20b)}$$

If a surface absorbs some of the energy, and reflects some of it, then $\Delta p = a\,\Delta U/c$, where a is a factor between 1 and 2.

Using Newton's second law we can calculate the force and the pressure exerted by radiation on the object. The force F is given by

$$F = \frac{dp}{dt}.$$

The average rate that energy is delivered to the object is related to the Poynting vector by

$$\frac{dU}{dt} = \overline{S}A,$$

where A is the cross-sectional area of the object which intercepts the radiation. The radiation pressure P (assuming full absorption) is given by (see Eq. 31–20a)

$$P = \frac{F}{A} = \frac{1}{A}\frac{dp}{dt} = \frac{1}{Ac}\frac{dU}{dt} = \frac{\overline{S}}{c}. \qquad \begin{bmatrix} \text{fully} \\ \text{absorbed} \end{bmatrix} \quad \textbf{(31–21a)}$$

If the light is fully reflected, the pressure is twice as great (Eq. 31–20b):

$$P = \frac{2\overline{S}}{c}. \qquad \begin{bmatrix} \text{fully} \\ \text{reflected} \end{bmatrix} \quad \textbf{(31–21b)}$$

EXAMPLE 31–7 | **ESTIMATE** | **Solar pressure.** Radiation from the Sun that reaches the Earth's surface (after passing through the atmosphere) transports energy at a rate of about 1000 W/m^2. Estimate the pressure and force exerted by the Sun on your outstretched hand.

APPROACH The radiation is partially reflected and partially absorbed, so let us estimate simply $P = \overline{S}/c$.

SOLUTION $\quad P \approx \dfrac{\overline{S}}{c} = \dfrac{1000 \text{ W/m}^2}{3 \times 10^8 \text{ m/s}} \approx 3 \times 10^{-6} \text{ N/m}^2.$

An estimate of the area of your outstretched hand might be about 10 cm by 20 cm, so $A = 0.02 \text{ m}^2$. Then the force is

$$F = PA \approx (3 \times 10^{-6} \text{ N/m}^2)(0.02 \text{ m}^2) \approx 6 \times 10^{-8} \text{ N}.$$

NOTE These numbers are tiny. The force of gravity on your hand, for comparison, is maybe a half pound, or with $m = 0.2 \text{ kg}$, $mg \approx (0.2 \text{ kg})(9.8 \text{ m/s}^2) \approx 2 \text{ N}$. The radiation pressure on your hand is imperceptible compared to gravity.

[†]Very roughly, if we think of light as particles (and we do—see Chapter 37), the force that would be needed to bring such a particle moving at speed c to "rest" (i.e. absorption) is $F = \Delta p/\Delta t$. But F is also related to energy by Eq. 8–7, $F = \Delta U/\Delta x$, so $\Delta p = F\,\Delta t = \Delta U/(\Delta x/\Delta t) = \Delta U/c$ where we identify $(\Delta x/\Delta t)$ with the speed of light c.

EXAMPLE 31–8 **ESTIMATE** **A solar sail.** Proposals have been made to use the radiation pressure from the Sun to help propel spacecraft around the solar system. (*a*) About how much force would be applied on a 1 km × 1 km highly reflective sail, and (*b*) by how much would this increase the speed of a 5000-kg spacecraft in one year? (*c*) If the spacecraft started from rest, about how far would it travel in a year?

APPROACH Pressure P is force per unit area, so $F = PA$. We use the estimate of Example 31–7, doubling it for a reflecting surface $P = 2\overline{S}/c$. We find the acceleration from Newton's second law, and assume it is constant, and then find the speed from $v = v_0 + at$. The distance traveled is given by $x = \frac{1}{2}at^2$.

SOLUTION (*a*) Doubling the result of Example 31–7, the solar pressure is $2\overline{S}/c = 6 \times 10^{-6}\,\text{N/m}^2$. Then the force is $F \approx PA = (6 \times 10^{-6}\,\text{N/m}^2)(10^6\,\text{m}^2) \approx 6\,\text{N}$. (*b*) The acceleration is $a \approx F/m \approx (6\,\text{N})/(5000\,\text{kg}) \approx 1.2 \times 10^{-3}\,\text{m/s}^2$. The speed increase is $v - v_0 = at = (1.2 \times 10^{-3}\,\text{m/s}^2)(365\,\text{days})(24\,\text{hr/day})(3600\,\text{s/hr}) \approx 4 \times 10^4\,\text{m/s}$ ($\approx 150,000\,\text{km/h}$!). (*c*) Starting from rest, this acceleration would result in a distance of about $\frac{1}{2}at^2 \approx 6 \times 10^{11}\,\text{m}$ in a year, about four times the Sun-Earth distance. The starting point should be far from the Earth so the Earth's gravitational force is small compared to 6 N.

NOTE A large sail providing a small force over a long time can result in a lot of motion.

Although you cannot directly feel the effects of radiation pressure, the phenomenon is quite dramatic when applied to atoms irradiated by a finely focused laser beam. An atom has a mass on the order of 10^{-27} kg, and a laser beam can deliver energy at a rate of $1000\,\text{W/m}^2$. This is the same intensity used in Example 31–7, but here a radiation pressure of $10^{-6}\,\text{N/m}^2$ would be very significant on a molecule whose mass might be 10^{-23} to 10^{-26} kg. It is possible to move atoms and molecules around by steering them with a laser beam, in a device called "optical tweezers." Optical tweezers have some remarkable applications. They are of great interest to biologists, especially since optical tweezers can manipulate live microorganisms, and components within a cell, without damaging them. Optical tweezers have been used to measure the elastic properties of DNA by pulling each end of the molecule with such a laser "tweezers."

PHYSICS APPLIED
Optical tweezers

31–10 Radio and Television; Wireless Communication

PHYSICS APPLIED
Wireless transmission

Electromagnetic waves offer the possibility of transmitting information over long distances. Among the first to realize this and put it into practice was Guglielmo Marconi (1874–1937) who, in the 1890s, invented and developed wireless communication. With it, messages could be sent at the speed of light without the use of wires. The first signals were merely long and short pulses that could be translated into words by a code, such as the "dots" and "dashes" of the Morse code: they were digital wireless, believe it or not. In 1895 Marconi sent wireless signals a kilometer or two in Italy. By 1901 he had sent test signals 3000 km across the ocean from Newfoundland, Canada, to Cornwall, England. In 1903 he sent the first practical commercial messages from Cape Cod, Massachusetts, to England: the London *Times* printed news items sent from its New York correspondent. 1903 was also the year of the first powered airplane flight by the Wright brothers. The hallmarks of the modern age—wireless communication and flight—date from the same year. Our modern world of wireless communication, including radio, television, cordless phones, cell phones, Bluetooth, wi-fi, and satellite communication, are simply modern applications of Marconi's pioneering work.

The next decade saw the development of vacuum tubes. Out of this early work radio and television were born. We now discuss briefly (1) how radio and TV signals are transmitted, and (2) how they are received at home.

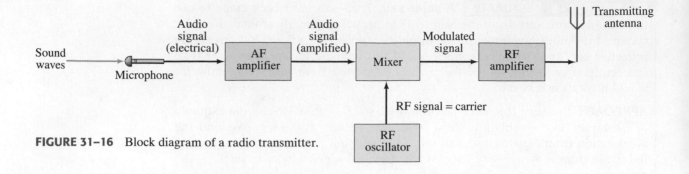

FIGURE 31–16 Block diagram of a radio transmitter.

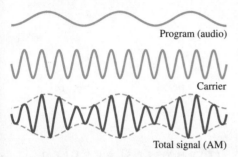

Program (audio)

Carrier

Total signal (AM)

FIGURE 31–17 In amplitude modulation (AM), the amplitude of the carrier signal is made to vary in proportion to the audio signal's amplitude.

AM and FM

FIGURE 31–18 In frequency modulation (FM), the frequency of the carrier signal is made to change in proportion to the audio signal's amplitude. This method is used by FM radio and television.

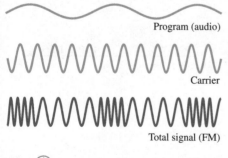

Program (audio)

Carrier

Total signal (FM)

PHYSICS APPLIED
Radio and TV receivers

The process by which a radio station transmits information (words and music) is outlined in Fig. 31–16. The audio (sound) information is changed into an electrical signal of the same frequencies by, say, a microphone or magnetic read/write head. This electrical signal is called an audiofrequency (AF) signal, since the frequencies are in the audio range (20 to 20,000 Hz). The signal is amplified electronically and is then mixed with a radio-frequency (RF) signal called its **carrier frequency**, which represents that station. AM radio stations have carrier frequencies from about 530 kHz to 1700 kHz. For example, "710 on your dial" means a station whose carrier frequency is 710 kHz. FM radio stations have much higher carrier frequencies, between 88 MHz and 108 MHz. The carrier frequencies for broadcast TV stations in the United States lie between 54 MHz and 72 MHz, between 76 MHz and 88 MHz, between 174 MHz and 216 MHz, and between 470 MHz and 698 MHz.

The mixing of the audio and carrier frequencies is done in two ways. In **amplitude modulation** (AM), the amplitude of the high-frequency carrier wave is made to vary in proportion to the amplitude of the audio signal, as shown in Fig. 31–17. It is called "amplitude modulation" because the *amplitude* of the carrier is altered ("modulate" means to change or alter). In **frequency modulation** (FM), the *frequency* of the carrier wave is made to change in proportion to the audio signal's amplitude, as shown in Fig. 31–18. The mixed signal is amplified further and sent to the transmitting antenna, where the complex mixture of frequencies is sent out in the form of EM waves. In digital communication, the signal is put into a digital form (Section 29–8) which modulates the carrier.

A television transmitter works in a similar way, using FM for audio and AM for video; both audio and video signals (see Section 23–9) are mixed with carrier frequencies.

Now let us look at the other end of the process, the reception of radio and TV programs at home. A simple radio receiver is diagrammed in Fig. 31–19. The EM waves sent out by all stations are received by the antenna. The signals the antenna detects and sends to the receiver are very small and contain frequencies from many different stations. The receiver selects out a particular RF frequency (actually a narrow range of frequencies) corresponding to a particular station using a resonant *LC* circuit (Sections 30–6 and 30–9).

FIGURE 31–19 Block diagram of a simple radio receiver.

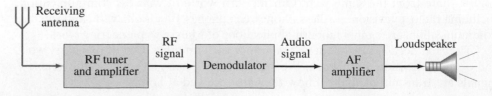

A simple way of tuning a station is shown in Fig. 31–20. A particular station is "tuned in" by adjusting C and/or L so that the resonant frequency of the circuit equals that of the station's carrier frequency. The signal, containing both audio and carrier frequencies, next goes to the *demodulator*, or *detector* (Fig. 31–19), where "demodulation" takes place—that is, the RF carrier frequency is separated from the audio signal. The audio signal is amplified and sent to a loudspeaker or headphones.

Modern receivers have more stages than those shown. Various means are used to increase the sensitivity and selectivity (ability to detect weak signals and distinguish them from other stations), and to minimize distortion of the original signal.[†]

A television receiver does similar things to both the audio and the video signals. The audio signal goes finally to the loudspeaker, and the video signal to the monitor, such as a *cathode ray tube* (CRT) or LCD screen (Sections 23–9 and 35–12).

One kind of antenna consists of one or more conducting rods; the electric field in the EM waves exerts a force on the electrons in the conductor, causing them to move back and forth at the frequencies of the waves (Fig. 31–21a). A second type of antenna consists of a tubular coil of wire which detects the magnetic field of the wave: the changing B field induces an emf in the coil (Fig. 31–21b).

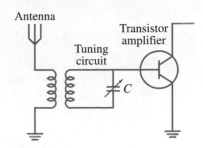

FIGURE 31–20 Simple tuning stage of a radio.

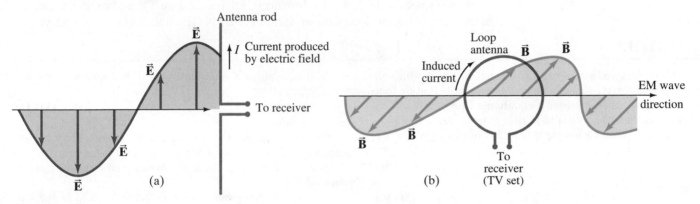

FIGURE 31–21 Antennas. (a) Electric field of EM wave produces a current in an antenna consisting of straight wire or rods. (b) Changing magnetic field induces an emf and current in a loop antenna.

A satellite dish (Fig. 31–22) consists of a parabolic reflector that focuses the EM waves onto a "horn," similar to a concave mirror telescope (Fig. 33–38).

FIGURE 31–22 A satellite dish.

EXAMPLE 31–9 Tuning a station. Calculate the transmitting wavelength of an FM radio station that transmits at 100 MHz.

APPROACH Radio is transmitted as an EM wave, so the speed is $c = 3.0 \times 10^8\,\text{m/s}$. The wavelength is found from Eq. 31–14, $\lambda = c/f$.

SOLUTION The carrier frequency is $f = 100\,\text{MHz} = 1.0 \times 10^8\,\text{s}^{-1}$, so

$$\lambda = \frac{c}{f} = \frac{(3.0 \times 10^8\,\text{m/s})}{(1.0 \times 10^8\,\text{s}^{-1})} = 3.0\,\text{m}.$$

NOTE The wavelengths of other FM signals (88 MHz to 108 MHz) are close to the 3.0-m wavelength of this station. FM antennas are typically 1.5 m long, or about a half wavelength. This length is chosen so that the antenna reacts in a resonant fashion and thus is more sensitive to FM frequencies. AM radio antennas would have to be very long to be either $\frac{1}{2}\lambda$ or $\frac{1}{4}\lambda$.

[†]For *FM stereo broadcasting*, two signals are carried by the carrier wave. One signal contains frequencies up to about 15 kHz, which includes most audio frequencies. The other signal includes the same range of frequencies, but 19 kHz is added to it. A stereo receiver subtracts this 19,000-Hz signal and distributes the two signals to the left and right channels. The first signal consists of the sum of left and right channels (L + R), so mono radios detect all the sound. The second signal is the difference between left and right (L − R). Hence the receiver must add and subtract the two signals to get pure left and right signals for each channel.

Other EM Wave Communications

The various regions of the radio-wave spectrum are assigned by governmental agencies for various purposes. Besides those mentioned above, there are "bands" assigned for use by ships, airplanes, police, military, amateurs, satellites and space, and radar. Cell phones, for example, are complete radio transmitters and receivers. In the U.S., CDMA cell phones function on two different bands: 800 MHz and 1900 MHz ($= 1.9$ GHz). Europe, Asia, and much of the rest of the world use a different system: the international standard called GSM (Global System for Mobile Communication), on 900-MHz and 1800-MHz bands. The U.S. now also has the GSM option (at 850 MHz and 1.9 GHz), as does much of the rest of the Americas. A 700-MHz band is now being made available for cell phones (it used to carry TV broadcast channels 52–69, now no longer used). Radio-controlled toys (cars, sailboats, robotic animals, etc.) can use various frequencies from 27 MHz to 75 MHz. Automobile remote (keyless) entry may operate around 300 MHz or 400 MHz.

Cable TV channels are carried as electromagnetic waves along a coaxial cable (see Fig. 31–13) rather than being broadcast and received through the "air." The channels are in the same part of the EM spectrum, hundreds of MHz, but some are at frequencies not available for TV broadcast. Digital satellite TV and radio are carried in the microwave portion of the spectrum (12 to 14 GHz and 2.3 GHz, respectively).

Summary

James Clerk Maxwell synthesized an elegant theory in which all electric and magnetic phenomena could be described using four equations, now called **Maxwell's equations**. They are based on earlier ideas, but Maxwell added one more—that a changing electric field produces a magnetic field. Maxwell's equations are

$$\oint \vec{E} \cdot d\vec{A} = \frac{Q}{\epsilon_0} \tag{31–5a}$$

$$\oint \vec{B} \cdot d\vec{A} = 0 \tag{31–5b}$$

$$\oint \vec{E} \cdot d\vec{\ell} = -\frac{d\Phi_B}{dt} \tag{31–5c}$$

$$\oint \vec{B} \cdot d\vec{\ell} = \mu_0 I + \mu_0 \epsilon_0 \frac{d\Phi_E}{dt}. \tag{31–5d}$$

The first two are Gauss's laws for electricity and for magnetism; the other two are Faraday's law and Ampère's law (as extended by Maxwell), respectively.

Maxwell's theory predicted that transverse **electromagnetic (EM) waves** would be produced by accelerating electric charges, and these waves would propagate through space at the speed of light c, given by

$$c = \frac{1}{\sqrt{\epsilon_0 \mu_0}} = 3.00 \times 10^8 \, \text{m/s}. \tag{31–12}$$

The wavelength λ and frequency f of EM waves are related to their speed c by

$$c = \lambda f, \tag{31–14}$$

just as for other waves.

The oscillating electric and magnetic fields in an EM wave are perpendicular to each other and to the direction of propagation. EM waves are waves of fields, not matter, and can propagate in empty space.

After EM waves were experimentally detected in the late 1800s, the idea that light is an EM wave (although of much higher frequency than those detected directly) became generally accepted. The **electromagnetic spectrum** includes EM waves of a wide variety of wavelengths, from microwaves and radio waves to visible light to X-rays and gamma rays, all of which travel through space at a speed $c = 3.00 \times 10^8$ m/s.

The energy carried by EM waves can be described by the **Poynting vector**

$$\vec{S} = \frac{1}{\mu_0} \vec{E} \times \vec{B} \tag{31–18}$$

which gives the rate energy is carried across unit area per unit time when the electric and magnetic fields in an EM wave in free space are $\vec{E}$ and $\vec{B}$.

EM waves carry momentum and exert a **radiation pressure** proportional to the intensity S of the wave.

Questions

1. An electric field $\vec{E}$ points away from you, and its magnitude is increasing. Will the induced magnetic field be clockwise or counterclockwise? What if $\vec{E}$ points toward you and is decreasing?

2. What is the direction of the displacement current in Fig. 31–3? (*Note*: The capacitor is discharging.)

3. Why is it that the magnetic field of a displacement current in a capacitor is so much harder to detect than the magnetic field of a conduction current?

4. Are there any good reasons for calling the term $\mu_0 \epsilon_0 \, d\Phi_E/dt$ in Eq. 31–1 as being due to a "current"? Explain.

5. The electric field in an EM wave traveling north oscillates in an east–west plane. Describe the direction of the magnetic field vector in this wave.

6. Is sound an electromagnetic wave? If not, what kind of wave is it?

7. Can EM waves travel through a perfect vacuum? Can sound waves?

8. When you flip a light switch, does the overhead light go on immediately? Explain.

9. Are the wavelengths of radio and television signals longer or shorter than those detectable by the human eye?

10. What does the wavelength calculated in Example 31–2 tell you about the phase of a 60-Hz ac current that starts at a power plant as compared to its phase at a house 200 km away?

11. When you connect two loudspeakers to the output of a stereo amplifier, should you be sure the lead wires are equal in length so that there will not be a time lag between speakers? Explain.

12. In the electromagnetic spectrum, what type of EM wave would have a wavelength of 10^3 km; 1 km; 1 m; 1 cm; 1 mm; 1 μm?

13. Can radio waves have the same frequencies as sound waves (20 Hz–20,000 Hz)?

14. Discuss how cordless telephones make use of EM waves. What about cellular telephones?

15. Can two radio or TV stations broadcast on the same carrier frequency? Explain.

16. If a radio transmitter has a vertical antenna, should a receiver's antenna (rod type) be vertical or horizontal to obtain best reception?

17. The carrier frequencies of FM broadcasts are much higher than for AM broadcasts. On the basis of what you learned about diffraction in Chapter 15, explain why AM signals can be detected more readily than FM signals behind low hills or buildings.

18. A lost person may signal by flashing a flashlight on and off using Morse code. This is actually a modulated EM wave. Is it AM or FM? What is the frequency of the carrier, approximately?

Problems

31–1 $\vec{B}$ Produced by Changing $\vec{E}$

1. (I) Determine the rate at which the electric field changes between the round plates of a capacitor, 6.0 cm in diameter, if the plates are spaced 1.1 mm apart and the voltage across them is changing at a rate of 120 V/s.

2. (I) Calculate the displacement current I_D between the square plates, 5.8 cm on a side, of a capacitor if the electric field is changing at a rate of 2.0×10^6 V/m·s.

3. (II) At a given instant, a 2.8-A current flows in the wires connected to a parallel-plate capacitor. What is the rate at which the electric field is changing between the plates if the square plates are 1.60 cm on a side?

4. (II) A 1500-nF capacitor with circular parallel plates 2.0 cm in diameter is accumulating charge at the rate of 38.0 mC/s at some instant in time. What will be the induced magnetic field strength 10.0 cm radially outward from the center of the plates? What will be the value of the field strength after the capacitor is fully charged?

5. (II) Show that the displacement current through a parallel-plate capacitor can be written $I_D = C\,dV/dt$, where V is the voltage across the capacitor at any instant.

6. (II) Suppose an air-gap capacitor has circular plates of radius $R = 2.5$ cm and separation $d = 1.6$ mm. A 76.0-Hz emf, $\mathcal{E} = \mathcal{E}_0 \cos \omega t$, is applied to the capacitor. The maximum displacement current is 35 μA. Determine (a) the maximum conduction current I, (b) the value of $\mathcal{E}_0$, (c) the maximum value of $d\Phi_E/dt$ between the plates. Neglect fringing.

7. (III) Suppose that a circular parallel-plate capacitor has radius $R_0 = 3.0$ cm and plate separation $d = 5.0$ mm. A sinusoidal potential difference $V = V_0 \sin(2\pi ft)$ is applied across the plates, where $V_0 = 150$ V and $f = 60$ Hz. (a) In the region between the plates, show that the magnitude of the induced magnetic field is given by $B = B_0(R) \cos(2\pi ft)$, where R is the radial distance from the capacitor's central axis. (b) Determine the expression for the amplitude $B_0(R)$ of this time-dependent (sinusoidal) field when $R \leq R_0$, and when $R > R_0$. (c) Plot $B_0(R)$ in tesla for the range $0 \leq R \leq 10$ cm.

31–5 EM Waves

8. (I) If the electric field in an EM wave has a peak magnitude of 0.57×10^{-4} V/m, what is the peak magnitude of the magnetic field strength?

9. (I) If the magnetic field in a traveling EM wave has a peak magnitude of 12.5 nT, what is the peak magnitude of the electric field?

10. (I) In an EM wave traveling west, the B field oscillates vertically and has a frequency of 80.0 kHz and an rms strength of 7.75×10^{-9} T. Determine the frequency and rms strength of the electric field. What is its direction?

11. (II) The electric field of a plane EM wave is given by $E_x = E_0 \cos(kz + \omega t)$, $E_y = E_z = 0$. Determine (a) the direction of propagation and (b) the magnitude and direction of $\vec{B}$.

12. (III) Consider two possible candidates $E(x, t)$ as solutions of the wave equation for an EM wave's electric field. Let A and α be constants. Show that (a) $E(x, t) = Ae^{-\alpha(x - vt)^2}$ satisfies the wave equation, and that (b) $E(x, t) = Ae^{-(\alpha x^2 - vt)}$ does not satisfy the wave equation.

31–6 Electromagnetic Spectrum

13. (I) What is the frequency of a microwave whose wavelength is 1.50 cm?

14. (I) (a) What is the wavelength of a 25.75×10^9 Hz radar signal? (b) What is the frequency of an X-ray with wavelength 0.12 nm?

15. (I) How long does it take light to reach us from the Sun, 1.50×10^8 km away?

16. (I) An EM wave has frequency 8.56×10^{14} Hz. What is its wavelength, and how would we classify it?

17. (I) Electromagnetic waves and sound waves can have the same frequency. (a) What is the wavelength of a 1.00-kHz electromagnetic wave? (b) What is the wavelength of a 1.00-kHz sound wave? (The speed of sound in air is 341 m/s.) (c) Can you hear a 1.00-kHz electromagnetic wave?

18. (II) Pulsed lasers used for science and medicine produce very brief bursts of electromagnetic energy. If the laser light wavelength is 1062 nm (Neodymium–YAG laser), and the pulse lasts for 38 picoseconds, how many wavelengths are found within the laser pulse? How brief would the pulse need to be to fit only one wavelength?

19. (II) How long would it take a message sent as radio waves from Earth to reach Mars when Mars is (a) nearest Earth, (b) farthest from Earth? Assume that Mars and Earth are in the same plane and that their orbits around the Sun are circles (Mars is 230×10^6 km from the Sun).

20. (II) An electromagnetic wave has an electric field given by
$$\vec{E} = \hat{i}(225 \text{ V/m}) \sin[(0.077 \text{ m}^{-1})z - (2.3 \times 10^7 \text{ rad/s})t].$$
(a) What are the wavelength and frequency of the wave? (b) Write down an expression for the magnetic field.

31–7 Speed of Light

21. (II) What is the minimum angular speed at which Michelson's eight-sided mirror would have had to rotate to reflect light into an observer's eye by succeeding mirror faces (1/8 of a revolution, Fig. 31–14)?

31–8 EM Wave Energy; Poynting Vector

22. (I) The $\vec{E}$ field in an EM wave has a peak of 26.5 mV/m. What is the average rate at which this wave carries energy across unit area per unit time?

23. (II) The magnetic field in a traveling EM wave has an rms strength of 22.5 nT. How long does it take to deliver 335 J of energy to $1.00\,cm^2$ of a wall that it hits perpendicularly?

24. (II) How much energy is transported across a $1.00\,cm^2$ area per hour by an EM wave whose E field has an rms strength of 32.8 mV/m?

25. (II) A spherically spreading EM wave comes from a 1500-W source. At a distance of 5.0 m, what is the intensity, and what is the rms value of the electric field?

26. (II) If the amplitude of the B field of an EM wave is $2.5 \times 10^{-7}\,T$, (a) what is the amplitude of the E field? (b) What is the average power per unit area of the EM wave?

27. (II) What is the energy contained in a $1.00\text{-}m^3$ volume near the Earth's surface due to radiant energy from the Sun? See Example 31–6.

28. (II) A 15.8-mW laser puts out a narrow beam 2.00 mm in diameter. What are the rms values of E and B in the beam?

29. (II) Estimate the average power output of the Sun, given that about $1350\,W/m^2$ reaches the upper atmosphere of the Earth.

30. (II) A high-energy pulsed laser emits a 1.0-ns-long pulse of average power $1.8 \times 10^{11}\,W$. The beam is $2.2 \times 10^{-3}\,m$ in radius. Determine (a) the energy delivered in each pulse, and (b) the rms value of the electric field.

31. (II) How practical is solar power for various devices? Assume that on a sunny day, sunlight has an intensity of $1000\,W/m^2$ at the surface of Earth and that, when illuminated by that sunlight, a solar-cell panel can convert 10% of the sunlight's energy into electric power. For each device given below, calculate the area A of solar panel needed to power it. (a) A calculator consumes 50 mW. Find A in cm^2. Is A small enough so that the solar panel can be mounted directly on the calculator that it is powering? (b) A hair dryer consumes 1500 W. Find A in m^2. Assuming no other electronic devices are operating within a house at the same time, is A small enough so that the hair dryer can be powered by a solar panel mounted on the house's roof? (c) A car requires 20 hp for highway driving at constant velocity (this car would perform poorly in situations requiring acceleration). Find A in m^2. Is A small enough so that this solar panel can be mounted directly on the car and power it in "real time"?

32. (III) (a) Show that the Poynting vector $\vec{S}$ points radially inward toward the center of a circular parallel-plate capacitor when it is being charged as in Example 31–1. (b) Integrate $\vec{S}$ over the cylindrical boundary of the capacitor gap to show that the rate at which energy enters the capacitor is equal to the rate at which electrostatic energy is being stored in the electric field of the capacitor (Section 24–4). Ignore fringing of $\vec{E}$.

33. (III) The Arecibo radio telescope in Puerto Rico can detect a radio wave with an intensity as low as $1 \times 10^{-23}\,W/m^2$. As a "best-case" scenario for communication with extraterrestrials, consider the following: suppose an advanced civilization located at point A, a distance x away from Earth, is somehow able to harness the entire power output of a Sun-like star, converting that power completely into a radio-wave signal which is transmitted uniformly in all directions from A. (a) In order for Arecibo to detect this radio signal, what is the maximum value for x in light-years $(1\,ly \approx 10^{16}\,m)$? (b) How does this maximum value compare with the 100,000-ly size of our Milky Way galaxy? The intensity of sunlight at Earth's orbital distance from the Sun is $1350\,W/m^2$.

31–9 Radiation Pressure

34. (II) Estimate the radiation pressure due to a 75-W bulb at a distance of 8.0 cm from the center of the bulb. Estimate the force exerted on your fingertip if you place it at this point.

35. (II) Laser light can be focused (at best) to a spot with a radius r equal to its wavelength λ. Suppose that a 1.0-W beam of green laser light $(\lambda = 5 \times 10^{-7}\,m)$ is used to form such a spot and that a cylindrical particle of about that size (let the radius and height equal r) is illuminated by the laser as shown in Fig. 31–23. Estimate the acceleration of the particle, if its density equals that of water and it absorbs the radiation. [This order-of-magnitude calculation convinced researchers of the feasibility of "optical tweezers," p. 829.]

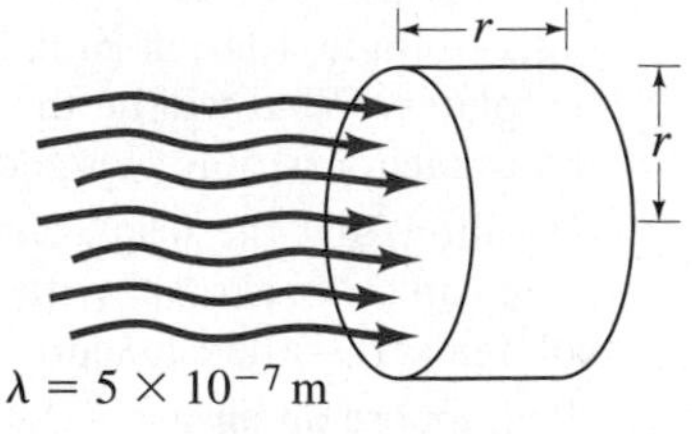

FIGURE 31–23
Problem 35.

$\lambda = 5 \times 10^{-7}\,m$

36. (II) The powerful laser used in a laser light show provides a 3-mm diameter beam of green light with a power of 3 W. When a space-walking astronaut is outside the Space Shuttle, her colleague inside the Shuttle playfully aims such a laser beam at the astronaut's space suit. The masses of the suited astronaut and the Space Shuttle are 120 kg and 103,000 kg, respectively. (a) Assuming the suit is perfectly reflecting, determine the "radiation-pressure" force exerted on the astronaut by the laser beam. (b) Assuming the astronaut is separated from the Shuttle's center of mass by 20 m, model the Shuttle as a sphere in order to estimate the gravitation force it exerts on the astronaut. (c) Which of the two forces is larger, and by what factor?

37. (II) What size should the solar panel on a satellite orbiting Jupiter be if it is to collect the same amount of radiation from the Sun as a $1.0\text{-}m^2$ solar panel on a satellite orbiting Earth?

38. (I) What is the range of wavelengths for (a) FM radio (88 MHz to 108 MHz) and (b) AM radio (535 kHz to 1700 kHz)?

39. (I) Estimate the wavelength for 1.9-GHz cell phone reception.

40. (I) The variable capacitor in the tuner of an AM radio has a capacitance of 2200 pF when the radio is tuned to a station at 550 kHz. What must the capacitance be for a station near the other end of the dial, 1610 kHz?

General Problems

43. A 1.60-m-long FM antenna is oriented parallel to the electric field of an EM wave. How large must the electric field be to produce a 1.00-mV (rms) voltage between the ends of the antenna? What is the rate of energy transport per square meter?

44. Who will hear the voice of a singer first: a person in the balcony 50.0 m away from the stage (see Fig. 31–24), or a person 1500 km away at home whose ear is next to the radio listening to a live broadcast? Roughly how much sooner? Assume the microphone is a few centimeters from the singer and the temperature is 20°C.

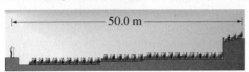

FIGURE 31–24 Problem 44.

45. Light is emitted from an ordinary lightbulb filament in wave-train bursts about 10^{-8} s in duration. What is the length in space of such wave trains?

46. Radio-controlled clocks throughout the United States receive a radio signal from a transmitter in Fort Collins, Colorado, that accurately (within a microsecond) marks the beginning of each minute. A slight delay, however, is introduced because this signal must travel from the transmitter to the clocks. Assuming Fort Collins is no more than 3000 km from any point in the U.S., what is the longest travel-time delay?

47. The voice from an astronaut on the Moon (Fig. 31–25) was beamed to a listening crowd on Earth. If you were standing 25 m from the loudspeaker on Earth, what was the total time lag between when you heard the sound and when the sound entered a microphone on the Moon? Explain whether the microphone was inside the space helmet, or outside, and why.

FIGURE 31–25
Problem 47.

41. (II) A certain FM radio tuning circuit has a fixed capacitor $C = 620$ pF. Tuning is done by a variable inductance. What range of values must the inductance have to tune stations from 88 MHz to 108 MHz?

42. (II) A satellite beams microwave radiation with a power of 12 kW toward the Earth's surface, 550 km away. When the beam strikes Earth, its circular diameter is about 1500 m. Find the rms electric field strength of the beam at the surface of the Earth.

48. Cosmic microwave background radiation fills all space with an average energy density of 4×10^{-14} J/m³. (a) Find the rms value of the electric field associated with this radiation. (b) How far from a 7.5-kW radio transmitter emitting uniformly in all directions would you find a comparable value?

49. What are E_0 and B_0 2.00 m from a 75-W light source? Assume the bulb emits radiation of a single frequency uniformly in all directions.

50. Estimate the rms electric field in the sunlight that hits Mars, knowing that the Earth receives about 1350 W/m² and that Mars is 1.52 times farther from the Sun (on average) than is the Earth.

51. At a given instant in time, a traveling EM wave is noted to have its maximum magnetic field pointing west and its maximum electric field pointing south. In which direction is the wave traveling? If the rate of energy flow is 560 W/m², what are the maximum values for the two fields?

52. How large an emf (rms) will be generated in an antenna that consists of a circular coil 2.2 cm in diameter having 320 turns of wire, when an EM wave of frequency 810 kHz transporting energy at an average rate of 1.0×10^{-4} W/m² passes through it? [Hint: you can use Eq. 29–4 for a generator, since it could be applied to an observer moving with the coil so that the magnetic field is oscillating with the frequency $f = \omega/2\pi$.]

53. The average intensity of a particular TV station's signal is 1.0×10^{-13} W/m² when it arrives at a 33-cm-diameter satellite TV antenna. (a) Calculate the total energy received by the antenna during 6.0 hours of viewing this station's programs. (b) What are the amplitudes of the E and B fields of the EM wave?

54. A radio station is allowed to broadcast at an average power not to exceed 25 kW. If an electric field amplitude of 0.020 V/m is considered to be acceptable for receiving the radio transmission, estimate how many kilometers away you might be able to hear this station.

55. A point source emits light energy uniformly in all directions at an average rate P_0 with a single frequency f. Show that the peak electric field in the wave is given by

$$E_0 = \sqrt{\frac{\mu_0 c P_0}{2\pi r^2}}.$$

56. Suppose a 35-kW radio station emits EM waves uniformly in all directions. (a) How much energy per second crosses a 1.0-m² area 1.0 km from the transmitting antenna? (b) What is the rms magnitude of the $\vec{E}$ field at this point, assuming the station is operating at full power? What is the rms voltage induced in a 1.0-m-long vertical car antenna (c) 1.0 km away, (d) 50 km away?

57. What is the maximum power level of a radio station so as to avoid electrical breakdown of air at a distance of 0.50 m from the transmitting antenna? Assume the antenna is a point source. Air breaks down in an electric field of about $3 \times 10^6 \, \text{V/m}$.

58. In free space ("vacuum"), where the net charge and current flow is zero, the speed of an EM wave is given by $v = 1/\sqrt{\epsilon_0 \mu_0}$. If, instead, an EM wave travels in a nonconducting ("dielectric") material with dielectric constant K, then $v = 1/\sqrt{K \epsilon_0 \mu_0}$. For frequencies corresponding to the visible spectrum (near $5 \times 10^{14} \, \text{Hz}$), the dielectric constant of water is 1.77. Predict the speed of light in water and compare this value (as a percentage) with the speed of light in a vacuum.

59. The metal walls of a microwave oven form a cavity of dimensions 37 cm $\times$ 37 cm $\times$ 20 cm. When 2.45-GHz microwaves are continuously introduced into this cavity, reflection of incident waves from the walls set up standing waves with nodes at the walls. Along the 37-cm dimension of the oven, how many nodes exist (excluding the nodes at the wall) and what is the distance between adjacent nodes? [Because no heating occurs at these nodes, most microwaves rotate food while operating.]

60. Imagine that a steady current I flows in a straight cylindrical wire of radius R_0 and resistivity ρ. (a) If the current is then changed at a rate dI/dt, show that a displacement current I_D exists in the wire of magnitude $\epsilon_0 \rho (dI/dt)$. (b) If the current in a copper wire is changed at the rate of 1.0 A/ms, determine the magnitude of I_D. (c) Determine the magnitude of the magnetic field B_D (T) created by I_D at the surface of a copper wire with $R_0 = 1.0 \, \text{mm}$. Compare (as a ratio) B_D with the field created at the surface of the wire by a steady current of 1.0 A.

61. The electric field of an EM wave pulse traveling along the x axis in free space is given by $E_y = E_0 \exp[-\alpha^2 x^2 - \beta^2 t^2 + 2\alpha\beta xt]$, where E_0, α, and β are positive constants. (a) Is the pulse moving in the $+x$ or $-x$ direction? (b) Express β in terms of α and c (speed of light in free space). (c) Determine the expression for the magnetic field of this EM wave.

62. Suppose that a right-moving EM wave overlaps with a left-moving EM wave so that, in a certain region of space, the total electric field in the y direction and magnetic field in the z direction are given by $E_y = E_0 \sin(kx - \omega t) + E_0 \sin(kx + \omega t)$ and $B_z = B_0 \sin(kx - \omega t) - B_0 \sin(kx + \omega t)$. (a) Find the mathematical expression that represents the standing electric and magnetic waves in the y and z directions, respectively. (b) Determine the Poynting vector and find the x locations at which it is zero at all times.

63. The electric and magnetic fields of a certain EM wave in free space are given by $\vec{E} = E_0 \sin(kx - \omega t)\hat{j} + E_0 \cos(kx - \omega t)\hat{k}$ and $\vec{B} = B_0 \cos(kx - \omega t)\hat{j} - B_0 \sin(kx - \omega t)\hat{k}$. (a) Show that $\vec{E}$ and $\vec{B}$ are perpendicular to each other at all times. (b) For this wave, $\vec{E}$ and $\vec{B}$ are in a plane parallel to the yz plane. Show that the wave moves in a direction perpendicular to both $\vec{E}$ and $\vec{B}$. (c) At any arbitrary choice of position x and time t, show that the magnitudes of $\vec{E}$ and $\vec{B}$ always equal E_0 and B_0, respectively. (d) At $x = 0$, draw the orientation of $\vec{E}$ and $\vec{B}$ in the yz plane at $t = 0$. Then qualitatively describe the motion of these vectors in the yz plane as time increases. [Note: The EM wave in this Problem is "circularly polarized."]

Answers to Exercises

A: (c).

B: (b).

C: (a) $3.8 \times 10^6 \, \text{Hz}$; (b) $5.5 \times 10^{18} \, \text{Hz}$.

D: 45 cm.

E: Yes; the signal still travels 72,000 km.

F: Over 4 hours.

Reflection from still water, as from a glass mirror, can be analyzed using the ray model of light.

Is this picture right side up? How can you tell? What are the clues? Notice the people and position of the Sun. Ray diagrams, which we will learn to draw in this Chapter, can provide the answer. See Example 32–3.

In this first Chapter on light and optics, we use the ray model of light to understand the formation of images by mirrors, both plane and curved (spherical). We also begin our study of refraction—how light rays bend when they go from one medium to another—which prepares us for our study in the next Chapter of lenses, which are the crucial part of so many optical instruments.

32

Light: Reflection and Refraction

CHAPTER-OPENING QUESTION—Guess now!

A 2.0-m-tall person is standing 2.0 m from a flat vertical mirror staring at her image. What minimum height must the mirror have if the person is to see her entire body, from the top of her head to her feet?

(a) 0.50 m.
(b) 1.0 m.
(c) 1.5 m.
(d) 2.0 m.
(e) 2.5 m.

The sense of sight is extremely important to us, for it provides us with a large part of our information about the world. How do we see? What is the something called *light* that enters our eyes and causes the sensation of sight? How does light behave so that we can see everything that we do? We saw in Chapter 31 that light can be considered a form of electromagnetic radiation. We now examine the subject of light in detail in the next four Chapters.

We see an object in one of two ways: (1) the object may be a *source* of light, such as a lightbulb, a flame, or a star, in which case we see the light emitted directly from the source; or, more commonly, (2) we see an object by light *reflected* from it. In the latter case, the light may have originated from the Sun, artificial lights, or a campfire. An understanding of how objects *emit* light was not achieved until the 1920s, and will be discussed in Chapter 37. How light is *reflected* from objects was understood earlier, and will be discussed here, in Section 32–2.

CONTENTS

32–1 The Ray Model of Light

A great deal of evidence suggests that *light travels in straight lines* under a wide variety of circumstances. For example, a source of light like the Sun casts distinct shadows, and the light from a laser pointer appears to be a straight line. In fact, we infer the positions of objects in our environment by assuming that light moves from the object to our eyes in straight-line paths. Our orientation to the physical world is based on this assumption.

This reasonable assumption is the basis of the **ray model** of light. This model assumes that light travels in straight-line paths called **light rays**. Actually, a ray is an idealization; it is meant to represent an extremely narrow beam of light. When we see an object, according to the ray model, light reaches our eyes from each point on the object. Although light rays leave each point in many different directions, normally only a small bundle of these rays can enter an observer's eye, as shown in Fig. 32–1. If the person's head moves to one side, a different bundle of rays will enter the eye from each point.

We saw in Chapter 31 that light can be considered as an electromagnetic wave. Although the ray model of light does not deal with this aspect of light (we discuss the wave nature of light in Chapters 34 and 35), the ray model has been very successful in describing many aspects of light such as reflection, refraction, and the formation of images by mirrors and lenses.[†] Because these explanations involve straight-line rays at various angles, this subject is referred to as **geometric optics**.

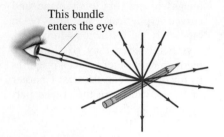

FIGURE 32–1 Light rays come from each single point on an object. A small bundle of rays leaving one point is shown entering a person's eye.

32–2 Reflection; Image Formation by a Plane Mirror

When light strikes the surface of an object, some of the light is reflected. The rest can be absorbed by the object (and transformed to thermal energy) or, if the object is transparent like glass or water, part can be transmitted through. For a very smooth shiny object such as a silvered mirror, over 95% of the light may be reflected.

When a narrow beam of light strikes a flat surface (Fig. 32–2), we define the **angle of incidence**, θ_i, to be the angle an incident ray makes with the normal (perpendicular) to the surface, and the **angle of reflection**, θ_r, to be the angle the reflected ray makes with the normal. It is found that the *incident and reflected rays lie in the same plane with the normal to the surface*, and that

the angle of reflection equals the angle of incidence, $\theta_r = \theta_i$.

This is the **law of reflection**, and it is depicted in Fig. 32–2. It was known to the ancient Greeks, and you can confirm it yourself by shining a narrow flashlight beam or a laser pointer at a mirror in a darkened room.

FIGURE 32–2 Law of reflection: (a) Shows a 3-D view of an incident ray being reflected at the top of a flat surface; (b) shows a side or "end-on" view, which we will usually use because of its clarity.

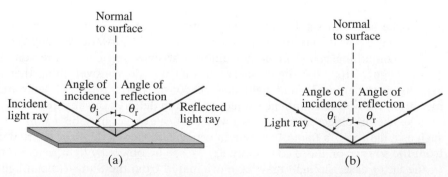

[†] In ignoring the wave properties of light we must be careful that when the light rays pass by objects or through apertures, these must be large compared to the wavelength of the light (so the wave phenomena of interference and diffraction, as discussed in Chapter 15, can be ignored), and we ignore what happens to the light at the edges of objects until we get to Chapters 34 and 35.

When light is incident upon a rough surface, even microscopically rough such as this page, it is reflected in many directions, as shown in Fig. 32–3. This is called **diffuse reflection**. The law of reflection still holds, however, at each small section of the surface. Because of diffuse reflection in all directions, an ordinary object can be seen at many different angles by the light reflected from it. When you move your head to the side, different reflected rays reach your eye from each point on the object (such as this page), Fig. 32–4a. Let us compare diffuse reflection to reflection from a mirror, which is known as **specular reflection**. ("Speculum" is Latin for mirror.) When a narrow beam of light shines on a mirror, the light will not reach your eye unless your eye is positioned at just the right place where the law of reflection is satisfied, as shown in Fig. 32–4b. This is what gives rise to the special image-forming properties of mirrors.

FIGURE 32–3 Diffuse reflection from a rough surface.

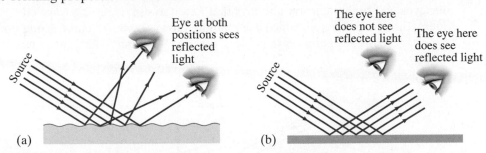

(a)

(b)

FIGURE 32–4 A narrow beam of light shines on (a) white paper, and (b) a mirror. In part (a), you can see with your eye the white light reflected at various positions because of diffuse reflection. But in part (b), you see the reflected light only when your eye is placed correctly $(\theta_r = \theta_i)$; mirror reflection is also known as specular reflection. (Galileo, using similar arguments, showed that the Moon must have a rough surface rather than a highly polished surface like a mirror, as some people thought.)

EXAMPLE 32–1 **Reflection from flat mirrors.** Two flat mirrors are perpendicular to each other. An incoming beam of light makes an angle of 15° with the first mirror as shown in Fig. 32–5a. What angle will the outgoing beam make with the second mirror?

APPROACH We sketch the path of the beam as it reflects off the two mirrors, and draw the two normals to the mirrors for the two reflections. We use geometry and the law of reflection to find the various angles.

FIGURE 32–5 Example 32–1.

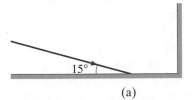

(a)

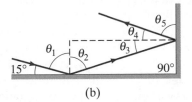

(b)

SOLUTION In Fig. 32–5b, $\theta_1 + 15° = 90°$, so $\theta_1 = 75°$; by the law of reflection $\theta_2 = \theta_1 = 75°$ too. The two normals to the two mirrors are perpendicular to each other, so $\theta_2 + \theta_3 + 90° = 180°$ as for any triangle. Thus $\theta_3 = 180° - 90° - 75° = 15°$. By the law of reflection, $\theta_4 = \theta_3 = 15°$, so $\theta_5 = 75°$ is the angle the reflected ray makes with the second mirror surface.

NOTE The outgoing ray is parallel to the incoming ray. Red reflectors on bicycles and cars use this principle.

When you look straight into a mirror, you see what appears to be yourself as well as various objects around and behind you, Fig. 32–6. Your face and the other objects look as if they are in front of you, beyond the mirror. But what you see in the mirror is an **image** of the objects, including yourself, that are in front of the mirror.

FIGURE 32–6 When you look in a mirror, you see an image of yourself and objects around you. You don't see yourself as others see you, because left and right appear reversed in the image.

A "plane" mirror is one with a smooth flat reflecting surface. Figure 32–7 shows how an image is formed by a plane mirror according to the ray model. We are viewing the mirror, on edge, in the diagram of Fig. 32–7, and the rays are shown reflecting from the front surface. (Good mirrors are generally made by putting a highly reflective metallic coating on one surface of a very flat piece of glass.) Rays from two different points on an object (the bottle on the left in Fig. 32–7) are shown: two rays are shown leaving from a point on the top of the bottle, and two more from a point on the bottom. Rays leave each point on the object going in many directions, but only those that enclose the bundle of rays that enter the eye from each of the two points are shown. Each set of diverging rays that reflect from the mirror and enter the eye *appear* to come from a single point (called the image point) behind the mirror, as shown by the dashed lines. That is, our eyes and brain interpret any rays that enter an eye as having traveled straight-line paths. The point from which each bundle of rays seems to come is one point on the image. For each point on the object, there is a corresponding image point.

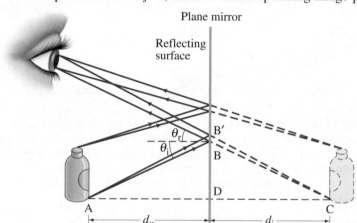

FIGURE 32–7 Formation of a virtual image by a plane mirror.

Let us concentrate on the two rays that leave point A on the object in Fig. 32–7, and strike the mirror at points B and B′. We use geometry for the rays at B. The angles ADB and CDB are right angles; and because of the law of reflection, $\theta_i = \theta_r$ at point B. Therefore, angles ABD and CBD are also equal. The two triangles ABD and CBD are thus congruent, and the length AD = CD. That is, the image appears as far behind the mirror as the object is in front. The **image distance**, d_i (perpendicular distance from mirror to image, Fig. 32–7), equals the **object distance**, d_o (perpendicular distance from object to mirror). From the geometry, we can also see that the height of the image is the same as that of the object.

The light rays do not actually pass through the image location itself in Fig. 32–7. (Note where the red lines are dashed to show they are our projections, not rays.) The image would not appear on paper or film placed at the location of the image. Therefore, it is called a **virtual image**. This is to distinguish it from a **real image** in which the light does pass through the image and which therefore could appear on film or in an electronic sensor, and even on a white sheet of paper or screen placed at the position of the image. Our eyes can see both real and virtual images, as long as the diverging rays enter our pupils. We will see that curved mirrors and lenses can form real images, as well as virtual. A movie projector lens, for example, produces a real image that is visible on the screen.

EXAMPLE 32–2 How tall must a full-length mirror be? A woman 1.60 m tall stands in front of a vertical plane mirror. What is the minimum height of the mirror, and how close must its lower edge be to the floor, if she is to be able to see her whole body? Assume her eyes are 10 cm below the top of her head.

APPROACH For her to see her whole body, light rays from the top of her head and from the bottom of her foot must reflect from the mirror and enter her eye: see Fig. 32–8. We don't show two rays diverging from each point as we did in Fig. 32–7, where we wanted to find where the image is. Now that we know the image is the same distance behind a plane mirror as the object is in front, we only need to show one ray leaving point G (top of head) and one ray leaving point A (her toe), and then use simple geometry.

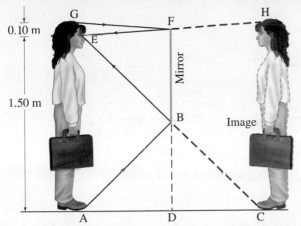

0.10 m

1.50 m

G

E

F

Mirror

H

B

Image

A   D   C

FIGURE 32–8 Seeing oneself in a mirror. Example 32–2.

SOLUTION First consider the ray that leaves her foot at A, reflects at B, and enters the eye at E. The mirror needs to extend no lower than B. The angle of reflection equals the angle of incidence, so the height BD is half of the height AE. Because AE = 1.60 m − 0.10 m = 1.50 m, then BD = 0.75 m. Similarly, if the woman is to see the top of her head, the top edge of the mirror only needs to reach point F, which is 5 cm below the top of her head (half of GE = 10 cm). Thus, DF = 1.55 m, and the mirror needs to have a vertical height of only (1.55 m − 0.75 m) = 0.80 m. The mirror's bottom edge must be 0.75 m above the floor.

NOTE We see that a mirror, if positioned well, need be only half as tall as a person for that person to see all of himself or herself.

EXERCISE A Does the result of Example 32–2 depend on your distance from the mirror? (Try it.)

EXERCISE B Return to the Chapter-Opening Question, page 837, and answer it again now. Try to explain why you may have answered differently the first time.

EXERCISE C Suppose you are standing about 3 m in front of a mirror in a hair salon. You can see yourself from your head to your waist, where the end of the mirror cuts off the rest of your image. If you walk closer to the mirror (*a*) you will not be able to see any more of your image; (*b*) you will be able to see more of your image, below your waist; (*c*) you will see less of your image, with the cutoff rising to be above your waist.

CONCEPTUAL EXAMPLE 32–3 Is the photo upside down? Close examination of the photograph on the first page of this Chapter reveals that in the top portion, the image of the Sun is seen clearly, whereas in the lower portion, the image of the Sun is partially blocked by the tree branches. Show why the reflection is not the same as the real scene by drawing a sketch of this situation, showing the Sun, the camera, the branch, and two rays going from the Sun to the camera (one direct and one reflected). Is the photograph right side up?

RESPONSE We need to draw two diagrams, one assuming the photo on p. 837 is right side up, and another assuming it is upside down. Figure 32–9 is drawn assuming the photo is upside down. In this case, the Sun blocked by the tree would be the direct view, and the full view of the Sun the reflection: the ray which reflects off the water and into the camera travels at an angle below the branch, whereas the ray that travels directly to the camera passes through the branches. This works. Try to draw a diagram assuming the photo is right side up (thus assuming that the image of the Sun in the reflection is higher above the horizon than it is as viewed directly). It won't work. The photo on p. 837 is upside down.

Also, what about the people in the photo? Try to draw a diagram showing why they don't appear in the reflection. [*Hint*: Assume they are not sitting on the edge of poolside, but back from the edge a bit.] Then try to draw a diagram of the reverse (i.e., assume the photo is right side up so the people are visible only in the reflection). Reflected images are not perfect replicas when different planes (distances) are involved.

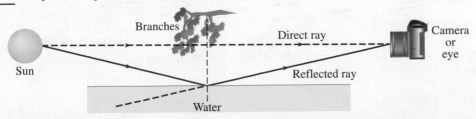

Branches

Sun

Direct ray

Camera or eye

Reflected ray

Water

FIGURE 32–9 Example 32–3.

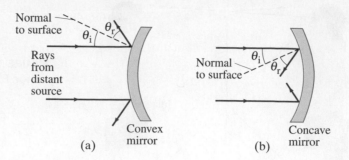

FIGURE 32–10 Mirrors with convex and concave spherical surfaces. Note that $\theta_r = \theta_i$ for each ray.

Normal to surface
θ_i θ_r
Rays from distant source
Convex mirror
(a)

Normal to surface
θ_i θ_r
Concave mirror
(b)

32–3 Formation of Images by Spherical Mirrors

Reflecting surfaces do not have to be flat. The most common *curved* mirrors are *spherical*, which means they form a section of a sphere. A spherical mirror is called **convex** if the reflection takes place on the outer surface of the spherical shape so that the center of the mirror surface bulges out toward the viewer (Fig. 32–10a). A mirror is called **concave** if the reflecting surface is on the inner surface of the sphere so that the mirror surface sinks away from the viewer (like a "cave"), Fig. 32–10b. Concave mirrors are used as shaving or cosmetic mirrors (Fig. 32–11a) because they magnify, and convex mirrors are sometimes used on cars and trucks (rearview mirrors) and in shops (to watch for thieves), because they take in a wide field of view (Fig. 32–11b).

FIGURE 32–11 (a) A concave cosmetic mirror gives a magnified image. (b) A convex mirror in a store reduces image size and so includes a wide field of view.

(a)

(b)

Focal Point and Focal Length

To see how spherical mirrors form images, we first consider an object that is very far from a concave mirror. For a distant object, as shown in Fig. 32–12, the rays from each point on the object that strike the mirror will be nearly parallel. *For an object infinitely far away* (the Sun and stars approach this), *the rays would be precisely parallel.*

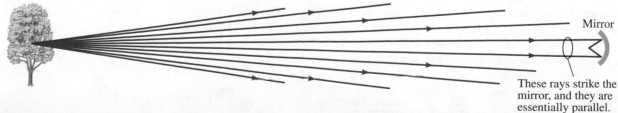

Mirror

These rays strike the mirror, and they are essentially parallel.

FIGURE 32–12 If the object's distance is large compared to the size of the mirror (or lens), the rays are nearly parallel. They are parallel for an object at infinity (∞).

Now consider such parallel rays falling on a concave mirror as in Fig. 32–13. The law of reflection holds for each of these rays at the point each strikes the mirror. As can be seen, they are not all brought to a single point. In order to form a sharp image, the rays must come to a point. Thus a spherical mirror will not make as sharp an image as a plane mirror will. However, as we show below, if the mirror is small compared to its radius of curvature, so that a reflected ray makes only a *small angle* with the incident ray (2θ in Fig. 32–14), then the rays will cross each other at very nearly a single point, or **focus**. In the case shown in Fig. 32–14, the incoming rays are parallel to the **principal axis**, which is defined as the straight line perpendicular to the curved surface at its center (line CA in Fig. 32–14). The point F, where incident parallel rays come to a focus after reflection, is called the **focal point** of the mirror. The distance between F and the center of the mirror, length FA, is called the **focal length**, f, of the mirror. The focal point is also the *image point for an object infinitely far away* along the principal axis. The image of the Sun, for example, would be at F.

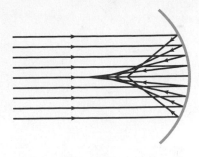

FIGURE 32–13 Parallel rays striking a concave spherical mirror do not intersect (or focus) at precisely a single point. (This "defect" is referred to as "spherical aberration.")

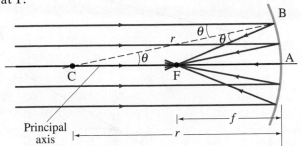

Principal axis

FIGURE 32–14 Rays parallel to the principal axis of a concave spherical mirror come to a focus at F, the focal point, as long as the mirror is small in width as compared to its radius of curvature, r, so that the rays are "paraxial"— that is, make only small angles with the horizontal axis.

Now we will show, for a mirror whose reflecting surface is small compared to its radius of curvature, that the rays very nearly meet at a common point, F, and we will also calculate the focal length f. In this approximation, we consider only rays that make a small angle with the principal axis; such rays are called **paraxial rays**, and their angles are exaggerated in Fig. 32–14 to make the labels clear. First we consider a ray that strikes the mirror at B in Fig. 32–14. The point C is the center of curvature of the mirror (the center of the sphere of which the mirror is a part). So the dashed line CB is equal to r, the radius of curvature, and CB is normal to the mirror's surface at B. The incoming ray that hits the mirror at B makes an angle θ with this normal, and hence the reflected ray, BF, also makes an angle θ with the normal (law of reflection). Note that angle BCF is also θ as shown. The triangle CBF is isosceles because two of its angles are equal. Thus we have length CF = BF. We assume the mirror surface is small compared to the mirror's radius of curvature, so the angles are small, and the length FB is nearly equal to length FA. In this approximation, FA = FC. But FA = f, the focal length, and CA = 2 × FA = r. Thus the focal length is half the radius of curvature:

$$f = \frac{r}{2}. \qquad \text{[spherical mirror]} \quad \textbf{(32–1)}$$

We assumed only that the angle θ was small, so this result applies for all other incident paraxial rays. Thus all paraxial rays pass through the same point F, the focal point.

Since it is only approximately true that the rays come to a perfect focus at F, the more curved the mirror, the worse the approximation (Fig. 32–13) and the more blurred the image. This "defect" of spherical mirrors is called **spherical aberration**; we will discuss it more with regard to lenses in Chapter 33. A **parabolic reflector**, on the other hand, will reflect the rays to a perfect focus. However, because parabolic shapes are much harder to make and thus much more expensive, spherical mirrors are used for most purposes. (Many astronomical telescopes use parabolic reflectors.) We consider here only spherical mirrors and we will assume that they are small compared to their radius of curvature so that the image is sharp and Eq. 32–1 holds.

Image Formation—Ray Diagrams

We saw that for an object at infinity, the image is located at the focal point of a concave spherical mirror, where $f = r/2$. But where does the image lie for an object not at infinity? First consider the object shown as an arrow in Fig. 32–15a, which is placed between F and C at point O (O for object). Let us determine where the image will be for a given point O′ at the top of the object.

(a) Ray 1 goes out from O′ parallel to the axis and reflects through F.

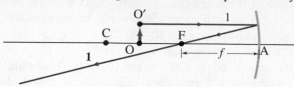

(b) Ray 2 goes through F and then reflects back parallel to the axis.

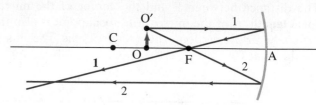

FIGURE 32–15 Rays leave point O′ on the object (an arrow). Shown are the three most useful rays for determining where the image I′ is formed. [Note that our mirror is not small compared to f, so our diagram will not give the precise position of the image.]

(c) Ray 3 is chosen perpendicular to mirror, and so must reflect back on itself and go through C (center of curvature).

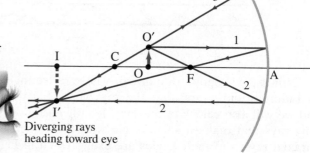

Diverging rays heading toward eye

To do this we can draw several rays and make sure these reflect from the mirror such that the angle of reflection equals the angle of incidence. Many rays could be drawn leaving any point on an object, but determining the image position is simplified if we deal with three particularly simple rays. These are the rays labeled 1, 2, and 3 in Fig. 32–15 and we draw them leaving object point O′ as follows:

Ray 1 is drawn parallel to the axis; therefore after reflection it must pass along a line through F (Fig. 32–15a).

Ray 2 leaves O′ and is made to pass through F (Fig. 32–15b); therefore it must reflect so it is parallel to the axis.

Ray 3 passes through C, the center of curvature (Fig. 32–15c); it is along a radius of the spherical surface and is perpendicular to the mirror, so it is reflected back on itself.

All three rays leave a single point O′ on the object. After reflection from a (small) mirror, the point at which these rays cross is the image point I′. All other rays from the same object point will also pass through this image point. To find the image point for any object point, only these three types of rays need to be drawn. Only two of these rays are needed, but the third serves as a check.

We have shown the image point in Fig. 32–15 only for a single point on the object. Other points on the object are imaged nearby, so a complete image of the object is formed, as shown by the dashed arrow in Fig. 32–15c. Because the light actually passes through the image itself, this is a **real image** that will appear on a piece of paper or film placed there. This can be compared to the virtual image formed by a plane mirror (the light does not actually pass through that image, Fig. 32–7).

The image in Fig. 32–15 can be seen by the eye when the eye is placed to the left of the image, so that some of the rays diverging from each point on the image (as point I′) can enter the eye as shown in Fig. 32–15c. (See also Figs. 32–1 and 32–7.)

Mirror Equation and Magnification

Image points can be determined, roughly, by drawing the three rays as just described, Fig. 32–15. But it is difficult to draw small angles for the "paraxial" rays as we assumed. For more accurate results, we now derive an equation that gives the image distance if the object distance and radius of curvature of the mirror are known. To do this, we refer to Fig. 32–16. The **object distance**, d_o, is the distance of the object (point O) from the center of the mirror. The **image distance**, d_i, is the distance of the image (point I) from the center of the mirror. The height of the object OO′ is called h_o and the height of the image, I′I, is h_i.

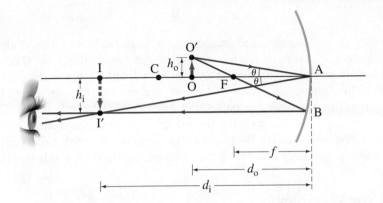

FIGURE 32–16 Diagram for deriving the mirror equation. For the derivation, we assume the mirror size is small compared to its radius of curvature.

Two rays leaving O′ are shown: O′FBI′ (same as ray 2 in Fig. 32–15) and O′AI′, which is a fourth type of ray that reflects at the center of the mirror and can also be used to find an image point. The ray O′AI′ obeys the law of reflection, so the two right triangles O′AO and I′AI are similar. Therefore, we have

$$\frac{h_o}{h_i} = \frac{d_o}{d_i}.$$

For the other ray shown, O′FBI′, the triangles O′FO and AFB are also similar because the angles are equal and we use the approximation AB = h_i (mirror small compared to its radius). Furthermore FA = f, the focal length of the mirror, so

$$\frac{h_o}{h_i} = \frac{OF}{FA} = \frac{d_o - f}{f}.$$

The left sides of the two preceding expressions are the same, so we can equate the right sides:

$$\frac{d_o}{d_i} = \frac{d_o - f}{f}.$$

We now divide both sides by d_o and rearrange to obtain

$$\frac{1}{d_o} + \frac{1}{d_i} = \frac{1}{f}. \qquad (32\text{–}2)$$

Mirror equation

This is the equation we were seeking. It is called the **mirror equation** and relates the object and image distances to the focal length f (where $f = r/2$).

The **lateral magnification**, m, of a mirror is defined as the height of the image divided by the height of the object. From our first set of similar triangles above, or the first equation on this page, we can write:

$$m = \frac{h_i}{h_o} = -\frac{d_i}{d_o}. \qquad (32\text{–}3)$$

The minus sign in Eq. 32–3 is inserted as a convention. Indeed, we must be careful about the signs of all quantities in Eqs. 32–2 and 32–3. Sign conventions are chosen so as to give the correct locations and orientations of images, as predicted by ray diagrams.

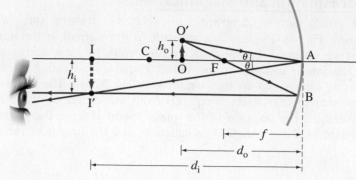

FIGURE 32–16 (Repeated from previous page.)

The sign conventions we use are: the image height h_i is positive if the image is upright, and negative if inverted, relative to the object (assuming h_o is taken as positive); d_i or d_o is positive if image or object is in front of the mirror (as in Fig. 32–16); if either image or object is behind the mirror, the corresponding distance is negative (an example can be seen in Fig. 32–17, Example 32–6). Thus the magnification (Eq. 32–3) is positive for an upright image and negative for an inverted image (upside down). We summarize sign conventions more fully in the Problem Solving Strategy following our discussion of convex mirrors later in this Section.

Concave Mirror Examples

EXAMPLE 32–4 **Image in a concave mirror.** A 1.50-cm-high diamond ring is placed 20.0 cm from a concave mirror with radius of curvature 30.0 cm. Determine (*a*) the position of the image, and (*b*) its size.

APPROACH We determine the focal length from the radius of curvature (Eq. 32–1), $f = r/2 = 15.0$ cm. The ray diagram is basically like that shown in Fig. 32–16 (repeated here on this page), since the object is between F and C. The position and size of the image are found from Eqs. 32–2 and 32–3.

SOLUTION Referring to Fig. 32–16, we have CA = r = 30.0 cm, FA = f = 15.0 cm, and OA = d_o = 20.0 cm.

(*a*) From Eq. 32–2,

$$\frac{1}{d_i} = \frac{1}{f} - \frac{1}{d_o}$$

$$= \frac{1}{15.0 \text{ cm}} - \frac{1}{20.0 \text{ cm}} = 0.0167 \text{ cm}^{-1}.$$

So $d_i = 1/(0.0167 \text{ cm}^{-1}) = 60.0$ cm. Because d_i is positive, the image is 60.0 cm in front of the mirror, on the same side as the object.

(*b*) From Eq. 32–3, the magnification is

$$m = -\frac{d_i}{d_o}$$

$$= -\frac{60.0 \text{ cm}}{20.0 \text{ cm}} = -3.00.$$

The image is 3.0 times larger than the object, and its height is

$$h_i = mh_o = (-3.00)(1.5 \text{ cm}) = -4.5 \text{ cm}.$$

The minus sign reminds us that the image is inverted, as in Fig. 32–16.

NOTE When an object is further from a concave mirror than the focal point, we can see from Fig. 32–15 or 32–16 that the image is always inverted and real.

CONCEPTUAL EXAMPLE 32–5 **Reversible rays.** If the object in Example 32–4 is placed instead where the image is (see Fig. 32–16), where will the new image be?

RESPONSE The mirror equation is *symmetric* in d_o and d_i. Thus the new image will be where the old object was. Indeed, in Fig. 32–16 we need only reverse the direction of the rays to get our new situation.

EXAMPLE 32–6 **Object closer to concave mirror.** A 1.00-cm-high object is placed 10.0 cm from a concave mirror whose radius of curvature is 30.0 cm. (*a*) Draw a ray diagram to locate (approximately) the position of the image. (*b*) Determine the position of the image and the magnification analytically.

APPROACH We draw the ray diagram using the rays as in Fig. 32–15, page 844. An analytic solution uses Eqs. 32–1, 32–2, and 32–3.

SOLUTION (*a*) Since $f = r/2 = 15.0$ cm, the object is between the mirror and the focal point. We draw the three rays as described earlier (Fig. 32–15); they are shown leaving the tip of the object in Fig. 32–17. Ray 1 leaves the tip of our object heading toward the mirror parallel to the axis, and reflects through F. Ray 2 cannot head toward F because it would not strike the mirror; so ray 2 must point as if it started at F (dashed line) and heads to the mirror, and then is reflected parallel to the principal axis. Ray 3 is perpendicular to the mirror, as before. The rays reflected from the mirror diverge and so never meet at a point. They appear, however, to be coming from a point behind the mirror. This point locates the image of the tip of the arrow. The image is thus behind the mirror and is *virtual*. (Why?)

(*b*) We use Eq. 32–2 to find d_i when $d_o = 10.0$ cm:

$$\frac{1}{d_i} = \frac{1}{f} - \frac{1}{d_o} = \frac{1}{15.0 \text{ cm}} - \frac{1}{10.0 \text{ cm}} = \frac{2 - 3}{30.0 \text{ cm}} = -\frac{1}{30.0 \text{ cm}}.$$

Therefore, $d_i = -30.0$ cm. The minus sign means the image is behind the mirror, which our diagram also told us. The magnification is $m = -d_i/d_o = -(-30.0 \text{ cm})/(10.0 \text{ cm}) = +3.00$. So the image is 3.00 times larger than the object. The plus sign indicates that the image is upright (same as object), which is consistent with the ray diagram, Fig. 32–17.

NOTE The image distance cannot be obtained accurately by measuring on Fig. 32–17, because our diagram violates the paraxial ray assumption (we draw rays at steeper angles to make them clearly visible).

NOTE When the object is located inside the focal point of a concave mirror ($d_o < f$), the image is always upright and virtual. And if the object O in Fig. 32–17 is you, you see yourself clearly, because the reflected rays at point O are diverging. Your image is upright and enlarged.

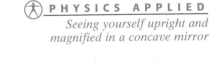

PHYSICS APPLIED
Seeing yourself upright and magnified in a concave mirror

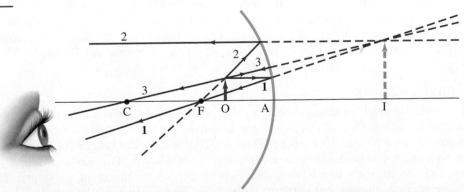

FIGURE 32–17 Object placed within the focal point F. The image is *behind* the mirror and is *virtual*, Example 32–6. [Note that the vertical scale (height of object = 1.0 cm) is different from the horizontal (OA = 10.0 cm) for ease of drawing, and reduces the precision of the drawing.]

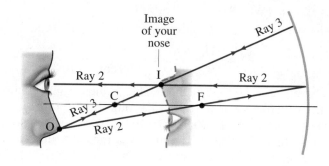

PHYSICS APPLIED

Magnifying mirror
(shaving/cosmetic)

It is useful to compare Figs. 32–15 and 32–17. We can see that if the object is within the focal point $(d_o < f)$, as in Fig. 32–17, the image is virtual, upright, and magnified. This is how a shaving or cosmetic mirror is used—you must place your head closer to the mirror than the focal point if you are to see yourself right-side up (see the photograph of Fig. 32–11a). If the object is *beyond* the focal point, as in Fig. 32–15, the image is real and inverted (upside down—and hard to use!). Whether the magnification has magnitude greater or less than 1.0 in the latter case depends on the position of the object relative to the center of curvature, point C. Practice making ray diagrams with various object distances.

The mirror equation also holds for a plane mirror: the focal length is $f = r/2 = \infty$, and Eq. 32–2 gives $d_i = -d_o$.

Seeing the Image

For a person's eye to see a sharp image, the eye must be at a place where it intercepts diverging rays from points on the image, as is the case for the eye's position in Figs. 32–15 and 32–16. Our eyes are made to see normal objects, which always means the rays are diverging toward the eye as shown in Fig. 32–1. (Or, for very distant objects like stars, the rays become essentially parallel, as in Fig. 32–12.) If you placed your eye between points O and I in Fig. 32–16, for example, *converging* rays from the object OO′ would enter your eye and the lens of your eye could not bring them to a focus; you would see a blurry image. [We will discuss the eye more in Chapter 33.]

FIGURE 32–18 You can see a clear inverted image of your face when you are beyond C $(d_o > 2f)$, because the rays that arrive at your eye are diverging. Standard rays 2 and 3 are shown leaving point O on your nose. Ray 2 (and other nearby rays) enters your eye. Notice that rays are diverging as they move to the left of image point I.

FIGURE 32–19 Convex mirror: (a) the focal point is at F, behind the mirror; (b) the image I of the object at O is virtual, upright, and smaller than the object. [Not to scale for Example 32–7.]

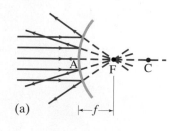

(a)

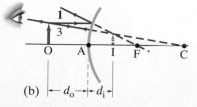

(b)

If *you* are the object OO′ in Fig. 32–16, situated between F and C, and are trying to see yourself in the mirror, you would see a blur; but the person whose eye is shown in Fig. 32–16 can see you clearly. You can see yourself clearly, but upside down, if you are to the left of C in Fig. 32–16, where $d_o > 2f$. Why? Because then the rays arriving from the image will be *diverging* at your position as demonstrated in Fig. 32–18, and your eye can then focus them. You can also see yourself clearly, and right-side up, if you are closer to the mirror than its focal point $(d_o < f)$, as we saw in Example 32–6, Fig. 32–17.

Convex Mirrors

The analysis used for concave mirrors can be applied to **convex** mirrors. Even the mirror equation (Eq. 32–2) holds for a convex mirror, although the quantities involved must be carefully defined. Figure 32–19a shows parallel rays falling on a convex mirror. Again spherical aberration is significant (Fig. 32–13), unless we assume the mirror's size is very small compared to its radius of curvature. The reflected rays diverge, but seem to come from point F behind the mirror. This is the **focal point**, and its distance from the center of the mirror (point A) is the **focal length**, f. It is easy to show that again $f = r/2$. We see that an object at infinity produces a virtual image in a convex mirror. Indeed, no matter where the object is placed on the reflecting side of a convex mirror, the image will be virtual and upright, as indicated in Fig. 32–19b. To find the image we draw rays 1 and 3 according to the rules used before on the concave mirror, as shown in Fig. 32–19b. Note that although rays 1 and 3 don't actually pass through points F and C, the line along which each is drawn does (shown dashed).

The mirror equation, Eq. 32–2, holds for convex mirrors but the focal length f must be considered negative, as must the radius of curvature. The proof is left as a Problem. It is also left as a Problem to show that Eq. 32–3 for the magnification is also valid.

Spherical Mirrors

1. Always **draw a ray diagram** even though you are going to make an analytic calculation—the diagram serves as a check, even if not precise. From one point on the object, draw at least two, preferably three, of the easy-to-draw rays using the rules described in Fig. 32–15. The image point is where the reflected rays intersect or appear to intersect.

2. Apply the **mirror equation**, Eq. 32–2, and the **magnification equation**, Eq. 32–3. It is crucially important to follow the sign conventions—see the next point.

3. **Sign Conventions**
   (a) When the object, image, or focal point is on the reflecting side of the mirror (on the left in our drawings), the corresponding distance is positive. If any of these points is behind the mirror (on the right) the corresponding distance is negative.[†]
   (b) The image height h_i is positive if the image is upright, and negative if inverted, relative to the object (h_o is always taken as positive).

4. **Check** that the analytical solution is consistent with the ray diagram.

[†]Object distances are positive for material objects, but can be negative in systems with more than one mirror or lens—see Section 33–3.

EXAMPLE 32–7 **Convex rearview mirror.** An external rearview car mirror is convex with a radius of curvature of 16.0 m (Fig. 32–20). Determine the location of the image and its magnification for an object 10.0 m from the mirror.

PHYSICS APPLIED
Convex rearview mirror

APPROACH We follow the steps of the Problem Solving Strategy explicitly.

SOLUTION

1. **Draw a ray diagram.** The ray diagram will be like Fig. 32–19b, but the large object distance $(d_o = 10.0\,\text{m})$ makes a precise drawing difficult. We have a convex mirror, so r is negative by convention.

2. **Mirror and magnification equations.** The center of curvature of a convex mirror is behind the mirror, as is its focal point, so we set $r = -16.0\,\text{m}$ so that the focal length is $f = r/2 = -8.0\,\text{m}$. The object is in front of the mirror, $d_o = 10.0\,\text{m}$. Solving the mirror equation, Eq. 32–2, for $1/d_i$ gives

$$\frac{1}{d_i} = \frac{1}{f} - \frac{1}{d_o} = \frac{1}{-8.0\,\text{m}} - \frac{1}{10.0\,\text{m}} = \frac{-10.0 - 8.0}{80.0\,\text{m}} = -\frac{18}{80.0\,\text{m}}.$$

FIGURE 32–20 Example 32–7.

Thus $d_i = -80.0\,\text{m}/18 = -4.4\,\text{m}$. Equation 32–3 gives the magnification

$$m = -\frac{d_i}{d_o} = -\frac{(-4.4\,\text{m})}{(10.0\,\text{m})} = +0.44.$$

3. **Sign conventions.** The image distance is negative, $-4.4\,\text{m}$, so the image is *behind* the mirror. The magnification is $m = +0.44$, so the image is *upright* (same orientation as object) and less than half as tall as the object.

4. **Check.** Our results are consistent with Fig. 32–19b.

Convex rearview mirrors on vehicles sometimes come with a warning that objects are closer than they appear in the mirror. The fact that d_i may be smaller than d_o (as in Example 32–7) seems to contradict this observation. The real reason the object seems farther away is that its image in the convex mirror is *smaller* than it would be in a plane mirror, and we judge distance of ordinary objects such as other cars mostly by their size.

TABLE 32–1 Indices of Refraction[†]	
Material	$n = \dfrac{c}{v}$
Vacuum	1.0000
Air (at STP)	1.0003
Water	1.33
Ethyl alcohol	1.36
Glass	
Fused quartz	1.46
Crown glass	1.52
Light flint	1.58
Lucite or Plexiglas	1.51
Sodium chloride	1.53
Diamond	2.42

[†] $\lambda = 589$ nm.

32–4 Index of Refraction

We saw in Chapter 31 that the speed of light in vacuum is

$$c = 2.99792458 \times 10^8 \, \text{m/s},$$

which is usually rounded off to

$$3.00 \times 10^8 \, \text{m/s}$$

when extremely precise results are not required.

In air, the speed is only slightly less. In other transparent materials such as glass and water, the speed is always less than that in vacuum. For example, in water light travels at about $\frac{3}{4}c$. The ratio of the speed of light in vacuum to the speed v in a given material is called the **index of refraction**, n, of that material:

$$n = \frac{c}{v}. \tag{32–4}$$

The index of refraction is never less than 1, and values for various materials are given in Table 32–1. For example, since $n = 2.42$ for diamond, the speed of light in diamond is

$$v = \frac{c}{n} = \frac{\left(3.00 \times 10^8 \, \text{m/s}\right)}{2.42} = 1.24 \times 10^8 \, \text{m/s}.$$

As we shall see later, n varies somewhat with the wavelength of the light—except in vacuum—so a particular wavelength is specified in Table 32–1, that of yellow light with wavelength $\lambda = 589$ nm.

That light travels more slowly in matter than in vacuum can be explained at the atomic level as being due to the absorption and reemission of light by atoms and molecules of the material.

32–5 Refraction: Snell's Law

When light passes from one transparent medium into another with a different index of refraction, part of the incident light is reflected at the boundary. The remainder passes into the new medium. If a ray of light is incident at an angle to the surface (other than perpendicular), the ray changes direction as it enters the new medium. This change in direction, or bending, is called **refraction**.

Figure 32–21a shows a ray passing from air into water. Angle θ_1 is the angle the incident ray makes with the normal (perpendicular) to the surface and is called the **angle of incidence**. Angle θ_2 is the **angle of refraction**, the angle the refracted ray makes with the normal to the surface. Notice that the ray bends toward the normal when entering the water. This is always the case when the ray enters a medium where the speed of light is *less* (and the index of refraction greater, Eq. 32–4). If light travels from one medium into a second where its speed is *greater*, the ray bends away from the normal; this is shown in Fig. 32–21b for a ray traveling from water to air.

⚠ **CAUTION**

Angles θ_1 and θ_2 are measured from the perpendicular, not from surface

FIGURE 32–21 Refraction.
(a) Light refracted when passing from air (n_1) into water (n_2): $n_2 > n_1$.
(b) Light refracted when passing from water (n_1) into air (n_2): $n_1 > n_2$.

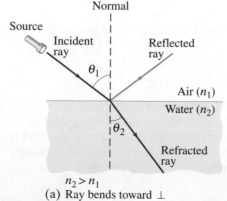

$n_2 > n_1$
(a) Ray bends toward ⊥

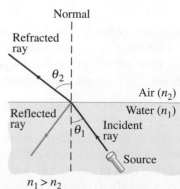

$n_1 > n_2$
(b) Ray bends away from ⊥

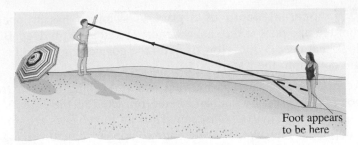

FIGURE 32–22 Ray diagram showing why a person's legs look shorter when standing in waist-deep water: the path of light traveling from the bather's foot to the observer's eye bends at the water's surface, and our brain interprets the light as having traveled in a straight line, from higher up (dashed line).

Foot appears to be here

Refraction is responsible for a number of common optical illusions. For example, a person standing in waist-deep water appears to have shortened legs. As shown in Fig. 32–22, the rays leaving the person's foot are bent at the surface. The observer's brain assumes the rays to have traveled a straight-line path (dashed red line), and so the feet appear to be higher than they really are. Similarly, when you put a straw in water, it appears to be bent (Fig. 32–23).

Snell's Law

The angle of refraction depends on the speed of light in the two media and on the incident angle. An analytical relation between θ_1 and θ_2 in Fig. 32–21 was arrived at experimentally about 1621 by Willebrord Snell (1591–1626). It is known as **Snell's law** and is written:

$$n_1 \sin \theta_1 = n_2 \sin \theta_2. \qquad (32\text{–}5)$$

θ_1 is the angle of incidence and θ_2 is the angle of refraction; n_1 and n_2 are the respective indices of refraction in the materials. See Fig. 32–21. The incident and refracted rays lie in the same plane, which also includes the perpendicular to the surface. Snell's law is the **law of refraction**. (Snell's law was derived in Section 15–10 where Eq. 15–19 is just a combination of Eqs. 32–5 and 32–4. We also derive it in Chapter 34 using the wave theory of light.)

It is clear from Snell's law that if $n_2 > n_1$, then $\theta_2 < \theta_1$. That is, if light enters a medium where n is greater (and its speed is less), then the ray is bent toward the normal. And if $n_2 < n_1$, then $\theta_2 > \theta_1$, so the ray bends away from the normal. This is what we saw in Fig. 32–21.

EXERCISE D Light passes from a medium with $n = 1.3$ into a medium with $n = 1.5$. Is the light bent toward or away from the perpendicular to the interface?

EXAMPLE 32–8 **Refraction through flat glass.** Light traveling in air strikes a flat piece of uniformly thick glass at an incident angle of 60°, as shown in Fig. 32–24. If the index of refraction of the glass is 1.50, (a) what is the angle of refraction θ_A in the glass; (b) what is the angle θ_B at which the ray emerges from the glass?

APPROACH We apply Snell's law at the first surface, where the light enters the glass, and again at the second surface where it leaves the glass and enters the air.

SOLUTION (a) The incident ray is in air, so $n_1 = 1.00$ and $n_2 = 1.50$. Applying Snell's law where the light enters the glass $(\theta_1 = 60°)$ gives

$$\sin \theta_A = \frac{1.00}{1.50} \sin 60° = 0.5774,$$

so $\theta_A = 35.3°$.

(b) Since the faces of the glass are parallel, the incident angle at the second surface is just θ_A (simple geometry), so $\sin \theta_A = 0.5774$. At this second interface, $n_1 = 1.50$ and $n_2 = 1.00$. Thus the ray re-enters the air at an angle $\theta_B \ (= \theta_2)$ given by

$$\sin \theta_B = \frac{1.50}{1.00} \sin \theta_A = 0.866,$$

and $\theta_B = 60°$. The direction of a light ray is thus unchanged by passing through a flat piece of glass of uniform thickness.

NOTE This result is valid for any angle of incidence. The ray is displaced slightly to one side, however. You can observe this by looking through a piece of glass (near its edge) at some object and then moving your head to the side slightly so that you see the object directly. It "jumps."

FIGURE 32–23 A straw in water looks bent even when it isn't.

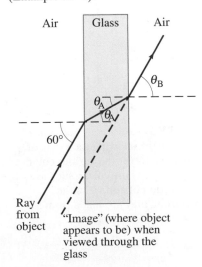

SNELL'S LAW
(LAW OF REFRACTION)

FIGURE 32–24 Light passing through a piece of glass (Example 32–8).

Air   Glass   Air

θ_B

θ_A
θ_A

60°

Ray from object

"Image" (where object appears to be) when viewed through the glass

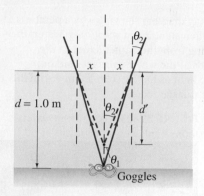

FIGURE 32–25 Example 32–9.

EXAMPLE 32–9 Apparent depth of a pool. A swimmer has dropped her goggles to the bottom of a pool at the shallow end, marked as 1.0 m deep. But the goggles don't look that deep. Why? How deep do the goggles appear to be when you look straight down into the water?

APPROACH We draw a ray diagram showing two rays going upward from a point on the goggles at a small angle, and being refracted at the water's (flat) surface, Fig. 32–25. The two rays traveling upward from the goggles are refracted *away* from the normal as they exit the water, and so appear to be diverging from a point above the goggles (dashed lines), which is why the water seems less deep than it actually is.

SOLUTION To calculate the apparent depth d' (Fig. 32–25), given a real depth $d = 1.0$ m, we use Snell's law with $n_1 = 1.33$ for water and $n_2 = 1.0$ for air:

$$\sin \theta_2 = n_1 \sin \theta_1.$$

We are considering only small angles, so $\sin \theta \approx \tan \theta \approx \theta$, with θ in radians. So Snell's law becomes

$$\theta_2 \approx n_1 \theta_1.$$

From Fig. 32–25, we see that

$$\theta_2 \approx \tan \theta_2 = \frac{x}{d'} \quad \text{and} \quad \theta_1 \approx \tan \theta_1 = \frac{x}{d}.$$

Putting these into Snell's law, $\theta_2 \approx n_1 \theta_1$, we get

$$\frac{x}{d'} \approx n_1 \frac{x}{d}$$

or

$$d' \approx \frac{d}{n_1} = \frac{1.0 \text{ m}}{1.33} = 0.75 \text{ m}.$$

The pool seems only three-fourths as deep as it actually is.

32–6 Visible Spectrum and Dispersion

An obvious property of visible light is its color. Color is related to the wavelengths or frequencies of the light. (How this was discovered will be discussed in Chapter 34.) Visible light—that to which our eyes are sensitive—has wavelengths in air in the range of about 400 nm to 750 nm.[†] This is known as the **visible spectrum**, and within it lie the different colors from violet to red, as shown in Fig. 32–26. Light with wavelength shorter than 400 nm (= violet) is called **ultraviolet** (UV), and light with wavelength greater than 750 nm (= red) is called **infrared** (IR).[‡] Although human eyes are not sensitive to UV or IR, some types of photographic film and digital cameras do respond to them.

A prism can separate white light into a rainbow of colors, as shown in Fig. 32–27. This happens because the index of refraction of a material depends on the wavelength, as shown for several materials in Fig. 32–28. White light is a

FIGURE 32–26 The spectrum of visible light, showing the range of wavelengths for the various colors as seen in air. Many colors, such as brown, do not appear in the spectrum; they are made from a mixture of wavelengths.

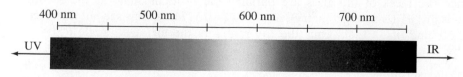

[†]Sometimes the angstrom (Å) unit is used when referring to light: 1 Å = 1×10^{-10} m. Then visible light falls in the wavelength range of 4000 Å to 7500 Å.

[‡]The complete electromagnetic spectrum is illustrated in Fig. 31–12.

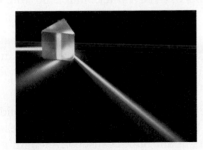

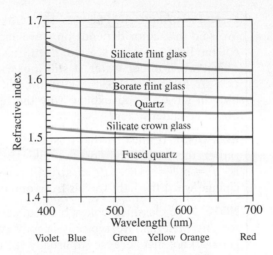

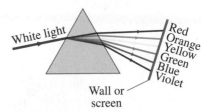

FIGURE 32–27 White light passing through a prism is broken down into its constituent colors.

FIGURE 32–28 Index of refraction as a function of wavelength for various transparent solids.

FIGURE 32–29 White light dispersed by a prism into the visible spectrum.

⊛ PHYSICS APPLIED
Rainbows

mixture of all visible wavelengths, and when incident on a prism, as in Fig. 32–29, the different wavelengths are bent to varying degrees. Because the index of refraction is greater for the shorter wavelengths, violet light is bent the most and red the least, as indicated. This spreading of white light into the full spectrum is called **dispersion**.

Rainbows are a spectacular example of dispersion—by drops of water. You can see rainbows when you look at falling water droplets with the Sun behind you. Figure 32–30 shows how red and violet rays are bent by spherical water droplets and are reflected off the back surface of the droplet. Red is bent the least and so reaches the observer's eyes from droplets higher in the sky, as shown in the diagram. Thus the top of the rainbow is red.

FIGURE 32–30 (a) Ray diagram explaining how a rainbow (b) is formed.

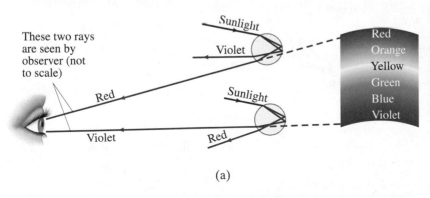

(a)

(b)

The visible spectrum, Fig. 32–26, does not show all the colors seen in nature. For example, there is no brown in Fig. 32–26. Many of the colors we see are a mixture of wavelengths. For practical purposes, most natural colors can be reproduced using three primary colors. They are red, green, and blue for direct source viewing such as TV and computer monitors. For inks used in printing, the primary colors are cyan (the color of the margin notes in this book), yellow, and magenta (the color we use for light rays in diagrams).

For any wave, its velocity v is related to its wavelength λ and frequency f by $v = f\lambda$ (Eq. 15–1 or 31–14). When a wave travels from one material into another, the frequency of the wave does not change across the boundary since a point (an atom) at the boundary oscillates at that frequency. Thus if light goes from air into a material with index of refraction n, the wavelength becomes (recall Eq. 32–4):

$$\lambda_n = \frac{v}{f} = \frac{c}{nf} = \frac{\lambda}{n} \tag{32–6}$$

where λ is the wavelength in vacuum or air and λ_n is the wavelength in the material with index of refraction n.

Observed color of light under water.
We said that color depends on wavelength. For example, for an object emitting 650 nm light in air, we see red. But this is true only in air. If we observe this same object when under water, it still looks red. But the wavelength in water λ_n is (Eq. 32–6) 650 nm/1.33 = 489 nm. Light with wavelength 489 nm would appear blue in air. Can you explain why the light appears red rather than blue when observed under water?

RESPONSE It must be that it is not the wavelength that the eye responds to, but rather the frequency. For example, the frequency of 650 nm red light in air is $f = c/\lambda = (3.0 \times 10^8 \, \text{m/s})/(650 \times 10^{-9} \, \text{m}) = 4.6 \times 10^{14} \, \text{Hz}$, and does not change when the light travels from one medium to another. Only λ changes.

NOTE If we classified colors by frequency, the color assignments would be valid for any material. We typically specify colors by *wavelength* in air (even if less general) not just because we usually see objects in air, but because wavelength is what is commonly measured (it is easier to measure than frequency).

32–7 Total Internal Reflection; Fiber Optics

When light passes from one material into a second material where the index of refraction is less (say, from water into air), the light bends away from the normal, as for rays I and J in Fig. 32–31. At a particular incident angle, the angle of refraction will be 90°, and the refracted ray would skim the surface (ray K) in this

FIGURE 32–31 Since $n_2 < n_1$, light rays are totally internally reflected if the incident angle $\theta_1 > \theta_C$, as for ray L. If $\theta_1 < \theta_C$, as for rays I and J, only a part of the light is reflected, and the rest is refracted.

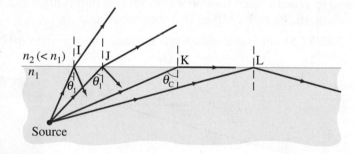

case. The incident angle at which this occurs is called the **critical angle**, θ_C. From Snell's law, θ_C is given by

$$\sin \theta_C = \frac{n_2}{n_1} \sin 90° = \frac{n_2}{n_1}. \tag{32–7}$$

For any incident angle less than θ_C, there will be a refracted ray, although part of the light will also be reflected at the boundary. However, for incident angles greater than θ_C, Snell's law would tell us that $\sin \theta_2$ is greater than 1.00. Yet the sine of an angle can never be greater than 1.00. In this case there is no refracted ray at all, and *all of the light is reflected*, as for ray L in Fig. 32–31. This effect is called **total internal reflection**. Total internal reflection can occur only when light strikes a boundary where the medium beyond has a lower index of refraction.

⚠ **CAUTION**

Total internal reflection (occurs only if refractive index is smaller beyond boundary)

EXERCISE E Fill a sink with water. Place a waterproof watch just below the surface with the watch's flat crystal parallel to the water surface. From above you can still see the watch reading. As you move your head to one side far enough what do you see, and why?

CONCEPTUAL EXAMPLE 32–11 **View up from under water.** Describe what a person would see who looked up at the world from beneath the perfectly smooth surface of a lake or swimming pool.

RESPONSE For an air–water interface, the critical angle is given by

$$\sin \theta_C = \frac{1.00}{1.33} = 0.750.$$

Therefore, $\theta_C = 49°$. Thus the person would see the outside world compressed into a circle whose edge makes a 49° angle with the vertical. Beyond this angle, the person would see reflections from the sides and bottom of the lake or pool (Fig. 32–32).

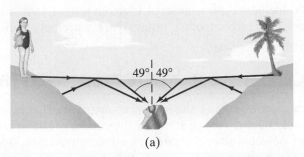

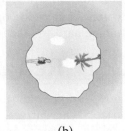

(a)

(b)

FIGURE 32–32 (a) Light rays, and (b) view looking upward from beneath the water (the surface of the water must be very smooth). Example 32–11.

Diamonds achieve their brilliance from a combination of dispersion and total internal reflection. Because diamonds have a very high index of refraction of about 2.4, the critical angle for total internal reflection is only 25°. The light dispersed into a spectrum inside the diamond therefore strikes many of the internal surfaces of the diamond before it strikes one at less than 25° and emerges.

After many such reflections, the light has traveled far enough that the colors have become sufficiently separated to be seen individually and brilliantly by the eye after leaving the diamond.

Many optical instruments, such as binoculars, use total internal reflection within a prism to reflect light. The advantage is that very nearly 100% of the light is reflected, whereas even the best mirrors reflect somewhat less than 100%. Thus the image is brighter, especially after several reflections. For glass with $n = 1.50$, $\theta_C = 41.8°$. Therefore, 45° prisms will reflect all the light internally, if oriented as shown in the binoculars of Fig. 32–33.

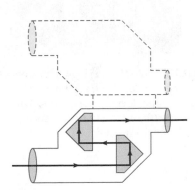

FIGURE 32–33 Total internal reflection of light by prisms in binoculars.

EXERCISE F If 45.0° plastic prisms are used in binoculars, what minimum index of refraction must the plastic have?

Fiber Optics

Total internal reflection is the principle behind **fiber optics**. Glass and plastic fibers as thin as a few micrometers in diameter are common. A bundle of such tiny fibers is called a **light pipe** or cable, and light[†] can be transmitted along it with almost no loss because of total internal reflection. Figure 32–34 shows how light traveling down a thin fiber makes only glancing collisions with the walls so that total internal reflection occurs. Even if the light pipe is bent into a complicated shape, the critical angle still won't be exceeded, so light is transmitted practically undiminished to the other end. Very small losses do occur, mainly by reflection at the ends and absorption within the fiber.

Important applications of fiber-optic cables are in communications and medicine. They are used in place of wire to carry telephone calls, video signals, and computer data. The signal is a modulated light beam (a light beam whose intensity can be varied) and data is transmitted at a much higher rate and with less loss and less interference than an electrical signal in a copper wire. Fibers have been developed that can support over one hundred separate wavelengths, each modulated to carry up to 10 gigabits (10^{10} bits) of information per second. That amounts to a terabit (10^{12} bits) per second for the full one hundred wavelengths.

FIGURE 32–34 Light reflected totally at the interior surface of a glass or transparent plastic fiber.

PHYSICS APPLIED
Fiber optics in communications

[†]Fiber optic devices can use not only visible light but also infrared light, ultraviolet light, and microwaves.

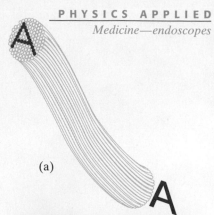

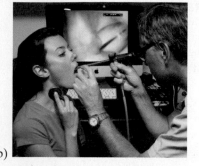

(a)

(b)

FIGURE 32–35 (a) How a fiber-optic image is made. (b) Example of a fiber-optic device inserted through the mouth to view the vocal cords, with image on screen.

The sophisticated use of fiber optics to transmit a clear picture is particularly useful in medicine, Fig. 32–35. For example, a patient's lungs can be examined by inserting a light pipe known as a bronchoscope through the mouth and down the bronchial tube. Light is sent down an outer set of fibers to illuminate the lungs. The reflected light returns up a central core set of fibers. Light directly in front of each fiber travels up that fiber. At the opposite end, a viewer sees a series of bright and dark spots, much like a TV screen—that is, a picture of what lies at the opposite end. Lenses are used at each end. The image may be viewed directly or on a monitor screen or film. The fibers must be optically insulated from one another, usually by a thin coating of material with index of refraction less than that of the fiber. The more fibers there are, and the smaller they are, the more detailed the picture. Such instruments, including bronchoscopes, colonoscopes (for viewing the colon), and endoscopes (stomach or other organs), are extremely useful for examining hard-to-reach places.

*32–8 Refraction at a Spherical Surface

We now examine the refraction of rays at the spherical surface of a transparent material. Such a surface could be one face of a lens or the cornea of the eye. To be general, let us consider an object which is located in a medium whose index of refraction is n_1, and rays from each point on the object can enter a medium whose index of refraction is n_2. The radius of curvature of the spherical boundary is R, and its center of curvature is at point C, Fig. 32–36. We now show that all rays leaving a point O on the object will be focused at a single point I, the image point, if we consider only paraxial rays: rays that make a small angle with the axis.

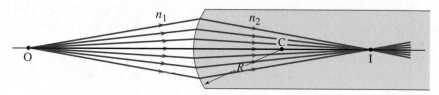

FIGURE 32–36 Rays from a point O on an object will be focused at a single image point I by a spherical boundary between two transparent materials $(n_2 > n_1)$, as long as the rays make small angles with the axis.

To do so, we consider a single ray that leaves point O as shown in Fig. 32–37. From Snell's law, Eq. 32–5, we have

$$n_1 \sin \theta_1 = n_2 \sin \theta_2.$$

We are assuming that angles θ_1, θ_2, α, β, and γ are small, so $\sin \theta \approx \theta$ (in radians), and Snell's law becomes, approximately,

$$n_1 \theta_1 = n_2 \theta_2.$$

Also, $\beta + \phi = 180°$ and $\theta_2 + \gamma + \phi = 180°$, so

$$\beta = \gamma + \theta_2.$$

Similarly for triangle OPC,

$$\theta_1 = \alpha + \beta.$$

FIGURE 32–37 Diagram for showing that all paraxial rays from O focus at the same point I $(n_2 > n_1)$.

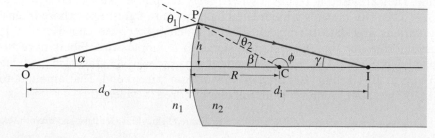

These three relations can be combined to yield

$$n_1 \alpha + n_2 \gamma = (n_2 - n_1)\beta.$$

Since we are considering only the case of small angles, we can write, approximately,

$$\alpha = \frac{h}{d_o}, \qquad \beta = \frac{h}{R}, \qquad \gamma = \frac{h}{d_i},$$

where d_o and d_i are the object and image distances and h is the height as shown in Fig. 32–37. We substitute these into the previous equation, divide through by h, and obtain

$$\frac{n_1}{d_o} + \frac{n_2}{d_i} = \frac{n_2 - n_1}{R}. \tag{32–8}$$

For a given object distance d_o, this equation tells us d_i, the image distance, does not depend on the angle of a ray. Hence all paraxial rays meet at the same point I. This is true only for rays that make small angles with the axis and with each other, and is equivalent to assuming that the width of the refracting spherical surface is small compared to its radius of curvature. If this assumption is not true, the rays will not converge to a point; there will be spherical aberration, just as for a mirror (see Fig. 32–13), and the image will be blurry. (Spherical aberration will be discussed further in Section 33–10.)

We derived Eq. 32–8 using Fig. 32–37 for which the spherical surface is convex (as viewed by the incoming ray). It is also valid for a concave surface—as can be seen using Fig. 32–38—if we use the following conventions:

1. If the surface is convex (so the center of curvature C is on the side of the surface opposite to that from which the light comes), R is positive; if the surface is concave (C on the same side from which the light comes) R is negative.

2. The image distance, d_i, follows the same convention: positive if on the opposite side from where the light comes, negative if on the same side.

3. The object distance is positive if on the same side from which the light comes (this is the normal case, although when several surfaces bend the light it may not be so), otherwise it is negative.

For the case shown in Fig. 32–38 with a concave surface, both R and d_i are negative when used in Eq. 32–8. Note, in this case, that the image is virtual.

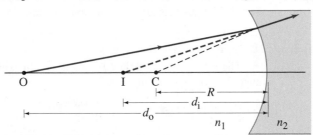

FIGURE 32–38 Rays from O refracted by a concave surface form a virtual image $(n_2 > n_1)$. Per our conventions, $R < 0$, $d_i < 0$, $d_o > 0$.

EXAMPLE 32–12 **Apparent depth II.** A person looks vertically down into a 1.0-m-deep pool. How deep does the water appear to be?

APPROACH Example 32–9 solved this problem using Snell's law. Here we use Eq. 32–8.

SOLUTION A ray diagram is shown in Fig. 32–39. Point O represents a point on the pool's bottom. The rays diverge and appear to come from point I, the image. We have $d_o = 1.0$ m and, for a flat surface, $R = \infty$. Then Eq. 32–8 becomes

$$\frac{1.33}{1.0 \text{ m}} + \frac{1.00}{d_i} = \frac{(1.00 - 1.33)}{\infty} = 0.$$

Hence $d_i = -(1.0 \text{ m})/(1.33) = -0.75$ m. So the pool appears to be only three-fourths as deep as it actually is, the same result we found in Example 32–9. The minus sign tells us the image point I is on the same side of the surface as O, and the image is virtual. At angles other than vertical, this conclusion must be modified.

FIGURE 32–39 Example 32–12.

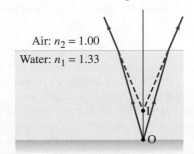

Air: $n_2 = 1.00$

Water: $n_1 = 1.33$

EXAMPLE 32–13 **A spherical "lens."** A point source of light is placed at a distance of 25.0 cm from the center of a glass sphere $(n = 1.5)$ of radius 10.0 cm, Fig. 32–40. Find the image of the source.

APPROACH As shown in Fig. 32–40, there are two refractions, and we treat them successively, one at a time. The light rays from the source first refract from the convex glass surface facing the source. We analyze this first refraction, treating it as in Fig. 32–36, ignoring the back side of the sphere.

SOLUTION Using Eq. 32–8 (assuming paraxial rays) with $n_1 = 1.0$, $n_2 = 1.5$, $R = 10.0$ cm, and $d_o = 25.0$ cm $- 10.0$ cm $= 15.0$ cm, we solve for the image distance as formed at surface 1, d_{i1}:

$$\frac{1}{d_{i1}} = \frac{1}{n_2}\left(\frac{(n_2 - n_1)}{R} - \frac{n_1}{d_o}\right) = \frac{1}{1.5}\left(\frac{1.5 - 1.0}{10.0\ \text{cm}} - \frac{1.0}{15.0\ \text{cm}}\right) = -\frac{1}{90.0\ \text{cm}}.$$

Thus, the image of the first refraction is located 90.0 cm *to the left* of the front surface. This image (I_1) now serves as the object for the refraction occurring at the back surface (surface 2) of the sphere. This surface is concave so $R = -10.0$ cm, and we consider a ray close to the axis. Then the object distance is $d_{o2} = 90.0$ cm $+ 2(10.0\ \text{cm}) = 110.0$ cm, and Eq. 32–8 yields, with $n_1 = 1.5$, $n_2 = 1.0$,

$$\frac{1}{d_{i2}} = \frac{1}{1.0}\left(\frac{1.0 - 1.5}{-10.0\ \text{cm}} - \frac{1.5}{110.0\ \text{cm}}\right) = \frac{4.0}{110.0\ \text{cm}}$$

so $d_{i2} = 28$ cm. Thus, the final image is located a distance 28 cm from the back side of the sphere.

FIGURE 32–40 Example 32–13.

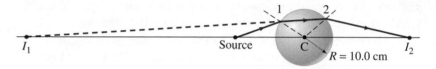

Summary

Light appears to travel along straight-line paths, called **rays**, at a speed v that depends on the **index of refraction**, n, of the material; that is

$$v = \frac{c}{n}, \tag{32–4}$$

where c is the speed of light in vacuum.

When light reflects from a flat surface, *the angle of reflection equals the angle of incidence*. This **law of reflection** explains why mirrors can form **images**.

In a **plane mirror**, the image is virtual, upright, the same size as the object, and is as far behind the mirror as the object is in front.

A **spherical mirror** can be concave or convex. A **concave** spherical mirror focuses parallel rays of light (light from a very distant object) to a point called the **focal point**. The distance of this point from the mirror is the **focal length** f of the mirror and

$$f = \frac{r}{2} \tag{32–1}$$

where r is the radius of curvature of the mirror.

Parallel rays falling on a **convex mirror** reflect from the mirror as if they diverged from a common point behind the mirror. The distance of this point from the mirror is the focal length and is considered negative for a convex mirror.

For a given object, the position and size of the image formed by a mirror can be found by ray tracing. Algebraically, the relation between image and object distances, d_i and d_o, and the focal length f, is given by the **mirror equation**:

$$\frac{1}{d_o} + \frac{1}{d_i} = \frac{1}{f}. \tag{32–2}$$

The ratio of image height to object height, which equals the magnification m of a mirror, is

$$m = \frac{h_i}{h_o} = -\frac{d_i}{d_o}. \tag{32–3}$$

If the rays that converge to form an image actually pass through the image, so the image would appear on film or a screen placed there, the image is said to be a **real image**. If the rays do not actually pass through the image, the image is a **virtual image**.

When light passes from one transparent medium into another, the rays bend or refract. The **law of refraction (Snell's law)** states that

$$n_1 \sin \theta_1 = n_2 \sin \theta_2, \tag{32–5}$$

where n_1 and θ_1 are the index of refraction and angle with the normal to the surface for the incident ray, and n_2 and θ_2 are for the refracted ray.

When light of wavelength λ enters a medium with index of refraction n, the wavelength is reduced to

$$\lambda_n = \frac{\lambda}{n}. \tag{32–6}$$

The frequency does not change.

The frequency or wavelength of light determines its color. The **visible spectrum** in air extends from about 400 nm (violet) to about 750 nm (red).

Glass prisms break white light down into its constituent colors because the index of refraction varies with wavelength, a phenomenon known as **dispersion**.

When light rays reach the boundary of a material where the index of refraction decreases, the rays will be **totally internally reflected** if the incident angle, θ_1, is such that Snell's law would predict $\sin \theta_2 > 1$. This occurs if θ_1 exceeds the critical angle θ_C given by

$$\sin \theta_C = \frac{n_2}{n_1}. \tag{32–7}$$

Questions

1. What would be the appearance of the Moon if it had (a) a rough surface; (b) a polished mirrorlike surface?

2. Archimedes is said to have burned the whole Roman fleet in the harbor of Syracuse by focusing the rays of the Sun with a huge spherical mirror. Is this[†] reasonable?

3. What is the focal length of a plane mirror? What is the magnification of a plane mirror?

4. An object is placed along the principal axis of a spherical mirror. The magnification of the object is -3.0. Is the image real or virtual, inverted or upright? Is the mirror concave or convex? On which side of the mirror is the image located?

5. Using the rules for the three rays discussed with reference to Fig. 32–15, draw ray 2 for Fig. 32–19b.

6. Does the mirror equation, Eq. 32–2, hold for a plane mirror? Explain.

7. If a concave mirror produces a real image, is the image necessarily inverted?

8. How might you determine the speed of light in a solid, rectangular, transparent object?

9. When you look at the Moon's reflection from a ripply sea, it appears elongated (Fig. 32–41). Explain.

FIGURE 32–41
Question 9.

10. How can a spherical mirror have a negative object distance?

11. What is the angle of refraction when a light ray is incident perpendicular to the boundary between two transparent materials?

12. When you look down into a swimming pool or a lake, are you likely to overestimate or underestimate its depth? Explain. How does the apparent depth vary with the viewing angle? (Use ray diagrams.)

13. Draw a ray diagram to show why a stick or straw looks bent when part of it is under water (Fig. 32–23).

†Students at MIT did a feasibility study. See www.mit.edu.

14. When a wide beam of parallel light enters water at an angle, the beam broadens. Explain.

15. You look into an aquarium and view a fish inside. One ray of light from the fish as it emerges from the tank is shown in Fig. 32–42. The apparent position of the fish is also shown. In the drawing, indicate the approximate position of the actual fish. Briefly justify your answer.

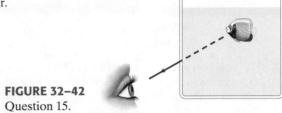

FIGURE 32–42
Question 15.

16. How can you "see" a round drop of water on a table even though the water is transparent and colorless?

17. A ray of light is refracted through three different materials (Fig. 32–43). Rank the materials according to their index of refraction, least to greatest.

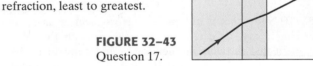

FIGURE 32–43
Question 17.

18. Can a light ray traveling in air be totally reflected when it strikes a smooth water surface if the incident angle is chosen correctly? Explain.

19. When you look up at an object in air from beneath the surface in a swimming pool, does the object appear to be the same size as when you see it directly in air? Explain.

20. What type of mirror is shown in Fig. 32–44?

FIGURE 32–44
Question 20.

21. Light rays from stars (including our Sun) always bend toward the vertical direction as they pass through the Earth's atmosphere. (a) Why does this make sense? (b) What can you conclude about the apparent positions of stars as viewed from Earth?

Problems

32–2 Reflection; Plane Mirrors

1. (I) When you look at yourself in a 60-cm-tall plane mirror, you see the same amount of your body whether you are close to the mirror or far away. (Try it and see.) Use ray diagrams to show why this should be true.

2. (I) Suppose that you want to take a photograph of yourself as you look at your image in a mirror 2.8 m away. For what distance should the camera lens be focused?

3. (II) Two plane mirrors meet at a 135° angle, Fig. 32–45. If light rays strike one mirror at 38° as shown, at what angle ϕ do they leave the second mirror?

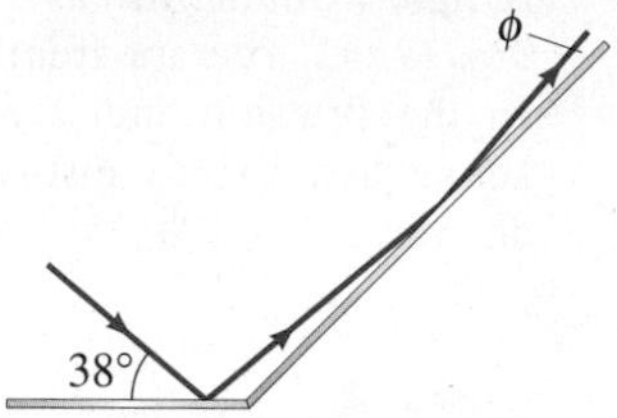

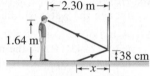

FIGURE 32–45
Problem 3.

38°

4. (II) A person whose eyes are 1.64 m above the floor stands 2.30 m in front of a vertical plane mirror whose bottom edge is 38 cm above the floor, Fig. 32–46. What is the horizontal distance x to the base of the wall supporting the mirror of the nearest point on the floor that can be seen reflected in the mirror?

|← 2.30 m →|

1.64 m

‡38 cm

|←x→|

FIGURE 32–46
Problem 4.

5. (II) Show that if two plane mirrors meet at an angle ϕ, a single ray reflected successively from both mirrors is deflected through an angle of 2ϕ independent of the incident angle. Assume $\phi < 90°$ and that only two reflections, one from each mirror, take place.

6. (II) Suppose you are 88 cm from a plane mirror. What area of the mirror is used to reflect the rays entering one eye from a point on the tip of your nose if your pupil diameter is 4.5 mm?

7. (II) Stand up two plane mirrors so they form a 90.0° angle as in Fig. 32–47. When you look into this double mirror, you see yourself as others see you, instead of reversed as in a single mirror. Make a ray diagram to show how this occurs.

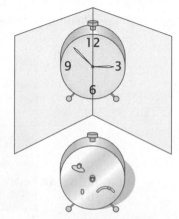

FIGURE 32–47
Problems 7 and 8.

8. (III) Suppose a third mirror is placed beneath the two shown in Fig. 32–47, so that all three are perpendicular to each other. (a) Show that for such a "corner reflector," any incident ray will return in its original direction after three reflections. (b) What happens if it makes only two reflections?

32–3 Spherical Mirrors

9. (I) A solar cooker, really a concave mirror pointed at the Sun, focuses the Sun's rays 18.8 cm in front of the mirror. What is the radius of the spherical surface from which the mirror was made?

10. (I) How far from a concave mirror (radius 24.0 cm) must an object be placed if its image is to be at infinity?

11. (I) When walking toward a concave mirror you notice that the image flips at a distance of 0.50 m. What is the radius of curvature of the mirror?

12. (II) A small candle is 35 cm from a concave mirror having a radius of curvature of 24 cm. (a) What is the focal length of the mirror? (b) Where will the image of the candle be located? (c) Will the image be upright or inverted?

13. (II) You look at yourself in a shiny 9.2-cm-diameter Christmas tree ball. If your face is 25.0 cm away from the ball's front surface, where is your image? Is it real or virtual? Is it upright or inverted?

14. (II) A mirror at an amusement park shows an upright image of any person who stands 1.7 m in front of it. If the image is three times the person's height, what is the radius of curvature of the mirror? (See Fig. 32–44.)

15. (II) A dentist wants a small mirror that, when 2.00 cm from a tooth, will produce a 4.0× upright image. What kind of mirror must be used and what must its radius of curvature be?

16. (II) Some rearview mirrors produce images of cars to your rear that are smaller than they would be if the mirror were flat. Are the mirrors concave or convex? What is a mirror's radius of curvature if cars 18.0 m away appear 0.33 their normal size?

17. (II) You are standing 3.0 m from a convex security mirror in a store. You estimate the height of your image to be half of your actual height. Estimate the radius of curvature of the mirror.

18. (II) An object 3.0 mm high is placed 18 cm from a convex mirror of radius of curvature 18 cm. (a) Show by ray tracing that the image is virtual, and estimate the image distance. (b) Show that the (negative) image distance can be computed from Eq. 32–2 using a focal length of −9.0 cm. (c) Compute the image size, using Eq. 32–3.

19. (II) The image of a distant tree is virtual and very small when viewed in a curved mirror. The image appears to be 16.0 cm behind the mirror. What kind of mirror is it, and what is its radius of curvature?

20. (II) Use two techniques, (a) a ray diagram, and (b) the mirror equation, to show that the magnitude of the magnification of a concave mirror is less than 1 if the object is beyond the center of curvature C $(d_o > r)$, and is greater than 1 if the object is within C $(d_o < r)$.

21. (II) Show, using a ray diagram, that the magnification m of a convex mirror is $m = -d_i/d_o$, just as for a concave mirror. [*Hint*: Consider a ray from the top of the object that reflects at the center of the mirror.]

22. (II) Use ray diagrams to show that the mirror equation, Eq. 32–2, is valid for a convex mirror as long as f is considered negative.

23. (II) The magnification of a convex mirror is +0.55× for objects 3.2 m from the mirror. What is the focal length of this mirror?

24. (II) (a) Where should an object be placed in front of a concave mirror so that it produces an image at the same location as the object? (b) Is the image real or virtual? (c) Is the image inverted or upright? (d) What is the magnification of the image?

25. (II) A 4.5-cm tall object is placed 26 cm in front of a spherical mirror. It is desired to produce a virtual image that is upright and 3.5 cm tall. (a) What type of mirror should be used? (b) Where is the image located? (c) What is the focal length of the mirror? (d) What is the radius of curvature of the mirror?

26. (II) A shaving or makeup mirror is designed to magnify your face by a factor of 1.35 when your face is placed 20.0 cm in front of it. (a) What type of mirror is it? (b) Describe the type of image that it makes of your face. (c) Calculate the required radius of curvature for the mirror.

27. (II) A concave mirror has focal length f. When an object is placed a distance $d_o > f$ from this mirror, a real image with magnification m is formed. (a) Show that $m = f/(f - d_o)$. (b) Sketch m vs. d_o over the range $f < d_o < +\infty$ where $f = 0.45$ m. (c) For what value of d_o will the real image have the same (lateral) size as the object? (d) To obtain a real image that is much larger than the object, in what general region should the object be placed relative to the mirror?

28. (II) Let the focal length of a convex mirror be written as $f = -|f|$. Show that the magnification m of an object a distance d_o from this mirror is given by $m = |f|/(d_o + |f|)$. Based on this relation, explain why your nose looks bigger than the rest of your face when looking into a convex mirror.

29. (II) A spherical mirror of focal length f produces an image of an object with magnification m. (a) Show that the object is a distance $d_o = f\left(1 - \dfrac{1}{m}\right)$ from the reflecting side of the mirror. (b) Use the relation in part (a) to show that, no matter where an object is placed in front of a convex mirror, its image will have a magnification in the range $0 \le m \le +1$.

30. (III) An object is placed a distance r in front of a wall, where r exactly equals the radius of curvature of a certain concave mirror. At what distance from the wall should this mirror be placed so that a real image of the object is formed on the wall? What is the magnification of the image?

31. (III) A short thin object (like a short length of wire) of length ℓ is placed along the axis of a spherical mirror (perpendicular to the glass surface). Show that its image has length $\ell' = m^2\ell$ so the longitudinal magnification is equal to $-m^2$ where m is the normal "lateral" magnification, Eq. 32–3. Why the minus sign? [Hint: Find the image positions for both ends of the wire, and assume ℓ is very small.]

32–4 Index of Refraction

32. (I) The speed of light in ice is 2.29×10^8 m/s. What is the index of refraction of ice?

33. (I) What is the speed of light in (a) ethyl alcohol, (b) lucite, (c) crown glass?

34. (I) Our nearest star (other than the Sun) is 4.2 light years away. That is, it takes 4.2 years for the light to reach Earth. How far away is it in meters?

35. (I) How long does it take light to reach us from the Sun, 1.50×10^8 km away?

36. (II) The speed of light in a certain substance is 88% of its value in water. What is the index of refraction of that substance?

37. (II) Light is emitted from an ordinary lightbulb filament in wave-train bursts of about 10^{-8} s in duration. What is the length in space of such wave trains?

32–5 Snell's Law

38. (I) A diver shines a flashlight upward from beneath the water at a 38.5° angle to the vertical. At what angle does the light leave the water?

39. (I) A flashlight beam strikes the surface of a pane of glass ($n = 1.56$) at a 63° angle to the normal. What is the angle of refraction?

40. (I) Rays of the Sun are seen to make a 33.0° angle to the vertical beneath the water. At what angle above the horizon is the Sun?

41. (I) A light beam coming from an underwater spotlight exits the water at an angle of 56.0°. At what angle of incidence did it hit the air–water interface from below the surface?

42. (II) A beam of light in air strikes a slab of glass ($n = 1.56$) and is partially reflected and partially refracted. Determine the angle of incidence if the angle of reflection is twice the angle of refraction.

43. (II) A light beam strikes a 2.0-cm-thick piece of plastic with a refractive index of 1.62 at a 45° angle. The plastic is on top of a 3.0-cm-thick piece of glass for which $n = 1.47$. What is the distance D in Fig. 32–48?

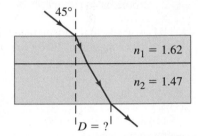

FIGURE 32–48
Problem 43.

44. (II) An aquarium filled with water has flat glass sides whose index of refraction is 1.56. A beam of light from outside the aquarium strikes the glass at a 43.5° angle to the perpendicular (Fig. 32–49). What is the angle of this light ray when it enters (a) the glass, and then (b) the water? (c) What would be the refracted angle if the ray entered the water directly?

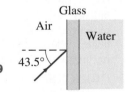

FIGURE 32–49
Problem 44.

45. (II) In searching the bottom of a pool at night, a watchman shines a narrow beam of light from his flashlight, 1.3 m above the water level, onto the surface of the water at a point 2.5 m from his foot at the edge of the pool (Fig. 32–50). Where does the spot of light hit the bottom of the pool, measured from the bottom of the wall beneath his foot, if the pool is 2.1 m deep?

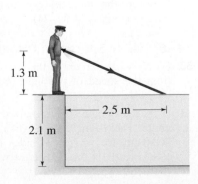

FIGURE 32–50
Problem 45.

46. (II) The block of glass ($n = 1.5$) shown in cross section in Fig. 32–51 is surrounded by air. A ray of light enters the block at its left-hand face with incident angle θ_1 and reemerges into the air from the right-hand face directed parallel to the block's base. Determine θ_1.

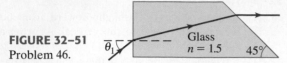

FIGURE 32–51
Problem 46.

Glass
$n = 1.5$ 45°
θ_1

47. (II) A laser beam of diameter $d_1 = 3.0$ mm in air has an incident angle $\theta_1 = 25°$ at a flat air–glass surface. If the index of refraction of the glass is $n = 1.5$, determine the diameter d_2 of the beam after it enters the glass.

48. (II) Light is incident on an equilateral glass prism at a 45.0° angle to one face, Fig. 32–52. Calculate the angle at which light emerges from the opposite face. Assume that $n = 1.54$.

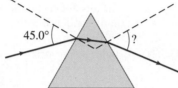

45.0°   ?

FIGURE 32–52
Problems 48 and 65.

49. (II) A triangular prism made of crown glass ($n = 1.52$) with base angles of 30.0° is surrounded by air. If parallel rays are incident normally on its base as shown in Fig. 32–53, what is the angle ϕ between the two emerging rays?

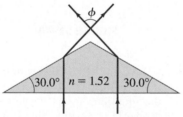

ϕ

30.0° $n = 1.52$ 30.0°

FIGURE 32–53
Problem 49.

50. (II) Show in general that for a light beam incident on a uniform layer of transparent material, as in Fig. 32–24, the direction of the emerging beam is parallel to the incident beam, independent of the incident angle θ. Assume the air on the two sides of the transparent material is the same.

51. (III) A light ray is incident on a flat piece of glass with index of refraction n as in Fig. 32–24. Show that if the incident angle θ is small, the emerging ray is displaced a distance $d = t\theta(n - 1)/n$, where t is the thickness of the glass, θ is in radians, and d is the perpendicular distance between the incident ray and the (dashed) line of the emerging ray (Fig. 32–24).

32–6 Visible Spectrum; Dispersion

52. (I) By what percent is the speed of blue light (450 nm) less than the speed of red light (680 nm), in silicate flint glass (see Fig. 32–28)?

53. (I) A light beam strikes a piece of glass at a 60.00° incident angle. The beam contains two wavelengths, 450.0 nm and 700.0 nm, for which the index of refraction of the glass is 1.4831 and 1.4754, respectively. What is the angle between the two refracted beams?

54. (II) A parallel beam of light containing two wavelengths, $\lambda_1 = 465$ nm and $\lambda_2 = 652$ nm, enters the silicate flint glass of an equilateral prism as shown in Fig. 32–54. At what angle does each beam leave the prism (give angle with normal to the face)? See Fig. 32–28.

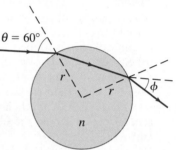

FIGURE 32–54
Problem 54.

55. (III) A ray of light with wavelength λ is incident from air at precisely 60° ($= \theta$) on a spherical water drop of radius r and index of refraction n (which depends on λ). When the ray reemerges into the air from the far side of the drop, it has been deflected an angle ϕ from its original direction as shown in Fig. 32–55. By how much does the value of ϕ for violet light ($n = 1.341$) differ from the value for red light ($n = 1.330$)?

$\theta = 60°$

r r ϕ

n

FIGURE 32–55
Problem 55.

56. (III) For visible light, the index of refraction n of glass is roughly 1.5, although this value varies by about 1% across the visible range. Consider a ray of white light incident from air at angle θ_1 onto a flat piece of glass. (a) Show that, upon entering the glass, the visible colors contained in this incident ray will be dispersed over a range $\Delta\theta_2$ of refracted angles given approximately by

$$\Delta\theta_2 \approx \frac{\sin\theta_1}{\sqrt{n^2 - \sin^2\theta_1}} \frac{\Delta n}{n}.$$

[*Hint:* For x in radians, $(d/dx)(\sin^{-1} x) = 1/\sqrt{1 - x^2}$.]
(b) If $\theta_1 = 0°$, what is $\Delta\theta_2$ in degrees? (c) If $\theta_1 = 90°$, what is $\Delta\theta_2$ in degrees?

32–7 Total Internal Reflection

57. (I) What is the critical angle for the interface between water and diamond? To be internally reflected, the light must start in which material?

58. (I) The critical angle for a certain liquid–air surface is 49.6°. What is the index of refraction of the liquid?

59. (II) A beam of light is emitted in a pool of water from a depth of 72.0 cm. Where must it strike the air–water interface, relative to the spot directly above it, in order that the light does *not* exit the water?

60. (II) A ray of light, after entering a light fiber, reflects at an angle of 14.5° with the long axis of the fiber, as in Fig. 32–56. Calculate the distance along the axis of the fiber that the light ray travels between successive reflections off the sides of the fiber. Assume that the fiber has an index of refraction of 1.55 and is 1.40×10^{-4} m in diameter.

14.5°

FIGURE 32–56 Problem 60.

61. (II) A beam of light is emitted 8.0 cm beneath the surface of a liquid and strikes the surface 7.6 cm from the point directly above the source. If total internal reflection occurs, what can you say about the index of refraction of the liquid?

62. (II) Figure 32–57 shows a liquid-detecting prism device that might be used inside a washing machine or other liquid-containing appliance. If no liquid covers the prism's hypotenuse, total internal reflection of the beam from the light source produces a large signal in the light sensor. If liquid covers the hypotenuse, some light escapes from the prism into the liquid and the light sensor's signal decreases. Thus a large signal from the light sensor indicates the absence of liquid in the reservoir. If this device is designed to detect the presence of water, determine the allowable range for the prism's index of refraction n. Will the device work properly if the prism is constructed from (inexpensive) lucite? For lucite, $n = 1.5$.

FIGURE 32–57 Problem 62.

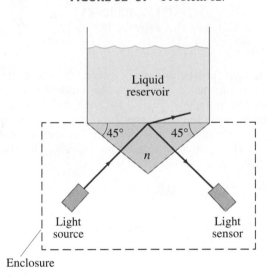

Liquid reservoir

45°   45°

n

Light source   Light sensor

Enclosure

63. (II) Two rays A and B travel down a cylindrical optical fiber of diameter $d = 75.0\,\mu\text{m}$, length $\ell = 1.0\,\text{km}$, and index of refraction $n_1 = 1.465$. Ray A travels a straight path down the fiber's axis, whereas ray B propagates down the fiber by repeated reflections at the critical angle each time it impinges on the fiber's boundary. Determine the extra time Δt it takes for ray B to travel down the entire fiber in comparison with ray A (Fig. 32–58), assuming (a) the fiber is surrounded by air, (b) the fiber is surrounded by a cylindrical glass "cladding" with index of refraction $n_2 = 1.460$.

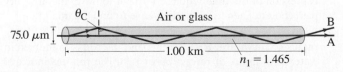

θ_C Air or glass   B

$75.0\,\mu\text{m}$ A

|— 1.00 km —| $n_1 = 1.465$

FIGURE 32–58 Problem 63.

64. (II) (a) What is the minimum index of refraction for a glass or plastic prism to be used in binoculars (Fig. 32–33) so that total internal reflection occurs at 45°? (b) Will binoculars work if their prisms (assume $n = 1.58$) are immersed in water? (c) What minimum n is needed if the prisms are immersed in water?

65. (III) Suppose a ray strikes the left face of the prism in Fig. 32 52 at 45.0° as shown, but is totally internally reflected at the opposite side. If the apex angle (at the top) is $\theta = 60.0°$, what can you say about the index of refraction of the prism?

66. (III) A beam of light enters the end of an optic fiber as shown in Fig. 32–59. (a) Show that we can guarantee total internal reflection at the side surface of the material (at point A), if the index of refraction is greater than about 1.42. In other words, regardless of the angle α, the light beam reflects back into the material at point A, assuming air outside.

FIGURE 32–59 Problem 66.

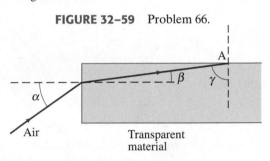

A

β γ

α

Air   Transparent material

*32–8 Refraction at Spherical Surface

*67. (II) A 13.0-cm-thick plane piece of glass ($n = 1.58$) lies on the surface of a 12.0-cm-deep pool of water. How far below the top of the glass does the bottom of the pool seem, as viewed from directly above?

*68. (II) A fish is swimming in water inside a thin spherical glass bowl of uniform thickness. Assuming the radius of curvature of the bowl is 28.0 cm, locate the image of the fish if the fish is located: (a) at the center of the bowl; (b) 20.0 cm from the side of the bowl between the observer and the center of the bowl. Assume the fish is small.

*69. (III) In Section 32–8, we derived Eq. 32–8 for a convex spherical surface with $n_2 > n_1$. Using the same conventions and using diagrams similar to Fig. 32–37, show that Eq. 32–8 is valid also for (a) a convex spherical surface with $n_2 < n_1$, (b) a concave spherical surface with $n_2 > n_1$, and (c) a concave spherical surface with $n_2 < n_1$.

*70. (III) A coin lies at the bottom of a 0.75-m-deep pool. If a viewer sees it at a 45° angle, where is the image of the coin, relative to the coin? [*Hint*: The image is found by tracing back to the intersection of two rays.]

General Problems

71. Two identical concave mirrors are set facing each other 1.0 m apart. A small lightbulb is placed halfway between the mirrors. A small piece of paper placed just to the left of the bulb prevents light from the bulb from directly shining on the left mirror, but light reflected from the right mirror still reaches the left mirror. A good image of the bulb appears on the left side of the piece of paper. What is the focal length of the mirrors?

72. A slab of thickness D, whose two faces are parallel, has index of refraction n. A ray of light incident from air onto one face of the slab at incident angle θ_1 splits into two rays A and B. Ray A reflects directly back into the air, while B travels a total distance ℓ within the slab before reemerging from the slab's face a distance d from its point of entry (Fig. 32–60). (a) Derive expressions for ℓ and d in terms of D, n, and θ_1. (b) For normal incidence (i.e., $\theta_1 = 0°$) show that your expressions yield the expected values for ℓ and d.

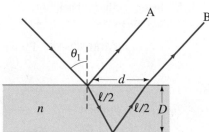

FIGURE 32–60
Problem 72.

73. Two plane mirrors are facing each other 2.2 m apart as in Fig. 32–61. You stand 1.5 m away from one of these mirrors and look into it. You will see multiple images of yourself. (a) How far away from you are the first three images of yourself in the mirror in front of you? (b) Are these first three images facing toward you or away from you?

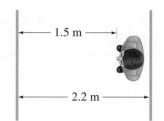

FIGURE 32–61
Problem 73.

74. We wish to determine the depth of a swimming pool filled with water by measuring the width $(x = 5.50\text{ m})$ and then noting that the bottom edge of the pool is just visible at an angle of 13.0° above the horizontal as shown in Fig. 32–62. Calculate the depth of the pool.

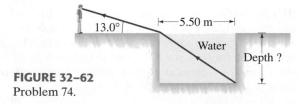

FIGURE 32–62
Problem 74.

75. A 1.80-m-tall person stands 3.80 m from a convex mirror and notices that he looks precisely half as tall as he does in a plane mirror placed at the same distance. What is the radius of curvature of the convex mirror? (Assume that $\sin\theta \approx \theta$.) [*Hint*: The viewing *angle* is half.]

76. The critical angle of a certain piece of plastic in air is $\theta_C = 39.3°$. What is the critical angle of the same plastic if it is immersed in water?

77. Each student in a physics lab is assigned to find the location where a bright object may be placed in order that a concave mirror, with radius of curvature $r = 46$ cm, will produce an image three times the size of the object. Two students complete the assignment at different times using identical equipment, but when they compare notes later, they discover that their answers for the object distance are not the same. Explain why they do not necessarily need to repeat the lab, and justify your response with a calculation.

78. A kaleidoscope makes symmetric patterns with two plane mirrors having a 60° angle between them as shown in Fig. 32–63. Draw the location of the images (some of them images of images) of the object placed between the mirrors.

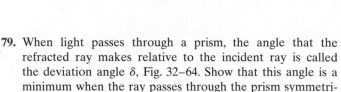

FIGURE 32–63
Problem 78.

79. When light passes through a prism, the angle that the refracted ray makes relative to the incident ray is called the deviation angle δ, Fig. 32–64. Show that this angle is a minimum when the ray passes through the prism symmetrically, perpendicular to the bisector of the apex angle ϕ, and show that the minimum deviation angle, δ_m, is related to the prism's index of refraction n by

$$n = \frac{\sin\frac{1}{2}(\phi + \delta_m)}{\sin\phi/2}.$$

[*Hint*: For θ in radians, $(d/d\theta)(\sin^{-1}\theta) = 1/\sqrt{1 - \theta^2}$.]

FIGURE 32–64
Problems 79 and 80.

80. If the apex angle of a prism is $\phi = 72°$ (see Fig. 32–64), what is the minimum incident angle for a ray if it is to emerge from the opposite side (i.e., not be totally internally reflected), given $n = 1.58$?

81. **Fermat's principle** states that "light travels between two points along the path that requires the least time, as compared to other nearby paths." From Fermat's principle derive (a) the law of reflection $(\theta_i = \theta_r)$ and (b) the law of refraction (Snell's law). [*Hint*: Choose two appropriate points so that a ray between them can undergo reflection or refraction. Draw a rough path for a ray between these points, and write down an expression of the time required for light to travel the arbitrary path chosen. Then take the derivative to find the minimum.]

*82. Suppose Fig. 32–36 shows a cylindrical rod whose end has a radius of curvature $R = 2.0$ cm, and the rod is immersed in water with index of refraction of 1.33. The rod has index of refraction 1.53. Find the location and height of the image of an object 2.0 mm high located 23 cm away from the rod.

83. An optical fiber is a long transparent cylinder of diameter d and index of refraction n. If this fiber is bent sharply, some light hitting the side of the cylinder may escape rather than reflect back into the fiber (Fig. 32–65). What is the smallest radius r at a short bent section for which total internal reflection will be assured for light initially travelling parallel to the axis of the fiber?

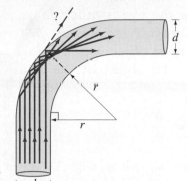

FIGURE 32–65
Problem 83.

84. An object is placed 15 cm from a certain mirror. The image is half the height of the object, inverted, and real. How far is the image from the mirror, and what is the radius of curvature of the mirror?

85. The end faces of a cylindrical glass rod $(n = 1.51)$ are perpendicular to the sides. Show that a light ray entering an end face at any angle will be totally internally reflected inside the rod when it strikes the sides. Assume the rod is in air. What if it were in water?

86. The paint used on highway signs often contains small transparent spheres which provide nighttime illumination of the sign's lettering by retro-reflecting vehicle headlight beams. Consider a light ray from air incident on one such sphere of radius r and index of refraction n. Let θ be its incident angle, and let the ray follow the path shown in Fig. 32–66, so that the ray exits the sphere in the direction exactly antiparallel to its incoming direction. Considering only rays for which $\sin\theta$ can be approximated as θ, determine the required value for n.

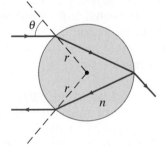

FIGURE 32–66
Problem 86.

*87. (II) The index of refraction, n, of crown flint glass at different wavelengths (λ) of light are given in the Table below.

λ (nm)	1060	546.1	365.0	312.5
n	1.50586	1.51978	1.54251	1.5600

Make a graph of n versus λ. The variation in index of refraction with wavelength is given by the Cauchy equation $n = A + B/\lambda^2$. Make another graph of n versus $1/\lambda^2$ and determine the constants A and B for the glass by fitting the data with a straight line.

*88. (III) Consider a ray of sunlight incident from air on a spherical raindrop of radius r and index of refraction n. Defining θ to be its incident angle, the ray then follows the path shown in Fig. 32–67, exiting the drop at a "scattering angle" ϕ compared with its original incoming direction. (a) Show that $\phi = 180° + 2\theta - 4\sin^{-1}(\sin\theta/n)$. (b) The parallel rays of sunlight illuminate a raindrop with rays of all possible incident angles from 0° to 90°. Plot ϕ vs. θ in the range $0° \le \theta \le 90°$, in 0.5° steps, assuming $n = 1.33$ as is appropriate for water at visible-light wavelengths. (c) From your plot, you should find that a fairly large fraction of the incident angles have nearly the same scattering angle. Approximately what fraction of the possible incident angles is within roughly 1° of $\phi = 139°$? [This subset of incident rays is what creates the rainbow. Wavelength-dependent variations in n cause the rainbow to form at slightly different ϕ for the various visible colors.]

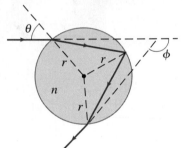

FIGURE 32–67
Problem 88.

Answers to Exercises

A: No.

B: (b).

C: (a).

D: Toward.

E: The face becomes shiny: total internal reflection.

F: 1.414.

Of the many optical devices we discuss in this Chapter, the magnifying glass is the simplest. Here it is magnifying part of page 886 of this Chapter, which describes how the magnifying glass works according to the ray model. In this Chapter we examine thin lenses in detail, seeing how to determine image position as a function of object position and the focal length of the lens, based on the ray model of light. We then examine optical devices including film and digital cameras, the human eye, eyeglasses, telescopes, and microscopes.

subtends a ~~~~~~~~~ As shown in Fig. 33–33a, the object is placed at the focal point ~ **magnifying** *glass* converging lens produces a virtual image, which ~~ the eye is to focus on it. If the eye is relaxed, ~ **comparison of part (a)** or ~ase the object is exactly at the focal point. ~iewed at the near point with ~f when you "focus" on the object ~ object subtends at the eye is much la~ part (b), in which **magnification** or **magnifying power**, M~ eye, rev~ angle subtended by an object when using ~~~ ~gular unaided eye, with the object at the near~ ~he ratio of the normal eye): ~e subtended using the ~ the eye ($N = 25$ cm for a

$$M = \frac{\theta'}{\theta},$$ **(33–5)**

where θ and θ' are shown in Fig. 33–33. ~ write M in terms of the focal length by noting that $\theta = h/N$ (Fig. 33–~ $\theta' = h/d_o$ (Fig. 33–33a), where h is the height of the object and we ass~ ~ngles are small so θ and θ' equal ~ their sines and tangents. If the eye is ~ ~r least eye strain), the image will ~ at infinity and the object will b~ ~ at the focal point; see Fig. 33–34. ~ $d_o = f$ and $\theta' = h/f$. Th~

CHAPTER 33

Lenses and Optical Instruments

CONTENTS

CHAPTER-OPENING QUESTION—Guess now!

A converging lens, like the type used in a magnifying glass,

(a) always produces a magnified image (image taller than the object).

(b) can also produce an image smaller than the object.

(c) always produces an upright image.

(d) can also produce an inverted image (upside down).

(e) None of these statements are true.

The laws of reflection and refraction, particularly the latter, are the basis for explaining the operation of many optical instruments. In this Chapter we discuss and analyze simple lenses using the model of ray optics discussed in the previous Chapter. We then analyze a number of optical instruments, from the magnifying glass and the human eye to telescopes and microscopes. The importance of lenses is that they form images of objects, as shown in Fig. 33–1.

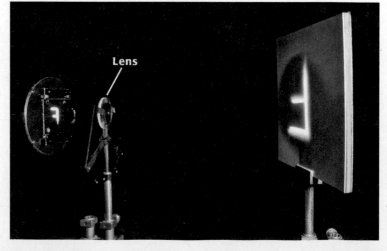

FIGURE 33–1 Converging lens (in holder) forms an image (large "F" on screen at right) of a bright object (illuminated "F" at the left).

Double convex   Planoconvex   Convex meniscus

(a) Converging lenses

Double concave   Planoconcave   Concave meniscus

(b) Diverging lenses

(c)

(d)

FIGURE 33–2 (a) Converging lenses and (b) diverging lenses, shown in cross section. Converging lenses are thicker in the center whereas diverging lenses are thinner in the center. (c) Photo of a converging lens (on the left) and a diverging lens (right). (d) Converging lenses (above), and diverging lenses (below), lying flat, and raised off the paper to form images.

33–1 Thin Lenses; Ray Tracing

The most important simple optical device is no doubt the thin lens. The development of optical devices using lenses dates to the sixteenth and seventeenth centuries, although the earliest record of eyeglasses dates from the late thirteenth century. Today we find lenses in eyeglasses, cameras, magnifying glasses, telescopes, binoculars, microscopes, and medical instruments. A thin lens is usually circular, and its two faces are portions of a sphere. (Cylindrical faces are also possible, but we will concentrate on spherical.) The two faces can be concave, convex, or plane. Several types are shown in Fig. 33–2 a and b in cross section.

Consider parallel rays striking the double convex lens shown in cross section in Fig. 33–3a. We assume the lens is made of material such as glass or transparent plastic, with index of refraction greater than that of the air outside. The **axis** of a lens is a straight line passing through the center of the lens and perpendicular to its two surfaces (Fig. 33–3). From Snell's law, we can see that each ray in Fig. 33–3a is bent toward the axis when the ray enters the lens and again when it leaves the lens at the back surface. (Note the dashed lines indicating the normals to each surface for the top ray.) If rays parallel to the axis fall on a thin lens, they will be focused to a point called the **focal point**, F. This will not be precisely true for a lens with spherical surfaces. But it will be very nearly true—that is, parallel rays will be focused to a tiny region that is nearly a point—if the diameter of the lens is small compared to the radii of curvature of the two lens surfaces. This criterion is satisfied by a **thin lens**, one that is very thin compared to its diameter, and we consider only thin lenses here.

The rays from a point on a distant object are essentially parallel—see Fig. 32–12. Therefore we can say that *the focal point is the image point for an object at infinity on the lens axis.* Thus, the focal point of a lens can be found by locating the point where the Sun's rays (or those of some other distant object) are brought to a sharp image, Fig. 33–4. The distance of the focal point from the center of the lens is called the **focal length**, f. A lens can be turned around so that light can pass through it from the opposite side. The focal length is *the same* on both sides, as we shall see later, even if the curvatures of the two lens surfaces are different. If parallel rays fall on a lens at an angle, as in Fig. 33–3b, they focus at a point F_a. The plane containing all focus points, such as F and F_a in Fig. 33–3b, is called the **focal plane** of the lens.

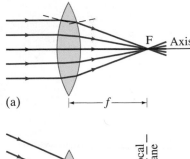

(a)

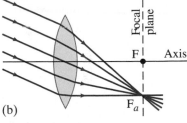

(b)

FIGURE 33–3 Parallel rays are brought to a focus by a converging thin lens.

FIGURE 33–4 Image of the Sun burning wood.

Any lens (in air) that is thicker in the center than at the edges will make parallel rays converge to a point, and is called a **converging lens** (see Fig. 33–2a). Lenses that are thinner in the center than at the edges (Fig. 33–2b) are called **diverging lenses** because they make parallel light diverge, as shown in Fig. 33–5. The focal point, F, of a diverging lens is defined as that point from which refracted rays, originating from parallel incident rays, seem to emerge as shown in Fig. 33–5. And the distance from F to the lens is called the **focal length**, f, just as for a converging lens.

FIGURE 33–5 Diverging lens.

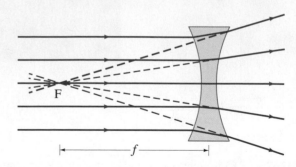

Optometrists and ophthalmologists, instead of using the focal length, use the reciprocal of the focal length to specify the strength of eyeglass (or contact) lenses. This is called the **power**, P, of a lens:

$$P = \frac{1}{f}. \tag{33–1}$$

The unit for lens power is the **diopter** (D), which is an inverse meter: $1\,D = 1\,m^{-1}$. For example, a 20-cm-focal-length lens has a power $P = 1/(0.20\,m) = 5.0\,D$. We will mainly use the focal length, but we will refer again to the power of a lens when we discuss eyeglass lenses in Section 33–6.

The most important parameter of a lens is its focal length f. For a converging lens, f is easily measured by finding the image point for the Sun or other distant objects. Once f is known, the image position can be calculated for any object. To find the image point by drawing rays would be difficult if we had to determine the refractive angles at the front surface of the lens and again at the back surface where the ray exits. We can save ourselves a lot of effort by making use of certain facts we already know, such as that a ray parallel to the axis of the lens passes (after refraction) through the focal point. To determine an image point, we can consider only the three rays indicated in Fig. 33–6, which uses an arrow (on the left) as the object, and a converging lens forming an image (dashed arrow) to the right. These rays, emanating from a single point on the object, are drawn as if the lens were infinitely thin, and we show only a single sharp bend at the center line of the lens instead of the refractions at each surface. These three rays are drawn as follows:

RAY DIAGRAM

Finding the image position formed by a thin lens

Ray 1 is drawn parallel to the axis, Fig. 33–6a; therefore it is refracted by the lens so that it passes along a line through the focal point F behind the lens. (See also Fig. 33–3a.)

Ray 2 is drawn on a line passing through the other focal point F′ (front side of lens in Fig. 33–6) and emerges from the lens parallel to the axis, Fig. 33–6b.

Ray 3 is directed toward the very center of the lens, where the two surfaces are essentially parallel to each other, Fig. 33–6c; this ray therefore emerges from the lens at the same angle as it entered; the ray would be displaced slightly to one side, as we saw in Example 32–8, but since we assume the lens is thin, we draw ray 3 straight through as shown.

The point where these three rays cross is the image point for that object point. Actually, any two of these rays will suffice to locate the image point, but drawing the third ray can serve as a check.

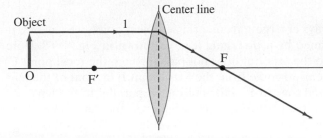

(a) Ray 1 leaves one point on object going parallel to the axis, then refracts through focal point behind the lens.

FIGURE 33–6 Finding the image by ray tracing for a converging lens. Rays are shown leaving one point on the object (an arrow). Shown are the three most useful rays, leaving the tip of the object, for determining where the image of that point is formed.

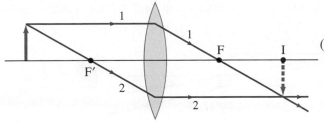

(b) Ray 2 passes through F′ in front of the lens; therefore it is parallel to the axis behind the lens.

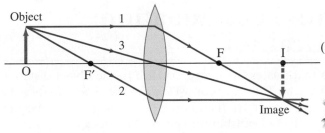

(c) Ray 3 passes straight through the center of the lens (assumed very thin).

Using these three rays for one object point, we can find the image point for that point of the object (the top of the arrow in Fig. 33–6). The image points for all other points on the object can be found similarly to determine the complete image of the object. Because the rays actually pass through the image for the case shown in Fig. 33–6, it is a **real image** (see page 840). The image could be detected by film or electronic sensor, or actually seen on a white surface or screen placed at the position of the image (Fig. 33–7a).

FIGURE 33–7 (a) A converging lens can form a real image (here of a distant building, upside down) on a screen. (b) That same real image is also directly visible to the eye. [Figure 33–2d shows images (graph paper) seen by the eye made by both diverging and converging lenses.]

| CONCEPTUAL EXAMPLE 33–1 | **Half-blocked lens.** What happens to the image of an object if the top half of a lens is covered by a piece of cardboard?

RESPONSE Let us look at the rays in Fig. 33–6. If the top half (or any half of the lens) is blocked, you might think that half the image is blocked. But in Fig. 33–6c, we see how the rays used to create the "top" of the image pass through both the top and the bottom of the lens. Only three of many rays are shown—many more rays pass through the lens, and they can form the image. You don't lose the image, but covering part of the lens cuts down on the total light received and reduces the brightness of the image.

NOTE If the lens is partially blocked by your thumb, you may notice an out of focus image of part of that thumb.

Seeing the Image

The image can also be seen directly by the eye when the eye is placed behind the image, as shown in Fig. 33–6c, so that some of the rays diverging from each point on the image can enter the eye. We can see a sharp image only for rays *diverging* from each point on the image, because we see normal objects when diverging rays from each point enter the eye as was shown in Fig. 32–1. Your eye cannot focus rays converging on it; if your eye was positioned between points F and I in Fig. 33–6c, it would not see a clear image. (More about our eyes in Section 33–6.) Figure 33–7 shows an image seen (a) on a screen and (b) directly by the eye (and a camera) placed behind the image. The eye can see both real and virtual images (see next page) as long as the eye is positioned so rays diverging from the image enter it.

(a)

(b)

Diverging Lens

By drawing the same three rays emerging from a single object point, we can determine the image position formed by a diverging lens, as shown in Fig. 33–8. Note that ray 1 is drawn parallel to the axis, but does not pass through the focal point F′ behind the lens. Instead it seems to come from the focal point F in front of the lens (dashed line). Ray 2 is directed toward F′ and is refracted parallel to the lens axis by the lens. Ray 3 passes directly through the center of the lens. The three refracted rays seem to emerge from a point on the left of the lens. This is the image point, I. Because the rays do not pass through the image, it is a **virtual image**. Note that the eye does not distinguish between real and virtual images—both are visible.

FIGURE 33–8 Finding the image by ray tracing for a diverging lens.

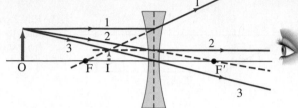

33–2 The Thin Lens Equation; Magnification

We now derive an equation that relates the image distance to the object distance and the focal length of a thin lens. This equation will make the determination of image position quicker and more accurate than doing ray tracing. Let d_o be the object distance, the distance of the object from the center of the lens, and d_i be the image distance, the distance of the image from the center of the lens.

FIGURE 33–9 Deriving the lens equation for a converging lens.

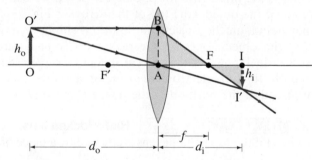

Let h_o and h_i refer to the heights of the object and image. Consider the two rays shown in Fig. 33–9 for a converging lens, assumed to be very thin. The right triangles FI′I and FBA (highlighted in yellow) are similar because angle AFB equals angle IFI′; so

$$\frac{h_i}{h_o} = \frac{d_i - f}{f},$$

since length AB = h_o. Triangles OAO′ and IAI′ are similar as well. Therefore,

$$\frac{h_i}{h_o} = \frac{d_i}{d_o}.$$

We equate the right sides of these two equations (the left sides are the same), and divide by d_i to obtain

$$\frac{1}{f} - \frac{1}{d_i} = \frac{1}{d_o}$$

or

$$\frac{1}{d_o} + \frac{1}{d_i} = \frac{1}{f}. \tag{33–2}$$

THIN LENS EQUATION

This is called the **thin lens equation**. It relates the image distance d_i to the object distance d_o and the focal length f. It is the most useful equation in geometric optics.

(Interestingly, it is exactly the same as the mirror equation, Eq. 32–2.) If the object is at infinity, then $1/d_o = 0$, so $d_i = f$. Thus the focal length is the image distance for an object at infinity, as mentioned earlier.

We can derive the lens equation for a diverging lens using Fig. 33–10. Triangles IAI′ and OAO′ are similar; and triangles IFI′ and AFB are similar. Thus (noting that length $AB = h_o$)

$$\frac{h_i}{h_o} = \frac{d_i}{d_o} \quad \text{and} \quad \frac{h_i}{h_o} = \frac{f - d_i}{f}.$$

When we equate the right sides of these two equations and simplify, we obtain

$$\frac{1}{d_o} - \frac{1}{d_i} = -\frac{1}{f}.$$

This equation becomes the same as Eq. 33–2 if we make f and d_i negative. That is, we take f to be *negative for a diverging lens*, and d_i negative when the image is on the same side of the lens as the light comes from. Thus Eq. 33–2 will be valid for both converging and diverging lenses, and for *all* situations, if we use the following **sign conventions**:

1. The focal length is positive for converging lenses and negative for diverging lenses.
2. The object distance is positive if the object is on the side of the lens from which the light is coming (this is usually the case, although when lenses are used in combination, it might not be so); otherwise, it is negative.
3. The image distance is positive if the image is on the opposite side of the lens from where the light is coming; if it is on the same side, d_i is negative. Equivalently, the image distance is positive for a real image and negative for a virtual image.
4. The height of the image, h_i, is positive if the image is upright, and negative if the image is inverted relative to the object. (h_o is always taken as upright and positive.)

The **lateral magnification**, m, of a lens is defined as the ratio of the image height to object height, $m = h_i/h_o$. From Figs. 33–9 and 33–10 and the conventions just stated (for which we'll need a minus sign below), we have

$$m = \frac{h_i}{h_o} = -\frac{d_i}{d_o}. \tag{33–3}$$

For an upright image the magnification is positive ($h_i > 0$ and $d_i < 0$), and for an inverted image the magnification is negative ($h_i < 0$ and $d_i > 0$).

From sign convention 1, it follows that the power (Eq. 33–1) of a converging lens, in diopters, is positive, whereas the power of a diverging lens is negative. A converging lens is sometimes referred to as a **positive lens**, and a diverging lens as a **negative lens**.

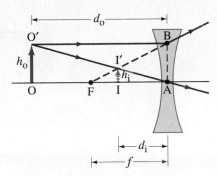

FIGURE 33–10 Deriving the lens equation for a diverging lens.

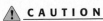

⚠ **CAUTION**

Focal length is negative for diverging lens

PROBLEM SOLVING

SIGN CONVENTIONS for lenses

P R O B L E M   S O L V I N G

Thin Lenses

1. Draw a **ray diagram**, as precise as possible, but even a rough one can serve as confirmation of analytic results. Choose one point on the object and draw at least two, or preferably three, of the easy-to-draw rays described in Figs. 33–6 and 33–8. The image point is where the rays intersect.

2. For analytic solutions, solve for unknowns in the **thin lens equation** (Eq. 33–2) and the **magnification equation** (Eq. 33–3). The thin lens equation involves reciprocals — don't forget to take the reciprocal.

3. Follow the **sign conventions** listed just above.

4. Check that your analytic answers are **consistent** with your ray diagram.

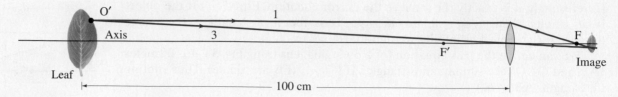

FIGURE 33–11 Example 33–2. (Not to scale.)

EXAMPLE 33–2 **Image formed by converging lens.** What is (*a*) the position, and (*b*) the size, of the image of a 7.6-cm-high leaf placed 1.00 m from a +50.0-mm-focal-length camera lens?

APPROACH We follow the steps of the Problem Solving Strategy explicitly.

SOLUTION

1. **Ray diagram.** Figure 33–11 is an approximate ray diagram, showing only rays 1 and 3 for a single point on the leaf. We see that the image ought to be a little behind the focal point F, to the right of the lens.

2. **Thin lens and magnification equations.** (*a*) We find the image position analytically using the thin lens equation, Eq. 33–2. The camera lens is converging, with $f = +5.00$ cm, and $d_o = 100$ cm, and so the thin lens equation gives

$$\frac{1}{d_i} = \frac{1}{f} - \frac{1}{d_o} = \frac{1}{5.00 \text{ cm}} - \frac{1}{100 \text{ cm}}$$

$$= \frac{20.0 - 1.0}{100 \text{ cm}} = \frac{19.0}{100 \text{ cm}}.$$

Then, taking the reciprocal,

$$d_i = \frac{100 \text{ cm}}{19.0} = 5.26 \text{ cm},$$

or 52.6 mm behind the lens.

(*b*) The magnification is

$$m = -\frac{d_i}{d_o} = -\frac{5.26 \text{ cm}}{100 \text{ cm}} = -0.0526,$$

so

$$h_i = m h_o = (-0.0526)(7.6 \text{ cm}) = -0.40 \text{ cm}.$$

The image is 4.0 mm high.

3. **Sign conventions.** The image distance d_i came out positive, so the image is behind the lens. The image height is $h_i = -0.40$ cm; the minus sign means the image is inverted.

4. **Consistency.** The analytic results of steps (2) and (3) are consistent with the ray diagram, Fig. 33–11: the image is behind the lens and inverted.

NOTE Part (*a*) tells us that the image is 2.6 mm farther from the lens than the image for an object at infinity, which would equal the focal length, 50.0 mm. Indeed, when focusing a camera lens, the closer the object is to the camera, the farther the lens must be from the sensor or film.

EXERCISE A If the leaf (object) of Example 33–2 is moved farther from the lens, does the image move closer to or farther from the lens? (Don't calculate!)

EXAMPLE 33–3 **Object close to converging lens.** An object is placed 10 cm from a 15-cm-focal-length converging lens. Determine the image position and size (a) analytically, and (b) using a ray diagram.

APPROACH We first use Eqs. 33–2 and 33–3 to obtain an analytic solution, and then confirm with a ray diagram using the special rays 1, 2, and 3 for a single object point.

SOLUTION (a) Given $f = 15$ cm and $d_o = 10$ cm, then

$$\frac{1}{d_i} = \frac{1}{15 \text{ cm}} - \frac{1}{10 \text{ cm}} = -\frac{1}{30 \text{ cm}},$$

and $d_i = -30$ cm. (Remember to take the reciprocal!) Because d_i is negative, the image must be virtual and on the same side of the lens as the object. The magnification

$$m = -\frac{d_i}{d_o} = -\frac{-30 \text{ cm}}{10 \text{ cm}} = 3.0.$$

⚠ **CAUTION**
Don't forget to take the reciprocal

The image is three times as large as the object and is upright. This lens is being used as a simple magnifying glass, which we discuss in more detail in Section 33–7. (b) The ray diagram is shown in Fig. 33–12 and confirms the result in part (a). We choose point O′ on the top of the object and draw ray 1, which is easy. But ray 2 may take some thought: if we draw it heading toward F′, it is going the wrong way—so we have to draw it as if coming from F′ (and so dashed), striking the lens, and then going out parallel to the lens axis. We project it back parallel, with a dashed line, as we must do also for ray 1, in order to find where they cross. Ray 3 is drawn through the lens center, and it crosses the other two rays at the image point, I′.

NOTE From Fig. 33–12 we can see that, whenever an object is placed between a converging lens and its focal point, the image is virtual.

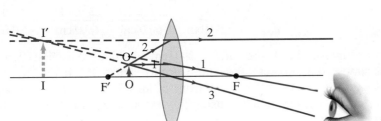

FIGURE 33–12 An object placed within the focal point of a converging lens produces a virtual image. Example 33–3.

EXAMPLE 33–4 **Diverging lens.** Where must a small insect be placed if a 25-cm-focal-length diverging lens is to form a virtual image 20 cm from the lens, on the same side as the object?

APPROACH The ray diagram is basically that of Fig. 33–10 because our lens here is diverging and our image is in front of the lens within the focal distance. (It would be a valuable exercise to draw the ray diagram to scale, precisely, now.) The insect's distance, d_o, can be calculated using the thin lens equation.

SOLUTION The lens is diverging, so f is negative: $f = -25$ cm. The image distance must be negative too because the image is in front of the lens (sign conventions), so $d_i = -20$ cm. Equation 33–2 gives

$$\frac{1}{d_o} = \frac{1}{f} - \frac{1}{d_i} = -\frac{1}{25 \text{ cm}} + \frac{1}{20 \text{ cm}} = \frac{-4 + 5}{100 \text{ cm}} = \frac{1}{100 \text{ cm}}.$$

So the object must be 100 cm in front of the lens.

FIGURE 33–13 Exercise C.

(a)

(b)

EXERCISE B Return to the Chapter-Opening Question, page 866, and answer it again now. Try to explain why you may have answered differently the first time.

EXERCISE C Figure 33–13 shows a converging lens held above three equal-sized letters A. In (a) the lens is 5 cm from the paper, and in (b) the lens is 15 cm from the paper. Estimate the focal length of the lens. What is the image position for each case?

33–3 Combinations of Lenses

Optical instruments typically use lenses in combination. When light passes through more than one lens, we find the image formed by the first lens as if it were alone. This image becomes the *object* for the second lens, and we find the image then formed by this second lens, which is the final image if there are only two lenses. The total magnification will be the product of the separate magnifications of each lens, as we shall see. Even if the second lens intercepts the light from the first lens before it forms an image, this technique still works.

EXAMPLE 33–5 **A two-lens system.** Two converging lenses, A and B, with focal lengths $f_A = 20.0$ cm and $f_B = 25.0$ cm, are placed 80.0 cm apart, as shown in Fig. 33–14a. An object is placed 60.0 cm in front of the first lens as shown in Fig. 33–14b. Determine (a) the position, and (b) the magnification, of the final image formed by the combination of the two lenses.

APPROACH Starting at the tip of our object O, we draw rays 1, 2, and 3 for the first lens, A, and also a ray 4 which, after passing through lens A, acts as "ray 3" (through the center) for the second lens, B. Ray 2 for lens A exits parallel, and so is ray 1 for lens B. To determine the position of the image I_A formed by lens A, we use Eq. 33–2 with $f_A = 20.0$ cm and $d_{oA} = 60.0$ cm. The distance of I_A (lens A's image) from lens B is the object distance d_{oB} for lens B. The final image is found using the thin lens equation, this time with all distances relative to lens B. For (b) the magnifications are found from Eq. 33–3 for each lens in turn.

SOLUTION (a) The object is a distance $d_{oA} = +60.0$ cm from the first lens, A, and this lens forms an image whose position can be calculated using the thin lens equation:

$$\frac{1}{d_{iA}} = \frac{1}{f_A} - \frac{1}{d_{oA}} = \frac{1}{20.0 \text{ cm}} - \frac{1}{60.0 \text{ cm}} = \frac{3-1}{60.0 \text{ cm}} = \frac{1}{30.0 \text{ cm}}.$$

> ⚠ **CAUTION**
> *Object distance for second lens is **not** equal to the image distance for first lens*

So the first image I_A is at $d_{iA} = 30.0$ cm behind the first lens. This image becomes the object for the second lens, B. It is a distance $d_{oB} = 80.0 \text{ cm} - 30.0 \text{ cm} = 50.0$ cm in front of lens B, as shown in Fig. 33–14b. The image formed by lens B, again using the thin lens equation, is at a distance d_{iB} from the lens B:

$$\frac{1}{d_{iB}} = \frac{1}{f_B} - \frac{1}{d_{oB}} = \frac{1}{25.0 \text{ cm}} - \frac{1}{50.0 \text{ cm}} = \frac{2-1}{50.0 \text{ cm}} = \frac{1}{50.0 \text{ cm}}.$$

Hence $d_{iB} = 50.0$ cm behind lens B. This is the final image—see Fig. 33–14b.

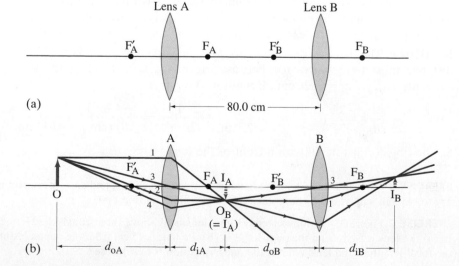

FIGURE 33–14 Two lenses, A and B, used in combination, Example 33–5. The small numbers refer to the easily drawn rays.

(b) Lens A has a magnification (Eq. 33–3)

$$m_A = -\frac{d_{iA}}{d_{oA}} = -\frac{30.0\ \text{cm}}{60.0\ \text{cm}} = -0.500.$$

Thus, the first image is inverted and is half as high as the object (again Eq. 33–3):

$$h_{iA} = m_A h_{oA} = -0.500 h_{oA}.$$

Lens B takes this image as object and changes its height by a factor

$$m_B = -\frac{d_{iB}}{d_{oB}} = \frac{50.0\ \text{cm}}{50.0\ \text{cm}} = -1.000$$

The second lens reinverts the image (the minus sign) but doesn't change its size. The final image height is (remember h_{oB} is the same as h_{iA})

$$h_{iB} = m_B h_{oB} = m_B h_{iA} = m_B m_A h_{oA} = (m_{\text{total}}) h_{oA}.$$

The total magnification is the product of m_A and m_B, which here equals $m_{\text{total}} = m_A m_B = (-1.000)(-0.500) = +0.500$, or half the original height, and the final image is upright.

EXAMPLE 33–6 **Measuring f for a diverging lens.** To measure the focal length of a diverging lens, a converging lens is placed in contact with it, as shown in Fig. 33–15. The Sun's rays are focused by this combination at a point 28.5 cm, behind the lenses as shown. If the converging lens has a focal length f_C of 16.0 cm, what is the focal length f_D of the diverging lens? Assume both lenses are thin and the space between them is negligible.

APPROACH The image distance for the first lens equals its focal length (16.0 cm) since the object distance is infinity (∞). The position of this image, even though it is never actually formed, acts as the object for the second (diverging) lens. We apply the thin lens equation to the diverging lens to find its focal length, given that the final image is at $d_i = 28.5\ \text{cm}$.

SOLUTION If the diverging lens was absent, the converging lens would form the image at its focal point—that is, at a distance $f_C = 16.0\ \text{cm}$ behind it (dashed lines in Fig. 33–15). When the diverging lens is placed next to the converging lens, we treat the image formed by the first lens as the *object* for the second lens. Since this object lies to the right of the diverging lens, this is a situation where d_o is negative (see the sign conventions, page 871). Thus, for the diverging lens, the object is virtual and $d_o = -16.0\ \text{cm}$. The diverging lens forms the image of this virtual object at a distance $d_i = 28.5\ \text{cm}$ away (given). Thus,

$$\frac{1}{f_D} = \frac{1}{d_o} + \frac{1}{d_i} = \frac{1}{-16.0\ \text{cm}} + \frac{1}{28.5\ \text{cm}} = -0.0274\ \text{cm}^{-1}.$$

We take the reciprocal to find $f_D = -1/(0.0274\ \text{cm}^{-1}) = -36.5\ \text{cm}$.

NOTE If this technique is to work, the converging lens must be "stronger" than the diverging lens—that is, it must have a focal length whose magnitude is less than that of the diverging lens. (Rays from the Sun are focused 28.5 cm behind the combination, so the focal length of the total combination is $f_T = 28.5\ \text{cm}$.)

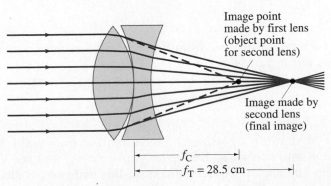

Image point made by first lens (object point for second lens)

Image made by second lens (final image)

f_C

$f_T = 28.5\ \text{cm}$

FIGURE 33–15 Determining the focal length of a diverging lens. Example 33–6.

*33–4 Lensmaker's Equation

In this Section, we will show that parallel rays are brought to a focus at a single point for a thin lens. At the same time, we will also derive an equation that relates the focal length of a lens to the radii of curvature of its two surfaces and its index of refraction, which is known as the lensmaker's equation.

In Fig. 33–16, a ray parallel to the axis of a lens is refracted at the front surface of the lens at point A_1 and is refracted at the back surface at point A_2. This ray then passes through point F, which we call the focal point for this ray. Point A_1 is a height h_1 above the axis, and point A_2 is height h_2 above the axis. C_1 and C_2 are the centers of curvature of the two lens surfaces; so the length $C_1 A_1 = R_1$, the radius of curvature of the front surface, and $C_2 A_2 = R_2$ is the radius of the second surface. We consider a double convex lens and by convention choose the radii of both lens surfaces as positive. The thickness of the lens has been grossly exaggerated so that the various angles would be clear. But we will assume that the lens is actually very thin and that angles between the rays and the axis are small. In this approximation, the sines and tangents of all the angles will be equal to the angles themselves in radians. For example, $\sin \theta_1 \approx \tan \theta_1 \approx \theta_1$ (radians).

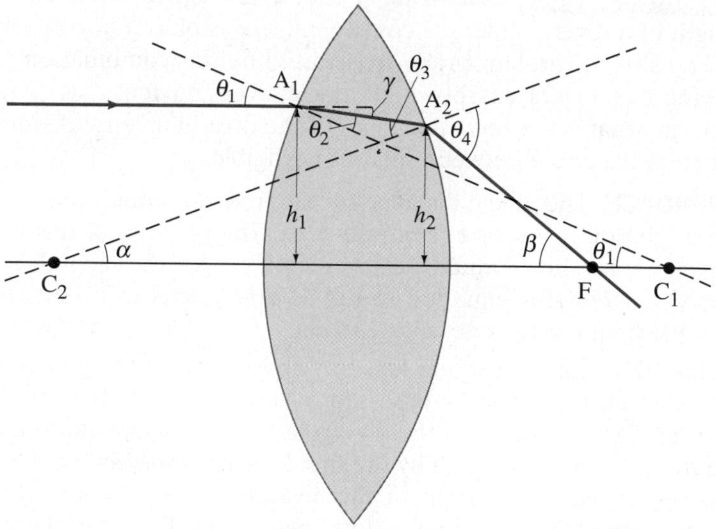

FIGURE 33–16 Diagram of a ray passing through a lens for derivation of the lensmaker's equation.

To this approximation, then, Snell's law tells us that

$$\theta_1 = n\theta_2$$
$$\theta_4 = n\theta_3$$

where n is the index of refraction of the glass, and we assume that the lens is surrounded by air $(n = 1)$. Notice also in Fig. 33–16 that

$$\theta_1 \approx \sin \theta_1 = \frac{h_1}{R_1}$$

$$\alpha \approx \frac{h_2}{R_2}$$

$$\beta \approx \frac{h_2}{f}.$$

The last follows because the distance from F to the lens (assumed very thin) is f. From the diagram, the angle γ is

$$\gamma = \theta_1 - \theta_2.$$

A careful examination of Fig. 33–16 shows also that

$$\alpha = \theta_3 - \gamma.$$

This can be seen by drawing a horizontal line to the left from point A_2, which divides the angle θ_3 into two parts. The upper part equals γ and the lower part equals α. (The opposite angles between an oblique line and two parallel lines are equal.) Thus, $\theta_3 = \gamma + \alpha$. Furthermore, by drawing a horizontal line to the right from point A_2, we divide θ_4 into two parts. The upper part is α and the lower is β.

Thus

$$\theta_4 = \alpha + \beta.$$

We now combine all these equations:

$$\alpha = \theta_3 - \gamma = \frac{\theta_4}{n} - (\theta_1 - \theta_2) = \frac{\alpha}{n} + \frac{\beta}{n} - \theta_1 + \theta_2,$$

or

$$\frac{h_2}{R_2} = \frac{h_2}{nR_2} + \frac{h_2}{nf} - \frac{h_1}{R_1} + \frac{h_1}{nR_1}.$$

Because the lens is thin, $h_1 \approx h_2$ and all h's can be canceled from all the numerators. We then multiply through by n and rearrange to find that

$$\frac{1}{f} = (n - 1)\left(\frac{1}{R_1} + \frac{1}{R_2}\right). \tag{33–4}$$

This is called the **lensmaker's equation**. It relates the focal length of a thin lens to the radii of curvature of its two surfaces and its index of refraction. Notice that f for a thin lens does not depend on h_1 or h_2. Thus the position of the point F does not depend on where the ray strikes the lens. Hence, all rays parallel to the axis of a thin lens will pass through the same point F, which we wished to prove.

In our derivation, both surfaces are convex and R_1 and R_2 are considered positive.[†] Equation 33–4 also works for lenses with one or both surfaces concave; but for a concave surface, the radius must be considered *negative*.

Notice in Eq. 33–4 that the equation is *symmetrical* in R_1 and R_2. Thus, if a lens is turned around so that light impinges on the other surface, the focal length is the same even if the two lens surfaces are different.

EXAMPLE 33–7 **Calculating f for a converging lens.** A convex meniscus lens (Figs. 33–2a and 33–17) is made from glass with $n = 1.50$. The radius of curvature of the convex surface is 22.4 cm and that of the concave surface is 46.2 cm. (*a*) What is the focal length? (*b*) Where will the image be for an object 2.00 m away?

APPROACH We use Eq. 33–4, noting that R_2 is negative because it refers to the concave surface.

SOLUTION (*a*) $R_1 = 22.4$ cm and $R_2 = -46.2$ cm. Then

$$\frac{1}{f} = (1.50 - 1.00)\left(\frac{1}{22.4\text{ cm}} - \frac{1}{46.2\text{ cm}}\right)$$

$$= 0.0115\text{ cm}^{-1}.$$

So

$$f = \frac{1}{0.0115\text{ cm}^{-1}} = 87.0\text{ cm}$$

and the lens is converging. Notice that if we turn the lens around so that $R_1 = -46.2$ cm and $R_2 = +22.4$ cm, we get the same result.

(*b*) From the lens equation, with $f = 0.870$ m and $d_o = 2.00$ m, we have

$$\frac{1}{d_i} = \frac{1}{f} - \frac{1}{d_o} = \frac{1}{0.870\text{ m}} - \frac{1}{2.00\text{ m}}$$

$$= 0.649\text{ m}^{-1},$$

so $d_i = 1/0.649\text{ m}^{-1} = 1.54$ m.

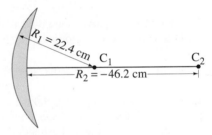

FIGURE 33–17 Example 33–7.

EXERCISE D A Lucite planoconcave lens (see Fig. 33–2b) has one flat surface and the other has $R = -18.4$ cm. What is the focal length? Is the lens converging or diverging?

[†]Some books use a different convention—for example, R_1 and R_2 are considered positive if their centers of curvature are to the right of the lens, in which case a minus sign appears in their equivalent of Eq. 33–4.

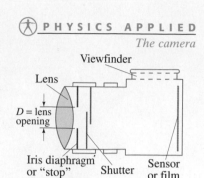

FIGURE 33–18 A simple camera.

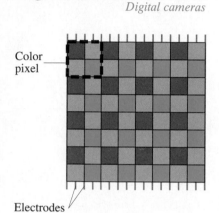

FIGURE 33–19 Portion of a typical CCD sensor. A square group of four pixels $^{RG}_{GB}$ is sometimes called a "color pixel."

FIGURE 33–20 Suppose we take a picture that includes a thin black line (our object) on a white background. The *image* of this black line has a colored halo (red above, blue below) due to the mosaic arrangement of color filter pixels, as shown by the colors transmitted. Computer averaging can minimize color problems such as this (the green at top and bottom of image can be averaged with nearby pixels to give white or nearly so) but the image is consequently "softened" or blurred. The layered color pixel described in the text would avoid this artifact.

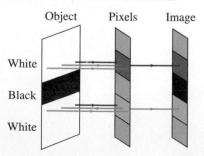

33–5 Cameras: Film and Digital

The basic elements of a **camera** are a lens, a light-tight box, a shutter to let light pass through the lens only briefly, and in a digital camera an electronic sensor or in a traditional camera a piece of film (Fig. 33–18). When the shutter is opened for a brief "exposure," light from external objects in the field of view is focused by the lens as an image on the film or sensor. Film contains light-sensitive chemicals that change when light strikes them. In the development process, chemical reactions cause the changed areas to turn opaque, so the image is recorded on the film.[†]

You can see the image yourself if you remove the back of a conventional camera and view through a piece of tissue paper (on which an image can form) placed where the film should be with the shutter open.

Digital Cameras, Electronic Sensors (CCD, CMOS)

In a **digital camera**, the film is replaced by a semiconductor sensor. Two types are in common use: **CCD** (*charge-coupled device*) and **CMOS** (*complementary metal oxide semiconductor*). A CCD sensor is made up of millions of tiny **pixels** ("picture elements")—see Figs. 35–42 and 33–19. A 6-MP (6-megapixel) sensor[‡] would contain about 2000 pixels vertically by 3000 pixels horizontally over an area of perhaps 4×6 mm or even 24×36 mm. Light reaching any pixel liberates electrons from the semiconductor. The more intense the light, the more charge accumulates during the brief exposure time. Conducting electrodes carry each pixel's charge (serially in time, row by row—hence the name "charge-coupled") to a central processor that stores the relative brightness of pixels, and allows reformation of the image later onto a computer screen or printer. A CCD is fully reusable. Once the pixel charges are transferred to memory, a new picture can be taken.

A CMOS sensor also uses a silicon semiconductor, and incorporates some electronics within each pixel, allowing parallel readout.

Color is achieved by red, green, and blue filters over alternating pixels as shown in Fig. 33–19, similar to a color CRT or LCD screen. The sensor type shown in Fig. 33–19 contains twice as many green pixels as red or blue (because green is claimed to have a stronger influence on the sensation of sharpness). The computer-analyzed color at each pixel is that pixel's intensity averaged with the intensities of the nearest-neighbor colors.

To reduce the amount of memory for each picture, compression programs average over pixels, but with a consequent loss of sharpness, or "resolution."

*Digital Artifacts

Digital cameras can produce image artifacts (artificial effects in the image not present in the original) resulting from the imaging process. One example using the "mosaic" of pixels (Fig. 33–19) is described in Fig. 33–20. An alternative technology uses a semitransparent silicon semiconductor layer system wherein different wavelengths of light penetrate silicon to different depths: each pixel is a sandwich of partly transparent layers, one for each color. The top layer can absorb blue light, allowing green and red light to pass through. The second layer absorbs green and the bottom layer detects the red. All three colors are detected by each pixel, resulting in better color resolution and fewer artifacts.

[†]This is called a *negative*, because the black areas correspond to bright objects and vice versa. The same process occurs during printing to produce a black-and-white "positive" picture from the negative. Color film has three emulsion layers (or dyes) corresponding to the three primary colors.

[‡]Each different color of pixel in a CCD is counted as a separate pixel. In contrast, in an LCD screen (Section 35–12), a group of three subpixels is counted as one pixel, a more conservative count.

Camera Adjustments

There are three main adjustments on good-quality cameras: shutter speed, f-stop, and focusing. Although most cameras today make these adjustments automatically, it is valuable to understand these adjustments to use a camera effectively. For special or top-quality work, a manual camera is indispensable (Fig. 33–21).

Exposure time or shutter speed This refers to how quickly the digital sensor can make an accurate reading, or how long the shutter of a camera is open and the film or sensor exposed. It could vary from a second or more ("time exposures"), to $\frac{1}{1000}$ s or less. To avoid blurring from camera movement, exposure times shorter than $\frac{1}{100}$ s are normally needed. If the object is moving, even shorter exposure times are needed to "stop" the action. If the exposure (or sampling) time is not fast enough, the image will be blurred by camera shake no matter how many pixels a digital camera claims. Blurring in low light conditions is more of a problem with cell-phone cameras whose sensors are not usually the most sophisticated. Digital still cameras or cell phones that take short videos must have a fast enough "sampling" time and fast "clearing" (of the charge) time so as to take pictures at least 15 frames per second, and preferably 30 fps.

FIGURE 33–21 On this camera, the f-stops and the focusing ring are on the camera lens. Shutter speeds are selected on the small wheel on top of the camera body.

f-stop The amount of light reaching the film must be carefully controlled to avoid **underexposure** (too little light so the picture is dark and only the brightest objects show up) or **overexposure** (too much light, so that all bright objects look the same, with a consequent lack of contrast and a "washed-out" appearance). Most cameras these days make f-stop and shutter speed adjustments automatically. A high quality camera controls the exposure with a "stop" or iris diaphragm, whose opening is of variable diameter, placed behind the lens (Fig. 33–18). The size of the opening is controlled (automatically or manually) to compensate for bright or dark lighting conditions, the sensitivity of the sensor or film,[†] and for different shutter speeds. The size of the opening is specified by the **f-number** or **f-stop**, defined as

$$f\text{-stop} = \frac{f}{D},$$

where f is the focal length of the lens and D is the diameter of the lens opening (Fig. 33–18). For example, when a 50-mm-focal-length lens has an opening $D = 25$ mm, we say it is set at $f/2$. When this lens is set at $f/8$, the opening is only $6\frac{1}{4}$ mm $\left(50/6\frac{1}{4} = 8\right)$. For faster shutter speeds, or low light conditions, a wider lens opening must be used to get a proper exposure, which corresponds to a smaller f-stop number. The smaller the f-stop number, the larger the opening and the more light passes through the lens to the sensor or film. The smallest f-number of a lens (largest opening) is referred to as the *speed* of the lens. The best lenses may have a speed of $f/2.0$, or even faster. The advantage of a fast lens is that it allows pictures to be taken under poor lighting conditions. Good quality lenses consist of several elements to reduce the defects present in simple thin lenses (Section 33–10). Standard f-stops are

$$1.0, \quad 1.4, \quad 2.0, \quad 2.8, \quad 4.0, \quad 5.6, \quad 8, \quad 11, \quad 16, \quad 22, \quad \text{and} \quad 32$$

(Fig. 33–21). Each of these stops corresponds to a diameter reduction by a factor of about $\sqrt{2} \approx 1.4$. Because the amount of light reaching the film is proportional to the *area* of the opening, and therefore proportional to the diameter squared, each standard f-stop corresponds to a factor of 2 in light intensity reaching the film.

Focusing Focusing is the operation of placing the lens at the correct position relative to the sensor or film for the sharpest image. The image distance is smallest for objects at infinity (the symbol ∞ is used for infinity) and is equal to the focal length. For closer objects, the image distance is greater than the focal length, as can be seen from the lens equation, $1/f = 1/d_o + 1/d_1$ (Eq. 33–2). To focus on nearby objects, the lens must therefore be moved away from the sensor or film, and this is usually done on a manual camera by turning a ring on the lens.

[†]Different films have different sensitivities to light, referred to as the "film speed" and specified as an "ISO (or ASA) number." A "faster" film is more sensitive and needs less light to produce a good image. Faster films are grainier so offer less sharpness (resolution) when enlarged. Digital cameras may have a "gain" or "ISO" adjustment for sensitivity. Adjusting a CCD to be "faster" for low light conditions results in "noise," the digital equivalent of graininess.

FIGURE 33–22 Photos taken with a camera lens (a) focused on a nearby object with distant object blurry, and (b) focused on a more distant object with nearby object blurry.

(a)

(b)

If the lens is focused on a nearby object, a sharp image of it will be formed, but the image of distant objects may be blurry (Fig. 33–22). The rays from a point on the distant object will be out of focus—they will form a circle on the sensor or film as shown (exaggerated) in Fig. 33–23. The distant object will thus produce an image consisting of overlapping circles and will be blurred. These circles are called **circles of confusion**. To include near and distant objects in the same photo, you (or the camera) can try setting the lens focus at an intermediate position. For a given distance setting, there is a range of distances over which the circles of confusion will be small enough that the images will be reasonably sharp. This is called the **depth of field**. The depth of field varies with the lens opening. If the lens opening is smaller, only rays through the central part of the lens are accepted, and these form smaller circles of confusion for a given object distance. Hence, at smaller lens openings, a greater range of object distances will fit within the circle of confusion criterion, so the depth of field is greater.[†] For a sensor or film width of 36 mm (including 35-mm film cameras), the depth of field is usually based on a maximum circle of confusion diameter of 0.003 mm.

FIGURE 33–23 When the lens is positioned to focus on a nearby object, points on a distant object produce circles and are therefore blurred. (The effect is shown greatly exaggerated.)

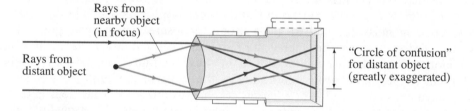

Rays from nearby object (in focus)

Rays from distant object

"Circle of confusion" for distant object (greatly exaggerated)

EXAMPLE 33–8 **Camera focus.** How far must a 50.0-mm-focal-length camera lens be moved from its infinity setting to sharply focus an object 3.00 m away?

APPROACH For an object at infinity, the image is at the focal point, by definition. For an object distance of 3.00 m, we use the thin lens equation, Eq. 33–2, to find the image distance (distance of lens to film or sensor).

SOLUTION When focused at infinity, the lens is 50.0 mm from the film. When focused at $d_o = 3.00$ m, the image distance is given by the lens equation,

$$\frac{1}{d_i} = \frac{1}{f} - \frac{1}{d_o} = \frac{1}{50.0 \text{ mm}} - \frac{1}{3000 \text{ mm}} = \frac{3000 - 50}{(3000)(50.0) \text{ mm}} = \frac{2950}{150,000 \text{ mm}}.$$

We solve for d_i and find $d_i = 50.8$ mm, so the lens needs to move 0.8 mm away from the film or digital sensor.

[†]Smaller lens openings, however, result in reduced resolution due to diffraction (discussed in Chapter 35). Best resolution is typically found around $f/8$.

CONCEPTUAL EXAMPLE 33–9 | **Shutter speed.** To improve the depth of field, you "stop down" your camera lens by two f-stops from $f/4$ to $f/8$. What should you do to the shutter speed to maintain the same exposure?

RESPONSE The amount of light admitted by the lens is proportional to the area of the lens opening. Reducing the lens opening by two f-stops reduces the diameter by a factor of 2, and the area by a factor of 4. To maintain the same exposure, the shutter must be open four times as long. If the shutter speed had been $\frac{1}{500}$ s, you would have to increase the exposure time to $\frac{1}{125}$ s.

*Picture Sharpness

The sharpness of a picture depends not only on accurate focusing and short exposure times, but also on the graininess of the film, or the number of pixels for a digital camera. Fine-grained films are "slower," meaning they require longer exposures for a given light level. Digital cameras have averaging (or "compression") programs, such as JPEG, which reduce memory size by averaging over pixels where little contrast is detected. Hence it is unusual to use all pixels available. They also average over pixels in low light conditions, resulting in a less sharp photo.

The quality of the lens strongly affects the image quality, and we discuss lens resolution and diffraction effects in Sections 33–10 and 35–4. The sharpness, or *resolution*, of a lens is often given as so many lines per millimeter, measured by photographing a standard set of parallel lines on fine-grain film or high quality sensor, or as so many dots per inch (dpi). The minimum spacing of distinguishable lines or dots gives the resolution; 50 lines/mm is reasonable, 100 lines/mm is quite good ($= 100$ dots/mm ≈ 2500 dpi on the sensor).

EXAMPLE 33–10 **Pixels and resolution.** A 6-MP (6-megapixel) digital camera offers a maximum resolution of 2000×3000 pixels on a 16-mm $\times$ 24-mm CCD sensor. How sharp should the lens be to make use of this resolution?

APPROACH We find the number of pixels per millimeter and require the lens to be at least that good.

SOLUTION We can either take the image height (2000 pixels in 16 mm) or the width (3000 pixels in 24 mm):

$$\frac{3000 \text{ pixels}}{24 \text{ mm}} = 125 \text{ pixels/mm}.$$

We would want the lens to be able to resolve at least 125 lines or dots per mm as well, which would be a very good lens. If the lens is not this good, fewer pixels and less memory could be used.

NOTE Increasing lens resolution is a tougher problem today than is squeezing more pixels on a CCD or CMOS. The sensor for high MP cameras must also be physically larger for greater light sensitivity (low light conditions).

EXAMPLE 33–11 **Blown-up photograph.** An enlarged photograph looks sharp at normal viewing distances if the dots or lines are resolved to about 10 dots/mm. Would an 8×10-inch enlargement of a photo taken by the camera in Example 33–10 seem sharp? To what maximum size could you enlarge this 2000×3000-pixel image?

APPROACH We assume the image is 2000×3000 pixels on a 16×24-mm CCD as in Example 33–10, or 125 pixels/mm. We make an enlarged photo 8×10 in. $= 20$ cm $\times$ 25 cm.

SOLUTION The short side of the CCD is 16 mm $= 1.6$ cm long, and that side of the photograph is 8 inches or 20 cm. Thus the size is increased by a factor of 20 cm/1.6 cm $= 12.5\times$ (or 25 cm/2.4 cm $\approx 10\times$). To fill the 8×10-in. paper, we assume the enlargement is $12.5\times$. The pixels are thus enlarged $12.5\times$; so the pixel count of 125/mm on the CCD becomes 10 per mm on the print. Hence an 8×10-inch print is just about the maximum possible for a sharp photograph with 6 megapixels. If you feel 7 dots per mm is good enough, you can enlarge to maybe 11×14 inches.

*Telephotos and Wide-angles

Camera lenses are categorized into normal, telephoto, and wide angle, according to focal length and film size. A **normal lens** covers the sensor or film with a field of view that corresponds approximately to that of normal vision. A normal lens for 35-mm film has a focal length in the vicinity of 50 mm. The best digital cameras aim for a sensor of the same size[†] (24 mm × 36mm). (If the sensor is smaller, digital cameras sometimes specify focal lengths to correspond with classic 35-mm cameras.) **Telephoto lenses** act like telescopes to magnify images. They have longer focal lengths than a normal lens: as we saw in Section 33–2 (Eq. 33–3), the height of the image for a given object distance is proportional to the image distance, and the image distance will be greater for a lens with longer focal length. For distant objects, the image height is very nearly proportional to the focal length. Thus a 200-mm telephoto lens for use with a 35-mm camera gives a 4× magnification over the normal 50-mm lens. A **wide-angle lens** has a shorter focal length than normal: a wider field of view is included, and objects appear smaller. A **zoom lens** is one whose focal length can be changed so that you seem to zoom up to, or away from, the subject as you change the focal length.

Digital cameras may have an **optical zoom** meaning the lens can change focal length and maintain resolution. But an "electronic" or **digital zoom** just enlarges the dots (pixels) with loss of sharpness.

Different types of viewing systems are used in cameras. In some cameras, you view through a small window just above the lens as in Fig. 33–18. In a **single-lens reflex** camera (SLR), you actually view through the lens with the use of prisms and mirrors (Fig. 33–24). A mirror hangs at a 45° angle behind the lens and flips up out of the way just before the shutter opens. SLRs have the advantage that you can see almost exactly what you will get. Digital cameras use an LCD display, and it too can show what you will get on the photo if it is carefully designed.

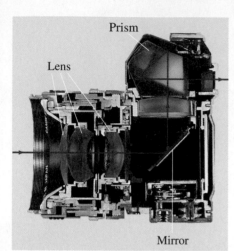

FIGURE 33–24 Single-lens reflex (SLR) camera, showing how the image is viewed through the lens with the help of a movable mirror and prism.

PHYSICS APPLIED

The eye

FIGURE 33–25 Diagram of a human eye.

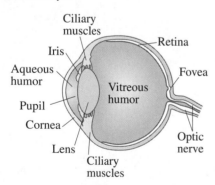

33–6 The Human Eye; Corrective Lenses

The human eye resembles a camera in its basic structure (Fig. 33–25), but is far more sophisticated. The interior of the eye is filled with a transparent gel-like substance called the *vitreous humor* with index of refraction $n = 1.337$. Light enters this enclosed volume through the cornea and lens. Between the cornea and lens is a watery fluid, the aqueous humor (*aqua* is "water" in Latin) with $n = 1.336$. A diaphragm, called the **iris** (the colored part of your eye), adjusts automatically to control the amount of light entering the eye, similar to a camera. The hole in the iris through which light passes (the **pupil**) is black because no light is reflected from it (it's a hole), and very little light is reflected back out from the interior of the eye. The **retina**, which plays the role of the film or sensor in a camera, is on the curved rear surface of the eye. The retina consists of a complex array of nerves and receptors known as *rods* and *cones* which act to change light energy into electrical signals that travel along the nerves. The reconstruction of the image from all these tiny receptors is done mainly in the brain, although some analysis may also be done in the complex interconnected nerve network at the retina itself. At the center of the retina is a small area called the **fovea**, about 0.25 mm in diameter, where the cones are very closely packed and the sharpest image and best color discrimination are found.

Unlike a camera, the eye contains no shutter. The equivalent operation is carried out by the nervous system, which analyzes the signals to form images at the rate of about 30 per second. This can be compared to motion picture or television cameras, which operate by taking a series of still pictures at a rate of 24 (movies) or 30 (U.S. television) per second; their rapid projection on the screen gives the appearance of motion.

[†]A "35-mm camera" uses film that is physically 35 mm wide; that 35 mm is not to be confused with a focal length. 35-mm film has sprocket holes, so only 24 mm of its height is used for the photo; the width is usually 36 mm for stills. Thus one frame is 24 mm × 36 mm. Movie frames are 18 mm × 24 mm.

The lens of the eye ($n = 1.386$ to 1.406) does little of the bending of the light rays. Most of the refraction is done at the front surface of the **cornea** ($n = 1.376$) at its interface with air ($n = 1.0$). The lens acts as a fine adjustment for focusing at different distances. This is accomplished by the ciliary muscles (Fig. 33–25), which change the curvature of the lens so that its focal length is changed. To focus on a distant object, the ciliary muscles of the eye are relaxed and the lens is thin, as shown in Fig. 33–26a, and parallel rays focus at the focal point (on the retina). To focus on a nearby object, the muscles contract, causing the center of the lens to thicken, Fig. 33–26b, thus shortening the focal length so that images of nearby objects can be focused on the retina, behind the new focal point. This focusing adjustment is called **accommodation**.

The closest distance at which the eye can focus clearly is called the **near point** of the eye. For young adults it is typically 25 cm, although younger children can often focus on objects as close as 10 cm. As people grow older, the ability to accommodate is reduced and the near point increases. A given person's **far point** is the farthest distance at which an object can be seen clearly. For some purposes it is useful to speak of a **normal eye** (a sort of average over the population), defined as an eye having a near point of 25 cm and a far point of infinity. To check your own near point, place this book close to your eye and slowly move it away until the type is sharp.

The "normal" eye is sort of an ideal. Many people have eyes that do not accommodate within the "normal" range of 25 cm to infinity, or have some other defect. Two common defects are nearsightedness and farsightedness. Both can be corrected to a large extent with lenses—either eyeglasses or contact lenses.

In **nearsightedness**, or *myopia*, the eye can focus only on nearby objects. The far point is not infinity but some shorter distance, so distant objects are not seen clearly. It is usually caused by an eyeball that is too long, although sometimes it is the curvature of the cornea that is too great. In either case, images of distant objects are focused in front of the retina. A diverging lens, because it causes parallel rays to diverge, allows the rays to be focused at the retina (Fig. 33–27a) and thus corrects this defect.

In **farsightedness**, or *hyperopia*, the eye cannot focus on nearby objects. Although distant objects are usually seen clearly, the near point is somewhat greater than the "normal" 25 cm, which makes reading difficult. This defect is caused by an eyeball that is too short or (less often) by a cornea that is not sufficiently curved. It is corrected by a converging lens, Fig. 33–27b. Similar to hyperopia is *presbyopia*, which refers to the lessening ability of the eye to accommodate as a person ages, and the near point moves out. Converging lenses also compensate for this.

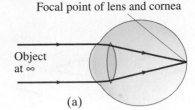

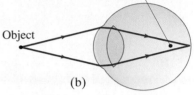

FIGURE 33–26 Accommodation by a normal eye: (a) lens relaxed, focused at infinity; (b) lens thickened, focused on a nearby object.

PHYSICS APPLIED
Corrective lenses

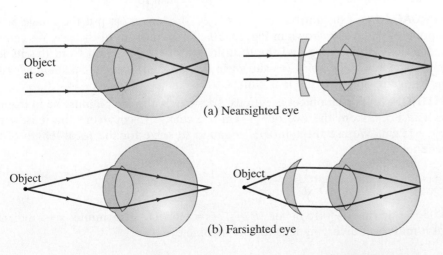

(a) Nearsighted eye

(b) Farsighted eye

FIGURE 33–27 Correcting eye defects with lenses: (a) a nearsighted eye, which cannot focus clearly on distant objects, can be corrected by use of a diverging lens; (b) a farsighted eye, which cannot focus clearly on nearby objects, can be corrected by use of a converging lens.

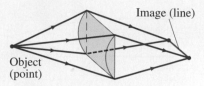

Image (line)

Object
(point)

FIGURE 33–28 A cylindrical lens forms a line image of a point object because it is converging in one plane only.

Astigmatism is usually caused by an out-of-round cornea or lens so that point objects are focused as short lines, which blurs the image. It is as if the cornea were spherical with a cylindrical section superimposed. As shown in Fig. 33–28, a cylindrical lens focuses a point into a line parallel to its axis. An astigmatic eye may focus rays in one plane, such as the vertical plane, at a shorter distance than it does for rays in a horizontal plane. Astigmatism is corrected with the use of a compensating cylindrical lens. Lenses for eyes that are nearsighted or farsighted as well as astigmatic are ground with superimposed spherical and cylindrical surfaces, so that the radius of curvature of the correcting lens is different in different planes.

EXAMPLE 33–12 **Farsighted eye.** Sue is farsighted with a near point of 100 cm. Reading glasses must have what lens power so that she can read a newspaper at a distance of 25 cm? Assume the lens is very close to the eye.

APPROACH When the object is placed 25 cm from the lens, we want the image to be 100 cm away on the *same* side of the lens (so the eye can focus it), and so the image is virtual, as shown in Fig. 33–29, and $d_i = -100$ cm will be negative. We use the thin lens equation (Eq. 33–2) to determine the needed focal length. Optometrists' prescriptions specify the power ($P = 1/f$, Eq. 33–1) given in diopters $(1 \, \text{D} = 1 \, \text{m}^{-1})$.

SOLUTION Given that $d_o = 25$ cm and $d_i = -100$ cm, the thin lens equation gives

$$\frac{1}{f} = \frac{1}{d_o} + \frac{1}{d_i} = \frac{1}{25 \, \text{cm}} + \frac{1}{-100 \, \text{cm}} = \frac{4-1}{100 \, \text{cm}} = \frac{1}{33 \, \text{cm}}.$$

So $f = 33$ cm $= 0.33$ m. The power P of the lens is $P = 1/f = +3.0$ D. The plus sign indicates that it is a converging lens.

NOTE We chose the image position to be where the eye can actually focus. The lens needs to put the image there, given the desired placement of the object (newspaper).

FIGURE 33–29 Lens of reading glasses (Example 33–12).

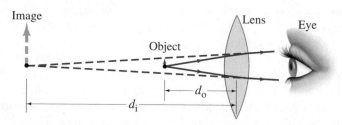

Image

Lens

Eye

Object

d_o

d_i

FIGURE 33–30 Example 33–13.

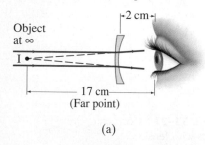

|-2 cm-|

Object
at ∞

I

17 cm
(Far point)

(a)

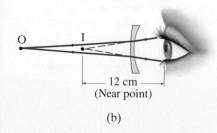

O   I

12 cm
(Near point)

(b)

EXAMPLE 33–13 **Nearsighted eye.** A nearsighted eye has near and far points of 12 cm and 17 cm, respectively. (*a*) What lens power is needed for this person to see distant objects clearly, and (*b*) what then will be the near point? Assume that the lens is 2.0 cm from the eye (typical for eyeglasses).

APPROACH For a distant object $(d_o = \infty)$, the lens must put the image at the far point of the eye as shown in Fig. 33–30a, 17 cm in front of the eye. We can use the thin lens equation to find the focal length of the lens, and from this its lens power. The new near point (as shown in Fig. 33–30b) can be calculated for the lens by again using the thin lens equation.

SOLUTION (*a*) For an object at infinity $(d_o = \infty)$, the image must be in front of the lens 17 cm from the eye or $(17 \, \text{cm} - 2 \, \text{cm}) = 15 \, \text{cm}$ from the lens; hence $d_i = -15$ cm. We use the thin lens equation to solve for the focal length of the needed lens:

$$\frac{1}{f} = \frac{1}{d_o} + \frac{1}{d_i} = \frac{1}{\infty} + \frac{1}{-15 \, \text{cm}} = -\frac{1}{15 \, \text{cm}}.$$

So $f = -15$ cm $= -0.15$ m or $P = 1/f = -6.7$ D. The minus sign indicates that it must be a diverging lens for the myopic eye.

(b) The near point when glasses are worn is where an object is placed (d_o) so that the lens forms an image at the "near point of the naked eye," namely 12 cm from the eye. That image point is $(12 \text{ cm} - 2 \text{ cm}) = 10 \text{ cm}$ in front of the lens, so $d_i = -0.10 \text{ m}$ and the thin lens equation gives

$$\frac{1}{d_o} = \frac{1}{f} - \frac{1}{d_i} = -\frac{1}{0.15 \text{ m}} + \frac{1}{0.10 \text{ m}} = \frac{-2 + 3}{0.30 \text{ m}} = \frac{1}{0.30 \text{ m}}.$$

So $d_o = 30 \text{ cm}$, which means the near point when the person is wearing glasses is 30 cm in front of the lens, or 32 cm from the eye.

Suppose contact lenses are used to correct the eye in Example 33–13. Since contacts are placed directly on the cornea, we would not subtract out the 2.0 cm for the image distances. That is, for distant objects $d_i = f = -17 \text{ cm}$, so $P = 1/f = -5.9 \text{ D}$. The new near point would be 41 cm. Thus we see that a contact lens and an eyeglass lens will require slightly different powers, or focal lengths, for the same eye because of their different placements relative to the eye. We also see that glasses in this case give a better near point than contacts.

PHYSICS APPLIED
Contact lenses

EXERCISE E What power contact lens is needed for an eye to see distant objects if its far point is 25 cm?

When your eyes are under water, distant underwater objects look blurry because at the water–cornea interface, the difference in indices of refraction is very small: $n = 1.33$ for water, 1.376 for the cornea. Hence light rays are bent very little and are focused far behind the retina, Fig. 33–31a. If you wear goggles or a face mask, you restore an air–cornea interface ($n = 1.0$ and 1.376, respectively) and the rays can be focused, Fig. 33–31b.

PHYSICS APPLIED
Underwater vision

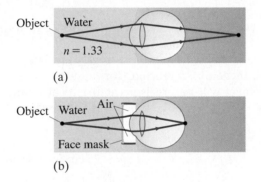

(a)

(b)

FIGURE 33–31 (a) Under water, we see a blurry image because light rays are bent much less than in air. (b) If we wear goggles, we again have an air–cornea interface and can see clearly.

33–7 Magnifying Glass

Much of the remainder of this Chapter will deal with optical devices that are used to produce magnified images of objects. We first discuss the **simple magnifier**, or **magnifying glass**, which is simply a converging lens (see Chapter-Opening Photo).

How large an object appears, and how much detail we can see on it, depends on the size of the image it makes on the retina. This, in turn, depends on the angle subtended by the object at the eye. For example, a penny held 30 cm from the eye looks twice as tall as one held 60 cm away because the angle it subtends is twice as great (Fig. 33–32). When we want to examine detail on an object, we bring it up close to our eyes so that it subtends a greater angle. However, our eyes can accommodate only up to a point (the near point), and we will assume a standard distance of $N = 25 \text{ cm}$ as the near point in what follows.

FIGURE 33–32 When the same object is viewed at a shorter distance, the image on the retina is greater, so the object appears larger and more detail can be seen. The angle θ that the object subtends in (a) is greater than in (b). *Note:* This is not a normal ray diagram because we are showing only one ray from each point.

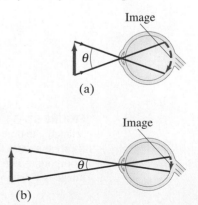

(a)

(b)

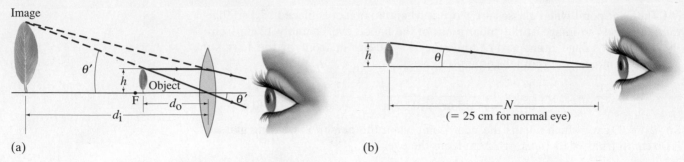

(a)

(b)

FIGURE 33–33 Leaf viewed (a) through a magnifying glass, and (b) with the unaided eye. The eye is focused at its near point in both cases.

A magnifying glass allows us to place the object closer to our eye so that it subtends a greater angle. As shown in Fig. 33–33a, the object is placed at the focal point or just within it. Then the converging lens produces a virtual image, which must be at least 25 cm from the eye if the eye is to focus on it. If the eye is relaxed, the image will be at infinity, and in this case the object is exactly at the focal point. (You make this slight adjustment yourself when you "focus" on the object by moving the magnifying glass.)

A comparison of part (a) of Fig. 33–33 with part (b), in which the same object is viewed at the near point with the unaided eye, reveals that the angle the object subtends at the eye is much larger when the magnifier is used. The **angular magnification** or **magnifying power**, M, of the lens is defined as the ratio of the angle subtended by an object when using the lens, to the angle subtended using the unaided eye, with the object at the near point N of the eye ($N = 25$ cm for a normal eye):

$$M = \frac{\theta'}{\theta}, \tag{33–5}$$

where θ and θ' are shown in Fig. 33–33. We can write M in terms of the focal length by noting that $\theta = h/N$ (Fig. 33–33b) and $\theta' = h/d_o$ (Fig. 33–33a), where h is the height of the object and we assume the angles are small so θ and θ' equal their sines and tangents. If the eye is relaxed (for least eye strain), the image will be at infinity and the object will be precisely at the focal point; see Fig. 33–34. Then $d_o = f$ and $\theta' = h/f$, whereas $\theta = h/N$ as before (Fig. 33–33b). Thus

$$M = \frac{\theta'}{\theta} = \frac{h/f}{h/N} = \frac{N}{f}. \quad \begin{bmatrix} \text{eye focused at } \infty; \\ N = 25 \text{ cm for normal eye} \end{bmatrix} \tag{33–6a}$$

We see that the shorter the focal length of the lens, the greater the magnification.[†]

The magnification of a given lens can be increased a bit by moving the lens and adjusting your eye so it focuses on the image at the eye's near point. In this case, $d_i = -N$ (see Fig. 33–33a) if your eye is very near the magnifier.

[†]Simple single-lens magnifiers are limited to about 2 or 3× because of blurring due to spherical aberration (Section 33–10).

FIGURE 33–34 With the eye relaxed, the object is placed at the focal point, and the image is at infinity. Compare to Fig. 33–33a where the image is at the eye's near point.

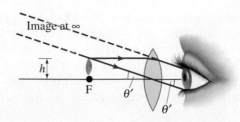

Then the object distance d_o is given by

$$\frac{1}{d_o} = \frac{1}{f} - \frac{1}{d_i} = \frac{1}{f} + \frac{1}{N}.$$

We see from this equation that $d_o = fN/(f + N) < f$, as shown in Fig. 33–33a, since $N/(f + N)$ must be less than 1. With $\theta' = h/d_o$ the magnification is

$$M = \frac{\theta'}{\theta} = \frac{h/d_o}{h/N} = \frac{N}{d_o} = N\left(\frac{1}{f} + \frac{1}{N}\right)$$

or

$$M = \frac{N}{f} + 1. \qquad \left[\begin{array}{l}\text{eye focused at near point, } N;\\ N = 25\text{ cm for normal eye}\end{array}\right] \quad \textbf{(33–6b)}$$

We see that the magnification is slightly greater when the eye is focused at its near point, as compared to when it is relaxed.

EXAMPLE 33–14 ESTIMATE A jeweler's "loupe." An 8-cm-focal-length converging lens is used as a "jeweler's loupe," which is a magnifying glass. Estimate (a) the magnification when the eye is relaxed, and (b) the magnification if the eye is focused at its near point $N = 25$ cm.

APPROACH The magnification when the eye is relaxed is given by Eq. 33–6a. When the eye is focused at its near point, we use Eq. 33–6b and we assume the lens is near the eye.

SOLUTION (a) With the relaxed eye focused at infinity,

$$M = \frac{N}{f} = \frac{25\text{ cm}}{8\text{ cm}} \approx 3\times.$$

(b) The magnification when the eye is focused at its near point ($N = 25$ cm), and the lens is near the eye, is

$$M = 1 + \frac{N}{f} = 1 + \frac{25}{8} \approx 4\times.$$

33–8 Telescopes

A telescope is used to magnify objects that are very far away. In most cases, the object can be considered to be at infinity.

Galileo, although he did not invent it,[†] developed the telescope into a usable and important instrument. He was the first to examine the heavens with the telescope (Fig. 33–35), and he made world-shaking discoveries, including the moons of Jupiter, the phases of Venus, sunspots, the structure of the Moon's surface, and that the Milky Way is made up of a huge number of individual stars.

[†]Galileo built his first telescope in 1609 after having heard of such an instrument existing in Holland. The first telescopes magnified only three to four times, but Galileo soon made a 30-power instrument. The first Dutch telescope seems to date from about 1604, but there is a reference suggesting it may have been copied from an Italian telescope built as early as 1590. Kepler (see Chapter 6) gave a ray description (in 1611) of the Keplerian telescope, which is named for him because he first described it, although he did not build it.

FIGURE 33–35 (a) Objective lens (mounted now in an ivory frame) from the telescope with which Galileo made his world-shaking discoveries, including the moons of Jupiter. (b) Later telescopes made by Galileo.

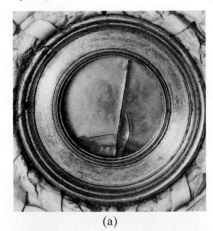

(a)

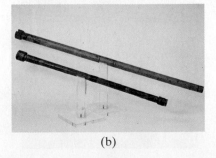

(b)

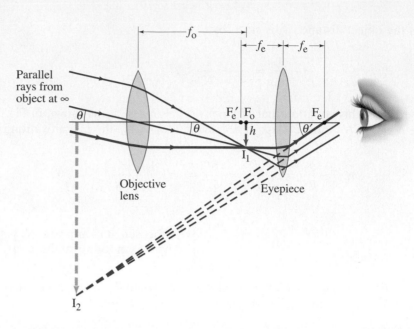

FIGURE 33–36 Astronomical telescope (refracting). Parallel light from one point on a distant object $(d_o = \infty)$ is brought to a focus by the objective lens in its focal plane. This image (I_1) is magnified by the eyepiece to form the final image I_2. Only two of the rays shown entering the objective are standard rays (2 and 3) as described in Fig. 33–6.

Several types of **astronomical telescope** exist. The common **refracting** type, sometimes called **Keplerian**, contains two converging lenses located at opposite ends of a long tube, as illustrated in Fig. 33–36. The lens closest to the object is called the **objective lens** (focal length f_o) and forms a real image I_1 of the distant object in the plane of its focal point F_o (or near it if the object is not at infinity). The second lens, called the **eyepiece** (focal length f_e), acts as a magnifier. That is, the eyepiece magnifies the image I_1 formed by the objective lens to produce a second, greatly magnified image, I_2, which is virtual and inverted. If the viewing eye is relaxed, the eyepiece is adjusted so the image I_2 is at infinity. Then the real image I_1 is at the focal point F'_e of the eyepiece, and the distance between the lenses is $f_o + f_e$ for an object at infinity.

To find the total angular magnification of this telescope, we note that the angle an object subtends as viewed by the unaided eye is just the angle θ subtended at the telescope objective. From Fig. 33–36 we can see that $\theta \approx h/f_o$, where h is the height of the image I_1 and we assume θ is small so that $\tan \theta \approx \theta$. Note, too, that the thickest of the three rays drawn in Fig. 33–36 is parallel to the axis before it strikes the eyepiece and therefore is refracted through the eyepiece focal point F_e on the far side. Thus, $\theta' \approx h/f_e$ and the **total magnifying power** (that is, angular magnification, which is what is always quoted) of this telescope is

$$M = \frac{\theta'}{\theta} = \frac{(h/f_e)}{(h/f_o)} = -\frac{f_o}{f_e}, \qquad \text{[telescope]} \quad \textbf{(33–7)}$$

where we have inserted a minus sign to indicate that the image is inverted. To achieve a large magnification, the objective lens should have a long focal length and the eyepiece a short focal length.

FIGURE 33–37 This large refracting telescope was built in 1897 and is housed at Yerkes Observatory in Wisconsin. The objective lens is 102 cm (40 inches) in diameter, and the telescope tube is about 19 m long. Example 33–15.

EXAMPLE 33–15 Telescope magnification. The largest optical refracting telescope in the world is located at the Yerkes Observatory in Wisconsin, Fig. 33–37. It is referred to as a "40-inch" telescope, meaning that the diameter of the objective is 40 in., or 102 cm. The objective lens has a focal length of 19 m, and the eyepiece has a focal length of 10 cm. (*a*) Calculate the total magnifying power of this telescope. (*b*) Estimate the length of the telescope.

APPROACH Equation 33–7 gives the magnification. The length of the telescope is the distance between the two lenses.

SOLUTION (*a*) From Eq. 33–7 we find

$$M = -\frac{f_o}{f_e} = -\frac{19 \text{ m}}{0.10 \text{ m}} = -190\times.$$

(*b*) For a relaxed eye, the image I_1 is at the focal point of both the eyepiece and the objective lenses. The distance between the two lenses is thus $f_o + f_e \approx 19 \text{ m}$, which is essentially the length of the telescope.

EXERCISE F A 40× telescope has a 1.2-cm focal length eyepiece. What is the focal length of the objective lens?

For an astronomical telescope to produce bright images of faint stars, the objective lens must be large to allow in as much light as possible. Indeed, the diameter of the objective lens (and hence its "light-gathering power") is an important parameter for an astronomical telescope, which is why the largest ones are specified by giving the objective diameter (such as the 10-meter Keck telescope in Hawaii). The construction and grinding of large lenses is very difficult. Therefore, the largest telescopes are **reflecting telescopes** which use a curved mirror as the objective, Fig. 33–38. A mirror has only one surface to be ground and can be supported along its entire surface[†] (a large lens, supported at its edges, would sag under its own weight). Often, the eyepiece lens or mirror (see Fig. 33–38) is removed so that the real image formed by the objective mirror can be recorded directly on film or on an electronic sensor (CCD or CMOS, Section 33–5).

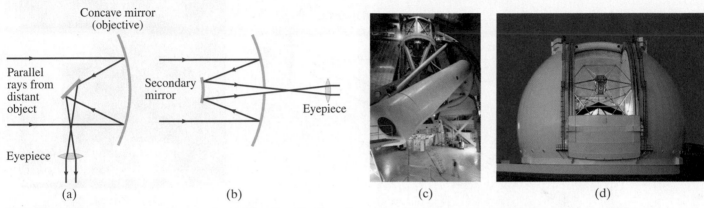

FIGURE 33–38 A concave mirror can be used as the objective of an astronomical telescope. Arrangement (a) is called the Newtonian focus, and (b) the Cassegrainian focus. Other arrangements are also possible. (c) The 200-inch (mirror diameter) Hale telescope on Palomar Mountain in California. (d) The 10-meter Keck telescope on Mauna Kea, Hawaii. The Keck combines thirty-six 1.8-meter six-sided mirrors into the equivalent of a very large single reflector, 10 m in diameter.

A **terrestrial telescope**, for viewing objects on Earth, must provide an upright image—seeing normal objects upside down would be difficult (much less important for viewing stars). Two designs are shown in Fig. 33–39. The **Galilean** type, which Galileo used for his great astronomical discoveries, has a diverging lens as eyepiece which intercepts the converging rays from the objective lens before they reach a focus, and acts to form a virtual upright image, Fig. 33–39a. This design is still used in opera glasses. The tube is reasonably short, but the field of view is small. The second type, shown in Fig. 33–39b, is often called a **spyglass** and makes use of a third convex lens that acts to make the image upright as shown. A spyglass must be quite long. The most practical design today is the **prism binocular** which was shown in Fig. 32–33. The objective and eyepiece are converging lenses. The prisms reflect the rays by total internal reflection and shorten the physical size of the device, and they also act to produce an upright image. One prism reinverts the image in the vertical plane, the other in the horizontal plane.

[†]Another advantage of mirrors is that they exhibit no chromatic aberration because the light doesn't pass through them; and they can be ground in a parabolic shape to correct for spherical aberration (Section 33–10). The reflecting telescope was first proposed by Newton.

FIGURE 33–39 Terrestrial telescopes that produce an upright image: (a) Galilean; (b) spyglass, or erector type.

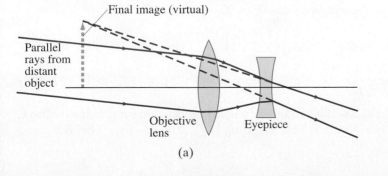

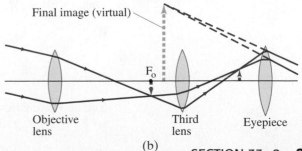

*33-9 Compound Microscope

(ⓧ) PHYSICS APPLIED
Microscopes

The compound **microscope**, like the telescope, has both objective and eyepiece (or ocular) lenses, Fig. 33–40. The design is different from that for a telescope because a microscope is used to view objects that are very close, so the object distance is very small. The object is placed just beyond the objective's focal point as shown in Fig. 33–40a. The image I_1 formed by the objective lens is real, quite far from the objective lens, and much enlarged. The eyepiece is positioned so that this image is near the eyepiece focal point F_e. The image I_1 is magnified by the eyepiece into a very large virtual image, I_2, which is seen by the eye and is inverted. Modern microscopes use a third "tube" lens behind the objective, but we will analyze the simpler arrangement shown in Fig. 33–40a.

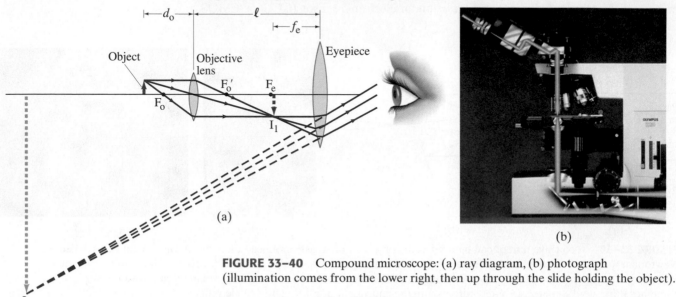

(a)

(b)

FIGURE 33–40 Compound microscope: (a) ray diagram, (b) photograph (illumination comes from the lower right, then up through the slide holding the object).

The overall magnification of a microscope is the product of the magnifications produced by the two lenses. The image I_1 formed by the objective lens is a factor m_o greater than the object itself. From Fig. 33–40a and Eq. 33–3 for the lateral magnification of a simple lens, we have

$$m_o = \frac{h_i}{h_o} = \frac{d_i}{d_o} = \frac{\ell - f_e}{d_o}, \qquad (33\text{–}8)$$

where d_o and d_i are the object and image distances for the objective lens, ℓ is the distance between the lenses (equal to the length of the barrel), and we ignored the minus sign in Eq. 33–3 which only tells us that the image is inverted. We set $d_i = \ell - f_e$, which is true only if the eye is relaxed, so that the image I_1 is at the eyepiece focal point F_e. The eyepiece acts like a simple magnifier. If we assume that the eye is relaxed, the eyepiece angular magnification M_e is (from Eq. 33–6a)

$$M_e = \frac{N}{f_e}, \qquad (33\text{–}9)$$

where the near point $N = 25$ cm for the normal eye. Since the eyepiece enlarges the image formed by the objective, the overall angular magnification M is the product of the lateral magnification of the objective lens, m_o, times the angular magnification, M_e, of the eyepiece lens (Eqs. 33–8 and 33–9):

$$M = M_e m_o = \left(\frac{N}{f_e}\right)\left(\frac{\ell - f_e}{d_o}\right) \qquad \text{[microscope]} \quad (33\text{–}10a)$$

$$\approx \frac{N\ell}{f_e f_o}. \qquad [f_o \text{ and } f_e \ll \ell] \quad (33\text{–}10b)$$

The approximation, Eq. 33–10b, is accurate when f_e and f_o are small compared to ℓ, so $\ell - f_e \approx \ell$, and the object is near F_o so $d_o \approx f_o$ (Fig. 33–40a). This is a good approximation for large magnifications, which are obtained when f_o and f_e are very small (they are in the denominator of Eq. 33–10b). To make lenses of very short focal length, compound lenses involving several elements must be used to avoid serious aberrations, as discussed in the next Section.

EXAMPLE 33–16 **Microscope.** A compound microscope consists of a $10\times$ eyepiece and a $50\times$ objective 17.0 cm apart. Determine (a) the overall magnification, (b) the focal length of each lens, and (c) the position of the object when the final image is in focus with the eye relaxed. Assume a normal eye, so $N = 25$ cm.

APPROACH The overall magnification is the product of the eyepiece magnification and the objective magnification. The focal length of the eyepiece is found from Eq. 33–6a or 33–9 for the magnification of a simple magnifier. For the objective lens, it is easier to next find d_o (part c) using Eq. 33–8 before we find f_o.

SOLUTION (a) The overall magnification is $(10\times)(50\times) = 500\times$.
(b) The eyepiece focal length is (Eq. 33–9) $f_e = N/M_e = 25 \text{ cm}/10 = 2.5 \text{ cm}$. Next we solve Eq. 33–8 for d_o, and find

$$d_o = \frac{\ell - f_e}{m_o} = \frac{(17.0 \text{ cm} - 2.5 \text{ cm})}{50} = 0.29 \text{ cm}.$$

Then, from the thin lens equation for the objective with $d_i = \ell - f_e = 14.5$ cm (see Fig. 33–40a),

$$\frac{1}{f_o} = \frac{1}{d_o} + \frac{1}{d_i} = \frac{1}{0.29 \text{ cm}} + \frac{1}{14.5 \text{ cm}} = 3.52 \text{ cm}^{-1};$$

so $f_o = 1/(3.52 \text{ cm}^{-1}) = 0.28$ cm.
(c) We just calculated $d_o = 0.29$ cm, which is very close to f_o.

---

*33–10 Aberrations of Lenses and Mirrors

Earlier in this Chapter we developed a theory of image formation by a thin lens. We found, for example, that all rays from each point on an object are brought to a single point as the image point. This result, and others, were based on approximations for a thin lens, mainly that all rays make small angles with the axis and that we can use $\sin\theta \approx \theta$. Because of these approximations, we expect deviations from the simple theory, which are referred to as **lens aberrations**. There are several types of aberration; we will briefly discuss each of them separately, but all may be present at one time.

Consider an object at any point (even at infinity) on the axis of a lens with spherical surfaces. Rays from this point that pass through the outer regions of the lens are brought to a focus at a different point from those that pass through the center of the lens. This is called **spherical aberration**, and is shown exaggerated in Fig. 33–41. Consequently, the image seen on a screen or film will not be a point but a tiny circular patch of light. If the sensor or film is placed at the point C, as indicated, the circle will have its smallest diameter, which is referred to as the **circle of least confusion**. Spherical aberration is present whenever spherical surfaces are used. It can be reduced by using nonspherical (= aspherical) lens surfaces, but grinding such lenses is difficult and expensive. Spherical aberration can be reduced by the use of several lenses in combination, and by using primarily the central part of lenses.

PHYSICS APPLIED
Lens aberrations

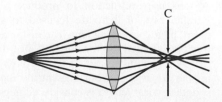
FIGURE 33–41 Spherical aberration (exaggerated). Circle of least confusion is at C.

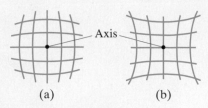

FIGURE 33–42 Distortion: lenses may image a square grid of perpendicular lines to produce (a) barrel distortion or (b) pincushion distortion. These distortions can be seen in the photograph of Fig. 33–2d.

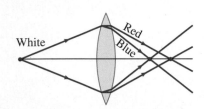

FIGURE 33–43 Chromatic aberration. Different colors are focused at different points.

FIGURE 33–44 Achromatic doublet.

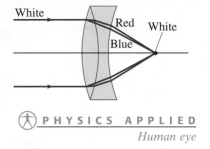

For object points off the lens axis, additional aberrations occur. Rays passing through the different parts of the lens cause spreading of the image that is noncircular. There are two effects: **coma** (because the image of a point is comet-shaped rather than a circle) and **off-axis astigmatism**.† Furthermore, the image points for objects off the axis but at the same distance from the lens do not fall on a flat plane but on a curved surface—that is, the focal plane is not flat. (We expect this because the points on a flat plane, such as the film in a camera, are not equidistant from the lens.) This aberration is known as **curvature of field** and is a problem in cameras and other devices where the film is placed in a flat plane. In the eye, however, the retina is curved, which compensates for this effect.

Another aberration, **distortion**, is a result of variation of magnification at different distances from the lens axis. Thus a straight-line object some distance from the axis may form a curved image. A square grid of lines may be distorted to produce "barrel distortion," or "pincushion distortion," Fig. 33–42. The former is common in extreme wide-angle lenses.

All the above aberrations occur for monochromatic light and hence are referred to as *monochromatic aberrations*. Normal light is not monochromatic, and there will also be **chromatic aberration**. This aberration arises because of dispersion—the variation of index of refraction of transparent materials with wavelength (Section 32–6). For example, blue light is bent more than red light by glass. So if white light is incident on a lens, the different colors are focused at different points, Fig. 33–43, and have slightly different magnifications resulting in colored fringes in the image. Chromatic aberration can be eliminated for any two colors (and reduced greatly for all others) by the use of two lenses made of different materials with different indices of refraction and dispersion. Normally one lens is converging and the other diverging, and they are often cemented together (Fig. 33–44). Such a lens combination is called an **achromatic doublet** (or "color-corrected" lens).

To reduce aberrations, high-quality lenses are **compound lenses** consisting of many simple lenses, referred to as **elements**. A typical high-quality camera lens may contain six to eight (or more) elements. For simplicity we will usually indicate lenses in diagrams as if they were simple lenses.

The human eye is also subject to aberrations, but they are minimal. Spherical aberration, for example, is minimized because (1) the cornea is less curved at the edges than at the center, and (2) the lens is less dense at the edges than at the center. Both effects cause rays at the outer edges to be bent less strongly, and thus help to reduce spherical aberration. Chromatic aberration is partially compensated for because the lens absorbs the shorter wavelengths appreciably and the retina is less sensitive to the blue and violet wavelengths. This is just the region of the spectrum where dispersion—and thus chromatic aberration—is greatest (Fig. 32–28).

Spherical mirrors (Section 32–3) also suffer aberrations including spherical aberration (see Fig. 32–13). Mirrors can be ground in a parabolic shape to correct for aberrations, but they are much harder to make and therefore very expensive. Spherical mirrors do not, however, exhibit chromatic aberration because the light does not pass through them (no refraction, no dispersion).

†Although the effect is the same as for astigmatism in the eye (Section 33–6), the cause is different. Off-axis astigmatism is no problem in the eye because objects are clearly seen only at the fovea, on the lens axis.

Summary

A lens uses refraction to produce a real or virtual image. Parallel rays of light are focused to a point, called the **focal point**, by a **converging** lens. The distance of the focal point from the lens is called the **focal length** f of the lens.

After parallel rays pass through a **diverging** lens, they appear to diverge from a point, its focal point; and the corresponding focal length is considered negative.

The **power** P of a lens, which is

$$P = \frac{1}{f} \tag{33–1}$$

is given in diopters, which are units of inverse meters (m^{-1}).

For a given object, the position and size of the image formed by a lens can be found approximately by ray tracing.

Algebraically, the relation between image and object distances, d_i and d_o, and the focal length f, is given by the **thin lens equation**:

$$\frac{1}{d_o} + \frac{1}{d_i} = \frac{1}{f}. \qquad (33-2)$$

The ratio of image height to object height, which equals the **lateral magnification** m, is

$$m = \frac{h_i}{h_o} = -\frac{d_i}{d_o}. \qquad (33-3)$$

When using the various equations of geometric optics, it is important to remember the **sign conventions** for all quantities involved: carefully review them (page 871) when doing Problems.

When two (or more) thin lenses are used in combination to produce an image, the thin lens equation can be used for each lens in sequence. The image produced by the first lens acts as the object for the second lens.

[*The **lensmaker's equation** relates the radii of curvature of the lens surfaces and the len's index of refraction to the focal length of the lens.]

A **camera** lens forms an image on film, or on an electronic sensor (CCD or CMOS) in a digital camera, by allowing light in through a shutter. The image is focused by moving the lens relative to the film, and the ***f*-stop** (or lens opening) must be adjusted for the brightness of the scene and the chosen shutter speed. The *f*-stop is defined as the ratio of the focal length to the diameter of the lens opening.

The human **eye** also adjusts for the available light—by opening and closing the iris. It focuses not by moving the lens, but by adjusting the shape of the lens to vary its focal length. The image is formed on the retina, which contains an array of receptors known as rods and cones.

Diverging eyeglass or contact lenses are used to correct the defect of a nearsighted eye, which cannot focus well on distant objects. Converging lenses are used to correct for defects in which the eye cannot focus on close objects.

A **simple magnifier** is a converging lens that forms a virtual image of an object placed at (or within) the focal point. The **angular magnification**, when viewed by a relaxed normal eye, is

$$M = \frac{N}{f}, \qquad (33-6a)$$

where f is the focal length of the lens and N is the near point of the eye (25 cm for a "normal" eye).

An **astronomical telescope** consists of an **objective** lens or mirror, and an **eyepiece** that magnifies the real image formed by the objective. The **magnification** is equal to the ratio of the objective and eyepiece focal lengths, and the image is inverted:

$$M = -\frac{f_o}{f_e}. \qquad (33-7)$$

[*A compound **microscope** also uses objective and eyepiece lenses, and the final image is inverted. The total magnification is the product of the magnifications of the two lenses and is approximately

$$M \approx \frac{N\ell}{f_e f_o}, \qquad (33-10b)$$

where ℓ is the distance between the lenses, N is the near point of the eye, and f_o and f_e are the focal lengths of objective and eyepiece, respectively.]

[*Microscopes, telescopes, and other optical instruments are limited in the formation of sharp images by **lens aberrations**. These include **spherical aberration**, in which rays passing through the edge of a lens are not focused at the same point as those that pass near the center; and **chromatic aberration**, in which different colors are focused at different points. Compound lenses, consisting of several elements, can largely correct for aberrations.]

Questions

1. Where must the film be placed if a camera lens is to make a sharp image of an object far away?

2. A photographer moves closer to his subject and then refocuses. Does the camera lens move farther away from or closer to the sensor? Explain.

3. Can a diverging lens form a real image under any circumstances? Explain.

4. Use ray diagrams to show that a real image formed by a thin lens is always inverted, whereas a virtual image is always upright if the object is real.

5. Light rays are said to be "reversible." Is this consistent with the thin lens equation? Explain.

6. Can real images be projected on a screen? Can virtual images? Can either be photographed? Discuss carefully.

7. A thin converging lens is moved closer to a nearby object. Does the real image formed change (*a*) in position, (*b*) in size? If yes, describe how.

8. Compare the mirror equation with the thin lens equation. Discuss similarities and differences, especially the sign conventions for the quantities involved.

9. A lens is made of a material with an index of refraction $n = 1.30$. In air, it is a converging lens. Will it still be a converging lens if placed in water? Explain, using a ray diagram.

10. Explain how you could have a virtual object.

11. A dog with its tail in the air stands facing a converging lens. If the nose and the tail are each focused on a screen in turn, which will have the greater magnification?

12. A cat with its tail in the air stands facing a converging lens. Under what circumstances (if any) would the image of the nose be virtual and the image of the tail be real? Where would the image of the rest of the cat be?

13. Why, in Example 33–6, must the converging lens have a shorter focal length than the diverging lens if the latter's focal length is to be determined by combining them?

14. The thicker a double convex lens is in the center as compared to its edges, the shorter its focal length for a given lens diameter. Explain.

15. Does the focal length of a lens depend on the fluid in which it is immersed? What about the focal length of a spherical mirror? Explain.

16. An underwater lens consists of a carefully shaped thin-walled plastic container filled with air. What shape should it have in order to be (a) converging (b) diverging? Use ray diagrams to support your answer.

17. Consider two converging lenses separated by some distance. An object is placed so that the image from the first lens lies exactly at the focal point of the second lens. Will this combination produce an image? If so, where? If not, why not?

18. Will a nearsighted person who wears corrective lenses in her glasses be able to see clearly underwater when wearing those glasses? Use a diagram to show why or why not.

19. You can tell whether people are nearsighted or farsighted by looking at the width of their face through their glasses. If a person's face appears narrower through the glasses, (Fig. 33–45), is the person farsighted or nearsighted?

FIGURE 33–45
Question 19.

20. The human eye is much like a camera—yet, when a camera shutter is left open and the camera is moved, the image will be blurred. But when you move your head with your eyes open, you still see clearly. Explain.

21. In attempting to discern distant details, people will sometimes squint. Why does this help?

22. Is the image formed on the retina of the human eye upright or inverted? Discuss the implications of this for our perception of objects.

23. Reading glasses use converging lenses. A simple magnifier is also a converging lens. Are reading glasses therefore magnifiers? Discuss the similarities and differences between converging lenses as used for these two different purposes.

24. Why must a camera lens be moved farther from the film to focus on a closer object?

*25. Spherical aberration in a thin lens is minimized if rays are bent equally by the two surfaces. If a planoconvex lens is used to form a real image of an object at infinity, which surface should face the object? Use ray diagrams to show why.

*26. For both converging and diverging lenses, discuss how the focal length for red light differs from that for violet light.

Problems

33–1 and 33–2 Thin Lenses

1. (I) A sharp image is located 373 mm behind a 215-mm-focal-length converging lens. Find the object distance (a) using a ray diagram, (b) by calculation.

2. (I) Sunlight is observed to focus at a point 18.5 cm behind a lens. (a) What kind of lens is it? (b) What is its power in diopters?

3. (I) (a) What is the power of a 23.5-cm-focal-length lens? (b) What is the focal length of a -6.75-D lens? Are these lenses converging or diverging?

4. (II) A certain lens focuses an object 1.85 m away as an image 48.3 cm on the other side of the lens. What type of lens is it and what is its focal length? Is the image real or virtual?

5. (II) A 105-mm-focal-length lens is used to focus an image on the sensor of a camera. The maximum distance allowed between the lens and the sensor plane is 132 mm. (a) How far ahead of the sensor should the lens be if the object to be photographed is 10.0 m away? (b) 3.0 m away? (c) 1.0 m away? (d) What is the closest object this lens could photograph sharply?

6. (II) A stamp collector uses a converging lens with focal length 28 cm to view a stamp 18 cm in front of the lens. (a) Where is the image located? (b) What is the magnification?

7. (II) It is desired to magnify reading material by a factor of $2.5\times$ when a book is placed 9.0 cm behind a lens. (a) Draw a ray diagram and describe the type of image this would be. (b) What type of lens is needed? (c) What is the power of the lens in diopters?

8. (II) A -8.00-D lens is held 12.5 cm from an ant 1.00 mm high. Describe the position, type, and height of the image.

9. (II) An object is located 1.50 m from an 8.0-D lens. By how much does the image move if the object is moved (a) 0.90 m closer to the lens, and (b) 0.90 m farther from the lens?

10. (II) (a) How far from a 50.0-mm-focal-length lens must an object be placed if its image is to be magnified $2.50\times$ and be real? (b) What if the image is to be virtual and magnified $2.50\times$?

11. (II) How far from a converging lens with a focal length of 25 cm should an object be placed to produce a real image which is the same size as the object?

12. (II) (a) A 2.80-cm-high insect is 1.30 m from a 135-mm-focal-length lens. Where is the image, how high is it, and what type is it? (b) What if $f = -135$ mm?

13. (II) A bright object and a viewing screen are separated by a distance of 86.0 cm. At what location(s) between the object and the screen should a lens of focal length 16.0 cm be placed in order to produce a sharp image on the screen? [*Hint*: first draw a diagram.]

14. (II) How far apart are an object and an image formed by an 85-cm-focal-length converging lens if the image is $2.95\times$ larger than the object and is real?

15. (II) Show analytically that the image formed by a converging lens (a) is real and inverted if the object is beyond the focal point ($d_o > f$), and (b) is virtual and upright if the object is within the focal point ($d_o < f$). Next, describe the image if the object is itself an image (formed by another lens), and its position is on the opposite side of the lens from the incoming light, (c) for $-d_o > f$, and (d) for $0 < -d_o < f$.

16. (II) A converging lens has focal length f. When an object is placed a distance $d_o > f$ from this lens, a real image with magnification m is formed. (a) Show that $m = f/(f - d_o)$. (b) Sketch m vs. d_o over the range $f < d_o < \infty$ where $f = 0.45$ cm. (c) For what value of d_o will the real image have the same (lateral) size as the object? (d) To obtain a real image that is much larger than the object, in what general region should the object be placed relative to the lens?

17. (II) In a slide or movie projector, the film acts as the object whose image is projected on a screen (Fig. 33–46). If a 105-mm-focal-length lens is to project an image on a screen 6.50 m away, how far from the lens should the slide be? If the slide is 36 mm wide, how wide will the picture be on the screen?

FIGURE 33–46
Slide projector, Problem 17.

Slide   Lens   Screen

18. (III) A bright object is placed on one side of a converging lens of focal length f, and a white screen for viewing the image is on the opposite side. The distance $d_T = d_i + d_o$ between the object and the screen is kept fixed, but the lens can be moved. (a) Show that if $d_T > 4f$, there will be two positions where the lens can be placed and a sharp image will be produced on the screen. (b) If $d_T < 4f$, show that there will be no lens position where a sharp image is formed. (c) Determine a formula for the distance between the two lens positions in part (a), and the ratio of the image sizes.

19. (III) (a) Show that the lens equation can be written in the *Newtonian form*:
$$xx' = f^2,$$
where x is the distance of the object from the focal point on the front side of the lens, and x' is the distance of the image to the focal point on the other side of the lens. Calculate the location of an image if the object is placed 48.0 cm in front of a convex lens with a focal length of 38.0 cm using (b) the standard form of the thin lens formula, and (c) the Newtonian form, derived above.

33–3 Lens Combinations

20. (II) A diverging lens with $f = -33.5$ cm is placed 14.0 cm behind a converging lens with $f = 20.0$ cm. Where will an object at infinity be focused?

21. (II) Two 25.0-cm-focal-length converging lenses are placed 16.5 cm apart. An object is placed 35.0 cm in front of one lens. Where will the final image formed by the second lens be located? What is the total magnification?

22. (II) A 34.0-cm-focal-length converging lens is 24.0 cm behind a diverging lens. Parallel light strikes the diverging lens. After passing through the converging lens, the light is again parallel. What is the focal length of the diverging lens? [*Hint*: first draw a ray diagram.]

23. (II) The two converging lenses of Example 33–5 are now placed only 20.0 cm apart. The object is still 60.0 cm in front of the first lens as in Fig. 33–14. In this case, determine (a) the position of the final image, and (b) the overall magnification. (c) Sketch the ray diagram for this system.

24. (II) A diverging lens with a focal length of -14 cm is placed 12 cm to the right of a converging lens with a focal length of 18 cm. An object is placed 33 cm to the left of the converging lens. (a) Where will the final image be located? (b) Where will the image be if the diverging lens is 38 cm from the converging lens?

25. (II) Two lenses, one converging with focal length 20.0 cm and one diverging with focal length -10.0 cm, are placed 25.0 cm apart. An object is placed 60.0 cm in front of the converging lens. Determine (a) the position and (b) the magnification of the final image formed. (c) Sketch a ray diagram for this system.

26. (II) A diverging lens is placed next to a converging lens of focal length f_C, as in Fig. 33–15. If f_T represents the focal length of the combination, show that the focal length of the diverging lens, f_D, is given by
$$\frac{1}{f_D} = \frac{1}{f_T} - \frac{1}{f_C}.$$

27. (II) A lighted candle is placed 36 cm in front of a converging lens of focal length $f_1 = 13$ cm, which in turn is 56 cm in front of another converging lens of focal length $f_2 = 16$ cm (see Fig. 33–47). (a) Draw a ray diagram and estimate the location and the relative size of the final image. (b) Calculate the position and relative size of the final image.

FIGURE 33–47
Problem 27.

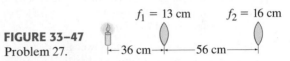

$f_1 = 13$ cm $f_2 = 16$ cm

⊢—36 cm—⊣—56 cm—⊣

*33–4 Lensmaker's Equation

*28. (I) A double concave lens has surface radii of 33.4 cm and 28.8 cm. What is the focal length if $n = 1.58$?

*29. (I) Both surfaces of a double convex lens have radii of 31.4 cm. If the focal length is 28.9 cm, what is the index of refraction of the lens material?

*30. (I) Show that if the lens of Example 33–7 is reversed, the focal length is unchanged.

*31. (I) A planoconvex lens (Fig. 33–2a) is to have a focal length of 18.7 cm. If made from fused quartz, what must be the radius of curvature of the convex surface?

*32. (II) An object is placed 90.0 cm from a glass lens ($n = 1.52$) with one concave surface of radius 22.0 cm and one convex surface of radius 18.5 cm. Where is the final image? What is the magnification?

*33. (II) A prescription for a corrective lens calls for $+3.50$ diopters. The lensmaker grinds the lens from a "blank" with $n = 1.56$ and convex front surface of radius of curvature of 30.0 cm. What should be the radius of curvature of the other surface?

33–5 Camera

34. (I) A properly exposed photograph is taken at $f/16$ and $\frac{1}{120}$ s. What lens opening is required if the shutter speed is $\frac{1}{1000}$ s?

35. (I) A television camera lens has a 17-cm focal length and a lens diameter of 6.0 cm. What is its f-number?

36. (II) A "pinhole" camera uses a tiny pinhole instead of a lens. Show, using ray diagrams, how reasonably sharp images can be formed using such a pinhole camera. In particular, consider two point objects 2.0 cm apart that are 1.0 m from a 1.0-mm-diameter pinhole. Show that on a piece of film 7.0 cm behind the pinhole the two objects produce two separate circles that do not overlap.

37. (II) Suppose that a correct exposure is $\frac{1}{250}$ s at $f/11$. Under the same conditions, what exposure time would be needed for a pinhole camera (Problem 36) if the pinhole diameter is 1.0 mm and the film is 7.0 cm from the hole?

38. (II) Human vision normally covers an angle of about 40° horizontally. A "normal" camera lens then is defined as follows: When focused on a distant horizontal object which subtends an angle of 40°, the lens produces an image that extends across the full horizontal extent of the camera's light-recording medium (film or electronic sensor). Determine the focal length f of the "normal" lens for the following types of cameras: (a) a 35-mm camera that records images on film 36 mm wide; (b) a digital camera that records images on a charge-coupled device (CCD) 1.00 cm wide.

39. (II) A nature photographer wishes to photograph a 38-m tall tree from a distance of 65 m. What focal-length lens should be used if the image is to fill the 24-mm height of the sensor?

33–6 Eye and Corrective Lenses

40. (I) A human eyeball is about 2.0 cm long and the pupil has a maximum diameter of about 8.0 mm. What is the "speed" of this lens?

41. (II) A person struggles to read by holding a book at arm's length, a distance of 55 cm away. What power of reading glasses should be prescribed for her, assuming they will be placed 2.0 cm from the eye and she wants to read at the "normal" near point of 25 cm?

42. (II) Reading glasses of what power are needed for a person whose near point is 105 cm, so that he can read a computer screen at 55 cm? Assume a lens–eye distance of 1.8 cm.

43. (II) If the nearsighted person in Example 33–13 wore contact lenses corrected for the far point ($= \infty$), what would be the near point? Would glasses be better in this case?

44. (II) An eye is corrected by a -4.50-D lens, 2.0 cm from the eye. (a) Is this eye near- or farsighted? (b) What is this eye's far point without glasses?

45. (II) A person's right eye can see objects clearly only if they are between 25 cm and 78 cm away. (a) What power of contact lens is required so that objects far away are sharp? (b) What will be the near point with the lens in place?

46. (II) A person has a far point of 14 cm. What power glasses would correct this vision if the glasses were placed 2.0 cm from the eye? What power contact lenses, placed on the eye, would the person need?

47. (II) One lens of a nearsighted person's eyeglasses has a focal length of -23.0 cm and the lens is 1.8 cm from the eye. If the person switches to contact lenses placed directly on the eye, what should be the focal length of the corresponding contact lens?

48. (II) What is the focal length of the eye lens system when viewing an object (a) at infinity, and (b) 38 cm from the eye? Assume that the lens–retina distance is 2.0 cm.

49. (II) A nearsighted person has near and far points of 10.6 and 20.0 cm respectively. If she puts on contact lenses with power $P = -4.00$ D, what are her new near and far points?

50. (II) The closely packed cones in the fovea of the eye have a diameter of about 2 μm. For the eye to discern two images on the fovea as distinct, assume that the images must be separated by at least one cone that is not excited. If these images are of two point-like objects at the eye's 25-cm near point, how far apart are these barely resolvable objects? Assume the diameter of the eye (cornea-to-fovea distance) is 2.0 cm.

33–7 Magnifying Glass

51. (I) What is the focal length of a magnifying glass of 3.8× magnification for a relaxed normal eye?

52. (I) What is the magnification of a lens used with a relaxed eye if its focal length is 13 cm?

53. (I) A magnifier is rated at 3.0× for a normal eye focusing on an image at the near point. (a) What is its focal length? (b) What is its focal length if the 3.0× refers to a relaxed eye?

54. (II) Sherlock Holmes is using an 8.80-cm-focal-length lens as his magnifying glass. To obtain maximum magnification, where must the object be placed (assume a normal eye), and what will be the magnification?

55. (II) A small insect is placed 5.85 cm from a $+6.00$-cm-focal-length lens. Calculate (a) the position of the image, and (b) the angular magnification.

56. (II) A 3.40-mm-wide bolt is viewed with a 9.60-cm-focal-length lens. A normal eye views the image at its near point. Calculate (a) the angular magnification, (b) the width of the image, and (c) the object distance from the lens.

57. (II) A magnifying glass with a focal length of 9.5 cm is used to read print placed at a distance of 8.3 cm. Calculate (a) the position of the image; (b) the angular magnification.

58. (II) A magnifying glass is rated at 3.0× for a normal eye that is relaxed. What would be the magnification for a relaxed eye whose near point is (a) 65 cm, and (b) 17 cm? Explain the differences.

59. (II) A converging lens of focal length $f = 12$ cm is being used by a writer as a magnifying glass to read some fine print on his book contract. Initially, the writer holds the lens above the fine print so that its image is at infinity. To get a better look, he then moves the lens so that the image is at his 25-cm near point. How far, and in what direction (toward or away from the fine print) did the writer move the lens? Assume the writer's eye is adjusted to remain always very near the magnifying glass.

33–8 Telescopes

60. (I) What is the magnification of an astronomical telescope whose objective lens has a focal length of 78 cm, and whose eyepiece has a focal length of 2.8 cm? What is the overall length of the telescope when adjusted for a relaxed eye?

61. (I) The overall magnification of an astronomical telescope is desired to be 35×. If an objective of 88 cm focal length is used, what must be the focal length of the eyepiece? What is the overall length of the telescope when adjusted for use by the relaxed eye?

62. (II) A 7.0× binocular has 3.0-cm-focal-length eyepieces. What is the focal length of the objective lenses?

63. (II) An astronomical telescope has an objective with focal length 75 cm and a $+35$ D eyepiece. What is the total magnification?

64. (II) An astronomical telescope has its two lenses spaced 78.0 cm apart. If the objective lens has a focal length of 75.5 cm, what is the magnification of this telescope? Assume a relaxed eye.

65. (II) A Galilean telescope adjusted for a relaxed eye is 33.8 cm long. If the objective lens has a focal length of 36.0 cm, what is the magnification?

66. (II) What is the magnifying power of an astronomical telescope using a reflecting mirror whose radius of curvature is 6.4 m and an eyepiece whose focal length is 2.8 cm?

67. (II) The Moon's image appears to be magnified 120× by a reflecting astronomical telescope with an eyepiece having a focal length of 3.1 cm. What are the focal length and radius of curvature of the main (objective) mirror?

68. (II) A 120× astronomical telescope is adjusted for a relaxed eye when the two lenses are 1.25 m apart. What is the focal length of each lens?

69. (II) An astronomical telescope longer than about 50 cm is not easy to hold by hand. Based on this fact, estimate the maximum angular magnification achievable for a telescope designed to be handheld. Assume its eyepiece lens, if used as a magnifying glass, provides a magnification of 5× for a relaxed eye with near point $N = 25$ cm.

70. (III) A reflecting telescope (Fig. 33–38b) has a radius of curvature of 3.00 m for its objective mirror and a radius of curvature of −1.50 m for its eyepiece mirror. If the distance between the two mirrors is 0.90 m, how far in front of the eyepiece should you place the electronic sensor to record the image of a star?

71. (III) A 7.5× pair of binoculars has an objective focal length of 26 cm. If the binoculars are focused on an object 4.0 m away (from the objective), what is the magnification? (The 7.5× refers to objects at infinity; Eq. 33–7 holds only for objects at infinity and not for nearby ones.)

*33–9 Microscopes

*72. (I) A microscope uses an eyepiece with a focal length of 1.50 cm. Using a normal eye with a final image at infinity, the barrel length is 17.5 cm and the focal length of the objective lens is 0.65 cm. What is the magnification of the microscope?

*73. (I) A 680× microscope uses a 0.40-cm-focal-length objective lens. If the barrel length is 17.5 cm, what is the focal length of the eyepiece? Assume a normal eye and that the final image is at infinity.

*74. (I) A 17-cm-long microscope has an eyepiece with a focal length of 2.5 cm and an objective with a focal length of 0.28 cm. What is the approximate magnification?

*75. (II) A microscope has a 13.0× eyepiece and a 58.0× objective lens 20.0 cm apart. Calculate (a) the total magnification, (b) the focal length of each lens, and (c) where the object must be for a normal relaxed eye to see it in focus.

*76. (II) Repeat Problem 75 assuming that the final image is located 25 cm from the eyepiece (near point of a normal eye).

*77. (II) A microscope has a 1.8-cm-focal-length eyepiece and a 0.80-cm objective. Assuming a relaxed normal eye, calculate (a) the position of the object if the distance between the lenses is 16.8 cm, and (b) the total magnification.

*78. (II) The eyepiece of a compound microscope has a focal length of 2.80 cm and the objective lens has $f = 0.740$ cm. If an object is placed 0.790 cm from the objective lens, calculate (a) the distance between the lenses when the microscope is adjusted for a relaxed eye, and (b) the total magnification.

*79. (II) An inexpensive instructional lab microscope allows the user to select its objective lens to have a focal length of 32 mm, 15 mm, or 3.9 mm. It also has two possible eyepieces with magnifications 5× and 10×. Each objective forms a real image 160 mm beyond its focal point. What are the largest and smallest overall magnifications obtainable with this instrument?

*80. (III) Given two 12-cm-focal-length lenses, you attempt to make a crude microscope using them. While holding these lenses a distance 55 cm apart, you position your microscope so that its objective lens is distance d_o from a small object. Assume your eye's near point $N = 25$ cm. (a) For your microscope to function properly, what should d_o be? (b) Assuming your eye is relaxed when using it, what magnification M does your microscope achieve? (c) Since the length of your microscope is not much greater than the focal lengths of its lenses, the approximation $M \approx N\ell/f_e f_o$ is not valid. If you apply this approximation to your microscope, what % error do you make in your microscope's true magnification?

*33–10 Lens Aberrations

*81. (II) A planoconvex lens (Fig. 33–2a) has one flat surface and the other has $R = 15.3$ cm. This lens is used to view a red and yellow object which is 66.0 cm away from the lens. The index of refraction of the glass is 1.5106 for red light and 1.5226 for yellow light. What are the locations of the red and yellow images formed by the lens?

*82. (II) An achromatic lens is made of two very thin lenses, placed in contact, that have focal lengths $f_1 = -28$ cm and $f_2 = +25$ cm. (a) Is the combination converging or diverging? (b) What is the net focal length?

General Problems

83. A 200-mm-focal-length lens can be adjusted so that it is 200.0 mm to 206.4 mm from the film. For what range of object distances can it be adjusted?

84. If a 135-mm telephoto lens is designed to cover object distances from 1.30 m to ∞, over what distance must the lens move relative to the plane of the sensor or film?

85. For a camera equipped with a 58-mm-focal-length lens, what is the object distance if the image height equals the object height? How far is the object from the image on the film?

86. Show that for objects very far away (assume infinity), the magnification of any camera lens is proportional to its focal length.

87. A small object is 25.0 cm from a diverging lens as shown in Fig. 33–48. A converging lens with a focal length of 12.0 cm is 30.0 cm to the right of the diverging lens. The two-lens system forms a real inverted image 17.0 cm to the right of the converging lens. What is the focal length of the diverging lens?

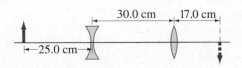

FIGURE 33–48 Problem 87.

88. A converging lens with focal length of 13.0 cm is placed in contact with a diverging lens with a focal length of −20.0 cm. What is the focal length of the combination, and is the combination converging or diverging?

89. An astronomical telescope has a magnification of 8.0×. If the two lenses are 28 cm apart, determine the focal length of each lens.

90. (a) Show that if two thin lenses of focal lengths f_1 and f_2 are placed in contact with each other, the focal length of the combination is given by $f_T = f_1 f_2/(f_1 + f_2)$. (b) Show that the power P of the combination of two lenses is the sum of their separate powers, $P = P_1 + P_2$.

91. How large is the image of the Sun on film used in a camera with (a) a 28-mm-focal-length lens, (b) a 50-mm-focal-length lens, and (c) a 135-mm-focal-length lens? (d) If the 50-mm lens is considered normal for this camera, what relative magnification does each of the other two lenses provide? The Sun has diameter 1.4×10^6 km, and it is 1.5×10^8 km away.

92. Two converging lenses are placed 30.0 cm apart. The focal length of the lens on the right is 20.0 cm, and the focal length of the lens on the left is 15.0 cm. An object is placed to the left of the 15.0-cm-focal-length lens. A final image from both lenses is inverted and located halfway between the two lenses. How far to the left of the 15.0-cm-focal-length lens is the original object?

93. When an object is placed 60.0 cm from a certain converging lens, it forms a real image. When the object is moved to 40.0 cm from the lens, the image moves 10.0 cm farther from the lens. Find the focal length of this lens.

94. Figure 33–49 was taken from the NIST Laboratory (National Institute of Standards and Technology) in Boulder, CO, 2 km from the hiker in the photo. The Sun's image was 15 mm across on the film. Estimate the focal length of the camera lens (actually a telescope). The Sun has diameter 1.4×10^6 km, and it is 1.5×10^8 km away.

FIGURE 33–49 Problem 94.

95. A movie star catches a reporter shooting pictures of her at home. She claims the reporter was trespassing. To prove her point, she gives as evidence the film she seized. Her 1.75-m height is 8.25 mm high on the film, and the focal length of the camera lens was 220 mm. How far away from the subject was the reporter standing?

96. As early morning passed toward midday, and the sunlight got more intense, a photographer noted that, if she kept her shutter speed constant, she had to change the f-number from $f/5.6$ to $f/16$. By what factor had the sunlight intensity increased during that time?

97. A child has a near point of 15 cm. What is the maximum magnification the child can obtain using an 8.5-cm-focal-length magnifier? What magnification can a normal eye obtain with the same lens? Which person sees more detail?

98. A woman can see clearly with her right eye only when objects are between 45 cm and 155 cm away. Prescription bifocals should have what powers so that she can see distant objects clearly (upper part) and be able to read a book 25 cm away (lower part) with her right eye? Assume that the glasses will be 2.0 cm from the eye.

99. What is the magnifying power of a +4.0-D lens used as a magnifier? Assume a relaxed normal eye.

100. A physicist lost in the mountains tries to make a telescope using the lenses from his reading glasses. They have powers of +2.0 D and +4.5 D, respectively. (a) What maximum magnification telescope is possible? (b) Which lens should be used as the eyepiece?

101. A 50-year-old man uses +2.5-D lenses to read a newspaper 25 cm away. Ten years later, he must hold the paper 32 cm away to see clearly with the same lenses. What power lenses does he need now in order to hold the paper 25 cm away? (Distances are measured from the lens.)

102. An object is moving toward a converging lens of focal length f with constant speed v_o such that its distance d_o from the lens is always greater than f. (a) Determine the velocity v_i of the image as a function of d_o. (b) Which direction (toward or away from the lens) does the image move? (c) For what d_o does the image's speed equal the object's speed?

103. The objective lens and the eyepiece of a telescope are spaced 85 cm apart. If the eyepiece is +23 D, what is the total magnification of the telescope?

* 104. Two converging lenses, one with $f = 4.0$ cm and the other with $f = 44$ cm, are made into a telescope. (a) What are the length and magnification? Which lens should be the eyepiece? (b) Assume these lenses are now combined to make a microscope; if the magnification needs to be 25×, how long would the microscope be?

105. Sam purchases +3.50-D eyeglasses which correct his faulty vision to put his near point at 25 cm. (Assume he wears the lenses 2.0 cm from his eyes.) (a) Calculate the focal length of Sam's glasses. (b) Calculate Sam's near point without glasses. (c) Pam, who has normal eyes with near point at 25 cm, puts on Sam's glasses. Calculate Pam's near point with Sam's glasses on.

106. The proper functioning of certain optical devices (e.g., optical fibers and spectrometers) requires that the input light be a collection of diverging rays within a cone of half-angle θ (Fig. 33–50). If the light initially exists as a collimated beam (i.e., parallel rays), show that a single lens of focal length f and diameter D can be used to create the required input light if $D/f = 2 \tan \theta$. If $\theta = 3.5°$ for a certain spectrometer, what focal length lens should be used if the lens diameter is 5.0 cm?

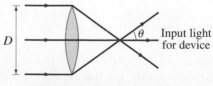

FIGURE 33–50 Problem 106.

107. In a science-fiction novel, an intelligent ocean-dwelling creature's eye functions underwater with a near point of 25 cm. This creature would like to create an underwater magnifier out of a thin plastic container filled with air. What shape should the air-filled plastic container have (i.e., determine radii of curvature of its surfaces) in order for it to be used by the creature as a $3.0\times$ magnifier? Assume the eye is focused at its near point.

108. A telephoto lens system obtains a large magnification in a compact package. A simple such system can be constructed out of two lenses, one converging and one diverging, of focal lengths f_1 and $f_2 = -\frac{1}{2}f_1$, respectively, separated by a distance $\ell = \frac{3}{4}f_1$ as shown in Fig. 33–51. (*a*) For a distant object located at distance d_o from the first lens, show that the first lens forms an image with magnification $m_1 \approx -f_1/d_o$ located very close to its focal point. Go on to show that the total magnification for the two-lens system is $m \approx -2f_1/d_o$. (*b*) For an object located at infinity, show that the two-lens system forms an image that is a distance $\frac{5}{4}f_1$ behind the first lens. (*c*) A single 250-mm-focal-length lens would have to be mounted about 250 mm from a camera's film in order to produce an image of a distant object at d_o with magnification $-(250 \text{ mm})/d_o$. To produce an image of this object with the same magnification using the two-lens system, what value of f_1 should be used and how far in front of the film should the first lens be placed? How much smaller is the "focusing length" (i.e., first lens-to-final image distance) of this two-lens system in comparison with the 250-mm "focusing length" of the equivalent single lens?

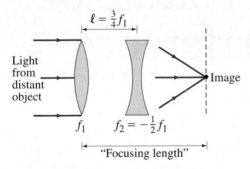

FIGURE 33–51 Problem 108.

*Numerical/Computer

*109. (III) In the "magnification" method, the focal length f of a converging lens is found by placing an object of known size at various locations in front of the lens and measuring the resulting real-image distances d_i and their associated magnifications m (minus sign indicates that image is inverted). The data taken in such an experiment are given here:

d_i (cm)	20	25	30	35	40
m	-0.43	-0.79	-1.14	-1.50	-1.89

(*a*) Show analytically that a graph of m vs. d_i should produce a straight line. What are the theoretically expected values for the slope and the *y*-intercept of this line? [*Hint:* d_o is not constant.] (*b*) Using the data above, graph m vs. d_i and show that a straight line does indeed result. Use the slope of this line to determine the focal length of the lens. Does the *y*-intercept of your plot have the expected value? (*c*) In performing such an experiment, one has the practical problem of locating the exact center of the lens since d_i must be measured from this point. Imagine, instead, that one measures the image distance d_i' from the back surface of the lens, which is a distance ℓ from the lens's center. Then, $d_i = d_i' + \ell$. Show that, when implementing the magnification method in this fashion, a plot of m vs. d_i' will still result in a straight line. How can f be determined from this straight line?

Answers to Exercises

A: Closer to it.

B: (*b*) and (*d*) are true.

C: $f = 10$ cm for both cases; $d_i = -10$ cm in (*a*) and 30 cm in (*b*).

D: -36 cm; diverging.

E: $P = -4.0$ D.

F: 48 cm.

The beautiful colors from the surface of this soap bubble can be nicely explained by the wave theory of light. A soap bubble is a very thin spherical film filled with air. Light reflected from the outer and inner surfaces of this thin film of soapy water interferes constructively to produce the bright colors. Which color we see at any point depends on the thickness of the soapy water film at that point and also on the viewing angle. Near the top of the bubble, we see a small black area surrounded by a silver or white area. The bubble's thickness is smallest at that black spot, perhaps only about 30 nm thick, and is fully transparent (we see the black background).

We cover fundamental aspects of the wave nature of light, including two-slit interference and interference in thin films.

CHAPTER 34

The Wave Nature of Light; Interference

CHAPTER-OPENING QUESTION—Guess now!

When a thin layer of oil lies on top of water or wet pavement, you can often see swirls of color. We also see swirls of color on the soap bubble shown above. What causes these colors?

(a) Additives in the oil or soap reflect various colors.

(b) Chemicals in the oil or soap absorb various colors.

(c) Dispersion due to differences in index of refraction in the oil or soap.

(d) The interactions of the light with a thin boundary layer where the oil (or soap) and the water have mixed irregularly.

(e) Light waves reflected from the top and bottom surfaces of the thin oil or soap film can add up constructively for particular wavelengths.

That light carries energy is obvious to anyone who has focused the Sun's rays with a magnifying glass on a piece of paper and burned a hole in it. But how does light travel, and in what form is this energy carried? In our discussion of waves in Chapter 15, we noted that energy can be carried from place to place in basically two ways: by particles or by waves. In the first case, material objects or particles can carry energy, such as an avalanche of rocks or rushing water. In the second case, water waves and sound waves, for example, can carry energy over long distances even though the oscillating particles of the medium do not travel these distances. In view of this, what can we say about the nature of light: does light travel as a stream of particles away from its source, or does light travel in the form of waves that spread outward from the source?

Historically, this question has turned out to be a difficult one. For one thing, light does not reveal itself in any obvious way as being made up of tiny particles; nor do we see tiny light waves passing by as we do water waves. The evidence seemed to favor first one side and then the other until about 1830, when most physicists had accepted the wave theory. By the end of the nineteenth century, light was considered to be an *electromagnetic wave* (Chapter 31). In the early twentieth century, light was shown to have a particle nature as well, as we shall discuss in Chapter 37. We now speak of the wave–particle duality of light. The wave theory of light remains valid and has proved very successful. We now investigate the evidence for the wave theory and how it has been used to explain a wide range of phenomena.

34–1 Waves versus Particles; Huygens' Principle and Diffraction

The Dutch scientist Christian Huygens (1629–1695), a contemporary of Newton, proposed a wave theory of light that had much merit. Still useful today is a technique Huygens developed for predicting the future position of a wave front when an earlier position is known. By a wave front, we mean all the points along a two- or three-dimensional wave that form a wave crest—what we simply call a "wave" as seen on the ocean. Wave fronts are perpendicular to rays as we already discussed in Chapter 15 (Fig. 15–20). **Huygens' principle** can be stated as follows: *Every point on a wave front can be considered as a source of tiny wavelets that spread out in the forward direction at the speed of the wave itself. The new wave front is the envelope of all the wavelets—that is, the tangent to all of them.*

As a simple example of the use of Huygens' principle, consider the wave front AB in Fig. 34–1, which is traveling away from a source S. We assume the medium is *isotropic*—that is, the speed v of the waves is the same in all directions. To find the wave front a short time t after it is at AB, tiny circles are drawn with radius $r = vt$. The centers of these tiny circles are blue dots on the original wave front AB, and the circles represent Huygens' (imaginary) wavelets. The tangent to all these wavelets, the curved line CD, is the new position of the wave front.

Huygens' principle is particularly useful for analyzing what happens when waves impinge on an obstacle and the wave fronts are partially interrupted. Huygens' principle predicts that waves bend in behind an obstacle, as shown in Fig. 34–2. This is just what water waves do, as we saw in Chapter 15 (Figs. 15–31 and 15–32). The bending of waves behind obstacles into the "shadow region" is known as **diffraction**. Since diffraction occurs for waves, but not for particles, it can serve as one means for distinguishing the nature of light.

Note, as shown in Fig. 34–2, that diffraction is most prominent when the size of the opening is on the order of the wavelength of the wave. If the opening is much larger than the wavelength, diffraction goes unnoticed.

Does light exhibit diffraction? In the mid-seventeenth century, the Jesuit priest Francesco Grimaldi (1618–1663) had observed that when sunlight entered a darkened room through a tiny hole in a screen, the spot on the opposite wall was larger than would be expected from geometric rays. He also observed that the border of the image was not clear but was surrounded by colored fringes. Grimaldi attributed this to the diffraction of light.

The wave model of light nicely accounts for diffraction, and we discuss diffraction in detail in the next Chapter. But the ray model (Chapter 32) cannot account for diffraction, and it is important to be aware of such limitations to the ray model. Geometric optics using rays is successful in a wide range of situations only because normal openings and obstacles are much larger than the wavelength of the light, and so relatively little diffraction or bending occurs.

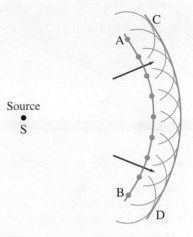

FIGURE 34–1 Huygens' principle, used to determine wave front CD when wave front AB is given.

FIGURE 34–2 Huygens' principle is consistent with diffraction (a) around the edge of an obstacle, (b) through a large hole, (c) through a small hole whose size is on the order of the wavelength of the wave.

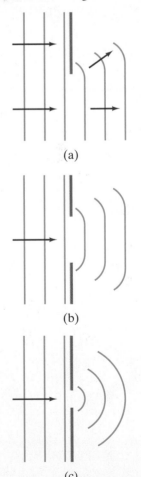

(a)

(b)

(c)

34–2 Huygens' Principle and the Law of Refraction

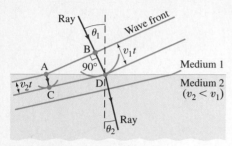

FIGURE 34–3 Refraction explained, using Huygens' principle. Wave fronts are perpendicular to the rays.

The laws of reflection and refraction were well known in Newton's time. The law of reflection could not distinguish between the two theories we just discussed: waves versus particles. For when waves reflect from an obstacle, the angle of incidence equals the angle of reflection (Fig. 15–21). The same is true of particles—think of a tennis ball without spin striking a flat surface.

The law of refraction is another matter. Consider a ray of light entering a medium where it is bent toward the normal, as when traveling from air into water. As shown in Fig. 34–3, this bending can be constructed using Huygens' principle if we assume the speed of light is less in the second medium $(v_2 < v_1)$. In time t, point B on wave front AB (perpendicular to the incoming ray) travels a distance $v_1 t$ to reach point D. Point A on the wave front, traveling in the second medium, goes a distance $v_2 t$ to reach point C, and $v_2 t < v_1 t$. Huygens' principle is applied to points A and B to obtain the curved wavelets shown at C and D. The wave front is tangent to these two wavelets, so the new wave front is the line CD. Hence the rays, which are perpendicular to the wave fronts, bend toward the normal if $v_2 < v_1$, as drawn.

Newton favored a particle theory of light which predicted the opposite result, that the speed of light would be greater in the second medium $(v_2 > v_1)$. Thus the wave theory predicts that the speed of light in water, for example, is less than in air; and Newton's particle theory predicts the reverse. An experiment to actually measure the speed of light in water was performed in 1850 by the French physicist Jean Foucault, and it confirmed the wave-theory prediction. By then, however, the wave theory was already fully accepted, as we shall see in the next Section.

Snell's law of refraction follows directly from Huygens' principle, given that the speed of light v in any medium is related to the speed in a vacuum, c, and the index of refraction, n, by Eq. 32–4: that is, $v = c/n$. From the Huygens' construction of Fig. 34–3, angle ADC is equal to θ_2 and angle BAD is equal to θ_1. Then for the two triangles that have the common side AD, we have

$$\sin \theta_1 = \frac{v_1 t}{AD}, \qquad \sin \theta_2 = \frac{v_2 t}{AD}.$$

We divide these two equations and obtain

$$\frac{\sin \theta_1}{\sin \theta_2} = \frac{v_1}{v_2}.$$

Then, by Eq. 32–4 $v_1 = c/n_1$ and $v_2 = c/n_2$, so we have

$$n_1 \sin \theta_1 = n_2 \sin \theta_2,$$

which is Snell's law of refraction, Eq. 32–5. (The law of reflection can be derived from Huygens' principle in a similar way: see Problem 1 at the end of this Chapter.)

When a light wave travels from one medium to another, its frequency does not change, but its wavelength does. This can be seen from Fig. 34–3, where each of the blue lines representing a wave front corresponds to a crest (peak) of the wave. Then

$$\frac{\lambda_2}{\lambda_1} = \frac{v_2 t}{v_1 t} = \frac{v_2}{v_1} = \frac{n_1}{n_2},$$

where, in the last step, we used Eq. 32–4, $v = c/n$. If medium 1 is a vacuum (or air), so $n_1 = 1, v_1 = c$, and we call λ_1 simply λ, then the wavelength in another medium of index of refraction $n \, (= n_2)$ will be

$$\lambda_n = \frac{\lambda}{n}. \tag{34–1}$$

This result is consistent with the frequency f being unchanged no matter what medium the wave is traveling in, since $c = f\lambda$.

EXERCISE A A light beam in air with wavelength = 500 nm, frequency = 6.0×10^{14} Hz, and speed = 3.0×10^8 m/s goes into glass which has an index of refraction = 1.5. What are the wavelength, frequency, and speed of the light in the glass?

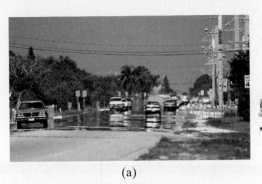

(a)

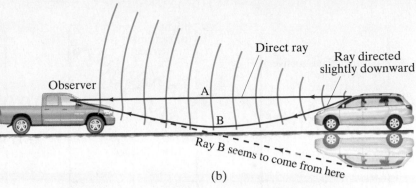

Direct ray

Ray directed
slightly downward

Observer

A

B

Ray B seems to come from here

(b)

Wave fronts can be used to explain how mirages are produced by refraction of light. For example, on a hot day motorists sometimes see a mirage of water on the highway ahead of them, with distant vehicles seemingly reflected in it (Fig. 34–4a). On a hot day, there can be a layer of very hot air next to the roadway (made hot by the Sun beating on the road). Hot air is less dense than cooler air, so the index of refraction is slightly lower in the hot air. In Fig. 34–4b, we see a diagram of light coming from one point on a distant car (on the right) heading left toward the observer. Wave fronts and two rays (perpendicular to the wave fronts) are shown. Ray A heads directly at the observer and follows a straight-line path, and represents the normal view of the distant car. Ray B is a ray initially directed slightly downward, but it bends slightly as it moves through layers of air of different index of refraction. The wave fronts, shown in blue in Fig. 34–4b, move slightly faster in the layers of air nearer the ground. Thus ray B is bent as shown, and seems to the observer to be coming from below (dashed line) as if reflected off the road. Hence the mirage.

FIGURE 34–4 (a) A highway mirage. (b) Drawing (greatly exaggerated) showing wave fronts and rays to explain highway mirages. Note how sections of the wave fronts near the ground move faster and so are farther apart.

PHYSICS APPLIED
Highway mirages

34–3 Interference—Young's Double-Slit Experiment

In 1801, the Englishman Thomas Young (1773–1829) obtained convincing evidence for the wave nature of light and was even able to measure wavelengths for visible light. Figure 34–5a shows a schematic diagram of Young's famous double-slit experiment.

FIGURE 34–5 (a) Young's double-slit experiment. (b) If light consists of particles, we would expect to see two bright lines on the screen behind the slits. (c) In fact, many lines are observed. The slits and their separation need to be very thin.

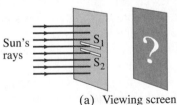

Sun's rays

S_1
S_2

(a) Viewing screen

(b) Viewing screen
(particle theory
prediction)

(c) Viewing screen
(actual)

To have light from a single source, Young used the Sun passing through a very narrow slit in a window covering. This beam of parallel rays falls on a screen containing two closely spaced slits, S_1 and S_2. (The slits and their separation are very narrow, not much larger than the wavelength of the light.) If light consists of tiny particles, we might expect to see two bright lines on a screen placed behind the slits as in (b). But instead a series of bright lines are seen, as in (c). Young was able to explain this result as a **wave-interference** phenomenon.

To understand why, we consider the simple situation of **plane waves**[†] of light of a single wavelength—called **monochromatic**, meaning "one color"—falling on the two slits as shown in Fig. 34–6. Because of diffraction, the waves leaving the two small slits spread out as shown. This is equivalent to the interference pattern produced when two rocks are thrown into a lake (Fig. 15–23), or when sound from two loudspeakers interferes (Fig. 16–15). Recall Section 15–8 on wave interference.

FIGURE 34–6 If light is a wave, light passing through one of two slits should interfere with light passing through the other slit.

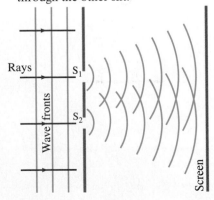

Rays

Wave fronts

S_1

S_2

Screen

[†]See pages 410 and 818.

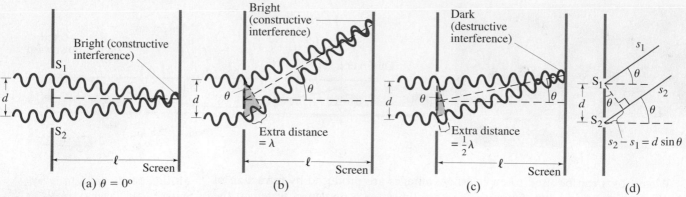

FIGURE 34–7 How the wave theory explains the pattern of lines seen in the double-slit experiment. (a) At the center of the screen the waves from each slit travel the same distance and are in phase. (b) At this angle θ, the lower wave travels an extra distance of one whole wavelength, and the waves are in phase; note from the shaded triangle that the path difference equals $d\sin\theta$. (c) For this angle θ, the lower wave travels an extra distance equal to one-half wavelength, so the two waves arrive at the screen fully out of phase. (d) A more detailed diagram showing the geometry for parts (b) and (c).

FIGURE 34–8 Two traveling waves are shown undergoing (a) constructive interference, (b) destructive interference. (See also Section 15–8.)

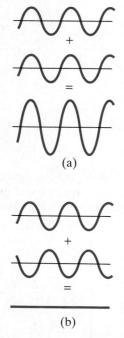

To see how an interference pattern is produced on the screen, we make use of Fig. 34–7. Waves of wavelength λ are shown entering the slits S_1 and S_2, which are a distance d apart. The waves spread out in all directions after passing through the slits (Fig. 34–6), but they are shown only for three different angles θ. In Fig. 34–7a, the waves reaching the center of the screen are shown ($\theta = 0°$). The waves from the two slits travel the same distance, so they are in phase: a crest of one wave arrives at the same time as a crest of the other wave. Hence the amplitudes of the two waves add to form a larger amplitude as shown in Fig. 34–8a. This is **constructive interference**, and there is a bright area at the center of the screen. Constructive interference also occurs when the paths of the two rays differ by one wavelength (or any whole number of wavelengths), as shown in Fig. 34–7b; also here there will be brightness on the screen. But if one ray travels an extra distance of one-half wavelength (or $\frac{3}{2}\lambda, \frac{5}{2}\lambda$, and so on), the two waves are exactly out of phase when they reach the screen: the crests of one wave arrive at the same time as the troughs of the other wave, and so they add to produce zero amplitude (Fig. 34–8b). This is **destructive interference**, and the screen is dark, Fig. 34–7c. Thus, there will be a series of bright and dark lines (or **fringes**) on the viewing screen.

To determine exactly where the bright lines fall, first note that Fig. 34–7 is somewhat exaggerated; in real situations, the distance d between the slits is very small compared to the distance ℓ to the screen. The rays from each slit for each case will therefore be essentially parallel, and θ is the angle they make with the horizontal as shown in Fig. 34–7d. From the shaded right triangles shown in Figs. 34–7b and c, we can see that the extra distance traveled by the lower ray is $d\sin\theta$ (seen more clearly in Fig. 34–7d). Constructive interference will occur, and a bright fringe will appear on the screen, when the *path difference*, $d\sin\theta$, equals a whole number of wavelengths:

$$d\sin\theta = m\lambda, \qquad m = 0, 1, 2, \cdots. \qquad \begin{bmatrix} \text{constructive} \\ \text{interference} \\ \text{(bright)} \end{bmatrix} \quad \textbf{(34–2a)}$$

The value of m is called the **order** of the interference fringe. The first order ($m = 1$), for example, is the first fringe on each side of the central fringe (which is at $\theta = 0, m = 0$). Destructive interference occurs when the path difference $d\sin\theta$ is $\frac{1}{2}\lambda, \frac{3}{2}\lambda$, and so on:

$$d\sin\theta = \left(m + \tfrac{1}{2}\right)\lambda, \qquad m = 0, 1, 2, \cdots. \qquad \begin{bmatrix} \text{destructive} \\ \text{interference} \\ \text{(dark)} \end{bmatrix} \quad \textbf{(34–2b)}$$

The bright fringes are peaks or maxima of light intensity, the dark fringes are minima.

The intensity of the bright fringes is greatest for the central fringe ($m = 0$) and decreases for higher orders, as shown in Fig. 34–9. How much the intensity decreases with increasing order depends on the width of the two slits.

CONCEPTUAL EXAMPLE 34–1 **Interference pattern lines.** (*a*) Will there be an infinite number of points on the viewing screen where constructive and destructive interference occur, or only a finite number of points? (*b*) Are neighboring points of constructive interference uniformly spaced, or is the spacing between neighboring points of constructive interference not uniform?

RESPONSE (*a*) When you look at Eqs. 34–2a and b you might be tempted to say, given the statement $m = 0, 1, 2, \cdots$ beside the equations, that there are an infinite number of points of constructive and destructive interference. However, recall that $\sin \theta$ cannot exceed 1. Thus, there is an upper limit to the values of m that can be used in these equations. For Eq. 34–2a, the maximum value of m is the integer closest in value but smaller than d/λ. So there are a *finite* number of points of constructive and destructive interference no matter how large the screen. (*b*) The spacing between neighboring points of constructive or destructive interference is not uniform: The spacing gets larger as θ gets larger, and you can verify this statement mathematically. For small values of θ the spacing is nearly uniform as you will see in Example 34–2.

EXAMPLE 34–2 **Line spacing for double-slit interference.** A screen containing two slits 0.100 mm apart is 1.20 m from the viewing screen. Light of wavelength $\lambda = 500$ nm falls on the slits from a distant source. Approximately how far apart will adjacent bright interference fringes be on the screen?

APPROACH The angular position of bright (constructive interference) fringes is found using Eq. 34–2a. The distance between the first two fringes (say) can be found using right triangles as shown in Fig. 34–10.

SOLUTION Given $d = 0.100$ mm $= 1.00 \times 10^{-4}$ m, $\lambda = 500 \times 10^{-9}$ m, and $\ell = 1.20$ m, the first-order fringe ($m = 1$) occurs at an angle θ given by

$$\sin \theta_1 = \frac{m\lambda}{d} = \frac{(1)(500 \times 10^{-9} \, \text{m})}{1.00 \times 10^{-4} \, \text{m}} = 5.00 \times 10^{-3}.$$

This is a very small angle, so we can take $\sin \theta \approx \theta$, with θ in radians. The first-order fringe will occur a distance x_1 above the center of the screen (see Fig. 34–10), given by $x_1/\ell = \tan \theta_1 \approx \theta_1$, so

$$x_1 \approx \ell\theta_1 = (1.20 \, \text{m})(5.00 \times 10^{-3}) = 6.00 \, \text{mm}.$$

The second-order fringe ($m = 2$) will occur at

$$x_2 \approx \ell\theta_2 = \ell\frac{2\lambda}{d} = 12.0 \, \text{mm}$$

above the center, and so on. Thus the lower order fringes are 6.00 mm apart.

NOTE The spacing between fringes is essentially uniform until the approximation $\sin \theta \approx \theta$ is no longer valid.

CONCEPTUAL EXAMPLE 34–3 **Changing the wavelength.** (*a*) What happens to the interference pattern shown in Fig. 34–10, Example 34–2, if the incident light (500 nm) is replaced by light of wavelength 700 nm? (*b*) What happens instead if the wavelength stays at 500 nm but the slits are moved farther apart?

RESPONSE (*a*) When λ increases in Eq. 34–2a but d stays the same, then the angle θ for bright fringes increases and the interference pattern spreads out. (*b*) Increasing the slit spacing d reduces θ for each order, so the lines are closer together.

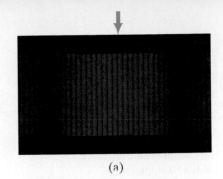

(a)

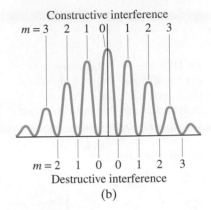

(b)

FIGURE 34–9 (a) Interference fringes produced by a double-slit experiment and detected by photographic film placed on the viewing screen. The arrow marks the central fringe. (b) Graph of the intensity of light in the interference pattern. Also shown are values of m for Eq. 34–2a (constructive interference) and Eq. 34–2b (destructive interference).

FIGURE 34–10 Examples 34–2 and 34–3. For small angles θ (give θ in radians), the interference fringes occur at distance $x = \theta\ell$ above the center fringe ($m = 0$); θ_1 and x_1 are for the first-order fringe ($m = 1$); θ_2 and x_2 are for $m = 2$.

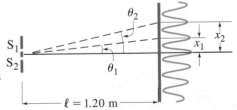

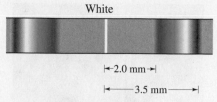

White

|←2.0 mm→|

|←——3.5 mm——→|

FIGURE 34–11 First-order fringes are a full spectrum, like a rainbow. Also Example 34–4.

FIGURE 34–10 (Repeated.) For small angles θ (give θ in radians), the interference fringes occur at distance $x = \theta\ell$ above the center fringe ($m = 0$); θ_1 and x_1 are for the first-order fringe ($m = 1$), θ_2 and x_2 are for $m = 2$.

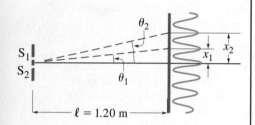

$\ell = 1.20$ m

From Eqs. 34–2 we can see that, except for the zeroth-order fringe at the center, the position of the fringes depends on wavelength. Consequently, when white light falls on the two slits, as Young found in his experiments, the central fringe is white, but the first- (and higher-) order fringes contain a spectrum of colors like a rainbow; θ was found to be smallest for violet light and largest for red (Fig. 34–11). By measuring the position of these fringes, Young was the first to determine the wavelengths of visible light (using Eqs. 34–2). In doing so, he showed that what distinguishes different colors physically is their wavelength (or frequency), an idea put forward earlier by Grimaldi in 1665.

EXAMPLE 34–4 **Wavelengths from double-slit interference.** White light passes through two slits 0.50 mm apart, and an interference pattern is observed on a screen 2.5 m away. The first-order fringe resembles a rainbow with violet and red light at opposite ends. The violet light is about 2.0 mm and the red 3.5 mm from the center of the central white fringe (Fig. 34–11). Estimate the wavelengths for the violet and red light.

APPROACH We find the angles for violet and red light from the distances given and the diagram of Fig. 34–10. Then we use Eq. 34–2a to obtain the wavelengths. Because 3.5 mm is much less than 2.5 m, we can use the small-angle approximation.

SOLUTION We use Eq. 34–2a with $m = 1$ and $\sin\theta \approx \tan\theta \approx \theta$. Then for violet light, $x = 2.0$ mm, so (see also Fig. 34–10)

$$\lambda = \frac{d\sin\theta}{m} \approx \frac{d\theta}{m} \approx \frac{d}{m}\frac{x}{\ell} = \left(\frac{5.0 \times 10^{-4}\,\text{m}}{1}\right)\left(\frac{2.0 \times 10^{-3}\,\text{m}}{2.5\,\text{m}}\right) = 4.0 \times 10^{-7}\,\text{m},$$

or 400 nm. For red light, $x = 3.5$ mm, so

$$\lambda = \frac{d}{m}\frac{x}{\ell} = \left(\frac{5.0 \times 10^{-4}\,\text{m}}{1}\right)\left(\frac{3.5 \times 10^{-3}\,\text{m}}{2.5\,\text{m}}\right) = 7.0 \times 10^{-7}\,\text{m} = 700\,\text{nm}.$$

Coherence

The two slits in Fig. 34–7 act as if they were two sources of radiation. They are called **coherent sources** because the waves leaving them have the same wavelength and frequency, and bear the same phase relationship to each other at all times. This happens because the waves come from a single source to the left of the two slits in Fig. 34–7, splitting the original beam into two. An interference pattern is observed only when the sources are coherent. If two tiny lightbulbs replaced the two slits, an interference pattern would not be seen. The light emitted by one lightbulb would have a random phase with respect to the second bulb, and the screen would be more or less uniformly illuminated. Two such sources, whose output waves have phases that bear no fixed relationship to each other over time, are called **incoherent sources**.

*34–4 Intensity in the Double-Slit Interference Pattern

We saw in Section 34–3 that the interference pattern produced by the coherent light from two slits, S_1 and S_2 (Figs. 34–7 and 34–9), produces a series of bright and dark fringes. If the two monochromatic waves of wavelength λ are in phase at the slits, the maxima (brightest points) occur at angles θ given by (Eqs. 34–2)

$$d\sin\theta = m\lambda,$$

and the minima (darkest points) when

$$d\sin\theta = \left(m + \tfrac{1}{2}\right)\lambda,$$

where m is an integer ($m = 0, 1, 2, \cdots$).

We now determine the intensity of the light at all points in the pattern, assuming that if either slit were covered, the light passing through the other would diffract sufficiently to illuminate a large portion of the screen uniformly. The intensity I of the light at any point is proportional to the square of its wave amplitude (Section 15–3). Treating light as an electromagnetic wave, I is proportional to the square of the electric field E (or to the magnetic field B, Section 31–8): $I \propto E^2$. The electric field $\vec{E}$ at any point P (see Fig. 34–12) will be the sum of the electric field vectors of the waves coming from each of the two slits, $\vec{E}_1$ and $\vec{E}_2$. Since $\vec{E}_1$ and $\vec{E}_2$ are essentially parallel (on a screen far away compared to the slit separation), the magnitude of the electric field at angle θ (that is, at point P) will be

$$E_\theta = E_1 + E_2.$$

Both E_1 and E_2 vary sinusoidally with frequency $f = c/\lambda$, but they differ in phase, depending on their different travel distances from the slits. The electric field at P can then be written for the light from each of the two slits, using $\omega = 2\pi f$, as

$$E_1 = E_{10} \sin \omega t$$
$$E_2 = E_{20} \sin(\omega t + \delta) \qquad \text{(34–3)}$$

where E_{10} and E_{20} are their respective amplitudes and δ is the phase difference. The value of δ depends on the angle θ, so let us now determine δ as a function of θ.

At the center of the screen (point 0), $\delta = 0$. If the difference in path length from P to S_1 and S_2 is $d \sin \theta = \lambda/2$, the two waves are exactly out of phase so $\delta = \pi$ (or 180°). If $d \sin \theta = \lambda$, the two waves differ in phase by $\delta = 2\pi$. In general, then, δ is related to θ by

$$\frac{\delta}{2\pi} = \frac{d \sin \theta}{\lambda}$$

or

$$\delta = \frac{2\pi}{\lambda} d \sin \theta. \qquad \text{(34–4)}$$

To determine $E_\theta = E_1 + E_2$, we add the two scalars E_1 and E_2 which are sine functions differing by the phase δ. One way to determine the sum of E_1 and E_2 is to use a **phasor diagram**. (We used this technique before, in Chapter 30.) As shown in Fig. 34–13, we draw an arrow of length E_{10} to represent the amplitude of E_1 (Eq. 34–3); and the arrow of length E_{20}, which we draw to make a fixed angle δ with E_{10}, represents the amplitude of E_2. When the diagram rotates at angular frequency ω about the origin, the projections of E_{10} and E_{20} on the vertical axis represent E_1 and E_2 as a function of time (see Eq. 34–3). We let $E_{\theta 0}$ be the "vector" sum[†] of E_{10} and E_{20}; $E_{\theta 0}$ is the amplitude of the sum $E_\theta = E_1 + E_2$, and the projection of $E_{\theta 0}$ on the vertical axis is just E_θ. If the two slits provide equal illumination, so that $E_{10} = E_{20} = E_0$, then from *symmetry* in Fig. 34–13, the angle $\phi = \delta/2$, and we can write

$$E_\theta = E_{\theta 0} \sin\left(\omega t + \frac{\delta}{2}\right). \qquad \text{(34–5a)}$$

From Fig. 34–13 we can also see that

$$E_{\theta 0} = 2E_0 \cos \phi = 2E_0 \cos \frac{\delta}{2}. \qquad \text{(34–5b)}$$

Combining Eqs. 34–5a and b, we obtain

$$E_\theta = 2E_0 \cos \frac{\delta}{2} \sin\left(\omega t + \frac{\delta}{2}\right), \qquad \text{(34–5c)}$$

where δ is given by Eq. 34–4.

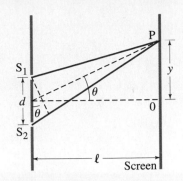

FIGURE 34–12 Determining the intensity in a double-slit interference pattern. Not to scale: in fact $\ell \gg d$, and the two rays become essentially parallel.

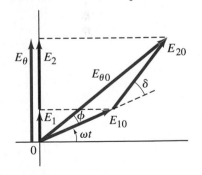

FIGURE 34–13 Phasor diagram for double-slit interference pattern.

[†]We are not adding the actual electric field vectors; instead we are using the "phasor" technique to add the amplitudes, taking into account the phase difference of the two waves.

We are not really interested in E_θ as a function of time, since for visible light the frequency (10^{14} to 10^{15} Hz) is much too high to be noticeable. We are interested in the average intensity, which is proportional to the amplitude squared, $E_{\theta 0}^2$. We now drop the word "average," and we let I_θ ($I_\theta \propto E_\theta^2$) be the intensity at any point P at an angle θ to the horizontal. We let I_0 be the intensity at point O, the center of the screen, where $\theta = \delta = 0$, so $I_0 \propto (E_{10} + E_{20})^2 = (2E_0)^2$. Then the ratio I_θ / I_0 is equal to the ratio of the squares of the electric-field amplitudes at these two points, so

$$\frac{I_\theta}{I_0} = \frac{E_{\theta 0}^2}{(2E_0)^2} = \cos^2 \frac{\delta}{2}$$

where we used Eq. 34–5b. Thus the intensity I_θ at any point is related to the maximum intensity at the center of the screen by

$$I_\theta = I_0 \cos^2 \frac{\delta}{2}$$

$$= I_0 \cos^2 \left(\frac{\pi d \sin \theta}{\lambda} \right) \tag{34–6}$$

where δ was given by Eq. 34–4. This is the relation we sought.

From Eq. 34–6 we see that maxima occur where $\cos \delta/2 = \pm 1$, which corresponds to $\delta = 0, 2\pi, 4\pi, \cdots$. From Eq. 34–4, δ has these values when

$$d \sin \theta = m\lambda, \qquad m = 0, 1, 2, \cdots.$$

Minima occur where $\delta = \pi, 3\pi, 5\pi, \cdots$, which corresponds to

$$d \sin \theta = \left(m + \tfrac{1}{2} \right)\lambda, \qquad m = 0, 1, 2, \cdots.$$

These are the same results we obtained in Section 34–3. But now we know not only the position of maxima and minima, but from Eq. 34–6 we can determine the intensity at all points.

In the usual situation where the distance ℓ to the screen from the slits is large compared to the slit separation d ($\ell \gg d$), if we consider only points P whose distance y from the center (point O) is small compared to ℓ ($y \ll \ell$)—see Fig. 34–12—then

$$\sin \theta = \frac{y}{\ell}.$$

From this it follows (see Eq. 34–4) that

$$\delta = \frac{2\pi}{\lambda} \frac{d}{\ell} y.$$

Equation 34–6 then becomes

$$I_\theta = I_0 \left[\cos \left(\frac{\pi d}{\lambda \ell} y \right) \right]^2. \qquad [y \ll \ell, \ d \ll \ell] \quad \textbf{(34–7)}$$

EXERCISE B What are the values for the intensity I_θ when (a) $y = 0$, (b) $y = \lambda\ell/4d$, and (c) $y = \lambda\ell/2d$?

The intensity I_θ as a function of the phase difference δ is plotted in Fig. 34–14. In the approximation of Eq. 34–7, the horizontal axis could as well be y, the position on the screen.

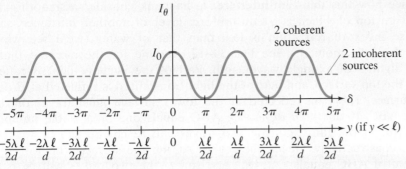

The intensity pattern expressed in Eqs. 34–6 and 34–7, and plotted in Fig. 34–14, shows a series of maxima of equal height, and is based on the assumption that each slit (alone) would illuminate the screen uniformly. This is never quite true, as we shall see when we discuss diffraction in the next Chapter. We will see that the center maximum is strongest and each succeeding maximum to each side is a little less strong.

EXAMPLE 34–5 **Antenna intensity.** Two radio antennas are located close to each other as shown in Fig. 34–15, separated by a distance d. The antennas radiate in phase with each other, emitting waves of intensity I_0 at wavelength λ. (a) Calculate the net intensity as a function of θ for points very far from the antennas. (b) For $d = \lambda$, determine I and find in which directions I is a maximum and a minimum. (c) Repeat part (b) when $d = \lambda/2$.

APPROACH This setup is similar to Young's double-slit experiment.

SOLUTION (a) Points of constructive and destructive interference are still given by Eqs. 34–2a and b, and the net intensity as a function of θ is given by Eq. 34–6. (b) We let $d = \lambda$ in Eq. 34–6, and find for the intensity,

$$I = I_0 \cos^2(\pi \sin \theta).$$

I is a maximum, equal to I_0, when $\sin \theta = 0$, 1, or -1, meaning $\theta = 0, 90°, 180°,$ and $270°$. I is zero when $\sin \theta = \frac{1}{2}$ and $-\frac{1}{2}$, for which $\theta = 30°, 150°, 210°,$ and $330°$. (c) For $d = \lambda/2$, I is maximized for $\theta = 0$ and $180°$, and minimized for $90°$ and $270°$.

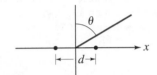

FIGURE 34–15 Example 34–5. The two dots represent the antennas.

34–5 Interference in Thin Films

Interference of light gives rise to many everyday phenomena such as the bright colors reflected from soap bubbles and from thin oil or gasoline films on water, Fig. 34–16. In these and other cases, the colors are a result of constructive interference between light reflected from the two surfaces of the thin film.

FIGURE 34–16 Thin film interference patterns seen in (a) a soap bubble, (b) a thin film of soapy water, and (c) a thin layer of oil on wet pavement.

(a)

(b)

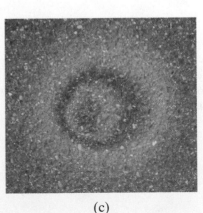

(c)

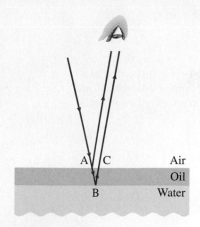

FIGURE 34–17 Light reflected from the upper and lower surfaces of a thin film of oil lying on water. This analysis assumes the light strikes the surface nearly perpendicularly, but is shown here at an angle so we can display each ray.

To see how this **thin-film interference** happens, consider a smooth surface of water on top of which is a thin uniform layer of another substance, say an oil whose index of refraction is less than that of water (we'll see why we assume this in a moment); see Fig. 34–17. Assume for the moment that the incident light is of a single wavelength. Part of the incident light is reflected at A on the top surface, and part of the light transmitted is reflected at B on the lower surface. The part reflected at the lower surface must travel the extra distance ABC. If this *path difference* ABC equals one or a whole number of wavelengths in the film (λ_n), the two waves will reach the eye in phase and interfere constructively. Hence the region AC on the surface film will appear bright. But if ABC equals $\frac{1}{2}\lambda_n, \frac{3}{2}\lambda_n$, and so on, the two waves will be exactly out of phase and destructive interference occurs: the area AC on the film will show no reflection—it will be dark (or better, transparent to the dark material below). The wavelength λ_n is *the wavelength in the film*: $\lambda_n = \lambda/n$, where n is the index of refraction in the film and λ is the wavelength in vacuum. See Eq. 34–1.

When white light falls on such a film, the path difference ABC will equal λ_n (or $m\lambda_n$, with m = an integer) for only one wavelength at a given viewing angle. The color corresponding to λ (λ in air) will be seen as very bright. For light viewed at a slightly different angle, the path difference ABC will be longer or shorter and a different color will undergo constructive interference. Thus, for an extended (nonpoint) source emitting white light, a series of bright colors will be seen next to one another. Variations in thickness of the film will also alter the path difference ABC and therefore affect the color of light that is most strongly reflected.

EXERCISE C Return to the Chapter-Opening Question, page 900, and answer it again now. Try to explain why you may have answered differently the first time.

When a curved glass surface is placed in contact with a flat glass surface, Fig. 34–18, a series of concentric rings is seen when illuminated from above by either white light (as shown) or by monochromatic light. These are called **Newton's rings**[†] and they are due to interference between waves reflected by the top and bottom surfaces of the very thin *air gap* between the two pieces of glass. Because this gap (which is equivalent to a thin film) increases in width from the central contact point out to the edges, the extra path length for the lower ray (equal to BCD) varies; where it equals $0, \frac{1}{2}\lambda, \lambda, \frac{3}{2}\lambda, 2\lambda$, and so on, it corresponds to constructive and destructive interference; and this gives rise to the series of bright colored circles seen in Fig. 34–18b. The color you see at a given radius corresponds to constructive interference; at that radius, other colors partially or fully destructively interfere. (If monochromatic light is used, the rings are alternately bright and dark.)

[†]Although Newton gave an elaborate description of them, they had been first observed and described by his contemporary, Robert Hooke.

FIGURE 34–18 Newton's rings. (a) Light rays reflected from upper and lower surfaces of the thin air gap can interfere. (b) Photograph of interference patterns using white light.

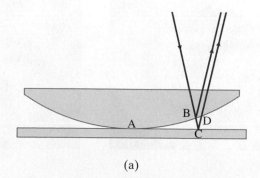

(a)

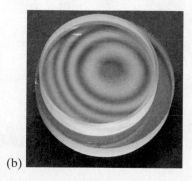

(b)

The point of contact of the two glass surfaces (A in Fig. 34–18a) is dark in Fig. 34–18b. Since the path difference is zero here, our previous analysis would suggest that the waves reflected from each surface are in phase and so this central area ought to be bright. But it is dark, which tells us something else is happening here: the two waves must be completely out of phase. This can happen if one of the waves, upon reflection, flips over—a crest becomes a trough—see Fig. 34–19. We say that the reflected wave has undergone a phase change of 180°, or of half a wave cycle. Indeed, this and other experiments reveal that, at normal incidence,

a beam of light reflected by a material with index of refraction greater than that of the material in which it is traveling, changes phase by 180° or $\frac{1}{2}$ cycle;

see Fig. 34–19. This phase change acts just like a path difference of $\frac{1}{2}\lambda$. If the index of refraction of the reflecting material is less than that of the material in which the light is traveling, no phase change occurs.[†]

Thus the wave reflected at the curved surface above the air gap in Fig. 34–18a undergoes no change in phase. But the wave reflected at the lower surface, where the beam in air strikes the glass, undergoes a $\frac{1}{2}$-cycle phase change, equivalent to a $\frac{1}{2}\lambda$ path difference. Thus the two waves reflected near the point of contact A of the two glass surfaces (where the air gap approaches zero thickness) will be a half cycle (or 180°) out of phase, and a dark spot occurs. Bright colored rings will occur when the path difference is $\frac{1}{2}\lambda, \frac{3}{2}\lambda$, and so on, because the phase change at one surface effectively adds a path difference of $\frac{1}{2}\lambda \left(=\frac{1}{2}\text{cycle}\right)$. (If monochromatic light is used, the bright Newton's rings will be separated by dark bands which occur when the path difference BCD in Fig. 34–18a is equal to an integral number of wavelengths.)

Returning for a moment to Fig. 34–17, the light reflecting at both interfaces, air–oil and oil–water, underwent a phase change of 180° equivalent to a path difference of $\frac{1}{2}\lambda$, since we assumed $n_{\text{water}} > n_{\text{oil}} > n_{\text{air}}$; since the phase changes were equal, they didn't affect our analysis.

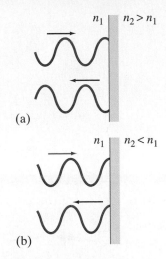

FIGURE 34–19 (a) Reflected ray changes phase by 180° or $\frac{1}{2}$ cycle if $n_2 > n_1$, but (b) does not if $n_2 < n_1$.

FIGURE 34–20 (a) Light rays reflected from the upper and lower surfaces of a thin wedge of air interfere to produce bright and dark bands. (b) Pattern observed when glass plates are optically flat; (c) pattern when plates are not so flat. See Example 34–6.

EXAMPLE 34–6 **Thin film of air, wedge-shaped.** A very fine wire 7.35×10^{-3} mm in diameter is placed between two flat glass plates as in Fig. 34–20a. Light whose wavelength in air is 600 nm falls (and is viewed) perpendicular to the plates and a series of bright and dark bands is seen, Fig. 34–20b. How many light and dark bands will there be in this case? Will the area next to the wire be bright or dark?

APPROACH We need to consider two effects: (1) path differences for rays reflecting from the two close surfaces (thin wedge of air between the two glass plates), and (2) the $\frac{1}{2}$-cycle phase change at the lower surface (point E in Fig. 34–20a), where rays in air can enter glass. Because of the phase change at the lower surface, there will be a dark band (no reflection) when the path difference is $0, \lambda, 2\lambda, 3\lambda$, and so on. Since the light rays are perpendicular to the plates, the extra path length equals $2t$, where t is the thickness of the air gap at any point.

SOLUTION Dark bands will occur where

$$2t = m\lambda, \qquad m = 0, 1, 2, \cdots.$$

Bright bands occur when $2t = \left(m + \frac{1}{2}\right)\lambda$, where m is an integer. At the position of the wire, $t = 7.35 \times 10^{-6}$ m. At this point there will be $2t/\lambda = (2)(7.35 \times 10^{-6}\,\text{m})/(6.00 \times 10^{-7}\,\text{m}) = 24.5$ wavelengths. This is a "half integer," so the area next to the wire will be bright. There will be a total of 25 dark lines along the plates, corresponding to path lengths of $0\lambda, 1\lambda, 2\lambda, 3\lambda, \ldots, 24\lambda$, including the one at the point of contact A ($m = 0$). Between them, there will be 24 bright lines plus the one at the end, or 25.

NOTE The bright and dark bands will be straight only if the glass plates are extremely flat. If they are not, the pattern is uneven, as in Fig. 34–20c. Thus we see a very precise way of testing a glass surface for flatness. Spherical lens surfaces can be tested for precision by placing the lens on a flat glass surface and observing Newton's rings (Fig. 34–18b) for perfect circularity.

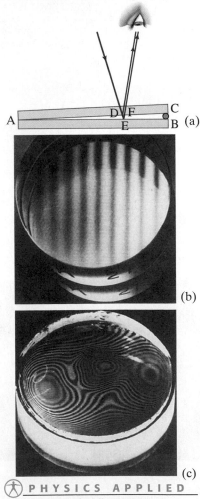

PHYSICS APPLIED
Testing glass for flatness

[†]This result can be derived from Maxwell's equations. It corresponds to the reflection of a wave traveling along a cord when it reaches the end; as we saw in Fig. 15–18, if the end is tied down, the wave changes phase and the pulse flips over, but if the end is free, no phase change occurs.

If the wedge between the two glass plates of Example 34–6 is filled with some transparent substance other than air—say, water—the pattern shifts because the wavelength of the light changes. In a material where the index of refraction is n, the wavelength is $\lambda_n = \lambda/n$, where λ is the wavelength in vacuum (Eq. 34–1). For instance, if the thin wedge of Example 34–6 were filled with water, then $\lambda_n = 600\,\text{nm}/1.33 = 450\,\text{nm}$; instead of 25 dark lines, there would be 33.

When white light (rather than monochromatic light) is incident on the thin wedge of air in Figs. 34–18a or 34–20a, a colorful series of fringes is seen because constructive interference occurs for different wavelengths in the reflected light at different thicknesses along the wedge.

A soap bubble (Fig. 34–16a and Chapter-Opening Photo) is a thin spherical shell (or film) with air inside. The variations in thickness of a soap bubble film gives rise to bright colors reflected from the soap bubble. (There is air on both sides of the bubble film.) Similar variations in film thickness produce the bright colors seen reflecting from a thin layer of oil or gasoline on a puddle or lake (Fig. 34–16c). Which wavelengths appear brightest also depends on the viewing angle.

EXAMPLE 34–7 **Thickness of soap bubble skin.** A soap bubble appears green ($\lambda = 540\,\text{nm}$) at the point on its front surface nearest the viewer. What is the smallest thickness the soap bubble film could have? Assume $n = 1.35$.

APPROACH Assume the light is reflected perpendicularly from the point on a spherical surface nearest the viewer, Fig. 34–21. The light rays also reflect from the inner surface of the soap bubble film as shown. The path difference of these two reflected rays is $2t$, where t is the thickness of the soap film. Light reflected from the first (outer) surface undergoes a 180° phase change (index of refraction of soap is greater than that of air), whereas reflection at the second (inner) surface does not. To determine the thickness t for an interference maximum, we must use the wavelength of light in the soap ($n = 1.35$).

SOLUTION The 180° phase change at only one surface is equivalent to a $\frac{1}{2}\lambda$ path difference. Therefore, green light is bright when the minimum path difference equals $\frac{1}{2}\lambda_n$. Thus, $2t = \lambda/2n$, so

$$t = \frac{\lambda}{4n} = \frac{(540\,\text{nm})}{(4)(1.35)} = 100\,\text{nm}.$$

This is the smallest thickness; but the green color is more likely to be seen at the *next* thickness that gives constructive interference, $2t = 3\lambda/2n$, because other colors would be more fully cancelled by destructive interference. The more likely thickness is $3\lambda/4n = 300\,\text{nm}$, or even $5\lambda/4n = 500\,\text{nm}$. Note that green is seen in air, so $\lambda = 540\,\text{nm}$ (not λ/n).

*Colors in a Thin Soap Film

The thin film of soapy water shown in Fig. 34–16b (repeated here) has stood vertically for a long time so that gravity has pulled much of the soapy water toward the bottom. The top section is so thin (perhaps 30 nm thick) that light reflected from the front and back surfaces have almost no path difference. Thus the 180° phase change at the front surface assures that the two reflected waves are 180° out of phase for all wavelengths of visible light. The white light incident on this thin film does not reflect at the top part of the film. Thus the top **is** transparent, and we see the background which is black.

Below the black area at the top, there is a thin blue line, and then a white band. The film thickness is perhaps 75 to 100 nm, so the shortest wavelength (blue) light begins to partially interfere constructively; but just below, where the thickness is slightly greater (100 nm), the path length is reasonably close to $\lambda/2$ for much of the spectrum and we see white or silver. (Why? Recall that red starts at 600 nm in air; so most colors in the spectrum lie between 450 nm and 600 nm in air; but in water the wavelengths are $n = 1.33$ times smaller, 340 nm to 450 nm, so a 100 nm thickness is a 200 nm path length, not far from $\lambda/2$ for most colors.) Immediately below the white band we see a brown band (around 200 nm in thickness) where selected wavelengths (not all) are close to exactly λ and those colors destructively interfere, leaving only a few colors to partially interfere constructively, giving us murky brown.

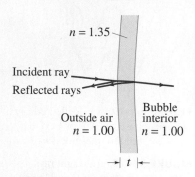

FIGURE 34–21 Example 34–7. The incident and reflected rays are assumed to be perpendicular to the bubble's surface. They are shown at a slight angle so we can distinguish them.

CAUTION

A formula is not enough: you must also check for phase changes at surfaces

FIGURE 34–16b (Repeated.)

Farther down, with increasing thickness t, a path length $2t = 510$ nm corresponds nicely to $\frac{3}{2}\lambda$ for blue, but not for other colors, so we see blue ($\frac{3}{2}\lambda$ path difference plus $\frac{1}{2}\lambda$ phase change = constructive interference). Other colors experience constructive interference (at $\frac{3}{2}\lambda$ and then at $\frac{5}{2}\lambda$) at still greater thicknesses, so we see a series of separated colors something like a rainbow.

In the soap bubble of our Chapter-Opening Photo (p. 900), similar things happen: at the top (where the film is thinnest) we see black and then silver-white, just as within the loop shown in Fig. 34–16b. And examine the oil film on wet pavement shown in Fig. 34–16c (repeated here); the oil film is thickest at the center and thins out toward the edges. Notice the whitish outer ring where most colors constructively interfere, which would suggest a thickness on the order of 100 nm as discussed above for the white band in the soap film. Beyond the outer white band of the oil film, Fig. 34–16c, there is still some oil, but the film is so thin that reflected light from upper and lower surfaces destructively interfere and you can see right through this very thin oil film.

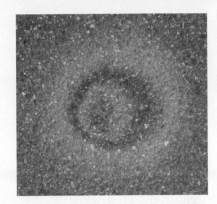

FIGURE 34–16c (Repeated.)

Lens Coatings

An important application of thin-film interference is in the coating of glass to make it "nonreflecting," particularly for lenses. A glass surface reflects about 4% of the light incident upon it. Good-quality cameras, microscopes, and other optical devices may contain six to ten thin lenses. Reflection from all these surfaces can reduce the light level considerably, and multiple reflections produce a background haze that reduces the quality of the image. By reducing reflection, transmission is increased. A very thin coating on the lens surfaces can reduce reflections considerably. The thickness of the film is chosen so that light (at least for one wavelength) reflecting from the front and rear surfaces of the film destructively interferes. The amount of reflection at a boundary depends on the difference in index of refraction between the two materials. Ideally, the coating material should have an index of refraction which is the geometric mean $\left(= \sqrt{n_1 n_2}\right)$ of those for air and glass, so that the amount of reflection at each surface is about equal. Then destructive interference can occur nearly completely for one particular wavelength depending on the thickness of the coating. Nearby wavelengths will at least partially destructively interfere, but a single coating cannot eliminate reflections for all wavelengths. Nonetheless, a single coating can reduce total reflection from 4% to 1% of the incident light. Often the coating is designed to eliminate the center of the reflected spectrum (around 550 nm). The extremes of the spectrum—red and violet—will not be reduced as much. Since a mixture of red and violet produces purple, the light seen reflected from such coated lenses is purple (Fig. 34–22). Lenses containing two or three separate coatings can more effectively reduce a wider range of reflecting wavelengths.

PHYSICS APPLIED
Lens coatings

FIGURE 34–22 A coated lens. Note color of light reflected from the front lens surface.

PROBLEM SOLVING

Interference

1. **Interference effects** depend on the simultaneous arrival of two or more waves at the same point in space.

2. **Constructive interference** occurs when the waves arrive in phase with each other: a crest of one wave arrives at the same time as a crest of the other wave. The amplitudes of the waves then add to form a larger amplitude. Constructive interference also occurs when the path difference is exactly one full wavelength or any integer multiple of a full wavelength: 1λ, 2λ, 3λ, $\cdots$.

3. **Destructive interference** occurs when a crest of one wave arrives at the same time as a trough of the other wave. The amplitudes add, but they are of opposite sign, so the total amplitude is reduced to zero if the two amplitudes are equal. Destructive interference occurs whenever the phase difference is half a wave cycle, or the path difference is a half-integral number of wavelengths. Thus, the total amplitude will be zero if two identical waves arrive one-half wavelength out of phase, or $\left(m + \frac{1}{2}\right)\lambda$ out of phase, where m is an integer.

4. For thin-film interference, an extra half-wavelength **phase shift** occurs when light **reflects** from an optically more dense medium (going from a material of lesser toward greater index of refraction).

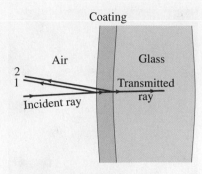

FIGURE 34–23 Example 34–8. Incident ray of light is partially reflected at the front surface of a lens coating (ray 1) and again partially reflected at the rear surface of the coating (ray 2), with most of the energy passing as the transmitted ray into the glass.

EXAMPLE 34–8 **Nonreflective coating.** What is the thickness of an optical coating of MgF_2 whose index of refraction is $n = 1.38$ and which is designed to eliminate reflected light at wavelengths (in air) around 550 nm when incident normally on glass for which $n = 1.50$?

APPROACH We explicitly follow the procedure outlined in the Problem Solving Strategy on page 913.

SOLUTION

1. **Interference effects.** Consider two rays reflected from the front and rear surfaces of the coating on the lens as shown in Fig. 34–23. The rays are drawn not quite perpendicular to the lens so we can see each of them. These two reflected rays will interfere with each other.
2. **Constructive interference.** We want to eliminate reflected light, so we do not consider constructive interference.
3. **Destructive interference.** To eliminate reflection, we want reflected rays 1 and 2 to be $\frac{1}{2}$ cycle out of phase with each other so that they destructively interfere. The phase difference is due to the path difference $2t$ traveled by ray 2, as well as any phase change in either ray due to reflection.
4. **Reflection phase shift.** Rays 1 and 2 *both* undergo a change of phase by $\frac{1}{2}$ cycle when they reflect from the coating's front and rear surfaces, respectively (at both surfaces the index of refraction increases). Thus there is no net change in phase due to the reflections. The net phase difference will be due to the extra path $2t$ taken by ray 2 in the coating, where $n = 1.38$. We want $2t$ to equal $\frac{1}{2}\lambda_n$ so that destructive interference occurs, where $\lambda_n = \lambda/n$ is the wavelength in the coating. With $2t = \lambda_n/2 = \lambda/2n$, then

$$t = \frac{\lambda_n}{4} = \frac{\lambda}{4n} = \frac{(550\,\text{nm})}{(4)(1.38)} = 99.6\,\text{nm}.$$

NOTE We could have set $2t = \left(m + \frac{1}{2}\right)\lambda_n$, where m is an integer. The smallest thickness ($m = 0$) is usually chosen because destructive interference will occur over the widest angle.

NOTE Complete destructive interference occurs only for the given wavelength of visible light. Longer and shorter wavelengths will have only partial cancellation.

*34–6 Michelson Interferometer

A useful instrument involving wave interference is the **Michelson interferometer** (Fig. 34–24),[†] invented by the American Albert A. Michelson (Section 31–7). Monochromatic light from a single point on an extended source is shown striking a half-silvered mirror M_S. This **beam splitter** mirror M_S has a thin layer of silver that reflects only half the light that hits it, so that half of the beam passes through to a fixed mirror M_2, where it is reflected back. The other half is reflected by M_S to a mirror M_1 that is movable (by a fine-thread screw), where it is also reflected back. Upon its return, part of beam 1 passes through M_S and reaches the eye; and part of beam 2, on its return, is reflected by M_S into the eye. If the two path lengths are identical, the two coherent beams entering the eye constructively interfere and brightness will be seen. If the movable mirror is moved a distance $\lambda/4$, one beam will travel an extra distance equal to $\lambda/2$ (because it travels back and forth over the distance $\lambda/4$). In this case, the two beams will destructively interfere and darkness will be seen. As M_1 is moved farther, brightness will recur (when the path difference is λ), then darkness, and so on.

Very precise length measurements can be made with an interferometer. The motion of mirror M_1 by only $\frac{1}{4}\lambda$ produces a clear difference between brightness and darkness. For $\lambda = 400\,\text{nm}$, this means a precision of 100 nm or $10^{-4}\,\text{mm}$! If mirror M_1 is tilted very slightly, the bright or dark spots are seen instead as a series of bright and dark lines or "fringes." By counting the number of fringes (or fractions thereof) that pass a reference line, extremely precise length measurements can be made.

FIGURE 34–24 Michelson interferometer.

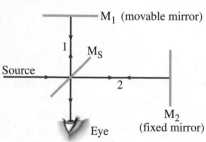

[†]There are other types of interferometer, but Michelson's is the best known.

*34–7 Luminous Intensity

The *intensity* of light, as for any electromagnetic wave, is measured by the Poynting vector in W/m², and the total power output of a source can be measured in watts (the *radiant flux*). But for measuring the visual sensation we call brightness, we must consider *only* the visible spectrum as well as the eye's sensitivity to different wavelengths—the eye is most sensitive in the central, 550-nm (green), portion of the spectrum.

These factors are taken into account in the quantity **luminous flux**, F_ℓ, whose unit is the **lumen** (lm). One lumen is equivalent to $\frac{1}{683}$ watts of 555-nm light.

Since the luminous flux from a source may not be uniform over all directions, we define the **luminous intensity** I_ℓ as the luminous flux per unit solid angle[†] (steradian). Its unit is the **candela** (cd) where $1\,\text{cd} = 1\,\text{lm/sr}$, and it is one of the seven basic quantities in the SI. (See Section 1–4 and Table 1–5.)

The **illuminance**, E_ℓ, is the luminous flux incident on a surface per unit area of the surface: $E_\ell = F_\ell/A$. Its unit is the lumen per square meter (lm/m^2) and is a measure of the illumination falling on a surface.[‡]

EXAMPLE 34–9 **Lightbulb illuminance.** The brightness of a particular type of 100-W lightbulb is rated at 1700 lm. Determine (*a*) the luminous intensity and (*b*) the illuminance at a distance of 2.0 m.

APPROACH Assume the light output is uniform in all directions.

SOLUTION (*a*) A full sphere corresponds to 4π sr. Hence, $I_\ell = 1700\,\text{lm}/4\pi\,\text{sr} = 135\,\text{cd}$. It does not depend on distance. (*b*) At $d = 2.0\,\text{m}$ from the source, the luminous flux per unit area is

$$E_\ell = \frac{F_\ell}{4\pi d^2} = \frac{1700\,\text{lm}}{(4\pi)(2.0\,\text{m})^2} = 34\,\text{lm/m}^2.$$

The illuminance decreases as the square of the distance.

[†] A solid angle is a sort of two-dimensional angle and is measured in steradians. Think of a solid angle starting at a point and intercepting an area ΔA on a sphere of radius r surrounding that point. The solid angle has magnitude $\Delta A/r^2$ steradians. A solid angle including all of space intercepts the full surface area of the sphere, $4\pi r^2$, and so has magnitude $4\pi r^2/r^2 = 4\pi$ steradians. (Compare to a normal angle, for which a full circle subtends 2π radians.)

[‡] The British unit is the foot-candle, or lumen per square foot.

Summary

The wave theory of light is strongly supported by the observations that light exhibits **interference** and **diffraction**. Wave theory also explains the refraction of light and the fact that light travels more slowly in transparent solids and liquids than it does in air.

An aid to predicting wave behavior is **Huygens' principle**, which states that every point on a wave front can be considered as a source of tiny wavelets that spread out in the forward direction at the speed of the wave itself. The new wave front is the envelope (the common tangent) of all the wavelets.

The wavelength of light in a medium with index of refraction n is

$$\lambda_n = \frac{\lambda}{n}, \tag{34–1}$$

where λ is the wavelength in vacuum; the frequency is not changed.

Young's double-slit experiment clearly demonstrated the interference of light. The observed bright spots of the interference pattern are explained as constructive interference between the beams coming through the two slits, where the beams differ in path length by an integral number of wavelengths. The dark areas in between are due to destructive interference when the path lengths differ by $\frac{1}{2}\lambda, \frac{3}{2}\lambda$, and so on. The angles θ at which **constructive interference** occurs are given by

$$\sin\theta = m\frac{\lambda}{d}, \tag{34–2a}$$

where λ is the wavelength of the light, d is the separation of the slits, and m is an integer $(0, 1, 2, \cdots)$. **Destructive interference** occurs at angles θ given by

$$\sin\theta = \left(m + \tfrac{1}{2}\right)\frac{\lambda}{d}, \tag{34–2b}$$

where m is an integer $(0, 1, 2, \cdots)$.

The light intensity I_θ at any point in a double-slit interference pattern can be calculated using a phasor diagram, which predicts that

$$I_\theta = I_0 \cos^2 \frac{\delta}{2} \qquad \text{(34–6)}$$

where I_0 is the intensity at $\theta = 0$ and the phase angle δ is

$$\delta = \frac{2\pi d}{\lambda} \sin \theta. \qquad \text{(34–4)}$$

Two sources of light are perfectly **coherent** if the waves leaving them are of the same single frequency and maintain the same phase relationship at all times. If the light waves from the two sources have a random phase with respect to each other over time (as for two incandescent lightbulbs) the two sources are **incoherent**.

Light reflected from the front and rear surfaces of a thin film of transparent material can interfere constructively or destructively, depending on the path difference. A phase change of 180° or $\frac{1}{2}\lambda$ occurs when the light reflects at a surface where the index of refraction increases. Such **thin-film interference** has many practical applications, such as lens coatings and using Newton's rings to check the uniformity of glass surfaces.

Questions

1. Does Huygens' principle apply to sound waves? To water waves?
2. What is the evidence that light is energy?
3. Why is light sometimes described as rays and sometimes as waves?
4. We can hear sounds around corners but we cannot see around corners; yet both sound and light are waves. Explain the difference.
5. Can the wavelength of light be determined from reflection or refraction measurements?
6. Two rays of light from the same source destructively interfere if their path lengths differ by how much?
7. Monochromatic red light is incident on a double slit and the interference pattern is viewed on a screen some distance away. Explain how the fringe pattern would change if the red light source is replaced by a blue light source.
8. If Young's double-slit experiment were submerged in water, how would the fringe pattern be changed?
9. Compare a double-slit experiment for sound waves to that for light waves. Discuss the similarities and differences.
10. Suppose white light falls on the two slits of Fig. 34–7, but one slit is covered by a red filter (700 nm) and the other by a blue filter (450 nm). Describe the pattern on the screen.
11. Why doesn't the light from the two headlights of a distant car produce an interference pattern?
12. Why are interference fringes noticeable only for a *thin* film like a soap bubble and not for a thick piece of glass, say?
13. Why are Newton's rings (Fig. 34–18) closer together farther from the center?
14. Some coated lenses appear greenish yellow when seen by reflected light. What wavelengths do you suppose the coating is designed to eliminate completely?
15. A drop of oil on a pond appears bright at its edges where its thickness is much less than the wavelengths of visible light. What can you say about the index of refraction of the oil compared to that of water?

Problems

34–2 Huygens' Principle

1. (II) Derive the law of reflection—namely, that the angle of incidence equals the angle of reflection from a flat surface—using Huygens' principle for waves.

34–3 Double-Slit Interference

2. (I) Monochromatic light falling on two slits 0.018 mm apart produces the fifth-order bright fringe at a 9.8° angle. What is the wavelength of the light used?
3. (I) The third-order bright fringe of 610 nm light is observed at an angle of 28° when the light falls on two narrow slits. How far apart are the slits?
4. (II) Monochromatic light falls on two very narrow slits 0.048 mm apart. Successive fringes on a screen 6.00 m away are 8.5 cm apart near the center of the pattern. Determine the wavelength and frequency of the light.
5. (II) If 720-nm and 660-nm light passes through two slits 0.68 mm apart, how far apart are the second-order fringes for these two wavelengths on a screen 1.0 m away?
6. (II) A red laser from the physics lab is marked as producing 632.8-nm light. When light from this laser falls on two closely spaced slits, an interference pattern formed on a wall several meters away has bright fringes spaced 5.00 mm apart near the center of the pattern. When the laser is replaced by a small laser pointer, the fringes are 5.14 mm apart. What is the wavelength of light produced by the pointer?

7. (II) Light of wavelength λ passes through a pair of slits separated by 0.17 mm, forming a double-slit interference pattern on a screen located a distance 35 cm away. Suppose that the image in Fig. 34–9a is an actual-size reproduction of this interference pattern. Use a ruler to measure a pertinent distance on this image; then utilize this measured value to determine λ (nm).
8. (II) Light of wavelength 680 nm falls on two slits and produces an interference pattern in which the third-order bright fringe is 38 mm from the central fringe on a screen 2.6 m away. What is the separation of the two slits?
9. (II) A parallel beam of light from a He–Ne laser, with a wavelength 633 nm, falls on two very narrow slits 0.068 mm apart. How far apart are the fringes in the center of the pattern on a screen 3.8 m away?
10. (II) A physics professor wants to perform a lecture demonstration of Young's double-slit experiment for her class using the 633-nm light from a He–Ne laser. Because the lecture hall is very large, the interference pattern will be projected on a wall that is 5.0 m from the slits. For easy viewing by all students in the class, the professor wants the distance between the $m = 0$ and $m = 1$ maxima to be 25 cm. What slit separation is required in order to produce the desired interference pattern?

11. (II) Suppose a thin piece of glass is placed in front of the lower slit in Fig. 34–7 so that the two waves enter the slits 180° out of phase (Fig. 34–25). Describe in detail the interference pattern on the screen.

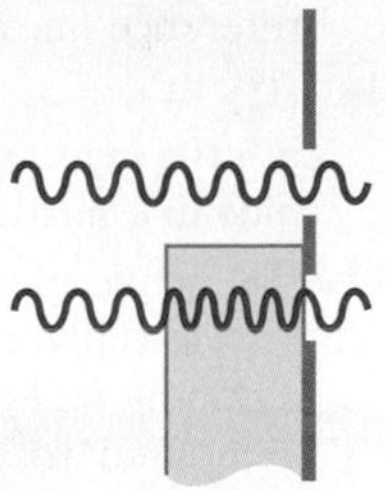

FIGURE 34–25
Problem 11.

12. (II) In a double-slit experiment it is found that blue light of wavelength 480 nm gives a second-order maximum at a certain location on the screen. What wavelength of visible light would have a minimum at the same location?

13. (II) Two narrow slits separated by 1.0 mm are illuminated by 544 nm light. Find the distance between adjacent bright fringes on a screen 5.0 m from the slits.

14. (II) In a double-slit experiment, the third-order maximum for light of wavelength 500 nm is located 12 mm from the central bright spot on a screen 1.6 m from the slits. Light of wavelength 650 nm is then projected through the same slits. How far from the central bright spot will the second-order maximum of this light be located?

15. (II) Light of wavelength 470 nm in air falls on two slits 6.00×10^{-2} mm apart. The slits are immersed in water, as is a viewing screen 50.0 cm away. How far apart are the fringes on the screen?

16. (II) A very thin sheet of plastic $(n = 1.60)$ covers one slit of a double-slit apparatus illuminated by 680-nm light. The center point on the screen, instead of being a maximum, is dark. What is the (minimum) thickness of the plastic?

34–4 Intensity in Two-Slit Interference

*17. (I) If one slit in Fig. 34–12 is covered, by what factor does the intensity at the center of the screen change?

*18. (II) Derive an expression similar to Eq. 34–2 which gives the angles for which the double-slit intensity is one-half its maximum value, $I_\theta = \frac{1}{2} I_0$.

*19. (II) Show that the angular full width at half maximum of the central peak in a double-slit interference pattern is given by $\Delta\theta = \lambda/2d$ if $\lambda \ll d$.

*20. (II) In a two-slit interference experiment, the path length to a certain point P on the screen differs for one slit in comparison with the other by 1.25λ. (a) What is the phase difference between the two waves arriving at point P? (b) Determine the intensity at P, expressed as a fraction of the maximum intensity I_0 on the screen.

*21. (III) Suppose that one slit of a double-slit apparatus is wider than the other so that the intensity of light passing through it is twice as great. Determine the intensity I as a function of position (θ) on the screen for coherent light.

*22. (III) (a) Consider three equally spaced and equal-intensity coherent sources of light (such as adding a third slit to the two slits of Fig. 34–12). Use the phasor method to obtain the intensity as a function of the phase difference δ (Eq. 34–4). (b) Determine the positions of maxima and minima.

34–5 Thin-Film Interference

23. (I) If a soap bubble is 120 nm thick, what wavelength is most strongly reflected at the center of the outer surface when illuminated normally by white light? Assume that $n = 1.32$.

24. (I) How far apart are the dark fringes in Example 34–6 if the glass plates are each 28.5 cm long?

25. (II) (a) What is the smallest thickness of a soap film $(n = 1.33)$ that would appear black if illuminated with 480-nm light? Assume there is air on both sides of the soap film. (b) What are two other possible thicknesses for the film to appear black? (c) If the thickness t was much less than λ, why would the film also appear black?

26. (II) A lens appears greenish yellow $(\lambda = 570 \, nm$ is strongest) when white light reflects from it. What minimum thickness of coating $(n = 1.25)$ do you think is used on such a glass $(n = 1.52)$ lens, and why?

27. (II) A thin film of oil $(n_o = 1.50)$ with varying thickness floats on water $(n_w = 1.33)$. When it is illuminated from above by white light, the reflected colors are as shown in Fig. 34–26. In air, the wavelength of yellow light is 580 nm. (a) Why are there no reflected colors at point A? (b) What is the oil's thickness t at point B?

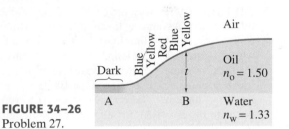

FIGURE 34–26
Problem 27.

28. (II) A thin oil slick $(n_o = 1.50)$ floats on water $(n_w = 1.33)$. When a beam of white light strikes this film at normal incidence from air, the only enhanced reflected colors are red (650 nm) and violet (390 nm). From this information, deduce the (minimum) thickness t of the oil slick.

29. (II) A total of 31 bright and 31 dark Newton's rings (not counting the dark spot at the center) are observed when 560-nm light falls normally on a planoconvex lens resting on a flat glass surface (Fig. 34–18). How much thicker is the center than the edges?

30. (II) A fine metal foil separates one end of two pieces of optically flat glass, as in Fig. 34–20. When light of wavelength 670 nm is incident normally, 28 dark lines are observed (with one at each end). How thick is the foil?

31. (II) How thick (minimum) should the air layer be between two flat glass surfaces if the glass is to appear bright when 450-nm light is incident normally? What if the glass is to appear dark?

32. (II) A uniform thin film of alcohol $(n = 1.36)$ lies on a flat glass plate $(n = 1.56)$. When monochromatic light, whose wavelength can be changed, is incident normally, the reflected light is a minimum for $\lambda = 512 \, nm$ and a maximum for $\lambda = 635 \, nm$. What is the minimum thickness of the film?

33. (II) Show that the radius r of the mth dark Newton's ring, as viewed from directly above (Fig. 34–18), is given by $r = \sqrt{m\lambda R}$ where R is the radius of curvature of the curved glass surface and λ is the wavelength of light used. Assume that the thickness of the air gap is much less than R at all points and that $r \ll R$. [Hint: Use the binomial expansion.]

34. (II) Use the result of Problem 33 to show that the distance between adjacent dark Newton's rings is

$$\Delta r \approx \sqrt{\frac{\lambda R}{4m}}$$

for the mth ring, assuming $m \gg 1$.

35. (II) When a Newton's ring apparatus (Fig. 34–18) is immersed in a liquid, the diameter of the eighth dark ring decreases from 2.92 cm to 2.54 cm. What is the refractive index of the liquid? [Hint: see Problem 33.]

36. (II) A planoconvex lucite lens 3.4 cm in diameter is placed on a flat piece of glass as in Fig. 34–18. When 580-nm light is incident normally, 44 bright rings are observed, the last one right at the edge. What is the radius of curvature of the lens surface, and the focal length of the lens? [Hint: see Problem 33.]

37. (II) Let's explore why only "thin" layers exhibit thin-film interference. Assume a layer of water, sitting atop a flat glass surface, is illuminated from the air above by white light (all wavelengths from 400 nm to 700 nm). Further, assume that the water layer's thickness t is much greater than a micron ($= 1000$ nm); in particular, let $t = 200\ \mu m$. Take the index of refraction for water to be $n = 1.33$ for all visible wavelengths. (a) Show that a visible color will be reflected from the water layer if its wavelength is $\lambda = 2nt/m$, where m is an integer. (b) Show that the two extremes in wavelengths (400 nm and 700 nm) of the incident light are both reflected from the water layer and determine the m-value associated with each. (c) How many other visible wavelengths, besides $\lambda = 400$ nm and 700 nm, are reflected from the "thick" layer of water? (d) How does this explain why such a thick layer does not reflect colorfully, but is white or grey?

38. (III) A single optical coating reduces reflection to zero for $\lambda = 550$ nm. By what factor is the intensity reduced by the coating for $\lambda = 430$ nm and $\lambda = 670$ nm as compared to no coating? Assume normal incidence.

*34–6 Michelson Interferometer

*39. (II) How far must the mirror M_1 in a Michelson interferometer be moved if 650 fringes of 589-nm light are to pass by a reference line?

*40. (II) What is the wavelength of the light entering an interferometer if 384 bright fringes are counted when the movable mirror moves 0.125 mm?

*41. (II) A micrometer is connected to the movable mirror of an interferometer. When the micrometer is tightened down on a thin metal foil, the net number of bright fringes that move, compared to the empty micrometer, is 272. What is the thickness of the foil? The wavelength of light used is 589 nm.

*42. (III) One of the beams of an interferometer (Fig. 34–27) passes through a small evacuated glass container 1.155 cm deep. When a gas is allowed to slowly fill the container, a total of 176 dark fringes are counted to move past a reference line. The light used has a wavelength of 632.8 nm. Calculate the index of refraction of the gas at its final density, assuming that the interferometer is in vacuum.

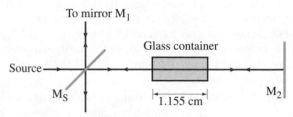

FIGURE 34–27 Problem 42.

*43. (III) The yellow sodium D lines have wavelengths of 589.0 and 589.6 nm. When they are used to illuminate a Michelson interferometer, it is noted that the interference fringes disappear and reappear periodically as the mirror M_1 is moved. Why does this happen? How far must the mirror move between one disappearance and the next?

*34–7 Luminous Intensity

*44. (II) The illuminance of direct sunlight on Earth is about 10^5 lm/m². Estimate the luminous flux and luminous intensity of the Sun.

*45. (II) The luminous efficiency of a lightbulb is the ratio of luminous flux to electric power input. (a) What is the luminous efficiency of a 100-W, 1700-lm bulb? (b) How many 40-W, 60-lm/W fluorescent lamps would be needed to provide an illuminance of 250 lm/m² on a factory floor of area 25 m × 30 m? Assume the lights are 10 m above the floor and that half their flux reaches the floor.

General Problems

46. Light of wavelength 5.0×10^{-7} m passes through two parallel slits and falls on a screen 4.0 m away. Adjacent bright bands of the interference pattern are 2.0 cm apart. (a) Find the distance between the slits. (b) The same two slits are next illuminated by light of a different wavelength, and the fifth-order minimum for this light occurs at the same point on the screen as the fourth-order minimum for the previous light. What is the wavelength of the second source of light?

47. Television and radio waves reflecting from mountains or airplanes can interfere with the direct signal from the station. (a) What kind of interference will occur when 75-MHz television signals arrive at a receiver directly from a distant station, and are reflected from a nearby airplane 122 m directly above the receiver? Assume $\frac{1}{2}\lambda$ change in phase of the signal upon reflection. (b) What kind of interference will occur if the plane is 22 m closer to the receiver?

48. A radio station operating at 88.5 MHz broadcasts from two identical antennas at the same elevation but separated by a 9.0-m horizontal distance d, Fig. 34–28. A maximum signal is found along the midline, perpendicular to d at its midpoint and extending horizontally in both directions. If the midline is taken as 0°, at what other angle(s) θ is a maximum signal detected? A minimum signal? Assume all measurements are made much farther than 9.0 m from the antenna towers.

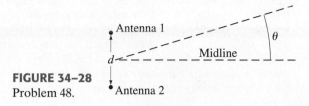

FIGURE 34–28
Problem 48.

49. Light of wavelength 690 nm passes through two narrow slits 0.66 mm apart. The screen is 1.60 m away. A second source of unknown wavelength produces its second-order fringe 1.23 mm closer to the central maximum than the 690-nm light. What is the wavelength of the unknown light?

50. Monochromatic light of variable wavelength is incident normally on a thin sheet of plastic film in air. The reflected light is a maximum only for $\lambda = 491.4$ nm and $\lambda = 688.0$ nm in the visible spectrum. What is the thickness of the film $(n = 1.58)$? [*Hint*: Assume successive values of m.]

*51. Suppose the mirrors in a Michelson interferometer are perfectly aligned and the path lengths to mirrors M_1 and M_2 are identical. With these initial conditions, an observer sees a bright maximum at the center of the viewing area. Now one of the mirrors is moved a distance x. Determine a formula for the intensity at the center of the viewing area as a function of x, the distance the movable mirror is moved from the initial position.

52. A highly reflective mirror can be made for a particular wavelength at normal incidence by using two thin layers of transparent materials of indices of refraction n_1 and n_2 $(1 < n_1 < n_2)$ on the surface of the glass $(n > n_2)$. What should be the minimum thicknesses d_1 and d_2 in Fig. 34–29 in terms of the incident wavelength λ, to maximize reflection?

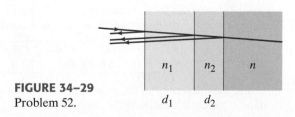

FIGURE 34–29
Problem 52.

53. Calculate the minimum thickness needed for an antireflective coating $(n = 1.38)$ applied to a glass lens in order to eliminate (*a*) blue (450 nm), or (*b*) red (720 nm) reflections for light at normal incidence.

54. Stealth aircraft are designed to not reflect radar, whose wavelength is typically 2 cm, by using an antireflecting coating. Ignoring any change in wavelength in the coating, estimate its thickness.

55. Light of wavelength λ strikes a screen containing two slits a distance d apart at an angle θ_i to the normal. Determine the angle θ_m at which the mth-order maximum occurs.

56. Consider two antennas radiating 6.0-MHz radio waves in phase with each other. They are located at points S_1 and S_2, separated by a distance $d = 175$ m, Fig. 34–30. Determine the points on the y axis where the signals from the two sources will be out of phase (crests of one meet troughs of the other).

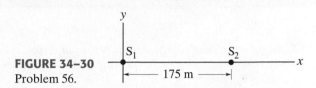

FIGURE 34–30
Problem 56.

57. What is the minimum (non-zero) thickness for the air layer between two flat glass surfaces if the glass is to appear dark when 680-nm light is incident normally? What if the glass is to appear bright?

58. *Lloyd's mirror* provides one way of obtaining a double-slit interference pattern from a single source so the light is coherent. As shown in Fig. 34–31, the light that reflects from the plane mirror appears to come from the virtual image of the slit. Describe in detail the interference pattern on the screen.

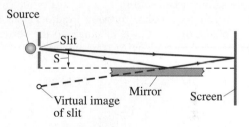

FIGURE 34–31   Problem 58.

59. Consider the antenna array of Example 34–5, Fig. 34–15. Let $d = \lambda/2$, and suppose that the two antennas are now 180° out of phase with each other. Find the directions for constructive and destructive interference, and compare with the case when the sources are in phase. (These results illustrate the basis for directional antennas.)

60. Suppose you viewed the light *transmitted* through a thin film layered on a flat piece of glass. Draw a diagram, similar to Fig. 34–17 or 34–23, and describe the conditions required for maxima and minima. Consider all possible values of index of refraction. Discuss the relative size of the minima compared to the maxima and to zero.

61. A thin film of soap $(n = 1.34)$ coats a piece of flat glass $(n = 1.52)$. How thick is the film if it reflects 643-nm red light most strongly when illuminated normally by white light?

62. Two identical sources S_1 and S_2, separated by distance d, coherently emit light of wavelength λ uniformly in all directions. Defining the x axis with its origin at S_1 as shown in Fig. 34–32, find the locations (expressed as multiples of λ) where the signals from the two sources are out of phase along this axis for $x > 0$, if $d = 3\lambda$.

FIGURE 34–32
Problem 62.

63. A two-slit interference set-up with slit separation $d = 0.10\,\text{mm}$ produces interference fringes at a particular set of angles θ_m (where $m = 0, 1, 2, \ldots$) for red light of frequency $f = 4.6 \times 10^{14}\,\text{Hz}$. If one wishes to construct an analogous two-slit interference set-up that produces interference fringes at the same set of angles θ_m for room-temperature sound of middle-C frequency $f_S = 262\,\text{Hz}$, what should the slit separation d_S be for this analogous set-up?

64. A radio telescope, whose two antennas are separated by 55 m, is designed to receive 3.0-MHz radio waves produced by astronomical objects. The received radio waves create 3.0-MHz electronic signals in the telescope's left and right antennas. These signals then travel by equal-length cables to a centrally located amplifier, where they are added together. The telescope can be "pointed" to a certain region of the sky by adding the instantaneous signal from the right antenna to a "time-delayed" signal received by the left antenna a time Δt ago. (This time delay of the left signal can be easily accomplished with the proper electronic circuit.) If a radio astronomer wishes to "view" radio signals arriving from an object oriented at a 12° angle to the vertical as in Fig. 34–33, what time delay Δt is necessary?

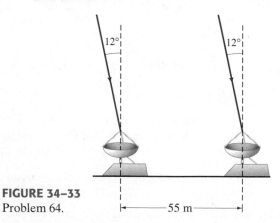

FIGURE 34–33
Problem 64.

65. In a compact disc (CD), digital information is stored as a sequence of raised surfaces called "pits" and recessed surfaces called "lands." Both pits and lands are highly reflective and are embedded in a thick plastic material with index of refraction $n = 1.55$ (Fig. 34–34). As a 780-nm wavelength (in air) laser scans across the pit–land sequence, the transition between a neighboring pit and land is sensed by monitoring the intensity of reflected laser light from the CD. At the moment when half the width of the laser beam is reflected from the pit and the other half from the land, we want the two reflected halves of the beam to be 180° out of phase with each other. What should be the (minimum) height difference t between a pit and land? [When this light enters a detector, cancellation of the two out-of-phase halves of the beam produces a minimum detector output.]

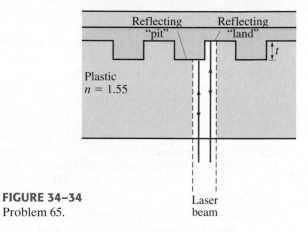

FIGURE 34–34
Problem 65.

*Numerical/Computer

*66. (II) A Michelson interferometer can be used to determine the index of refraction of a glass plate. A glass plate (thickness t) is placed on a platform that can rotate. The plate is placed in the light's path between the beam splitter and either the fixed or movable mirror, so that its thickness is in the direction of the laser beam. The platform is rotated to various angles, and the number of fringes shifted is counted. It can be shown that if N is the number of fringes shifted when the angle of rotation changes by θ, the index of refraction is $n = (2t - N\lambda)(1 - \cos\theta)/[2t(1 - \cos\theta) - N\lambda]$ where t is the thickness of the plate. The accompanying Table shows the data collected by a student in determining the index of refraction of a transparent plate by a Michelson interferometer.

N	25	50	75	100	125	150
θ (degree)	5.5	6.9	8.6	10.0	11.3	12.5

In the experiment $\lambda = 632.8\,\text{nm}$ and $t = 4.0\,\text{mm}$. Determine n for each θ and find the average n.

Answers to Exercises

A: 333 nm; $6.0 \times 10^{14}\,\text{Hz}$; $2.0 \times 10^8\,\text{m/s}$.

B: (a) I_0; (b) $0.50I_0$; (c) 0.

C: (e).

Parallel coherent light from a laser, which acts as nearly a point source, illuminates these shears. Instead of a clean shadow, there is a dramatic diffraction pattern, which is a strong confirmation of the wave theory of light. Diffraction patterns are washed out when typical extended sources of light are used, and hence are not seen, although a careful examination of shadows will reveal fuzziness. We will examine diffraction by a single slit, and how it affects the double-slit pattern. We also discuss diffraction gratings and diffraction of X-rays by crystals. We will see how diffraction affects the resolution of optical instruments, and that the ultimate resolution can never be greater than the wavelength of the radiation used. Finally we study the polarization of light.

Diffraction and Polarization

35

CONTENTS

CHAPTER-OPENING QUESTION—Guess now!

Because of diffraction, a light microscope has a maximum useful magnification of about

(a) 50×;
(b) 100×;
(c) 500×;
(d) 2000×;
(e) 5000×;

and the smallest objects it can resolve have a size of about

(a) 10 nm;
(b) 100 nm;
(c) 500 nm;
(d) 2500 nm;
(e) 5500 nm.

Young's double-slit experiment put the wave theory of light on a firm footing. But full acceptance came only with studies on diffraction more than a decade later, in the 1810s and 1820s.

We have already discussed diffraction briefly with regard to water waves (Section 15–11) as well as for light (Section 34–1). We have seen that it refers to the spreading or bending of waves around edges. Now we look at diffraction in more detail, including its important practical effects of limiting the amount of detail, or *resolution*, that can be obtained with any optical instrument such as telescopes, cameras, and the eye.

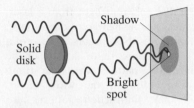

FIGURE 35–1 If light is a wave, a bright spot will appear at the center of the shadow of a solid disk illuminated by a point source of monochromatic light.

In 1819, Augustin Fresnel (1788–1827) presented to the French Academy a wave theory of light that predicted and explained interference and diffraction effects. Almost immediately Siméon Poisson (1781–1840) pointed out a counter-intuitive inference: according to Fresnel's wave theory, if light from a point source were to fall on a solid disk, part of the incident light would be diffracted around the edges and would constructively interfere at the center of the shadow (Fig. 35–1). That prediction seemed very unlikely. But when the experiment was actually carried out by François Arago, the bright spot was seen at the very center of the shadow (Fig. 35–2a). This was strong evidence for the wave theory.

Figure 35–2a is a photograph of the shadow cast by a coin using a coherent point source of light, a laser in this case. The bright spot is clearly present at the center. Note also the bright and dark fringes beyond the shadow. These resemble the interference fringes of a double slit. Indeed, they are due to interference of waves diffracted around the disk, and the whole is referred to as a **diffraction pattern**. A diffraction pattern exists around any sharp-edged object illuminated by a point source, as shown in Fig. 35–2b and c. We are not always aware of diffraction because most sources of light in everyday life are not points, so light from different parts of the source washes out the pattern.

FIGURE 35–2 Diffraction pattern of (a) a circular disk (a coin), (b) razorblade, (c) a single slit, each illuminated by a coherent point source of monochromatic light, such as a laser.

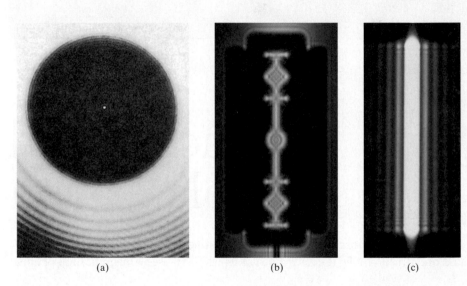

(a)   (b)   (c)

35–1 Diffraction by a Single Slit or Disk

To see how a diffraction pattern arises, we will analyze the important case of monochromatic light passing through a narrow slit. We will assume that parallel rays (plane waves) of light fall on the slit of width D, and pass through to a viewing screen very far away. If the viewing screen is not far away, lenses can be used to make the rays parallel.[†] As we know from studying water waves and from Huygens' principle, the waves passing through the slit spread out in all directions. We will now examine how the waves passing through different parts of the slit interfere with each other.

Parallel rays of monochromatic light pass through the narrow slit as shown in Fig. 35–3a. The slit width D is on the order of the wavelength λ of the light, but the slit's length (in and out of page) is large compared to λ. The light falls on a screen which is assumed to be very far away, so the rays heading for any point are very nearly parallel before they meet at the screen.

[†]Such a diffraction pattern, involving parallel rays, is called *Fraunhofer diffraction*. If the screen is close and no lenses are used, it is called *Fresnel diffraction*. The analysis in the latter case is rather involved, so we consider only the limiting case of Fraunhofer diffraction.

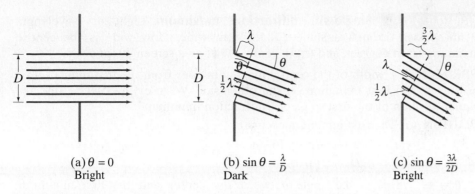
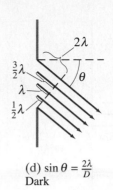

(a) $\theta = 0$
Bright

(b) $\sin\theta = \dfrac{\lambda}{D}$
Dark

(c) $\sin\theta = \dfrac{3\lambda}{2D}$
Bright

(d) $\sin\theta = \dfrac{2\lambda}{D}$
Dark

FIGURE 35–3 Analysis of diffraction pattern formed by light passing through a narrow slit of width D.

First we consider rays that pass straight through as in Fig. 35–3a. They are all in phase, so there will be a central bright spot on the screen (see Fig. 35–2c). In Fig. 35–3b, we consider rays moving at an angle θ such that the ray from the top of the slit travels exactly one wavelength farther than the ray from the bottom edge to reach the screen. The ray passing through the very center of the slit will travel one-half wavelength farther than the ray at the bottom of the slit. These two rays will be exactly out of phase with one another and so will destructively interfere when they overlap at the screen. Similarly, a ray slightly above the bottom one will cancel a ray that is the same distance above the central one. Indeed, each ray passing through the lower half of the slit will cancel with a corresponding ray passing through the upper half. Thus, all the rays destructively interfere in pairs, and so the light intensity will be zero on the viewing screen at this angle. The angle θ at which this takes place can be seen from Fig. 35–3b to occur when $\lambda = D\sin\theta$, so

$$\sin\theta = \frac{\lambda}{D}. \qquad\qquad \text{[first minimum]} \quad \textbf{(35–1)}$$

The light intensity is a maximum at $\theta = 0°$ and decreases to a minimum (intensity = zero) at the angle θ given by Eq. 35–1.

Now consider a larger angle θ such that the top ray travels $\frac{3}{2}\lambda$ farther than the bottom ray, as in Fig. 35–3c. In this case, the rays from the bottom third of the slit will cancel in pairs with those in the middle third because they will be $\lambda/2$ out of phase. However, light from the top third of the slit will still reach the screen, so there will be a bright spot centered near $\sin\theta \approx 3\lambda/2D$, but it will not be nearly as bright as the central spot at $\theta = 0°$. For an even larger angle θ such that the top ray travels 2λ farther than the bottom ray, Fig. 35–3d, rays from the bottom quarter of the slit will cancel with those in the quarter just above it because the path lengths differ by $\lambda/2$. And the rays through the quarter of the slit just above center will cancel with those through the top quarter. At this angle there will again be a minimum of zero intensity in the diffraction pattern. A plot of the intensity as a function of angle is shown in Fig. 35–4. This corresponds well with the photo of Fig. 35–2c. Notice that minima (zero intensity) occur on both sides of center at

$$D\sin\theta = m\lambda, \qquad m = \pm1, \pm2, \pm3, \cdots, \qquad \text{[minima]} \quad \textbf{(35–2)}$$

but *not* at $m = 0$ where there is the strongest maximum. Between the minima, smaller intensity maxima occur at approximately (not exactly) $m \approx \frac{3}{2}, \frac{5}{2}, \cdots$.

Note that the *minima* for a diffraction pattern, Eq. 35–2, satisfy a criterion that looks very similar to that for the *maxima* (bright fringes) for double-slit interference, Eq. 34–2a. Also note that D is a single slit width, whereas d in Eq. 34–2 is the distance between two slits.

FIGURE 35–4 Intensity in the diffraction pattern of a single slit as a function of $\sin\theta$. Note that the central maximum is not only much higher than the maxima to each side, but it is also twice as wide ($2\lambda/D$ wide) as any of the others (only λ/D wide each).

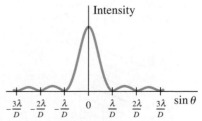

⚠ **CAUTION**

Don't confuse Eqs. 34–2 for interference with Eq. 35–1 for diffraction: note the differences

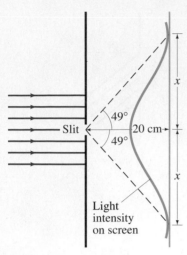

FIGURE 35–5 Example 35–1.

EXAMPLE 35–1 **Single-slit diffraction maximum.** Light of wavelength 750 nm passes through a slit 1.0×10^{-3} mm wide. How wide is the central maximum (a) in degrees, and (b) in centimeters, on a screen 20 cm away?

APPROACH The width of the central maximum goes from the first minimum on one side to the first minimum on the other side. We use Eq. 35–1 to find the angular position of the first single-slit diffraction minimum.

SOLUTION (a) The first minimum occurs at

$$\sin \theta = \frac{\lambda}{D} = \frac{7.5 \times 10^{-7}\, \text{m}}{1.0 \times 10^{-6}\, \text{m}} = 0.75.$$

So $\theta = 49°$. This is the angle between the center and the first minimum, Fig. 35–5. The angle subtended by the whole central maximum, between the minima above and below the center, is twice this, or 98°.

(b) The width of the central maximum is $2x$, where $\tan \theta = x/20$ cm. So $2x = 2(20\, \text{cm})(\tan 49°) = 46\, \text{cm}$.

NOTE A large width of the screen will be illuminated, but it will not normally be very bright since the amount of light that passes through such a small slit will be small and it is spread over a large area. Note also that we *cannot* use the small-angle approximation here ($\theta \approx \sin \theta \approx \tan \theta$) because θ is large.

EXERCISE A In Example 35–1, red light ($\lambda = 750$ nm) was used. If instead yellow light at 575 nm had been used, would the central maximum be wider or narrower?

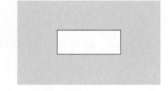

FIGURE 35–6 Example 35–2.

CONCEPTUAL EXAMPLE 35–2 **Diffraction spreads.** Light shines through a rectangular hole that is narrower in the vertical direction than the horizontal, Fig. 35–6. (a) Would you expect the diffraction pattern to be more spread out in the vertical direction or in the horizontal direction? (b) Should a rectangular loudspeaker horn at a stadium be high and narrow, or wide and flat?

RESPONSE (a) From Eq. 35–1 we can see that if we make the slit (width D) narrower, the pattern spreads out more. This is consistent with our study of waves in Chapter 15. The diffraction through the rectangular hole will be wider vertically, since the opening is smaller in that direction. (b) For a loudspeaker, the sound pattern desired is one spread out horizontally, so the horn should be tall and narrow (rotate Fig. 35–6 by 90°).

*35–2 Intensity in Single-Slit Diffraction Pattern

We have determined the positions of the minima in the diffraction pattern produced by light passing through a single slit, Eq. 35–2. We now discuss a method for predicting the amplitude and intensity at any point in the pattern using the phasor technique already discussed in Section 34–4.

Let us consider the slit divided into N very thin strips of width Δy as indicated in Fig. 35–7. Each strip sends light in all directions toward a screen on the right. Again we take the rays heading for any particular point on the distant screen to be parallel, all making an angle θ with the horizontal as shown. We choose the strip width Δy to be much smaller than the wavelength λ of the monochromatic light falling on the slit, so all the light from a given strip is in phase. The strips are of equal size, and if the whole slit is uniformly illuminated, we can take the electric field wave amplitudes ΔE_0 from each thin strip to be equal as long as θ is not too large. However, the separate amplitudes from the different strips will differ in phase. The phase difference in the light coming from adjacent strips will be (see Section 34–4, Eq. 34–4)

$$\Delta \beta = \frac{2\pi}{\lambda} \Delta y \sin \theta \qquad (35–3)$$

since the difference in path length is $\Delta y \sin \theta$.

FIGURE 35–7 Slit of width D divided into N strips of width Δy.

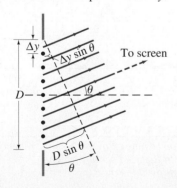

The total amplitude on the screen at any angle θ will be the sum of the separate wave amplitudes due to each strip. These wavelets have the same amplitude ΔE_0 but differ in phase. To obtain the total amplitude, we can use a phasor diagram as we did in Section 34–4 (Fig. 34–13). The phasor diagrams for four different angles θ are shown in Fig. 35–8. At the center of the screen, $\theta = 0$, the waves from each strip are all in phase ($\Delta \beta = 0$, Eq. 35–3), so the arrows representing each ΔE_0 line up as shown in Fig. 35–8a. The total amplitude of the light arriving at the center of the screen is then $E_0 = N \, \Delta E_0$.

FIGURE 35–8 Phasor diagram for single-slit diffraction, giving the total amplitude E_θ at four different angles θ.

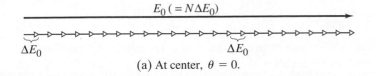

(a) At center, $\theta = 0$.

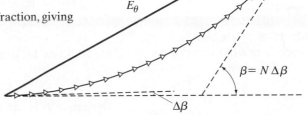

(b) Between center and first minimum.

(c) First minimum, $E_\theta = 0$ ($\beta = 2\pi = 360°$).

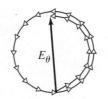

(d) Near secondary maximum.

At a small angle θ, for a point on the distant screen not far from the center, Fig. 35–8b shows how the wavelets of amplitude ΔE_0 add up to give E_θ, the total amplitude on the screen at this angle θ. Note that each wavelet differs in phase from the adjacent one by $\Delta \beta$. The phase difference between the wavelets from the top and bottom edges of the slit is

$$\beta = N \, \Delta \beta = \frac{2\pi}{\lambda} N \, \Delta y \sin \theta = \frac{2\pi}{\lambda} D \sin \theta \qquad \textbf{(35–4)}$$

where $D = N \, \Delta y$ is the total width of the slit. Although the "arc" in Fig. 35–8b has length $N \, \Delta E_0$, and so would equal E_0 (total amplitude at $\theta = 0$), the amplitude of the total wave E_θ at angle θ is the *vector* sum of each wavelet amplitude and so is equal to the length of the chord as shown. The chord is shorter than the arc, so $E_\theta < E_0$.

For greater θ, we eventually come to the case, illustrated in Fig. 35–8c, where the chain of arrows closes on itself. In this case the vector sum is zero, so $E_\theta = 0$ for this angle θ. This corresponds to the first minimum. Since $\beta = N \, \Delta \beta$ is 360° or 2π in this case, we have from Eq. 35–3,

$$2\pi = N \, \Delta \beta = N \left(\frac{2\pi}{\lambda} \Delta y \sin \theta \right)$$

or, since the slit width $D = N \, \Delta y$,

$$\sin \theta = \frac{\lambda}{D}.$$

Thus the first minimum $\left(E_\theta = 0 \right)$ occurs where $\sin \theta = \lambda/D$, which is the same result we obtained in the previous Section, Eq. 35–1.

For even greater values of θ, the chain of arrows spirals beyond 360°. Figure 35–8d shows the case near the secondary maximum next to the first minimum. Here $\beta = N \, \Delta \beta \approx 360° + 180° = 540°$ or 3π. When greater angles θ are considered, new maxima and minima occur. But since the total length of the coil remains constant, equal to $N \, \Delta E_0 \left(= E_0 \right)$, each succeeding maximum is smaller and smaller as the coil winds in on itself.

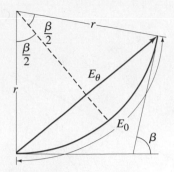

FIGURE 35–9 Determining amplitude E_θ as a function of θ for single-slit diffraction.

To obtain a quantitative expression for the amplitude (and intensity) for any point on the screen (that is, for any angle θ), we now consider the limit $N \to \infty$ so Δy becomes the infinitesimal width dy. In this case, the diagrams of Fig. 35–8 become smooth curves, one of which is shown in Fig. 35–9. For any angle θ, the wave amplitude on the screen is E_θ, equal to the chord in Fig. 35–9. The length of the arc is E_0, as before. If r is the radius of curvature of the arc, then

$$\frac{E_\theta}{2} = r \sin \frac{\beta}{2}.$$

Using radian measure for $\beta/2$, we also have

$$\frac{E_0}{2} = r \frac{\beta}{2}.$$

We combine these to obtain

$$E_\theta = E_0 \frac{\sin \beta/2}{\beta/2}. \tag{35–5}$$

The angle β is the phase difference between the waves from the top and bottom edges of the slit. The path difference for these two rays is $D \sin \theta$ (see Fig. 35–7 as well as Eq. 35–4), so

$$\beta = \frac{2\pi}{\lambda} D \sin \theta. \tag{35–6}$$

Intensity is proportional to the square of the wave amplitude, so the intensity I_θ at any angle θ is, from Eq. 35–5,

$$I_\theta = I_0 \left(\frac{\sin \beta/2}{\beta/2} \right)^2 \tag{35–7}$$

where $I_0 \, (\propto E_0^2)$ is the intensity at $\theta = 0$ (the central maximum). We can combine Eqs. 35–7 and 35–6 (although it is often simpler to leave them as separate equations) to obtain

$$I_\theta = I_0 \left[\frac{\sin\left(\dfrac{\pi D \sin \theta}{\lambda} \right)}{\left(\dfrac{\pi D \sin \theta}{\lambda} \right)} \right]^2. \tag{35–8}$$

According to Eq. 35–8, minima $(I_\theta = 0)$ occur where $\sin(\pi D \sin \theta/\lambda) = 0$, which means $\pi D \sin \theta/\lambda$ must be $\pi, 2\pi, 3\pi$, and so on, or

$$D \sin \theta = m\lambda, \qquad m = 1, 2, 3, \cdots \qquad \text{[minima]}$$

which is what we have obtained previously, Eq. 35–2. Notice that m cannot be zero: when $\beta/2 = \pi D \sin \theta/\lambda = 0$, the denominator as well as the numerator in Eqs. 35–7 or 35–8 vanishes. We can evaluate the intensity in this case by taking the limit as $\theta \to 0$ (or $\beta \to 0$); for very small angles, $\sin \beta/2 \approx \beta/2$, so $(\sin \beta/2)/(\beta/2) \to 1$ and $I_\theta = I_0$, the *maximum* at the center of the pattern.

The intensity I_θ as a function of θ, as given by Eq. 35–8, corresponds to the diagram of Fig. 35–4.

EXAMPLE 35–3 **ESTIMATE** **Intensity at secondary maxima.** Estimate the intensities of the first two secondary maxima to either side of the central maximum.

APPROACH The secondary maxima occur close to halfway between the minima, at about

$$\frac{\beta}{2} = \frac{\pi D \sin \theta}{\lambda} \approx (m + \tfrac{1}{2})\pi. \qquad m = 1, 2, 3, \cdots$$

The actual maxima are not quite at these points—their positions can be determined by differentiating Eq. 35–7 (see Problem 14)—but we are only seeking an estimate.

SOLUTION Using these values for β in Eq. 35–7 or 35–8, with $\sin(m + \tfrac{1}{2})\pi = 1$, gives

$$I_\theta = \frac{I_0}{(m + \tfrac{1}{2})^2 \pi^2}. \qquad m = 1, 2, 3, \cdots$$

For $m = 1$ and 2, we get

$$I_\theta = \frac{I_0}{22.2} = 0.045I_0 \qquad\qquad [m = 1]$$

$$I_\theta = \frac{I_0}{61.7} = 0.016I_0. \qquad\qquad [m = 2]$$

The first maximum to the side of the central peak has only 1/22, or 4.5%, the intensity of the central peak, and succeeding ones are smaller still, just as we can see in Fig. 35–4 and the photo of Fig. 35–2c.

Diffraction by a circular opening produces a similar pattern (though circular rather than rectangular) and is of great practical importance, since lenses are essentially circular apertures through which light passes. We will discuss this in Section 35–4 and see how diffraction limits the resolution (or sharpness) of images.

*35–3 Diffraction in the Double-Slit Experiment

When we analyzed Young's double-slit experiment in Section 34–4, we assumed that the central portion of the screen was uniformly illuminated. This is equivalent to assuming the slits are infinitesimally narrow, so that the central diffraction peak is spread out over the whole screen. This can never be the case for real slits; diffraction reduces the intensity of the bright interference fringes to the side of center so they are not all of the same height as they were shown in Fig. 34–14. (They were shown more correctly in Fig. 34–9b.)

To calculate the intensity in a double-slit interference pattern, including diffraction, let us assume the slits have equal widths D and their centers are separated by a distance d. Since the distance to the screen is large compared to the slit separation d, the wave amplitude due to each slit is essentially the same at each point on the screen. Then the total wave amplitude at any angle θ will no longer be

$$E_{\theta 0} = 2E_0 \cos\frac{\delta}{2},$$

as was given by Eq. 34–5b. Rather, it must be modified, because of diffraction, by Eq. 35–5, so that

$$E_{\theta 0} = 2E_0 \left(\frac{\sin\beta/2}{\beta/2}\right) \cos\frac{\delta}{2}.$$

Thus the intensity will be given by

$$I_\theta = I_0 \left(\frac{\sin\beta/2}{\beta/2}\right)^2 \left(\cos\frac{\delta}{2}\right)^2 \qquad\qquad \textbf{(35–9)}$$

where $I_0 = 4E_0^2$, and from Eqs. 35–6 and 34–4 we have

$$\frac{\beta}{2} = \frac{\pi}{\lambda}D\sin\theta \qquad \text{and} \qquad \frac{\delta}{2} = \frac{\pi}{\lambda}d\sin\theta.$$

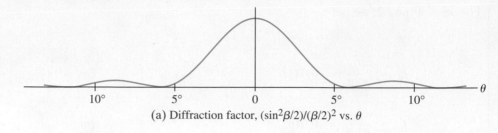

(a) Diffraction factor, $(\sin^2\beta/2)/(\beta/2)^2$ vs. θ

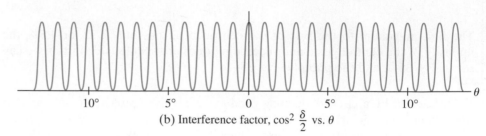

(b) Interference factor, $\cos^2\dfrac{\delta}{2}$ vs. θ

FIGURE 35–10
(a) Diffraction factor,
(b) interference factor, and
(c) the resultant intensity I_θ,
plotted as a function of θ
for $d = 6D = 60\lambda$.

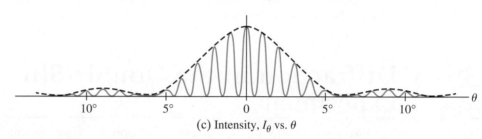

(c) Intensity, I_θ vs. θ

Equation 35–9 for the intensity in a double-slit pattern, as we just saw, is

$$I_\theta = I_0\left(\frac{\sin \beta/2}{\beta/2}\right)^2\left(\cos\frac{\delta}{2}\right)^2. \tag{35–9}$$

The first term in parentheses is sometimes called the "diffraction factor" and the second one the "interference factor." These two factors are plotted in Fig. 35–10a and b for the case when $d = 6D$, and $D = 10\lambda$. (Figure 35–10b is essentially the same as Fig. 34–14.) Figure 35–10c shows the product of these two curves (times I_0) which is the actual intensity as a function of θ (or as a function of position on the screen for θ not too large) as given by Eq. 35–9. As indicated by the dashed lines in Fig. 35–10c, the diffraction factor acts as a sort of envelope that limits the interference peaks.

FIGURE 35–11 Photographs of a double-slit interference pattern using a laser beam showing effects of diffraction. In both cases $d = 0.50$ mm, whereas $D = 0.040$ mm in (a) and 0.080 mm in (b).

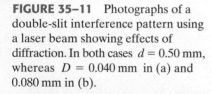

(a)

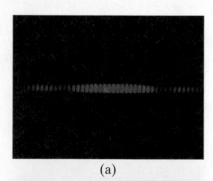

(b)

EXAMPLE 35–4 **Diffraction plus interference.** Show why the central diffraction peak in Fig. 35–10c contains 11 interference fringes.

APPROACH The first minimum in the diffraction pattern occurs where

$$\sin \theta = \frac{\lambda}{D}.$$

Since $d = 6D$,

$$d \sin \theta = 6D\left(\frac{\lambda}{D}\right) = 6\lambda.$$

SOLUTION From Eq. 34–2a, interference peaks (maxima) occur for $d \sin \theta = m\lambda$ where m can be $0, 1, \cdots$ or any integer. Thus the diffraction minimum ($d \sin \theta = 6\lambda$) coincides with $m = 6$ in the interference pattern, so the $m = 6$ peak won't appear. Hence the central diffraction peak encloses the central interference peak ($m = 0$) and five peaks ($m = 1$ to 5) on each side for a total of 11. Since the sixth order doesn't appear, it is said to be a "missing order."

Notice from Example 35–4 that the number of interference fringes in the central diffraction peak depends only on the ratio d/D. It does not depend on wavelength λ. The actual spacing (in angle, or in position on the screen) does depend on λ. For the case illustrated, $D = 10\lambda$, and so the first diffraction minimum occurs at $\sin \theta = \lambda/D = 0.10$ or about 6°.

The decrease in intensity of the interference fringes away from the center, as graphed in Fig. 35–10, is shown in Fig. 35–11.

Interference vs. Diffraction

The patterns due to interference and diffraction arise from the same phenomenon—the superposition of coherent waves of different phase. The distinction between them is thus not so much physical as for convenience of description, as in this Section where we analyzed the two-slit pattern in terms of interference and diffraction separately. In general, we use the word "diffraction" when referring to an analysis by superposition of many infinitesimal and usually contiguous sources, such as when we subdivide a source into infinitesimal parts. We use the term "interference" when we superpose the wave from a finite (and usually small) number of coherent sources.

35–4 Limits of Resolution; Circular Apertures

The ability of a lens to produce distinct images of two point objects very close together is called the **resolution** of the lens. The closer the two images can be and still be seen as distinct (rather than overlapping blobs), the higher the resolution. The resolution of a camera lens, for example, is often specified as so many dots or lines per millimeter, as mentioned in Section 33–5.

Two principal factors limit the resolution of a lens. The first is lens aberrations. As we saw in Chapter 33, because of spherical and other aberrations, a point object is not a point on the image but a tiny blob. Careful design of compound lenses can reduce aberrations significantly, but they cannot be eliminated entirely. The second factor that limits resolution is *diffraction*, which cannot be corrected for because it is a natural result of the wave nature of light. We discuss it now.

In Section 35–1, we saw that because light travels as a wave, light from a point source passing through a slit is spread out into a diffraction pattern (Figs. 35–2 and 35–4). A lens, because it has edges, acts like a round slit. When a lens forms the image of a point object, the image is actually a tiny diffraction pattern. Thus *an image would be blurred even if aberrations were absent.*

In the analysis that follows, we assume that the lens is free of aberrations, so we can concentrate on diffraction effects and how much they limit the resolution of a lens. In Fig. 35–4 we saw that the diffraction pattern produced by light passing through a rectangular slit has a central maximum in which most of the light falls. This central peak falls to a minimum on either side of its center at an angle θ given by $\sin \theta = \lambda/D$ (this is Eq. 35–1), where D is the slit width and λ the wavelength of light used. θ is the angular half-width of the central maximum, and for small angles can be written

$$\theta \approx \sin \theta = \frac{\lambda}{D}.$$

There are also low-intensity fringes beyond.

For a lens, or any circular hole, the image of a point object will consist of a *circular* central peak (called the *diffraction spot* or *Airy disk*) surrounded by faint circular fringes, as shown in Fig. 35–12a. The central maximum has an angular half width given by

$$\theta = \frac{1.22\lambda}{D},$$

where D is the diameter of the circular opening. This is a theoretical result for a perfect circle or lens. For real lenses or circles, the factor is on the order of 1 to 2. This formula differs from that for a slit (Eq. 35–1) by the factor 1.22. This factor appears because the width of a circular hole is not uniform (like a rectangular slit) but varies from its diameter D to zero. A mathematical analysis shows that the "average" width is $D/1.22$. Hence we get the equation above rather than Eq. 35–1. The intensity of light in the diffraction pattern from a point source of light passing through a circular opening is shown in Fig. 35–13. The image for a non-point source is a superposition of such patterns. For most purposes we need consider only the central spot, since the concentric rings are so much dimmer.

If two point objects are very close, the diffraction patterns of their images will overlap as shown in Fig. 35–12b. As the objects are moved closer, a separation is

(a)

(b)

FIGURE 35–12 Photographs of images (greatly magnified) formed by a lens, showing the diffraction pattern of an image for: (a) a single point object; (b) two point objects whose images are barely resolved.

FIGURE 35–13 Intensity of light across the diffraction pattern of a circular hole.

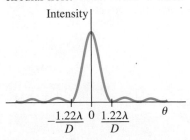

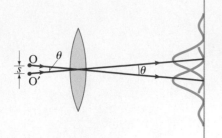

FIGURE 35–14 The *Rayleigh criterion*. Two images are just resolvable when the center of the diffraction peak of one is directly over the first minimum in the diffraction pattern of the other. The two point objects O and O' subtend an angle θ at the lens; only one ray (it passes through the center of the lens) is drawn for each object, to indicate the center of the diffraction pattern of its image.

FIGURE 35–15 Hubble Space Telescope, with Earth in the background. The flat orange panels are solar cells that collect energy from the Sun.

(🏃) PHYSICS APPLIED
How well the eye can see

reached where you can't tell if there are two overlapping images or a single image. The separation at which this happens may be judged differently by different observers. However, a generally accepted criterion is that proposed by Lord Rayleigh (1842–1919). This **Rayleigh criterion** states that *two images are just resolvable when the center of the diffraction disk of one image is directly over the first minimum in the diffraction pattern of the other*. This is shown in Fig. 35–14. Since the first minimum is at an angle $\theta = 1.22\lambda/D$ from the central maximum, Fig. 35–14 shows that two objects can be considered *just resolvable* if they are separated by at least an angle θ given by

$$\theta = \frac{1.22\lambda}{D}. \qquad \text{[}\theta \text{ in radians]} \quad \textbf{(35–10)}$$

In this equation, D is the diameter of the lens, and applies also to a mirror diameter. This is the limit on resolution set by the wave nature of light due to diffraction. A smaller angle means better resolution: you can make out closer objects. We see from Eq. 35–10 that using a shorter wavelength λ can reduce θ and thus increase resolution.

EXERCISE B Green light (550 nm) passes through a 25-mm-diameter camera lens. What is the angular half-width of the resulting diffraction pattern? (*a*) 2.7×10^{-5} degrees, (*b*) 1.5×10^{-3} degrees, (*c*) $3.2°$, (*d*) $27°$, (*e*) 1.5×10^3 degrees.

EXAMPLE 35–5 **Hubble Space Telescope.** The Hubble Space Telescope (HST) is a reflecting telescope that was placed in orbit above the Earth's atmosphere, so its resolution would not be limited by turbulence in the atmosphere (Fig. 35–15). Its objective diameter is 2.4 m. For visible light, say $\lambda = 550$ nm, estimate the improvement in resolution the Hubble offers over Earth-bound telescopes, which are limited in resolution by movement of the Earth's atmosphere to about half an arc second. (Each degree is divided into 60 minutes each containing 60 seconds, so $1° = 3600$ arc seconds.)

APPROACH Angular resolution for the Hubble is given (in radians) by Eq. 35–10. The resolution for Earth telescopes is given, and we first convert it to radians so we can compare.

SOLUTION Earth-bound telescopes are limited to an angular resolution of

$$\theta = \frac{1}{2}\left(\frac{1}{3600}\right)^{°}\left(\frac{2\pi \text{ rad}}{360°}\right) = 2.4 \times 10^{-6} \text{ rad}.$$

The Hubble, on the other hand, is limited by diffraction (Eq. 35–10) which for $\lambda = 550$ nm is

$$\theta = \frac{1.22\lambda}{D} = \frac{1.22(550 \times 10^{-9} \text{ m})}{2.4 \text{ m}} = 2.8 \times 10^{-7} \text{ rad},$$

thus giving almost ten times better resolution ($2.4 \times 10^{-6} \text{ rad}/2.8 \times 10^{-7} \text{ rad} \approx 9\times$).

EXAMPLE 35–6 **ESTIMATE** **Eye resolution.** You are in an airplane at an altitude of 10,000 m. If you look down at the ground, estimate the minimum separation s between objects that you could distinguish. Could you count cars in a parking lot? Consider only diffraction, and assume your pupil is about 3.0 mm in diameter and $\lambda = 550$ nm.

APPROACH We use the Rayleigh criterion, Eq. 35–10, to estimate θ. The separation s of objects is $s = \ell\theta$, where $\ell = 10^4$ m and θ is in radians.

SOLUTION In Eq. 35–10, we set $D = 3.0$ mm for the opening of the eye:

$$s = \ell\theta = \ell\frac{1.22\lambda}{D} = \frac{(10^4 \text{ m})(1.22)(550 \times 10^{-9} \text{ m})}{3.0 \times 10^{-3} \text{ m}} = 2.2 \text{ m}.$$

Yes, you could just resolve a car (roughly 2 m wide by 3 or 4 m long) and count them.

35–5 Resolution of Telescopes and Microscopes; the λ Limit

You might think that a microscope or telescope could be designed to produce any desired magnification, depending on the choice of focal lengths and quality of the lenses. But this is not possible, because of diffraction. An increase in magnification above a certain point merely results in magnification of the diffraction patterns. This can be highly misleading since we might think we are seeing details of an object when we are really seeing details of the diffraction pattern. To examine this problem, we apply the Rayleigh criterion: two objects (or two nearby points on one object) are just resolvable if they are separated by an angle θ (Fig. 35–14) given by Eq. 35–10:

$$\theta = \frac{1.22\lambda}{D}.$$

This formula is valid for either a microscope or a telescope, where D is the diameter of the objective lens or mirror. For a telescope, the resolution is specified by stating θ as given by this equation.[†]

FIGURE 35–16 The 300-meter radiotelescope in Arecibo, Puerto Rico, uses radio waves (Fig. 31–12) instead of visible light.

EXAMPLE 35–7 **Telescope resolution (radio wave vs. visible light).** What is the theoretical minimum angular separation of two stars that can just be resolved by (a) the 200-inch telescope on Palomar Mountain (Fig. 33–38c); and (b) the Arecibo radiotelescope (Fig. 35–16), whose diameter is 300 m and whose radius of curvature is also 300 m. Assume $\lambda = 550\ nm$ for the visible-light telescope in part (a), and $\lambda = 4\ cm$ (the shortest wavelength at which the radiotelescope has been operated) in part (b).

APPROACH We apply the Rayleigh criterion (Eq. 35–10) for each telescope.

SOLUTION (a) Since $D = 200\ in. = 5.1\ m,$ we have from Eq. 35–10 that

$$\theta = \frac{1.22\lambda}{D} = \frac{(1.22)(5.50 \times 10^{-7}\ m)}{(5.1\ m)} = 1.3 \times 10^{-7}\ rad,$$

or 0.75×10^{-5} deg. (Note that this is equivalent to resolving two points less than 1 cm apart from a distance of 100 km!)

(b) For radio waves with $\lambda = 0.04\ m$ emitted by stars, the resolution is

$$\theta = \frac{(1.22)(0.04\ m)}{(300\ m)} = 1.6 \times 10^{-4}\ rad.$$

The resolution is less because the wavelength is so much larger, but the larger objective collects more radiation and thus detects fainter objects.

NOTE In both cases, we determined the limit set by diffraction. The resolution for a visible-light Earth-bound telescope is not this good because of aberrations and, more importantly, turbulence in the atmosphere. In fact, large-diameter objectives are not justified by increased resolution, but by their greater light-gathering ability—they allow more light in, so fainter objects can be seen. Radiotelescopes are not hindered by atmospheric turbulence, and the resolution found in (b) is a good estimate.

[†]Earth-bound telescopes with large-diameter objectives are usually limited not by diffraction but by other effects such as turbulence in the atmosphere. The resolution of a high-quality microscope, on the other hand, normally *is* limited by diffraction; microscope objectives are complex compound lenses containing many elements of small diameter (since f is small), thus reducing aberrations.

For a microscope, it is more convenient to specify the actual distance, s, between two points that are just barely resolvable: see Fig. 35–14. Since objects are normally placed near the focal point of the microscope objective, the angle subtended by two objects is $\theta = s/f$, so $s = f\theta$. If we combine this with Eq. 35–10, we obtain the **resolving power** (**RP**) of a microscope

$$\text{RP} = s = f\theta = \frac{1.22\lambda f}{D}, \tag{35–11}$$

where f is the objective lens' focal length (not frequency). This distance s is called the resolving power of the lens because it is the minimum separation of two object points that can just be resolved — assuming the highest quality lens since this limit is imposed by the wave nature of light. A smaller RP means better resolution, better detail.

EXERCISE C What is the resolving power of a microscope with a 5-mm-diameter objective which has $f = 9$ mm? (a) 550 nm, (b) 750 nm, (c) 1200 nm, (d) 0.05 nm, (e) 0.005 nm.

Diffraction sets an ultimate limit on the detail that can be seen on any object. In Eq. 35–11 for resolving power of a microscope, the focal length of the lens cannot practically be made less than (approximately) the radius of the lens, and even that is very difficult (see the lensmaker's equation, Eq. 33–4). In this best case, Eq. 35–11 gives, with $f \approx D/2$,

$$\text{RP} \approx \frac{\lambda}{2}. \tag{35–12}$$

Thus we can say, to within a factor of 2 or so, that

it is not possible to resolve detail of objects smaller than the wavelength of the radiation being used.

This is an important and useful rule of thumb.

Compound lenses in microscopes are now designed so well that the actual limit on resolution is often set by diffraction — that is, by the wavelength of the light used. To obtain greater detail, one must use radiation of shorter wavelength. The use of UV radiation can increase the resolution by a factor of perhaps 2. Far more important, however, was the discovery in the early twentieth century that electrons have wave properties (Chapter 37) and that their wavelengths can be very small. The wave nature of electrons is utilized in the electron microscope (Section 37–8), which can magnify 100 to 1000 times more than a visible-light microscope because of the much shorter wavelengths. X-rays, too, have very short wavelengths and are often used to study objects in great detail (Section 35–10).

*35–6 Resolution of the Human Eye and Useful Magnification

The resolution of the human eye is limited by several factors, all of roughly the same order of magnitude. The resolution is best at the fovea, where the cone spacing is smallest, about $3\,\mu\text{m}$ ($= 3000$ nm). The diameter of the pupil varies from about 0.1 cm to about 0.8 cm. So for $\lambda = 550$ nm (where the eye's sensitivity is greatest), the diffraction limit is about $\theta \approx 1.22\lambda/D \approx 8 \times 10^{-5}$ rad to 6×10^{-4} rad. The eye is about 2 cm long, giving a resolving power (Eq. 35–11) of $s \approx (2 \times 10^{-2}\,\text{m})(8 \times 10^{-5}\,\text{rad}) \approx 2\,\mu\text{m}$ at best, to about $10\,\mu\text{m}$ at worst (pupil small). Spherical and chromatic aberration also limit the resolution to about $10\,\mu\text{m}$. The net result is that the eye can just resolve objects whose angular separation is around

$$5 \times 10^{-4}\,\text{rad}. \qquad \left[\begin{array}{c}\text{best eye}\\\text{resolution}\end{array}\right]$$

This corresponds to objects separated by 1 cm at a distance of about 20 m.

The typical near point of a human eye is about 25 cm. At this distance, the eye can just resolve objects that are $(25\,\text{cm})(5 \times 10^{-4}\,\text{rad}) \approx 10^{-4}\,\text{m} = \frac{1}{10}$ mm apart. Since the best light microscopes can resolve objects no smaller than about 200 nm at best (Eq. 35–12 for violet light, $\lambda = 400$ nm), the useful magnification

[= (resolution by naked eye)/(resolution by microscope)] is limited to about

$$\frac{10^{-4}\,\text{m}}{200 \times 10^{-9}\,\text{m}} \approx 500\times. \qquad \left[\begin{array}{l}\text{maximum useful}\\ \text{microscope magnification}\end{array}\right]$$

In practice, magnifications of about 1000× are often used to minimize eyestrain. Any greater magnification would simply make visible the diffraction pattern produced by the microscope objective lens.

Now you have the answers to the Chapter-Opening Question: (c), by the equation above, and (c) by the λ rule.

35–7 Diffraction Grating

A large number of equally spaced parallel slits is called a **diffraction grating**, although the term "interference grating" might be as appropriate. Gratings can be made by precision machining of very fine parallel lines on a glass plate. The untouched spaces between the lines serve as the slits. Photographic transparencies of an original grating serve as inexpensive gratings. Gratings containing 10,000 lines per centimeter are common, and are very useful for precise measurements of wavelengths. A diffraction grating containing slits is called a **transmission grating**. Another type of diffraction grating is the **reflection grating**, made by ruling fine lines on a metallic or glass surface from which light is reflected and analyzed. The analysis is basically the same as for a transmission grating, which we now discuss.

The analysis of a diffraction grating is much like that of Young's double-slit experiment. We assume parallel rays of light are incident on the grating as shown in Fig. 35–17. We also assume that the slits are narrow enough so that diffraction by each of them spreads light over a very wide angle on a distant screen beyond the grating, and interference can occur with light from all the other slits. Light rays that pass through each slit without deviation ($\theta = 0°$) interfere constructively to produce a bright line at the center of the screen. Constructive interference also occurs at an angle θ such that rays from adjacent slits travel an extra distance of $\Delta\ell = m\lambda$, where m is an integer. If d is the distance between slits, then we see from Fig. 35–17 that $\Delta\ell = d\sin\theta$, and

$$\sin\theta = \frac{m\lambda}{d}, \qquad m = 0, 1, 2, \cdots \qquad \left[\begin{array}{l}\text{diffraction grating,}\\ \text{principal maxima}\end{array}\right] \quad \textbf{(35–13)}$$

is the criterion to have a brightness maximum. This is the same equation as for the double-slit situation, and again m is called the order of the pattern.

There is an important difference between a double-slit and a multiple-slit pattern. The bright maxima are much *sharper* and *narrower* for a grating. Why? Suppose that the angle θ is increased just slightly beyond that required for a maximum. In the case of only two slits, the two waves will be only slightly out of phase, so nearly full constructive interference occurs. This means the maxima are wide (see Fig. 34–9). For a grating, the waves from two adjacent slits will also not be significantly out of phase. But waves from one slit and those from a second one a few hundred slits away may be exactly out of phase; all or nearly all the light can cancel in pairs in this way. For example, suppose the angle θ is very slightly different from its first-order maximum, so that the extra path length for a pair of adjacent slits is not exactly λ but rather 1.0010λ. The wave through one slit and another one 500 slits below will have a path difference of $1\lambda + (500)(0.0010\lambda) = 1.5000\lambda$, or $1\frac{1}{2}$ wavelengths, so the two will cancel. A pair of slits, one below each of these, will also cancel. That is, the light from slit 1 cancels with that from slit 501; light from slit 2 cancels with that from slit 502, and so on. Thus even for a tiny angle[†] corresponding to an extra path length of $\frac{1}{1000}\lambda$, there is much destructive interference, and so the maxima are very narrow. The more lines there are in a grating, the sharper will be the peaks (see Fig. 35–18). Because a grating produces much sharper lines than two slits alone can (and much brighter lines because there are many more slits), a grating is a far more precise device for measuring wavelengths.

[†]Depending on the total number of slits, there may or may not be complete cancellation for such an angle, so there will be very tiny peaks between the main maxima (see Fig. 35–18b), but they are usually much too small to be seen.

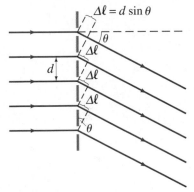

FIGURE 35–17 Diffraction grating.

⚠ **CAUTION**

Diffraction grating is analyzed using interference formulas, not diffraction formulas

FIGURE 35–18 Intensity as a function of viewing angle θ (or position on the screen) for (a) two slits, (b) six slits. For a diffraction grating, the number of slits is very large ($\approx 10^4$) and the peaks are narrower still.

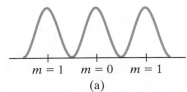

$m = 1$ $m = 0$ $m = 1$
(a)

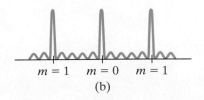

$m = 1$ $m = 0$ $m = 1$
(b)

Suppose the light striking a diffraction grating is not monochromatic, but consists of two or more distinct wavelengths. Then for all orders other than $m = 0$, each wavelength will produce a maximum at a different angle (Fig. 35–19a), just as for a double slit. If white light strikes a grating, the central ($m = 0$) maximum will be a sharp white peak. But for all other orders, there will be a distinct spectrum of colors spread out over a certain angular width, Fig. 35–19b. Because a diffraction grating spreads out light into its component wavelengths, the resulting pattern is called a **spectrum**.

FIGURE 35–19 Spectra produced by a grating: (a) two wavelengths, 400 nm and 700 nm; (b) white light. The second order will normally be dimmer than the first order. (Higher orders are not shown.) If grating spacing is small enough, the second and higher orders will be missing.

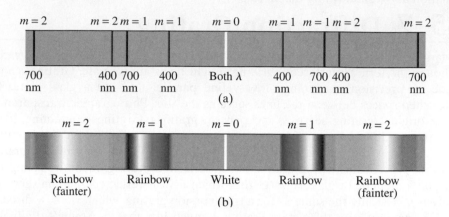

EXAMPLE 35–8 **Diffraction grating: lines.** Determine the angular positions of the first- and second-order maxima for light of wavelength 400 nm and 700 nm incident on a grating containing 10,000 lines/cm.

APPROACH First we find the distance d between grating lines: if the grating has N lines in 1 m, then the distance between lines must be $d = 1/N$ meters. Then we use Eq. 35–13, $\sin \theta = m\lambda/d$, to find the angles for the two wavelengths for $m = 1$ and 2.

SOLUTION The grating contains 1.00×10^4 lines/cm $= 1.00 \times 10^6$ lines/m, which means the distance between lines is $d = (1/1.00 \times 10^6)$ m $= 1.00 \times 10^{-6}$ m $= 1.00 \, \mu$m. In first order ($m = 1$), the angles are

$$\sin \theta_{400} = \frac{m\lambda}{d} = \frac{(1)(4.00 \times 10^{-7} \, \text{m})}{1.00 \times 10^{-6} \, \text{m}} = 0.400$$

$$\sin \theta_{700} = \frac{(1)(7.00 \times 10^{-7} \, \text{m})}{1.00 \times 10^{-6} \, \text{m}} = 0.700$$

so $\theta_{400} = 23.6°$ and $\theta_{700} = 44.4°$. In second order,

$$\sin \theta_{400} = \frac{2\lambda}{d} = \frac{(2)(4.00 \times 10^{-7} \, \text{m})}{1.00 \times 10^{-6} \, \text{m}} = 0.800$$

$$\sin \theta_{700} = \frac{(2)(7.00 \times 10^{-7} \, \text{m})}{1.00 \times 10^{-6} \, \text{m}} = 1.40$$

so $\theta_{400} = 53.1°$. But the second order does not exist for $\lambda = 700$ nm because $\sin \theta$ cannot exceed 1. No higher orders will appear.

EXAMPLE 35–9 **Spectra overlap.** White light containing wavelengths from 400 nm to 750 nm strikes a grating containing 4000 lines/cm. Show that the blue at $\lambda = 450$ nm of the third-order spectrum overlaps the red at 700 nm of the second order.

APPROACH We use $\sin \theta = m\lambda/d$ to calculate the angular positions of the $m = 3$ blue maximum and the $m = 2$ red one.

SOLUTION The grating spacing is $d = (1/4000)$ cm $= 2.50 \times 10^{-6}$ m. The blue of the third order occurs at an angle θ given by

$$\sin\theta = \frac{m\lambda}{d} = \frac{(3)(4.50 \times 10^{-7}\,\text{m})}{(2.50 \times 10^{-6}\,\text{m})} = 0.540.$$

Red in second order occurs at

$$\sin\theta = \frac{(2)(7.00 \times 10^{-7}\,\text{m})}{(2.50 \times 10^{-6}\,\text{m})} = 0.560,$$

which is a greater angle; so the second order overlaps into the beginning of the third-order spectrum.

CONCEPTUAL EXAMPLE 35–10 **Compact disk.** When you look at the surface of a music CD (Fig. 35–20), you see the colors of a rainbow. (*a*) Estimate the distance between the curved lines (they are read by a laser). (*b*) Estimate the distance between lines, noting that a CD contains at most 80 min of music, that it rotates at speeds from 200 to 500 rev/min, and that $\frac{2}{3}$ of its 6 cm radius contains the lines.

RESPONSE (*a*) The CD acts like a reflection diffraction grating. To satisfy Eq. 35–13, we might estimate the line spacing as one or a few (2 or 3) wavelengths ($\lambda \approx 550$ nm) or 0.5 to 1.5 μm. (*b*) Average rotation speed of 350 rev/min times 80 min gives 28,000 total rotations or 28,000 lines, which are spread over $(\frac{2}{3})(6\,\text{cm}) = 4$ cm. So we have a sort of reflection diffraction grating with about $(28{,}000\,\text{lines})/(4\,\text{cm}) = 7000$ lines/cm. The distance d between lines is roughly $1\,\text{cm}/7000\,\text{lines} \approx 1.4 \times 10^{-6}$ m $= 1.4\,\mu$m. Our results in (*a*) and (*b*) agree.

FIGURE 35–20 A compact disk, Example 35–10.

35–8 The Spectrometer and Spectroscopy

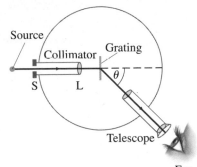

FIGURE 35–21 Spectrometer or spectroscope.

A **spectrometer** or **spectroscope**, Fig. 35–21, is a device to measure wavelengths accurately using a diffraction grating (or a prism) to separate different wavelengths of light. Light from a source passes through a narrow slit S in the "collimator." The slit is at the focal point of the lens L, so parallel light falls on the grating. The movable telescope can bring the rays to a focus. Nothing will be seen in the viewing telescope unless it is positioned at an angle θ that corresponds to a diffraction peak (first order is usually used) of a wavelength emitted by the source. The angle θ can be measured to very high accuracy, so the wavelength of a line can be determined to high accuracy using Eq. 35–13:

$$\lambda = \frac{d}{m}\sin\theta,$$

where m is an integer representing the order, and d is the distance between grating lines. The line you see in a spectrometer corresponding to each wavelength is actually an image of the slit S. A narrower slit results in dimmer light but we can measure the angular positions more precisely. If the light contains a continuous range of wavelengths, then a continuous spectrum is seen in the spectroscope.

The spectrometer in Fig. 35–21 uses a transmission grating. Others may use a reflection grating, or sometimes a prism. A prism works because of dispersion (Section 32–6), bending light of different wavelengths into different angles. (A prism is not a linear device and must be calibrated.)

An important use of a spectrometer is for the identification of atoms or molecules. When a gas is heated or an electric current is passed through it, the gas emits a characteristic **line spectrum**. That is, only certain discrete wavelengths of light are emitted, and these are different for different elements and compounds.[†] Figure 35–22 shows the line spectra for a number of elements in the gas state. Line spectra occur only for gases at high temperatures and low pressure and density. The light from heated solids, such as a lightbulb filament, and even from a dense gaseous object such as the Sun, produces a **continuous spectrum** including a wide range of wavelengths.

[†]Why atoms and molecules emit line spectra was a great mystery for many years and played a central role in the development of modern quantum theory, as we shall see in Chapter 37.

Atomic hydrogen

Mercury

Sodium

Solar absorption spectrum

FIGURE 35–22 Line spectra for the gases indicated, and the spectrum from the Sun showing absorption lines.

Figure 35–22 also shows the Sun's "continuous spectrum," which contains a number of *dark* lines (only the most prominent are shown), called **absorption lines**. Atoms and molecules can absorb light at the same wavelengths at which they emit light. The Sun's absorption lines are due to absorption by atoms and molecules in the cooler outer atmosphere of the Sun, as well as by atoms and molecules in the Earth's atmosphere. A careful analysis of all these thousands of lines reveals that at least two-thirds of all elements are present in the Sun's atmosphere. The presence of elements in the atmosphere of other planets, in interstellar space, and in stars, is also determined by spectroscopy.

Spectroscopy is useful for determining the presence of certain types of molecules in laboratory specimens where chemical analysis would be difficult. For example, biological DNA and different types of protein absorb light in particular regions of the spectrum (such as in the UV). The material to be examined, which is often in solution, is placed in a monochromatic light beam whose wavelength is selected by the placement angle of a diffraction grating or prism. The amount of absorption, as compared to a standard solution without the specimen, can reveal not only the presence of a particular type of molecule, but also its concentration.

Light emission and absorption also occur outside the visible part of the spectrum, such as in the UV and IR regions. Glass absorbs light in these regions, so reflection gratings and mirrors (in place of lenses) are used. Special types of film or sensors are used for detection.

(⊛) PHYSICS APPLIED
Chemical and biochemical analysis by spectroscopy

EXAMPLE 35–11 **Hydrogen spectrum.** Light emitted by hot hydrogen gas is observed with a spectroscope using a diffraction grating having 1.00×10^4 lines/cm. The spectral lines nearest to the center $(0°)$ are a violet line at 24.2°, a blue line at 25.7°, a blue-green line at 29.1°, and a red line at 41.0° from the center. What are the wavelengths of these spectral lines of hydrogen?

APPROACH The wavelengths can be determined from the angles by using $\lambda = (d/m) \sin\theta$ where d is the spacing between slits, and m is the order of the spectrum (Eq. 35–13).

SOLUTION Since these are the closest lines to $\theta = 0°$, this is the first-order spectrum $(m = 1)$. The slit spacing is $d = 1/(1.00 \times 10^4\,\text{cm}^{-1}) = 1.00 \times 10^{-6}\,\text{m}$. The wavelength of the violet line is

$$\lambda = \left(\frac{d}{m}\right)\sin\theta = \left(\frac{1.00 \times 10^{-6}\,\text{m}}{1}\right)\sin 24.2° = 4.10 \times 10^{-7}\,\text{m} = 410\,\text{nm}.$$

The other wavelengths are:

blue: $\quad\quad \lambda = (1.00 \times 10^{-6}\,\text{m})\sin 25.7° = 434\,\text{nm},$

blue-green: $\quad \lambda = (1.00 \times 10^{-6}\,\text{m})\sin 29.1° = 486\,\text{nm},$

red: $\quad\quad\, \lambda = (1.00 \times 10^{-6}\,\text{m})\sin 41.0° = 656\,\text{nm}.$

NOTE In an unknown mixture of gases, these four spectral lines need to be seen to identify that the mixture contains hydrogen.

*35–9 Peak Widths and Resolving Power for a Diffraction Grating

We now look at the pattern of maxima produced by a multiple-slit grating using phasor diagrams. We can determine a formula for the width of each peak, and we will see why there are tiny maxima between the principal maxima, as indicated in Fig. 35–18b. First of all, it should be noted that the two-slit and six-slit patterns shown in Fig. 35–18 were drawn assuming very narrow slits so that diffraction does not limit the height of the peaks. For real diffraction gratings, this is not normally the case: the slit width D is often not much smaller than the slit separation d, and diffraction thus limits the intensity of the peaks so the central peak $(m = 0)$ is brighter than the side peaks. We won't worry about this effect on intensity except to note that if a diffraction minimum coincides with a particular order of the interference pattern, that order will not appear. (For example, if $d = 2D$, all the even orders, $m = 2, 4, \cdots$, will be missing. Can you see why? Hint: See Example 35–4.)

Figures 35–23 and 35–24 show phasor diagrams for a two-slit and a six-slit grating, respectively. Each short arrow represents the amplitude of a wave from a single slit, and their vector sum (as phasors) represents the total amplitude for a given viewing angle θ. Part (a) of each Figure shows the phasor diagram at $\theta = 0°$, at the center of the pattern, which is the central maximum $(m = 0)$. Part (b) of each Figure shows the condition for the adjacent minimum: where the arrows first close on themselves (add to zero) so the amplitude E_θ is zero. For two slits, this occurs when the two separate amplitudes are 180° out of phase. For six slits, it occurs when each amplitude makes a 60° angle with its neighbor. For two slits, the minimum occurs when the phase between slits is $\delta = 2\pi/2$ (in radians); for six slits it occurs when the phase δ is $2\pi/6$; and in the general case of N slits, the minimum occurs for a phase difference between adjacent slits of

$$\delta = \frac{2\pi}{N}. \qquad (35–14)$$

What does this correspond to in θ? First note that δ is related to θ by (Eq. 34–4)

$$\frac{\delta}{2\pi} = \frac{d \sin \theta}{\lambda} \qquad \text{or} \qquad \delta = \frac{2\pi}{\lambda} d \sin \theta. \qquad (35–15)$$

Let us call $\Delta\theta_0$ the angular position of the minimum next to the peak at $\theta = 0$. Then

$$\frac{\delta}{2\pi} = \frac{d \sin \Delta\theta_0}{\lambda}.$$

We insert Eq. 35–14 for δ and find

$$\sin \Delta\theta_0 = \frac{\lambda}{Nd}. \qquad (35–16a)$$

Since $\Delta\theta_0$ is usually small (N is usually very large for a grating), $\sin \Delta\theta_0 \approx \Delta\theta_0$, so in the small angle limit we can write

$$\Delta\theta_0 = \frac{\lambda}{Nd}. \qquad (35–16b)$$

It is clear from either of the last two relations that the larger N is, the narrower will be the central peak. (For $N = 2$, $\sin \Delta\theta_0 = \lambda/2d$, which is what we obtained earlier for the double slit, Eq. 34–2b with $m = 0$.)

Either of Eqs. 35–16 shows why the peaks become narrower for larger N. The origin of the small secondary maxima between the principal peaks (see Fig. 35–18b) can be deduced from the diagram of Fig. 35–25. This is just a continuation of Fig. 35–24b (where $\delta = 60°$); but now the phase δ has been increased to almost 90°, where E_θ is a relative maximum. Note that E_θ is much less than E_0 (Fig. 35–24a), so the intensity in this secondary maximum is much smaller than in a principal peak. As δ (and θ) is increased further, E_θ again decreases to zero (a "double circle"), then reaches another tiny maximum, and so on. Eventually the diagram unfolds again and when $\delta = 360°$, all the amplitudes again lie in a straight line (as in Fig. 35–24a) corresponding to the next principal maximum ($m = 1$ in Eq. 35–13).

(a) Central maximum: $\theta = 0$, $\delta = 0$

$E = 0$

$\delta = 180°$

(b) Minimum: $\delta = 180°$

FIGURE 35–23 Phasor diagram for two slits (a) at the central maximum, (b) at the nearest minimum.

FIGURE 35–24 Phasor diagram for six slits (a) at the central maximum, (b) at the nearest minimum.

E_0

(a) Central maximum: $\theta = 0$, $\delta = 0$

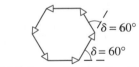

$\delta = 60°$

$\delta = 60°$

(b) Minimum: $\delta = 60°$, $E_\theta = 0$

FIGURE 35–25 Phasor diagram for the secondary peak.

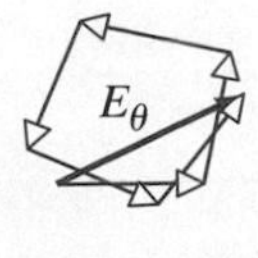

E_θ

Equation 35–16 b gives the half-width of the central $(m = 0)$ peak. To determine the half-width of higher order peaks, $\Delta\theta_m$ for order m, we differentiate Eq. 35–15 so as to relate the change $\Delta\delta$ in δ, to the change $\Delta\theta$ in the angle θ:

$$\Delta\delta \approx \frac{d\delta}{d\theta}\Delta\theta = \frac{2\pi d}{\lambda}\cos\theta\,\Delta\theta.$$

If $\Delta\theta_m$ represents the half-width of a peak of order m $(m = 1, 2, \cdots)$—that is, the angle between the peak maximum and the minimum to either side—then $\Delta\delta = 2\pi/N$ as given by Eq. 35–14. We insert this into the above relation and find

$$\Delta\theta_m = \frac{\lambda}{Nd\cos\theta_m}, \tag{35–17}$$

where θ_m is the angular position of the mth peak as given by Eq. 35–13. This derivation is valid, of course, only for small $\Delta\delta$ $(= 2\pi/N)$ which is indeed the case for real gratings since N is on the order of 10^4 or more.

An important property of any diffraction grating used in a spectrometer is its ability to resolve two very closely spaced wavelengths (wavelength difference $= \Delta\lambda$). The **resolving power** R of a grating is defined as

$$R = \frac{\lambda}{\Delta\lambda}. \tag{35–18}$$

With a little work, using Eq. 35–17, we can show that $\Delta\lambda = \lambda/Nm$ where N is the total number of grating lines and m is the order. Then we have

$$R = \frac{\lambda}{\Delta\lambda} = Nm. \tag{35–19}$$

The larger the value of R, the closer two wavelengths can be resolvable. If R is given, the minimum separation $\Delta\lambda$ between two wavelengths near λ, is (by Eq. 35–18)

$$\Delta\lambda = \frac{\lambda}{R}.$$

EXAMPLE 35–12 **Resolving two close lines.** Yellow sodium light, which consists of two wavelengths, $\lambda_1 = 589.00\,\text{nm}$ and $\lambda_2 = 589.59\,\text{nm}$, falls on a 7500-line/cm diffraction grating. Determine (a) the maximum order m that will be present for sodium light, (b) the width of grating necessary to resolve the two sodium lines.

APPROACH We first find $d = 1\,\text{cm}/7500 = 1.33 \times 10^{-6}\,\text{m}$, and then use Eq. 35–13 to find m. For (b) we use Eqs. 35–18 and 35–19.

SOLUTION (a) The maximum value of m at $\lambda = 589\,\text{nm}$, using Eq. 35–13 with $\sin\theta \leq 1$, is

$$m = \frac{d}{\lambda}\sin\theta \leq \frac{d}{\lambda} = \frac{1.33 \times 10^{-6}\,\text{m}}{5.89 \times 10^{-7}\,\text{m}} = 2.26,$$

so $m = 2$ is the maximum order present.

(b) The resolving power needed is

$$R = \frac{\lambda}{\Delta\lambda} = \frac{589\,\text{nm}}{0.59\,\text{nm}} = 1000.$$

From Eq. 35–19, the total number N of lines needed for the $m = 2$ order is $N = R/m = 1000/2 = 500$, so the grating need only be $500/7500\,\text{cm}^{-1} = 0.0667\,\text{cm}$ wide. A typical grating is a few centimeters wide, and so will easily resolve the two lines.

35–10 X-Rays and X-Ray Diffraction

In 1895, W. C. Roentgen (1845–1923) discovered that when electrons were accelerated by a high voltage in a vacuum tube and allowed to strike a glass or metal surface inside the tube, fluorescent minerals some distance away would glow, and photographic film would become exposed. Roentgen attributed these effects to a new type of radiation (different from cathode rays). They were given the name **X-rays** after the algebraic symbol x, meaning an unknown quantity. He soon found that X-rays penetrated through some materials better than through others, and within a few weeks he presented the first X-ray photograph (of his wife's hand). The production of X-rays today is usually done in a tube (Fig. 35–26) similar to Roentgen's, using voltages of typically 30 kV to 150 kV.

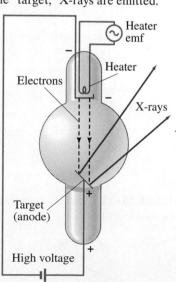

FIGURE 35–26 X-ray tube. Electrons emitted by a heated filament in a vacuum tube are accelerated by a high voltage. When they strike the surface of the anode, the "target," X-rays are emitted.

Investigations into the nature of X-rays indicated they were not charged particles (such as electrons) since they could not be deflected by electric or magnetic fields. It was suggested that they might be a form of invisible light. However, they showed no diffraction or interference effects using ordinary gratings. Indeed, if their wavelengths were much smaller than the typical grating spacing of 10^{-6} m $(= 10^3$ nm), no effects would be expected. Around 1912, Max von Laue (1879–1960) suggested that if the atoms in a crystal were arranged in a regular array (see Fig. 17–2a), such a crystal might serve as a diffraction grating for very short wavelengths on the order of the spacing between atoms, estimated to be about 10^{-10} m $(= 10^{-1}$ nm). Experiments soon showed that X-rays scattered from a crystal did indeed show the peaks and valleys of a diffraction pattern (Fig. 35–27). Thus it was shown, in a single blow, that X-rays have a wave nature and that atoms are arranged in a regular way in crystals. Today, X-rays are recognized as electromagnetic radiation with wavelengths in the range of about 10^{-2} nm to 10 nm, the range readily produced in an X-ray tube.

We saw in Section 35–5 that light of shorter wavelength provides greater resolution when we are examining an object microscopically. Since X-rays have much shorter wavelengths than visible light, they should in principle offer much greater resolution. However, there seems to be no effective material to use as lenses for the very short wavelengths of X-rays. Instead, the clever but complicated technique of **X-ray diffraction** (or **crystallography**) has proved very effective for examining the microscopic world of atoms and molecules. In a simple crystal such as NaCl, the atoms are arranged in an orderly cubical fashion, Fig. 35–28, with atoms spaced a distance d apart. Suppose that a beam of X-rays is incident on the crystal at an angle ϕ to the surface, and that the two rays shown are reflected from two subsequent planes of atoms as shown. The two rays will constructively interfere if the extra distance ray I travels is a whole number of wavelengths farther than the distance ray II travels. This extra distance is $2d \sin \phi$. Therefore, constructive interference will occur when

$$m\lambda = 2d \sin \phi, \qquad m = 1, 2, 3, \cdots, \qquad (35\text{–}20)$$

where m can be any integer. (Notice that ϕ is *not* the angle with respect to the normal to the surface.) This is called the **Bragg equation** after W. L. Bragg (1890–1971), who derived it and who, together with his father W. H. Bragg (1862–1942), developed the theory and technique of X-ray diffraction by crystals in 1912–1913. If the X-ray wavelength is known and the angle ϕ is measured, the distance d between atoms can be obtained. This is the basis for X-ray crystallography.

EXERCISE D When X-rays of wavelength 0.10×10^{-9} m are scattered from a sodium chloride crystal, a second-order diffraction peak is observed at 21°. What is the spacing between the planes of atoms for this scattering?

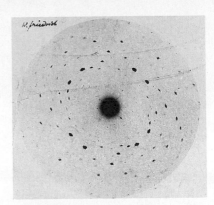

FIGURE 35–27 This X-ray diffraction pattern is one of the first observed by Max von Laue in 1912 when he aimed a beam of X-rays at a zinc sulfide crystal. The diffraction pattern was detected directly on a photographic plate.

FIGURE 35–28 X-ray diffraction by a crystal.

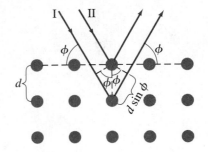

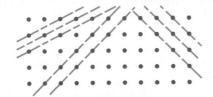

FIGURE 35–29 X-rays can be diffracted from many possible planes within a crystal.

Actual X-ray diffraction patterns are quite complicated. First of all, a crystal is a three-dimensional object, and X-rays can be diffracted from different planes at different angles within the crystal, as shown in Fig. 35–29. Although the analysis is complex, a great deal can be learned about any substance that can be put in crystalline form.

X-ray diffraction has also been very useful in determining the structure of biologically important molecules, such as the double helix structure of DNA, worked out by James Watson and Francis Crick in 1953. See Fig. 35–30, and for models of the double helix, Figs. 21–47a and 21–48. Around 1960, the first detailed structure of a protein molecule, myoglobin, was elucidated with the aid of X-ray diffraction. Soon the structure of an important constituent of blood, hemoglobin, was worked out, and since then the structures of a great many molecules have been determined with the help of X-rays.

FIGURE 35–30 X-ray diffraction photo of DNA molecules taken by Rosalind Franklin in the early 1950s. The cross of spots suggested that DNA is a helix.

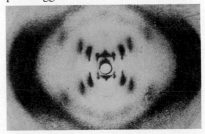

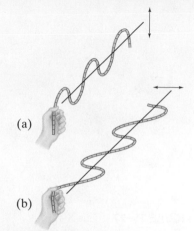

(a)

(b)

FIGURE 35–31 Transverse waves on a rope polarized (a) in a vertical plane and (b) in a horizontal plane.

FIGURE 35–32 (a) Vertically polarized wave passes through a vertical slit, but (b) a horizontally polarized wave will not.

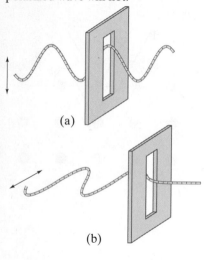

(a)

(b)

FIGURE 35–33 (below) (a) Oscillation of the electric field vectors in unpolarized light. The light is traveling into or out of the page. (b) Electric field in linear polarized light.

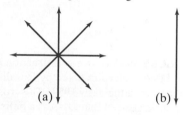

(a)

(b)

FIGURE 35–34 (right) Vertical Polaroid transmits only the vertical component of a wave (electric field) incident upon it.

35–11 Polarization

An important and useful property of light is that it can be *polarized*. To see what this means, let us examine waves traveling on a rope. A rope can be set into oscillation in a vertical plane as in Fig. 35–31a, or in a horizontal plane as in Fig. 35–31b. In either case, the wave is said to be **linearly polarized** or **plane-polarized**—that is, the oscillations are in a plane.

If we now place an obstacle containing a vertical slit in the path of the wave, Fig. 35–32, a vertically polarized wave passes through the vertical slit, but a horizontally polarized wave will not. If a horizontal slit were used, the vertically polarized wave would be stopped. If both types of slit were used, both types of wave would be stopped by one slit or the other. Note that polarization can exist *only* for *transverse waves*, and not for longitudinal waves such as sound. The latter oscillate only along the direction of motion, and neither orientation of slit would stop them.

Light is not necessarily polarized. It can also be **unpolarized**, which means that the source has oscillations in many planes at once, as shown in Fig. 35–33. An ordinary incandescent lightbulb emits unpolarized light, as does the Sun.

Polaroids (Polarization by Absorption)

Plane-polarized light can be obtained from unpolarized light using certain crystals such as tourmaline. Or, more commonly, we use a **Polaroid sheet**. (Polaroid materials were invented in 1929 by Edwin Land.) A Polaroid sheet consists of long complex molecules arranged parallel to one another. Such a Polaroid acts like a series of parallel slits to allow one orientation of polarization to pass through nearly undiminished. This direction is called the *transmission axis* of the Polaroid. Polarization perpendicular to this direction is absorbed almost completely by the Polaroid.

Absorption by a Polaroid can be explained at the molecular level. An electric field $\vec{E}$ that oscillates parallel to the long molecules can set electrons into motion along the molecules, thus doing work on them and transferring energy. Hence, if $\vec{E}$ is parallel to the molecules, it gets absorbed. An electric field $\vec{E}$ perpendicular to the long molecules does not have this possibility of doing work and transferring its energy, and so passes through freely. When we speak of the *transmission axis* of a Polaroid, we mean the direction for which $\vec{E}$ is passed, so a Polaroid axis is *perpendicular* to the long molecules. If we want to think of there being slits between the parallel molecules in the sense of Fig. 35–32, then Fig. 35–32 would apply for the $\vec{B}$ field in the EM wave, not the $\vec{E}$ field.

If a beam of plane-polarized light strikes a Polaroid whose transmission axis is at an angle θ to the incident polarization direction, the beam will emerge plane-polarized parallel to the Polaroid transmission axis, and the amplitude of E will be reduced to $E \cos \theta$, Fig. 35–34. Thus, a Polaroid passes only that component of polarization (the electric field vector, $\vec{E}$) that is parallel to its transmission axis. Because the intensity of a light beam is proportional to the square of the amplitude (Sections 15–3 and 31–8), we see that the intensity of a plane-polarized beam transmitted by a polarizer is

$$I = I_0 \cos^2 \theta, \qquad \left[\begin{array}{c}\text{intensity of plane polarized} \\ \text{wave reduced by polarizer}\end{array}\right] \quad \textbf{(35–21)}$$

where I_0 is the incoming intensity and θ is the angle between the polarizer transmission axis and the plane of polarization of the incoming wave.

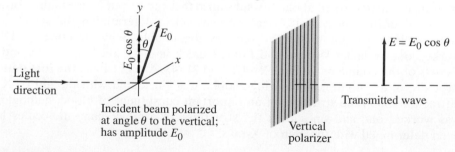

A Polaroid can be used as a **polarizer** to produce plane-polarized light from unpolarized light, since only the component of light parallel to the axis is transmitted. A Polaroid can also be used as an **analyzer** to determine (1) if light is polarized and (2) the plane of polarization. A Polaroid acting as an analyzer will pass the same amount of light independent of the orientation of its axis if the light is unpolarized; try rotating one lens of a pair of Polaroid sunglasses while looking through it at a lightbulb. If the light is polarized, however, when you rotate the Polaroid the transmitted light will be a maximum when the plane of polarization is parallel to the Polaroid's axis, and a minimum when perpendicular to it. If you do this while looking at the sky, preferably at right angles to the Sun's direction, you will see that skylight is polarized. (Direct sunlight is unpolarized, but don't look directly at the Sun, even through a polarizer, for damage to the eye may occur.) If the light transmitted by an analyzer Polaroid falls to zero at one orientation, then the light is 100% plane-polarized. If it merely reaches a minimum, the light is *partially polarized*.

Unpolarized light consists of light with random directions of polarization. Each of these polarization directions can be resolved into components along two mutually perpendicular directions. On average, an unpolarized beam can be thought of as two plane-polarized beams of equal magnitude perpendicular to one another. When unpolarized light passes through a polarizer, one component is eliminated. So the intensity of the light passing through is reduced by half since half the light is eliminated: $I = \frac{1}{2}I_0$ (Fig. 35–35).

When two Polaroids are *crossed*—that is, their polarizing axes are perpendicular to one another—unpolarized light can be entirely stopped. As shown in Fig. 35–36, unpolarized light is made plane-polarized by the first Polaroid (the polarizer).

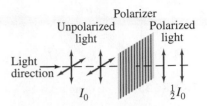

FIGURE 35–35 Unpolarized light has equal intensity vertical and horizontal components. After passing through a polarizer, one of these components is eliminated. The intensity of the light is reduced to half.

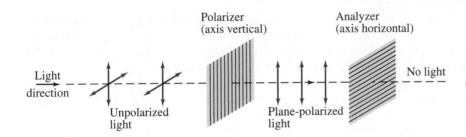

FIGURE 35–36 Crossed Polaroids completely eliminate light.

The second Polaroid, the analyzer, then eliminates this component since its transmission axis is perpendicular to the first. You can try this with Polaroid sunglasses (Fig. 35–37). Note that Polaroid sunglasses eliminate 50% of unpolarized light because of their polarizing property; they absorb even more because they are colored.

EXAMPLE 35–13 **Two Polaroids at 60°.** Unpolarized light passes through two Polaroids; the axis of one is vertical and that of the other is at 60° to the vertical. Describe the orientation and intensity of the transmitted light.

APPROACH Half of the unpolarized light is absorbed by the first Polaroid, and the remaining light emerges plane polarized. When that light passes through the second Polaroid, the intensity is further reduced according to Eq. 35–21, and the plane of polarization is then along the axis of the second Polaroid.

SOLUTION The first Polaroid eliminates half the light, so the intensity is reduced by half: $I_1 = \frac{1}{2}I_0$. The light reaching the second polarizer is vertically polarized and so is reduced in intensity (Eq. 35–21) to

$$I_2 = I_1(\cos 60°)^2 = \frac{1}{4}I_1.$$

Thus, $I_2 = \frac{1}{8}I_0$. The transmitted light has an intensity one-eighth that of the original and is plane-polarized at a 60° angle to the vertical.

FIGURE 35–37 Crossed Polaroids. When the two polarized sunglass lenses overlap, with axes perpendicular, almost no light passes through.

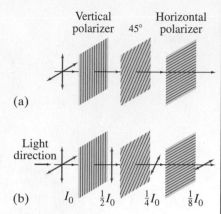

Vertical polarizer 45° Horizontal polarizer

(a)

Light direction

(b) I_0 $\frac{1}{2}I_0$ $\frac{1}{4}I_0$ $\frac{1}{8}I_0$

FIGURE 35–38 Example 35–14.

CONCEPTUAL EXAMPLE 35–14 **Three Polaroids.** We saw in Fig. 35–36 that when unpolarized light falls on two crossed Polaroids (axes at 90°), no light passes through. What happens if a third Polaroid, with axis at 45° to each of the other two, is placed between them (Fig. 35–38a)?

RESPONSE We start just as in Example 35–13 and recall again that light emerging from each Polaroid is polarized parallel to that Polaroid's axis. Thus the angle in Eq. 35–21 is that between the transmission axes of each pair of Polaroids taken in turn. The first Polaroid changes the unpolarized light to plane-polarized and reduces the intensity from I_0 to $I_1 = \frac{1}{2}I_0$. The second polarizer further reduces the intensity by $(\cos 45°)^2$, Eq. 35–21:

$$I_2 = I_1(\cos 45°)^2 = \frac{1}{2}I_1 = \frac{1}{4}I_0.$$

The light leaving the second polarizer is plane polarized at 45° (Fig. 35–38b) relative to the third polarizer, so the third one reduces the intensity to

$$I_3 = I_2(\cos 45°)^2 = \frac{1}{2}I_2,$$

or $I_3 = \frac{1}{8}I_0$. Thus $\frac{1}{8}$ of the original intensity gets transmitted.

NOTE If we don't insert the 45° Polaroid, zero intensity results (Fig. 35–36).

EXERCISE E How much light would pass through if the 45° polarizer in Example 35–14 was placed not between the other two polarizers but (*a*) before the vertical (first) polarizer, or (*b*) after the horizontal polarizer?

Polarization by Reflection

Another means of producing polarized light from unpolarized light is by reflection. When light strikes a nonmetallic surface at any angle other than perpendicular, the reflected beam is polarized preferentially in the plane parallel to the surface, Fig. 35–39. In other words, the component with polarization in the plane perpendicular to the surface is preferentially transmitted or absorbed. You can check this by rotating Polaroid sunglasses while looking through them at a flat surface of a lake or road. Since most outdoor surfaces are horizontal, Polaroid sunglasses are made with their axes vertical to eliminate the more strongly reflected horizontal component, and thus reduce glare. People who go fishing wear Polaroids to eliminate reflected glare from the surface of a lake or stream and thus see beneath the water more clearly (Fig. 35–40).

FIGURE 35–39 Light reflected from a nonmetallic surface, such as the smooth surface of water in a lake, is partially polarized parallel to the surface.

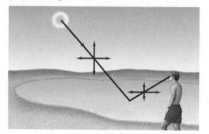

FIGURE 35–40 Photographs of a river, (a) allowing all light into the camera lens, and (b) using a polarizer. The polarizer is adjusted to absorb most of the (polarized) light reflected from the water's surface, allowing the dimmer light from the bottom of the river, and any fish lying there, to be seen more readily.

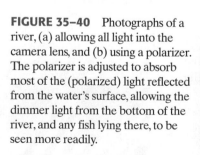

(a)

(b)

The amount of polarization in the reflected beam depends on the angle, varying from no polarization at normal incidence to 100% polarization at an angle known as the **polarizing angle** θ_p.[†] This angle is related to the index of refraction of the two materials on either side of the boundary by the equation

$$\tan \theta_p = \frac{n_2}{n_1}, \qquad \text{(35–22a)}$$

where n_1 is the index of refraction of the material in which the beam is traveling, and n_2 is that of the medium beyond the reflecting boundary. If the beam is traveling in air, $n_1 = 1$, and Eq. 35–22a becomes

$$\tan \theta_p = n. \qquad \text{(35–22b)}$$

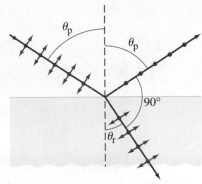

FIGURE 35–41 At θ_p the reflected light is plane-polarized parallel to the surface, and $\theta_p + \theta_r = 90°$, where θ_r is the refraction angle. (The large dots represent vibrations perpendicular to the page.)

The polarizing angle θ_p is also called **Brewster's angle**, and Eqs. 35–22 *Brewster's law*, after the Scottish physicist David Brewster (1781–1868), who worked it out experimentally in 1812. Equations 35–22 can be derived from the electromagnetic wave theory of light. It is interesting that at Brewster's angle, the reflected ray and the transmitted (refracted) ray make a 90° angle to each other; that is, $\theta_p + \theta_r = 90°$, where θ_r is the refraction angle (Fig. 35–41). This can be seen by substituting Eq. 35–22a, $n_2 = n_1 \tan \theta_p = n_1 \sin \theta_p / \cos \theta_p$, into Snell's law, $n_1 \sin \theta_p = n_2 \sin \theta_r$, which gives $\cos \theta_p = \sin \theta_r$ which can only hold if $\theta_p = 90° - \theta_r$.

EXAMPLE 35–15 **Polarizing angle.** (*a*) At what incident angle is sunlight reflected from a lake plane-polarized? (*b*) What is the refraction angle?

APPROACH The polarizing angle at the surface is Brewster's angle, Eq. 35–22b. We find the angle of refraction from Snell's law.

SOLUTION (*a*) We use Eq. 35–22b with $n = 1.33$, so $\tan \theta_p = 1.33$ giving $\theta_p = 53.1°$. (*b*) From Snell's law, $\sin \theta_r = \sin \theta_p / n = \sin 53.1° / 1.33 = 0.601$ giving $\theta_r = 36.9°$.

NOTE $\theta_p + \theta_r = 53.1° + 36.9° = 90.0°$, as expected.

*35–12 Liquid Crystal Displays (LCD)

A wonderful use of polarization is in a **liquid crystal display** (LCD). LCDs are used as the display in hand-held calculators, digital wrist watches, cell phones, and in beautiful color flat-panel computer and television screens.

A liquid crystal display is made up of many tiny rectangles called **pixels**, or "picture elements." The picture you see depends on which pixels are dark or light and of what color, as suggested in Fig. 35–42 for a simple black and white picture.

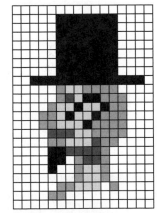

FIGURE 35–42 Example of an image made up of many small squares or *pixels* (picture elements). This one has rather poor resolution.

Liquid crystals are organic materials that at room temperature exist in a phase that is neither fully solid nor fully liquid. They are sort of gooey, and their molecules display a randomness of position characteristic of liquids, as we discussed in Section 17–1 and Fig. 17–2. They also show some of the orderliness of a solid crystal (Fig. 17–2a), but only in one dimension. The liquid crystals we find useful are made up of relatively rigid rod-like molecules that interact weakly with each other and tend to align parallel to each other, as shown in Fig. 35–43.

FIGURE 35–43 Liquid crystal molecules tend to align in one dimension (parallel to each other) but have random positions (left-right, up-down).

[†]Only a fraction of the incident light is reflected at the surface of a transparent medium. Although this reflected light is 100% polarized (if $\theta = \theta_p$), the remainder of the light, which is transmitted into the new medium, is only partially polarized.

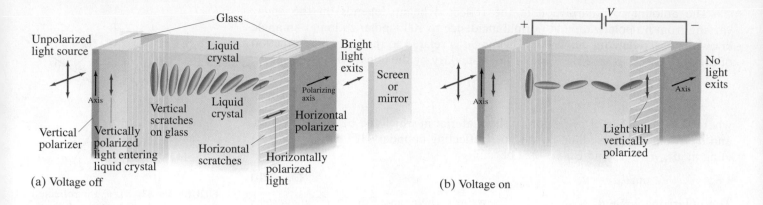

(a) Voltage off   (b) Voltage on

FIGURE 35–44 (a) "Twisted" form of liquid crystal. Light polarization plane is rotated 90°, and so is transmitted by the horizontal polarizer. Only one line of molecules is shown. (b) Molecules disoriented by electric field. Plane of polarization is not changed, so light does not pass through the horizontal polarizer. (The transparent electrodes are not shown.)

FIGURE 35–45 Calculator LCD display. The black segments or pixels have a voltage applied to them. Note that the 8 uses all seven segments (pixels), whereas other numbers use fewer.

In a simple LCD, each pixel (picture element) contains a liquid crystal sandwiched between two glass plates whose inner surfaces have been brushed to form nanometer-wide parallel scratches. The rod-like liquid crystal molecules in contact with the scratches tend to line up along the scratches. The two plates typically have their scratches at 90° to each other, and the weak forces between the rod-like molecules tend to keep them nearly aligned with their nearest neighbors, resulting in the twisted pattern shown in Fig. 35–44a.

The outer surfaces of the glass plates each have a thin film polarizer, they too oriented at 90° to each other. Unpolarized light incident from the left becomes plane-polarized and the liquid crystal molecules keep this polarization aligned with their rod-like shape. That is, the plane of polarization of the light rotates with the molecules as the light passes through the liquid crystal. The light emerges with its plane of polarization rotated by 90°, and passes through the second polarizer readily (Fig. 35–44a). A tiny LCD pixel in this situation will appear bright.

Now suppose a voltage is applied to transparent electrodes on each glass plate of the pixel. The rod-like molecules are polar (or can acquire an internal separation of charge due to the applied electric field). The applied voltage tends to align the molecules and they no longer follow the twisted pattern shown in Fig. 35–44a, with the end molecules always lying in a plane parallel to the glass plates. Instead the applied electric field tends to align the molecules flat, left to right (perpendicular to the glass plates), and they don't affect the light polarization significantly. The entering plane-polarized light no longer has its plane of polarization rotated as it passes through, and no light can exit through the second (horizontal) polarizer (Fig. 35–44b). With the voltage on, the pixel appears dark.[†]

The simple display screens of watches and calculators use ambient light as the source (you can't see the display in the dark), and a mirror behind the LCD to reflect the light back. There are only a few pixels, corresponding to the elongated segments needed to form the numbers from 0 to 9 (and letters in some displays), as seen in Fig. 35–45. Any pixels to which a voltage is applied appear dark and form part of a number. With no voltage, pixels pass light through the polarizers to the mirror and back out, which forms a bright background to the dark numbers on the display.

Color television and computer LCDs are more sophisticated. A color pixel consists of three cells, or subpixels, each covered with a red, green, or blue filter. Varying brightnesses of these three primary colors can yield almost any natural color. A good-quality screen consists of a million or more pixels. Behind this array of pixels is a light source, often thin fluorescent tubes the diameter of a straw. The light passes through the pixels, or not, depending on the voltage applied to each subpixel, as in Fig. 35–44a and b.

[†]In some displays, the polarizers are parallel to each other (the scratches remain at 90° to maintain the twist). Then voltage off results in black (no light), and voltage on results in bright light.

*35–13 Scattering of Light by the Atmosphere

Sunsets are red, the sky is blue, and skylight is polarized (at least partially). These phenomena can be explained on the basis of the *scattering* of light by the molecules of the atmosphere. In Fig. 35–46 we see unpolarized light from the Sun impinging on a molecule of the Earth's atmosphere. The electric field of the EM wave sets the electric charges within the molecule into oscillation, and the molecule absorbs some of the incident radiation. But the molecule quickly reemits this light since the charges are oscillating. As discussed in Section 31–4, oscillating electric charges produce EM waves. The intensity is strongest along the direction perpendicular to the oscillation, and drops to zero along the line of oscillation (Section 31–4). In Fig. 35–46 the motion of the charges is resolved into two components. An observer at right angles to the direction of the sunlight, as shown, will see plane-polarized light because no light is emitted along the line of the other component of the oscillation. (When viewing along the line of an oscillation, you don't see that oscillation, and hence see no waves made by it.) At other viewing angles, both components will be present; one will be stronger, however, so the light appears partially polarized. Thus, the process of scattering explains the polarization of skylight.

Scattering of light by the Earth's atmosphere depends on wavelength λ. For particles much smaller than the wavelength of light (such as molecules of air), the particles will be less of an obstruction to long wavelengths than to short ones. The scattering decreases, in fact, as $1/\lambda^4$. Blue and violet light are thus scattered much more than red and orange, which is why the sky looks blue. At sunset, the Sun's rays pass through a maximum length of atmosphere. Much of the blue has been taken out by scattering. The light that reaches the surface of the Earth, and reflects off clouds and haze, is thus lacking in blue. That is why sunsets appear reddish.

The dependence of scattering on $1/\lambda^4$ is valid only if the scattering objects are much smaller than the wavelength of the light. This is valid for oxygen and nitrogen molecules whose diameters are about 0.2 nm. Clouds, however, contain water droplets or crystals that are much larger than λ. They scatter all frequencies of light nearly uniformly. Hence clouds appear white (or gray, if shadowed).

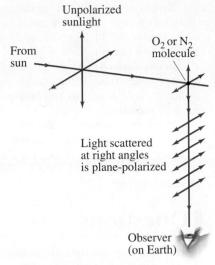

FIGURE 35–46 Unpolarized sunlight scattered by molecules of the air. An observer at right angles sees plane-polarized light, since the component of oscillation along the line of sight emits no light along that line.

PHYSICS APPLIED

Why the sky is blue
Why sunsets are red

PHYSICS APPLIED

Why clouds are white

Summary

Diffraction refers to the fact that light, like other waves, bends around objects it passes, and spreads out after passing through narrow slits. This bending gives rise to a **diffraction pattern** due to interference between rays of light that travel different distances.

Light passing through a very narrow slit of width D (on the order of the wavelength λ) will produce a pattern with a bright central maximum of half-width θ given by

$$\sin\theta = \frac{\lambda}{D}, \tag{35–1}$$

flanked by fainter lines to either side.

The minima in the diffraction pattern occur at

$$D\sin\theta = m\lambda \tag{35–2}$$

where $m = 1, 2, 3, \cdots$, but not $m = 0$ (for which the pattern has its strongest maximum).

The **intensity** at any point in the single-slit diffraction pattern can be calculated using **phasor** diagrams. The same technique can be used to determine the intensity of the pattern produced by two slits.

The pattern for two-slit interference can be described as a series of maxima due to interference of light from the two slits, modified by an "envelope" due to diffraction at each slit.

The wave nature of light limits the sharpness or **resolution** of images. Because of diffraction, it is not possible to *discern details smaller than the wavelength* of the radiation being used. The useful magnification of a light microscope is limited by diffraction to about 500×.

A **diffraction grating** consists of many parallel slits or lines, each separated from its neighbors by a distance d. The peaks of constructive interference occur at angles θ given by

$$\sin\theta = \frac{m\lambda}{d}, \tag{35–13}$$

where $m = 0, 1, 2, \cdots$. The peaks of constructive interference are much brighter and sharper for a diffraction grating than for a simple two-slit apparatus. [*Peak width is inversely proportional to the total number of lines in the grating.]

[*A diffraction grating (or a prism) is used in a **spectrometer** to separate different colors or to observe **line spectra**. For a given order m, θ depends on λ. Precise determination of wavelength can be done with a spectrometer by careful measurement of θ.]

X-rays are a form of electromagnetic radiation of very short wavelength. They are produced when high-speed electrons, accelerated by high voltage in an evacuated tube, strike a glass or metal target.

In **unpolarized light**, the electric field vectors oscillate in all transverse directions. If the electric vector oscillates only in one plane, the light is said to be **plane-polarized**. Light can also be partially polarized.

When an unpolarized light beam passes through a **Polaroid** sheet, the emerging beam is plane-polarized. When a light beam is polarized and passes through a Polaroid, the intensity varies as the Polaroid is rotated. Thus a Polaroid can act as a **polarizer** or as an **analyzer**.

The intensity I of a plane-polarized light beam incident on a Polaroid is reduced by the factor

$$I = I_0 \cos^2 \theta \qquad (35-21)$$

where θ is the angle between the axis of the Polaroid and the initial plane of polarization.

Light can also be partially or fully **polarized by reflection**. If light traveling in air is reflected from a medium of index of refraction n, the reflected beam will be *completely* plane-polarized if the incident angle θ_p is given by

$$\tan \theta_p = n. \qquad (35-22b)$$

The fact that light can be polarized shows that it must be a transverse wave.

Questions

1. Radio waves and light are both electromagnetic waves. Why can a radio receive a signal behind a hill when we cannot see the transmitting antenna?

2. Hold one hand close to your eye and focus on a distant light source through a narrow slit between two fingers. (Adjust your fingers to obtain the best pattern.) Describe the pattern that you see.

3. Explain why diffraction patterns are more difficult to observe with an extended light source than for a point source. Compare also a monochromatic source to white light.

4. For diffraction by a single slit, what is the effect of increasing (a) the slit width, (b) the wavelength?

5. Describe the single-slit diffraction pattern produced when white light falls on a slit having a width of (a) 50 nm, (b) 50,000 nm.

6. What happens to the diffraction pattern of a single slit if the whole apparatus is immersed in (a) water, (b) a vacuum, instead of in air.

7. In the single-slit diffraction pattern, why does the first off-center maximum not occur at exactly $\sin \theta = \frac{3}{2}\lambda/D$?

8. Discuss the similarities, and differences, of double-slit interference and single-slit diffraction.

*9. Figure 35–10 shows a two-slit interference pattern for the case when d is larger than D. Can the reverse case occur, when d is less than D?

*10. When both diffraction and interference are taken into account in the double-slit experiment, discuss the effect of increasing (a) the wavelength, (b) the slit separation, (c) the slit width.

11. Does diffraction limit the resolution of images formed by (a) spherical mirrors, (b) plane mirrors?

12. Do diffraction effects occur for virtual as well as real images?

13. Give at least two advantages for the use of large reflecting mirrors in astronomical telescopes.

14. Atoms have diameters of about 10^{-8} cm. Can visible light be used to "see" an atom? Explain.

15. Which color of visible light would give the best resolution in a microscope? Explain.

16. Could a diffraction grating just as well be called an interference grating? Discuss.

17. Suppose light consisting of wavelengths between 400 nm and 700 nm is incident normally on a diffraction grating. For what orders (if any) would there be overlap in the observed spectrum? Does your answer depend on the slit spacing?

18. What is the difference in the interference patterns formed by two slits 10^{-4} cm apart as compared to a diffraction grating containing 10^4 lines/cm?

19. White light strikes (a) a diffraction grating and (b) a prism. A rainbow appears on a wall just below the direction of the horizontal incident beam in each case. What is the color of the top of the rainbow in each case? Explain.

20. Explain why there are tiny peaks between the main peaks produced by a diffraction grating illuminated with monochromatic light. Why are the peaks so tiny?

21. What does polarization tell us about the nature of light?

22. How can you tell if a pair of sunglasses is polarizing or not?

*23. What would be the color of the sky if the Earth had no atmosphere?

Problems

[Note: Assume light passing through slits is in phase, unless stated otherwise.]

35–1 Single-Slit Diffraction

1. (I) If 680-nm light falls on a slit 0.0365 mm wide, what is the angular width of the central diffraction peak?

2. (I) Monochromatic light falls on a slit that is 2.60×10^{-3} mm wide. If the angle between the first dark fringes on either side of the central maximum is 32.0° (dark fringe to dark fringe), what is the wavelength of the light used?

3. (II) Light of wavelength 580 nm falls on a slit that is 3.80×10^{-3} mm wide. Estimate how far the first brightest diffraction fringe is from the strong central maximum if the screen is 10.0 m away.

4. (II) Consider microwaves which are incident perpendicular to a metal plate which has a 1.6-cm slit in it. Discuss the angles at which there are diffraction minima for wavelengths of (a) 0.50 cm, (b) 1.0 cm, and (c) 3.0 cm.

5. (II) If parallel light falls on a single slit of width D at a 23.0° angle to the normal, describe the diffraction pattern.

6. (II) Monochromatic light of wavelength 633 nm falls on a slit. If the angle between the first bright fringes on either side of the central maximum is 35°, estimate the slit width.

7. (II) If a slit diffracts 580-nm light so that the diffraction maximum is 6.0 cm wide on a screen 2.20 m away, what will be the width of the diffraction maximum for light with a wavelength of 460 nm?

8. (II) (a) For a given wavelength λ, what is the minimum slit width for which there will be no diffraction minima? (b) What is the minimum slit width so that no visible light exhibits a diffraction minimum?

9. (II) When blue light of wavelength 440 nm falls on a single slit, the first dark bands on either side of center are separated by 55.0°. Determine the width of the slit.

10. (II) A single slit 1.0 mm wide is illuminated by 450-nm light. What is the width of the central maximum (in cm) in the diffraction pattern on a screen 5.0 m away?

11. (II) Coherent light from a laser diode is emitted through a rectangular area $3.0\,\mu m \times 1.5\,\mu m$ (horizontal-by-vertical). If the laser light has a wavelength of 780 nm, determine the angle between the first diffraction minima (a) above and below the central maximum, (b) to the left and right of the central maximum.

*35–2 Intensity, Single-Slit Diffraction Pattern

*12. (II) If you double the width of a single slit, the intensity of the light passing through the slit is doubled. (a) Show, however, that the intensity at the center of the screen increases by a factor of 4. (b) Explain why this does not violate conservation of energy.

*13. (II) Light of wavelength 750 nm passes through a slit $1.0\,\mu m$ wide and a single-slit diffraction pattern is formed vertically on a screen 25 cm away. Determine the light intensity I 15 cm above the central maximum, expressed as a fraction of the central maximum's intensity I_0.

*14. (III) (a) Explain why the secondary maxima in the single-slit diffraction pattern do not occur precisely at $\beta/2 = (m + \frac{1}{2})\pi$ where $m = 1, 2, 3, \cdots$. (b) By differentiating Eq. 35–7 with respect to β show that the secondary maxima occur when $\beta/2$ satisfies the relation $\tan(\beta/2) = \beta/2$. (c) Carefully and precisely plot the curves $y = \beta/2$ and $y = \tan \beta/2$. From their intersections, determine the values of β for the first and second secondary maxima. What is the percent difference from $\beta/2 = (m + \frac{1}{2})\pi$?

*35–3 Intensity, Double-Slit Diffraction

*15. (II) If a double-slit pattern contains exactly nine fringes in the central diffraction peak, what can you say about the slit width and separation? Assume the first diffraction minimum occurs at an interference minimum.

*16. (II) Design a double-slit apparatus so that the central diffraction peak contains precisely seventeen fringes. Assume the first diffraction minimum occurs at (a) an interference miminum, (b) an interference maximum.

*17. (II) 605-nm light passes through a pair of slits and creates an interference pattern on a screen 2.0 m behind the slits. The slits are separated by 0.120 mm and each slit is 0.040 mm wide. How many constructive interference fringes are formed on the screen? [Many of these fringes will be of very low intensity.]

*18. (II) In a double-slit experiment, if the central diffraction peak contains 13 interference fringes, how many fringes are contained within each secondary diffraction peak (between $m = +1$ and $+2$ in Eq. 35–2). Assume the first diffraction minimum occurs at an interference minimum.

*19. (II) Two 0.010-mm-wide slits are 0.030 mm apart (center to center). Determine (a) the spacing between interference fringes for 580 nm light on a screen 1.0 m away and (b) the distance between the two diffraction minima on either side of the central maximum of the envelope.

*20. (II) Suppose $d = D$ in a double-slit apparatus, so that the two slits merge into one slit of width $2D$. Show that Eq. 35–9 reduces to the correct equation for single-slit diffraction.

*21. (II) In a double-slit experiment, let $d = 5.00 D = 40.0 \lambda$. Compare (as a ratio) the intensity of the third-order interference maximum with that of the zero-order maximum.

*22. (II) How many fringes are contained in the central diffraction peak for a double-slit pattern if (a) $d = 2.00D$, (b) $d = 12.0D$, (c) $d = 4.50D$, (d) $d = 7.20D$.

*23. (III) (a) Derive an expression for the intensity in the interference pattern for three equally spaced slits. Express in terms of $\delta = 2\pi d \sin \theta/\lambda$ where d is the distance between adjacent slits and assume the slit width $D \approx \lambda$. (b) Show that there is only one secondary maximum between principal peaks.

35–4 and 35–5 Resolution Limits

24. (I) What is the angular resolution limit (degrees) set by diffraction for the 100-inch (254-cm mirror diameter) Mt. Wilson telescope ($\lambda = 560$ nm)?

25. (II) Two stars 16 light-years away are barely resolved by a 66-cm (mirror diameter) telescope. How far apart are the stars? Assume $\lambda = 550$ nm and that the resolution is limited by diffraction.

26. (II) The nearest neighboring star to the Sun is about 4 light-years away. If a planet happened to be orbiting this star at an orbital radius equal to that of the Earth–Sun distance, what minimum diameter would an Earth-based telescope's aperture have to be in order to obtain an image that resolved this star–planet system? Assume the light emitted by the star and planet has a wavelength of 550 nm.

27. (II) If you shine a flashlight beam toward the Moon, estimate the diameter of the beam when it reaches the Moon. Assume that the beam leaves the flashlight through a 5.0-cm aperture, that its white light has an average wavelength of 550 nm, and that the beam spreads due to diffraction only.

28. (II) Suppose that you wish to construct a telescope that can resolve features 7.5 km across on the moon, 384,000 km away. You have a 2.0-m-focal-length objective lens whose diameter is 11.0 cm. What focal-length eyepiece is needed if your eye can resolve objects 0.10 mm apart at a distance of 25 cm? What is the resolution limit set by the size of the objective lens (that is, by diffraction)? Use $\lambda = 560$ nm.

29. (II) The normal lens on a 35-mm camera has a focal length of 50.0 mm. Its aperture diameter varies from a maximum of 25 mm ($f/2$) to a minimum of 3.0 mm ($f/16$). Determine the resolution limit set by diffraction for ($f/2$) and ($f/16$). Specify as the number of lines per millimeter resolved on the detector or film. Take $\lambda = 560$ nm.

35–7 and 35–8 Diffraction Grating, Spectroscopy

30. (I) At what angle will 480-nm light produce a second-order maximum when falling on a grating whose slits are 1.35×10^{-3} cm apart?

31. (I) A source produces first-order lines when incident normally on a 12,000-line/cm diffraction grating at angles 28.8°, 36.7°, 38.6°, and 47.9°. What are the wavelengths?

32. (I) A 3500-line/cm grating produces a third-order fringe at a 26.0° angle. What wavelength of light is being used?

33. (I) A grating has 6800 lines/cm. How many spectral orders can be seen (400 to 700 nm) when it is illuminated by white light?

34. (II) How many lines per centimeter does a grating have if the third order occurs at a 15.0° angle for 650-nm light?

35. (II) Red laser light from a He–Ne laser ($\lambda = 632.8$ nm) is used to calibrate a diffraction grating. If this light creates a second-order fringe at 53.2° after passing through the grating, and light of an unknown wavelength λ creates a first-order fringe at 20.6°, find λ.

36. (II) White light containing wavelengths from 410 nm to 750 nm falls on a grating with 7800 lines/cm. How wide is the first-order spectrum on a screen 2.80 m away?

37. (II) A diffraction grating has 6.0×10^5 lines/m. Find the angular spread in the second-order spectrum between red light of wavelength 7.0×10^{-7} m and blue light of wavelength 4.5×10^{-7} m.

38. (II) A tungsten–halogen bulb emits a continuous spectrum of ultraviolet, visible, and infrared light in the wavelength range 360 nm to 2000 nm. Assume that the light from a tungsten–halogen bulb is incident on a diffraction grating with slit spacing d and that the first-order brightness maximum for the wavelength of 1200 nm occurs at angle θ. What other wavelengths within the spectrum of incident light will produce a brightness maximum at this same angle θ? [Optical filters are used to deal with this bothersome effect when a continuous spectrum of light is measured by a spectrometer.]

39. (II) Show that the second- and third-order spectra of white light produced by a diffraction grating always overlap. What wavelengths overlap?

40. (II) Two first-order spectrum lines are measured by a 9650 line/cm spectroscope at angles, on each side of center, of +26°38′, +41°02′ and −26°18′, −40°27′. Calculate the wavelengths based on these data.

41. (II) Suppose the angles measured in Problem 40 were produced when the spectrometer (but not the source) was submerged in water. What then would be the wavelengths (in air)?

42. (II) The first-order line of 589-nm light falling on a diffraction grating is observed at a 16.5° angle. How far apart are the slits? At what angle will the third order be observed?

43. (II) White light passes through a 610-line/mm diffraction grating. First-order and second-order visible spectra ("rainbows") appear on the wall 32 cm away as shown in Fig. 35–47. Determine the widths ℓ_1 and ℓ_2 of the two "rainbows" (400 nm to 700 nm). In which order is the "rainbow" dispersed over a larger distance?

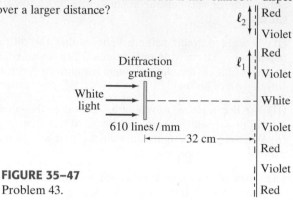

FIGURE 35–47
Problem 43.

44. (II) **Missing orders** occur for a diffraction grating when a diffraction minimum coincides with an interference maximum. Let D be the width of each slit and d the separation of slits. (a) Show that if $d = 2D$, all even orders ($m = 2, 4, 6, \cdots$) are missing. (b) Show there will be missing orders whenever

$$\frac{d}{D} = \frac{m_1}{m_2}$$

where m_1 and m_2 are integers. (c) Discusss the case $d = D$, the limit in which the space between slits becomes negligible.

45. (II) Monochromatic light falls on a transmission diffraction grating at an angle ϕ to the normal. (a) Show that Eq. 35–13 for diffraction maxima must be replaced by

$$d(\sin\phi + \sin\theta) = \pm m\lambda. \qquad m = 0, 1, 2, \cdots.$$

(b) Explain the $\pm$ sign. (c) Green light with a wavelength of 550 nm is incident at an angle of 15° to the normal on a diffraction grating with 5000 lines/cm. Find the angles at which the first-order maxima occur.

*35–9 Grating, Peak Widths, Resolving Power

***46.** (II) A 6500-line/cm diffraction grating is 3.18 cm wide. If light with wavelengths near 624 nm falls on the grating, how close can two wavelengths be if they are to be resolved in any order? What order gives the best resolution?

***47.** (II) A diffraction grating has 16,000 rulings in its 1.9 cm width. Determine (a) its resolving power in first and second orders, and (b) the minimum wavelength resolution ($\Delta\lambda$) it can yield for $\lambda = 410$ nm.

***48.** (II) Let 580-nm light be incident normally on a diffraction grating for which $d = 3.00D = 1050$ nm. (a) How many orders (principal maxima) are present? (b) If the grating is 1.80 cm wide, what is the full angular width of each principal maximum?

35–10 X-Ray Diffraction

49. (II) X-rays of wavelength 0.138 nm fall on a crystal whose atoms, lying in planes, are spaced 0.285 nm apart. At what angle ϕ (relative to the surface, Fig. 35–28) must the X-rays be directed if the first diffraction maximum is to be observed?

50. (II) First-order Bragg diffraction is observed at 26.8° relative to the crystal surface, with spacing between atoms of 0.24 nm. (a) At what angle will second order be observed? (b) What is the wavelength of the X-rays?

51. (II) If X-ray diffraction peaks corresponding to the first three orders ($m = 1, 2,$ and 3) are measured, can both the X-ray wavelength λ and lattice spacing d be determined? Prove your answer.

35-11 Polarization

52. (I) Two polarizers are oriented at $65°$ to one another. Unpolarized light falls on them. What fraction of the light intensity is transmitted?

53. (I) Two Polaroids are aligned so that the light passing through them is a maximum. At what angle should one of them be placed so the intensity is subsequently reduced by half?

54. (I) What is Brewster's angle for an air–glass ($n = 1.58$) surface?

55. (I) What is Brewster's angle for a diamond submerged in water if the light is hitting the diamond ($n = 2.42$) while traveling in the water?

56. (II) The critical angle for total internal reflection at a boundary between two materials is $55°$. What is Brewster's angle at this boundary? Give two answers, one for each material.

57. (II) At what angle should the axes of two Polaroids be placed so as to reduce the intensity of the incident unpolarized light to (a) $\frac{1}{3}$, (b) $\frac{1}{10}$?

58. (II) Two polarizers are oriented at $36.0°$ to one another. Light polarized at an $18.0°$ angle to each polarizer passes through both. What is the transmitted intensity (%)?

59. (II) What would Brewster's angle be for reflections off the surface of water for light coming from beneath the surface? Compare to the angle for total internal reflection, and to Brewster's angle from above the surface.

60. (II) Unpolarized light passes through six successive Polaroid sheets each of whose axis makes a $45°$ angle with the previous one. What is the intensity of the transmitted beam?

61. (II) Two polarizers A and B are aligned so that their transmission axes are vertical and horizontal, respectively. A third polarizer is placed between these two with its axis aligned at angle θ with respect to the vertical. Assuming vertically polarized light of intensity I_0 is incident upon polarizer A, find an expression for the light intensity I transmitted through this three-polarizer sequence. Calculate the derivative $dI/d\theta$; then use it to find the angle θ that maximizes I.

62. (III) The percent polarization P of a partially polarized beam of light is defined as

$$ P = \frac{I_{max} - I_{min}}{I_{max} + I_{min}} \times 100 $$

where I_{max} and I_{min} are the maximum and minimum intensities that are obtained when the light passes through a polarizer that is slowly rotated. Such light can be considered as the sum of two unequal plane-polarized beams of intensities I_{max} and I_{min} perpendicular to each other. Show that the light transmitted by a polarizer, whose axis makes an angle ϕ to the direction in which I_{max} is obtained, has intensity

$$ \frac{1 + p\cos 2\phi}{1 + p} I_{max} $$

where $p = P/100$ is the "fractional polarization."

General Problems

63. When violet light of wavelength 415 nm falls on a single slit, it creates a central diffraction peak that is 8.20 cm wide on a screen that is 2.85 m away. How wide is the slit?

64. A series of polarizers are each placed at a $10°$ interval from the previous polarizer. Unpolarized light is incident on this series of polarizers. How many polarizers does the light have to go through before it is $\frac{1}{4}$ of its original intensity?

65. The wings of a certain beetle have a series of parallel lines across them. When normally incident 480-nm light is reflected from the wing, the wing appears bright when viewed at an angle of $56°$. How far apart are the lines?

66. A teacher stands well back from an outside doorway 0.88 m wide, and blows a whistle of frequency 850 Hz. Ignoring reflections, estimate at what angle(s) it is *not* possible to hear the whistle clearly on the playground outside the doorway. Assume 340 m/s for the speed of sound.

67. Light is incident on a diffraction grating with 7600 lines/cm and the pattern is viewed on a screen located 2.5 m from the grating. The incident light beam consists of two wavelengths, $\lambda_1 = 4.4 \times 10^{-7}$ m and $\lambda_2 = 6.8 \times 10^{-7}$ m. Calculate the linear distance between the first-order bright fringes of these two wavelengths on the screen.

68. How many lines per centimeter must a grating have if there is to be no second-order spectrum for any visible wavelength?

69. When yellow sodium light, $\lambda = 589$ nm, falls on a diffraction grating, its first-order peak on a screen 66.0 cm away falls 3.32 cm from the central peak. Another source produces a line 3.71 cm from the central peak. What is its wavelength? How many lines/cm are on the grating?

70. Two of the lines of the atomic hydrogen spectrum have wavelengths of 656 nm and 410 nm. If these fall at normal incidence on a grating with 8100 lines/cm, what will be the angular separation of the two wavelengths in the first-order spectrum?

71. (a) How far away can a human eye distinguish two car headlights 2.0 m apart? Consider only diffraction effects and assume an eye diameter of 6.0 mm and a wavelength of 560 nm. (b) What is the minimum angular separation an eye could resolve when viewing two stars, considering only diffraction effects? In reality, it is about $1'$ of arc. Why is it not equal to your answer in (b)?

72. A laser beam passes through a slit of width 1.0 cm and is pointed at the Moon, which is approximately 380,000 km from the Earth. Assume the laser emits waves of wavelength 633 nm (the red light of a He–Ne laser). Estimate the width of the beam when it reaches the Moon.

73. A He–Ne gas laser which produces monochromatic light of wavelength $\lambda = 6.328 \times 10^{-7}$ m is used to calibrate a reflection grating in a spectroscope. The first-order diffraction line is found at an angle of $21.5°$ to the incident beam. How many lines per meter are there on the grating?

74. The entrance to a boy's bedroom consists of two doorways, each 1.0 m wide, which are separated by a distance of 3.0 m. The boy's mother yells at him through the two doors as shown in Fig. 35–48, telling him to clean up his room. Her voice has a frequency of 400 Hz. Later, when the mother discovers the room is still a mess, the boy says he never heard her telling him to clean his room. The velocity of sound is 340 m/s. (a) Find all of the angles θ (Fig. 35–48) at which no sound will be heard within the bedroom when the mother yells. Assume sound is completely absorbed when it strikes a bedroom wall. (b) If the boy was at the position shown when his mother yelled, does he have a good explanation for not having heard her? Explain.

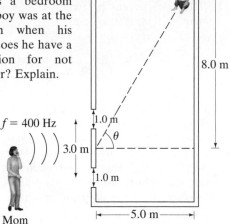

FIGURE 35–48
Problem 74.

75. At what angle above the horizon is the Sun when light reflecting off a smooth lake is polarized most strongly?

76. Unpolarized light falls on two polarizer sheets whose axes are at right angles. (a) What fraction of the incident light intensity is transmitted? (b) What fraction is transmitted if a third polarizer is placed between the first two so that its axis makes a 66° angle with the axis of the first polarizer? (c) What if the third polarizer is in front of the other two?

77. At what angle should the axes of two Polaroids be placed so as to reduce the intensity of the incident unpolarized light by an additional factor (after the first Polaroid cuts it in half) of (a) 4, (b) 10, (c) 100?

78. Four polarizers are placed in succession with their axes vertical, at 30.0° to the vertical, at 60.0° to the vertical, and at 90.0° to the vertical. (a) Calculate what fraction of the incident unpolarized light is transmitted by the four polarizers. (b) Can the transmitted light be *decreased* by removing one of the polarizers? If so, which one? (c) Can the transmitted light intensity be extinguished by removing polarizers? If so, which one(s)?

79. Spy planes fly at extremely high altitudes (25 km) to avoid interception. Their cameras are reportedly able to discern features as small as 5 cm. What must be the minimum aperture of the camera lens to afford this resolution? (Use $\lambda = 580$ nm.)

80. Two polarizers are oriented at 48° to each other and plane-polarized light is incident on them. If only 25% of the light gets through both of them, what was the initial polarization direction of the incident light?

81. X-rays of wavelength 0.0973 nm are directed at an unknown crystal. The second diffraction maximum is recorded when the X-rays are directed at an angle of 23.4° relative to the crystal surface. What is the spacing between crystal planes?

82. X-rays of wavelength 0.10 nm fall on a microcrystalline powder sample. The sample is located 12 cm from the photographic film. The crystal structure of the sample has an atomic spacing of 0.22 nm. Calculate the radii of the diffraction rings corresponding to first- and second-order scattering. Note in Fig. 35–28 that the X-ray beam is deflected through an angle 2ϕ.

83. The Hubble Space Telescope with an objective diameter of 2.4 m, is viewing the Moon. Estimate the minimum distance between two objects on the Moon that the Hubble can distinguish. Consider diffraction of light with wavelength 550 nm. Assume the Hubble is near the Earth.

84. The Earth and Moon are separated by about 400×10^6 m. When Mars is 8×10^{10} m from Earth, could a person standing on Mars resolve the Earth and its Moon as two separate objects without a telescope? Assume a pupil diameter of 5 mm and $\lambda = 550$ nm.

85. A slit of width $D = 22 \ \mu$m is cut through a thin aluminum plate. Light with wavelength $\lambda = 650$ nm passes through this slit and forms a single-slit diffraction pattern on a screen a distance $\ell = 2.0$ m away. Defining x to be the distance between the first minima ($m = +1$ and $m = -1$) in this diffraction pattern, find the change Δx in this distance when the temperature T of the metal plate is changed by amount $\Delta T = 55$ C°. [Hint: Since $\lambda \ll D$, the first minima occur at a small angle.]

*Numerical/Computer

*86. (II) A student shined a laser light onto a single slit of width 0.04000 mm. He placed a screen at a distance of 1.490 m from the slit to observe the diffraction pattern of the laser light. The accompanying Table shows the distances of the dark fringes from the center of the central bright fringe for different orders.

Order number, m:	1	2	3	4	5	6	7	8
Distance (m)	0.0225	0.0445	0.0655	0.0870	0.1105	0.1320	0.1540	0.1775

Determine the angle of diffraction, θ, and $\sin \theta$ for each order. Make a graph of $\sin \theta$ vs. order number, m, and find the wavelength, λ, of the laser from the best-fit straight line.

*87. (III) Describe how to rotate the plane of polarization of a plane-polarized beam of light by 90° and produce only a 10% loss in intensity, using polarizers. Let N be the number of polarizers and θ be the (same) angle between successive polarizers.

88. (III) The "full-width at half-maximum" (FWHM) of the central peak for single-slit diffraction is defined as the angle $\Delta \theta$ between the two points on either side of center where the intensity is $\frac{1}{2} I_0$. (a) Determine $\Delta \theta$ in terms of (λ/D). Use graphs or a spreadsheet to solve $\sin \alpha = \alpha/\sqrt{2}$. (b) Determine $\Delta \theta$ (in degrees) for $D = \lambda$ and for $D = 100\lambda$.

Answers to Exercises

A: Narrower.

B: (b).

C: (c).

D: 0.28 nm.

E: Zero for both (a) and (b), because the two successive polarizers at 90° cancel all light. The 45° Polaroid must be inserted *between* the other two if any transmission is to occur.

Mathematical Formulas

A–1 Quadratic Formula

If $ax^2 + bx + c = 0$

then $x = \dfrac{-b \pm \sqrt{b^2 - 4ac}}{2a}$

A–2 Binomial Expansion

$$(1 \pm x)^n = 1 \pm nx + \frac{n(n-1)}{2!}x^2 \pm \frac{n(n-1)(n-2)}{3!}x^3 + \cdots$$

$$(x + y)^n = x^n\left(1 + \frac{y}{x}\right)^n = x^n\left(1 + n\frac{y}{x} + \frac{n(n-1)}{2!}\frac{y^2}{x^2} + \cdots\right)$$

A–3 Other Expansions

$$e^x = 1 + x + \frac{x^2}{2!} + \frac{x^3}{3!} + \cdots$$

$$\ln(1 + x) = x - \frac{x^2}{2} + \frac{x^3}{3} - \frac{x^4}{4} + \cdots$$

$$\sin\theta = \theta - \frac{\theta^3}{3!} + \frac{\theta^5}{5!} - \cdots$$

$$\cos\theta = 1 - \frac{\theta^2}{2!} + \frac{\theta^4}{4!} - \cdots$$

$$\tan\theta = \theta + \frac{\theta^3}{3} + \frac{2}{15}\theta^5 + \cdots \qquad |\theta| < \frac{\pi}{2}$$

In general: $f(x) = f(0) + \left(\dfrac{df}{dx}\right)_0 x + \left(\dfrac{d^2f}{dx^2}\right)_0 \dfrac{x^2}{2!} + \cdots$

A–4 Exponents

$$(a^n)(a^m) = a^{n+m}$$
$$(a^n)(b^n) = (ab)^n$$
$$(a^n)^m = a^{nm}$$

$$\frac{1}{a^n} = a^{-n}$$
$$a^n a^{-n} = a^0 = 1$$
$$a^{\frac{1}{2}} = \sqrt{a}$$

A–5 Areas and Volumes

Object	Surface area	Volume
Circle, radius r	πr^2	—
Sphere, radius r	$4\pi r^2$	$\frac{4}{3}\pi r^3$
Right circular cylinder, radius r, height h	$2\pi r^2 + 2\pi rh$	$\pi r^2 h$
Right circular cone, radius r, height h	$\pi r^2 + \pi r\sqrt{r^2 + h^2}$	$\frac{1}{3}\pi r^2 h$

A–6 Plane Geometry

1. *Equal angles:*

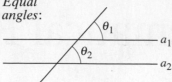

FIGURE A–1 If line a_1 is parallel to line a_2, then $\theta_1 = \theta_2$.

2. *Equal angles:*

FIGURE A–2 If $a_1 \perp a_2$ and $b_1 \perp b_2$, then $\theta_1 = \theta_2$.

3. The sum of the angles in any plane triangle is 180°.

4. *Pythagorean theorem:*

FIGURE A–3

In any right triangle (one angle = 90°) of sides a, b, and c:

$$a^2 + b^2 = c^2$$

where c is the length of the hypotenuse (opposite the 90° angle).

5. *Similar triangles:* Two triangles are said to be similar if all three of their angles are equal (in Fig. A–4, $\theta_1 = \phi_1$, $\theta_2 = \phi_2$, and $\theta_3 = \phi_3$). Similar triangles can have different sizes and different orientations.

(*a*) Two triangles are similar if any two of their angles are equal. (This follows because the third angles must also be equal since the sum of the angles of a triangle is 180°.)

(*b*) The ratios of corresponding sides of two similar triangles are equal (Fig. A–4):

$$\frac{a_1}{b_1} = \frac{a_2}{b_2} = \frac{a_3}{b_3}.$$

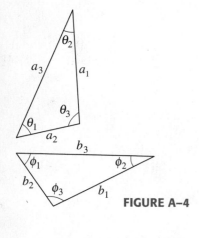

FIGURE A–4

6. *Congruent triangles:* Two triangles are congruent if one can be placed precisely on top of the other. That is, they are similar triangles and they have the same size. Two triangles are congruent if any of the following holds:

(*a*) The three corresponding sides are equal.

(*b*) Two sides and the enclosed angle are equal ("side-angle-side").

(*c*) Two angles and the enclosed side are equal ("angle-side-angle").

A–7 Logarithms

Logarithms are defined in the following way:

$$\text{if} \quad y = A^x, \quad \text{then} \quad x = \log_A y.$$

That is, the logarithm of a number y to the base A is that number which, as the exponent of A, gives back the number y. For **common logarithms**, the base is 10, so

$$\text{if} \quad y = 10^x, \quad \text{then} \quad x = \log y.$$

The subscript 10 on $\log_{10}$ is usually omitted when dealing with common logs. Another important base is the exponential base $e = 2.718\ldots$, a natural number. Such logarithms are called **natural logarithms** and are written ln. Thus,

$$\text{if} \quad y = e^x, \quad \text{then} \quad x = \ln y.$$

For any number y, the two types of logarithm are related by

$$\ln y = 2.3026 \log y.$$

Some simple rules for logarithms are as follows:

$$\log(ab) = \log a + \log b, \tag{i}$$

which is true because if $a = 10^n$ and $b = 10^m$, then $ab = 10^{n+m}$. From the definition of logarithm, $\log a = n$, $\log b = m$, and $\log(ab) = n + m$; hence, $\log(ab) = n + m = \log a + \log b$. In a similar way, we can show that

$$\log\left(\frac{a}{b}\right) = \log a - \log b \qquad \text{(ii)}$$

and

$$\log a^n = n \log a. \qquad \text{(iii)}$$

These three rules apply to any kind of logarithm.

If you do not have a calculator that calculates logs, you can easily use a **log table**, such as the small one shown here (Table A–1): the number N whose log we want is given to two digits. The first digit is in the vertical column to the left, the second digit is in the horizontal row across the top. For example, Table A–1 tells us that $\log 1.0 = 0.000$, $\log 1.1 = 0.041$, and $\log 4.1 = 0.613$. Table A–1 does not include the decimal point. The Table gives logs for numbers between 1.0 and 9.9. For larger or smaller numbers, we use rule (i) above, $\log(ab) = \log a + \log b$. For example, $\log(380) = \log(3.8 \times 10^2) = \log(3.8) + \log(10^2)$. From the Table, $\log 3.8 = 0.580$; and from rule (iii) above $\log(10^2) = 2\log(10) = 2$, since $\log(10) = 1$. [This follows from the definition of the logarithm: if $10 = 10^1$, then $1 = \log(10)$.] Thus,

$$
\begin{aligned}
\log(380) &= \log(3.8) + \log(10^2) \\
&= 0.580 + 2 \\
&= 2.580.
\end{aligned}
$$

Similarly,

$$
\begin{aligned}
\log(0.081) &= \log(8.1) + \log(10^{-2}) \\
&= 0.908 - 2 = -1.092.
\end{aligned}
$$

The reverse process of finding the number N whose log is, say, 2.670, is called "taking the **antilogarithm**." To do so, we separate our number 2.670 into two parts, making the separation at the decimal point:

$$
\begin{aligned}
\log N &= 2.670 = 2 + 0.670 \\
&= \log 10^2 + 0.670.
\end{aligned}
$$

We now look at Table A–1 to see what number has its log equal to 0.670; none does, so we must **interpolate**: we see that $\log 4.6 = 0.663$ and $\log 4.7 = 0.672$. So the number we want is between 4.6 and 4.7, and closer to the latter by $\frac{7}{9}$. Approximately we can say that $\log 4.68 = 0.670$. Thus

$$
\begin{aligned}
\log N &= 2 + 0.670 \\
&= \log(10^2) + \log(4.68) = \log(4.68 \times 10^2),
\end{aligned}
$$

so $N = 4.68 \times 10^2 = 468$.

If the given logarithm is negative, say, -2.180, we proceed as follows:

$$
\begin{aligned}
\log N &= -2.180 = -3 + 0.820 \\
&= \log 10^{-3} + \log 6.6 = \log 6.6 \times 10^{-3},
\end{aligned}
$$

so $N = 6.6 \times 10^{-3}$. Notice that we added to our given logarithm the next largest integer (3 in this case) so that we have an integer, plus a decimal number between 0 and 1.0 whose antilogarithm can be looked up in the Table.

TABLE A–1 Short Table of Common Logarithms

N	0.0	0.1	0.2	0.3	0.4	0.5	0.6	0.7	0.8	0.9
1	000	041	079	114	146	176	204	230	255	279
2	301	322	342	362	380	398	415	431	447	462
3	477	491	505	519	531	544	556	568	580	591
4	602	613	623	633	643	653	663	672	681	690
5	699	708	716	724	732	740	748	756	763	771
6	778	785	792	799	806	813	820	826	833	839
7	845	851	857	863	869	875	881	886	892	898
8	903	908	914	919	924	929	935	940	944	949
9	954	959	964	968	973	978	982	987	991	996

A–8 Vectors

Vector addition is covered in Sections 3–2 to 3–5.
Vector multiplication is covered in Sections 3–3, 7–2, and 11–2.

A–9 Trigonometric Functions and Identities

The trigonometric functions are defined as follows (see Fig. A–5, o = side opposite, a = side adjacent, h = hypotenuse. Values are given in Table A–2):

$$\sin \theta = \frac{o}{h} \qquad\qquad \csc \theta = \frac{1}{\sin \theta} = \frac{h}{o}$$

$$\cos \theta = \frac{a}{h} \qquad\qquad \sec \theta = \frac{1}{\cos \theta} = \frac{h}{a}$$

$$\tan \theta = \frac{o}{a} = \frac{\sin \theta}{\cos \theta} \qquad\qquad \cot \theta = \frac{1}{\tan \theta} = \frac{a}{o}$$

and recall that

$$a^2 + o^2 = h^2 \qquad\qquad \text{[Pythagorean theorem]}.$$

Figure A–6 shows the signs (+ or −) that cosine, sine, and tangent take on for angles θ in the four quadrants (0° to 360°). Note that angles are measured counterclockwise from the x axis as shown; negative angles are measured from *below* the x axis, clockwise: for example, $-30° = +330°$, and so on.

The following are some useful identities among the trigonometric functions:

$$\sin^2 \theta + \cos^2 \theta = 1$$

$$\sec^2 \theta - \tan^2 \theta = 1, \quad \csc^2 \theta - \cot^2 \theta = 1$$

$$\sin 2\theta = 2 \sin \theta \cos \theta$$

$$\cos 2\theta = \cos^2 \theta - \sin^2 \theta = 2 \cos^2 \theta - 1 = 1 - 2 \sin^2 \theta$$

$$\tan 2\theta = \frac{2 \tan \theta}{1 - \tan^2 \theta}$$

$$\sin(A \pm B) = \sin A \cos B \pm \cos A \sin B$$

$$\cos(A \pm B) = \cos A \cos B \mp \sin A \sin B$$

$$\tan(A \pm B) = \frac{\tan A \pm \tan B}{1 \mp \tan A \tan B}$$

$$\sin(180° - \theta) = \sin \theta$$

$$\cos(180° - \theta) = -\cos \theta$$

$$\sin(90° - \theta) = \cos \theta$$

$$\cos(90° - \theta) = \sin \theta$$

$$\sin(-\theta) = -\sin \theta$$

$$\cos(-\theta) = \cos \theta$$

$$\tan(-\theta) = -\tan \theta$$

$$\sin\tfrac{1}{2}\theta = \sqrt{\frac{1 - \cos \theta}{2}}, \quad \cos\tfrac{1}{2}\theta = \sqrt{\frac{1 + \cos \theta}{2}}, \quad \tan\tfrac{1}{2}\theta = \sqrt{\frac{1 - \cos \theta}{1 + \cos \theta}}$$

$$\sin A \pm \sin B = 2 \sin\left(\frac{A \pm B}{2}\right) \cos\left(\frac{A \mp B}{2}\right).$$

For any triangle (see Fig. A–7):

$$\frac{\sin \alpha}{a} = \frac{\sin \beta}{b} = \frac{\sin \gamma}{c} \qquad\qquad \text{[Law of sines]}$$

$$c^2 = a^2 + b^2 - 2ab \cos \gamma. \qquad\qquad \text{[Law of cosines]}$$

Values of sine, cosine, tangent are given in Table A–2.

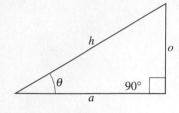

FIGURE A–5

FIGURE A–6

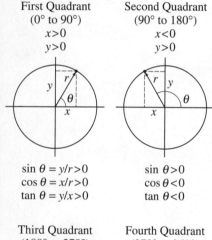

First Quadrant
(0° to 90°)
$x > 0$
$y > 0$

$\sin \theta = y/r > 0$
$\cos \theta = x/r > 0$
$\tan \theta = y/x > 0$

Second Quadrant
(90° to 180°)
$x < 0$
$y > 0$

$\sin \theta > 0$
$\cos \theta < 0$
$\tan \theta < 0$

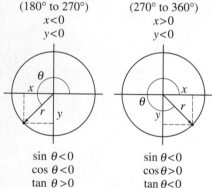

Third Quadrant
(180° to 270°)
$x < 0$
$y < 0$

$\sin \theta < 0$
$\cos \theta < 0$
$\tan \theta > 0$

Fourth Quadrant
(270° to 360°)
$x > 0$
$y < 0$

$\sin \theta < 0$
$\cos \theta > 0$
$\tan \theta < 0$

FIGURE A–7

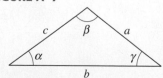

TABLE A–2 Trigonometric Table: Numerical Values of Sin, Cos, Tan

Angle in Degrees	Angle in Radians	Sine	Cosine	Tangent	Angle in Degrees	Angle in Radians	Sine	Cosine	Tangent
0°	0.000	0.000	1.000	0.000					
1°	0.017	0.017	1.000	0.017	46°	0.803	0.719	0.695	1.036
2°	0.035	0.035	0.999	0.035	47°	0.820	0.731	0.682	1.072
3°	0.052	0.052	0.999	0.052	48°	0.838	0.743	0.669	1.111
4°	0.070	0.070	0.998	0.070	49°	0.855	0.755	0.656	1.150
5°	0.087	0.087	0.996	0.087	50°	0.873	0.766	0.643	1.192
6°	0.105	0.105	0.995	0.105	51°	0.890	0.777	0.629	1.235
7°	0.122	0.122	0.993	0.123	52°	0.908	0.788	0.616	1.280
8°	0.140	0.139	0.990	0.141	53°	0.925	0.799	0.602	1.327
9°	0.157	0.156	0.988	0.158	54°	0.942	0.809	0.588	1.376
10°	0.175	0.174	0.985	0.176	55°	0.960	0.819	0.574	1.428
11°	0.192	0.191	0.982	0.194	56°	0.977	0.829	0.559	1.483
12°	0.209	0.208	0.978	0.213	57°	0.995	0.839	0.545	1.540
13°	0.227	0.225	0.974	0.231	58°	1.012	0.848	0.530	1.600
14°	0.244	0.242	0.970	0.249	59°	1.030	0.857	0.515	1.664
15°	0.262	0.259	0.966	0.268	60°	1.047	0.866	0.500	1.732
16°	0.279	0.276	0.961	0.287	61°	1.065	0.875	0.485	1.804
17°	0.297	0.292	0.956	0.306	62°	1.082	0.883	0.469	1.881
18°	0.314	0.309	0.951	0.325	63°	1.100	0.891	0.454	1.963
19°	0.332	0.326	0.946	0.344	64°	1.117	0.899	0.438	2.050
20°	0.349	0.342	0.940	0.364	65°	1.134	0.906	0.423	2.145
21°	0.367	0.358	0.934	0.384	66°	1.152	0.914	0.407	2.246
22°	0.384	0.375	0.927	0.404	67°	1.169	0.921	0.391	2.356
23°	0.401	0.391	0.921	0.424	68°	1.187	0.927	0.375	2.475
24°	0.419	0.407	0.914	0.445	69°	1.204	0.934	0.358	2.605
25°	0.436	0.423	0.906	0.466	70°	1.222	0.940	0.342	2.747
26°	0.454	0.438	0.899	0.488	71°	1.239	0.946	0.326	2.904
27°	0.471	0.454	0.891	0.510	72°	1.257	0.951	0.309	3.078
28°	0.489	0.469	0.883	0.532	73°	1.274	0.956	0.292	3.271
29°	0.506	0.485	0.875	0.554	74°	1.292	0.961	0.276	3.487
30°	0.524	0.500	0.866	0.577	75°	1.309	0.966	0.259	3.732
31°	0.541	0.515	0.857	0.601	76°	1.326	0.970	0.242	4.011
32°	0.559	0.530	0.848	0.625	77°	1.344	0.974	0.225	4.331
33°	0.576	0.545	0.839	0.649	78°	1.361	0.978	0.208	4.705
34°	0.593	0.559	0.829	0.675	79°	1.379	0.982	0.191	5.145
35°	0.611	0.574	0.819	0.700	80°	1.396	0.985	0.174	5.671
36°	0.628	0.588	0.809	0.727	81°	1.414	0.988	0.156	6.314
37°	0.646	0.602	0.799	0.754	82°	1.431	0.990	0.139	7.115
38°	0.663	0.616	0.788	0.781	83°	1.449	0.993	0.122	8.144
39°	0.681	0.629	0.777	0.810	84°	1.466	0.995	0.105	9.514
40°	0.698	0.643	0.766	0.839	85°	1.484	0.996	0.087	11.43
41°	0.716	0.656	0.755	0.869	86°	1.501	0.998	0.070	14.301
42°	0.733	0.669	0.743	0.900	87°	1.518	0.999	0.052	19.081
43°	0.750	0.682	0.731	0.933	88°	1.536	0.999	0.035	28.636
44°	0.768	0.695	0.719	0.966	89°	1.553	1.000	0.017	57.290
45°	0.785	0.707	0.707	1.000	90°	1.571	1.000	0.000	∞

Derivatives and Integrals

B–1 Derivatives: General Rules

(See also Section 2–3.)

$$\frac{dx}{dx} = 1$$

$$\frac{d}{dx}[af(x)] = a\frac{df}{dx} \qquad [a = \text{constant}]$$

$$\frac{d}{dx}[f(x) + g(x)] = \frac{df}{dx} + \frac{dg}{dx}$$

$$\frac{d}{dx}[f(x)g(x)] = \frac{df}{dx}g + f\frac{dg}{dx}$$

$$\frac{d}{dx}[f(y)] = \frac{df}{dy}\frac{dy}{dx} \qquad [\text{chain rule}]$$

$$\frac{dx}{dy} = \frac{1}{\left(\dfrac{dy}{dx}\right)} \qquad \text{if } \frac{dy}{dx} \neq 0.$$

B–2 Derivatives: Particular Functions

$$\frac{da}{dx} = 0 \qquad [a = \text{constant}]$$

$$\frac{d}{dx}x^n = nx^{n-1}$$

$$\frac{d}{dx}\sin ax = a\cos ax$$

$$\frac{d}{dx}\cos ax = -a\sin ax$$

$$\frac{d}{dx}\tan ax = a\sec^2 ax$$

$$\frac{d}{dx}\ln ax = \frac{1}{x}$$

$$\frac{d}{dx}e^{ax} = ae^{ax}$$

B–3 Indefinite Integrals: General Rules

(See also Section 7–3.)

$$\int dx = x$$

$$\int af(x)\,dx = a\int f(x)\,dx \qquad [a = \text{constant}]$$

$$\int [f(x) + g(x)]\,dx = \int f(x)\,dx + \int g(x)\,dx$$

$$\int u\,dv = uv - \int v\,du \qquad [\text{integration by parts: see also B–6}]$$

B–4 Indefinite Integrals: Particular Functions

(An arbitrary constant can be added to the right side of each equation.)

$$\int a\, dx = ax \quad [a = \text{constant}]$$

$$\int x^m\, dx = \frac{1}{m+1} x^{m+1} \quad [m \neq -1]$$

$$\int \sin ax\, dx = -\frac{1}{a}\cos ax$$

$$\int \cos ax\, dx = \frac{1}{a}\sin ax$$

$$\int \tan ax\, dx = \frac{1}{a}\ln|\sec ax|$$

$$\int \frac{1}{x}\, dx = \ln x$$

$$\int e^{ax}\, dx = \frac{1}{a}e^{ax}$$

$$\int \frac{dx}{\sqrt{x^2 \pm a^2}} = \ln(x + \sqrt{x^2 \pm a^2})$$

$$\int \frac{dx}{\sqrt{a^2 - x^2}} = \sin^{-1}\left(\frac{x}{a}\right) = -\cos^{-1}\left(\frac{x}{a}\right) \quad [\text{if } x^2 \le a^2]$$

$$\int \frac{dx}{(x^2 \pm a^2)^{\frac{3}{2}}} = \frac{\pm x}{a^2\sqrt{x^2 \pm a^2}}$$

$$\int \frac{x\, dx}{(x^2 \pm a^2)^{\frac{3}{2}}} = \frac{-1}{\sqrt{x^2 \pm a^2}}$$

$$\int \sin^2 ax\, dx = \frac{x}{2} - \frac{\sin 2ax}{4a}$$

$$\int xe^{-ax}\, dx = -\frac{e^{-ax}}{a^2}(ax + 1)$$

$$\int x^2 e^{-ax}\, dx = -\frac{e^{-ax}}{a^3}(a^2 x^2 + 2ax + 2)$$

$$\int \frac{dx}{x^2 + a^2} = \frac{1}{a}\tan^{-1}\frac{x}{a}$$

$$\int \frac{dx}{x^2 - a^2} = \frac{1}{2a}\ln\left(\frac{x-a}{x+a}\right) \quad [x^2 > a^2]$$

$$= -\frac{1}{2a}\ln\left(\frac{a+x}{a-x}\right) \quad [x^2 < a^2]$$

B–5 A Few Definite Integrals

$$\int_0^\infty x^n e^{-ax}\, dx = \frac{n!}{a^{n+1}}$$

$$\int_0^\infty e^{-ax^2}\, dx = \sqrt{\frac{\pi}{4a}}$$

$$\int_0^\infty xe^{-ax^2}\, dx = \frac{1}{2a}$$

$$\int_0^\infty x^2 e^{-ax^2}\, dx = \sqrt{\frac{\pi}{16a^3}}$$

$$\int_0^\infty x^3 e^{-ax^2}\, dx = \frac{1}{2a^2}$$

$$\int_0^\infty x^{2n} e^{-ax^2}\, dx = \frac{1\cdot 3\cdot 5\cdots(2n-1)}{2^{n+1}a^n}\sqrt{\frac{\pi}{a}}$$

B–6 Integration by Parts

Sometimes a difficult integral can be simplified by carefully choosing the functions u and v in the identity:

$$\int u\, dv = uv - \int v\, du. \qquad [\text{Integration by parts}]$$

This identity follows from the property of derivatives

$$\frac{d}{dx}(uv) = u\frac{dv}{dx} + v\frac{du}{dx}$$

or as differentials: $d(uv) = u\, dv + v\, du$.

For example $\int xe^{-x}\, dx$ can be integrated by choosing $u = x$ and $dv = e^{-x} dx$ in the "integration by parts" equation above:

$$\int xe^{-x}\, dx = (x)(-e^{-x}) + \int e^{-x}\, dx$$

$$= -xe^{-x} - e^{-x} = -(x+1)e^{-x}.$$

More on Dimensional Analysis

An important use of dimensional analysis (Section 1–7) is to obtain the *form* of an equation: how one quantity depends on others. To take a concrete example, let us try to find an expression for the period T of a simple pendulum. First, we try to figure out what T could depend on, and make a list of these variables. It might depend on its length ℓ, on the mass m of the bob, on the angle of swing θ, and on the acceleration due to gravity, g. It might also depend on air resistance (we would use the viscosity of air), the gravitational pull of the Moon, and so on; but everyday experience suggests that the Earth's gravity is the major force involved, so we ignore the other possible forces. So let us assume that T is a function of ℓ, m, θ, and g, and that each of these factors is present to some power:

$$T = C\ell^w m^x \theta^y g^z.$$

C is a dimensionless constant, and w, x, y, and z are exponents we want to solve for. We now write down the dimensional equation (Section 1–7) for this relationship:

$$[T] = [L]^w [M]^x [L/T^2]^z.$$

Because θ has no dimensions (a radian is a length divided by a length—see Eq. 10–1a), it does not appear. We simplify and obtain

$$[T] = [L]^{w+z} [M]^x [T]^{-2z}$$

To have dimensional consistency, we must have

$$1 = -2z$$
$$0 = w + z$$
$$0 = x.$$

We solve these equations and find that $z = -\frac{1}{2}$, $w = \frac{1}{2}$, and $x = 0$. Thus our desired equation must be

$$T = C\sqrt{\ell/g}\, f(\theta), \tag{C–1}$$

where $f(\theta)$ is some function of θ that we cannot determine using this technique. Nor can we determine in this way the dimensionless constant C. (To obtain C and f, we would have to do an analysis such as that in Chapter 14 using Newton's laws, which reveals that $C = 2\pi$ and $f \approx 1$ for small θ). But look what we *have* found, using only dimensional consistency. We obtained the form of the expression that relates the period of a simple pendulum to the major variables of the situation, ℓ and g (see Eq. 14–12c), and saw that it does not depend on the mass m.

How did we do it? And how useful is this technique? Basically, we had to use our intuition as to which variables were important and which were not. This is not always easy, and often requires a lot of insight. As to usefulness, the final result in our example could have been obtained from Newton's laws, as in Chapter 14. But in many physical situations, such a derivation from other laws cannot be done. In those situations, dimensional analysis can be a powerful tool.

In the end, any expression derived by the use of dimensional analysis (or by any other means, for that matter) must be checked against experiment. For example, in our derivation of Eq. C–1, we can compare the periods of two pendulums of different lengths, ℓ_1 and ℓ_2, whose amplitudes (θ) are the same. For, using Eq. C–1, we would have

$$\frac{T_1}{T_2} = \frac{C\sqrt{\ell_1/g}\, f(\theta)}{C\sqrt{\ell_2/g}\, f(\theta)} = \sqrt{\frac{\ell_1}{\ell_2}}.$$

Because C and $f(\theta)$ are the same for both pendula, they cancel out, so we can experimentally determine if the ratio of the periods varies as the ratio of the square roots of the lengths. This comparison to experiment checks our derivation, at least in part; C and $f(\theta)$ could be determined by further experiments.

D Gravitational Force due to a Spherical Mass Distribution

In Chapter 6 we stated that the gravitational force exerted by or on a uniform sphere acts as if all the mass of the sphere were concentrated at its center, if the other object (exerting or feeling the force) is outside the sphere. In other words, the gravitational force that a uniform sphere exerts on a particle outside it is

$$ F = G\frac{mM}{r^2}, \qquad [m \text{ outside sphere of mass } M] $$

where m is the mass of the particle, M the mass of the sphere, and r the distance of m from the center of the sphere. Now we will derive this result. We will use the concepts of infinitesimally small quantities and integration.

First we consider a very thin, uniform spherical shell (like a thin-walled basketball) of mass M whose thickness t is small compared to its radius R (Fig. D–1). The force on a particle of mass m at a distance r from the center of the shell can be calculated as the vector sum of the forces due to all the particles of the shell. We imagine the shell divided up into thin (infinitesimal) circular strips so that all points on a strip are equidistant from our particle m. One of these circular strips, labeled AB, is shown in Fig. D–1. It is $R\,d\theta$ wide, t thick, and has a radius $R\sin\theta$. The force on our particle m due to a tiny piece of the strip at point A is represented by the vector $\vec{F}_A$ shown. The force due to a tiny piece of the strip at point B, which is diametrically opposite A, is the force $\vec{F}_B$. We take the two pieces at A and B to be of equal mass, so $F_A = F_B$. The horizontal components of $\vec{F}_A$ and $\vec{F}_B$ are each equal to

$$ F_A \cos\phi $$

and point toward the center of the shell. The vertical components of $\vec{F}_A$ and $\vec{F}_B$ are of equal magnitude and point in opposite directions, and so cancel. Since for every point on the strip there is a corresponding point diametrically opposite (as with A and B), we see that the net force due to the entire strip points toward the center of the shell. Its magnitude will be

$$ dF = G\frac{m\,dM}{\ell^2}\cos\phi, $$

where dM is the mass of the entire circular strip and ℓ is the distance from all points on the strip to m, as shown. We write dM in terms of the density ρ; by density we mean the mass per unit volume (Section 13–2). Hence, $dM = \rho\,dV$, where dV is the volume of the strip and equals $(2\pi R\sin\theta)(t)(R\,d\theta)$. Then the force dF due to the circular strip shown is

$$ dF = G\frac{m\rho 2\pi R^2 t \sin\theta\,d\theta}{\ell^2}\cos\phi. \qquad \textbf{(D–1)} $$

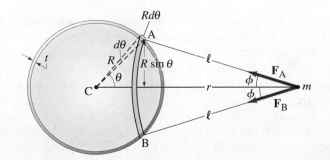

FIGURE D–1 Calculating the gravitational force on a particle of mass m due to a uniform spherical shell of radius R and mass M.

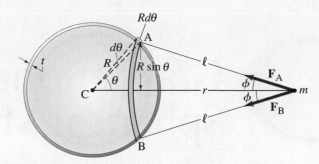

FIGURE D–1 (repeated)
Calculating the gravitational force on a particle of mass m due to a uniform spherical shell of radius R and mass M.

To get the total force F that the entire shell exerts on the particle m, we must integrate over all the circular strips: that is, we integrate

$$dF = G \frac{m\rho 2\pi R^2 t \sin\theta \, d\theta}{\ell^2} \cos\phi \qquad \textbf{(D–1)}$$

from $\theta = 0°$ to $\theta = 180°$. But our expression for dF contains ℓ and ϕ, which are functions of θ. From Fig. D–1 we can see that

$$\ell \cos\phi = r - R\cos\theta.$$

Furthermore, we can write the law of cosines for triangle CmA:

$$\cos\theta = \frac{r^2 + R^2 - \ell^2}{2rR}. \qquad \textbf{(D–2)}$$

With these two expressions we can reduce our three variables (ℓ, θ, ϕ) to only one, which we take to be ℓ. We do two things with Eq. D–2: (1) We put it into the equation for $\ell \cos\phi$ above:

$$\cos\phi = \frac{1}{\ell}(r - R\cos\theta) = \frac{r^2 + \ell^2 - R^2}{2r\ell}.$$

and (2) we take the differential of both sides of Eq. D–2 (because $\sin\theta \, d\theta$ appears in the expression for dF, Eq. D–1), considering r and R to be constants when summing over the strips:

$$-\sin\theta \, d\theta = -\frac{2\ell \, d\ell}{2rR} \quad \text{or} \quad \sin\theta \, d\theta = \frac{\ell \, d\ell}{rR}.$$

We insert these into Eq. D–1 for dF and find

$$dF = Gm\rho\pi t \frac{R}{r^2}\left(1 + \frac{r^2 - R^2}{\ell^2}\right) d\ell.$$

Now we integrate to get the net force on our thin shell of radius R. To integrate over all the strips ($\theta = 0°$ to $180°$), we must go from $\ell = r - R$ to $\ell = r + R$ (see Fig. D–1). Thus,

$$F = Gm\rho\pi t \frac{R}{r^2}\left[\ell - \frac{r^2 - R^2}{\ell}\right]_{\ell = r - R}^{\ell = r + R}$$

$$= Gm\rho\pi t \frac{R}{r^2}(4R).$$

The volume V of the spherical shell is its area $(4\pi R^2)$ times the thickness t. Hence the mass $M = \rho V = \rho 4\pi R^2 t$, and finally

$$F = G\frac{mM}{r^2}. \qquad \left[\begin{array}{c}\text{particle of mass } m \text{ outside a} \\ \text{thin uniform spherical shell of mass } M\end{array}\right]$$

This result gives us the force a thin shell exerts on a particle of mass m a distance r from the center of the shell, and *outside* the shell. We see that the force is the same as that between m and a particle of mass M at the center of the shell. In other words, for purposes of calculating the gravitational force exerted on or by a uniform spherical shell, we can consider all its mass concentrated at its center.

What we have derived for a shell holds also for a solid sphere, since a solid sphere can be considered as made up of many concentric shells, from $R = 0$ to $R = R_0$, where R_0 is the radius of the solid sphere. Why? Because if each shell has mass dM, we write for each shell, $dF = Gm\,dM/r^2$, where r is the distance from the center C to mass m and is the same for all shells. Then the total force equals the sum or integral over dM, which gives the total mass M. Thus the result

$$F = G\frac{mM}{r^2} \qquad \left[\begin{array}{c}\text{particle of mass } m \text{ outside}\\ \text{solid sphere of mass } M\end{array}\right] \quad \textbf{(D–3)}$$

is valid for a solid sphere of mass M even if the density varies with distance from the center. (It is not valid if the density varies within each shell—that is, depends not only on R.) Thus the gravitational force exerted on or by spherical objects, including nearly spherical objects like the Earth, Sun, and Moon, can be considered to act as if the objects were point particles.

This result, Eq. D–3, is true only if the mass m is outside the sphere. Let us next consider a point mass m that is located inside the spherical shell of Fig. D–1. Here, r would be less than R, and the integration over ℓ would be from $\ell = R - r$ to $\ell = R + r$, so

$$\left[\ell - \frac{r^2 - R^2}{\ell}\right]_{R-r}^{R+r} = 0.$$

Thus the force on any mass inside the shell would be zero. This result has particular importance for the electrostatic force, which is also an inverse square law. For the gravitational situation, we see that at points within a solid sphere, say 1000 km below the Earth's surface, only the mass up to that radius contributes to the net force. The outer shells beyond the point in question contribute zero net gravitational effect.

The results we have obtained here can also be reached using the gravitational analog of Gauss's law for electrostatics (Chapter 22).

E Differential Form of Maxwell's Equations

Maxwell's equations can be written in another form that is often more convenient than Eqs. 31–5. This material is usually covered in more advanced courses, and is included here simply for completeness.

We quote here two theorems, without proof, that are derived in vector analysis textbooks. The first is called **Gauss's theorem** or the **divergence theorem**. It relates the integral over a surface of any vector function $\vec{\mathbf{F}}$ to a volume integral over the volume enclosed by the surface:

$$\oint_{\text{Area } A} \vec{\mathbf{F}} \cdot d\vec{\mathbf{A}} = \int_{\text{Volume } V} \vec{\boldsymbol{\nabla}} \cdot \vec{\mathbf{F}} \, dV.$$

The operator $\vec{\boldsymbol{\nabla}}$ is the **del operator**, defined in Cartesian coordinates as

$$\vec{\boldsymbol{\nabla}} = \hat{\mathbf{i}} \frac{\partial}{\partial x} + \hat{\mathbf{j}} \frac{\partial}{\partial y} + \hat{\mathbf{k}} \frac{\partial}{\partial z}.$$

The quantity

$$\vec{\boldsymbol{\nabla}} \cdot \vec{\mathbf{F}} = \frac{\partial F_x}{\partial x} + \frac{\partial F_y}{\partial y} + \frac{\partial F_z}{\partial z}$$

is called the **divergence** of $\vec{\mathbf{F}}$. The second theorem is **Stokes's theorem**, and relates a line integral around a closed path to a surface integral over any surface enclosed by that path:

$$\oint_{\text{Line}} \vec{\mathbf{F}} \cdot d\vec{\boldsymbol{\ell}} = \int_{\text{Area } A} \vec{\boldsymbol{\nabla}} \times \vec{\mathbf{F}} \cdot d\vec{\mathbf{A}}.$$

The quantity $\vec{\boldsymbol{\nabla}} \times \vec{\mathbf{F}}$ is called the **curl** of $\vec{\mathbf{F}}$. (See Section 11–2 on the vector product.)

We now use these two theorems to obtain the differential form of Maxwell's equations in free space. We apply Gauss's theorem to Eq. 31–5a (Gauss's law):

$$\oint_A \vec{\mathbf{E}} \cdot d\vec{\mathbf{A}} = \int \vec{\boldsymbol{\nabla}} \cdot \vec{\mathbf{E}} \, dV = \frac{Q}{\epsilon_0}.$$

Now the charge Q can be written as a volume integral over the charge density ρ: $Q = \int \rho \, dV$. Then

$$\int \vec{\boldsymbol{\nabla}} \cdot \vec{\mathbf{E}} \, dV = \frac{1}{\epsilon_0} \int \rho \, dV.$$

Both sides contain volume integrals over the same volume, and for this to be true over *any* volume, whatever its size or shape, the integrands must be equal:

$$\vec{\boldsymbol{\nabla}} \cdot \vec{\mathbf{E}} = \frac{\rho}{\epsilon_0}. \qquad \qquad \textbf{(E–1)}$$

This is the differential form of Gauss's law. The second of Maxwell's equations, $\oint \vec{\mathbf{B}} \cdot d\vec{\mathbf{A}} = 0$, is treated in the same way, and we obtain

$$\vec{\boldsymbol{\nabla}} \cdot \vec{\mathbf{B}} = 0. \qquad \qquad \textbf{(E–2)}$$

Next, we apply Stokes's theorem to the third of Maxwell's equations,

$$\oint \vec{\mathbf{E}} \cdot d\vec{\boldsymbol{\ell}} = \int \vec{\boldsymbol{\nabla}} \times \vec{\mathbf{E}} \cdot d\vec{\mathbf{A}} = -\frac{d\Phi_B}{dt}.$$

Since the magnetic flux $\Phi_B = \int \vec{\mathbf{B}} \cdot d\vec{\mathbf{A}}$, we have

$$\int \vec{\boldsymbol{\nabla}} \times \vec{\mathbf{E}} \cdot d\vec{\mathbf{A}} = -\frac{\partial}{\partial t} \int \vec{\mathbf{B}} \cdot d\vec{\mathbf{A}}$$

where we use the partial derivative, $\partial \vec{\mathbf{B}}/\partial t$, since B may also depend on position. These are surface integrals over the same area, and to be true over any area, even a very small one, we must have

$$\vec{\boldsymbol{\nabla}} \times \vec{\mathbf{E}} = -\frac{\partial \vec{\mathbf{B}}}{\partial t}. \qquad \text{(E–3)}$$

This is the third of Maxwell's equations in differential form. Finally, to the last of Maxwell's equations,

$$\oint \vec{\mathbf{B}} \cdot d\vec{\boldsymbol{\ell}} = \mu_0 I + \mu_0 \epsilon_0 \frac{d\Phi_E}{dt},$$

we apply Stokes's theorem and write $\Phi_E = \int \vec{\mathbf{E}} \cdot d\vec{\mathbf{A}}$:

$$\int \vec{\boldsymbol{\nabla}} \times \vec{\mathbf{B}} \cdot d\vec{\mathbf{A}} = \mu_0 I + \mu_0 \epsilon_0 \frac{\partial}{\partial t} \int \vec{\mathbf{E}} \cdot d\vec{\mathbf{A}}.$$

The conduction current I can be written in terms of the current density $\vec{\mathbf{j}}$, using Eq. 25–12:

$$I = \int \vec{\mathbf{j}} \cdot d\vec{\mathbf{A}}.$$

Then Maxwell's fourth equation becomes:

$$\int \vec{\boldsymbol{\nabla}} \times \vec{\mathbf{B}} \cdot d\vec{\mathbf{A}} = \mu_0 \int \vec{\mathbf{j}} \cdot d\vec{\mathbf{A}} + \mu_0 \epsilon_0 \frac{\partial}{\partial t} \int \vec{\mathbf{E}} \cdot d\vec{\mathbf{A}}.$$

For this to be true over any area A, whatever its size or shape, the integrands on each side of the equation must be equal:

$$\vec{\boldsymbol{\nabla}} \times \vec{\mathbf{B}} = \mu_0 \vec{\mathbf{j}} + \mu_0 \epsilon_0 \frac{\partial \vec{\mathbf{E}}}{\partial t}. \qquad \text{(E–4)}$$

Equations E–1, 2, 3, and 4 are Maxwell's equations in differential form for free space. They are summarized in Table E–1.

TABLE E–1 Maxwell's Equations in Free Space[†]

Integral form	Differential form
$\oint \vec{\mathbf{E}} \cdot d\vec{\mathbf{A}} = \dfrac{Q}{\epsilon_0}$	$\vec{\boldsymbol{\nabla}} \cdot \vec{\mathbf{E}} = \dfrac{\rho}{\epsilon_0}$
$\oint \vec{\mathbf{B}} \cdot d\vec{\mathbf{A}} = 0$	$\vec{\boldsymbol{\nabla}} \cdot \vec{\mathbf{B}} = 0$
$\oint \vec{\mathbf{E}} \cdot d\vec{\boldsymbol{\ell}} = -\dfrac{d\Phi_B}{dt}$	$\vec{\boldsymbol{\nabla}} \times \vec{\mathbf{E}} = -\dfrac{\partial \vec{\mathbf{B}}}{\partial t}$
$\oint \vec{\mathbf{B}} \cdot d\vec{\boldsymbol{\ell}} = \mu_0 I + \mu_0 \epsilon_0 \dfrac{d\Phi_E}{dt}$	$\vec{\boldsymbol{\nabla}} \times \vec{\mathbf{B}} = \mu_0 \vec{\mathbf{j}} + \mu_0 \epsilon_0 \dfrac{\partial \vec{\mathbf{E}}}{\partial t}$

[†]$\vec{\boldsymbol{\nabla}}$ stands for the *del operator* $\vec{\boldsymbol{\nabla}} = \hat{\mathbf{i}}\dfrac{\partial}{\partial x} + \hat{\mathbf{j}}\dfrac{\partial}{\partial y} + \hat{\mathbf{k}}\dfrac{\partial}{\partial z}$ in Cartesian coordinates.

Selected Isotopes

(1) Atomic Number Z	(2) Element	(3) Symbol	(4) Mass Number A	(5) Atomic Mass†	(6) % Abundance (or Radioactive Decay‡ Mode)	(7) Half-life (if radioactive)
0	(Neutron)	n	1	1.008665	β^-	10.23 min
1	Hydrogen	H	1	1.007825	99.9885%	
	Deuterium	d or D	2	2.014082	0.0115%	
	Tritium	t or T	3	3.016049	β^-	12.312 yr
2	Helium	He	3	3.016029	0.000137%	
			4	4.002603	99.999863%	
3	Lithium	Li	6	6.015123	7.59%	
			7	7.016005	92.41%	
4	Beryllium	Be	7	7.016930	EC, γ	53.22 days
			9	9.012182	100%	
5	Boron	B	10	10.012937	19.9%	
			11	11.009305	80.1%	
6	Carbon	C	11	11.011434	β^+, EC	20.370 min
			12	12.000000	98.93%	
			13	13.003355	1.07%	
			14	14.003242	β^-	5730 yr
7	Nitrogen	N	13	13.005739	β^+, EC	9.9670 min
			14	14.003074	99.632%	
			15	15.000109	0.368%	
8	Oxygen	O	15	15.003066	β^+, EC	122.5 min
			16	15.994915	99.757%	
			18	17.999161	0.205%	
9	Fluorine	F	19	18.998403	100%	
10	Neon	Ne	20	19.992440	90.48%	
			22	21.991385	9.25%	
11	Sodium	Na	22	21.994436	β^+, EC, γ	2.6027 yr
			23	22.989769	100%	
			24	23.990963	β^-, γ	14.9574 h
12	Magnesium	Mg	24	23.985042	78.99%	
13	Aluminum	Al	27	26.981539	100%	
14	Silicon	Si	28	27.976927	92.2297%	
			31	30.975363	β^-, γ	157.3 min
15	Phosphorus	P	31	30.973762	100%	
			32	31.973907	β^-	14.284 days

† The masses given in column (5) are those for the neutral atom, including the Z electrons.

‡ Chapter 41; EC = electron capture.

(1) Atomic Number Z	(2) Element	(3) Symbol	(4) Mass Number A	(5) Atomic Mass	(6) % Abundance (or Radioactive Decay Mode)	(7) Half-life (if radioactive)
16	Sulfur	S	32	31.972071	94.9%	
			35	34.969032	β^-	87.32 days
17	Chlorine	Cl	35	34.968853	75.78%	
			37	36.965903	24.22%	
18	Argon	Ar	40	39.962383	99.600%	
19	Potassium	K	39	38.963707	93.258%	
			40	39.963998	0.0117%	
					β^-, EC, γ, β^+	1.265×10^9 yr
20	Calcium	Ca	40	39.962591	96.94%	
21	Scandium	Sc	45	44.955912	100%	
22	Titanium	Ti	48	47.947946	73.72%	
23	Vanadium	V	51	50.943960	99.750%	
24	Chromium	Cr	52	51.940508	83.789%	
25	Manganese	Mn	55	54.938045	100%	
26	Iron	Fe	56	55.934938	91.75%	
27	Cobalt	Co	59	58.933195	100%	
			60	59.933817	β^-, γ	5.2710 yr
28	Nickel	Ni	58	57.935343	68.077%	
			60	59.930786	26.223%	
29	Copper	Cu	63	62.929598	69.17%	
			65	64.927790	30.83%	
30	Zinc	Zn	64	63.929142	48.6%	
			66	65.926033	27.9%	
31	Gallium	Ga	69	68.925574	60.108%	
32	Germanium	Ge	72	71.922076	27.5%	
			74	73.921178	36.3%	
33	Arsenic	As	75	74.921596	100%	
34	Selenium	Se	80	79.916521	49.6%	
35	Bromine	Br	79	78.918337	50.69%	
36	Krypton	Kr	84	83.911507	57.00%	
37	Rubidium	Rb	85	84.911790	72.17%	
38	Strontium	Sr	86	85.909260	9.86%	
			88	87.905612	82.58%	
			90	89.907738	β^-	28.80 yr
39	Yttrium	Y	89	88.905848	100%	
40	Zirconium	Zr	90	89.904704	51.4%	
41	Niobium	Nb	93	92.906378	100%	
42	Molybdenum	Mo	98	97.905408	24.1%	
43	Technetium	Tc	98	97.907216	β^-, γ	4.2×10^6 yr
44	Ruthenium	Ru	102	101.904349	31.55%	
45	Rhodium	Rh	103	102.905504	100%	
46	Palladium	Pd	106	105.903486	27.33%	
47	Silver	Ag	107	106.905097	51.839%	
			109	108.904752	48.161%	
48	Cadmium	Cd	114	113.903359	28.7%	
49	Indium	In	115	114.903878	95.71%; β^-	4.41×10^{14} yr
50	Tin	Sn	120	119.902195	32.58%	
51	Antimony	Sb	121	120.903816	57.21%	

(1) Atomic Number Z	(2) Element	(3) Symbol	(4) Mass Number A	(5) Atomic Mass	(6) % Abundance (or Radioactive Decay Mode)	(7) Half-life (if radioactive)
52	Tellurium	Te	130	129.906224	34.1%; $\beta^-\beta^-$	$>9.7 \times 10^{22}$ yr
53	Iodine	I	127	126.904473	100%	
			131	130.906125	β^-, γ	8.0233 days
54	Xenon	Xe	132	131.904154	26.89%	
			136	135.907219	8.87%; $\beta^-\beta^-$	$>8.5 \times 10^{21}$ yr
55	Cesium	Cs	133	132.905452	100%	
56	Barium	Ba	137	136.905827	11.232%	
			138	137.905247	71.70%	
57	Lanthanum	La	139	138.906353	99.910%	
58	Cerium	Ce	140	139.905439	88.45%	
59	Praseodymium	Pr	141	140.907653	100%	
60	Neodymium	Nd	142	141.907723	27.2%	
61	Promethium	Pm	145	144.912749	EC, α	17.7 yr
62	Samarium	Sm	152	151.919732	26.75%	
63	Europium	Eu	153	152.921230	52.19%	
64	Gadolinium	Gd	158	157.924104	24.84%	
65	Terbium	Tb	159	158.925347	100%	
66	Dysprosium	Dy	164	163.929175	28.2%	
67	Holmium	Ho	165	164.930322	100%	
68	Erbium	Er	166	165.930293	33.6%	
69	Thulium	Tm	169	168.934213	100%	
70	Ytterbium	Yb	174	173.938862	31.8%	
71	Lutetium	Lu	175	174.940772	97.41%	
72	Hafnium	Hf	180	179.946550	35.08%	
73	Tantalum	Ta	181	180.947996	99.988%	
74	Tungsten (wolfram)	W	184	183.950931	30.64%; α	$>8.9 \times 10^{21}$ yr
75	Rhenium	Re	187	186.955753	62.60%; β^-	4.35×10^{10} yr
76	Osmium	Os	191	190.960930	β^-, γ	15.4 days
			192	191.961481	40.78%	
77	Iridium	Ir	191	190.960594	37.3%	
			193	192.962926	62.7%	
78	Platinum	Pt	195	194.964791	33.832%	
79	Gold	Au	197	196.966569	100%	
80	Mercury	Hg	199	198.968280	16.87%	
			202	201.970643	29.9%	
81	Thallium	Tl	205	204.974428	70.476%	
82	Lead	Pb	206	205.974465	24.1%	
			207	206.975897	22.1%	
			208	207.976652	52.4%	
			210	209.984188	β^-, γ, α	22.23 yr
			211	210.988737	β^-, γ	36.1 min
			212	211.991898	β^-, γ	10.64 h
			214	213.999805	β^-, γ	26.8 min
83	Bismuth	Bi	209	208.980399	100%	
			211	210.987269	α, γ, β^-	2.14 min
84	Polonium	Po	210	209.982874	α, γ, EC	138.376 days
			214	213.995201	α, γ	162.3 μs
85	Astatine	At	218	218.008694	α, β^-	1.4 s

(1) Atomic Number Z	(2) Element	(3) Symbol	(4) Mass Number A	(5) Atomic Mass	(6) % Abundance (or Radioactive Decay Mode)	(7) Half-life (if radioactive)
86	Radon	Rn	222	222.017578	α, γ	3.8235 days
87	Francium	Fr	223	223.019736	β^-, γ, α	22.00 min
88	Radium	Ra	226	226.025410	α, γ	1600 yr
89	Actinium	Ac	227	227.027752	β^-, γ, α	21.772 yr
90	Thorium	Th	228	228.028741	α, γ	1.9116 yr
			232	232.038055	100%; α, γ	1.405×10^{10} yr
91	Protactinium	Pa	231	231.035884	α, γ	3.276×10^4 yr
92	Uranium	U	232	232.037156	α, γ	68.9 yr
			233	233.039635	α, γ	1.592×10^5 yr
			235	235.043930	0.720%; α, γ	7.04×10^8 yr
			236	236.045568	α, γ	2.342×10^7 yr
			238	238.050788	99.274%; α, γ	4.468×10^9 yr
			239	239.054293	β^-, γ	23.45 min
93	Neptunium	Np	237	237.048173	α, γ	2.144×10^6 yr
			239	239.052939	β^-, γ	2.356 days
94	Plutonium	Pu	239	239.052163	α, γ	24,110 yr
			244	244.064204	α	8.00×10^7 yr
95	Americium	Am	243	243.061381	α, γ	7370 yr
96	Curium	Cm	247	247.070354	α, γ	1.56×10^7 yr
97	Berkelium	Bk	247	247.070307	α, γ	1380 yr
98	Californium	Cf	251	251.079587	α, γ	898 yr
99	Einsteinium	Es	252	252.082980	α, EC, γ	471.7 days
100	Fermium	Fm	257	257.095105	α, γ	100.5 days
101	Mendelevium	Md	258	258.098431	α, γ	51.5 days
102	Nobelium	No	259	259.10103	α, EC	58 min
103	Lawrencium	Lr	262	262.10963	$\alpha, EC,$ fission	≈ 4 h
104	Rutherfordium	Rf	263	263.11255	fission	10 min
105	Dubnium	Db	262	262.11408	$\alpha,$ fission, EC	35 s
106	Seaborgium	Sg	266	266.12210	$\alpha,$ fission	≈ 21 s
107	Bohrium	Bh	264	264.12460	α	≈ 0.44 s
108	Hassium	Hs	269	269.13406	α	≈ 10 s
109	Meitnerium	Mt	268	268.13870	α	21 ms
110	Darmstadtium	Ds	271	271.14606	α	≈ 70 ms
111	Roentgenium	Rg	272	272.15360	α	3.8 ms
112		Uub	277	277.16394	α	≈ 0.7 ms

Preliminary evidence (unconfirmed) has been reported for elements 113, 114, 115, 116 and 118.

Answers to Odd-Numbered Problems

CHAPTER 21

1. 2.7×10^{-3} N.

3. 7200 N.

5. $(4.9 \times 10^{-14})\%$.

7. 4.88 cm.

9. -5.8×10^8 C, 0.

11. (a) $q_1 = q_2 = \frac{1}{2}Q_T$;

    (b) $q_1 = 0$, $q_2 = Q_T$.

13. $F_1 = 0.53$ N at $265°$,

    $F_2 = 0.33$ N at $112°$,

    $F_3 = 0.26$ N at $53°$.

15. $F = 2.96 \times 10^7$ N, away from

    center of square.

17. 1.0×10^{12} electrons.

19. (a) $\pi\sqrt{\dfrac{md^3}{kQq}}$;

    (b) 0.2 ps.

21. 3.08×10^{-16} N west.

23. 1.10×10^7 N/C up.

25. $(172\,\hat{\mathbf{j}})$ N/C.

27. 1.01×10^{14} m/s^2, opposite to the

    field.

29.

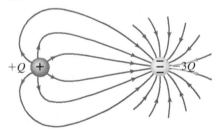

31. $(-4.7 \times 10^{11}\,\hat{\mathbf{i}})$ N/C

    $- (1.6 \times 10^{11}\,\hat{\mathbf{j}})$ N/C;

    or

    5.0×10^{11} N/C at $199°$.

33. $E = 2.60 \times 10^4$ N/C, away from

    the center.

35. $\dfrac{4kQxa}{(x^2 - a^2)^2}$, left.

37. $\dfrac{\lambda}{2\pi\varepsilon_0}\sqrt{\dfrac{1}{x^2} + \dfrac{1}{y^2}}$, $\tan^{-1}\dfrac{x}{y}$.

39.

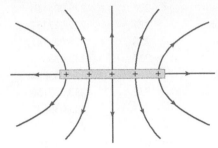

41. $\frac{1}{4}$.

43. (a) $\dfrac{Qy}{2\pi\varepsilon_0(y^2 + \ell^2)^{3/2}}$.

45. 1.8×10^6 N/C, away from the wire.

47. $\dfrac{8\lambda\ell z}{\pi\varepsilon_0(\ell^2 + 4z^2)\sqrt{4z^2 + 2\ell^2}}$, vertical.

49. $-\dfrac{2\lambda\sin\theta_0}{4\pi\varepsilon_0 R}\hat{\mathbf{i}}$.

51. (a) $\dfrac{\lambda}{4\pi\varepsilon_0\,x(x^2 + \ell^2)^{1/2}}$

    $\times (\ell\hat{\mathbf{i}} + [x - (x^2 + \ell^2)^{1/2}]\hat{\mathbf{j}})$.

53. $\dfrac{Q}{4\pi\varepsilon_0\,x(x + \ell)}$.

55. $\dfrac{Q(x\hat{\mathbf{i}} - \frac{2a}{\pi}\hat{\mathbf{j}})}{4\pi\varepsilon_0(x^2 + a^2)^{3/2}}$.

57. (a) $(-3.5 \times 10^{15}\,\text{m/s}^2)\,\hat{\mathbf{i}}$

    $- (1.41 \times 10^{16}\,\text{m/s}^2)\,\hat{\mathbf{j}}$;

    (b) $166°$ counterclockwise from the

    initial direction.

59. $-23°$.

61. (b) $2\pi\sqrt{\dfrac{4\pi\varepsilon_0\,mR^3}{qQ}}$.

63. (a) 3.4×10^{-20} C;

    (b) no;

    (c) 8.5×10^{-26} m·N;

    (d) 2.5×10^{-26} J.

65. (a) θ very small;

    (b) $\dfrac{1}{2\pi}\sqrt{\dfrac{pE}{I}}$.

67. (a) In the direction of the dipole.

69. 3.5×10^9 C.

71. 6.8×10^5 C, negative.

73. 1.0×10^7 electrons.

75. 5.71×10^{13} C.

77. 1.6 m from Q_2, 3.6 m from Q_1.

79. $\dfrac{1.08 \times 10^7}{[3.00 - \cos(13.9t)]^2}$ N/C (upwards).

81. 5×10^{-9} C.

83. 8.0×10^{-9} C.

85. $18°$.

87. $E_A = 3.4 \times 10^4$ N/C, to the right;

    $E_B = 2.3 \times 10^4$ N/C, to the left;

    $E_C = 5.6 \times 10^3$ N/C, to the right;

    $E_D = 3.4 \times 10^3$ N/C, to the left.

89. -7.66×10^{-6} C, unstable.

91. (a) 9.18×10^6 N/C, down;

    (b) 1.63×10^{-4} C/m^2.

93. (a) $\dfrac{a}{\sqrt{2}} = 7.07$ cm;

    (b) yes;

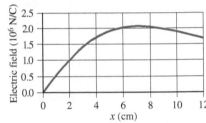

    (c) and (d)

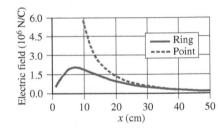

    (e) 37 cm.

CHAPTER 22

1. (a) 31 N·m^2/C;

    (b) 22 N·m^2/C;

    (c) 0.

3. (a) 0;

    (b) 0, 0, 0, 0, $E_0\ell^2$, $-E_0\ell^2$.

5. 1.63×10^{-8} C.

7. (a) -1.1×10^5 N·m^2/C;

    (b) 0.

9. -8.3×10^{-7} C.

11. 4.3×10^{-5} C/m.

13. -8.52×10^{-11} C.

15. (a) -2.6×10^4 N/C (toward wire);

    (b) -8.6×10^4 N/C (toward wire).

17. (a) $-(1.9 \times 10^{11} \text{ N/C·m})r$;
(b) $-(1.1 \times 10^8 \text{ N·m}^2/\text{C})/r^2$
$+ (3.0 \times 10^{11} \text{ N/C·m})r$;
(c) $(4.1 \times 10^8 \text{ N·m}^2/\text{C})/r^2$;
(d) yes.

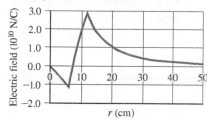

19.

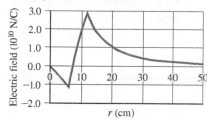

21. (a) 5.5×10^7 N/C (outward);
(b) 0;
(c) 5.5×10^5 N/C (outward).

23. (a) $-8.00 \, \mu\text{C}$;
(b) $+1.90 \, \mu\text{C}$.

25. (a) 0;
(b) $\dfrac{\sigma}{\varepsilon_0}$ (outward, if both plates are positive);
(c) same.

27. (a) 0;
(b) $\dfrac{r_1^2 \sigma_1}{\varepsilon_0 r^2}$;
(c) $\dfrac{(r_1^2 \sigma_1 + r_2^2 \sigma_2)}{\varepsilon_0 r^2}$;
(d) $\sigma_1 = -\left(\dfrac{r_2}{r_1}\right)^2 \sigma_2$;
(e) $\sigma_1 = 0$, or place $Q = -4\pi\sigma_1 r_1^2$ inside r_1.

29. (a) 0;
(b) $\dfrac{Q}{4\pi\varepsilon_0}\left(\dfrac{1}{r_0^3 - r_1^3}\right)\left(\dfrac{r^3 - r_1^3}{r^2}\right)$;
(c) $\dfrac{kQ}{r^2}$.

31. (a) $-q$;
(b) $Q + q$;
(c) $\dfrac{kq}{r^2}$;
(d) 0;
(e) $\dfrac{k(q + Q)}{r^2}$.

33. (a) $\dfrac{\sigma R_0}{\varepsilon_0 R}$, radially outward;
(b) 0;
(c) same for $R > R_0$ if $\lambda = 2\pi R_0 \sigma$.

35. (a) 0;
(b) $\dfrac{1}{2\pi\varepsilon_0}\dfrac{(Q/\ell)}{r}$;
(c) 0;
(d) $\dfrac{e}{4\pi\varepsilon_0}\left(\dfrac{Q}{\ell}\right)$.

37. (a) 1.9×10^7 m/s;
(b) 5.5×10^5 m/s.

39. (a) $\dfrac{\rho_E r}{3\varepsilon_0}$;
(b) $\dfrac{\rho_E r_0^3}{3\varepsilon_0 r^2}$;
(c) 0;
(d) $\left(\dfrac{\rho_E r_0^3}{3\varepsilon_0} + \dfrac{Q}{4\pi\varepsilon_0}\right)\dfrac{1}{r^2}$.

41. (a) 0;
(b) $\dfrac{Q}{2500\pi\varepsilon_0 R_0^2}$.

43. (a) $\dfrac{\rho_E d}{2\varepsilon_0}$ away from surface.

45. (a) 13 N (attractive);
(b) 0.064 J.

47. (a) 0;
(b) $-\dfrac{\rho_0(d - x)}{\varepsilon_0}\hat{\mathbf{i}}$;
(c) $-\dfrac{\rho_0(d + x)}{\varepsilon_0}\hat{\mathbf{i}}$.

49. $\dfrac{Q}{4\pi\varepsilon_0}\dfrac{r^2}{r_0^4}$, radially outward.

51. $\Phi = \oint \vec{g} \cdot d\vec{A} = -4\pi G M_{\text{enc}}$.

53. $a\ell^3 \varepsilon_0$.

55. 475 N·m²/C, 475 N·m²/C.

57. (a) 0;
(b) $E_{\text{max}} = \dfrac{Q}{\pi\varepsilon_0 r_0^2}$, $E_{\text{min}} = \dfrac{Q}{25\pi\varepsilon_0 r_0^2}$;
(c) no;
(d) no.

59. (a) 1.1×10^{-19} C;
(b) 3.5×10^{11} N/C.

61. (a) $\dfrac{\rho_E r_0}{6\varepsilon_0}$, right;
(b) $\dfrac{17}{54}\dfrac{\rho_E r_0}{\varepsilon_0}$, left.

63. (a) 0;
(b) 5.65×10^5 N/C, right;
(c) 5.65×10^5 N/C, right;
(d) -5.00×10^{-6} C/m³;
(e) $+5.00 \times 10^{-6}$ C/m³.

65. (a) On inside surface of shell.
(b) $r < 0.10$ m,
$$E = \left(\dfrac{2.7 \times 10^4}{r^2}\right) \text{N/C};$$
$r > 0.10$ m, $E = 0$.

67. -46 N·m²/C, -4.0×10^{-10} C.

CHAPTER 23

1. -0.71 V.

3. 3280 V, plate B has a higher potential.

5. 30 m.

7. $1.4 \, \mu\text{C}$.

9. 1.2 cm, 46 nC.

11. (a) 0;
(b) -29.4 V;
(c) -29.4 V.

13. (a) -9.6×10^8 V;
(b) 9.6×10^8 V.

15. (a) They are equal;
(b) $Q\left(\dfrac{r_2}{r_1 + r_2}\right)$.

17. (a) 10–20 kV;
(b) $30 \, \mu\text{C/m}^2$.

19. (a) $\dfrac{Q}{4\pi\varepsilon_0 r}$;
(b) $\dfrac{Q}{8\pi\varepsilon_0 r_0}\left(3 - \dfrac{r^2}{r_0^2}\right)$;
(c) Let $V_0 = V$ at $r = r_0$, and $E_0 = E$ at $r = r_0$:

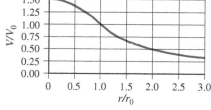

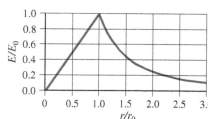

21. $\dfrac{\rho_0}{\varepsilon_0}\left(\dfrac{r_0^2}{4} - \dfrac{r^2}{6} + \dfrac{r^4}{20 r_0^2}\right)$.

23. (a) $\dfrac{R_0 \sigma}{\varepsilon_0}\ln\left(\dfrac{R_0}{R}\right) + V_0$;
(b) V_0;
(c) no, from part (a) $V \to -\infty$ due to length of wire.

25. (a) 29 V;
(b) -4.6×10^{-18} J.

27. 0.34 J.

29. 4.2 MV.

31. 9.64×10^5 m/s.

33. (a) 0;

 (b) $E_x = 0$,

$$E_y = \frac{Q}{4\pi\varepsilon_0}\frac{R}{(x^2 + R^2)^{3/2}}, \text{ looks}$$

 like a dipole.

35. $\frac{\sigma}{2\varepsilon_0}(\sqrt{R_2^2 + x^2} - \sqrt{R_1^2 + x^2})$.

37. 29 m/s.

39. $\frac{Q}{8\pi\varepsilon_0\ell}\ln\left(\frac{x + \ell}{x - \ell}\right)$.

41. $\frac{a}{6\varepsilon_0}(R^2 - 2x^2)\sqrt{R^2 + x^2} + \frac{a|x|^3}{3\varepsilon_0}$.

43. 2 mm.

45. (a) 2.6 mV;

 (b) 1.8 mV;

 (c) −1.8 mV.

49. -7.1×10^{-11} C/m² on $x = 0$ plate, 7.1×10^{-11} C/m² on other plate.

51. $(-2.5y + 3.5yz)\hat{\mathbf{i}}$
  $+ (-2y - 2.5x + 3.5xz)\hat{\mathbf{j}}$
  $+ (3.5xy)\hat{\mathbf{k}}$.

53. (a) $\frac{Q}{4\pi\varepsilon_0}\left(\frac{1}{y\sqrt{\ell^2 + y^2}}\right)\hat{\mathbf{j}}$;

 (b) $\frac{Q}{4\pi\varepsilon_0}\left(\frac{1}{x^2 - \ell^2}\right)\hat{\mathbf{i}}$.

55. −62.5 kV.

57. 1.3 eV.

59. (a) $\frac{1}{4\pi\varepsilon_0}\left(\frac{Q_1Q_2}{r_{12}} + \frac{Q_1Q_3}{r_{13}} + \frac{Q_1Q_4}{r_{14}}\right.$

 $\left. + \frac{Q_2Q_3}{r_{23}} + \frac{Q_2Q_4}{r_{24}} + \frac{Q_3Q_4}{r_{34}}\right)$;

 (b) $\frac{1}{4\pi\varepsilon_0}\left(\frac{Q_1Q_2}{r_{12}} + \frac{Q_1Q_3}{r_{13}} + \frac{Q_1Q_4}{r_{14}}\right.$

 $+ \frac{Q_1Q_5}{r_{15}} + \frac{Q_2Q_3}{r_{23}} + \frac{Q_2Q_4}{r_{24}}$

 $+ \frac{Q_2Q_5}{r_{25}} + \frac{Q_3Q_4}{r_{34}} + \frac{Q_3Q_5}{r_{35}}$

 $\left. + \frac{Q_4Q_5}{r_{45}}\right)$.

61. (a) 1.33 keV;

 (b) $v_e/v_p = 42.8$.

63. 250 MeV, same order of magnitude as observed values.

65. 1.11×10^5 m/s, 3.5×10^5 m/s.

67. 0.26 MV/m.

69. 600 V.

71. 1.5 J.

73. Yes, 2.0 pV.

75. 1.03×10^6 m/s.

77. $-\frac{\sqrt{3}Q}{2\pi\varepsilon_0\ell}, \frac{Q}{\pi\varepsilon_0\ell}\left(\frac{\sqrt{3}}{6} - 2\right),$

 $-\frac{Q}{\pi\varepsilon_0\ell}\left(1 + \frac{\sqrt{3}}{6}\right)$.

79. (a) 1.2 MV;

 (b) 1.8 kg.

81. (a) $\frac{\rho_E(r_2^3 - r_1^3)}{3\varepsilon_0 r}$;

 (b) $\frac{\rho_E}{\varepsilon_0}\left(\frac{r_2^2}{2} - \frac{r^2}{6} - \frac{r_1^3}{3r}\right)$;

 (c) $\frac{\rho_E}{2\varepsilon_0}(r_2^2 - r_1^2)$; yes.

83. $\vec{\mathbf{E}} = \frac{\lambda}{2\pi\varepsilon_0 R}$, radially outward.

85. (a) 23 kV;

 (b) $\frac{4Bx\hat{\mathbf{i}}}{(x^2 + R^2)^3}$;

 (c) $(2.3 \times 10^5 \text{ N/C})\hat{\mathbf{i}}$.

87. (a) and (b):

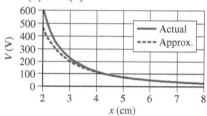

89. (a) Point charge;

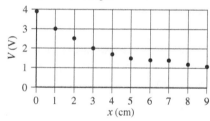

 (b) 1.5×10^{-11} C;

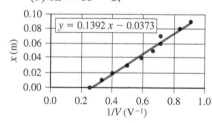

 (c) $x = -3.7$ cm.

CHAPTER 24

1. 3.0 μF.

3. 3.1 pF.

5. 56 μF.

7. 1.1 C.

9. 83 days.

11. 130 m².

13. 7.10×10^{-4} F.

15. 18 nC.

17. 5.8×10^4 V/m.

19. (a) $0.22 \ \mu\text{m} \le x \le 220 \ \mu\text{m}$;

 (b) $\frac{x^2 \Delta C}{\varepsilon_0 A}$;

 (c) 0.01%, 10%.

21. 3600 pF, yes.

23. 1.5 μF in series with the parallel combination of 2.0 μF and 3.0 μF, 2.8 V.

25. Add 11 μF connected in parallel.

27. $C_{max} = 1.94 \times 10^{-8}$ F, all in parallel, $C_{min} = 1.8 \times 10^{-9}$ F, all in series.

29. (a) $\frac{3}{5}C$;

 (b) $Q_1 = Q_2 = \frac{1}{5}CV, Q_3 = \frac{2}{5}CV,$

 $Q_4 = \frac{3}{5}CV, V_1 = V_2 = \frac{1}{5}V,$

 $V_3 = \frac{2}{5}V, V_4 = \frac{3}{5}V.$

31. $Q_1 = \frac{C_1C_2}{C_1 + C_2}V_0, Q_2 = \frac{C_2^2}{C_1 + C_2}V_0.$

33. (a) $Q_1 = 23 \ \mu\text{C}, Q_2 = Q_4 = 46 \ \mu\text{C}$;

 (b) $V_1 = V_2 = V_3 = V_4 = 2.9$ V;

 (c) 5.8 V.

35. 2.4 μF.

37. (a) $C_1 + \frac{C_2C_3}{C_2 + C_3}$;

 (b) $Q_1 = 8.40 \times 10^{-4}$ C,

 $Q_2 = Q_3 = 2.80 \times 10^{-4}$ C.

39. $C = \frac{\varepsilon_0 A}{d}\left(1 - \frac{\theta\sqrt{A}}{2d}\right).$

41. 6.8×10^{-3} J.

43. 2.0×10^3 J.

45. 1.70×10^{-3} J.

47. (a) $\frac{U_f}{U_i} = \frac{\ln\left(\frac{3R_a}{R_b}\right)}{\ln\left(\frac{R_a}{R_b}\right)} > 1,$

 work done to enlarge cylinder;

 (b) $\frac{U_f}{U_i} = \frac{\ln\left(\frac{R_a}{R_b}\right)}{\ln\left(\frac{3R_a}{R_b}\right)} < 1,$

 charge moved to battery.

49. (a) $-\frac{\varepsilon_0 A\ell V_0^2}{2d(d - \ell)}$;

 (b) $\frac{\varepsilon_0 A\ell V_0^2}{2(d - \ell)^2}$.

53. 2200 batteries, no.

55. 1.1×10^{-4} J.

57. (a) $0.32 \, \mu m^2$;

(b) 59 megabytes.

59. $\dfrac{\varepsilon_0 A}{2d}(K_1 + K_2)$.

61. $\dfrac{\varepsilon_0 A K_1 K_2}{(d_1 K_2 + d_2 K_1)}$.

63. (a) $\dfrac{\varepsilon_0 \ell^2}{d}\left[1 + (K - 1)\dfrac{x}{\ell}\right]$;

(b) $\dfrac{V_0^2 \varepsilon_0 \ell^2}{2d}\left[1 + (K - 1)\dfrac{x}{\ell}\right]$;

(c) $\dfrac{V_0^2 \varepsilon_0 \ell}{2d}(K - 1)$, left.

67. $\dfrac{\varepsilon_0 A}{d - \ell + \dfrac{\ell}{K}}$.

69. $E_{air} = 2.69 \times 10^4$ V/m,

$E_{glass} = 4.64 \times 10^3$ V/m,

$Q_{free} = 0.345 \, \mu C, Q_{ind} = 0.286 \, \mu C.$

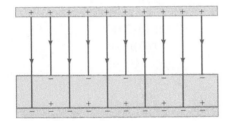

71. $43 \, \mu F$.

73. 15 V.

75. 840 V.

77. 3.76×10^{-9} F, 0.221 m^2.

79. $\dfrac{1}{2K}$, work done by the electric

field, $\dfrac{1}{K}$.

81. 1.2.

83. (a) 25 J;

(b) 940 kW.

85. (a) Parallel;

(b) 7.7 pF to 35 pF.

87. 5.15 pF.

89. $Q_1 = 11 \, \mu C, Q_2 = 13 \, \mu C,$

$Q_3 = 13 \, \mu C, V_1 = 11$ V,

$V_2 = 6.3$ V, $V_3 = 5.2$ V.

91. $\dfrac{Q^2 x}{2\varepsilon_0 A}$.

93. 9×10^{-16} m, no.

95. (a) $0.27 \, \mu C$, 15 kV/m, 5.9 nF,

6.0 μJ;

(b) $0.85 \, \mu C$, 15 kV/m, 19 nF, 19 μJ.

97. (a) 32 nF;

(b) $14 \, \mu C$;

(c) 7.0 mm;

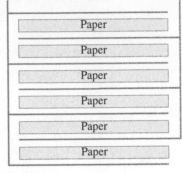

(d) 450 V.

CHAPTER 25

1. 8.13×10^{18} electrons/s.

3. 5.5×10^{-11} A.

5. (a) 28 A;

(b) 8.4×10^4 C.

7. 1.1×10^{21} electrons/min.

9. (a) $2.0 \times 10^1 \, \Omega$;

(b) 430 J.

11. 0.47 mm.

13. 0.64.

15. (a) Slope $= 1/R$, y-intercept $= 0$;

(b) yes, $R = 1.39 \, \Omega$;

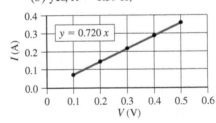

(c) $1.0 \times 10^{-6} \, \Omega \cdot$m, nichrome.

17. At $1/5.0$ of its length, $2.0 \, \Omega$, $8.0 \, \Omega$.

19. 2400°C.

21. $\sqrt{2}$.

23. 44.1°C.

25. One-quarter of the original.

27. $\dfrac{1}{4\pi\sigma}\left(\dfrac{1}{r_1} - \dfrac{1}{r_2}\right)$.

29. (a) $0.14 \, \Omega$;

(b) 0.60 A;

(c) $V_{Al} = 52$ mV, $V_{Cu} = 33$ mV.

31. 0.81 W.

33. 29 V.

35. (b) As large as possible.

37. (a) 0.83 A;

(b) $140 \, \Omega$.

39. 0.055 kWh, 7.9 cents/month.

41. 0.90 kWh $= 3.2 \times 10^6$ J.

43. 24 lightbulbs.

45. 11 kW.

47. 0.15 kg/s $= 150$ mL/s.

49. 0.12 A.

51. (a) ∞;

(b) $96 \, \Omega$.

53. (a) 930 V;

(b) 3.9 A.

55. (a) 1.3 kW;

(b) max $= 2.6$ kW, min $= 0$.

57. (a) 5.1×10^{-10} m/s;

(b) 6.9 A/m^2;

(c) 1.2×10^{-7} V/m.

59. 2.5 A/m^2, north.

61. 35 m/s, delay time from stimulus to action.

63. 11 hr.

65. 1.8 m, it would generate 540 W of heat and could start a fire.

67. 0.16 S.

69. (a) $35/month;

(b) 1300 kg/year.

71. (a) -19% change;

(b) % change would be slightly less.

73. (a) $190 \, \Omega$;

(b) $15 \, \Omega$.

75. (a) 1500 W;

(b) 12 A.

77. 2:1.

79. (a) $21 \, \Omega$;

(b) 2.0×10^1 s.

(c) 0.17 cents.

81. 36.0 m, 0.248 mm.

83. (a) 1200 W;

(b) 100 W.

85. 1.4×10^{12} protons.

87. (a) 3.1 kW;

(b) 24 W;

(c) 15 W;

(d) 38 cents/month.

89. (a) $55/kWh;

(b) $280/kWh, D-cells and AA-cells are $550\times$ and $2800\times$, respectively, more expensive.

91. $1.34 \times 10^{-4} \, \Omega$.

93. $\dfrac{4\ell\rho}{ab\pi}$.

95. $f = 1 - \dfrac{V}{V_0}$.

CHAPTER 26

1. (*a*) 5.93 V;
   (*b*) 5.99 V.

3. 0.060 Ω.

5. 9.3 V.

7. (*a*) 2.60 kΩ;
   (*b*) 270 Ω.

9. Connect nine 1.0-Ω resistors in
   series with battery; then connect
   output voltage circuit across four
   consecutive resistors.

11. 0.3 Ω.

13. 450 Ω, 0.024.

15. Solder a 1.6-kΩ resistor in parallel
   with 480-Ω resistor.

17. 120 Ω.

19. $\frac{13}{8}R$.

21. $R = r$.

23. (*a*) V_{left} decreases,
      V_{middle} increases,
      $V_{\text{right}} = 0$;
   (*b*) I_{left} decreases,
      I_{middle} increases,
      $I_{\text{right}} = 0$;
   (*c*) terminal voltage increases;
   (*d*) 8.5 V;
   (*e*) 8.6 V.

25. (*a*) V_1 and V_2 increase, V_3 and V_4
      decrease;
   (*b*) I_1 and I_2 increase, I_3 and I_4
      decrease;
   (*c*) increases;
   (*d*) before: $I_1 = 117\,\text{mA}$, $I_2 = 0$,
      $I_3 = I_4 = 59\,\text{mA}$;
      after: $I_1 = 132\,\text{mA}$,
      $I_2 = I_3 = I_4 = 44\,\text{mA}$, yes.

27. 0.38 A.

29. 0.

31. (*a*) 29 V;
   (*b*) 43 V, 73 V.

33. $I_1 = 0.68$ A left, $I_2 = 0.33$ A left.

37. 0.70 A.

39. 0.17 A.

41. (*a*) $\dfrac{R(5R' + 3R)}{8(R' + R)}$;
   (*b*) $\dfrac{R}{2}$.

43. $1 - 15$ MΩ.

45. 5.0 ms.

47. 44 s.

49. (*a*) $I_1 = \dfrac{2\mathscr{E}}{3R}$, $I_2 = I_3 = \dfrac{\mathscr{E}}{3R}$;
   (*b*) $I_1 = I_2 = \dfrac{\mathscr{E}}{2R}$, $I_3 = 0$;
   (*c*) $\dfrac{\mathscr{E}}{2}$.

51. (*a*) 8.0 V;
   (*b*) 14 V;
   (*c*) 8.0 V;
   (*d*) 4.8 μC.

53. 29 μA.

55. (*a*) Place in parallel with 0.22-mΩ
      shunt resistor;
   (*b*) place in series with 45-kΩ
      resistor.

57. 100 kΩ.

59. $V_{44} = 24$ V, $V_{27} = 15$ V;
   -15%, -15%.

61. 0.960 mA, 4.8 V.

63. 12 V.

65. Connect a 9.0-kΩ resistor in series
   with human body and battery.

67. 2.5 V, 117 V.

69. 92 kΩ.

71. (*a*) $\dfrac{R_2 R_3}{R_1}$;
   (*b*) 121 Ω.

73. Terminal voltage of mercury
   cell (3.99 V) is closer to 4.0 V
   than terminal voltage of dry cell
   (3.84 V).

75. 150 cells, 0.54 m^2, connect in series;
   connect four such sets in parallel to
   total 600 cells and deliver 120 V.

77. Counterclockwise current: -24 V,
   clockwise current: $+48$ V.

79. 10.7 V.

83. 9.0 Ω.

85. (*b*) 1.39 V;
   (*c*) 0.42 mV;
   (*d*) no current from "working"
      battery is needed to "power"
      galvanometer.

87. 1.0 mV, 2.0 mV, 4.0 mV, 10.0 mV.

89. (*a*) 6.8 V, 15 μC;
   (*b*) 48 μs.

91. 200 MΩ.

93. 4.5 ms.

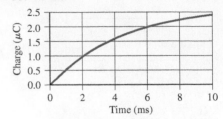

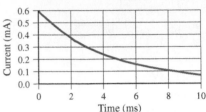

CHAPTER 27

1. (*a*) 8.5 N/m;
   (*b*) 4.9 N/m.

3. 2.6×10^{-4} N.

5. (*a*) South pole;
   (*b*) 3.41 A;
   (*c*) 7.39×10^{-2} N.

7. 2.13 N, 41.8° below negative y axis.

9. $(-2IrB_0 \sin\theta_0)\hat{\mathbf{j}}$.

13. 6.3×10^{-14} N, north.

15. 1.8 T.

17. (*a*) Downward;
   (*b*) into page;
   (*c*) right.

19. (*a*) 0.031 m;
   (*b*) 3.8×10^{-7} s.

23. 1.8 m.

25. $(0.78\hat{\mathbf{i}} - 1.0\hat{\mathbf{j}} + 0.1\hat{\mathbf{k}}) \times 10^{-15}$ N.

27. $L_{\text{final}} = \frac{1}{2}L_{\text{initial}}$.

29. (*a*) Negative;
   (*b*) $qB_0\left(\dfrac{\ell^2 + d^2}{2d}\right)$.

31. 1.3×10^8 m/s, yes.

33. (*a*) 45°;
   (*b*) 2.3×10^{-3} m.

35. (*a*) $2NIAB$;
   (*b*) 0.

37. (*a*) 4.85×10^{-5} m $\cdot$ N;
   (*b*) north.

39. (*a*) $(-4.3\,\hat{\mathbf{k}})$ A $\cdot$ m^2;
   (*b*) $(2.6\hat{\mathbf{i}} - 2.4\hat{\mathbf{j}})$ m $\cdot$ N;
   (*c*) -2.8 J.

41. 12%.

43. 39 μA.

45. 6 electrons.

47. (b) 0.05 nm, about $\frac{1}{6}$ the size of a typical metal atom;
(c) 10 mV.

49. 0.820 T.

51. 70 u, 72 u, 73 u, and 74 u.

53. 1.5 mm, 1.5 mm, 0.77 mm, 0.77 mm.

55. $^2_1$H, $^4_2$He.

57. 2.4 T, upwards.

59. (a) $\dfrac{IBd}{m}t$;
(b) $\left(\dfrac{IBd}{m} - \mu_k g\right)t$;
(c) east.

61. 1.1×10^{-6} m/s, west.

63. 3.8×10^{-4} m·N.

65. $\pi\left[\dfrac{mb(3a + b)}{3NIBa(a + b)}\right]^{1/2}$.

67. They do not enter second tube, 12°.

69. 1.1 A, down.

71. 7.3×10^{-3} T.

73. -6.9×10^{-20} J.

75. 0.083 N, northerly and 68° above the horizontal.

77. (a) Downward;
(b) 28 mT;
(c) 0.12 T.

CHAPTER 28

1. 0.37 mT, 7.4 times larger.

3. 0.15 N, toward other wire.

7. 0.12 mT, 82° above directly right.

9. 3.8×10^{-5} T, 17° below the horizontal to north.

11. (a) $(2.0 \times 10^{-5})(25 - I)$ T;
(b) $(2.0 \times 10^{-5})(25 + I)$ T.

15. Closer wire: 0.050 N/m, attractive, farther wire: 0.025 N/m, repulsive.

17. 17 A, downward.

19. $\dfrac{\mu_0 I}{2\pi}\left(\dfrac{d - 2x}{x(d - x)}\right)\hat{\mathbf{j}}$.

21. 46.6 μT.

23. (b) $\dfrac{\mu_0 I}{2\pi y}$, yes, looks like B from long straight wire.

25. 0.160 A.

27. (a) 5.3 mT;
(b) 3.2 mT;
(c) 1.8 mT.

29. (a) 0.554 m;
(b) 10.5 mT.

31. (a) $\dfrac{\mu_0 I_0 R}{2\pi R_1^2}$;

(b) $\dfrac{\mu_0 I_0}{2\pi R}$;

(c) $\dfrac{\mu_0 I_0}{2\pi R}\left(\dfrac{R_3^2 - R^2}{R_3^2 - R_2^2}\right)$;

(d) 0;

(e)

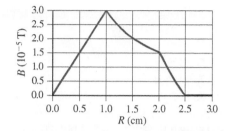

33. 3.6×10^{-6} T.

35. $0.075\,\mu_0 I/R$.

37. (a) $\dfrac{\mu_0 I}{4}\left(\dfrac{1}{R_1} + \dfrac{1}{R_2}\right)$, into the page;

(b) $\dfrac{\pi I(R_1^2 + R_2^2)}{2}$, into the page.

39. (a) $\dfrac{Q\omega R^2}{4}\hat{\mathbf{i}}$;

(b) $\dfrac{\mu_0 Q\omega}{2\pi R^2}\left(\dfrac{R^2 + 2x^2}{\sqrt{R^2 + x^2}} - 2x\right)\hat{\mathbf{i}}$;

(c) yes.

41. (b) $\dfrac{\mu_0 I}{4\pi y}\left(\dfrac{d}{\sqrt{d^2 + y^2}}\right)\hat{\mathbf{k}}$.

43. (a) $\dfrac{n\mu_0 I \tan(\pi/n)}{2\pi R}$, into the page.

45. $\dfrac{\mu_0 I}{4\pi}\left[\dfrac{\sqrt{x^2 + y^2}}{xy} + \dfrac{\sqrt{y^2 + (b - x)^2}}{(b - x)y}\right.$
$+ \dfrac{\sqrt{(a - y)^2 + (b - x)^2}}{(a - y)(b - x)}$
$\left.+ \dfrac{\sqrt{(a - y)^2 + x^2}}{x(a - y)}\right]$,
out of page.

47. (a) 16 A·m²;
(b) 13 m·N.

49. 2.4 T.

51. $(\vec{\mathbf{F}}/\ell)_M = 6.3 \times 10^{-4}$ N/m at 90°,
$(\vec{\mathbf{F}}/\ell)_N = 3.7 \times 10^{-4}$ N/m at 300°,
$(\vec{\mathbf{F}}/\ell)_P = 3.7 \times 10^{-4}$ N/m at 240°.

53. 170 A.

55. (a) 2.7×10^{-6} T;
(b) 5.3×10^{-6} T;
(c) no, no Newton's third-law-type of relationship;
(d) both 1.1×10^{-5} N/m, yes, Newton's third law holds.

57. $\dfrac{\mu_0 tj}{2}$, to the left above sheet (with current coming toward you).

61. (a) $\dfrac{N\mu_0 IR^2}{2}$

$\times \left(\dfrac{1}{(R^2 + x^2)^{3/2}} + \dfrac{1}{(R^2 + (x - R)^2)^{3/2}}\right)$;

(b) 4.5 mT.

63. 3×10^9 A.

65. (a) 46 turns;
(b) 0.83 mT;
(c) no.

67. $\dfrac{\mu_0 I\sqrt{5}}{2\pi a}$, into the page.

69. 0.10 N, south.

71. $\frac{2}{3}$.

73. (c) 1.5 A.

75.

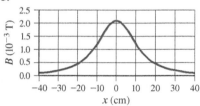

CHAPTER 29

1. -460 V.

3. Counterclockwise.

5. 1.2 mm/s.

7. (a) 0.010 Wb;
(b) 55°;
(c) 5.8 mWb.

9. Counterclockwise.

11. (a) Clockwise;
(b) 43 mV;
(c) 17 mA.

13. (a) 8.1 mJ;
(b) 4.2×10^{-3} C°.

15. (a) 0.15 A;
(b) 1.4 mW.

17. 8.81 C.

19. 21 μJ.

21. 23 mV, 26 mV.

23. (a) 0;
(b) 0.99 A, counterclockwise.

25. (a) $\dfrac{\mu_0 I a}{2\pi} \ln\left(1 + \dfrac{a}{b}\right)$;

(b) $\dfrac{\mu_0 I a^2 v}{2\pi b(a + b)}$;

(c) clockwise;

(d) $\dfrac{\mu_0^2 I^2 a^4 v}{4\pi^2 b^2 (a + b)^2 R}$.

27. 1.0 m/s.

29. (a) 0.11 V;

(b) 4.1 mA;

(c) 0.36 mN.

31. 0.39 m/s.

33. (a) Yes;

(b) $v_0\, e^{-B^2 \ell^2 t / mR}$.

35. (a) $\dfrac{v\mu_0 I}{2\pi} \ln\left(1 + \dfrac{a}{b}\right)$;

(b) $-\dfrac{v\mu_0 I}{2\pi} \ln\left(1 + \dfrac{a}{b}\right)$.

37. 57.2 loops.

41. 150 V.

43. 13 A.

45. (a) 2.4 kV;

(b) 190 V.

47. 50, 4.8 V.

49. (a) Step-up;

(b) 3.5.

51. (a) R;

(b) $\left(\dfrac{N_P}{N_S}\right)^2 R$.

53. 98 kW.

55. (b) Clockwise;

(c) increase.

57. (a) $\dfrac{IR}{\ell}$;

(b) $\dfrac{\mathcal{E}_0}{\ell}\, e^{-B^2 \ell^2 t / mR}$.

59. 10.1 mJ.

61. 0.6 nC.

63. (a) 41 kV;

(b) 31 MW;

(c) 0.88 MW;

(d) 3.0×10^7 W.

65. (a) Step-down;

(b) 2.9 A;

(c) 0.29 A;

(d) 4.1 Ω.

67. 46 mA, left to right through resistor.

69. 2.3×10^{17} electrons.

71. (a) 25 A;

(b) 98 V;

(c) 600 W;

(d) 81%.

73. $\frac{1}{2} B\omega\ell^2$.

77. $B\omega R$, radially in toward axis.

79. (a) $\dfrac{\pi d^2 B^2 \ell v}{16\rho}$;

(b) $16\rho\rho_m g / B^2$;

(c) 3.7 cm/s.

CHAPTER 30

1. (a) 31.0 mH;

(b) 3.79 V.

3. $\dfrac{\mu_0 N_1 N_2 A_2 \sin\theta}{\ell}$.

5. 12 V.

7. 0.566 H.

9. 11.3 V.

11. 46 m, 21 km, 0.70 kΩ.

15. 18.9 J.

17. 1.06×10^{-3} J/m³.

19. $\dfrac{\mu_0 N^2 I^2}{8\pi^2 r^2}$, $\dfrac{\mu_0 N^2 I^2 h}{4\pi} \ln\left(\dfrac{r_2}{r_1}\right)$.

21. $\dfrac{\mu_0 I^2}{16\pi}$.

23. 3.5 time constants.

25. (a) $\dfrac{L V_0^2}{2R^2}\left(1 - e^{-t/\tau}\right)^2$;

(b) 7.6 time constants.

27. (b) 6600 V.

29. $(12\text{ V})e^{-t/8.2\,\mu s}$, 0, 12 V.

31. (a) 0.16 nF;

(b) 62 μH.

33. (c) $(2 \times 10^{-4})\%$.

35. (a) $\dfrac{Q_0}{\sqrt{2}}$;

(b) $\frac{1}{8} T$.

37. $\dfrac{L}{R}\ln\left(\frac{4}{3}\right) = (0.29)\dfrac{L}{R}$.

39. 3300 Hz.

41.

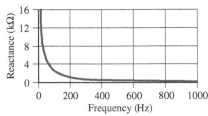

43. (a) $R + R'$;

(b) R'.

45. (a) 2800 Ω;

(b) 660 Hz, 11 A.

47. 2190 W.

49. (a) 0.40 kΩ;

(b) 75 Ω.

51. 1600 Hz.

53. 240 Hz, voltages are out of phase.

55. (a) 0.124 A;

(b) 5.02°;

(c) 14.8 W;

(d) 0.120 kV, 10.5 V.

57. 7.8 μF.

59. $I_0 V_0 \sin\omega t \sin(\omega t + \phi)$.

61. 130 Ω, 0.91.

63. 265 Hz, 324 W.

65. (b) 130 Ω.

67. (a) $\dfrac{V_0^2 R}{2\left[R^2 + \left(\omega L - \dfrac{1}{\omega C}\right)^2\right]}$;

(b) $\dfrac{1}{2\pi}\sqrt{\dfrac{1}{LC}}$;

(c) $\dfrac{R}{L}$.

69. 37 loops.

71. (a) 0.040 H;

(b) 28 mA;

(c) 16 μJ.

73. 2.4 mA, 0, 2.4 mA.

77. (a) $\dfrac{Q_0^2}{2C}\, e^{-Rt/L}$;

(b) $\dfrac{dU}{dt} = -I^2 R$.

79.

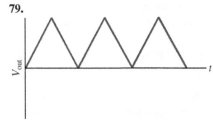

81. (a) 0;

(b) 0, 90° out of phase.

83. 2.2 kHz.

85. 69 mH, 18 Ω.

89. (a)

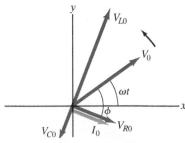

(b) $\dfrac{V_0}{\sqrt{R^2 + \left(\omega L - \dfrac{1}{\omega C}\right)^2}}\sin(\omega t - \phi)$,

$\phi = \tan^{-1}\dfrac{\omega L - \dfrac{1}{\omega C}}{R}$.

91. (a) $\left(\dfrac{V_{20}}{\omega L - \dfrac{1}{\omega C}}\right)\sin\left(\omega t - \tfrac{1}{2}\pi\right);$

(b) $\left(\dfrac{V_{20}}{\omega^2 LC - 1}\right)\sin(\omega t - \pi);$

(c) $\dfrac{1}{\omega^2 LC};$

(d) $V_{1\,\text{out}} = V_1.$

93. (a) $\dfrac{V_0}{R}\sin\omega t;$

(b) $\dfrac{V_0}{X_L}\sin\left(\omega t - \tfrac{1}{2}\pi\right);$

(c) $\dfrac{V_0}{X_C}\sin\left(\omega t + \tfrac{1}{2}\pi\right)$

(d) $\dfrac{V_0}{R}\sqrt{1 + \left(R\omega C - \dfrac{R}{\omega L}\right)^2}\sin(\omega t + \phi),$

$\phi = \tan^{-1}\left(R\omega C - \dfrac{R}{\omega L}\right);$

(e) $\dfrac{R}{\sqrt{1 + \left(R\omega C - \dfrac{R}{\omega L}\right)^2}};$

(f) $\dfrac{1}{\sqrt{1 + \left(R\omega C - \dfrac{R}{\omega L}\right)^2}}.$

95. 0.14 H.

97. 54 mH, 22 Ω.

99. $\sqrt{6.0}\, f_0 = 2.4 f_0.$

101. (a) 7.1 kHz, V_{rms};

(b) 0.90.

103. (b) For $f \to 0\; A \to 1$;
for $f \to \infty,\; A \to 0$;

(c) f is in s^{-1}:

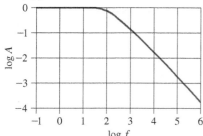

105.

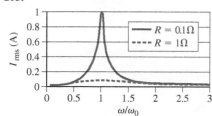

CHAPTER 31

1. 110 kV/m·s.

3. 1.2×10^{15} V/m·s.

7. (b) With R in meters, for $R \le R_0$,
$B_0 = (6.3 \times 10^{-11}\,\text{T/m})R$;

for $R > R_0$, $B_0 = \dfrac{5.7 \times 10^{-14}\,\text{T·m}}{R}.$

(c)

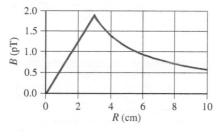

9. 3.75 V/m.

11. (a) $-\hat{\mathbf{k}}$;

(b) $\dfrac{E_0}{c}, -\hat{\mathbf{j}}.$

13. 2.00×10^{10} Hz.

15. 5.00×10^2 s = 8.33 min.

17. (a) 3.00×10^5 m;

(b) 34.1 cm;

(c) no.

19. (a) 261 s;

(b) 1260 s.

21. 3.4 krad/s.

23. 2.77×10^7 s.

25. 4.8 W/m^2, 42 V/m.

27. 4.50 μJ.

29. 3.80×10^{26} W.

31. (a) 5 cm^2, yes;

(b) 20 m^2, yes;

(c) 100 m^2, no.

33. (a) 2×10^8 ly;

(b) 2000 times larger.

35. 8×10^6 m/s^2.

37. 27 m^2.

39. 16 cm.

41. 3.5 nH to 5.3 nH.

43. 6.25×10^{-4} V/m;

1.04×10^{-9} W/m^2.

45. 3 m.

47. 1.35 s.

49. 34 V/m, 0.11 μT.

51. Down, 2.2 μT, 650 V/m.

53. (a) 0.18 nJ;

(b) 8.7 μV/m, 2.9×10^{-14} T.

57. 4×10^{10} W.

59. 5 nodes, 6.1 cm.

61. (a) $+x$;

(b) $\beta = \alpha c$;

(c) $\dfrac{E_0}{c}\, e^{-(\alpha x - \beta t)^2}.$

63. (d) Both $\vec{\mathbf{E}}$ and $\vec{\mathbf{B}}$ rotate
counterclockwise.

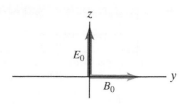

CHAPTER 32

1.

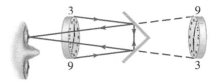

3. 7°.

7.

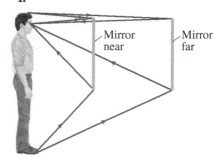

9. 37.6 cm.

11. 1.0 m.

13. 2.1 cm behind front surface of ball;
virtual, upright.

15. Concave, 5.3 cm.

17. -6.0 m.

19. Convex, -32.0 cm.

21.

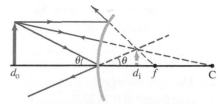

23. −3.9 m.

25. (a) Convex;

    (b) 20 cm behind mirror;

    (c) −91 cm;

    (d) −1.8 m.

27. (b)

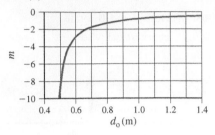

    (c) 0.90 m;

    (d) just beyond focal point.

31. Because the image is inverted.

33. (a) 2.21×10^8 m/s;

    (b) 1.99×10^8 m/s;

    (c) 1.97×10^8 m/s.

35. 8.33 min.

37. 3 m.

39. 35°.

41. 38.6°.

43. 2.6 cm.

45. 4.4 m.

47. 3.2 mm.

49. 38.9°.

53. 0.22°.

55. 0.80°.

57. 33.3°, diamond.

59. 82.1 cm.

61. $n \geq 1.5$.

63. (a) 2.3 μs;

    (b) 17 ns.

65. $n \geq 1.72$.

67. 17.3 cm.

71. 0.25 m, 0.50 m.

73. (a) 3.0 m, 4.4 m, 7.4 m;

    (b) toward, away, toward.

75. 3.80 m.

77. 31 cm for real image, 15 cm for virtual image.

83. $\dfrac{d}{n-1}$.

85. The light would totally internally reflect only if $\theta_i \leq 32.5°$.

---

87. $A = 1.5005$, $B = 5740 \text{ nm}^2$.

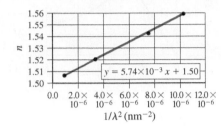

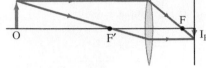

CHAPTER 33

1. (a)

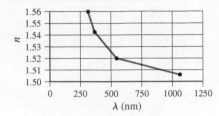

    (b) 508 mm.

3. (a) 4.26 D, converging;

    (b) −14.8 cm, diverging.

5. (a) 106 mm;

    (b) 109 mm;

    (c) 117 mm;

    (d) an object 0.513 m away.

7. (a) Virtual, upright, magnified;

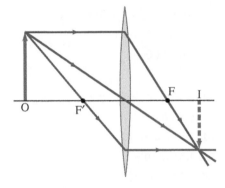

    (b) converging;

    (c) 6.7 D.

9. (a) 0.02 m;

    (b) 0.004 m.

11. 50 cm.

---

13. 21.3 cm, 64.7 cm.

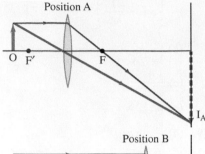

15. (c) Real, upright; (d) real, upright.

17. 0.107 m, 2.2 m.

19. (b) 182 cm; (c) 182 cm.

21. 18.5 cm beyond second lens, $-0.651\times$.

23. (a) 7.14 cm beyond second lens;

    (b) $-0.357\times$; (c)

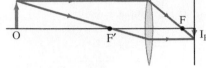

25. (a) 0.10 m to right of diverging lens;

    (b) $-1.0\times$;

    (c)

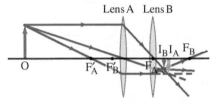

27. (a) 30 cm beyond second lens, half the size of object;

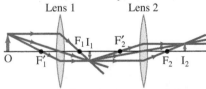

    (b) 29 cm beyond second lens, 0.46 times the size of object.

29. 1.54.

31. 8.6 cm.

33. 34 cm.

35. $f/2.8$.

37. $\frac{1}{6}$ s.

39. 41 mm.

41. +2.5 D.

43. 41 cm, yes.

45. (a) −1.3 D;
   (b) 37 cm.
47. −24.8 cm.
49. 18.4 cm, 1.00 m.
51. 6.6 cm.
53. (a) 13 cm;
   (b) 8.3 cm.
55. (a) −234 cm;
   (b) 4.17×.
57. (a) −66 cm;
   (b) 3.0×.
59. 4 cm, toward.
61. 2.5 cm, 91 cm.
63. −26×.
65. 16×.
67. 3.7 m, 7.4 m.
69. −9×.
71. 8.0×.
73. 1.6 cm.
75. (a) 754×;
   (b) 1.92 cm, 0.307 cm;
   (c) 0.312 cm.
77. (a) 0.85 cm;
   (b) 250×.
79. 410×, 25×.
81. 79.4 cm, 75.5 cm.
83. 6.450 m ≤ d_0 ≤ ∞.
85. 116 mm, 232 mm.
87. −19.0 cm.
89. 3.1 cm, 25 cm.
91. (a) 0.26 mm;
   (b) 0.47 mm;
   (c) 1.3 mm;
   (d) 0.56×, 2.7×.
93. 20.0 cm.
95. 47 m.
97. 2.8×, 3.9×, person with normal eye.
99. 1.0×.
101. +3.4 D.
103. −19×.
105. (a) 28.6 cm;
   (b) 120 cm;
   (c) 15 cm.
107. −6.2 cm.
109. (a) $-1/f$, 1;
   (b) 14 cm, yes,
    y-intercept = 1.03;

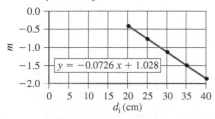

   (c) $f = -1/\text{slope}$.

CHAPTER 34

3. 3.9 μm.
5. 0.2 mm.
7. 660 nm.
9. 3.5 cm.
11. Inverted, starts with central dark line, and every place there was bright fringe before is now dark fringe and vice versa.
13. 2.7 mm.
15. 2.94 mm.
17. $\frac{1}{4}$.
21. $I_0 \left[\dfrac{3 + 2\sqrt{2} \cos\left(\dfrac{2\pi d \sin\theta}{\lambda} \right)}{3 + 2\sqrt{2}} \right]$.
23. 634 nm.
25. (a) 180 nm;
   (b) 361 nm, 541 nm.
27. (b) 290 nm.
29. 8.68 μm.
31. 113 nm, 225 nm.
35. 1.32.
37. (c) 571 nm.
39. 0.191 mm.
41. 80.1 μm.
43. 0.3 mm.
45. (a) 17 lm/W;
   (b) 160 lamps.
47. (a) Constructive;
   (b) destructive.
49. 440 nm.
51. $I_0 \cos^2\left(\dfrac{2\pi x}{\lambda} \right)$.
53. (a) 81.5 nm;
   (b) 0.130 μm.
55. $\theta = \sin^{-1}\left(\sin\theta_i \pm \dfrac{m\lambda}{d} \right)$.
57. 340 nm, 170 nm.
59. Constructive: 90°, 270°; destructive: 0°, 180°; exactly switched.
61. 240 nm.
63. 0.20 km.
65. 126 nm.

CHAPTER 35

1. 37.3 mrad = 2.13°.
3. 2.35 m.
5. Entire pattern is shifted, with central maximum at 23° to the normal.
7. 4.8 cm.
9. 953 nm.
11. (a) 63°;
   (b) 30°.

13. 0.15.
15. $d = 5D$.
17. 265 fringes.
19. (a) 1.9 cm;
   (b) 12 cm.
21. 0.255.
23. (a) $I_\theta = I_0 \left(\dfrac{1 + 2\cos\delta}{3} \right)^2$.
25. 1.5×10^{11} m.
27. 1.0×10^4 m.
29. 730 lines/mm, 88 lines/mm.
31. 0.40 μm, 0.50 μm, 0.52 μm, 0.62 μm.
33. Two full orders, plus part of a third order.
35. 556 nm.
37. 24°.
39. $\lambda_2 > 600$ nm overlap with $\lambda_3 < 467$ nm.
41. $\lambda_1 = 614$ nm, $\lambda_2 = 899$ nm.
43. 7 cm, 35 cm, second order.
45. (c) −32°, 0.9°.
47. (a) 16,000 and 32,000;
   (b) 26 pm, 13 pm.
49. 14.0°.
51. No.
53. 45°.
55. 61.2°.
57. (a) 35.3°;
   (b) 63.4°.
59. 36.9°, smaller than both angles.
61. $I = \dfrac{I_0}{4} \sin^2(2\theta)$, 45°.
63. 28.8 μm.
65. 580 nm.
67. 0.6 m.
69. 658 nm, 853 lines/cm.
71. (a) 18 km;
   (b) 23″, atmospheric distortions make it worse.
73. 5.79×10^5 lines/m.
75. 36.9°.
77. (a) 60°;
   (b) 71.6°;
   (c) 84.3°.
79. 0.4 m.
81. 0.245 nm.
83. 110 m.
85. −0.17 mm.
87. Use 24 polarizers, each rotated 3.75° from previous axis.

Index

Note: The abbreviation *defn* means the page cited gives the definition of the term; *fn* means the reference is in a footnote; *pr* means it is found in a Problem or Question; *ff* means "also the following pages."

A (atomic mass number), 1105
Aberration:
  chromatic, 889 *fn*, 892, 932
  of lenses, 891–92, 929, 931
  spherical, 843, 857, 891, 892, 932
Absolute pressure, 345
Absolute space, 953, 957
Absolute temperature scale, 457, 464, 469–70
Absolute time, 953
Absolute zero, 464, 549
Absorbed dose, 1148
Absorption lines, 936, 1002, 1081, 1084–85
Absorption spectra, 936, 1002, 1084
Absorption wavelength, 1008
Abundances, natural, 1105
Ac circuits, 664–65, 677 *fn*, 790–803
Ac generator, 766–67
Ac motor, 720
Accelerating reference frames, 85, 88, 155–56, 300–2
Acceleration, 24–42, 60–62
  angular, 251–56, 258–63
  average, 24–26
  centripetal, 120 *ff*
  constant, 28–29, 62
  constant angular, 255
  Coriolis, 301–2
  cosmic, 1223
  in *g*'s, 37
  due to gravity, 34–39, 87 *fn*, 92, 143–45
  instantaneous, 27–28, 60–61
  of the Moon, 121, 140
  motion at constant, 28–39, 62–71
  radial, 120 *ff*, 128
  related to force, 86–88
  tangential, 128–29, 251–52
  uniform, 28–39, 62–71
  variable, 39–43
Accelerators, particle, 1165–71
Accelerometer, 100
Acceptor level, 1094
Accommodation of eye, 883
Accuracy, 3–5
  precision vs., 5
Achromatic doublet, 892
Achromatic lens, 892
Actinides, 1054
Action at a distance, 154, 568
Action potential, 670
Action–reaction (Newton's third law), 89–91
Activation energy, 481, 1075, 1077
Active galactic nuclei (AGN), 1197

Active solar heating, 550
Activity, 1118
  and half-life, 1120
  source, 1147
Addition of vectors, 52–58
Addition of velocities:
  classical, 71–74
  relativistic, 970–71
Adhesion, 360
Adiabatic lapse rate, 525 *pr*
Adiabatic processes, 508, 514–15
ADP, 1076–77
AFM, 1039
AGN, 1197
Air bags, 31
Air cleaner, electrostatic, 645 *pr*
Air columns, vibrations of, 434–36
Air conditioners, 537–38
Air parcel, 525 *pr*
Air pollution, 551
Air resistance, 34–35, 129–30
Airplane wing, 356–57
Airy disk, 929
Alkali metals, 1054
Allowed transitions, 1048–49, 1080–81, 1083, 1084
Alpha decay, 1111–14, 1117
  and tunneling, 1038, 1113
Alpha particle (or ray), 1038, 1111–14
Alternating current (ac), 664–65, 677 *fn*, 796–803
Alternators, 768
AM radio, 830
Amino acids, 1079
Ammeter, 695–97, 721
  digital, 695, 697
Amorphous solids, 1085
Ampère, André, 654, 737
Ampere (A) (unit), 654, 736
  operational definition of, 736
Ampère's law, 737–43, 813–17
Amplifiers, 1097
Amplitude, 371, 397, 404
  intensity related to, 430
  pressure, 427
  of vibration, 371
  of wave, 371, 397, 402, 404, 426, 430, 1019
Amplitude modulation (AM), 830
Analog information, 775
Analog meters, 695–97, 721
Analyzer (of polarized light), 941
Anderson, Carl, 1174
Andromeda, 1196
Aneroid barometer, 347
Aneroid gauge, 347
Angle, 7 *fn*, 249
  attack, 356
  Brewster's, 943, 949 *pr*
  critical, 854
  of dip, 709
  of incidence, 410, 415, 838, 850
  phase, 373, 405, 800

  polarizing, 943–44
  radian measure of, 249
  of reflection, 410, 838
  of refraction, 415, 850
  solid, 7 *fn*, 915 *fn*
Angstrom (Å) (unit), 17 *pr*, 852 *fn*
Angular acceleration, 251–56, 258–63
  constant, 255
Angular displacement, 250, 381
Angular frequency, 373
Angular magnification, 886
Angular momentum, 285–89, 291–300, 1003
  in atoms, 1004, 1046–49, 1057–60
  conservation, law of, 285–89, 297–98, 1117
  directional nature of, 288–89, 291 *ff*
  nuclear, 1107
  of a particle, 291–92
  quantized in atoms, 1046–47
  quantized in molecules, 1080–81
  relation between torque and, 292–97
  total, 1059
  and uncertainty principle, 1023
  vector, 288, 291
Angular position, 249, 1023
Angular quantities, 249 *ff*
  vector nature, 254
Angular velocity, 250–55
  of precession, 299–300
Anisotropy of CMB, 1214, 1220
Annihilation (e^-e^+, particle–antiparticle), 996, 1175, 1217
Anode, 620
Antenna, 812, 817, 824, 831, 909
Anthropic principle, 1225
Anticodon, 1079
Antilogarithm, A-3
Antimatter, 1175, 1188, 1190 *pr* (*see also* Antiparticle)
Antineutrino, 1115–16, 1179
Antineutron, 1175
Antinodes, 412, 433, 434, 435
Antiparticle, 1116, 1174–76, 1179 (*see also* Antimatter)
Antiproton, 1164, 1174–75
Antiquark, 1179, 1183
Apparent brightness, 1197–98
Apparent magnitude, 1228 *pr*
Apparent weight, 148–49, 350
Apparent weightlessness, 148–49
Approximations, 9–12
Arago, F., 922
Arches, 327–28
Archimedes, 349–50
Archimedes' principle, 348–52
  and geology, 351
Area, 9, A-1, inside back cover
  under a curve or graph, 169–71
Arecibo, 931
Aristotle, 2, 84
Armature, 720, 766
Arteriosclerosis, 359

Photo Credits

Useful Geometry Formulas—Areas, Volumes

Circumference of circle $\quad C = \pi d = 2\pi r$

Area of circle $\quad A = \pi r^2 = \dfrac{\pi d^2}{4}$

Area of rectangle $\quad A = \ell w$

Area of parallelogram $\quad A = bh$

Area of triangle $\quad A = \frac{1}{2}hb$

Right triangle
(Pythagoras) $\quad c^2 = a^2 + b^2$

Sphere: surface area $\quad A = 4\pi r^2$
volume $\quad V = \frac{4}{3}\pi r^3$

Rectangular solid:
volume $\quad V = \ell w h$

Cylinder (right):
surface area $\quad A = 2\pi r \ell + 2\pi r^2$
volume $\quad V = \pi r^2 \ell$

Right circular cone:
surface area $\quad A = \pi r^2 + \pi r\sqrt{r^2 + h^2}$
volume $\quad V = \frac{1}{3}\pi r^2 h$

Exponents

$(a^n)(a^m) = a^{n+m}$ [Example: $(a^3)(a^2) = a^5$]
$(a^n)(b^n) = (ab)^n$ [Example: $(a^3)(b^3) = (ab)^3$]
$(a^n)^m = a^{nm}$
$\begin{bmatrix}\text{Example: } (a^3)^2 = a^6 \\ \text{Example: } (a^{\frac{1}{4}})^4 = a\end{bmatrix}$

$a^{-1} = \dfrac{1}{a} \qquad a^{-n} = \dfrac{1}{a^n} \qquad a^0 = 1$

$a^{\frac{1}{2}} = \sqrt{a} \qquad a^{\frac{1}{4}} = \sqrt{\sqrt{a}}$

$(a^n)(a^{-m}) = \dfrac{a^n}{a^m} = a^{n-m}$ [Ex.: $(a^5)(a^{-2}) = a^3$]

$\dfrac{a^n}{b^n} = \left(\dfrac{a}{b}\right)^n$

Logarithms [Appendix A–7; Table A–1]

If $y = 10^x$, then $x = \log_{10} y = \log y$.
If $y = e^x$, then $x = \log_e y = \ln y$.

$\log(ab) = \log a + \log b$

$\log\left(\dfrac{a}{b}\right) = \log a - \log b$

$\log a^n = n \log a$

Some Derivatives and Integrals†

$\dfrac{d}{dx}x^n = nx^{n-1} \qquad \int \sin ax \, dx = -\dfrac{1}{a}\cos ax$

$\dfrac{d}{dx}\sin ax = a \cos ax \qquad \int \cos ax \, dx = \dfrac{1}{a}\sin ax$

$\dfrac{d}{dx}\cos ax = -a \sin ax \qquad \int \dfrac{1}{x} dx = \ln x$

$\int x^m \, dx = \dfrac{1}{m+1}x^{m+1} \qquad \int e^{ax} \, dx = \dfrac{1}{a}e^{ax}$

†See Appendix B for more.

Quadratic Formula

Equation with unknown x, in the form
$$ax^2 + bx + c = 0,$$
has solutions
$$x = \dfrac{-b \pm \sqrt{b^2 - 4ac}}{2a}.$$

Binomial Expansion

$$(1 \pm x)^n = 1 \pm nx + \dfrac{n(n-1)}{2\cdot 1}x^2 \pm \dfrac{n(n-1)(n-2)}{3\cdot 2\cdot 1}x^3 + \cdots \quad \text{[for } x^2 < 1\text{]}$$
$$\approx 1 \pm nx \quad \text{[for } x \ll 1\text{]}$$

Trigonometric Formulas [Appendix A–9]

$\sin \theta = \dfrac{\text{opp}}{\text{hyp}}$

$\cos \theta = \dfrac{\text{adj}}{\text{hyp}}$

$\tan \theta = \dfrac{\text{opp}}{\text{adj}}$

$\text{adj}^2 + \text{opp}^2 = \text{hyp}^2 \quad$ (Pythagorean theorem)

$\tan \theta = \dfrac{\sin \theta}{\cos \theta}$

$\sin^2 \theta + \cos^2 \theta = 1$

$\sin 2\theta = 2 \sin \theta \cos \theta$

$\cos 2\theta = (\cos^2 \theta - \sin^2 \theta) = (1 - 2\sin^2 \theta) = (2\cos^2 \theta - 1)$

$\sin(180° - \theta) = \sin \theta \qquad \cos(180° - \theta) = -\cos \theta$
$\sin(90° - \theta) = \cos \theta$
$\cos(90° - \theta) = \sin \theta$
$\sin \frac{1}{2}\theta = \sqrt{(1 - \cos \theta)/2} \qquad \cos \frac{1}{2}\theta = \sqrt{(1 + \cos \theta)/2}$
$\sin \theta \approx \theta \quad$ [for small $\theta \lesssim 0.2$ rad]
$\cos \theta \approx 1 - \dfrac{\theta^2}{2} \quad$ [for small $\theta \lesssim 0.2$ rad]
$\sin(A \pm B) = \sin A \cos B \pm \cos A \sin B$
$\cos(A \pm B) = \cos A \cos B \mp \sin A \sin B$

For any triangle:
$c^2 = a^2 + b^2 - 2ab \cos \gamma \quad$ (law of cosines)
$\dfrac{\sin \alpha}{a} = \dfrac{\sin \beta}{b} = \dfrac{\sin \gamma}{c} \quad$ (law of sines)